Springer Complexity

Springer Complexity is an interdisciplinary program publishing the best research and academic-level teaching on both fundamental and applied aspects of complex systems—cutting across all traditional disciplines of the natural and life sciences, engineering, economics, medicine, neuroscience, social and computer science.

Complex Systems are systems that comprise many interacting parts with the ability to generate a new quality of macroscopic collective behavior the manifestations of which are the spontaneous formation of distinctive temporal, spatial or functional structures. Models of such systems can be successfully mapped onto quite diverse "real-life" situations like the climate, the coherent emission of light from lasers, chemical reaction-diffusion systems, biological cellular networks, the dynamics of stock markets and of the internet, earthquake statistics and prediction, freeway traffic, the human brain, or the formation of opinions in social systems, to name just some of the popular applications.

Although their scope and methodologies overlap somewhat, one can distinguish the following main concepts and tools: self-organization, nonlinear dynamics, synergetics, turbulence, dynamical systems, catastrophes, instabilities, stochastic processes, chaos, graphs and networks, cellular automata, adaptive systems, genetic algorithms and computational intelligence.

The three major book publication platforms of the Springer Complexity program are the monograph series "Understanding Complex Systems" focusing on the various applications of complexity, the "Springer Series in Synergetics", which is devoted to the quantitative theoretical and methodological foundations, and the "Springer Briefs in Complexity" which are concise and topical working reports, case studies, surveys, essays and lecture notes of relevance to the field. In addition to the books in these two core series, the program also incorporates individual titles ranging from textbooks to major reference works.

Series Editors

Understanding Complex Systems

Founding Editor: S. Kelso

Future scientific and technological developments in many fields will necessarily depend upon coming to grips with complex systems. Such systems are complex in both their composition – typically many different kinds of components interacting simultaneously and nonlinearly with each other and their environments on multiple levels – and in the rich diversity of behavior of which they are capable.

The Springer Series in Understanding Complex Systems series (UCS) promotes new strategies and paradigms for understanding and realizing applications of complex systems research in a wide variety of fields and endeavors. UCS is explicitly transdisciplinary. It has three main goals: First, to elaborate the concepts, methods and tools of complex systems at all levels of description and in all scientific fields, especially newly emerging areas within the life, social, behavioral, economic, neuro- and cognitive sciences (and derivatives thereof); second, to encourage novel applications of these ideas in various fields of engineering and computation such as robotics, nano-technology and informatics; third, to provide a single forum within which commonalities and differences in the workings of complex systems may be discerned, hence leading to deeper insight and understanding.

UCS will publish monographs, lecture notes and selected edited contributions aimed at communicating new findings to a large multidisciplinary audience.

More information about this series at http://www.springer.com/series/5394

Victor A. Sadovnichiy · Michael Z. Zgurovsky
Editors

Contemporary Approaches and Methods in Fundamental Mathematics and Mechanics

Springer

Editors
Victor A. Sadovnichiy
Lomonosov Moscow State University
Moscow, Russia

Michael Z. Zgurovsky
National Technical University of Ukraine
"Igor Sikorsky Kyiv Polytechnic Institute"
Kyiv, Ukraine

ISSN 1860-0832 ISSN 1860-0840 (electronic)
Understanding Complex Systems
ISBN 978-3-030-50301-7 ISBN 978-3-030-50302-4 (eBook)
https://doi.org/10.1007/978-3-030-50302-4

This Springer imprint is published by the registered company Springer Nature Switzerland AG
The registered company address is: Gewerbestrasse 11, 6330 Cham, Switzerland

International Editorial Board of this Volume

- D. Korkin, University of Missouri, Columbia, USA
- Pedro Marín-Rubio, Universidad de Sevilla, Spain
- Francisco Morillas, Universidad de Valencia, Spain.

This volume is dedicated to the distinguished mathematician Prof. Michael Z. Zgurovsky on the occasion of his 70th birthday

Preface

Given collection of papers have been organized as a result of regular open joint academic panels of researchers from Faculty of Mechanics and Mathematics of Lomonosov Moscow State University and Institute for Applied Systems Analysis of the National Technical University of Ukraine “Igor Sikorsky Kyiv Polytechnic Institute”. This volume devoted to the contemporary approaches and methods in fundamental mathematics and mechanics. It attracted attention of researchers from leading scientific schools of Australia, Brazil, China, Columbia, France, Germany, Italy, Poland, Russian Federation, Spain, Switzerland, Mexico, Ukraine, USA, Vietnam, and other countries.

Modern technological applications require development and synthesis of fundamental and applied scientific areas, with a view to reducing the gap that may still exist between theoretical basis used for solving complicated technical problems and implementation of obtained innovations. To solve these problems mathematicians, mechanics, and engineers from wide research and scientific centers have been worked together. Results of their joint efforts, including modern mechanics and its applications and advances in control and optimization as well as determinism and stochasticity in modeling in real phenomena are partially presented here. In fact, serial publication of such collected papers to similar seminars is planned.

This is the sequel of earlier volumes:

- Zgurovsky, Mikhail Z.; Sadovnichiy, Victor A. (Eds.) Continuous and Distributed Systems: Theory and Applications Series: Solid Mechanics and Its Applications, Vol. 211, 2014, XIX, 333 p. 33 illus., 14 illus. in color.
- Victor A. Sadovnichiy and Michael Z. Zgurovsky (Eds.), Continuous and Distributed Systems: Theory and Applications, Volume II, Studies in Systems, Decision and Control, Volume 30, 2015, Springer, Heidelberg xxiv+375 pp.
- Victor A. Sadovnichiy and Michael Z. Zgurovsky (Eds.), Advances in Dynamical Systems and Control, Studies in Systems, Decision and Control, Volume 69, 2016, Springer, Heidelberg xxii+471 pp.

...ovnichiy and Michael Z. Zgurovsky (Eds.), Modern Mathematics ...anics: Fundamentals, Problems and Challenges; Understanding ...x Systems, Springer International Publishing, Switzerland, 2019, ...+557 pp.

...nis volume we are planning to focus on the contemporary approaches and methods in fundamental mathematics and mechanics. We consider research and review papers based on recent advances in deterministic and stochastic modeling in real problems. The combination of new elements to describe complex situations in modeling allows us to get a more comprehensive analysis of factors, as in the rate of attraction (decay) of models, of the behavior of human brain with networks, of fractal dimensions of objects (upper and lower bounds), of the use of switching systems, Markov chains, ergodic properties and so on. Indeed, we aim to consider a wide range of papers concerning all these topics and beyond. The abstract mathematical approaches, such as differential geometry, differential and difference equations are applied to the practical applications in solid mechanics, hydro-, aerodynamics, optimization, decision-making theory, and control theory. In particular, it is shown that under small forcing intensity the global attractor of the 2D Navier–Stokes equations is a singleton, while when endowed with additive or multiplicative white noise no sufficient evidence was found that the random attractor keeps the singleton structure. We study Input-to-State Stability (ISS) property for nonlinear control systems with impulsive jumps at fixed moments. Sufficient conditions for the ISS are formulated in terms of a candidate ISS-Lyapunov function equipped with nonlinear rate functions which characterize its evolution along the discontinuous trajectories of the system. We will also introduce informational structures and informational fields related to the dynamics on complex structural networks representing a parcellation of a real brain, etc.

The book is addressed to a wide circle of mathematical, mechanical, and engineering readers.

Moscow, Russia
Kyiv, Ukraine
April 2020

Victor A. Sadovnichiy
Michael Z. Zgurovsky

Contents

Contributors

A. V. Aksenov Lomonosov Moscow State University, Moscow, Russian Federation

Oleg V. Barabash State University of Telecommunications, Kyiv, Ukraine

Lesia V. Baranovska Institute for Applied System Analysis, National Technical University of Ukraine "Kyiv Polytechnic Institute", Kyiv, Ukraine

Carles Bretó Departament d'Anàlisi Econòmica, Universitat de València, Valencia, Spain

Jose S. Cánovas Departamento de Matemática Aplicada y Estadística, Universidad Politécnica de Cartagena, Cartagena, Spain

Jan W. Cholewa Institute of Mathematics, University of Silesia in Katowice, Katowice, Poland

Marina Chugunova Institute of Mathematical Sciences, Claremont Graduate University, Claremont, CA, USA

Hongyong Cui School of Mathematics and Statistics, Huazhong University of Science & Technology, Wuhan, China

Radosław Czaja Institute of Mathematics, University of Silesia in Katowice, Katowice, Poland

Ciro D'Apice Dipartimento di Science Aziendali-Management e Innovation Systems, University of Salerno, Fisciano, SA, Italy

Umberto De Maio Dipartimento di Matematica e Applicazioni "R. Caccioppoli", Università degli Studi di Napoli "Federico II", Napoli, Italy

K. P. Druzhkov Lomonosov Moscow State University, Moscow, Russian Federation

Petro Feketa Kiel University, Kiel, Germany

Anatoly T. Fomenko Lomonosov Moscow State University, Moscow, Russian Federation

D. V. Georgievskii Lomonosov Moscow State University, Moscow, Russian Federation

I. M. Gorban Institute of Hydromechanics, National Academy of Sciences of Ukraine, Kyiv, Ukraine

Nataliia V. Gorban Institute for Applied System Analysis, National Technical University of Ukraine "Igor Sikorsky Kyiv Polytechnic Institute", Kyiv, Ukraine

Jianhua Huang College of Liberal Arts and Science, National University of Defense Technology, Changsha, China

Casey Johnson Institute of Mathematical Sciences, Claremont Graduate University, Claremont, CA, USA

D. O. Kapustian Taras Shevchenko National University of Kyiv, Kyiv, Ukraine

Olena Kapustian Taras Shevchenko National University of Kyiv, Kyiv, Ukraine

Oleksiy V. Kapustyan Taras Shevchenko National University of Kyiv, Institute for Applied System Analysis, National Technical University of Ukraine "Igor Sikorsky Kyiv Polytechnic Institute", Kyiv, Ukraine

Pavlo O. Kasyanov Institute for Applied System Analysis, National Technical University of Ukraine "Igor Sikorsky Kyiv Polytechnic Institute", Kyiv, Ukraine

Olha V. Khomenko Institute for Applied System Analysis, National Technical University of Ukraine "Igor Sikorsky Kyiv Polytechnic Institute", Kyiv, Ukraine

Vladislav A. Kibkalo Lomonosov Moscow State University, Moscow, Russian Federation;
Moscow Center for Fundamental and Applied Mathematics, Moscow, Russia

I. F. Kirichok Timoshenko Institute of Mechanics, National Academy of Sciences of Ukraine, Kyiv, Ukraine

Peter E. Kloeden Mathematisches Institut, Universität Tübingen, Tübingen, Germany

Peter I. Kogut Department of Differential Equations, Oles Honchar Dnipro National University, Dnipro, Ukraine

S. Yu. Kruts Taras Shevchenko National University of Kyiv, Kyiv, Ukraine

Olha P. Kupenko Department of Differential Equations, Oles Honchar Dnipro National University, Dnipro, Ukraine

O. Lavrova Taras Shevchenko National University of Kyiv, Kyiv, Ukraine

Yuhong Li School of Hydropower and Information Engineer, Huazhong University of Science and Technology, Wuhan, China

Ya. M. Linder Taras Shevchenko National University of Kyiv, Kyiv, Ukraine

Nataliia V. Lukova-Chuiko Taras Shevchenko National University of Kyiv, Kyiv, Ukraine

S. O. Mashchenko Taras Shevchenko National University of Kyiv, Kyiv, Ukraine

Thomas Meurer Kiel University, Kiel, Germany

V. Mogylova National Technical University of Ukraine "Igor Sikorsky Kyiv Polytechnic Institute", Kyiv, Ukraine

Andrii P. Musienko State University of Telecommunications, Kyiv, Ukraine

V. V. Pichkur Taras Shevchenko National University of Kyiv, Kyiv, Ukraine

Alexander Schaum Kiel University, Kiel, Germany

Tianlong Shen National Innovation Institute of Defense Technology, Academy of Military Sciences, Beijing, China

Valentyn V. Sobchuk State University of Telecommunications, Kyiv, Ukraine

O. Stanzhytskyi Taras Shevchenko National University of Kyiv, Kyiv, Ukraine

Olga V. Svynchuk State University of Telecommunications, Kyiv, Ukraine

Roman M. Taranets Institute of Applied Mathematics and Mechanics of the NASU, Sloviansk, Ukraine

L. N. Tereshchenko Institute of Hydromechanics, National Academy of Sciences of Ukraine, Kyiv, Ukraine

Nina V. Kasimova Taras Shevchenko National University of Kyiv, Kyiv, Ukraine

José Valero Centro de Investigación Operativa, Universidad Miguel Hernandez de Elche, Elche, Spain

Nataliya Vasylyeva Institute of Applied Mathematics and Mechanics of the NASU, Sloviansk, Ukraine

A. A. Voskoboinick Institute of Hydromechanics, National Academy of Sciences of Ukraine, Kyiv, Ukraine

A. V. Voskoboinick Institute of Hydromechanics, National Academy of Sciences of Ukraine, Kyiv, Ukraine

V. A. Voskoboinick Institute of Hydromechanics, National Academy of Sciences of Ukraine, Kyiv, Ukraine

E. B. Zavoychinskaya Lomonosov Moscow State University, Moscow, Russian Federation

Michael Z. Zgurovsky National Technical University of Ukraine "Igor Sikorsky Kyiv Politechnic Institute", Kyiv, Ukraine

Y. A. Zhuk Taras Shevchenko National University of Kyiv, Kyiv, Ukraine

Part I
Modern Mechanics and Its Applications

Chapter 1
Saddle Singularities in Integrable Hamiltonian Systems: Examples and Algorithms

Anatoly T. Fomenko and Vladislav A. Kibkalo

Abstract Saddle or hyperbolic singularities of Liouville foliations of integrable Hamiltonian systems are discussed. We observe new and classical results on their classification, representation and invariants with respect to topological equivalence depending on number of degrees of freedom. Then criterion of their component-wise stability by A.A. Oshemkov and its application are reminded. At last, we discuss saddle singularities of famous dynamical and physical systems, particularly problem of realization (modeling) of Liouville foliations and their singularities (A.T. Fomenko billiard conjecture) by integrable billiards. New result is obtained: loop molecules of all saddle-saddle singularities with one equilibrium are modeled by billiard books, i.e. integrable billiards on CW-complexes introduced by V.V. Vedyushkina.

1.1 Introduction

Integrability of a dynamical system, i.e. the presence of sufficient quantity of its independent first integrals, takes place in a lot of well-known Hamiltonian systems of mechanics, geometry and mathematical physics. Properties of integrable Hamiltonian systems (IHS) can be successfully described analyzing their phase topology, i.e. foliation of their phase space on common level surfaces of their fist integrals.

Liouville foliation (which is a Lagrangian fibration with singularities) and dynamics near a regular fiber can be completely described according to Liouville theorem (e.g., see book by A.V. Bolsinov and A.T. Fomenko, in Russian [1] and its English

A. T. Fomenko · V. A. Kibkalo (✉)
Lomonosov Moscow State University, GSP -1, Leninskie Gory, Moscow 119991, Russian Federation
e-mail: slava.kibkalo@gmail.com

A. T. Fomenko
e-mail: atfomenko@mail.ru

V. A. Kibkalo
Moscow Center for Fundamental and Applied Mathematics, Moscow, Russia

V. A. Sadovnichiy and M. Z. Zgurovsky (eds.), *Contemporary Approaches and Methods in Fundamental Mathematics and Mechanics*, Understanding Complex Systems,
https://doi.org/10.1007/978-3-030-50302-4_1

translation [2]). Studying Liouville foliations on the whole phase space is, usually, a non-trivial and rather complicated problem.

Firstly, momentum map can has critical points (where a fiber is not a smooth n-dimensional manifold). Secondly, even a system without such singularities can has non-trivial "global" invariants connected with topological properties of high-dimensional phase manifold M^{2n} (e.g. see [3]).

Obtained results and open problems of theory of local and semi-local singularities of IHS, Liouville foliations of IHS and their properties are observed and discussed in great detail in the book [2] (by A.V. Bolsinov and A.T. Fomenko) and in paper [8] by A.V. Bolsinov and A.A. Oshemkov.

In this paper we consider *only non-degenerate singularities*. This class is a generalization of critical points of Morse type on a manifold. In Sect. 1.2 invariants and classification of singularities are discussed both in *local* sense (in a small neighbourhood of a singular point in M^{2n}) and *semi-local* sense, i.e. in an invariant saturated neighbourhood of the fiber that contains this singular point (such fiber is called *singular*). We also consider *global* invariants of Liouville foliations of IHS with 2 degrees of freedom on invariant 3-submanifolds of 4-dimensional phase space.

This paper is mainly devoted to *saddle semi-local* singularities, i.e. local type of every its singular point should contain only saddle (hyperbolic) components and may be regular foliation of lower dimension.

In Hamiltonian systems with 1 degree of freedom we have an effective invariant of saddle semi-local singularities (saddle *2-atom*) that is *f-graph*, introduced by A.A. Oshemkov in [4].

For systems with 2 degrees of freedom non-degenerate $\text{rk} = 1$ (of momentum map) singularities of the *Bott type* usually are studied. Bott property in IHS is a natural analog of a Morse property of a critical point for a function on manifold and was introduced in integrable case by A.T. Fomenko in [5]. According to Fomenko theorem non-degenerate Bott saddle singularity of IHS with 2 degrees of freedom has a structure either $V \times S^1$ or $(V \times S^1)/\mathbb{Z}_2$ where V is a saddle 2-atom [5].

Singularities of $\text{rk} = 0$ of IHS with 2 degrees of freedom are classified by *Cl*-invariant, introduced by A.V. Bolsinov and studied in [6] by V.S. Matveev.

To study singularities $\text{rk} = 0$ in the case of arbitrary many degrees of freedom two different invariants are used. The first one is representation of a singularity as an *almost direct product* ([7], see also [8]). That is a direct product of 2-atoms quotient by action of a finite group G. The other invariant is a f_n-graphs introduced by A.A. Oshemkov in [9]. This graph is a generalization of f-graph of a 1 degree of freedom singularity. It turns out that f_n-graph construction is closely related to one independently appeared construction in integrable billiards (see [10], and Sect. 1.7.3).

Central role in the classification of Liouville foliations on 3-dimensional submanifolds of constant energy play famous Fomenko–Zieschang invariant (finite graph equipped by numerical marks) and Fomenko invariant (same graph without numerical marks, i.e. Reeb graph equipped by 3-atoms of bifurcations), particularly, such invariants of loop molecules of Liouville foliations on a 3-boundary of 4-neighbourhood of a non-degenerate semi-local singularity. Fomenko invariant is also called a coarse molecule. It turns out that properties of foliation in a 4-neighbourhood of such singularity depend to a large extent on its loop molecule (A.T. Fomenko conjecture, see [2, 11, 12]). In the case of singularity complexity 1 and 2 this conjecture is true, i.e. Liouville foliation on neighbourhood completely determines Liouville type of this singularity, [2].

Later non-equivalent saddle 4-singularities of complexity more than 2 were discovered, such that their loop molecules coincide. Recent results by M.A. Tuzhilin on this problem are discussed in Sect. 1.4. In Sect. 1.5 we recall give a brief account of singularity stability problem, including A.A. Oshemkov criterion on stability of a general saddle singularity with respect to component-wise integrable perturbation and its application by M.A. Tuzhilin to the case of complexity 2 singularities (their singular fiber contains two points of the $\mathrm{rk} = 0$) of systems with two degrees of freedom.

Then in Sect. 1.6 we list non-degenerate singularities discovered in various integrable systems of physics, mechanics and geometry and note which of them splits in these systems.

In the final Sect. 1.7 we discuss an up-to-date approach to realization of Liouville foliation via integrable billiards on CW-complexes, recently introduced by V.V. Vedyushkina ([14], also see [15, 17, 18]).

It turns out that adding a reflecting Hooke potential to billiard Hamiltonian essentially expands class of Liouville foliations modeled by integrable billiards ([19, 20]). It makes possible to realize not only arbitrary non-degenerate $\mathrm{rk} = 1$ singularities (3-atoms), but also some semi-local singularities of $\mathrm{rk} = 0$ of IHS with 2 degrees of freedom, [10].

In our paper new result is proved. Loop molecules with r-marks of every saddle complexity 1 singularities *are realized* by appropriate integrable billiard with potential. Recall that marked loop molecule uniquely (up to Liouville equivalence) determine a non-degenerate singularity of complexity 1 for the case of smooth IHS.

This paper was written with the support of the Russian Foundation for Basic Research (grant no. 19-01-00775-a).

1.2 Basic Notions

1.2.1 Liouville Theorem

On a symplectic manifold M^{2n} with form ω function $H \in C^{\infty}(M^{2n})$ generates a Hamiltonian vector field $(\mathrm{sgrad}\, H)^i = \omega^{ij}\,(\mathrm{grad}\, H)_j$. Recall that $u(H) = \omega(\mathrm{sgrad}\, H, u)$ for every tangent vector u. Such system $\dot{x} = \mathrm{sgrad}\, H$ is called a Hamiltonian system and function H is the Hamiltonian or the energy of this system. *First integral* of vector field sgrad H is a function $F \in C^{\infty}(M)$ such that $\mathrm{sgrad}\, H(F) = 0$.

Let us remind definition of a *completely integrable in Liouville sense* Hamiltonian system with n degrees of freedom (for brevity, also called an *integrable* system or

IHS with n degrees of freedom). Such systems have n first integrals $F_1, \dots, F_n$ of vector field sgrad H that

- are in involution: $\omega(\text{sgrad } F_i, \text{sgrad } F_j) = 0$ for all $0 \le i, j \le n$),
- $\mathrm{d}F_1, \dots, \mathrm{d}F_n$ are linearly independent almost everywhere on M^{2n},
- vector fields sgrad F_i are complete, i.e. a shift on every $t \in \mathbb{R}$ is well-defined on its trajectories.

Notice that involutive condition for F_i, F_j is equivalent to their commuting $\{F_i, F_j\} = 0$ w.r.t. the following Poisson bracket $\{f, g\} := (\omega^{-1})^{kl}(\frac{\mathrm{d}f}{\mathrm{d}x^k})(\frac{\mathrm{d}f}{\mathrm{d}x^l})$.

Map $\mathfrak{F} = (F_1, \dots, F_n)$ is called a *momentum map*. Liouville foliation (which is a Lagrangian fibration with singularities) is a decomposition of M^{2n} into connected components $\mathfrak{F}^{-1}(\xi)$ for $\xi \in \mathbb{R}^n$, i.e. $\{x \in M^{2n} | F_i(x) = \xi_i, i = 1, \dots, n\}$.

Consider a regular fiber $T_\xi = \{\mathfrak{F} = \xi \in \mathbb{R}^n\}$ and its rather narrow *saturated* neighbourhood $U(T_\xi)$. It means that if a point $x \in \xi$ from a fiber ξ belongs to U then the whole fiber $\xi \subset U$ is a subset of U. Notice that such a neighbourhood of a regular fiber is invariant w.r.t. the action of $\mathbb{R}^n$ on M^{2n}. Homomorphism $\mathbb{R}^n \longrightarrow \mathrm{Diff}\, M$ generates an action of $(\lambda_1, \dots, \lambda_n)$ on M^{2n} that is shift by unit along vector field sgrad $(\sum_i \lambda_i F_i))$. This action $\mathbb{R}^n$ usualy is called a Hamiltonian one or a Poisson one.

Liouville theorem completely fully describes Liouville foliation and dynamics of IHS in saturated neighbourhood $U(T_\xi)$ of a regular fiber T_ξ:

Theorem 1.1 (Liouville, [2]) *Let $v =$ sgrad H on (M^{2n}, ω) be a completely integrable Hamiltonian system with momentum map $\mathfrak{F} = (F_1, \dots, F_n)$ and $F_1 = H$. If gradients of F_i are linearly independent everywhere on a fiber T_ξ then*

- *fiber T_ξ is an n-dimensional Lagrangian submanifold invariant on vector fields* sgrad $F_i, i = 1, \dots, n$,
- *fiber T_ξ is diffeomorphic to $T^{n-k} \times \mathbb{R}^k$; if T_ξ is compact then it is diffeomorphic to n-dimensional torus T^n, called a Liouville torus,*
- *Liouville foliation in saturated neighbourhood $U(T_\xi)$ has structure of $D^n \times T^n$,*
- *such action-angle variables exist $(s_1, \dots, s_n)$ on D^n and $(\phi_1, \dots, \phi_n)$ on T^n that symplectic form has canonical form $\omega = \sum_i \mathrm{d}\phi_i \wedge \mathrm{d}s_i$, , integrals are functions on action variables $\mathfrak{F} = \mathfrak{F}(s_1, \dots, s_n)$ and the flow v is straightened: $\dot{s_i} = 0, \dots \dot{\phi_i} = q_i(s_1, \dots, s_n)$.*

Trajectories of the system are (quasi-)periodic on regular tori and are rational or irrational rectilinear winding on them.

1.2.2 *Equivalences of Non-degenerate Singularities*

The set of critical points of momentum map $\mathfrak{F}$ naturally is stratified by its rank. For every value of rank following notion of a *non-degenerate* singular point of such rank is defined.

At first, consider a singular point of rank 0 for IHS with n degrees of freedom. Such point is called *non-degenerate*, if linearizations $A_{F_i} = \Omega^{-1}\mathrm{d}^2 F_i$ of first integrals F_i generate a Cartan (i.e. maximal commutative) subalgebra in Lie algebra $sp(2n, \mathbb{R})$ of symplectic group. It means that linearizations are linearly independent and some of their linear combination $A = \sum_i c_i A_{F_i}$ has simple spectrum with $2n$ nonzero eigenvalues.

In the case of a point of the rank $k > 0$ for IHS with $n + k$ degrees of freedom it is possible to make symplectic reduction by Hamiltonian action first integrals $f_1, \ldots, f_k$ such that $\mathrm{d}f_1 \wedge \mathrm{d}f_2 \wedge \cdots \wedge \mathrm{d}f_k \neq 0$. If the resulting critical point of the rank 0 of the quotient system with n degrees of freedom is non-degenerate then the initial point will be called non-degenerate (see more in [2]).

Note two significant aspects. Firstly, functions that determine IHS (its phase manifold M, symplectic structure on it and first integrals) can be either real-analytic or only smooth enough. Methods of proofs and results of classification differs in these cases.

Secondly, singularities of IHS can be classified w.r.t. various equivalences. Choice of equivalence usually is determined by two following questions: 1) which subsets X_i (that contain critical points x_i) of phase space M_i are compared; 2) map $\phi : X_1 \longrightarrow X_2$ of which class is used for this comparison.

Usually there are three different answers on the first question (i.e. three types of equivalence): *local*, *semi-local* or *global* equivalences. In these cases one consider Liouville foliation either in a small neighbourhood of the critical point x or in a neighbourhood $U(L)$ that contains singular fiber $L(x)$ (recall that U should be *saturated*) or on an invariant (w.r.t. Hamiltonian action $\mathfrak{F}$) submanifold, that can contain several singularities "far" from each other. For example, such manifold can be M^{2n} or an isoenergy submanifold $Q_h^{2n-1} : H = h$).

Map $\phi : X_1 \longrightarrow X_2$ from the definition can belong to classes of homeomorphisms, diffeomorphisms or symplectomorphisms (it means that ϕ is a diffeomorphism such that $\phi * (\omega_2) = \omega_1$). Equivalences and their invariants corresponding these three classes are called *continuous* (*topological*), *smooth* or *symplectic* respectively. It is obvious that map ϕ should be *fiber-wise*, i.e. transform fiber into another fiber with necessary smoothness.

Remark 1 Topology of IHS can also be studied in a tubular neighbourhood of a singular orbit of the action of $\mathfrak{F}$. Resulting equivalence is a priori more strong than local equivalence and more weak than semi-local one.

Having fiber-wise homeomorphism of Liouville foliations of two IHS we will say that these systems are Liouville equivalent. For restriction on a non-special 3-dimensional submanifold $Q^3 \subset M^4$ (discussed in Sect. 1.2.5) of non-degenerate IHS (M^4, ω, H, F) we also demand preserving of orientation of Q^3 and directions of vector field sgrad H on circles that consist of critical points (more in [2], v. 1, Chap. 4).

1.2.3 Classification of Local Singularities

Simplest examples of local singularities of the rank 0 are small neighbourhoods of the origin at spaces $\mathbb{R}^2(p_1, q_1)$ with symplectic form $\omega = \mathrm{d}p_1 \wedge \mathrm{d}q_1$ and function $F_1 = p_1^2 + q_1^2$ (elliptic case, also called a center) or function $F_1 = p_1^2 - q_1^2$ (hyperbolic case, also called a saddle). Note that in the case of 1 degrees of freedom a critical point is non-degenerate if and only if Morse condition: $\det \mathrm{d}^2 F_1 \neq 0$ is true.

Significantly new singularity (focus type, or a focus-focus) appears only in system with 2 or more degrees of freedom, e.g. in space $\mathbb{R}^4(p_1, p_2, q_1, q_2)$ equipped by form $\omega = \sum_i \mathrm{d}p_i \wedge \mathrm{d}q_i$ and pair of functions $F_1 = p_1q_2 - p_2q_1$, $F_2 = p_1q_2 + p_2q_1$. The only critical point of this system has rank 0 and belongs to a 2-dimensional fiber.

In this section class of equivalent sufficiently small neighbourhoods of a critical point of a smooth IHS with n degrees of freedom where rank of $\mathrm{d}\mathfrak{F} = r$ is called a (local) singularity of rank r. In this case all three equivalences described earlier (topological, smooth and symplectic) generates the same classification. For every n it has only finite number of singularities.

Theorem 1.2 (L. Elliason, [21]) *Local singularity of rank r of a smooth IHS with n degrees of freedom is fiber-wise symplectomorphic to a direct product of model components: a regular foliation with r degrees of freedom and $k_e, k_h, k_f \geq 0$ items of elliptic, saddle and focus components respectively. Here $k_e + k_h + 2k_f = n - r$.*

This theorem is discussed in detail in [8]. Note that integrals F_i are lifted to the direct product and corresponding symplectic form is a sum of the components' forms: $\omega = \sum_i \mathrm{d}p_i \wedge \mathrm{d}q_i$.

Remark 2 *Hereinafter we do not consider singularities of focus type.* Then every local singularity for $n = 2$ is a product of foliations of two systems with 1 degree of freedom. It is either product of a Morse singularity (a center or a saddle) and a regular foliation on 2-manifolds (thus the singularity has rank 1) or product of two Morse singularities on 2-manifolds. Resulting singularity is of the rank 0 and is called center-center, center-saddle or saddle-saddle depending on components' types.

1.2.4 Semi-local Singularities of IHS

In the elliptic case local type of a rank 0 point coincides with the semi-local one. Singular fiber contains only point (that is critical of rank 0). In the case of saddle singularities of rank 0 of IHS semi-local problem is nontrivial because singular fiber always contains both critical and regular points.

Complexity of a saddle singularity of rank r is the quantity k of rank r orbits (of the Poisson action of $\mathfrak{F}$) on the singular fiber. For $r = 0$ such orbit is a point of rank 0 and is fixed under the $\mathfrak{F}$-action.

If $n = 1$ then every semi-local class if saddle singularities of complexity k (hereinafter called a *2-atom*) can be realized as a gluing of k saddle local singularities (saddles, or crosses).

In the case of $n = 2$ non-degeneracy of a rank 1 point is equivalent to Bott property: critical points should be united in non-degenerate critical submanifolds [2]. Classification of such *3-atoms* (in $Q^3 \subset M^4$) is obtained by A.T. Fomenko theorem [5]. Namely, it has the structure of either a direct product of appropriate 2-atom and a circle S^1 or its quotient by the action of appropriate involution $\mathbb{Z}_2$. It turns out that 3-atom of IHS also has is a Seifert manifold, may be with special 1-fibers of (2, 1) type ([22], see also [2, 23] and its translation [24]).

In addition to non-degeneracy of a singular point 0 N.T. Zung condition of non-splittability also is demanded in the saddle case. For $n = 2$ it means that intersection of the critical set and an invariant neighbourhood of a singular fiber should be mapped by $\mathfrak{F}$ in a pair of transversely intersecting smooth intervals of two curves of the bifurcation diagram Σ. Note that this nonsplittability differs from topological stability and splittability discussed in 1.5.

1.2.5 Topology of Liouville Foliations on Invariant Submanifolds

Liouville foliations of IHS with $n = 2$ restricted on an invariant non-special 3-dimensional submanifold can be described via Fomenko–Zieschang invariants, i.e. graphs with numerical marks. Their vertices correspond to rank 1 bifurcations and are equipped by their types, i.e. 3-atoms ([25, 26] see also [2]).

Recall that *loop 3-manifold* Q^3_γ of a singularity is the 3-dimensional boundary of its invariant 4-neighbourhood. Its $\mathfrak{F}$-image $\gamma \subset \mathbb{R}^2$ should transversely intersect arcs of the bifurcation diagram Σ, i.e. be of the general case.

Fomenko–Zieschang invariant of Liouville foliation on Q^3_γ is called a marked *loop molecule* of the singularity. Squared distance from a point on γ to the singular point of Σ we consider as local Hamiltonian. Some periodic coordinate on γ can be considered as additional Bott first integral.

Fomenko–Zieschang invariant classifies IHS restricted on Q^3 w.r.t. *Liouville equivalence*, i.e. to via fiber-wise homeomorphism that preserves orientation of Q^3 and direction of vector field sgrad H on critical circles (submanifolds).

1.3 Invariants of Semi-local Equivalence

1.3.1 f-Graph or Pair of Permutations Determine 2-Atom

Notions of 2-atoms and 3-atoms necessary for IHS classification were introduced by A.T. Fomenko [22, 27]. Every 2-atom P with fixed direction of function f growth on P (pair (P^2, f) is called f-atom) in paper by A.A. Oshemkov [4] was canonically corresponded to its *f-graph*. Notice that every *atom* corresponds to *two f-atoms* and *two their f-graphs*, a priori non-isomorphic.

Every vertex of f-graph has degree 3 (its in-degree is 1, out-degree is 1, and one edge is not directed). In the case of Liouville foliations on orientable 2-surfaces (in particular, for Hamiltonian systems with 1 degree of freedom) isomorphism of f-graphs means the equivalence of corresponding f-atoms (i.e. fiber-wise diffeomorphism of their foliations preserving directions of function growth).

f-graph also can be determined by a pair of permutations (σ, τ) acting on its vertices preserving direction of edges. Permutation σ determines the shift along cycles that consist of directed edges. Permutation τ consists of transpositions of pairs of vertices that are connected by non-directed edges of the f-graph. Thus τ is an involution without fixed points. Simmetries of 2-atoms and their f-graphs and related topics is discussed in [28].

Adding several vertices of degree 2 (with in-degree and out-degree equal to 1) in f-graphs allows to determine *f-graph with stars* for 2-atom with stars of an arbitrary 3-atom which is not of direct product type $V \times S^1$).

1.3.2 General Saddle Singularities

Two general questions of semi-local topological classification are:

- which invariants can uniquely determine this class,
- how algorithmically to enlist all different classes of singularities with fixed rank and degrees of freedom?

For $n = 2$ following Cl-invariant (introduced by A.V. Bolsinov) is a complete invariant and generates an effective algorithm of comparison of singularities (V.S. Matveev, [6]). Cl-invariant is a triple (V_1, V_2, L) with two embeddings of singular fibers L_i of 2-atoms V_i (may be not connected) in singular fiber L of the singularity. Singular fiber L is a 2-dimensional CW-complex and can be determined by finite numerical data. It is unknown [9] if this invariant is complete for $n > 2$.

In the paper [7] Zung formulated his condition of nonsplittability for singularities of IHS. In the case of $\mathrm{rk} = 0$ saddle singularities it means that "local" bifurcation diagrams of singular points of $\mathrm{rk} = 0$ should coincide. We formulate Zung's results only for saddle $\mathrm{rk} = 0$ singularities.

For compact saddle 2-atoms $V_1, \ldots, V_n$ with symplectic forms $\omega_1, \ldots, \omega_n$ and Morse functions $f_1, \ldots, f_n$ their *almost direct product* was defined by the following way. Suppose that finite group G acts on every atom V_i, and each of these actions ϕ_i preserves both ω_i and f_i.

If the action $\phi(g)(x_1, \ldots, x_n) = (\phi_1(g)x_1, \ldots, \phi_n(g)x_n)$ is free then quotient symplectic manifold $(V_1 \times \cdots \times V_n)/G$ and corresponding IHS with n degrees of freedom are well-defined. Liouville foliation on this manifold has a structure of a neighbourhood $U(L)$ of a connected singular n-fiber L with non-degenerate singular points of rk $= 0$.

We say that an almost direct product $(V_1 \times \cdots \times V_n)/G$ is *irreducible* if the group G has no elements acting nontrivially oFn only one of the factors of the direct product $V_1 \times \cdots \times V_n$.

Theorem 1.3 (N.T. Zung, [7]). *Every Zung-nonsplittable non-degenerate compact saddle singularity $U(L)$ of rank 0 is of almost direct product type, and the representation of the singularity $U(L)$ in the form of an irreducible almost direct product $(V_1 \times \cdots \times V_n)/G$ is unique up to Liouville equivalence.*

In the case of saddle rk $= 0$ singularities such representation is unique (Zung [7], some comments in [9, 29]) for *minimal models*. It means that every nontrivial element of G acts nontrivially at least on two components of the product $V_1 \times \cdots \times V_n$. Thus such invariant is complete, but not very convenient for algorithmic enlisting of singularities.

More effective algorithm was discovered by A.A. Oshemkov who introduced new invariant called *f_n-graph*.

Definition 1.1 Finite graph $\Theta = (V, E)$ is called an f_n-graph (here V and are sets of vertices and edges of Θ respectively) if

- $E = \bigsqcup_1^n E_i$, i.e. every edge belongs to one of n classes E_i
- some of edges are oriented and other non-oriented,
- subgraphs $\Theta_i = (V(E_i), E_i)$ are f-graphs,
- their permutations (σ_i, τ_i) as f-graphs commute by following ways:

$$\tau_i \circ \tau_j = \tau_j \circ \tau_i, \quad \tau_i \circ \sigma_j = \sigma_i \circ \tau_j \quad \sigma_i \circ \sigma_j = \sigma_i \circ \sigma_j \qquad \text{for } i \neq j$$

- $\mathbb{Z}_2^n$ action on the set of vertices V generated by $\tau_1, \ldots, \tau_n$ is free (i.e. every their combination is an involution without fixed points).

In the initial paper [9] permutations σ_i we denoted by μ_i.

Two such graphs are called *equivalent* if they correspond to the same foliation. It does not depend on growth direction of the growth of integrals F_i. It is known that in similar situation f-graphs are called equivalent if they correspond to f-atoms of the same atom. Class of equivalent f_n-graphs uniquely determines a minimal model, i.e. Liouville class of saddle singularity.

Applying this structure to the case of IHS with three degrees of freedom, A.A. Oshemkov effectively listed in [13] all 32 different singularities of rk $= 0$ and complexity 1.

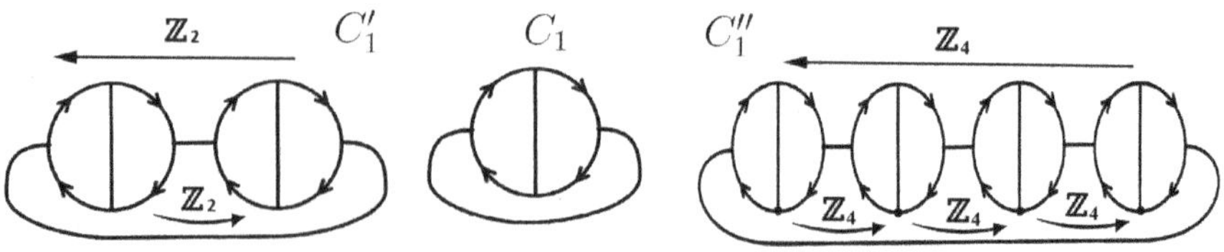

Fig. 1.1 Atoms C_1, C_1' and C_1'' from Grabezhnoy examples with group action

1.4 A.T. Fomenko Conjecture on Loop Molecules of Non-Degenerate Singularities

The concept of a loop molecule as an invariant of Liouville foliation was widely used in studying of topology of concrete, often rather complicated integrable systems: Kovalevskaya [30], Sokolov [31], Steklov [31], Kovalevskaya-Yehia ([33–35]), Kovalevskaya on Lie algebras $so(4)$ and $so(3, 1)$ ([36–38]).

Firstly, to describe loop molecule it is sufficient to know 3-dimensional 3-atoms (bifurcations of Liouville foliations) on arcs of bifurcation diagram that often help avoid more difficult studying of 4-dimensional singularities (even in the case of degenerate one). Secondly, singularities of studied IHS of physics and mechanics typically have relatively "simple" construction, thus calculation of its loop molecule significantly help "predict" which singularities are possible in this system.

A.T. Fomenko [2] formulated a problem if Liouville foliation on 3-boundary of a 4-singularity uniquely determine its type.

Conjecture 1 (A.T. Fomenko) Loop molecule uniquely determines semi-local fiber-wise homeomorphism class of non-degenerate 4-singularity of rank 0 in non-degenerate IHS with 2 degrees of freedom.

It turns out that the conjecture is true for all singularities of center-center, center-saddle and focus-focus types. It is also true for all saddle singularities of complexity 1 and 2. However counterexamples were constructed for saddle singularities of complexity 4 and more. Three first pair-wise nonequivalent 4-singularities with the same loop molecules were discovered by A.V. Grabezhnoy in his master thesis [11]. They are of complexity 4. Later in the paper [39] M.A. Tuzhilin constructed two infinite series of counterexamples (pair-wise nonequivalent with the same loop molecule).

Key idea of A.V. Grabezhnoy counterexample is to consider coverings of a 2-atom C_1 by another 2-atom. f-graphs of these 2-atoms C_1, C_1' and C_2'' with group $\mathbb{Z}_2$ and $\mathbb{Z}_4$ actions are shown on Fig. 1.1. These groups act on atom C_1 by rotation along the boundary circle and change sheet of the covering.

Proposition 1.1 *Irreducible non-equivalent 4-singularities* $C_1 \times C_1$, $(C_1 \times C_1')/\mathbb{Z}_2$, $(C_1 \times C_1'')/\mathbb{Z}_4$ *of complexity* 4 *have the same marked loop molecules, i.e. their Liouville foliations on the loop 3-manifolds are Liouville equivalent.*

In the paper [39] M.A. Tuzhilin constructed two infinite series of pair-wise nonequivalent 4-singularities with coinciding loop molecules.

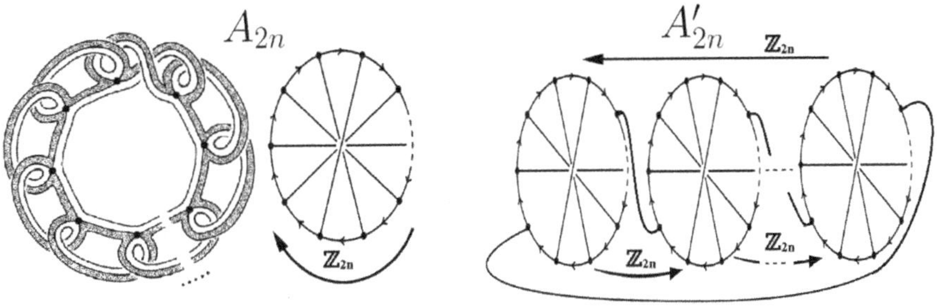

Fig. 1.2 Atom A_{2n}, its f-graph and $2n$-covering A'_{2n} with $\mathbb{Z}_{2n}$ actions

The first series consists of 4-singularities $(C_1^n \times C_1^n)/\mathbb{Z}_n$ with group $\mathbb{Z}_n$ acting on components by $2\pi/n$ rotation.

Proposition 1.2 *Saddle-saddle singularities* $(C_1^n \times C_1^n)/\mathbb{Z}_n$ *are pair-wise topologically non-equivalent but their loop molecules (without marks) coincide.*

The second series is based on consideration of $2n$-coverings A'_{2n} with atom A_{2n} as their base. Atom A_{2n}, its f-graph and f-graph of the covering A'_{2n} with $\mathbb{Z}_{2n}$ action (by a cyclic permutation of sheets) are shown on Fig. 1.2.

Describe the component-wise action of $\mathbb{Z}_{2n}$ the direct product $(A_{2n} \times A'_{2n})$. Firstly, it should be free. This action on A_{2n} is a rotation along each of boundary circles A_{2n} on angle multiple of $2\pi/n$. The action on A'_{2n} was described earlier. Thus these singularities are of saddle-saddle type.

Theorem 1.4 (M.A. Tuzhilin, [39]) *Semi-local saddle-saddle singularities* $A_{2n} \times A_{2n}$, $(A_{2n} \times A'_{2n})/\mathbb{Z}_{2n}$ *are topologically non-equivalent. Their marked loop molecules coincide for some choice of orientations of loop molecules and edge directions.*

1.5 Stability and Splittability of Singularities

First we recall the notion of an integrable perturbation. Family of IHSs $(M^{2n}, \omega_\alpha, F_{1,\alpha}, \dots, F_{n,\alpha})$ is an integrable perturbation of IHS $(M^{2n}, \omega, F_1, \dots, F_n)$ if $\omega_\alpha = \omega_{x_1,\dots x_{2n},\alpha}$ and $F_{i,\alpha} = F_i(x_1, \dots x_{2n}, \alpha)$ are smooth and coincide with ω, F_i when $\alpha = 0$.

Definition 1 IHS $(M_{2n}, \omega, F_1, \dots, F_n)$ is called *stable* if for any integrable perturbation $(M_{2n}, \omega, F_{1,\alpha}, \dots, F_{n,\alpha})$, there exists $\varepsilon > 0$ such that initial IHS is topologically equivalent to all perturbed systems for all $\alpha \in [0, \varepsilon]$.

This general definition of stability can be restricted on invariant neighbourhoods of semi-local saddle singularities of rk $= 0$. They can be stable (nonsplittable) or unstable (splittable). Precise definitions can be found in [12].

In the case of Hamiltonian system with 1 degree of freedom it is known that every saddle 2-atom with complexity $k > 1$ is splittable. Problem of a splittability of such atoms under integrable perturbations that preserve some symmetry group becomes non-trivial.

In the case of many degrees of freedom stability problem is much more complicated. Let such atom V be included in an almost direct product $(V_1 \times \cdots \times V)/G$ where action ϕ_n on atom V preserves both F and every F_α. Then the action ϕ of G on the direct product $(V_1 \times \cdots \times V')/G$ is well-defined and free. It means that for resulting system the functions $F_{1,\alpha}, \ldots, F_{n,\alpha}$ are integrals. This system is a splitting of initial singularity.

On the first stage another notion of component-wise stability was studied. Of course, a stable singularity is component-wise stable. Also a component-wise splittable singularity is generally splittable.

Let us define a *fiber-wise perturbation* $F_{1,\alpha}, \ldots, F_{n,\alpha}$ on $(V_1 \times \cdots \times V_n)/G$ of IHS with integrals $F_1, \ldots, F_n$ (on 2-atoms $V_1, V_2 \ldots, V_n$ respectively) as perturbation with two following conditions. Firstly, it should be invariant under the actions $\phi_1, \phi_2, \ldots, \phi_n$ of the group G (on corresponding 2-atoms respectively) and, secondly, this perturbation should be constant on the boundary circles of these 2-atoms.

Definition 2 Saddle singularity $(V_1 \times \cdots \times V_n)/G$ is called *component-wise stable* if for any component-wise integrable perturbation some positive ε can be chosen such that the resulting IHS is Liouville equivalent to the initial one for all $\alpha \in [0, \varepsilon]$.

Other singularities are called *component-wise splittable.*

Let us recall the Oshemkov's criterion of component-wise splittability for non-degenerate saddle singularities of $\mathrm{rk} = 0$. This result was applied by M.A. Tuzhilin for analysis of saddle singularities with 2 degrees of freedom, particularly in the case of complexity $k = 2$.

Theorem 1.5 (A.A. Oshemkov, [12]) *A singularity $(V_1 \times \cdots \times V_n)/G$ is component-wise stable (nonsplittable) if and only if the group G acts transitively on the vertices of each atom $V_1, \ldots, V_n$ (that is, on their saddle critical points).*

The following result was obtained from Oshemkov's criterion.

Theorem 1.6 (M.A. Tuzhilin, [12]) *Among the all 39 singularities of complexity 2 of saddle-saddle type, 28 singularities are component-wise splittable, and another 11 are component-wise nonsplittable: namely the singularities 15, 16, 17, 18, 27, 28, 35, 36, 37, 38 in the Table 9.2 [2].*

Note that the result of splitting of a complexity 2 singularity is a pair of saddle singularities of complexity 1. There are four such non-equivalent singularities $B \times B$, $(B \times C_2)/\mathbb{Z}_2$, $(B \times D_1)/\mathbb{Z}_2$, $(C_2 \times C_2)/(\mathbb{Z}_2 \oplus \mathbb{Z}_2)$. Their loop molecules are shown on the Fig. 1.3.

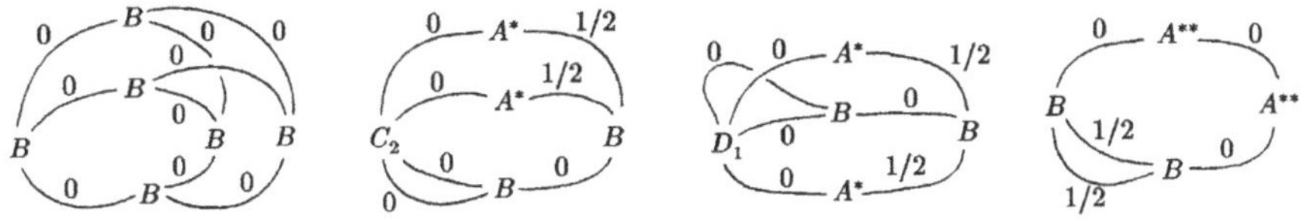

Fig. 1.3 Loop molecules of the all (namely, four) saddle-saddle singularities of complexity 1: $B \times B$, $(B \times C_2)/\mathbb{Z}_2$, $(B \times D_1)/\mathbb{Z}_2$, $(C_2 \times C_2)/(\mathbb{Z}_2 \oplus \mathbb{Z}_2)$

1.6 Saddle-Saddle Singularities in Physical, Mechanical and Geometrical Systems

In recent years Liouville foliations and their singularities were described for famous cases of integrability of physics and mechanics. A lot of them have structure (may be after reduction) of dynamical systems on dual space (co-algebra) of Lie algebra. These Lie groups usually are $e(3)$, $so(4)$, $so(3, 1)$, i.e. algebras of Lie groups of isometries of Euclidean 3-space $\mathbb{R}^3$ and rotation groups of Euclidean 3-space $\mathbb{R}^4$ and pseudo-Euclidean $\mathbb{R}^4_1$ with index equal to 1.

In Table 1.1 we list non-degenerate singularities of rank 0 that appear in following cases of integrability: Euler, Lagrange (discussed in [2]), Kovalevskaya (case of Lie algebras $so(3, 1)$: [38, 45], $e(3)$: [30, 40, 41], $so(4)$: [42]), Goryachev-Chaplygin-Sretensky [2, 43], Clebsch [32], Steklov [31], Sokolov [31, 44], Kovalevskaya-Yehia [34] and Goryachev-Chaplygin-Yehia [46].

Here $c - c$, $c - s$, $s - s$, $f - f$ are notations for singularity types: center-center, center-saddle, saddle-saddle, focus-focus respectively. Group actions on 2-atoms of singularities of saddle-saddle type are denoted according to [2]. Symmetry α is a central rotation on π for a flat representation of 2-atom as in [2]. Singularities $(B \times P_4)/\mathbb{Z}_2$ and $(C_2 \times C_2)/\mathbb{Z}_2$ of complexity 2 have numbers 12 and 17 respectively in the Table 9.1 [2] and in the paper [12].

Among them following singularities $A \times C_2$, $A \times P_4$, $B \times C_2$, $(B \times P_4)/\mathbb{Z}_2$ are splittable and singularity $(C_2 \times C_2)/\mathbb{Z}_2$ is component-wise non-splittable. Important question is which of them can be splitted varying parameters of this integrable dynamical system or by explicit perturbation of the system (in the class of IHS). Now it is known that singularity $A \times C_2$ is splitted by appropriate perturbation of Euler integrable case. Resulting system is Zhukovsky case with two saddle-saddle singularities $A \times B$. In Kovalevskaya system on $so(3, 1)$ varying of values of Casimir functions allows to split singularities $A \times C_2$ and $B \times C_2$ to pairs of singularities $A \times B$ and $B \times B$ respectively.

Table 1.1 Non-degenerate singularities of physical and mechanical integrable systems

Sing	Action	Type	IHS
$A \times A$		center-center	majority of considered IHS
$A \times B$		center-saddle	Zhukovsky, Kovalevskaya $so(3, 1) - e(3) - so(4)$, Steklov, Sokolov $e(3)$, Kovalevskaya-Yehia
$A \times C_2$		center-saddle	Euler, Kovalevskaya $so(3, 1)$ and $so(4)$, Steklov, Sokolov $e(3)$
$A \times P_4$		center-saddle	Goryachev-Chaplygin-Yehia
$B \times B$		saddle-saddle	Kovalevskaya $so(3, 1) - e(3) - so(4)$, Sokolov $e(3)$, Kovalevskaya-Yehia
$(B \times C_2)/\mathbb{Z}_2$	$\alpha(B), \alpha(C_2)$	saddle-saddle	Kovalevskaya $so(3, 1) - e(3) - so(4)$, Sokolov $e(3)$, Kovalevskaya-Yehia
$(B \times D_1)/\mathbb{Z}_2$	$\alpha(B), \alpha(D_1)$	saddle-saddle	Sretensky
$B \times C_2$		saddle-saddle	Kovalevskaya $so(3, 1)$, Steklov, Sokolov $e(3)$
$(B \times P_4)/\mathbb{Z}_2$	$\alpha(B), \alpha(P_4)$	saddle-saddle	Goryachev-Chaplygin-Yehia
$(C_2 \times C_2)/\mathbb{Z}_2$	$\alpha(C_2), \alpha(C_2)$	saddle-saddle	Clebsch $e(3)$
focus compl. 1		focus-focus	Lagrange, Steklov $e(3)$

1.7 Foliations and Singularities of Integrable Billiards

Wide class of studied mechanical and physical systems are smooth and even real-analytic. Their special regimes of motion that correspond to singularities of Liouville foliations often are rather complicated to describe them explicitly or compare with singularities of other systems.

It turns out that in the class of integrable billiards (which is a class of a priori not smooth but piece-wise smooth systems) a lot of foliations were obtained that are topologically equivalent to well-known singularities and Liouville foliations on submanifolds of constant energy from smooth and real-analytic IHS.

As the first example consider billiard system in ellipse that is integrable according to G.D. Birkhoff, [47]. *Elementary billiard table* is a flat compact domain with smooth or piece-wise smooth boundary consisting of arcs of confocal quadrics (ellipses for $\lambda < b$, hyperbolas for $b < \lambda < a$, two coordinate axes for $\lambda = b, a$) with a finite number of angles equal to $\pi/2$. These confocal quadrics belong to the family

$$(b - \lambda)x^2 + (a - \lambda)y^2 = (a - \lambda)(b - \lambda). \tag{1.1}$$

Singular trajectories of a particle in a billiard correspond to intervals of coordinate axes of the domain or to gluing arcs. In papers by V.V Vedyushkina, A.T. Fomenko and others the class of integrable billiards was dramatically extended and new equivalences between billiards and smooth IHS of physics, mechanics and geometry were obtained.

Particularly, V.V. Vedyushkina introduced new wide class of integrable *billiard books*, i.e. integrable billiards on CW-complexes [14]. Their 2-cells are elementary flat billiard tables. Their 1-cells are equipped by permutations and are gluing arcs of 2-cells along their common boundary arcs (spines of the book). Permutation σ on a gluing arc (which is a common boundary arc of several elementary billiards) determine the motion of a particle. Trajectory straight-line segment (which belongs to elementary billiard i) that intersects this boundary arc will be continued by another elementary billiard $\sigma(i)$.

Based on recent results, A.T. Fomenko formulated conjecture consisting of several items which is devoted to realization of wide class of IHS with 2 degrees of freedom by integrable billiards.

Conjecture A (atoms). Every non-degenerate bifurcation of 2-dimensional Liouville tori in isoenergy submanifolds of every non-degenerate IHS with 2 degrees of freedom can be realized (modeled) by integrable billiards. Liouville foliations of billiards and of these IHS will be fiber-wise homeomorphic.

Conjecture B (coarse molecules). Every coarse molecule (Fomenko invariants) i.e. base of every Liouville foliations on isoenergy submanifold of IHS with 2 degrees of freedom, can be realized (modeled) by integrable billiards w.r.t. coarse Liouville equivalence.

Conjecture C (marked molecules). Every marked molecule (Fomenko–Zieschang invariant) can be modeled by integrable billiards. Recall that Fomenko–Zieschang invariant determines a non-degenerate system on a 3-manifold w.r.t. Liouville equivalence. Thus every such Liouville foliation is fiber-wise homeomorphic to the one of an appropriate integrable billiard.

Conjecture D (isoenergy 3-manifolds). Every 3-dimensional closed compact isoenergy manifold of every non-degenerate IHS with 2 degrees of freedom is an isoenergy submanifold of phase space of an appropriate integrable billiard. This statement also is a corollary of the item C. Recall that according to A.V. Brailov and A.T. Fomenko theorem the class of isoenergy manifolds of non-degenerate IHS coincides with the class of graph-manifolds (Waldhausen manifolds, see [50, 51]).

Later problem of realization of rank 0 singularity by billiards was also formulated.

Let us describe obtained results on Fomenko conjecture:

- **Conjecture A** is completely proved. In other words, every Bott 3-atom can be realized as an atom of some billiard book: V.V. Vedyushkina, I.S. Kharcheva [14].
- **Conjecture B** is completely true, V.V. Vedyushkina, I.S. Kharcheva [15].
- **Conjecture C** is still open but a wide class of Fomenko–Zieschang invariants have already be realized by billiards, including following IHS:
  - Euler and Lagrange integrable cases: for all energy zones ([17, 48]),
  - Kovalevskaya, Goryachev–Chaplygin, Steklov, Kovalevskaya on $so(4)$ integrable cases: in several zones of energy ([17, 49]);
  - all integrable geodesic flows on oriented 2-surfaces with a linear or quadratic first integral [52],

- **local conjecture C** (A.T Fomenko) is true: all values of numerical marks (of Fomenko–Zieschang invariants) appear in billiards, i.e. arbitrary pair of edge marks (r, ε) re realized (V.V. Vedyushkina); arbitrary value of the Euler class (of corresponding) Seifert submanifold, i.e. arbitrary value of mark n is realized on an appropriate family of saddle atoms (V.V. Vedyushkina, V.A. Kibkalo);

- **Conjecture D** is still open, but class of billiard isoenergy manifolds is not a subclass of Seifert manifolds. For example, arbitrary connected sums of lens spaces $L(p, q)$ (including projective 3-space $\mathbb{R}P^3$) and products $S^1 \times S^2$ are realized (V.V. Vedyushkina). Note that integral billiards also realizes one spherical Seifert manifold not diffeomorphic to such a connected sum.

Sections 1.7.3 and 1.7.4 are devoted to realization of rank 0 singularities.

1.7.1 Brief Description of Integrable Billiards

For an elementary table $\Omega \subset \mathbb{R}^2(x, y)$ consider $T^*\Omega$ with the canonical projection on Ω. Let in coordinates (x, y, v_x, v_y) the symplectic form ω is standart and metrics on Ω is flat. Define motion via Hamiltonian field sgrad H with energy $H = v_x^2 + v_y^2$.

Standard elastic reflection from boundary determines factorization of $T^*\Omega$ on the set $\pi^{-1}(\partial\Omega)$. Equivalent pairs $(x, v) \in T^*\Omega$ for a point x of boundary smoothness $\partial\Omega$ are defined by following condition: $|v_1| = |v_2|$ and $v_1 - v_2 \perp T_x(\partial\Omega)$. In intersection point of two smooth boundary arcs (with $\pi/2$ angle) this rule is applied to each of the arcs. Connection between reflecton law and (local) integrability is discussed in great detail in [53]

Such a priori piece-wise smooth Hamiltonian system is integrable. Value of the billiard integral Λ on a trajectory is the value λ of the caustic (of this trajectory) as a curve of family 1.1.

Phase space of a system on billiard book is glued from phase spaces of its sheets or "pages" (that are elementary billiards) with factorization in the pre-image of their boundaries according to permutations. Condition $\sigma(i) = j$ on a gluing arc γ means that "inward" (in this arc of the table i) velocity vector should be identified with appropriate "outward" one on the sheet j.

Construction is well-defined (i.e. trajectory is continuous after passing through a boundary non-smoothness point) because of commutative condition $\sigma_1 \circ \sigma_2 = \sigma_1 \circ \sigma_2$ for every permutations on intersection gluing arcs.

1.7.2 Realization of 3-Atoms (rk = 1 Singularities)

For billiard without a potential Liouville foliations on Q_h^3 for all $h > 0$ are equivalent. Elliptic atoms A correspond to convex gluing arcs of the table Ω. Saddle atoms correspond to intervals of the focal line (x-axis) and non-convex arcs of gluing (both elliptic or hyperbolic arcs). Foliations of every table have singularity A, i.e. exactly saddle atom realization is nontrivial.

Hereinafter consider following standard table (billiard domain) Ω_0 which boundary consists of a two straight intervals (horizontal and vertical) and two arcs (one elliptic and one hyperbolic). Let correspond permutations $\tau_1, \tau_2, \sigma_1, \sigma_2 \in S_n$ to them respectively. Note that a permutation can be trivial.

In the paper [14] V.V. Vedyushkina and I.S. Kharcheva suggested how to build algorithmically a billiard book Ω_V for a fixed 3-atom (with stars or without them). Such book should consist of standard domains Ω_0 and have Liouville foliation in a neighbourhood of the level $\Lambda = b$ should have singularity of such atom type. Number of domains ("pages" of the book) depends on the complexity of atom and on the number of atom's stars

We formulate their construction in terms of f-graphs (in [14] it is described via atom's V doubles $\hat{V}$ and so-called semi-crosses).

If $l = 0$ (atom without stars) then 3-atom $V = V^3 \cong V^2 \times S^1$, where V^2 is a 2-atom of complexity k. Let its f-graph Γ (which directed cycles correspond to subcritical function levels) is determined by pair of permutations (σ, τ).

Theorem 1.7 (V.V. Vedyushkina, I.S. Kharcheva) *Billiard book consisting of* $2k$ *domains* Ω_0 *with permutations* $\sigma_2 = \tau_2 = id, \sigma_1 = \sigma, \tau_1 = \tau \in S_{2k}$ *will contain in its* Q^3_h *semi-local singularity* V *in a neighbourhood of the level* $\Lambda = b$*, i.e. models this 3-atom* V*.*

Remark 1.1 Independent cycle of length $s > 2$ in permutation τ_1 corresponds to a degenerate (non-Bott) singular circle of 3-atom V.

Now 3-atom V of complexity $k + l$ has l stars, i.e. it has l special fibers of (2, 1) type as a Seifert manifold. Then 2-atom with l marked points (stars) on the singular fiber corresponds to it.

Consider two items of its f-graph with stars Γ (a graph with $2k$ vertices of degree 3 and l vertices of degree 2) with the isomorphism ϕ of these graphs which preserves edge direction.

Construct a graph $\hat{\Gamma}$ adding non-directed edges (with mark $+1$, in general case, see [4]) between two vertices-stars $i, \phi(i)$. Resulting f-graph is realized by appropriate billiard book (according to previous theorem) that consists of $4k + 2l$ sheets and has non-trivial permutations σ_1, τ_1.

Isomorphism ϕ generates an involution $\hat{\tau}$ of the graph $\hat{\Gamma}$. It commutes with both τ and σ. Permutation ϕ maps independent cycles σ to each other (and will be an involution on orbits of σ). Every saddle-star corresponds to the same transpositions (j_1, j_2) in decompositions of both $\hat{\tau}$ and τ (into independent transpositions). Pairs $x, \phi(x)$ of the rest saddles correspond to pairs $(i_1, i_2)(i_3, i_4)$ and $(i_1, i_3)(i_2, i_4)$ in these decompositions of τ and $\hat{\tau}$ respectively.

Theorem 1.8 (V.V. Vedyushkina, I.S. Kharcheva) *Billiard book consisting of* $4k + 2L$ *sheets* Ω_0 *with permutations* $\sigma_2 = id, \sigma_1 = \sigma, \tau_1 = \tau, \tau_2 = \hat{\tau}$ *from* S_{4k+2l} *will contain in* Q^3_h *a semi-local singularity near the level* $\lambda = b$ *that is fiber-wise homeomorphic to 3-atom* V*, i.e. such book models 3-atom* V*.*

1.7.3 *Realization of 4-Dimensional Singularities of* **rk 0** *by Integrable Billiards with Hooke Potential*

In this subsection we discuss the realization problem of singularities of rank 0 using billiards formulated in [15]. Let us add a central elastic potential $k(x^2 + y^2)$ (later Hook potential) to the initial Hamiltonian $H_0 = v_x^2 + v_y^2$. In the paper [54] V.V. Kozlov proved integrability of such billiard in an ellipse for all real $k \in \mathbb{R}$ and discovered several other integrable generalizations. Easy to see that billiards on other elementary domains also remain integrable after adding such Hook potential.

In recent papers by I.F. Kobtsev [19] and S.E. Pustovoytov [20] phase topology was studied for billiards with Hooke potential for two domains: ellipse or an annulus between two confocal ellipses. Notice that bifurcation diagram without studying of phase topology for such system in ellipse with attracting potential was described in earlier in the paper [55].

In elliptic coordinates λ_1, λ_2 and conjugate momenta μ_1, μ_2 the Hamiltonian H and fisrt integral F are the following:

$$H = \frac{2(\lambda_1 + a)(\lambda_1 + b)}{\lambda_2 - \lambda_1}\mu_1^2 + \frac{2(\lambda_2 + a)(\lambda_2 + b)}{\lambda_1 - \lambda_2}\mu_2^2 + \frac{k}{2}(\lambda_1 + \lambda_2),$$

$$F = 2(\lambda_1 + a)(\lambda_1 + b)\mu_1^2 + \frac{k}{2}\lambda_1^2 - \lambda_1 H.$$

On the Fig. 1.4 neighbourhood of one singular point of Σ is shown. Projections (on the billiard table) of Liouville foliation fibers are shaded depending on values (h, f), close to the singular one. Arcs of Σ, denoted by a, b, c, d, decompose the neighbourhood of the singular point on four open sets I, II, III, IV. Also show projections of special trajectories for each of arcs a, b, c, d. Notice that trajectory is tangent two one or two caustics and reflects from boundary of the domain.

In another paper [56] I.S. Kobtsev constructed bifuracation diagram and calculated Fomenko–Zieschang invariant for the geodesic flow on a tree-axial ellipsoid in the repelling central Hooke potential in $\mathbb{R}^3$. Bifurcation diagram in $\mathbb{R}^2$ has the same structure (as for billiard in previous problem) but phase topology differs. Particularly, loop molecule of the point S will coincide with the loop molecule of $B \times C_2$ singularity (differs from $(B \times C_2)/\mathbb{Z}_2$ in the case of billiard inside an epllipse).

In the paper [10] V.A. Kibkalo noticed that non-degeneracy of rank 0 points of the geodesic flow on ellipsoid with a potential entails that non-degenerate singularities of this system are equivalent to corresponding singularities of a billiard system on the table glued by two ellipses along their common boundary.

Thus fiber-wise homeomorphism is true for both local and semi-local senses for singularities of center-center, center-saddle and saddle-saddle types. Notice that V.V. Vedyushkina and S.E. Pustovoytov later realized singularity of the rest focus-focus type using billiard bounded by a circle and equipped with repelling Hook potential.

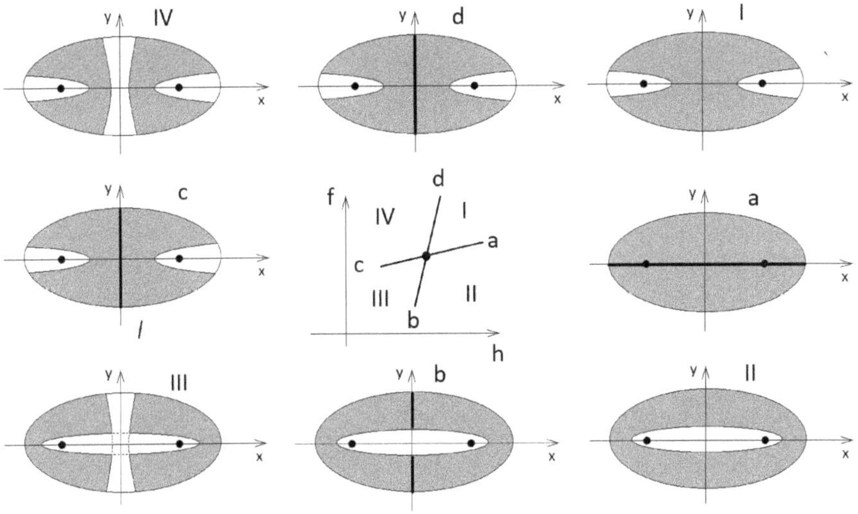

Fig. 1.4 Billiard in an ellipse with a repelling central Hook potential: projections of Liouville tori and critical trajectories for values (h, f), close to the singular point of Σ. Expected that the singularity in its pre-image has the type of saddle-saddle $(B \times C_2)/\mathbb{Z}_2$

These results follows that the problem of modeling of an arbitrary non-degenerate 4-singularity of the rank 0 by billiards with a potential deserves serious study. As discussed earlier, center-center singularity has already been realized on that two-ellipse table. Modeling an arbitrary singularity of center-saddle or saddle-saddle types is expected in the class of billiard books upon Ω_0 domain, similar of identical to the applied in Vedyushkina-Kharcheva algorithm. They will consist of $4k$ domains Ω_0 and be equipped by permutations $\tau_1, \tau_2, \sigma_1, \sigma_2$ (see Fig. 1.5). General statement on such modeling is not proved now but in the next subsection we show that in the case of complexity $k = 1$ it is possible to model all loop molecules of saddle-saddle singularities of complexity 1 (which in the smooth case determines such singularity uniquely).

1.7.4 Loop Molecules of Complexity 1 Saddle Singularities Are Realized in Billiards

Theorem 1.9 (V.A. Kibkalo) *Following billiard books $\Omega_1, \Omega_2, \Omega_3, \Omega_4$ upon domain Ω_0 with repelling Hooke potential realize singularities with the same loop molecules as saddle-saddle singularities of complexity 1: $B \times B, (B \times C_2)/\mathbb{Z}_2, (B \times D_1)/\mathbb{Z}_2, (C_2 \times C_2)/(\mathbb{Z}_2 \oplus \mathbb{Z}_2)$ respectively. Notice that for complexity 1 singularities of smooth IHS coincidence of loop molecules means semi-local equivalence of 4-singularities.*

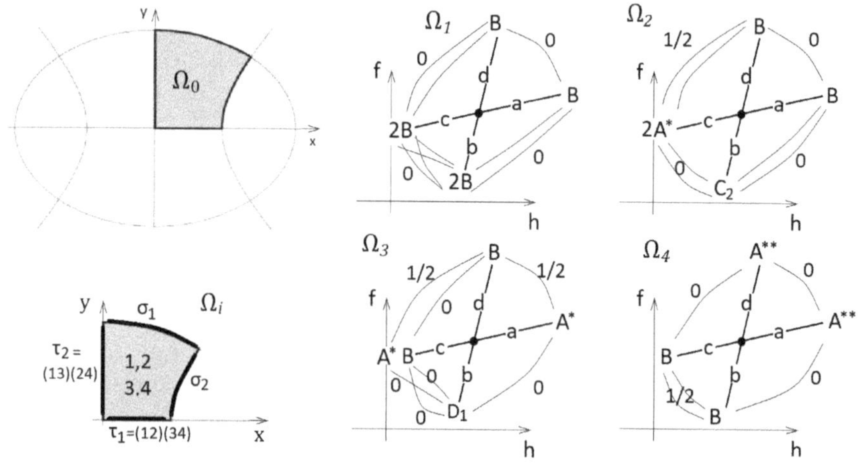

Fig. 1.5 Domain Ω_0, billiard books Ω_i, $1 \leq i \leq 4$ consisting of 4 sheets Ω_0 and loop molecules with r-marks of saddle-saddle singularities of their Liouville foliation

Constructed billiard books consist of four domains-sheets and their permutations $\tau_1 = (12)(34)$, $\tau_2 = (13)(24)$ *and* σ_1, σ_2 *are the following:*

- $\Omega_1 :\ B \times B :\quad \sigma_1 = \sigma_2 = id,$
- $\Omega_2 :\ (B \times C_2)/\mathbb{Z}_2 :\quad \sigma_1 = id,\ \sigma_2 = (12)(34),$
- $\Omega_3 :\ (B \times D_1)/\mathbb{Z}_2 :\quad \sigma_1 = id,\ \sigma_2 = (12),$
- $\Omega_4 :\ (C_2 \times C_2)/(\mathbb{Z}_2 \oplus \mathbb{Z}_2) :\quad \sigma_1 = \sigma_2 = (14)(23).$

2. Coarse Liouville equivalence can be expanded to the coincidence of marks r on edges of corresponding molecules. In this case it does not depend on edge directions and orientation of the loop manifold. Configuration of the singularity in space in projection, 3-atoms and marks r on edges connect them are shown on the Fig. 1.5.

Proof 1 1. Billiard system Ω_i with Hooke potential has the same 3-atom on the arc $s \in \{a, b, c, d\}$ of bifurcation diagram as the billiard without potential on the following book $\tilde{\Omega}_i$ near the level $\lambda = b$.
For the arc a book $\tilde{\Omega}_i = \Omega_i$. For the arc c they also coincide excepting $\tilde{\tau_2} = \text{id}$. For the arc d the book $\tilde{\Omega}_i$ is the result of "reflection" of $\Omega_i : \tilde{\sigma_1} = \sigma_2, \tilde{\tau_2} = \tau_1$, etc. For the arc b we need the same reflection with change $\tilde{\tau_2} = \text{id}$. So we can consider $\tilde{\Omega}_i$ instead of Ω_i for simplifying the proof.

2. Consider a book $\tilde{\Omega}_0$ and neighbourhood $X = \Lambda^{-1}([b - \varepsilon, b + \varepsilon])$ in Q^3_h that is a 3-atom of Liouville foliations. Pre-image of a hyperbola that intersects $\text{Int}(\Omega_0)$ in X consists of two same 2-atoms W_l, W_r equipped with velocity vectors directed to the "left" or "right" (this approach [57]) respectively. Generally speaking, 2-atoms W_l, W_r can be not connected, i.e. be disjoint unions of several 2-atoms. Their f-graph is determined by following pairs of permutations $\tilde{\sigma_1}, \tilde{\tau_1}$. For four books Ω_i they can have one of the following types: $B, 2B, C_2, D_1$.

For example, in the cases of Ω_1, Ω_4 every arc corresponds to 2-atoms $2B$ and C_2 respectively. In the case of Ω_2 such 2-atom is either C_2 for arcs b, d or $2B$ for arcs a, c.

3. Special trajectories of the particle, i.e. saddle singular circles of 3-atoms, are projected on an interval of the focal (horizontal) axis from $\partial\Omega_0$. These circles are glued from arcs that correspond to independent circles of $\tilde{\tau_1}$ equipped with "left" or "right" velocity vectors. Their gluing is determined by permutations $\tilde{\sigma_2}, \tilde{\tau_2}$.
4. Each of permutations $\tilde{\sigma_2}, \tilde{\tau_2}$ generates map from W_l to W_r and vice versa. Consider their composition $\tilde{\sigma_2} \circ \tilde{\tau_2}$ that 2-atom to itself. It uniquely determines 3-atom as a Seifert fibration upon 2-atom. It corresponds to the motion of a particle reflecting once from the vertical interval and once from the hyperbolic arc of $\partial\Omega_1$. Resulting 3-atoms have types of $B, 2B, C_2, D_1, A^*, 2A^*, A^* \sqcup B, A^{**}$. For example, for the arc a and for the book Ω_4 the composition $\tilde{\sigma_2} \circ \tilde{\tau_2} = (12)(34)$, i.e. group Z_2 acts on 2-atom C_2 such that neighbourhood of the singular circle is nontrivially mapped to itself. It means that the 3-atom has type of A^{**} with 2 stars, i.e. 3-dimensional Seifert manifold with 2 singular fibers of (2, 1) type.
5. Denominator q of the mark $r = p/q$ on an edge of a loop molecule coincides with the index of intersection of λ-cycles of two incidental 3-atoms. If $q \leq 2$ and cycles do not homologically coincide then the mark is equal to either 0 or $1/2$, particularly it does not depend on the direction of the edge of loop molecule.

   Cycle λ in a 3-atom is a regular fiber of its Seifert fibration. Thus the projection of cycle λ of 3-atom (i.e. a cycle on a boundary torus of this atom) is parallel (in the sense of the book Ω_i) to either vertical or horizontal axis depending on the projection of the critical circle of the 3-atom. These cycles consist of pairs "point-arrow" of a point of the billiard book Ω_i and an arrow $\uparrow, \downarrow, \rightarrow, \leftarrow$ that denotes type of projection on axis and the direction of motion.

Let us equip the cycles that are parallel to some axis with an arrow in a positive direction in the sense of the other coordinate axis. As the result we have triples "point-arrow-arrow" that determine a curve on a 4-sheet covering of the book Ω_i. Let us denote them by arrows $\nearrow, \nwarrow, \searrow, \swarrow$. Intersection of two cycles takes place if two pairs "sheet-arrow" coincide.

In the case of arcs a and d of the book Ω_4 the cycles λ_a and λ_d are coded as follows: $(1, \leftarrow), (4, \rightarrow), (2, \leftarrow), (3, \rightarrow)$ and $(1, \uparrow), (4, \downarrow), (3, \uparrow), (2, \downarrow)$ respectively. After second equipping we have cycles:

$$\lambda_a = ((1, \nwarrow), (4, \nearrow), (2, \nwarrow), (3, \nearrow)), \qquad \lambda_d = ((1, \nearrow), (4, \searrow), (3, \nearrow), (2, \searrow)).$$

They have the only intersection, i.e. index is equal to 1, and the mark $r = 0$.

In the case of arcs b and c it is possible to take any of two following cycles that pass by sheets 1, 4 or 2, 3 of the book (permutation $\tilde{\sigma_1} = (14)(23)$ maps them to each other). After second equipping we have

$$\lambda_b = ((1, \nearrow), (4, \searrow), (4, \nearrow), (3, \searrow)), \qquad \lambda_c = ((1, \nwarrow), (1, \nearrow), (4, \nwarrow), (4, \nearrow)).$$

They have exactly two intersections, the index is equal to 2, i.e. the mark $r = 1/2$.

Remark 3 It is known that two f-graphs (i.e. two pairs of permutations) correspond to every 2-atom. Permutations σ_1 or σ_2 in the books Ω_i can be not only reflected but also changed to another permutations of the same 2-atoms. Such process for f_n-graphs and saddle-saddle singularities of IHS was described by A.A. Oshemkov in great detail in [9]. Particularly, it will change the correspondence between 3-atoms and arcs of bifurcation diagram.

References

1. Bolsinov, A.V., Fomenko A.T.: Integrable Hamiltonian systems. Geometry, topology, classification. vol. I, II. Publishing House Udmurtia University, Izhevsk (1999) (in Russian)
2. Bolsinov, A.V., Fomenko A.T.: Integrable Hamiltonian systems. Geometry, Topology, Classification. Chapman & Hall/CRC, Boca Raton, FL (2004)
3. Duistermaat, J.J.: On global action-angle coordinates. Comm. Pure Appl. Math. **33**(6), 687–706 (1980)
4. Oshemkov, A.A.: Morse functions on two-dimensional surfaces. Encoding of singularities, Proc. Steklov Inst. Math. **205**, 119–127 (1995)
5. Fomenko, A.T.: The symplectic topology of completely integrable Hamiltonian systems, Russian Math. Surveys **44**(1) (265), 181–219 (1989)
6. Matveev, V.S.: Integrable Hamiltonian system with two degrees of freedom. The topological structure of saturated neighbourhoods of points of focus-focus and saddle-saddle type, Sb. Math. **187**(4), 495–524 (1996)
7. Zung, N.T.: Symplectic topology of integrable Hamiltonian systems. I: Arnold-Liouville with singularities. Compos. Math. **101**(2), 179–215 (1996)
8. Bolsinov, A.V., Oshemkov, A.A.: Singularities of integrable Hamiltonian systems, Topological methods in the theory of integrable systems, pp. 1–67. Cambridge Scientific Publishing, Cambridge (2006)
9. Oshemkov, A.A.: Classification of hyperbolic singularities of rank zero of integrable Hamiltonian systems. Sb. Math. **201**(8), 1153–1191 (2010)
10. Kibkalo, V.A.: Billiards with potential model four-dimensional singularities of integrable systems, Contemporary problems of mathematics and mechanics. In: International scientific conference Contemporary problems of mathematics and mechanics, dedicated to the 80th anniversary of academician V.A. Sadovnichii. Books of abstracts, vol. 2, pp. 563–566, Moscow (2019) (Russian)
11. Grabezhnoi, A.V.: Invariants of the Liouville foliation for 4-dimensional singularities of saddle-saddle type. Graduate Thesis, Moscow University, Moscow (2005) (Russian)
12. Oshemkov, A.A., Tuzhilin, M.A.: Integrable perturbations of saddle singularities of rank 0 of integrable Hamiltonian systems. Sb. Math. **209**(8), 1351–1375 (2018)
13. Oshemkov, A.A.: Saddle singularities of complexity 1 of integrable Hamiltonian systems. Moscow Univ. Math. Bull. **66**(2), 60–69 (2011)
14. Vediushkina, V.V., Kharcheva, I.S.: Billiard books model all three-dimensional bifurcations of integrable Hamiltonian systems. Sb. Math. **209**(12), 1690–1727 (2018)
15. Fomenko, A.T., Vediushkina, V.V.: Billiards and integrability in geometry and physics. New scope and new potential, Moscow Univ. Math. Bull. **74**(3), 98–107 (2019)
16. Kharcheva, I.E.: Isoenergy manifold of integrable billiard books, Moscow University Mathematics Bulletin (in press)

17. Fokicheva, V.V., Fomenko, A.T.: Integrable billiards model important integrable cases of rigid body dynamics. Dokl. Math. **92**(3), 682–684 (2015)
18. Fomenko, A.T., Vedyushkina (Fokicheva), V.V.: Integrable topological billiards and equivalent dynamical systems. Izv. Math. **81**(4), 688–733 (2017)
19. Kobtsev, I.F.: Elliptic billiard in field of potential forces: classification of motions, topological analysis. Sb. Math. 211, 93–120 (in press)
20. Pustovoytov, S.E.: Topological analysis of billiard in elliptic ring in field of potential forces. Fund. Appl. Math. **22**, 201–225 (2019) (in Russian) translation: in Journal of Mathematical Sciences
21. Eliasson, L.H.: Normal forms for Hamiltonian systems with Poisson commuting integrals - elliptic case. Comm. Math. Helv. **65**, 4–35 (1990)
22. Fomenko, A.T.: The topology of surfaces of constant energy in integrable Hamiltonian systems, and obstructions to integrability. Math. USSR-Izv. **29**(3), 629–658 (1987)
23. Fomenko, A.T., Matveev, S.V.: Algorithmical and computer methods in three-manifolds, p. 303. MSU Publishing, Moscow (1991) (In Russian)
24. Fomenko, A.T., Matveev, S.V.: Algorithmical and computer methods in three-manifolds. Kluwer Academic Publishers, The Netherlands (1997) (translation)
25. Fomenko, A.T., Zieschang, H.: On typical topological properties of integrable Hamiltonian systems. Math. USSR-Izv. **32**(2), 385–412 (1989)
26. Fomenko, A.T., Zieschang, H.: A topological invariant and a criterion for the equivalence of integrable Hamiltonian systems with two degrees of freedom. Math. USSR-Izv. **36**(3), 567–596 (1991)
27. Fomenko, A.T.: Morse theory of integrable Hamiltonian systems. Soviet Math. Dokl. **33**(2), 502–506 (1986)
28. Kudryavtseva, E.A., Nikonov, I.M., Fomenko, A.T.: Maximally symmetric cell decompositions of surfaces and their coverings. Sb. Math. **199**(9), 1263–1353 (2008)
29. Zung, N.T.: Symplectic topology of integrable Hamiltonian systems. I : Arnold-Liouville with singularities (2001). arXiv:math/0106013
30. Bolsinov, A.V., Richter, P., Fomenko, A.V.: The method of loop molecules and the topology of the Kovalevskaya top. Sb. Math. **191**(2), 151–188 (2000)
31. Morozov, P.V.: Topology of Liouville foliations in the Steklov and the Sokolov integrable cases of Kirchhoff's equations. Sb. Math. **195**(3), 369–412 (2004)
32. Morozov, P.V.: The Liouville classification of integrable systems of the Clebsch case. Sb. Math. **193**(10), 1507–1533 (2002)
33. Morozov, P.V.: Calculation of the Fomenko-Zieschang invariants in the Kovalevskaya-Yehia integrable case. Sb. Math. **198**(8), 1119–1143 (2007)
34. Logacheva, N.S.: Classification of nondegenerate equilibria and degenerate 1-dimensional orbits of the Kovalevskaya-Yehia integrable system. Sb. Math. **203**(1), 28–59 (2012)
35. Slavina, N.S.: Topological classification of systems of Kovalevskaya-Yehia type. Sb. Math. **205**(1), 101–155 (2014)
36. Kibkalo, V.A.: Topological analysis of the Liouville foliation for the Kovalevskaya integrable case on the Lie algebra so(4). Lobachevskii J. Math. **39**(9), 1396–1399 (2018)
37. Kibkalo, V.A.: Topological classification of Liouville foliations for the Kovalevskaya integrable case on the Lie algebra so(4). Sb. Math. **210**(5), 625–662 (2019)
38. Kibkalo, V.A.: Topological classification of Liouville foliations for the Kovalevskaya integrable case on the Lie algebra so(3, 1). Topol. Appl. (2020). https://doi.org/10.1016/j.topol.2019.107028
39. Tuzhilin, M.A.: Singularities of integrable Hamiltonian systems with the same boundary foliation. An infinite series. Moscow Univ. Math. Bull. **71**(5), 185–190 (2016)
40. Kharlamov, M.P.: Bifurcation of common levels of first integrals of the Kovalevskaya problem, J. Appl. Math. Mech. **47**(6), 737–743 (1983)
41. Kharlamov, M.P.: Topological analysis of classical integrable systems in the dynamics of the rigid body. Soviet Math. Dokl. **28**(3), 802–805 (1983)

42. Kozlov, I.K.: The topology of the Liouville foliation for the Kovalevskaya integrable case on the Lie algebra so(4). Sb. Math. **205**(4), 532–572 (2014)
43. Oshemkov, A.A.: Fomenko invariants for the main integrable cases of rigid body motion equations. AMS **4**, 67–146 (1991)
44. Ryabov, P.E.: Bifurcations of first integrals in the Sokolov case. Theor. Math. Phys. **134**, 181–197 (2003)
45. Kharlamov, M.P., Ryabov, P.E., Savushkin, AYu.: Topological Atlas of the Kowalevski-Sokolov top. Regul. Chaotic Dyn. **21**(1), 24–65 (2016)
46. Ryabov, P.E.: Bifurcation sets in an integrable problem on motion of a rigid body in fluid. Regul. Chaot. Dyn. **4**(4), 59–76 (1999)
47. Birkhoff, G.D.: Dynamical systems. American Mathematical Society Colloquium Publications, vol. 9. AMS, New York (1927)
48. Fomenko, A.T., Vedyushkina, V.V.: Topological obstacles to the realizability of integrable Hamiltonian systems by billiards. Dokl. Math. **488**, 471–475 (2019)
49. Vedyushkina, V.V.: Liouville foliation of billiard book modeling Goryachev-Chaplygin case. Moscow Univ. Math. Bull. **75**, 1 (2020)
50. Waldhausen, F.: Eine Klasse von 3-dimensionalen Mannigfaltighkeiten. I. Invent. Math. **3**(4), 308–333 (1967)
51. Waldhausen, F.: Eine Klasse von 3-dimensionalen Mannigfaltighkeiten. II. Invent. Math. **4**(2), 88–117 (1967)
52. Fomenko, A.T., Vedyushkina, V.V.: Integrable geodesic flows on orientable two-dimensional surfaces and topological billiards. Izv. RAN. Ser. Mat. **83**, 1137–1173 (2019)
53. Kudryavtseva, E.A.: Liouville integrable generalized billiard flows and Poncelet type theorems. J. Math. Sci. **225**(4), 611–638 (2017)
54. Kozlov, V.V.: Some integrable generalizations of the Jacobi problem on geodesics on an ellipsoid. J. Appl. Math. Mech. **59**, 1–7 (1995)
55. Il'inskaya, N.N.: Geometric analysis of a problem on a harmonic oscillator in an ellipse. Moscow Univ. Math. Bull. **1991**(1), 88–92 (1991)
56. Kobtsev, I.F: Geodesic flow of a 2D ellipsoid in an elastic stress field: topological classification of solutions. Moscow Univ. Math. Bull. **73**(2), 64–70 (2018)
57. Fokicheva (Vedyushkina), V.V.: A topological classification of billiards in locally planar domains bounded by arcs of confocal quadrics. Sb. Math. **206**(10), 1463–1507 (2015)

Chapter 2
Reduction of the Lamé Tensor Equations to the System of Non-Coupled Tetraharmonic Equations

D. V. Georgievskii

Abstract Reductions of the systems of equations in terms of displacements in 3D elasticity to the systems of higher orders based on the operators which are suitable for both numerical and analytical investigation better than the Lamé operator, are called representations of the solution. The Galerkin representation is one of such typical procedures in classical theory of elasticity. Below the Galerkin procedure is generalized in conformity to the systems generated by linear symmetric tensor (of the second rank) differential operator of the forth order acting on symmetric tensor field. Reduction of such systems to the systems of non-coupled tetraharmonic equations is realized. The fundamental solutions of the derived tetraharmonic equations in many-dimensional spaces are given.

2.1 Introduction

In tensor analysis all linear differential operators by coordinates in n-dimensional space are representable as superpositions of the following first order simplest operators: *divergence* div (or Div if it is applied to a tensor of rank $m > 1$), *curl* rot (or Rot if $m > 1$) and *gradient* grad (or Grad if $m > 0$). Each of them corresponds to some kind of multiplication (scalar, vector, or dyad) of the operator nabla by vector or tensor. The operator divergence decreases by one a rank of the applied object, the operator gradient increases it by one, whereas the operator curl does it equal to $n - m - 1$. The second order superpositions of the simplest operators such as the Laplace operator div grad (or Div Grad), grad div (or Grad Div), rot rot (or Rot Rot) do not change a rank.

D. V. Georgievskii (✉)
Lomonosov Moscow State University, GSP-1, Leninskie Gory, Moscow 119991, Russian Federation
e-mail: georgiev@mech.math.msu.su

V. A. Sadovnichiy and M. Z. Zgurovsky (eds.), *Contemporary Approaches and Methods in Fundamental Mathematics and Mechanics*, Understanding Complex Systems,
https://doi.org/10.1007/978-3-030-50302-4_2

2.2 The Lamé Vector Equations and the Galerkin Representation in Isotropic Elasticity

General form of a linear isotropic differential operator of the second order acting on a vector field $\mathbf{u}(\mathbf{x})$, $\mathbf{x} \in R^n$, $n \geq 2$, is the following

$$\check{\mathbf{L}}_{(a_1;a_2)}\mathbf{u} = a_1 \operatorname{grad} \operatorname{div} \mathbf{u} + a_2 \operatorname{Div} \operatorname{Grad} \mathbf{u}, \quad a_1, a_2 = \text{const}. \tag{2.1}$$

So, for example rot rot $= \check{\mathbf{L}}_{(1;-1)}$. In Cartesian components the equality (2.1) may be written as

$$\check{L}_{(a_1;a_2)i}\, \mathbf{u} = a_1\, u_{j,ji} + a_2\, u_{i,jj}. \tag{2.2}$$

Here and further all subscripts vary from 1 to n; if two subscripts are repeated in one monomial then the summation by them is taken within the same limits. Comma in subscript means partial differentiating with respect to corresponding coordinates.

Let we have not uniform system of n PDE

$$\check{L}_{(a_1;a_2)i}\, \mathbf{u} + f_i = 0 \tag{2.3}$$

with the known functions $f_i(\mathbf{x}) \in C_2$. We realize the following sequence of transforms on the system (2.3).

1°. Partial differentiating of (2.3) with respect to x_i and summation by i

$$(a_1 + a_2)\Delta\theta = -f_{i,i}, \quad \theta = \operatorname{div} \mathbf{u}. \tag{2.4}$$

2°. Double partial differentiating of (2.3) with respect to x_k and summation by k. With regard for (2.4) we receive the system of n non-coupled biharmonic equations

$$u_{i,jjkk} = \frac{1}{a_2}\Big(\frac{a_1}{a_1 + a_2} f_{j,ji} - f_{i,jj}\Big) \tag{2.5}$$

or in the indexless form

$$\Delta^2\mathbf{u} = \check{\mathbf{L}}_{(\alpha_1;\alpha_2)}\mathbf{f}, \quad \alpha_1 = \frac{a_1}{a_2(a_1 + a_2)}, \quad \alpha_2 = -\frac{1}{a_2}. \tag{2.6}$$

Search of solution of (2.6) as

$$\mathbf{u} = \check{\mathbf{L}}_{(\alpha_1;\alpha_2)}\varGamma \tag{2.7}$$

with respect to the new unknown vector $\varGamma(\mathbf{x})$ results in the following not uniform biharmonic equations

$$\Delta^2\varGamma = \mathbf{f} \tag{2.8}$$

being the sufficient conditions for fulfilment of (2.6) by virtue of the operators Δ and $\check{\mathbf{L}}_{(\alpha_1;\alpha_2)}$. As is obvious that above procedure of reduction to (2.8) may be realized if

$$a_2 \neq 0, \quad a_1 + a_2 \neq 0. \tag{2.9}$$

In 3D isotropic elasticity the vector $\mathbf{u}(\mathbf{x})$ has meaning displacements, the operator $\check{\mathbf{L}}_{(a_1;a_2)}$ when $a_1 = \lambda + \mu, a_2 = \mu$, makes sense the Lamé operator (λ and μ are Lamé constants), and the Eq. (2.3) are the equations of equilibrium in terms of displacements with the volume forces $\mathbf{f}$. The operator $\check{\mathbf{L}}_{(\alpha_1;\alpha_2)}$ where $\alpha_1 = 1/[2\mu(1-\nu)]$, $\alpha_2 = -1/\mu$, is called the Galerkin vector operator; a search of solution in the form (2.7) is said to be the Galerkin representation; the vector $\Gamma(\mathbf{x})$ is called the Galerkin vector (see, for example [1]). The generalization of the Galerkin representation in relation to transversal isotropic elasticity was created in [2].

Without loss of generality by any constants a_1 and a_2 complying with (2.9) we should call $\check{\mathbf{L}}_{(a_1;a_2)}$ and $\check{\mathbf{L}}_{(\alpha_1;\alpha_2)}$ (here α_1 and α_2 are connected with a_1 and a_2 by (2.6)) the Lamé vector operator and the Galerkin vector operator corresponding to $\check{\mathbf{L}}_{(a_1;a_2)}$.

2.3 Linear Differential Systems of the Forth Order for the Tensors of the Second Rank

Let us consider now the general view of a linear symmetric tensor (of the second rank) differential operator of the forth order acting on some symmetric tensor field $\mathbf{u}(\mathbf{x})$, $\mathbf{x} \in R^n$, $n \geq 2$:

$$\check{L}_{(b_1;b_2;b_3;b_4)ij}\,\mathbf{u} \equiv \check{L}_{(\mathbf{b})ij}\,\mathbf{u} = b_1 u_{kl,lkij} + b_2 u_{kk,llij} +$$

$$+ b_3(u_{il,lkkj} + u_{jl,lkki}) + b_4 u_{ij,kkll}, \quad b_1, b_2, b_3, b_4 = \text{const} \tag{2.10}$$

or in the indexless form

$$\check{\mathbf{L}}_{(\mathbf{b})} = \text{Grad grad}\,(b_1 \text{div Div} + b_2\, \Delta \text{tr}) +$$

$$+ \text{Div Grad}\,[b_3 \,\text{Grad Div} + b_3 (\text{Grad Div})^T + b_4 \,\text{Div Grad}\,]. \tag{2.11}$$

Notation in (2.10) and (2.11) and further are chosen for reasons of analogy with the previous section. So we should call $\check{\mathbf{L}}_{(\mathbf{b})}$ the Lamé tensor operator.

Let us write the following not uniform system of $n(n+1)/2$ differential equations

$$\check{L}_{(\mathbf{b})ij}\,\mathbf{u} + f_{ij} = 0, \quad f_{ij} \in C_4 \tag{2.12}$$

and realize the certain sequence of transformations on it which seems to be much more bulky than before.

1°. The operation of trace in (2.12).

$$(b_1 + 2b_3)\Delta\vartheta + (b_2 + b_4)\Delta^2 u + f = 0, \tag{2.13}$$

where $u = \operatorname{tr}\mathbf{u}$; $\vartheta = \operatorname{div}\operatorname{Div}\mathbf{u} = u_{kl,lk}$; $f = \operatorname{tr}\mathbf{f}$; $F = \operatorname{div}\operatorname{Div}\mathbf{f} = f_{kl,lk}$. It should be noted that one more scalar relation

$$(b_1 + 2b_3)\Delta^2\vartheta + (b_2 + b_4)\Delta^3 u + \Delta f = 0 \tag{2.14}$$

follows from (2.13).

2°. Double partial differentiating of (2.12) with respect to x_i, x_j and summation by i and j

$$(b_1 + 2b_3 + b_4)\Delta^2\vartheta + b_2\Delta^3 u + F = 0. \tag{2.15}$$

The invariants $\Delta^2\vartheta$ and $\Delta^3 u$ may be determined from the system of Eqs. (2.14) and (2.15):

$$\Delta^2\vartheta = \frac{1}{b_4 B}[b_2\Delta f - (b_2 + b_4)F], \quad \Delta^3 u = \frac{1}{b_4 B}[(b_1 + 2b_3)F - (b_1 + 2b_3 + b_4)\Delta f], \tag{2.16}$$

where $B = b_1 + b_2 + 2b_3 + b_4$.

3°. Let us: (a) assume in (2.12) $j = m$ and differentiate with respect to x_m, x_j with summation by m; (b) assume in (2.12) $i = m$ and differentiate with respect to x_m, x_i with summation by m; (c) add the obtained two tensor relations:

$$2(b_1 + b_3)\Delta\vartheta_{,ij} + 2b_2\Delta^2 u_{,ij} +$$

$$+ (b_3 + b_4)(u_{im,mj} + u_{jm,mi})_{,kkll} + f_{im,mj} + f_{jm,mi} = 0. \tag{2.17}$$

Applying the Laplace operator to both the hands of (2.17) one can express the invariants $\Delta^2\vartheta$ and $\Delta^3 u$ from (2.16):

$$(u_{im,mj} + u_{jm,mi})_{,kkllpp} = \frac{2}{b_4(b_3 + b_4)B}\Big[b_2(b_3 + b_4)\Delta f_{,ij} -$$

$$- (b_1 b_4 + b_3 b_4 - b_2 b_3)F_{,ij}\Big] - \frac{1}{b_3 + b_4}(f_{im,mj} + f_{jm,mi})_{,pp}. \tag{2.18}$$

4°. Double partial differentiating of (2.12) with respect to x_m and then one more double partial differentiating of (2.12) with respect to x_p with summation by m and p. The expressions by the coefficients b_1, b_2 and b_3 in the resulting relation are already known from (2.16) and (2.18). This allows to write $\Delta^4 u_{ij}$:

$$\Delta^4 u_{ij} = \beta_1 F_{,ij} + \beta_2\Delta f_{,ij} + \beta_3(f_{il,lj} f_{jl,li})_{,kk} + \beta_4 f_{ij,kkll} \equiv \check{L}_{(\beta)ij}\,\mathbf{u}, \tag{2.19}$$

where

$$\beta_1 = \frac{b_1(b_4-b_3) - 2b_3(b_2+b_3)}{b_4(b_3+b_4)B}, \quad \beta_2 = \frac{b_2}{b_4 B}, \quad \beta_3 = \frac{b_3}{b_4(b_3+b_4)}, \quad \beta_4 = -\frac{1}{b_4} \tag{2.20}$$

$$b_4 \neq 0, \quad b_3 + b_4 \neq 0, \quad b_1 + b_2 + 2b_3 + b_4 \neq 0. \tag{2.21}$$

In this way we have established reduction of the system (2.12) to the system (2.19) containing the same number of not uniform non-coupled tetraharmonic equations relative to the components $u_{ij}(\mathbf{x})$. The known right hands in (2.19) represent the components of the Galerkin tensor operator $\check{\mathbf{L}}_{(\beta)}$ which is applied to $\mathbf{f}(\mathbf{x})$. This operator is called to be corresponding to the Lam'e tensor operator $\check{\mathbf{L}}_{(\mathbf{b})}$. Connections of the constants b_1, b_2, b_3, b_4 and β_1, β_2, β_3, β_4 is defined by the formulae (2.20) with conditions (2.21).

Various aspects concerning general representation for solutions of elliptic equations of high orders, in particular, polyharmonic equations as well as their elementary solutions and the Green function, are contained in extensive literature including classical monographs [3–5] and articles of recent years [6–9].

By analogy with (2.7) search of solution of (2.19) as

$$\mathbf{u} = \check{\mathbf{L}}_{(\beta)} \Gamma \tag{2.22}$$

relative to new unknown Galerkin tensor $\Gamma(\mathbf{x})$ leads to the following not uniform tetraharmonic equations

$$\Delta^4 \Gamma = \mathbf{f} \tag{2.23}$$

which are sufficient conditions for compliance of the Eq. (2.19).

We note here one particular case when the tensor $\mathbf{f}(\mathbf{x})$ is spherical i.e. $\mathbf{f}(\mathbf{x}) = f_0(\mathbf{x})\mathbf{I}$ where $\mathbf{I}$ is the identity tensor of the second rank. Then the Galerkin tensor should be sought as $\Gamma(\mathbf{x}) = \Gamma_0(\mathbf{x})\mathbf{I}$ and also $\Delta^4\Gamma_0 = f_0$ on the basis of (2.23). The required tensor $\mathbf{u}(\mathbf{x})$ is connected with scalar $\Gamma_0(\mathbf{x})$ in the following way (see (2.22))

$$u_{ij} = (\beta_1 + n\beta_2 + 2\beta_3)\Delta\Gamma_{0,ij} + \beta_4\Delta^2\Gamma_0\delta_{ij}. \tag{2.24}$$

2.4 Fundamental Solutions of Tetraharmonic Equations in Many-Dimensional Spaces

Let us assume that $f_0(\mathbf{x}) = p\delta(\mathbf{x})$ where $\delta(\mathbf{x})$ is delta-function, p is some constant value. Using that the functions r^{2-n} when $n \geq 3$ and $\ln r$ when $r = 2$ ($r = \sqrt{x_j x_j}$) are the fundamental solutions of the Poisson equation in R^n we construct solutions $\Gamma_0(r)$ of not uniform tetraharmonic equation and after that obtain the functions u_{ij}

according to (2.24). These functions may be considered in some way as analogs of the Kelvin solution in isotropic elasticity. *1°*. $n \neq 2, 4, 6, 8$.

$$\Gamma_0(r) = \frac{pr^{8-n}}{48(8-n)(6-n)(4-n)(2-n)S_n}$$

$$u_{ij} = \frac{p}{8(4-n)(2-n)S_n}\big[(\beta_1 + n\beta_2 + 2\beta_3)\big((4-n)r^{2-n}x_i x_j + r^{4-n}\delta_{ij}\big) + 4\beta_4 r^{4-n}\delta_{ij}\big]$$

where S_n is the surface area of the unit n-dimensional sphere. It is obvious that $u_{ij}|_{r\to 0} \to 0$ in R^3 and $u_{ij}|_{r\to 0} \to \infty$ for other n under consideration in this item.

1°. $n = 2$.

$$\Gamma_0(r) = \frac{pr^6}{4608\pi}\Big(\ln r - \frac{11}{6}\Big)$$

$$u_{ij} = \frac{p}{128\pi}\big[(\beta_1 + 2\beta_2 + 2\beta_3)\big(2(4\ln r - 3)x_i x_j + r^2(4\ln r - 5)\delta_{ij}\big) +$$

$$+16\beta_4 r^2(\ln r - 1)\delta_{ij}\big].$$

2°. $n = 4$.

$$\Gamma_0(r) = -\frac{pr^4}{768 S_4}\Big(\ln r - \frac{7}{6}\Big)$$

$$u_{ij} = -\frac{p}{32 S_4}\Big[(\beta_1 + 4\beta_2 + 2\beta_3)\Big(\frac{2}{r^2}x_i x_j + \Big(2\ln r - \frac{1}{2}\Big)\delta_{ij}\Big) + 8\beta_4 \ln r\,\delta_{ij}\Big].$$

3°. $n = 6$.

$$\Gamma_0(r) = \frac{pr^2}{768 S_6}\Big(\ln r - \frac{2}{3}\Big)$$

$$u_{ij} = \frac{p}{64 S_6 r^2}\Big[(\beta_1 + 6\beta_2 + 2\beta_3)\Big(-\frac{2}{r^2}x_i x_j + \delta_{ij}\Big) + 4\beta_4\delta_{ij}\Big].$$

4°. $n = 8$.

$$\Gamma_0(r) = -\frac{p}{2304 S_8}\ln r$$

$$u_{ij} = \frac{p}{192 S_8 r^4}\Big[(\beta_1 + 8\beta_2 + 2\beta_3)\Big(-\frac{4}{r^2}x_i x_j + \delta_{ij}\Big) + 4\beta_4\delta_{ij}\Big].$$

Acknowledgements The work has been supported by Russian Fund for Fundamental Research (projects 18-29-10085 mk and 19-01-00016 a).

References

1. Nowacki, W.: Teoria Sprezystosci. PAN, Warszawa (1970)
2. Georgievskii, D.V.: Procedure of the Galerkin representation in transversely isotropic elasticity. In: Sadovnichiy, V.A., Zgurovsky, M.Z. (eds.) Modern Mathematics and Mechanics. Series Understanding Complex Systems, pp. 117–124. Springer, Berlin (2019)
3. Vekua, I.N.: New Methods for Solving Elliptic Equations. North-Holland Series in Applied Mathematics and Mechanics (1967)
4. Sobolev, S.L.: Cubature Formulas and Modern Analysis: An Introduction. Gordon and Breach Science Publishers, Montreux (1992)
5. Vladimirov, V.S.: Equations of Mathematical Physics. Mir Publisher, Moscow (1984)
6. Yarmukhamedov, S.: The Cauchy problem for polyharmonic equations. Dokl. Math. **67**(1), 27–30 (2003)
7. Karachik, V.V., Antropova, N.A.: On the solution of the inhomogeneous polyharmonic equation and the inhomogeneous Helmholtz equation. Differ. Equ. **46**(3), 387–399 (2010)
8. Kalmenov, TSh, Suragan, D.: Boundary conditions for the volume potential for the polyharmonic equation. Differ. Equ. **48**(4), 604–608 (2012)
9. Georgievskii, D.V.: The Galerkin tensor operator, reduction to tetraharmonic equations, and their fundamental solutions. Dokl. Phys. **60**(8), 364–367 (2015)

Chapter 3
Junction Flow Around Cylinder Group on Flat Platee

V. A. Voskoboinick, I. M. Gorban, A. A. Voskoboinick, L. N. Tereshchenko, and A. V. Voskoboinick

Abstract Results of numerical and physical simulations of a junction flow around a three-row cylinder group mounted on the rigid, hydraulically smooth flat surface are presented. Numerical model is based on the vortex method used to integrate 2D Navier–Stokes equations. Experimental researches are carried out in the open hydraulic channel where color inks and contrasting water-soluble coating are applied for visualization of the jet flows and vortex structures generating inside this group. Derived velocity fields demonstrate that the most intense lateral flows form between the first and second as well as between the penultimate and midday cylinders. The currents give rise to developing of large-scale coherent horseshoe vortices at the foot of the cylinders. The size, rotation frequency and angular velocity of the vortices were measured. The estimations show that the vortex generated between the first and second cylinders are more intense than the vortex locating before the group.

V. A. Voskoboinick · I. M. Gorban (✉) · A. A. Voskoboinick · L. N. Tereshchenko · A. V. Voskoboinick
Institute of Hydromechanics, National Academy of Sciences of Ukraine, Maria Kapnist St. 8/4, Kyiv 03057, Ukraine
e-mail: ivgorban@gmail.com

V. A. Voskoboinick
e-mail: vlad.vsk@gmail.com

A. A. Voskoboinick
e-mail: alexandr.vsk@gmail.com

L. N. Tereshchenko
e-mail: litere70@yahoo.com

A. V. Voskoboinick
e-mail: andrey.vsk@gmail.com

V. A. Sadovnichiy and M. Z. Zgurovsky (eds.), *Contemporary Approaches and Methods in Fundamental Mathematics and Mechanics*, Understanding Complex Systems,
https://doi.org/10.1007/978-3-030-50302-4_3

3.1 Introduction

Groups of bluff bodies mounted either on flat or curved surface and streamlined by a fluid flow are often observed in both natural environment and practical human activity. Ensembles of piles, which make up the piers of bridges or of other hydraulic and marine constructions, belong to them. The complex junction flows are generated in and around a group of bodies. They consist of jet and vortex flows including coherent vortex structures of various shapes and scales whose interactions lead to flow turbulization in the system. So, the junction flows are always of turbulent nature [1–3]. It should be noted that evolution of vortex structures in the junction flow around a complex group of piles has much in common with development of horseshoe or necklace-like vortices near a single pile or pier [4, 5]. The separation of the boundary layer and the subsequent formation of a stagnant zone in the lower part of pier is the result of an unfavorable pressure gradient due to the flow deceleration before the pier [6, 7]. In addition, an unfavorable longitudinal pressure gradient causes generation of horseshoe vortices in front of the pier [1, 3]. The horseshoe vortices arise due to the reorganization of vorticity in the boundary layer, when the flow approaches to the pier.

The tandem of two or more bodies is a clear example of their mutual influence [8, 9]. In this case, the aft body is located in the wake of the upstream body and, therefore, interacts with an unstable vortex flow [10, 11]. Theoretical and experimental studies have shown a significant dependence of the flow pattern in tandem on both the distance between the bodies and the Reynolds number [10–14]. At a fixed Reynolds number, depending on the gap the flow pattern may be symmetrical, when a stable pare of vortices develops between the cylinders; non-symmetrical, with separation of large vortices from the upstream cylinder; and bifurcation, when a sudden jump from one flow regime to another is possible. It should be noted that the modes were observed as for a circular [10, 13] as for a square cylinder [11], therefore, their realization is independent of the shape of the bodies that make up the tandem.

In recent years, much attention has been devoted to investigation of the flows that develop around the pile systems operating in open channels with erosion bed. First of all, the composite piers of hydraulic structures such as bridges, oil platforms, offshore wind power equipment are considered [15–18]. These constructions consisting of a large number of piles of small diameter are preferred over solid impenetrable piers because reducing the scour depth at the foot of the structure and savings construction materials.

It is good known that scour is the main cause of bridge failure. Therefore, understanding the scour mechanisms in each individual case is very important to prevent bridge accidents. The successes and results of the scientific research dealing with flows and scour around the pile structures have been reviewed in papers [19–21]. It has been noted in [22, 23], specific generation of vortex and jet flows in and around a pile group causes soil erosion which differentiates essentially from that around a solid impenetrable pier. According to conclusions of papers [24–26], two kinds of scour pattern are formed in a pile group. The local scour arises around an individual pile. It is caused by the generation of horseshoe and lee-wake vortices as well as by the contraction of streamlines and down flow. The global scour develops around the entire grillage. Its main mechanisms are the increase in the mean-flow velocity, formation of a large-scale horseshoe vortex structure surrounding the grillage as well as the turbulence generated by the individual piles. The depth of global scour increases with the number of piles N, but with N ≥ 4–5 it doesn't change [24].

If one considers the total scour as combination of local and global processes, four main mechanisms causing the total scour can be distinguished, namely, reinforcing, sheltering, shedding of lee-wake vortices as well as compressing of horseshoe vortex [20, 22, 27]. The strength of reinforcing most depends on the pile spacing; in particular, it reduces when increasing the gap between the piles. Both the distance between the upstream and downstream piles and the flow skew angle also effects on the strength of reinforcing [28, 29]. Sheltering in a pile group influences the scour depth at the rear piles owing to reduction of velocity and strength of the horseshoe vortex born around them. The separation of lee-wake vortices may arise if the rear pile is so close to the forward pile that influences the vortices that origin from the first [30, 31]. As the gap between two adjacent piles is decreased, the horseshoe vortices surrounding the piles compress, as a result, the flow velocity growths in this place that causes the increase of sediment transport rate and the depth of scoured hole. The effect of compressed horseshoe vortex becomes apparent when changing the gap between horizontal rows of the pile system. When increasing the flow skew angle, both the effect of sheltering and the intense of wake and compressed horseshoe vortices strengthen [20, 29].

Although the scour mechanisms in the pile system are well known, each individual device requires a separate analysis because it has a different configuration and operates under specific hydrological and morphological regimes. In the present study, a model of the three-row cylindrical pier, which has been designed for the Darnytsky bridge in Kyiv, is considered. Our purpose is to determine flow patterns generated inside the construction and identify the most dangerous areas with regard to bottom erosion. The paper includes the results of numerical and physical simulations of the junction flow developing inside the grillage and estimations of the shear stresses on the channel bed. The jet and vortex systems generated in and around the construction are identified. The most intense lateral jets are obtained to form between the first and second as well as between the penultimate and last vertical rows of the grillage. The scale, location and kinematic characteristics of horseshoe vortices generated in this pile system are estimated. The results point out a decrease in the intensity and scale of the horseshoe vortex formed before the grillage compared to that developing in front of an analogous solid impenetrable pier. The most intense and large vortices are detected before the lateral piles of the second row. It follows from these results that the most dangerous zones of bottom erosion in the grillage are located in this region.

3.2 Problem Statement, Experimental Setup and Numerical Procedure

Flow of viscous incompressible fluid in the system of staggered cylindrical piles mounted on the channel bed is considered. The pile system under consideration is the three-row grillage used as a bridge pier. The geometry of interest of the present study is depicted in Fig. 3.1. The cylinders have the same diameter d and the gaps between adjacent cylinders in horizontal and vertical rows are denoted as l and h, respectively. The flow approaches the grillage in the longitudinal direction, from left to right. The flow velocity is assumed to be constant. The channel top is bounded by an unperturbed free surface.

The sizes in Fig. 3.1 are given in mm and related to the laboratory experiments. The experimental investigations were carried out in the hydrodynamic flume (Fig. 3.2) of 14 m in length, 0.8 m in width and 0.8 m in depth with a free surface of water [32, 33]. Water in the hydrodynamic flume was fed by pumps through a settling chamber, a confuser, a honeycomb and turbulizing grids. The measuring section of the flume was fitted with the test equipment and means of visual recording of flow characteristics, lighting equipment and other auxiliary tools, which are necessary for experimental researches. In the measuring section, the hydraulically smooth plate of length 2 m was installed at the height of 0.1 m above the flume bottom. A three-row grillage with 31 cylindrical piles of diameter $d = 0.027$ m was mounted on this plate (Fig. 3.1). The grillage length was about 0.6 m, width = 0.1 m and height = 0.2 m.

The flow depth H was constant and equal to 0.2 m in all the experiments, and the free-stream velocity U was varied from 0.06 m/s to 0.4 m/s, which corresponds to the following ranges of the Reynolds and Froude numbers: $Re_x = Ux/\nu = 60000/400000$, $Re_d = Ud/\nu = 1620/10800$ and $Fr = U/\sqrt{gH} = 0.04/0.29$, where x is the longitudinal distance from the beginning of the bottom plate to the first central pile; g is the acceleration of gravity, ν is the kinematic viscosity of fluid.

A qualitative assessment of spatial and temporal characteristics of the vortex motion in the grillage and its interaction with the streamlined surface were carried out by visual experiments with color ink added to the water. The surface of the

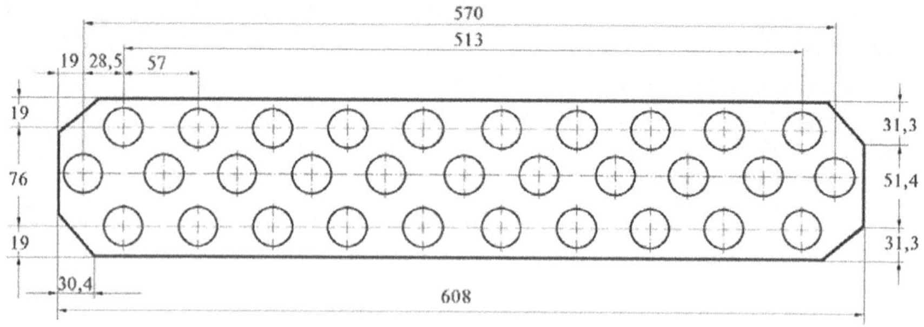

Fig. 3.1 Sketch of the free-row cylindrical grillage (top view)

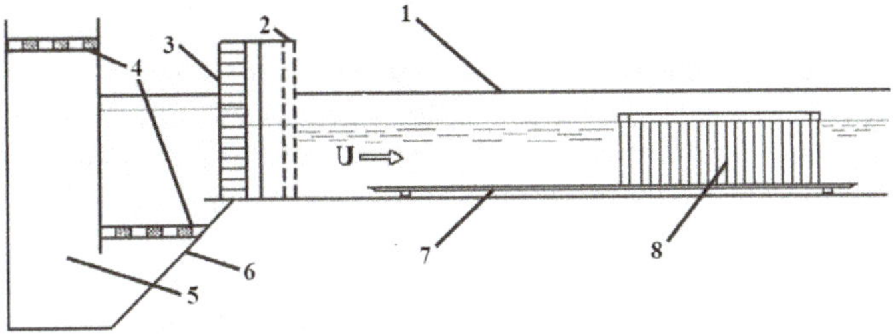

Fig. 3.2 Scheme of the experimental setup: **1**— hydrodynamic flume, **2**—turbulizing grids, **3**—honeycomb, **4**—grate, **5**— settling chamber, **6**—confuser, **7**—flat plate, **8**—three-row cylindrical grillage

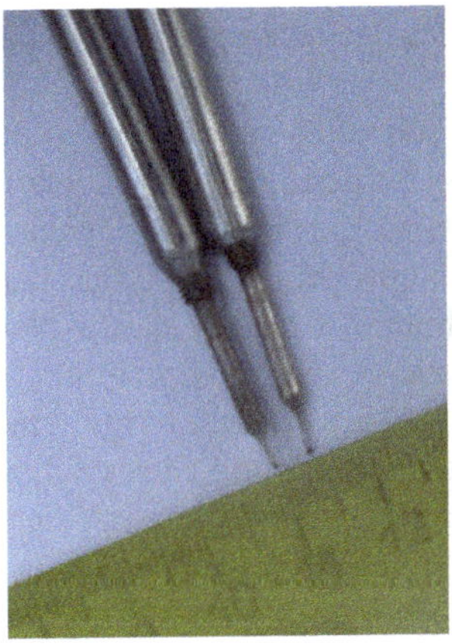

a

b

Fig. 3.3 Thermistor velocity sensors (**a**) and location of the correlation block near the grillage (**b**)

plate, on which the model of the grillage had been installed, was covered with water-soluble contrast medium. The erosion of this coating indicated the localization of the vortex and jet flows inside the grillage. In the places where vortex structures or jet currents had been detected, instrumental measurements of the velocity field as well as cross spectral and correlation dependencies of the velocity fluctuations were performed. The flow kinematic characteristics were measured using miniature thermistor velocity sensors and piezosensitive velocity head sensors [34, 35].

Thermistor velocity sensors of sensitivity diameter $8 \cdot 10^{-4}$ m (Fig. 3.3a), were mounted with the help of special holders in the correlation block, with a fixed distance between two adjacent sensors. The location of a pair of thermistor sensors near the explored model of the cylindrical group is shown in Fig. 3.3b. The control of the free-stream velocity was carried out using a piezoresistive velocity head sensor of thc Pito tube type.

The electric signals generated by the sensors were sent to the amplifier and control and measuring equipment. To register the information on the personal computer, the corresponding analogue-to-digital converters were applied. The obtained data were processed and analyzed with use of standard and developed specially codes based on the theory of probability and mathematical statistics on personal computers and on two-channel analyzers of the Bruel and Kjaer Company.

The measuring equipment was tested and calibrated in accordance with the passport data and the experimental method before, during and after the control experiment. The sensors were certified for absolute and relative accuracy on special stands and the corresponding equipment. The relative error in the measurements of the mean flow velocity did not exceed 4% (with a reliability of 0.95 or 2σ). The error in fluctuation velocity measurements did not exceed 6%. Correlation characteristics were obtained with an error of up to 8%, and spectral dependencies up to 2dB in the frequency range from 0.2 to 1000 Hz.

To improve the understanding of the hydrodynamic processes in the body system under consideration, a numerical simulation of the flow was performed. The numerical model used is based on 2-D fluid dynamics equations, so, the flow is considered in the cross section Oxy under the condition of an infinite flow depth. This means that the influence of both the free surface and the channel bottom is neglected. The flow evolution in the region is governed by the continuity equation and the Navier–Stokes equations with the no-slip condition on the boundary of the cylinders forming the grillage. To integrate the equations, the vortex method belonging to high-resolution Lagrangian-type schemes is used [36]. The vortex numerical scheme developed has been described in details in paper [37] therefore only its general principles are considered here.

The scheme is based on the idea of transition in the mathematical model from natural variables, pressure and velocity, to vorticity and focuses on its creation, transport and diffusion. The velocity field is recovered from the vorticity field with use of the Biot–Savart integral. The vorticity transfer equation, which now describes the flow, is split into convective and diffusive operators, which are integrated separately (see Cottet and Koumoutsakos [36]). In the present realization of the vortex method, the spatial derivative in the diffusion equation is approximated by the finite-difference scheme on the orthogonal grid put on the calculation domain. The convection of vorticity is simulated by the finite volume method, which controls vorticity flows across boundaries of elementary volumes. These volumes surround the nodes of the orthogonal grid. The vorticity is assumed to distribute evenly inside the elementary volume. Note that the cylinder boundary is approximated by the step function in this approach. To integrate the process in time, the explicit scheme of the second order with correction of all parameters after each operator performed is applied.

The body boundary is modeled by the attached vortex sheet whose intensity is determined based on the method of boundary integral equations. To redistribute the vorticity from the body boundary to adjacent fluid, the diffusion equation with the Neumann boundary condition is used. This procedure has been described in details by Ploumhans and Winckelmans [38] and successfully applied in paper [37] when modeling the viscous flow past a wing profile.

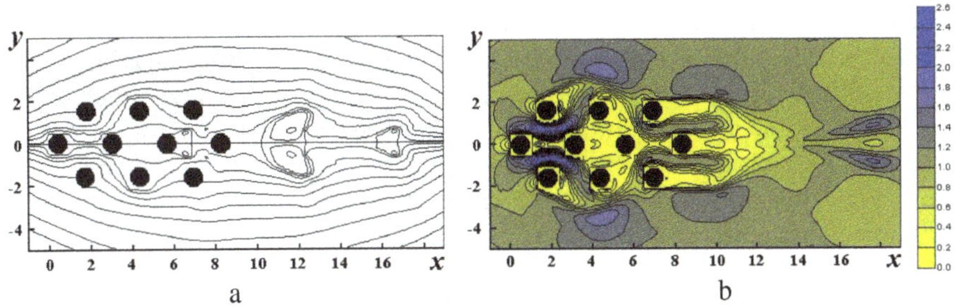

Fig. 3.4 The streamline pattern (**a**) and the longitudinal velocity field (**b**) around the group of ten staggered cylinders

It should be noted that the flow field around the body group is multiply connected [23, 39], so the present numerical scheme has features related to the fulfillment of the closed-loop vorticity theorem. In this regard, one has to evaluate separately the vorticity generated by each cylinder of the system. The corresponding calculation scheme has been described by the authors in paper [11]. Taking into account that the flow simulation in a multiply connected region is quite time-consuming, the calculations were performed on a three-level grid. The grid resolution Δ_1 in the domain containing the grillage depends on the step of discretization of the cylinder boundary. The cell size of each next grid is doubled compared with the previous. The present calculations were carried out at $N = 100$, where N is the number of the discrete panels simulating the boundary of a cylinder. The normalized time step Δt was chosen from the Courant–Friedrichs–Lewy condition. Estimates of the maximum flow velocity generated in the region under consideration have shown that $\Delta t = 0.01$ describes adequately the process.

3.3 Research Results

The flow pattern near a group of bluff bodies is known to be different from that around a similar impenetrable construction due to generation of vortex and jet flows in the gaps between the bodies [32, 34, 40]. The flow structure in the system of bodies depends on their shape and relative position and varies significantly in different designs. Therefore, each case requires special study and analysis.

The present numerical calculations deal with a three-row system of 10 staggered circular cylinders with normalized gaps in horizontal and vertical directions $h/d = 3$ and $l/d = 2.75$, respectively. This simplification compared to the body system shown in Fig. 3.1 allows to reduce computer costs and, at the same time, to reveal large-scale flow characteristics inherent for a three-row grillage. Figure 3.4 demonstrates the picture of streamlines (Fig. 3.4a) and distribution of the longitudinal velocity (Fig. 3.4b) in the system under consideration obtained with $Re = Ud/\nu = 1000$ at $tU/d = 50$.

These results indicate the formation of intense jet flows between the first and second, as well as between the penultimate and last vertical rows of the grillage. The strongest jet develops in the first case, since the flow velocity here is 2.5 times higher than the free-stream velocity. Obviously, this phenomenon will cause an intense erosion of the bottom in this local area. When the jets go outside, they interact with the external flow due to which they are washed out. In the middle of the grillage, the flow velocity is close to zero that indicates the formation of a stagnant zone here. The fluid that left the grillage with the first jet enters the area after the stagnant zone, as a result, the flow velocity between the penultimate and last rows increases again. Because of the flow energy is spent to overcome the individual obstacles being the elements of the construction the vortices generated in the wake of the grillage are weak and quickly attenuate. It should be noted that, although artificial asymmetry was introduced into the flow field at the initial stage of the calculations, the grillage flow quickly stabilized to the symmetrical form. This fact indicates the permeability is an important factor that can prevent high dynamic structural loads arising due to vortex shedding.

The conclusions of the numerical modeling are confirmed by the experimental data. The results of physical simulation presented below were obtained with the free-stream velocity $U = 0.1$ m/s that corresponds to $Re_d = 2.7 \cdot 10^3$, which is close to the Reynolds number of the numerical calculations. Figure 3.5 captures the pattern of coating erosion in the pile system (Fig. 3.5a, b—top view, Fig. 3.5c—side view).

One can see that the contrast medium is most washed out between the first and second as well as between the penultimate and last vertical rows of the grillage. This indicates the formation of intense vortex and jet flows in these regions, which is in accordance with the numerical results. In Fig. 3.5b, the snapshot of the flow in front of the grillage taken with use of color ink is shown. The distribution of the coloring substance indicates the formation of a large horseshoe vortex structure around the grillage, although, as data analysis shows, its intensity is much lower than that of the horseshoe vortex generated in front of a similar impenetrable body [41]. In the middle of the grillage, between the third and sixth rows, decreasing the contrast coating erosion is observed, which points out the formation of a stagnation zone in the middle of grillage. It is consistent with the observational data from paper [24]. The results presented in Fig. 3.5 demonstrate a decrease in global erosion in front of the grillage and the appearance of local erosion around its individual piles caused by jet flows and horseshoe vortices.

The flow visualization also shown that vortex structures with vertical axis, wake vortices, are generated near the grillage piles along with jet flows. The most intense wake vortices being form on the surface of front cylinder and of lateral cylinders of the second row (Fig. 3.6). Due to the lateral jet flows the symmetry of the oppositely rotating wake vortices violated. The vortex formed on the inside of the lateral cylinder is taken away by the jet to the outside, and a pair of oppositely rotating vortices is here localized, which is clearly seen in Fig. 3.6b. Note this process is in phase with shedding of the vortex generated at the front cylinder. The vortex pair is separated from the surface of the cylinder, which it was born on, and travels downstream along the outer side of the grillage forming a street of oppositely rotating vortices. The pro-

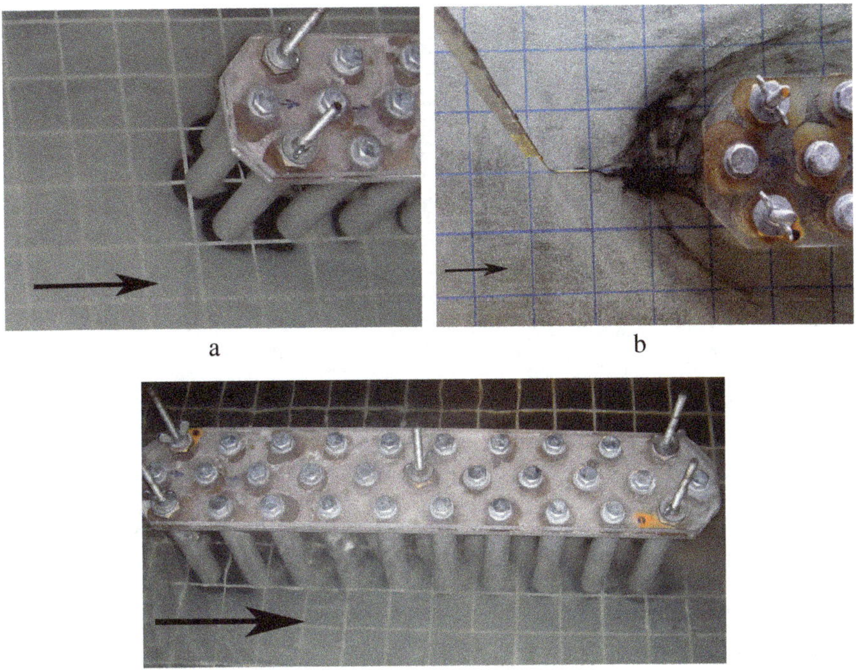

Fig. 3.5 Erosion of contrast substance on the flat plate with the three-row grillage (a, c) and distribution of color ink in the flow (b): a, b—top view, c—side view

cess occurs periodically with a frequency of approximately 0.5 Hz that corresponds to the Strouhal number $St = fd/U = 0.14$. It was observed that the vortex streets developing along the right and left outer sides of the grillage are asymmetric. They dissipate fast enough but new similar systems of wake vortices originate behind the lateral piles of the penultimate row of the grillage. In the end, the wake vortices and lateral jet flows combine with large-scale horseshoe vortex going around the grillage and form together an unsteady flow pattern along the outer sides of the construction.

The greatest threat to the stability of bridge piers is carried from large-scale vortices of horizontal axis formed when the incident flow approaches the pier. Then a portion of the approach flow moves down the front surface of the pier and when the portion reaches the bed a horseshoe vortex is generated at the base of the pier. It is seen that the process occurs differently in a pile grillage and around an impenetrable bridge pier. Results of the physical simulation of the bed erosion in the three-row pile grillage have been presented by the authors in papers [32–34]. In the present research, the vortical flow pattern forming around the grillage is reconstructed by detailed measurements of the velocity field in the vertical section that coincides with the coordinate plane Oxz. Note that the origin of the coordinate plane here coincides with the front point of the first cylinder. These measurements revealed a complex

a b

Fig. 3.6 Scour of a contrast coating in the three-row grillage: a—front view, b—side view

pattern of longitudinal and transversal flows in the gaps between the lateral piles, especially close to the second and penultimate rows. It is the result of the interaction between jet flows, large-scale vortex structures of horizontal axis surrounding the pile and wake vortices generated by an upstream pile.

Integral characteristics of the velocity field near the central and lateral piles of the three-row grillage are shown in Fig. 3.7. Figure 3.7a demonstrates the profiles of the mean longitudinal velocity $\overline{u}$ derived in front of the first central pile, where the numbers denote the percentage of the measured value of $\overline{u}$ to the free-stream velocity U. As seen in Fig. 3.7a, the system of horizontal-axis vortices is formed before the central pile of grillage. In paper [42], those were identified as primary vortices. The intensity of the vortex, which is closer to the pile, is much higher than that of another vortex. The vortex is created by a down flow propagating along the frontal surface of the cylinder; it is oval and stretched along the pile in vertical direction. Set of the vortices forms a horseshoe-like vortex structure surrounding the pile. According to Fig. 3.7a, the velocity in the core of the horseshoe vortex born around the grillage frontal pile is approximately 40% of the free-stream velocity, and the core is localized at a distance of (0.15–0.17)d forward from the pile and (0.1–0.12)d from the channel bottom. The second primary vortex in Fig. 3.7a is generated due to separation of the boundary layer formed above the channel bed; it is elongated along the bed in the horizontal direction. Its core lies at a distance of (0.3–0.33)d forward from the pile and (0.08–0.1)d from the bottom.

Similar horseshoe vortices form around other piles of the grillage. Since these vortices are the main factor of the scour, their presence is clearly visible on the snapshot of bed erosion presented in Fig. 3.5. It can be seen that the most intense horseshoe vortices are located in front of the grillage, namely, around the first central pile and lateral piles of the second row. Figure 3.7b demonstrates the time evolution of the mean longitudinal velocity measured before the frontal pile (curves **1**, **2**) and lateral pile of the second row (curves **3**, **4**). Note that curves **1**, **3** were taken at the core of a horseshoe vortex, and curves **2**, **4** at its periphery. It can be seen that the flow velocity inside the grillage is subject to strong oscillations, which is due

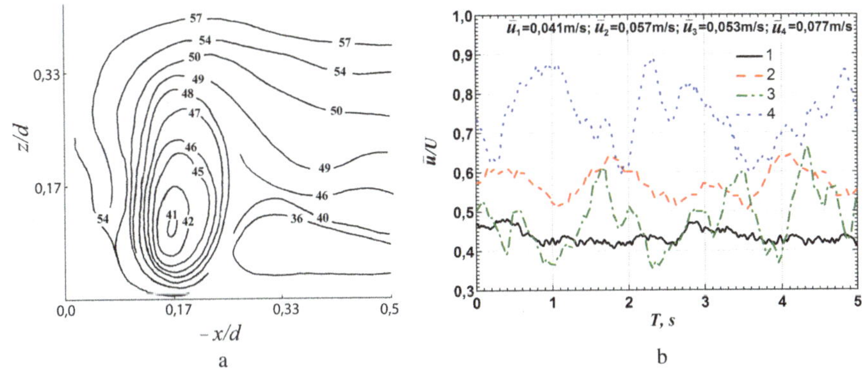

Fig. 3.7 The profiles of the mean longitudinal velocity $\overline{u}$ before the frontal cylinder (**a**) and the evolution of $\overline{u}$ before the frontal and second lateral cylinders of the grillage (**b**)

to the formation of a complex flow pattern here. Recall that the flow in this pile group is affected by the periodic generation of jets and vortices as of horizontal as of vertical axis, which interact with each other and the streamlined surface. These developing factors lead to significant non-stationarity and non-uniformity of the junction flow inside the body group. It follows from Fig. 3.7b, the velocity oscillations are most intense at the periphery of the horseshoe vortex, especially before the second lateral pile, which is consistent with the visual studies discussed above. The mean longitudinal velocity before the second lateral pile are almost (20–25)% higher than in front of the first central pile of the three-row grillage.

Similar data for the fluctuation velocity u' are presented in Fig. 3.8. Figure 3.8a demonstrates the longitudinal velocity fluctuations before the frontal pile (curves **1**, **2**) in comparison with those before the second lateral pile (curves **3**, **4**). The corresponding power spectral densities of the velocity fluctuations are presented in Fig. 3.8b. As before, curves **1**, **3** in Fig. 3.8a, b depict the appropriate values of velocity in the vortex core and curves **2**, **4** at the vortex periphery. As already reported in paper [34], the round horseshoe vortex of diameter $0.3d$ forms in front of the central pile. According to the equal velocity lines derived at the present measurements, the horseshoe vortex located before the lateral pile of the second row has an egg-like shape. The relative dimensions of the vortex are $l_x = 0.5d$ in longitudinal direction and $l_z = 0.4d$ in vertical direction and its core has the relative coordinates $x \approx 0.28d$; $z \approx 0.17d$. The more sharp part of the horseshoe vortex is closer to the aft of the pile. The velocity fluctuations observed before the piles of the second row are more significant compared with those which are generated in the core of the vortex formed before the frontal cylinder of the grillage. As the results presented in Fig. 3.8b shown, the cores of the horseshoe vortices developed around the lateral cylinders of the second row fluctuate intensively. This is due to the interaction of these vortices with the lateral jet flows, which itself have an oscillating character, and the wake vortices generated in this region.

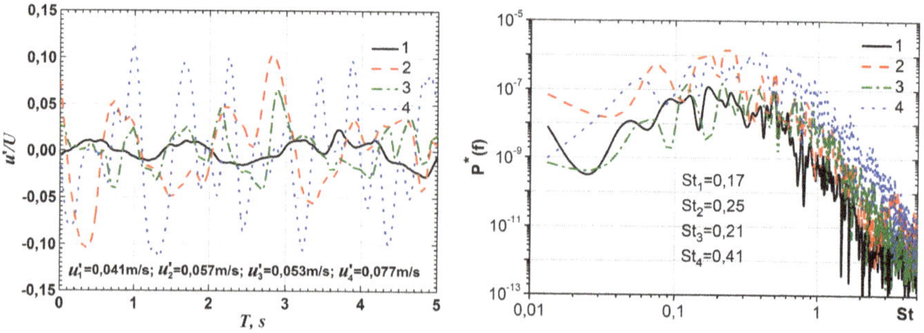

a b

Fig. 3.8 Time changes (**a**) and power spectral densities (**b**) of the longitudinal velocity fluctuations near the central cylinder and lateral cylinder of the second vertical row of the grillage

As follows from Fig. 3.8a, the velocity fluctuations at the periphery of the horseshoe vortices are higher than in their core, and the most intense fluctuations are observed before the lateral piles of the second row. At the same time, the intensity of turbulence, calculated as the ratio of the mean quadratic value of velocity fluctuations to the mean local velocity, is higher before the frontal pile than before the piles of the second row. From the results presented in Figs. 3.7b and 3.8a, one obtains that the turbulence intensity is 12 and 50%, respectively in the core and at the periphery of the vortex generated before the frontal pile and these values are equal to 7 and 29% for the lateral pile of the second row. This reduction is achieved due to significant increasing the mean velocity of vortex flow in the latter case, as shown in Fig. 3.7b.

The spectral characteristics of the field of velocity fluctuations in front of the second lateral piles are higher than those observed before the first pile of the grillage. It is seen in Fig. 3.8b, where spectra of dimensionless densities of velocity fluctuations $P^*(f) = P^2(f)d/(u'^2)U$ are presented against the Strouhal number $St = fd/U$, where f is the dimensional frequency. The maximum dimensional frequency is obtained to be 25 Hz before the lateral pile of the second row and 10 Hz before the frontal pile. This is the result of intense interactions between the horseshoe vortices, lee-wake vortices and jet flows when the piles are located in the near field one from other. The presence of high-frequency components in these spectra indicates the existence of small-scale sources of velocity fluctuations, which can be small vortices formed as a result of the destruction of the primary vortices [43, 44]. Maxima in the power spectral density of the velocity fluctuations in Fig. 3.8 are concentrated at frequencies close to (0.6–1.0) Hz and (0.9–1.5) Hz in front of the first central and second lateral cylinders, respectively. At the peripheries of horseshoe vortices, the frequency of oscillations is higher than in their cores. These high frequencies correspond to the most intense oscillations of the velocity field in the three-row grillage, which are caused by both the rotation and oscillations of the large-scale horseshoe vortex born in the gap between the frontal and second lateral piles.

When the free-stream-velocity increases, the dependencies of the spectral power densities of the longitudinal velocity fluctuations expand to the high frequency region and the levels of the spectral components of the high-frequency velocity fluctuations growth. This indicates the growth of the intensity of small-scale vortices in the vortex flow near the piles.

The obtained data indicate that both the scale and the rotational frequency of the horseshoe vortex structure formed near the second lateral pile are larger than those of the vortex developed around the frontal pile of the three-row grillage. Intensive jet flows between the first and second rows are the factor that has the greatest influence on the evolution of the horseshoe vortices in this region. In addition, the grillage as a whole should be considered as a bluff body that blocks the cross section of the flow, so, the regions of increased velocity are formed at its outer boundaries. These high-velocity flows lead to further stretching of the legs of the horseshoe vortex around the lateral cylinder that causes them to accelerate and rotate with a higher frequency. It should be noted also that the horseshoe vortices are of a quasiperiodic nature. They grow, pulsate and are carried out into the external flow during the life, giving way to new vortex formations.

3.4 Conclusion

Numerical and physical simulations of the junction flow inside the three-row grillage situated in the plane channel were carried out. The grillage was assumed to be a system of staggered cylindrical piles mounted on the channel bed. The purpose of the study was to determine the flow pattern generated by the uniform free stream inside the construction and identify the most dangerous areas with regard to bottom erosion. The vortex numerical scheme for simulation of viscous flow evolution in a multiply connected region was developed on base of 2-D fluid dynamics equations under the assumption of an infinite channel depth. It was applied for calculations of the flow inside the system of ten staggered cylinders at $Re = 10^3$. The experimental investigations were carried out in the hydrodynamic flume where the measuring section was fitted with the test equipment and means of visual recording of flow. For flow visualization, the channel bed was covered with water-soluble contrast medium whose erosion to be indicated the location of high shear stress zones caused either by vortices or by jet flows. Instrumental measurements of the longitudinal velocity field were performed in the places of localization of the most intense vortex and jet flows.

Both numerical and physical visualization revealed that the most intense jet flows develop between the first and second, as well as between the penultimate and last vertical rows of the pile grillage. At the same time, a stagnant zone forms in the middle of the grillage. A detailed study of the flow structure in the frontal part of the grillage showed that it is composed with the lateral jet flows developing between the first and second rows, wake vortices of vertical axis generated by cylinder walls and large-scale horseshoe vortices that surround the central pile and the lateral piles of the second row. The interaction of all these factors determines the pulsating nature of the flow between the piles, when large-scale formations are periodically thrown into the external flow and new ones are born in their place.

Detailed measurements of the longitudinal velocity in the vicinity of the front pile and lateral piles of the second row of grillage, which correspond to the developed turbulent flow regime with $Re_x = 100000$, $Re_d = 2700$, $Fr = 0.07$, were performed. This made it possible to establish the location, shape and scale of the horseshoe vortices generated in these regions. It was found that the round horseshoe vortex of diameter $0.3d$ forms in front of the central pile and egg-like vortex structure of extent $l_x = 0.5d$ and $l_z = 0.4d$ develop around the lateral pile of the second row. The rotational frequency and velocity of the horseshoe vortex placed in the second row are 1.5 Hz (or $St = 0.41$) and $0.7U$, respectively, against 1 Hz (or $St = 0.27$) and $0.56U$ for the frontal vortex. It is also established that the mean and fluctuating velocity in the core of a horseshoe vortex are (1.5–2) times less than at its periphery. Moreover, these characteristics for the horseshoe vortex located before the second row pile are 20–25% higher than for the frontal horseshoe vortex.

The spectral characteristics of the field of velocity fluctuations before the second lateral cylinder are filled to 25 Hz, or $St = 6.8$, and before the first central cylinder—up to 10 Hz, or $St = 2.7$. The peaks of the spectral densities of velocity fluctuations before the first pile are concentrated at the frequencies close to (0.6–1.0) Hz, or St = (0.16–0.27), and before the second lateral piles those are (0.9–1.5) Hz, or St = (0.24–0.41). At the periphery of the horseshoe vortices, the frequency of oscillations is higher than in their core. These frequencies correspond to the most intense oscillations of the velocity field due to the simultaneous rotation and transverse vibrations of a horseshoe vortex structure.

The obtained qualitative and quantitative characteristics of the junction flow in the three-row grillage allow us to conclude that the region between the first and second vertical rows of the grillage is most dangerous as regard to bed erosion.

References

1. Simpson, R.L.: Junction flows. Annu. Rev. Fluid Mech. **33**, 415–443 (2001). https://doi.org/10.1146/annurev.fluid.33.1.415
2. Paik J., Escauriaza C., Sotiropoulos F.: On the bimodal dynamics of the turbulent horseshoe vortex system in a wing-body junction. Phys. Fluids. **19**(4), 045107 (2007). https://doi.org/10.1063/1.2716813
3. Devenport, W.J., Simpson, R.L.: Time-dependent and time-averaged turbulence structure near the nose of a wing-body junction. J. Fluid Mech. **210**, 23–55 (1990). https://doi.org/10.1017/S0022112090001215
4. Gautam, P., Eldho, T.I., Mazumder, B.S., Behera, M.R.: Experimental study of flow and turbulence characteristics around simple and complex piers using PIV. Exp. Thermal Fluid Sci. **100**, 193–206 (2019). https://doi.org/10.1016/j.expthermflusci.2018.09.010
5. Launay G., Mignot E., Riviere N.: Laminar free-surface flow around emerging obstacles: Role of the obstacle elongation on the horseshoe vortex. Eur. J. Mech. B. Fluids **77**, 71–78 (2019). https://doi.org/10.1016/j.euromechflu.2019.04.006

6. Guan, D., Chiew, Y.-M., Wei, M., Hsieh, S.-C.: Characterization of horseshoe vortex in a developing scour hole at a cylindrical bridge pier. Int. J. Sedim. Res. **34**(2), 118–124 (2019). https://doi.org/10.1016/j.ijsrc.2018.07.001
7. Lange, E.A., Elahi, S.S., Lynch, S.P.: Time-resolved PIV measurements of the effect of freestream turbulence on horseshoe vortex dynamics. AIAA Pap. **2018–0584**, 1–12 (2018). https://doi.org/10.2514/6.2018-0584
8. Hamed A.M., Peterlein A.M., Randle L.V.: Turbulent boundary layer perturbation by two wall-mounted cylindrical roughness elements arranged in tandem: Effects of spacing and height ratio. Phys. Fluids **31**, 065110 (2019). https://doi.org/10.1063/1.5099493
9. Das, S., Mazumdar, A.: Evaluation of hydrodynamic consequences for horseshoe vortex system developing around two eccentrically arranged identical piers of diverse shapes. KSCE J. Civ. Eng. **22**(7), 2300–2314 (2018). https://doi.org/10.1007/s12205-017-1842-9
10. Tasaka Y., Kon S., Schouveiler L., Le Gal P.: Hysteretic mode exchange in the wake of two circular cylinder in tandem. J. Phys. Fluids. **18**(8), 084104 (2006). https://doi.org/10.1063/1.2227045
11. Gorban, V.O., Gorban, I.M.: Study of the interaction of square cylinders located in tandem. Appl. Hydromech. **10**(2), 36–47 (2008). (in Ukrainian)
12. Gowthaman, S., Nasvi, M.C.M.: Three-dimensional numerical simulation and validation of load-settlement behaviour of a pile group under compressive loading. Engineer **51**(1), 9–11 (2018). https://doi.org/10.4038/engineer.v51i1.7283
13. Mizushima J., Suehiro N.: Instability and transition of flow past two tandem circular cylinders. J. Phys. Fluids. **17**(10), 104107 (2005). https://doi.org/10.1063/1.2104689
14. Yang Y., Qi M., Li J., Ma X.: Evolution of hydrodynamic characteristics with scour hole developing around a pile group. Water **10**, 1632 (2018). https://doi.org/10.3390/w10111632
15. Ataie-Ashtiani, B., Beheshti, A.A.: Experimental investigation of clear-water local scour at pile groups. J. Hydr. Eng. **132**(10), 1100–1104 (2006). https://doi.org/10.1061/(ASCE)0733-9429(2006)132:10(1100)
16. Ji, C., Zhang, J.-F., Zhang, Q.-H., Li, M.-X., Chen, T.-Q.: Experimental investigation of local scour around a new pile-group foundation for offshore wind turbines in bi-directional current. Chin. Ocean Eng. **32**(6), 737–745 (2018). https://doi.org/10.1007/s13344-018-0075-0
17. Ma, L., Wang, L., Guo, Z., Jiang, H., Gao, Y.: Time development of scour around pile groups in tidal currents. Ocean Eng. **163**, 400–418 (2018). https://doi.org/10.1016/j.oceaneng.2018.06.035
18. Lanca, R., Fael, C., Maia, R., Pego, J., Cardoso, A.: Clear-water scour at pile groups. J. Hydr. Eng. **139**(10), 1089–1098 (2013). https://doi.org/10.1061/(ASCE)HY.1943-7900.0000770
19. Deng, L., Cai, C.S.: Bridge Scour: Prediction, modeling, monitoring, and countermeasures. Rev. Pract. Period. Struct. Des. Constr. **15**(2), 125–134 (2010). https://doi.org/10.1061/(asce)sc.1943-5576.0000041
20. Hosseini, R., Amini, A.: Scour depth estimation methods around pile groups. KSCE J. Civ. Eng. **19**(7), 2144–2156 (2015). https://doi.org/10.1007/s12205-015-0594-7
21. Kumar, A., Kothyari, U.C., Ranga Raju, K.G.: Flow structure and scour around circular compound bridge piers—a review. J. Hydro-Envir. Res. **6**, 251–265 (2012). https://doi.org/10.1016/j.jher.2012.05.006
22. Amini, A., Solaimani, N.: The effects of uniform and nonuniform pile spacing variations on local scour at pile groups. Mar. Georesour. Geotechnol. **36**(7), 861–866 (2018). https://doi.org/10.1080/1064119X.2017.1392658
23. Choi, S.-U., Choi, B., Lee, S.: Prediction of local scour around bridge piers using the ANFIS method. Neural Comput. Appl. **28**(2), 335–344 (2017). https://doi.org/10.1007/s00521-015-2062-1
24. Sumer B.M., Bundgaard K., Fredsoe J.: Global and local scour at pile group. In: Proceedings of the 15th International Offshore and Polar Engineering Conference, Seoul, Korea, 577–583 (2005)
25. Voskoboinick, A.V., Voskoboinick, V.A., Voskoboinick, A.A., Gorban, V.O., Gorban, I.M.: Features of vortex motion in the junction flow between pile group of three-row bridge pier. Appl. Hydromech. **11**(2), 16–29 (2009). (in Ukrainian)

26. Voskobijnyk A.V., Voskoboinick V.A., Voskoboinyk O.A., Tereshchenko L.M., Khizha I.A.: Feature of the vortex and the jet flows around and inside the three-row pile group. In: Proceeding of the 8th International Conference on Scour and Erosion (ICSE 2016) 12–15 Sep. 2016 Oxford, UK, pp. 897–903 (2016). https://doi.org/10.1201/9781315375045-114
27. Hannah C.R.: Scour at pile groups. MSc Thesis, University of Canterbury, 255p
28. Memar S., Zounemat-Kermani M., Beheshti A.-A., De Cesare G., Schleiss A.J.: Investigation of local scour around tandem piers for different skew-angles. E3S Web of Conferences 40, 03008 (2018). https://doi.org/10.1051/e3sconf/20184003008
29. Yang Y., Melville B.W., Sheppard D.M., Shamseldin A.Y.: Clear-water local scour at skewed complex bridge piers. J. Hydr. Eng. **144**(6), 04018019 (2018). https://doi.org/10.1061/(ASCE)HY.1943-7900.0001458
30. Amini, A., Melville, B., Mohamad, T.A., Ghazali, A.: Clear-water local scour around pile groups in shallow-water flow. J. Hydr. Eng. **138**(2), 177–185 (2012). https://doi.org/10.1061/(asce)hy.1943-7900.0000488
31. Chreties C., Teixeira L., Simarro G.: Influence of flow conditions on scour hole shape for pier groups. Proc. Inst. Civ. Eng.-Water Manag. **166**(3), 111–119 (2013). https://doi.org/10.1680/wama.11.00054
32. Voskoboinick A.V., Voskoboinick V.A., Voskoboinick A.A.: Junction flow of the three-row grillage on the flat surface. Part 1. Formation of the horseshoe vortices. Appl. Hydromech. **10**(3), 28–39 (2008). (in Ukrainian)
33. Voskoboinick A.A., Voskoboinick V.A., Voskoboinick A.V.: Vortex and jet flow about cylindrical group on the plate. In: Proceedings of the Acoustic Symposium Consonans-2011 Kyiv, IHM NASU, pp. 83–88 (2011). (in Russian)
34. Voskoboinick A.V., Voskoboinick V.A., Voskoboinick A.A.: Junction flow of the three-row grillage on the flat surface. Part 2. Space-time correlations and spectra. Appl. Hydromech. **10**(4), 13–25 (2008). (in Ukrainian)
35. Voskoboinick, V., Kornev, N., Turnow, J.: Study of near wall coherent flow structures on dimpled surfaces using unsteady pressure measurements. Flow Turbul. Combust. **90**(2), 86–99 (2013). https://doi.org/10.1007/s10494-012-9433-9
36. Cottet, G.-H., Koumoutsakos, P.: Vortex methods: Theory and Practice. Cambridge University Press, London (2000)
37. Gorban I.M., Lebid O.G.: Numerical modeling of the wing aerodynamics at angle-of-attack at low Reynolds numbers. In: Sadovnichiy V., Zgurovsky M. (eds) Modern Mathematics and Mechanics. Understanding Complex Systems, pp. 159–179. Springer, Cham, (2019). https://doi.org/10.1007/978-3-319-96755-4-10
38. Ploumhans, P., Winckelmans, G.S.: Vortex methods for high-resolution simulations of viscous flow past bluff bodies of general geometry. J. Comput. Phys. **165**, 354–406 (2000). https://doi.org/10.1006/jcph.2000.6614
39. Hassanzadeh, Y., Jafari-Bavil-Olyaei, A., Aalami, M.-T., Kardan, N.: Experimental and numerical investigation of bridge pier scour estimation using ANFIS and teaching-learning-based optimization methods. Eng. Comput. **35**(3), 1103–1120 (2019). https://doi.org/10.1007/s00366-018-0653-z
40. Olcmen, S.M., Simpson, R.L.: Experimental transport-rate budgets in complex 3-D turbulent flow near a wing/body junction. Int. J. Heat Fluid Flow **29**(4), 874–890 (2008). https://doi.org/10.1016/j.ijheatfluidflow.2007.12.004
41. Voskoboinick, A.A., Voskoboinick, A.V., Voskoboinick, V.A., Nikishov, V.I.: Pressure pulsations on the surface of soil erosion. Appl. Hydromech. **16**(2), 27–35 (2014). (in Russian)
42. Ballio, F., Franzetti, S.: Topological analysis of a junction vortex flow. Adv. Fluid Mech. **III**(29), 255–264 (2000)
43. Bendat, J.S., Piersol, A.G.: Random data: Analysis and measurement procedures. Willey, New York (1986)
44. Geoga, C.J., Haley, C.L., Siegel, A.R., Anitescu, M.: Frequency-wavenumber spectral analysis of spatio-temporal flows. J. Fluid Mech. **848**, 545–559 (2018). https://doi.org/10.1017/jfm.2018.366

Chapter 4
Accounting for Shear Deformation in the Problem of Vibrations and Dissipative Heating of Flexible Viscoelastic Structural Element with Piezoelectric Sensor and Actuator

I. F. Kirichok, Y. A. Zhuk, and S. Yu. Kruts

Abstract Statement of the problem on forced resonance vibration, active control and dissipative heating of flexible viscoelastic beam containing piezoelectric sensor and actuator with taking account for shear deformation and quadratic geometrical nonlinearity is elaborated on the base of coupled electro-thermomechanics theory. Procedure of numerical solution of the problem is developed as well. For the most energy intensive first bending mode of beam vibration, influence of mechanical end fixing conditions, heat exchange, shear deformation and accounting for geometrical nonlinearity onto frequency characteristics of maximum deflection amplitude, level of dissipative heating temperature and electric parameter of sensor are investigated under monoharmonic loading. Possibility of active damping of beam vibration by means of piezoactuator with making use of electric parameter of sensor is studied in the case of unknown loading amplitude.

4.1 Introduction

Thin wall viscoelastic structural elements are widely used nowadays in modern engineering such as space and aircraft technology, automotive and ship construction industry, mechanical engineering, microelectromechanical systems etc. The elements

I. F. Kirichok
Timoshenko Institute of Mechanics, National Academy of Sciences of Ukraine,
P. Nesterov str., 3, Kyiv 03057, Ukraine
e-mail: term@inmech.kiev.ua

Y. A. Zhuk (✉) · S. Yu. Kruts
Taras Shevchenko National University of Kyiv, Volodymyrska str., 60,
Kyiv 01601, Ukraine
e-mail: y.zhuk@i.ua

S. Yu. Kruts
e-mail: sofiya1402@ukr.net

V. A. Sadovnichiy and M. Z. Zgurovsky (eds.), *Contemporary Approaches and Methods in Fundamental Mathematics and Mechanics*, Understanding Complex Systems,
https://doi.org/10.1007/978-3-030-50302-4_4

are often subjected to different kinds of loadings, the nonstationary and harmonic excitations are among the most common of them. Under specific conditions, induced intensive unsteady vibrations as well as stationary vibrations at the frequency lying in the vicinity of the natural frequency of the element can be especially dangerous. To avoid the undesirable regimes and to prevent the structural damage and failure, the damping of vibrations of thin wall structural elements becomes an important issue.

On the one hand, traditional way to address this problem is to apply the passive damping technique (usage of elements fabricated of materials possessing high hysteresis losses as components within the structure of interest or as additional coatings) for vibration control. The amount of publications devoted to this approach is really vast (see reviews [6, 15] for the particular emphasis on the damping of thin wall structures).

On the other hand, rapid development of modern technology necessitates a change from the traditional methods of vibration control of thin wall structures to methods of active control that allow for the implementation of more complex and highly effective operating modes, while taking proper account of numerous life and reliability criteria. The recent trend is to use an active damping approach which lies in the embedding of an active (for instance piezoelectric) inclusions (usually distributed ones) into a passive (without the piezoelectric effect) thin wall structure made of a metallic, polymeric, or composite material [4, 21]. Some of these embedded elements function as sensors to provide information on the mechanical state of the body, while the others usually referred to as actuators are designed to deform or excite the structure. The state of the art in active damping of the vibrations of thin-wall elements in the isothermal case is discussed in the publications [19, 21–23].

There are two basic techniques for active damping of thin wall element vibrations. One employs actuators only which have a voltage (electric potential difference) applied to electrodes with the aim to balance the mechanical load. In the frame of the other technique, both sensors and actuators are used to provide the voltage that is proportional to the sensor's output (voltage itself and its first or second time derivative) is being applied to the actuator. The sensor–actuator relationship is described by feedback equations (see works [7, 8, 19, 21]). As a result of application of the both techniques mentioned, stiffness, dissipative, and inertial characteristics of the structure are being varied, which makes it possible to manipulate the amplitude of vibration of the element.

The efficiency of active damping of plates depends on many factors: geometrical and electromechanical characteristics of the sensors and actuators, the geometrical and mechanical characteristics of the passive element, the mechanical and electric boundary conditions as well as on temperature. Thermal effects are among the most profound ones and were previously discussed in [5–8, 10, 24]. It is well known that the inelastic deformation of a material is accompanied by the release of heat due to dissipation of mechanical energy. The heating effect can be particularly significant for the partial case of cyclic loading due to considerable hysteresis losses, low heat conductivity, and temperature dependence of the electromechanical properties for many active and passive materials. Under these conditions, the small temperature advance during one cycle can result in significant heating levels for a large number of cycles. As a result, the phenomenon of dissipative heating may significantly affect the performance of thin wall elements for several reasons.

First of all, the temperature dependence of the electromechanical properties may have a profound effect on the performance of the sensors and actuators and, consequently, the efficiency of active damping itself. Secondly, the piezoactive material loses the piezoelectric effect as soon as the heating temperature reaches the Curie point. Hence, the whole structure losses its electromechanical functionality though it preserves the integrity as a passive structure. Thirdly, disbalance between the heat input due to the dissipative heating and the heat losses due to heat transfer into the environment may cause the phenomenon usually referred to as thermal breakdown in both active and passive elements, i.e. catastrophic rise in temperature resulting in the absence of steady thermal state (for this well-known effect see [3, 10]). Obviously, neglecting the thermomechanical coupling in this case can lead to incorrect prediction of the system response.

To address the issue of active damping of the forced vibration, to simulate and study the thermo-electro-mechanical behavior of a layered thin wall structure and analyze the interplay between different factors, the theories and models were developed in [2, 5–8, 10, 13, 14, 23] based on the Kirchhoff–Love hypotheses mainly supplemented with correspondent assumptions regarding electric and temperature fields. However, if the structural element is thick enough and material properties of the passive and piezoactive layers are very different, it is necessary to use refined mechanical hypotheses and relevant assumptions on the electric and thermal variables (see [3, 7, 15] for details). The detailed reviews of the publications devoted to the investigation of thermomechanical response of the structures containing physically nonlinear both passive and active layers under monoharmonic mechanical or electrical loading are presented in [6, 7]. A review [8] is devoted to the study of dissipative heating influence on the working capacity of piezoelectric sensors and actuators and their applicability for controlling and damping of forced vibration of thin wall rods, beams, plates and shells composed of passive materials in the frame of classical and shear deformation theories.

In the present paper, the forced resonant vibrations and dissipative heating of the simplest structural element—a sandwich beam—containing passive viscoelastic central layer and piezoactive outer layers (sensors and actuators) are studied in the frame of refined formulation taking account of thermomechanical coupling, geometrical nonlinearity for flexible elements, physical nonlinearity of material properties as well as shear deformation and rotation inertia of the beam cross section. Interplay between factors mentioned as well as the influence of boundary conditions on the amplitude-frequency characteristics of the vibrations are investigated in details.

4.2 Problem Statement

General theory of thermoviscoelastic plates and shells with distributed sensors and actuators under monoharmonic mechanical or electrical loading was developed in [6–8]. Particular case of controlling the axisymmetric resonant vibrations and self-

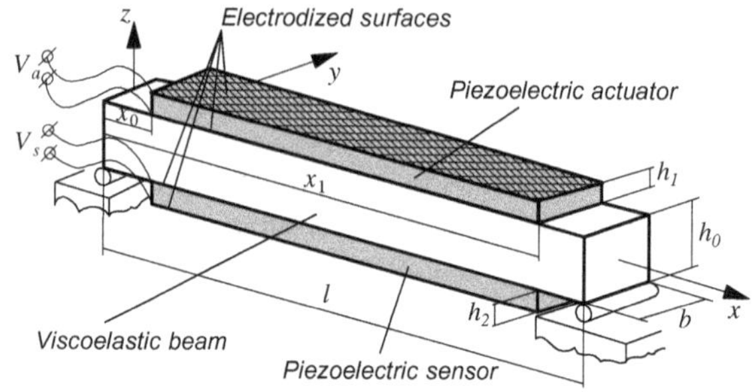

Fig. 4.1 Beam lay-up

heating of shells of revolution with piezoelectric sensors and actuators is addressed in [9, 12]. The problem of resonant vibration and heating of ring plates with piezoactuators under electromechanical loading is investigated in [11] in details.

In this paper, the three layer flexible beam of length l and width b is under consideration. The internal layer of the thickness h_0 is fabricated from electrically passive transversely isotropic viscoelastic material. The upper layer of thickness h_1 and the lower layer of thickness h_2 are made of the same viscoelastic piezoceramics polarized over the thickness in the opposite (to each other) directions. Let us assume the upper piezolayer of thickness h_1 is used as actuator and characterized by the piezoelectric constant $+d_{31}$ meanwhile the lower piezolayer of thickness h_2 functions as sensor with piezoelectric coefficient $-d_{31}$.

The beam is described by Cartesian coordinates x, y, z, so that $0 \le x \le l$, $|y| \le b/2$, $|z| \le h_0/2$ (see Fig. 4.1). The inner faces of the piezolayers are covered with solid continuous infinitely thin electrodes and are kept at zero electric potential $\varphi_{1,2}(h_0/2) = 0$. The outer faces have the part of the area $s = b\Delta_x$, $\Delta_x = x_1 - x_0$, $x_0 \ge 0$, $x_1 \le l$ covered with electrodes only. Under these assumptions, the electrical boundary conditions at the outer surface of the piezosensor h_2 can be written in the form

$$\iint\limits_S {}^2D_z ds = 0 \;\; if \;\; x_0 \le x \le x_1; \quad {}^2D_z = 0 \;\; if \;\; 0 \le x \le x_0, \; x_1 \le x \le l, \quad (4.1)$$

where D_z is the normal component of the electric-flux density vector in piezolayer, superscript stands for the piezolayer number.

The beam is loaded by the surface pressure $q_z = q_0' \cos \omega t$ that varies harmonically in time t with amplitude q_0' at frequency ω which lies in the vicinity of flexural vibration resonance of the beam.

The voltage of the same frequency and amplitude $V_a = \varphi_1(-h_0/2 - h_1) - \varphi_1(-h_0/2)$ is applied to the actuator (electrode patch of the area s of piezolayer) in-phase or in-antiphase to amplify or suppress the mechanically induced vibrations, correspondingly. As the beam is deformed harmonically, the voltage of amplitude $V_s = \varphi_2(h_0/2 + h_2) - \varphi_2(h_0/2)$ is induced across the open-circuited electrodes (electroded part of the area s of the piezolayer) of the sensor. This voltage can be either measured or calculated as the solution of problem statement of electromechanics under boundary conditions (4.1). The beam ends are assumed to be either hinged or rigidly fixed which leads to symmetric first bending mode of vibration under geometry considered. It is also supposed that heat exchange with the environment of temperature T_c takes place on the beam surface.

To simulate the electromechanical response of the beam, Timoshenko hypotheses are assumed to be valid for the mechanical variables of the problem through the total thickness of the laminate (shear deformation and rotation inertia of the cross section are taken into account as it was done in [16]). As for the electrical variables, let us assume the components ${}^{1,2}D_x$ and ${}^{1,2}D_y$ of electric-flux density vector can be neglected while normal component $D_z = \text{const}$ does not depend on the thickness coordinate (see [6] for details). At this, the equations of electrostatics are satisfied identically and components ${}^{1,2}E_x$, ${}^{1,2}E_y$ of the electric field vector can be determined using material constitutive equations ${}^{1,2}D_x = 0$, ${}^{1,2}D_y = 0$. To take into account the special case of geometric nonlinearity (i.e., finite deflections but strains are still small), both the squared angles of rotation of beam cross sections and nonlinear terms are retained in the Cauchy strain-displacement relations and in the equation of motion correspondingly.

Viscoelastic material properties are described by the integral operators of linear viscoelasticity (see [6]) which are reduced to the multiplication of the complex-value quantities under particular case of harmonic deformation of the material

$$a * b = (a' + ia'')(b' + ib''),$$

where operator itself is denoted by the asterisk, real and imaginary parts of quantities are marked as $(\cdot)'$ and $(\cdot)''$ correspondently; $i = \sqrt{-1}$. Temperature is considered to be constant through the beam thickness.

Under the assumptions adopted, the three-dimensional constitutive equations for the piezoceramics polarized along z direction for the piezolayers of actuator (of thickness h_1) and sensor (of thickness h_2) take the form [6, 9]

$$\begin{aligned} {}^m\sigma_x &= c_{11}^E * \varepsilon_x \mp b_{31} * {}^mE_z; \quad {}^m\sigma_{xz} = c_{44}^E * e_{xz}; \\ {}^mD_z &= \mp b_{31} * \varepsilon_x + b_{33} * {}^mE_z; \quad {}^mE_z = -d\varphi_m/dz, \end{aligned} \tag{4.2}$$

where

$$\begin{aligned} c_{11}^E &= 1/s_{11}^E; \quad b_{31} = d_{31}/s_{11}^E; \\ b_{33} &= \varepsilon_{33}^T - d_{31}^2/s_{11}^E; \quad c_{44}^E = 1/(s_{44}^E - d_{15}^2/\varepsilon_{11}^T); \quad m = 1, 2, \end{aligned} \tag{4.3}$$

and $s_{kk}^E = s'_{kk}\left(1 - i\delta_{kk}^s\right)$, $d_{ik} = d'_{ik}(1 - i\delta_{ik}^d)$, $\varepsilon_{kk}^T = \varepsilon'_{kk}(1 - i\delta_{kk}^\varepsilon)$ are viscoelastic compliances, piezoelectric coefficients and dielectric permeabilities for piezoceramics respectively; ${}^mE_z$ is normal component of the electric field vector in the piezolayers. For the viscoelastic material of the electrically passive layer (of thickness h_0), the first and second equations in (4.3) are applicable with the substitution $c_{11}^E = E$,

$c_{44}^E = G_{13}$, $b_{31} = 0$, where E and G_{13} are viscoelastic Young modulus and shear modulus correspondingly. The asterisk sign in the equations will be omitted further on. Hereafter the standard notation for complex variable is used: $a = a' + ia''$, $|a| = \left(a'^2 + a''^2\right)^{1/2}$.

Strain amplitudes in the Eq. (4.2) are expressed in terms of the complex amplitudes of longitudinal displacement u, beam deflection w and rotation angle of initially undeformed normal ψ_x (nonlinear Cauchy relations) as follows:

$$\varepsilon_x = e_x + z\kappa_x,\quad e_x = \frac{\partial u}{\partial x} + \frac{\vartheta_x^2}{2},\quad \kappa_x = \frac{\partial \psi_x}{\partial x},\quad e_{xz} = \psi_x - \vartheta_x,\quad \vartheta_x = -\frac{\partial w}{\partial x}. \tag{4.4}$$

Integration of third and fourth relations from expressions (4.2) with respect to z and accounting for electrostatic boundary conditions adopted yields

$$\begin{aligned}
{}^m\sigma_x &= \left(c_{11}^E + \gamma_{33}\right)\varepsilon_x \mp \gamma_{31}{}^m D_z,\\
{}^m\sigma_{xz} &= c_{44}^E\left(\psi_x - \vartheta_x\right),\\
{}^m D_z &= -\frac{b_{33} V_{a,s}}{h_{1,2}} \pm b_{31}\left(e_x \mp \tilde{h}_{1,2}\kappa_x\right);\\
\gamma_{31} &= \frac{b_{31}}{b_{33}},\quad \gamma_{33} = \gamma_{31} b_{31},\quad \tilde{h}_{1,2} = \frac{h_0 + h_{1,2}}{2};\quad m = 1, 2
\end{aligned} \tag{4.5}$$

Substituting ${}^2D_z$ from Eq. (4.5) into the integral electric condition (4.1) at the outer surface of the piezosensor, one can obtain the amplitude of the electric potentials difference, V_s,

$$V_s = -\frac{\gamma_{31} h_2}{\Delta_x} \int_{x_0}^{x_1} \left(e_x + \tilde{h}_2\kappa_x\right) dx. \tag{4.6}$$

Substitution of V_s from (4.6) into the first expression from (4.5) for sensor yields the integro-differential equation for ${}^2\sigma_x$ that complicates greatly the derivation of the relations for forces and moments for the beam under consideration. To avoid this, let us use the second expression from conditions (4.1) for ${}^2\sigma_x$ in the relations (4.5). It must be emphasize here that second expression from (4.1) is satisfied approximately on the segment Δ_x and is exact in the case of unelectroded surface of the piezolayer. This assumption is similar to that made to design piezoelectric transducers: to calculate the output voltage of the generating section of the idling transducer, the electric-flux density along the polarization of the piezoelectric element is assumed zero as it was shown in [2].

Replacing the components of the stress tensor bystatically equivalent integral characteristics

$$\left(N_x, Q_x, M_x\right) = b \int \left(\sigma_x, \sigma_{xz}, z\sigma_x\right) dz,$$

through the total thickness of the beam accounting for relations (4.4) and (4.5), one can obtain the following formulas for the forces and moments:

$$N_x = C_{11}e_x + K_{11}\kappa_x + N_E;\ Q_x = k_s C_{44}(\psi_x - \vartheta_x);\ M_x = K_{11}e_x + D_{11}\kappa_x + M_E, \tag{4.7}$$

where

$$\begin{aligned}
C_{11} &= bh_0\left[E + c_{11}^E(\delta_1 + \delta_2) + \gamma_{33}\delta_2\right];\\
C_{44} &= bh_0\left[G_{13} + c_{44}^E(\delta_1 + \delta_2)\right];\\
K_{11} &= \frac{bh_0}{2}\left[\left(c_{11}^E + \gamma_{33}\right)(1 + \delta_2)\delta_2 - c_{11}^E(1 + \delta_1)\delta_1\right];\\
D_{11} &= \frac{bh_0^3}{12}\left[E + c_{11}^E(\delta_{13} + \delta_{23}) + \gamma_{33}\left(\delta_{23} + \delta_1^3\right)\right];\\
N_E &= bb_{31}V_a;\quad M_E = -\frac{bh_0(1 + \delta_1)b_{31}V_a}{2};\\
\delta_k &= \frac{h_k}{h_0};\quad \delta_{k3} = 4\delta_k^3 + 6\delta_k^2 + 3\delta_k,\quad k = 1, 2
\end{aligned}$$

and k_s is shear coefficient (see [16]).

The equations of harmonic bending vibrations of the beam in terms of amplitudes have the following form:

$$\frac{\partial N_x}{\partial x} - \bar{\rho}\frac{\partial^2 u}{\partial t^2} = 0;\ \frac{\partial \tilde{Q}_x}{\partial x} + Fq_z - \bar{\rho}\frac{\partial^2 w}{\partial t^2} = 0;\ \frac{\partial M_x}{\partial x} - \tilde{Q}_x - N_x\vartheta_x - \hat{\rho}\frac{\partial^2 \psi_x}{\partial t^2} = 0 \tag{4.8}$$

with the following mechanical boundary conditions: (i) for the case of hinged ends

$$u = 0,\quad w = 0,\quad M_x = 0,\quad x = 0, l; \tag{4.9}$$

(ii) for the case of rigidly fixed ends

$$u = 0,\quad w = 0,\quad \psi_x = 0,\quad x = 0, l. \tag{4.10}$$

Next notations are used in the Eq. (4.8):

$$\begin{aligned}
&\tilde{Q}_x = Q_x - N_x\vartheta_x,\ F = bH,\ H = h_0(1 + \delta_1 + \delta_2),\\
&\bar{\rho} = bh_0(\rho_0 + \rho_1\delta_1 + \rho_1\delta_2),\ \hat{\rho} = \frac{bh_0^3(\rho_0 + \rho_1\delta_{13} + \rho_1\delta_{23})}{12};
\end{aligned}$$

ρ_0 and ρ_1 are specific density of the beam materials.

As it was mentioned in the Introduction, forced vibrations of viscoelastic elements are accompanied by dissipative heating due to hysteresis losses in the material. Therefore the Eqs. (4.4)–(4.10) should be supplemented with the energy equation [11, 12]. The energy equation averaged over a cycle of vibration and over the thickness of the beam yields the heat conduction equation

$$\frac{1}{a}\frac{\partial T}{\partial t} = \frac{\partial^2 T}{\partial x^2} - \frac{2\alpha_s (b + H)}{\lambda F}(T - T_c) + \langle W \rangle , \tag{4.11}$$

with initial and boundary conditions as follows:

$$T = T_0, \quad t = 0;$$
$$\lambda \frac{\partial T}{\partial x} = \pm\alpha_{0,l}\,(T - T_c)\,, \quad x = 0, l.$$

Here $\langle W \rangle$ is averaged dissipation rate

$$\langle W \rangle = \frac{\omega b}{\lambda F}\int\limits_{(H)} \left(\sigma''_x \varepsilon'_x - \sigma'_x \varepsilon''_x + \sigma''_{xz} e'_{xz} - \sigma'_{xz} e''_{xz} + {}^1D'_z {}^1E''_z - {}^1D''_z {}^1E'_z\right) dz, \tag{4.12}$$

$\alpha_s = (\alpha_+ \alpha_-)/2$, $\alpha_\pm$ and $\alpha_{0,l}$ are heat exchange coefficients on the beam surfaces and at the ends respectively; λ and a are averaged heat conductivity and thermal diffusivity coefficients correspondently; T_0 is initial temperature of the beam.

4.3 Solution of the Problem

This section is devoted to the description of the technique used to solve the nonlinear coupled problem of electro-thermoviscoelasticity formulated above. Let us rewrite the resolving Eqs. (4.4), (4.7) and (4.8) in terms of decision variables $u, w, \psi_x, N_x, \tilde{Q}_x$ and M_x accurate within second order components

$$\begin{aligned}
\frac{\partial N_x}{\partial x} &= \bar{\rho}\frac{\partial^2 u}{\partial t^2},\\
\frac{\partial \tilde{Q}_x}{\partial x} &= -Fq_z + \bar{\rho}\frac{\partial^2 w}{\partial t^2},\\
\frac{\partial M_x}{\partial x} &= \tilde{Q}_x + N_x\psi_x - J_{SD}\left(\tilde{Q}_x N_x\right) + \hat{\rho}\frac{\partial^2 \psi_x}{\partial t^2},\\
\frac{\partial u}{\partial x} &= J_{CK}\,(N_x - N_E) - J_{DK}\,(M_x - M_E) - \frac{1}{2}\psi_x^2 + J_{SD}\left(\psi_x \tilde{Q}_x\right),\\
\frac{\partial \psi_x}{\partial x} &= -J_{DK}\,(N_x - N_E) + J_D\,(M_x - M_E)\,,\\
\frac{\partial w}{\partial x} &= -\psi_x + J_{SD}\tilde{Q}_x,
\end{aligned} \tag{4.13}$$

where

$$J_C = \frac{1}{C_{11}}, \quad J_D = \frac{1}{D_{11}\left(1 - \nu_{KC}\nu_{KD}\right)}, \quad J_{CK} = J_C + J_D \nu_{KC}^2,$$

$$J_{DK} = J_D \nu_{KC}, \quad \nu_{KC} = \frac{K_{11}}{C_{11}}, \quad \nu_{KD} = \frac{K_{11}}{D_{11}}, \quad J_{SD} = \frac{1}{k_s C_{44}}.$$

Neglecting of transient processes under harmonic loading of the form $q_z = q_0' \cos\omega t - q_0'' \sin\omega t$ (in our particular case $q_0'' = 0$), approximate solution of the system of nonlinear equations (4.13) can be expanded into harmonic series with respect to time. Then, to construct the solution, the single frequency approximation is used according with [6–8]. To take account of both geometrical nonlinearity and shear deformation, single frequency approximation applied to the set of variables $A = \{w,\ \psi_x,\ \tilde{Q}_x,\ M_x\}$ which characterizes bending of the beam while first and second harmonics should be retained in the expansion for the set of the variables $B = \{u,\ N_x\}$ that describes plane deformation of the beam

$$A = \overset{1}{A}{}' \cos\omega t - \overset{1}{A}{}'' \sin\omega t, \quad B = \overset{0}{B} + \sum_{k=1}^{2} (\overset{k}{B}{}' \cos k\omega t - \overset{k}{B}{}'' \sin k\omega t). \tag{4.14}$$

Let us use the single frequency approximation approach developed in [6–8] along with the assumptions (4.14). Then the system of partial differential equations (4.13) yields the system of ordinary differential equations of order $m = 18$ with respect to coefficients of the expansions (4.14). Linearization of the obtained system by the application of quasilinearization technique (see [6]) with account of boundary conditions (4.9) and (4.10) leads to the sequence of linear boundary-value problems

$$\frac{d\mathbf{Y}^{n+1}}{dx} = \Re\mathbf{Y}^{n+1} + \mathbf{F}\left(q_z,\ N_E,\ M_E,\ \mathbf{Y}^n\right),$$

$$B_1\mathbf{Y}^{n+1}(0) = 0,\ B_2\mathbf{Y}^{n+1}(l) = 0,\ k = 1, 2,\ n = 0, 1, 2, ..., \tag{4.15}$$

where $(\mathbf{Y})^T = \{\overset{1}{u}{}',\ \overset{1}{u}{}'',\ \overset{2}{u}{}',\ \overset{2}{u}{}'',\ \overset{1}{w}{}',\ \overset{1}{w}{}'',\ \overset{1}{\psi}{}_x',\ \overset{1}{\psi}{}_x'',\ \overset{1}{N}{}_x',\ \overset{1}{N}{}_x'',\ \overset{2}{N}{}_x',\ \overset{2}{N}{}_x'',\ \overset{1}{\tilde{Q}}{}_x',\ \overset{1}{\tilde{Q}}{}_x'',\ \overset{1}{M}{}_x',\ \overset{1}{M}{}_x'',\ \overset{0}{u},\ \overset{0}{N}{}_x\}$ is a vector-column of unknown functions; $\Re$ is a square $m \times m$ matrix with components that depend on material parameters (which are frequency dependent in turn) and values of the unknowns at the previous iteration, $\mathbf{Y}^n$, $\mathbf{F}$ is a vector-column of free terms of the system. Expressions for the elements of $\Re$ and $\mathbf{F}$ are very cumbersome and are omitted here to save space. Rectangular matrices B_1 and B_2 are determined by the boundary conditions (4.9) and (4.10) respectively.

In terms of sought quantities, dissipative function (4.12) can be written as follows:

$$\langle W\rangle = \frac{\omega}{2\lambda F}\left\{\sum_{k=1}^{2} k\left(\overset{k}{N''_x}\overset{k}{e'_x} - \overset{k}{N'_x}\overset{k}{e''_x}\right)\right.$$
$$\left. + \overset{1}{M''_x}\overset{1}{\kappa'_x} - \overset{1}{M'_x}\overset{1}{\kappa''_x} + \overset{1}{Q''_x}\overset{1}{e'_{xz}} - \overset{1}{Q'_x}\overset{1}{e''_{xz}} + {}^1D''_z V'_a - {}^1D'_z V''_a\right\}. \quad (4.16)$$

The nonlinear system of normal form differential equations subjected to correspondent boundary conditions is solved numerically with application of a time-marching integration scheme [1, 21]. At each iteration, solution of the linear boundary value problem (4.15) is obtained by the stable discrete-orthogonalization method with application of the parameter extension procedure with respect to frequency along with standard routine for solving the system of ordinary differential equations. Solution of the geometrically linear problem is used as the initial approximation ($n = 0$). The heat conduction problem (4.11) with account of expression (4.16) is solved using explicit finite difference method.

There are two ways to determine the amplitude of compensating electric potential (actuator parameter), V_a, that should be applied to actuator for active damping of a forced vibration of the beam [8]. According to the first one, if pressure of a constant amplitude $q'_0 = q_0$ is known then

$$|V_a| = k_a(\Delta_x)\, q_0. \quad (4.17)$$

When mechanical loading is unknown, actuator factor, V_a, can be found with making use of sensor parameter, V_s, and feed-back relation

$$|V_a| = G_{as}\,|V_s|\,. \quad (4.18)$$

In the relations (4.17) and (4.18), k_a and G_{as} are control ratios. Values of k_a and G_{as} corresponding to maximal damping are determined by formulas $k_a = \left|w^1_{q\,\max}\right| / \left|w^1_{E\,\max}\right|$ and $G_{as} = \left|V^1_a\right| / \left|V^1_s\right|$ where $\left|w^1_{q\,\max}\right|$ and $\left|w^1_{E\,\max}\right|$ are maximal amplitudes of deflection which are calculated as solutions of reference (calibration) problems at the linear resonance frequency for $q_0 = 1$ Pa, $V_a = 0$ V and $q_0 = 0$ Pa, $V_a = 1$ V respectively. Values of $\left|V^1_a\right|$ and $\left|V^1_s\right|$ are determined as solutions of the problem for $q_0 = 1$ Pa with making use of formulas (4.17) and (4.6). Antiphase of electric potential supplied to actuator with the aim of the beam vibration damping is ensured by the law $V_a \cos(\omega t + \pi) = -V_a \cos \omega t$.

4.4 Results and Discussion

Calculations were performed for the beam made of electrically passive polymer as a material of a core layer possessing properties as follows [18]: $\overset{k}{E} = \overset{k}{E'} + i\,\overset{k}{E''}$; $\overset{k}{E'} = \overset{0}{E}\,(k\omega)^p$, $\overset{k}{E''} = \overset{k}{E'}\,\beta(k\omega)^q$, $G_{13} = 0.025\,\overset{k}{E}$, $(k = 1, 2)$; $\overset{0}{E} = 0.308 \cdot 10^{10}$ N/m^2,

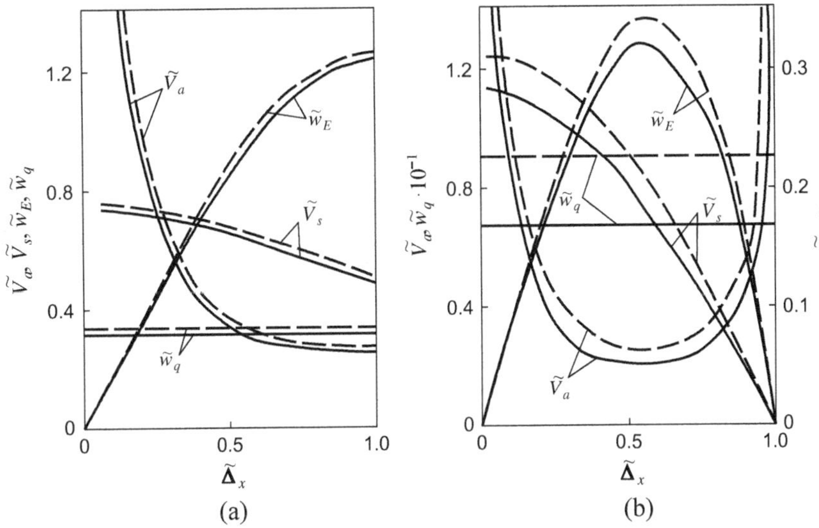

Fig. 4.2 Dependence of thermomechanical parameters on the actuator electroded region ratio for beam with hinged (**a**) and rigidly fixed ends (**b**)

$\beta = 0.16$, $q = -0.145$, $p = 0.076$, $\rho_0 = 2770$ kg/m^2, $\lambda = 0.45$ W/(m^2·K). Piezolayers of sensor and actuator are fabricated from the PZT piezoceramics with the following characteristics [2]: $s'_{11} = 12.5 \cdot 10^{-12}$ m^2/N, $s'_{44} = 39.7 \cdot 10^{-12}$ m^2/N, $d'_{31} = -1.6 \cdot 10^{-10}$ C/m, $\varepsilon'_{33} = 2100\varepsilon_0$, $d'_{15} = 4.5 \cdot 10^{-10}$ C/m, $\varepsilon_0 = 8.854 \cdot 10^{-12}$ F/m, $\varepsilon'_{11} = 18.5 \cdot 10^2\varepsilon_0$, $\rho_1 = \rho_2 = 7520$ kg/m^3, $\delta^s_{11} = 0.0016$, $\delta^s_{44} = 0.0014$, $\delta^d_{31} = 0.004$, $\delta^d_{15} = 0.0035$, $\delta^\varepsilon_{33} = 0.0035$, $\delta^\varepsilon_{11} = 0.005$, $\lambda = 0.47$ W/(m^2· K).

Calculations were performed for the wide variety of the geometrical parameters of the beam. Typical results are shown below for the beam dimensions $l = 0.04$ m, $h_0 = b = 0.02$ m, with the thicknesses of piezolayers are chosen to be equal $h_1 = h_2 = 0.5 \cdot 10^{-4}$ m. The shear coefficient, k_s, for this case is equal to 5/6 [16].

Due to structural symmetry, type of piezolayers polarization, and loading conditions, the beam undergoes purely bending vibration. Therefore, the excitation frequencies were chosen to be close to the first bending frequency of the beam. Dependencies of maximum deflection amplitude, $\tilde{w}_q = \left|w^1_{q\,\max}\right| \cdot 10^7$ m, electric parameter of sensor, $\tilde{V}_s = \left|V^1_s\right| \cdot 10^3$ V, for external pressure amplitude $q_0 = 1$ Pa (for voltage free actuator $V_a = 0$) as well as deflection amplitude $\tilde{w}_E = \left|w^1_{E\,\max}\right| \cdot 10^5$ m for supplied voltage to actuator $V_a = 1$ V (under pressure free condition $q_0 = 0$) and actuator parameter $\tilde{V}_a = \left|V^1_a\right| \cdot 10^2$ V which compensate the unit mechanical loading of $q_0 = 1$ Pa on the electroded region ratio $\tilde{\Delta}_x = \Delta_x / l$ (dimensionless parameter characterizing the ratio of the area covered with electrode over the total area of the piezolayer) are shown in Fig. 4.2a, b for beams with hinged and rigidly fixed ends respectively. It should be mentioned here that piezoactive region of both sensor and actuator are considered to be located symmetrically, so its center coincides

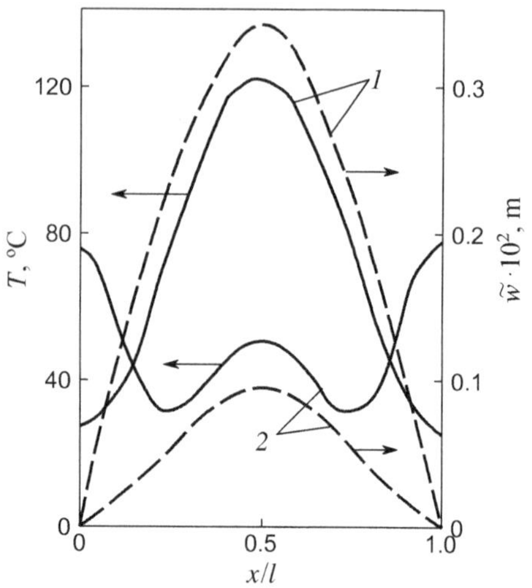

Fig. 4.3 Distribution of deflection amplitude and temperature of dissipative heating along the beam

with the point of maximal deflection, $x = 0.5l$. In Fig. 4.2, solid lines correspond to the classical problem statement solution at resonance frequencies $\omega_r = 532$ rad/s (Fig. 4.2a) and $\omega_r = 1236$ rad/s (Fig. 4.2b) while dashed lines are referred to the case of accounting for shear deformation at $\omega_r = 518$ rad/s and $\omega_r = 1096$ rad/s respectively.

Analysis of the curves in Fig. 4.2 shows that accounting for shear deformation leads to the decrease in the first bending resonance frequency of the beam and increase of the reference dynamic characteristic under mechanical ($q_0 = 1$ Pa, $V_a = 0$ V) and electrical ($q_0 = 0$ Pa, $V_a = 1$ V) loading. Beam fixing conditions transform qualitatively the behavior of the characteristics shown in Fig. 4.2. Boundary conditions modification leads to significant alteration of optimal dimensions of the piezoactuator designed for providing either largest beam deflection or the most effective damping of the beam vibration under electric loading. In this sense, the most effective actuator in the case of hinge ends is one with electrode region ratio $\tilde{\Delta}_x = 1$ (see Fig. 4.2a) and with $\tilde{\Delta}_x = 0.55$ for the case of the fixed ends (Fig. 4.2b). Accounting for shear deformation has little effect on the parameter $\tilde{\Delta}_x$.

Distribution of deflection amplitude, w, and temperature of dissipative heating, T, along the beam are shown in Fig. 4.3 by dashed and solid lines respectively for hinged (curves 1) and fixed (curves 2) ends under $q_0 = 0.1 \cdot 10^6$ Pa and $\alpha_s = \alpha_{0,l} = 5$ W/(m^2· K). Calculations of the curves 1 and 2 were performed with taking the shear deformation into account at the frequency $\omega = 518$ rad/s and $\omega = 1096$ rad/s respectively. It is clearly observable that influence of the boundary conditions under consideration manifests itself in variation of the deflection amplitude (dashed lines) but does not affect the shape of the amplitude distribution along the beam axis. At the same time, temperature distribution along the beam length demonstrates irregular behavior having maximum in the center for the case of the hinged ends and two maxima in the vicinity of the both ends which are rigidly fixed.

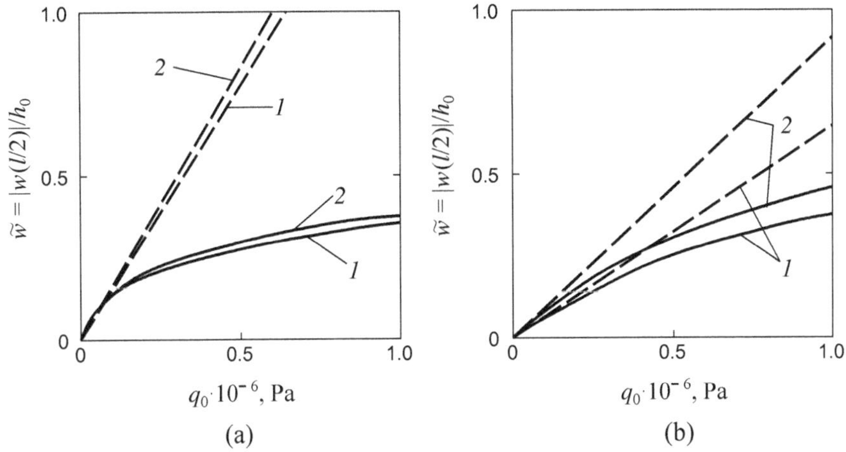

Fig. 4.4 Deflection amplitude versus amplitude of surface pressure for beam with hinged (**a**) and rigidly fixed ends (**b**)

Dependence of the calculated dimensionless maximum deflection amplitude, $\tilde{w} = |w\,(l/2)|\big/h_0$, on the amplitude of the surface pressure, q_0, calculated at the linear resonance frequency $\omega = 532$ rad/s (without accounting for shear deformation) and $\omega = 518$ rad/s (with taking it into account) are shown in Fig. 4.4a by the curves 1 and 2 correspondingly for the case of hinged beam. The same curves for rigidly fixed ends of the beam at the frequencies $\omega = 1236$ rad/s and $\omega = 1096$ rad/s are presented in Fig. 4.4b. Dashed and solid lines correspond to the solutions obtained in the frame of linear and nonlinear problem statements further on. For the flexible beam possessing the above listed physical and mechanical characteristics, comparison of the Fig. 4.4a, b shows that contribution of geometrical nonlinearity becomes noticeable at the loading amplitudes q_0 inducing relative deflection amplitudes $\tilde{w} \geq 0.1$ for both hinged and rigidly fixed ends of the beam. But in the latter case, the higher loading amplitudes q_0 are needed to reach this level of deflection.

In Figs. 4.5 and 4.6, the graphs illustrate dependence of maximum deflection amplitude, $\tilde{w}$, and electric parameter of sensor $|V_s|$ on excitation frequency for hinged (Figs. 4.5a and 4.6a) and rigidly fixed (Figs. 4.5b and 4.6b) ends of the beam of dimensions $l = 0.2$ m, $h_0 = b = 0.01$ m at the most effective values of the parameter Δ_x of sensor and actuator. Dashed and solid lines correspond to the solutions obtained in the frame of linear and nonlinear problem statements. Curves 1 correspond to the values $q_0 = 0.25 \cdot 10^6$ Pa, $V_a = 165.3$ V while curves 2 represent the case of values $q_0 = 0.5 \cdot 10^6$ Pa, $V_a = 322.3$ V. The dashed and solid lines 2 in Fig. 4.6b are very close. The dash-dot curves are obtained for the case of combined anti-phase action of mechanical loading with amplitude q_0 and compensating it electrical loading V_a. Here the compensating values of V_a were calculated according to formula (4.17) and (4.18). The dash-dot lines are the same for the cases of both linear and nonlinear problems. Their coincidence confirms the correctness of the formula (4.18)

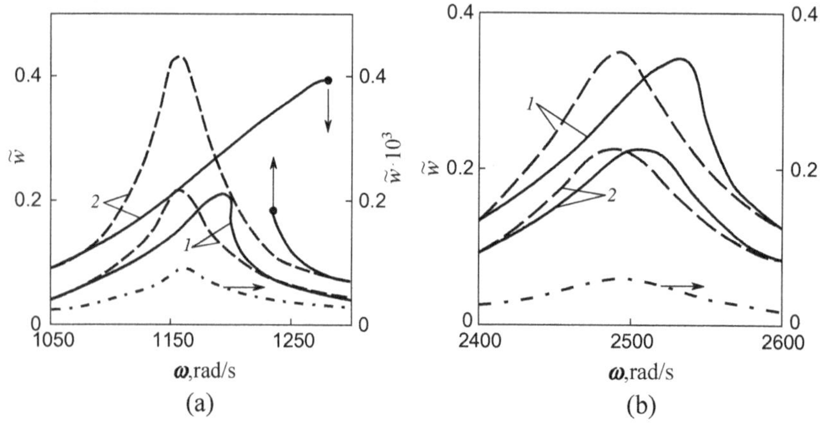

Fig. 4.5 Amplitude–frequency characteristics for beam with hinged (**a**) and rigidly fixed ends (**b**)

application for the case of unknown mechanical loading. Frequency dependencies of maximum temperature of dissipative heating over the beam volume at the steady-state stage under loading conditions considered are completely analogous to the ones shown in Fig. 4.5.

Analysis of the solid lines in Figs. 4.5 and 4.6 reveals that accounting for geometrical nonlinearity (that usually leads to formation of amplitude–frequency characteristics of a stiff type) manifests itself more vividly in the case of hinged beam (Figs. 4.5a and 4.6a) then in the case of rigidly fixed ends of the beam (Figs. 4.5b and 4.6b). For instance, it follows from the comparison of curves 2 under the loading parameter $q_0 = 0.5 \cdot 10^6$ Pa that accounting for geometrical nonlinearity leads to transformation of the linear frequency characteristics into nonlinear for hinged beam but does not change them in the case of rigidly fixed beam. The dash-dot lines in Fig. 4.5 shows that amplitudes of mechanically induced vibration being damped by application of compensating voltage to the actuator electrodes can be reduced by three orders of magnitude. Under these conditions, the influence of geometrical nonlinearity does not occur and temperature of the beam remains close to the initial value.

Under certain amplitudes of mechanical or electrical harmonic loading along with some conditions of heat transfer into environment, temperature of dissipative heating of complex thin wall structures composed of inelastic electrically passive and piezoactive materials can reach a critical value, T_{cr}, referred to as degradation point when thermal destruction of the system occurs due to either thermal softening of the passive material or due to depolarization of the piezoactive constituent (Curie point). At this, the critical amplitude of mechanical or electric loading (q_{cr} and V_{cr} respectively) which correspond to the critical value of steady-state temperature of dissipative heating, T_{cr}, should be determined. In addition, it is necessary to find the critical fatigue-life time interval τ_{cr} before the structure workability is being exhausted under the loading amplitude that exceeds the correspondent critical value.

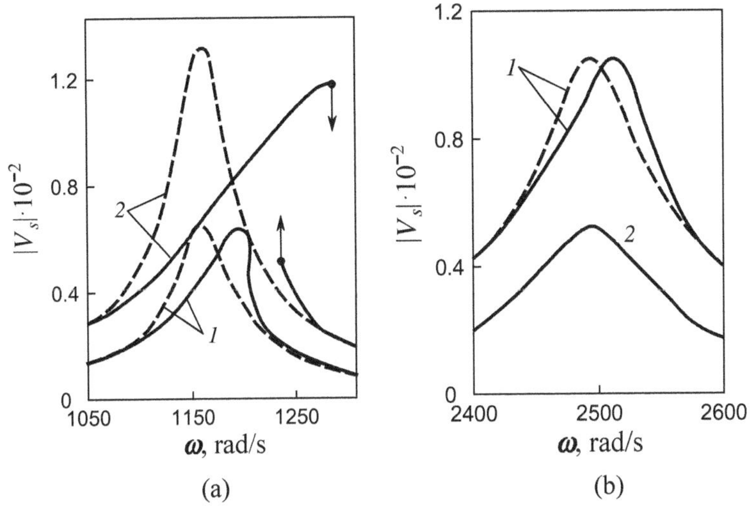

Fig. 4.6 Electric parameter of sensor versus frequency for beam with hinged (**a**) and rigidly fixed ends (**b**) in the vicinity of first resonance

Fig. 4.7 Dependence of maximum temperature on loading pressure amplitude

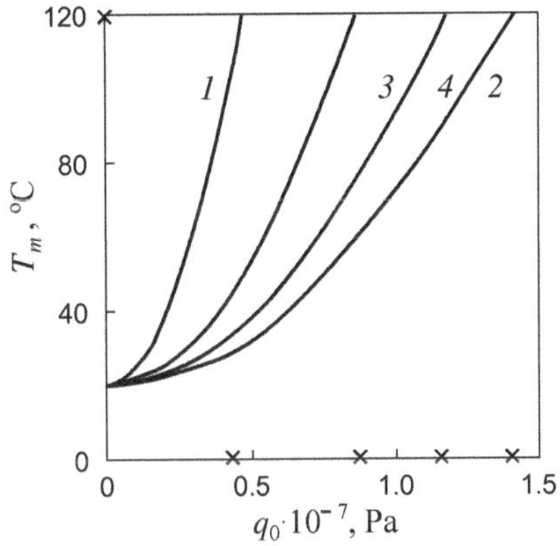

Let us assume the system degradation occurs if critical temperature $T_m = T_{cr} = 120\,°C$ is reached at one point of the beam at least. Curves 1–4 in Fig. 4.7 represent dependence of maximum steady-state temperature of dissipative heating, T_m, in the vicinity of the ends of the rigidly fixed beam on the amplitude of mechanical loading at the adjusted frequency $\omega = 1096$ rad/s for heat exchange coefficients $\alpha_s = \alpha_{0,l} = 1,\ 5,\ 10,\ 15$ W/(m^2·K) respectively. The critical values of mechanical loading amplitude, q_{cr}, corresponding to the critical temperature, T_{cr}, are marked with crosses on abscissa axis. It can be seen that increase in the heat exchange coefficient lead to the rise of the critical mechanical loading, q_{cr}, at which the heating temperature reaches the system degradation temperature. It is worth mentioning

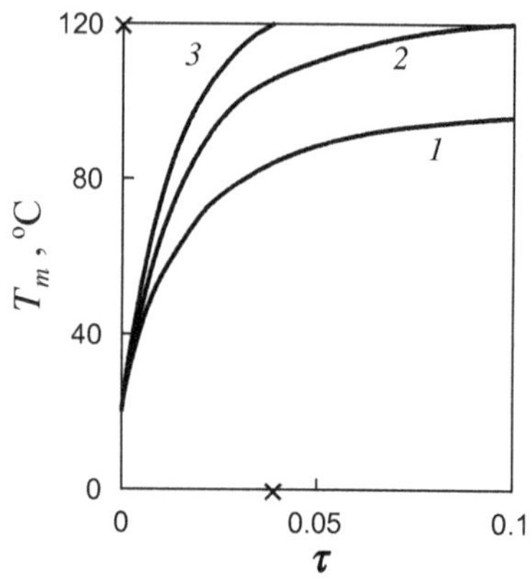

Fig. 4.8 Evolution of maximum temperature

here that geometrical nonlinearity does not affect the beam response in the excitation frequency range under consideration.

Let us emphasize that both steady-state and transient response of the beam were investigated. Mechanical part of the problem is calculated as stabilized in time solution while temperature is considered to be varying with cycles. This scheme is completely applicable because specific stabilization time for temperature processes is much longer that one for mechanical variables. Evolution of the maximum temperature of dissipative heating with respect to normalized time, $\tau = at/l^2$, is shown in Fig. 4.8 by the curves 1–3 calculated at frequency $\omega = 1096$ rad/s and heat exchange coefficients $\alpha_s = \alpha_{0,l} = 1$ W/(m^2·K) for values of mechanical loading amplitude $q_0 = 0.4,\ 0.46,\ 0.5$ Pa respectively. Degradation temperature, T_{cr}, and critical time interval, τ_{cr}, are marked with crosses on the ordinate and abscissa axis correspondingly. It is seen that if loading amplitude $q_0 \leq q_{cr} = 0.462 \cdot 10^7$ Pa then temperature of dissipative heating does not reach T_{cr} (see curves 1 and 2) and therefore the thermal break-down of the structure does not occur. For $q_0 \geq q_{cr}$, there is a time moment $\tau \geq \tau_{cr}$ when heating temperature becomes equal to the degradation point of material.

Relationship between critical loading amplitude, q_{cr}, and critical time parameter, τ_{cr}, is presented in Fig. 4.9 for two values of heat transfer coefficients $\alpha_s = \alpha_{0,l} = 1.0,\ 1.5$ W/(m^2· K) by curves 1 and 2 respectively at frequency $\omega = 1096$ rad/s. Safe value of loading amplitude can be found as the intersection point of horizontal asymptote with ordinate axis. These values are marked with crosses for the heat transfer coefficients mentioned above. Below these critical values, the thermal break-down of the beam never occurs. The curves presented in Fig. 4.9 are in fact the analogues of the S-N curves usually referred to as Wöhler diagrams. They are used to represent a relationship between applied stress versus number of cycles to break-down. In our case, the curves separate the region of safe functioning (below the curves) from the regimes that lead to break-down (above the curves). If the loading amplitude is specified as well as vibration frequency and thermal boundary conditions then the safe life-time for the structure can be found before the critical time, τ_{cr}, can be

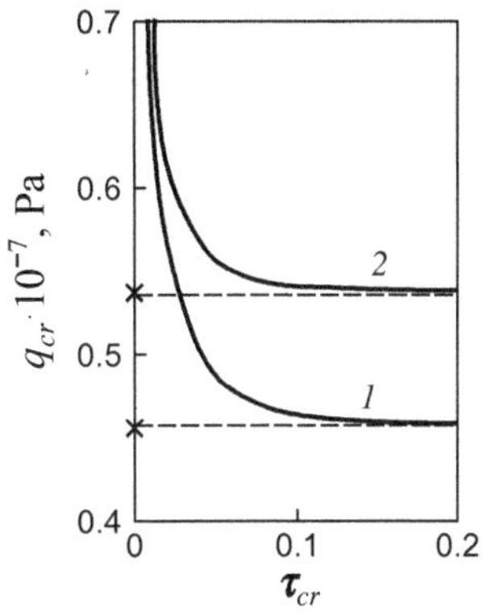

Fig. 4.9 Critical loading amplitude versus critical time parameter

reached and thermal break-down occurs due to thermal degradation of the material. It is clear as well that behavior of the curves depends significantly on the heat exchange conditions on the beam surface.

4.5 Conclusions

Statement of the problem on forced resonance vibration and dissipative heating of viscoelastic beam containing piezoelectric sensor and actuator with taking account for shear deformation and quadratic geometrical nonlinearity is developed. With application of monoharmonic approximation technique and making use of complex moduli concept, the problem statement is reduced to the normal system of ordinary differential equations supplemented with boundary conditions corresponding to the cases of hinged or rigidly fixed beam ends. The nonlinear boundary value problem is solved numerically. For the most energy intensive first bending mode of beam vibration, influence of the mechanical end fixing conditions, heat exchange, shear deformation and geometrical nonlinearity accounting onto frequency characteristics of maximum deflection amplitude, level of dissipative heating temperature and electric parameter of sensor are investigated under monoharmonic loading. It is shown that fixing conditions have significant effect on the temperature distribution along the beam length and on the dimensions of the piezoactuator providing the most effective active damping of the mechanically induced vibration. For example, in the partial case of hinged beam, the maximum heating temperature occurs in its middle section and the most effective damping takes place if the beam surface is covered with piezoactuator layer completely. Meanwhile, in the case of rigidly fixed beam, the maximum temperature of dissipative heating occurs in the vicinity of the beam ends and for partially covered surfaces by the piezoactuator layer. Accounting for shear deformation predicts lower resonance frequency of the beam and does not affect dimensions of the optimal effective actuator. Geometrical nonlinearity manifests itself under the loadings inducing relative deflections in the region $\tilde{w} \geq 0.1$ and leads to transformation of linear amplitude–frequency characteristics into nonlinear characteristics of the stiff type. Possibility of active control (damping) of the beam vibration by means of piezoactuatoris shown with making use of electric parameter (factor) of sensor in the case of unknown loading amplitude.

References

1. Blanguernon, A., Lene, F., Bernadou, M.: Active control a beam a piezoceramic element. Smart Mater. Struct. **8**, 116–124 (1999)
2. Bolkisev, A.M., Karlash, V.L., Shul'ga, N.A.: Temperature dependence of the properties of piezoelectric ceramics. Int. Appl. Mech. **20**, 650–653 (1984)
3. Franz, W.: Dielektrischer Durchschlag. Handbuch der Physik. Springer, Berlin-Gottingen-Heidelberg (1965)
4. Gabbert, U., Tzou, H.S.: Smart Structures and Structronic Systems. Kluwer Academic Publishers, Dordrecht (2001)
5. Guz, I.A., Zhuk, Y.A., Kashtalyan, M.: Dissipative heating and thermal fatigue life prediction for structures containing piezoactive layer. Technische Mechanik **32**, 238–250 (2012)
6. Karnaukhov, V.G., Kirichok, I.F.: Forced harmonic vibrations and dissipative heating-up of viscoelastic thin-walled elements (review). Int. Appl. Mech. **36**, 174–195 (2000)
7. Karnaukhov, V.G., Kirichok, I.F., Kozlov, V.I.: Electromechanical vibrations and dissipative heating of viscoelastic thin-walled piezoelements (review). Int. Appl. Mech. **37**, 182–212 (2001)
8. Karnaukhov, V.G., Kirichok, I.F., Kozlov, V.I.: Thermomechanics of inelastic thin-wall structural members with piezoelectric sensors and actuators under harmonic loading (review). Int. Appl. Mech. **53**, 6–58 (2017)
9. Karnaukhova, T.V., Pyatetskaya, E.V.: Basic equations for thermoviscoelastic plates with distributed actuators under monoharmonic loading. Int. Appl. Mech. **45**, 200–214 (2009)
10. Katunin, A., Wronkowicz, A., et al.: Criticality of self-heating in degradation processes of polymeric composites subjected to cyclic loading: A multiphysical approach. Arch. Civ. Mech. Eng. **17**, 806–815 (2017)
11. Kirichok, I.F.: Resonant vibration and heating of ring plates with piezoactuators under electromechanical loading and shear deformation. Int. Appl. Mech. **45**, 215–222 (2009)
12. Kirichok, I.F.: Control of axisymmetric resonant vibrations and self-heating of shells of revolution with piezoelectric sensors and actuators. Int. Appl. Mech. **46**, 890–901 (2011)
13. Lu, X., Hanagud, S.V.: Extended irreversible thermodynamics modeling for self-heating and dissipation in piezoelectric ceramics. IEEE Trans. on Ultrasonics. Ferroelectrics, and Frequency Control **31**, 1582–1592 (2004)
14. Mauk, L.D., Lynch, C.S.: Thermo-electro-mechanical behavior of ferroelectric materials. Part I: Computational micromechanical model versus experimental results. J. Int. Mat. Sys. Struct. **14**, 587–602 (2003)
15. Nashif, A.D., Johnes, D.J., Henderson, J.P.: Vibration Damping. John Wiley & Sons, New York (1985)
16. Reddy, J.N.: Theory and Analysis of Elastic Plates and Shells. CRC Press, Boca Raton-London-New York (2007)
17. Schwartz, M.M.: Encyclopedia of Smart Materials. Wiley, New York (2002)
18. Stevens, K.K.: Transverse vibration of a viscoelastic column with initial curvature under periodic axial load. J. Appl. Mech. **36**, 814–818 (1969)
19. Tzou, H.S.: Piezoelectric Shells (Distributed Sensing and Control of Continua). Kluwer Academic Publishers, Dordrecht-Boston-London (1993)

20. Tzou, H.S., Anderson, G.L.: Intelligent Structural Systems. Kluwer, Dordrecht (1992)
21. Tzou, H.S., Bergman, L.A.: Dynamics and Control of Distributed Systems. Cambridge University Press, New York (1998)
22. Zgurovsky, M.Z., Kasyanov, P.O., Paliichuk, L.S., Tkachuk, A.M.: Dynamics of solutions for controlled piezoelectric fields with multivalued "reaction-displaceme" law. Stud. Syst. Decis. Control **30**, 267–276 (2015)
23. Zhuk, Y.A., Guz, I.A.: Active damping of the forced vibration of a hinged beam with piezoelectric layers, geometrical and physical nonlinearities taken into account. Int. Appl. Mech. **45**, 94–108 (2009)
24. Zhuk, Y.A., Guz, I.A., Sands, C.M.: Analysis of the vibrationally induced dissipative heating of thin-wall structures containing piezoactive layers. Int. J. Non-Lin. Mech. **47**, 105–116 (2012)

Chapter 5
A Stochastic Theory of Scale-Structural Fatigue and Structure Durability at Operational Loading

E. B. Zavoychinskaya

Abstract Here are discussed the results of numerous experimental and theoretical investigations of multicyclic fatigue multilevel processes at complex stress state of metals and alloys on solid state physics, metal science and solid mechanics. Proposed by the author on their basis the scale-structural fatigue theory describing the evolution of fatigue defects and allowing to find the metal durability on a certain level of accumulated defects are presented. A new method for evaluation the durability of long structures, including criteria of structural reliability and technogenic safety of operation and calculation of structural element durability on the offered theory, is briefly considered.

5.1 Introduction

According to modern phenomenological approaches to fatigue estimation at long-term and cyclic loading the existence of a determinate damage function (tensor) is supposed. Failure is determined when this function (or some measure of the damage tensor) is reached unity. On the results of numerous experiments at various loading spectra there were calculated the values of these functions at failure. It was established that its values are in the range from 0.1 to 10. A statistical analysis of this interval shows that the damage function is a random function with an average value of unity and its probability distribution density on the truncated normal (log-normal) law. So the range of applicability of these approaches is the loading processes that are close to the processes of the material functions and parameters determination.

The proposed approach considers the failure process as a random steady-state process for which the multilevel failure probability function at loading in each time

E. B. Zavoychinskaya (✉)
Lomonosov Moscow State University, GSP-1, Leninskie Gory,
Moscow 119991, Russian Federation
e-mail: elen@velesgroup.com

V. A. Sadovnichiy and M. Z. Zgurovsky (eds.), *Contemporary Approaches and Methods in Fundamental Mathematics and Mechanics*, Understanding Complex Systems,
https://doi.org/10.1007/978-3-030-50302-4_5

moment is introduced. At arbitrary multiaxial loading the random processes of brittle and viscous failures can be described by linear functionals whose kernels are random time functions.

In this work within the framework of this approach it is presented a model of brittle failure evolution in metals and alloys at multicycle and gigacycle fatigue, based on the physical relations of the process sequentially at the nano- and microlevel of defect substructures and microstructures, at the mesoscale level of grain, and the macrolevel of grain ensembles up to leader brittle macrocrack of the characteristic length. The leader cracking is described by well-known methods and approaches of fracture mechanics.

Here briefly presents (in Sect. 5.2) the results of experimental explorations of the main properties of fatigue failure hierarchical processes at the micro-, meso-, and macrolevels [1–12]. Their analysis leads to the need to describe of the fatigue failure process in the framework of multilevel models. In Sect. 5.3 the critical overview of the modern theoretical approaches to metal fatigue description of solid state physics, metal science and solid mechanics [2–4, 7, 8, 10, 13–24] in the interpretation of its author.

Section 5.4 presents the main hypotheses of the proposed scale-structural fatigue theory and its structure [25–31]. From the entire hierarchy of scale levels, six levels are considered in accordance with the metal structure evolution by various physical mechanisms. There are formulated constitutive relations for failure probability at each level at three-dimensional proportional loading processes. There are plotted fatigue curves on defect levels.

In the last section, within the framework of the proposed approach and on the methodological base of famous works [32–36], it is considered the failure probability of long structure sections and proposed criterion of structural reliability [23, 37–39]. There was suggested the method for durability estimation, including a system of criteria of operation technogenic safety [37–39]. On their basis there was received the expression for the design durability of the structure operation safety as a random function of the material properties, structural scheme, operational loading, distribution of the failure flow at a similar structural scheme system operation, taking into account anthropogenic and technogenic factors.

5.2 The Main Properties of Metal Fatigue at Microscopic, Mesoscopic and Macroscopic Scale-Structural Levels

The solution of structure durability estimation problem is based on wide systematic experimental investigations of the multilevel failure process at long-term and cyclic loading of metals and alloys, which are carried out in solid state physics, metal science and solid mechanics by methods of the corresponding scientific directions. In known publications of recent years (e.g. [1–10]) the stochastic relations of the failure evolution at multi- and gigacyclic fatigue, related to the physical and mechanical material properties, the type of stress-strain state, temperature conditions, etc. are explored; the physical mechanisms of brittle and viscous failure at various scale-structural levels are considered; self-similar relations of the defect accumulation kinetics at different levels and macrocrack growing are investigated; scale-invariant characteristics of the defect areas and their relations with the mechanisms and levels of the failure process are determined; correlation between fracture characteristics and the material structure are established.

According to the results of experimental investigations of multicycle and gigacycle fatigue, associated with a wide variety of physical mechanisms of these processes, which are realized at various scale and structural levels, the process of fatigue failure is the result of the evolution of metal defects (physical heterogeneities of metallurgical and technological origin) during loading process. Various mechanisms of fatigue failure are determined by the structural heterogeneity of the metal, random size distributions (lying in a wide range, on average $d \in \left(10^{-4}, 5\right)$ mm), the shape and orientation of the crystal lattice of grains, grain boundaries, and there are associated with the occurrence of microstresses caused by random and inhomogeneous nucleation of crystallization centers, random processes of growing crystals interactions with each other and the distribution of particles of refractory non-metallic inclusions, oxides, impurities, etc. [2, 6–12].

On modern concepts, the processes of brittle fracture of metals and alloys are characterized by the obligatory following levels of its evolution.

Typical **nanoscale structures** include point defects of characteristic sizes, on average, $l_1 \leq 10^{-5}$ mm (vacancies, bivacancies, Schottky defects, Frenkel defects, impurity and doping atoms, electron-hole defects), linear defects (one-dimensional growth dislocations formed during crystal growth, disclinations, chains of vacancies or interstitial atoms, vacancy clusters), plane defects of size l_1 in two directions (grain, subgrain and twin boundaries, packing defects, phase interfaces, twinning planes, two-dimensional dislocations) and volumetric defects of size, on average, l_1 in three directions (pores, inclusions, particles of alloying, impurity or other elements, three-dimensional dislocations), formed as a result of microstresses caused by random and inhomogeneous nucleation of crystallization centers and the distribution of nonmetallic inclusions, oxides, impurities, etc., by random processes of interaction of growing crystals with each other at the influence of thermal, mechanical, electrical effects. At a certain stress level with an increase of cycle number the redistribution of existing defects occurs, their density q_1 increases, especially near boundaries and inclusions, in accordance with the principle of minimizing free energy; the dislocations growing begins, they are combined in slip lines and in slip bands (inelastic microstrain takes place). Changes in the metal substructure are observed by various physical methods: by measuring microhardness in volumes measured with grain sizes, of density, and of volume changes during hydrostatic weighing. In solid state physics and metal science work the authors establish that at this failure level there are formed ordered self-organizing substructures of two categories: a strip ("mesh") structure associated with the accumulation of submicrodefects (submicrodisconti nuities) of characteristic sizes l_1 and density q_1 and a cellular structure associated with multiple microshears along slip planes determined by the Burgers vector. At multi-cycle fatigue of some materials, only a strip structure is observed, in others (plastic materials) there is a stress level of beginning and growing of predominantly cellular structure. According to experiments at uniaxial loading stress amplitudes below the sensitivity to cyclic stresses limit $\sigma^c = \sigma^c(N, \omega)$ (N is the number of cycles, ω is the loading frequency) for a large class of materials, nanolevel defects only accumulate, do not saturate and go to the next structural level. In these cases, the microstress field created by the defects is in equilibrium with the microstress field arising as a reaction of the material to the submicrodefects origin. Nanoscale defects are reversible ones and can be "cured", for example, at recrystallization firing.

At a certain stress level as a result of the evolution of strip structures at increasing cycle number a certain critical value of the density of submicrodefects $q_{f,1}$ is achieved (for example, for BCC-metals in the slip band $q_{f,1} = 10^8\ \text{mm}^{-2}$) in the regions of critical curvature of the crystal lattice (called failure centers), and by their merging, a phase transition occurs, its consequence is the formation of a new, more stable phase of **microlevel structures** of characteristic sizes, on average, $l_2 \in \left(10^{-3}, 10^{-2}\right] d$, where d is the average metal grain size. Microdefects became sinks for nanodefects, their density decreases increasing the microdefect size. L. R. Botvina believes that the rate of microdefect density change determines the kinetic microfailure process. Brittle fatigue grooves resulting from the growth of brittle microcracks with an increase of the cycle number are observed on microphotographs. Researchers believe that the source of microcracks is the concentration of microstresses due to the possible development of submicrodefects in the most weakened grain or at the grain boundary, interface, nonmetallic inclusions, etc., depending on the stress state. As a result of the phase transition, microstresses relax, and the stored internal energy of the strip structures is converted to the surface energy of the formed microcracks. In this case, microdefects are formed "impressive" (compared with the atom size) sizes. The size of the observed microcracks in metals and alloys is, on average, $l_2 \in \left(10^{-4}, 5 * 10^{-3}\right)$ mm and their maximum density $q_{f,2}$ is unevenly distributed, its magnitude decreases by 2–3 orders moving into the sample. At this level, the physical and mechanical properties of metals and alloys change (electrical resistivity, microhardness, elastic moduli, magnetic characteristics), microdefects are irreversible ones.

The kinetics of the fracture process consists in the monotonous accumulation of defects and their merging when failure states are reached at the time with varying degrees of probability and the formation of the next level defects. Each scale-structural level is the result of processes of all previous levels. To change the level is to change in the mechanisms of failure and it is characterized by the defect density limits.

According results of many investigations depending on the stress level the microdefects birth and growing occurs by various physical mechanisms, namely from the body surface, when the surface outpaces the internal volumes in the microdefect accumulations at multicycle fatigue, and when microdefects reach limit states in the internal volumes earlier than on the surface at gigacyclic fatigue. So an increase of gigacyclic strength is associated with changes in the metal structure, an increase of purity due to a decrease in the percentage of residual gases, a decrease in the size of inclusions, etc.

The stability of microcrack dimensions makes it possible to increase their density in accordance with the principle of minimizing free energy. At a certain level of stresses during the interaction of neighboring microdefects the distances between them become comparable with the characteristic sizes of the defects themselves, which leads to their coalescence and nucleation of the following **microscopic level structures, namely short non-propagating cracks** inside the grain and along the grain boundaries of characteristic sizes $l_3 \in \left(10^{-2}, 10^{-1}\right] d$. Non-propagating cracks of limit density $q_{f,3} = 10^6 \mathrm{mm}^{-2}$ were observed in experiments.

As a result of the short non-propagating cracks merger failure reaches the grain level, characteristic **structures of short propagating cracks** on average, $l_4 \in \left(10^{-1}, 1\right] d$, are formed **at the mesoscopic level**. These structures texture a peculiar relief of the surface named "river pattern" (the reverse movement of the slip strip blocks during cyclic loading creates "wavy" relief in the form of extrusions and intrusions at the places of slip strips exit to the surface). The lines of the "river pattern" are oriented almost perpendicular to the brittle grooves. The bases of steps and troughs and intrusions bottom are concentrators from which the new microdefects develop.

In the areas of multicycle fatigue there is observed the surface grain or some grain size short cracks brittle macrofracture at elastic macrodeformation. For many materials under uniaxial symmetric loading, at the stress amplitude of short cracks (on average $l_{f,4} \in \left[0.01, 0.1\right]$ mm) failure is defined as the material endurance σ_{-1}. A brittle chip is formed by relatively flat facets of a brittle transcrystalline chip along the grain body or an intergranular chip along the grain boundaries. Defects of all levels are observed on the chip plane. As a rule, one facet is a chip of one grain. At low temperatures and high strain rates, the probability of brittle failure on short cracks increases. The type of metal structure also affects on endurance.

At gigacyclic fatigue macrofailure of many metallic materials (high-strength and hardened steels, aluminum and nickel alloys) occurs on surface short cracks of grain size or several grains originating from the "fish eye" center in the body volume (in most cases, from inclusions, fine precipitates, pores) at elastic macrodeformation [3–5]. The crystalline chip resulting from the propagation of brittle macrocracks has an atomically smooth surface. The fatigue limits in this area decrease with increasing number of cycles. In the study of a number of such materials it is experimentally discovered a stress interval, called the bifurcation region, between the regions of multicycle fatigue (on durability, on average, $N_f \in \left[5 * 10^3, 5 * 10^6\right]$ cycles) and gigacyclic fatigue ($N_f \in \left[5 * 10^6, 10^{12}\right]$ cycles), within which at the same stress state one part of the samples is destroyed from a failure center on the surface, in another part the defect nucleation occurs in the body volume, and so the energy dissipation methods change. It is noted a bifurcation region also takes place in the region of the transition from low-cycle fatigue (the area of durability, on average, $N_f \in \left[10^2, 5 * 10^3\right]$ cycles) to re-static failure ($N_f < 100$ cycles) and failure at monotonous loading with the subsurface defect nucleation.

To describe the crack growth at the mesoscale, the Paris equation and the Herzberg equation are used, as well as the approaches of Zhurkov, Botvina, Betekhtin and others. The French lines of irreversible damage corresponding to the end of the mesofailure within the grain and the beginning of the macrofailure are constructed.

In the literature, there are conventionally distinguished the characteristic **structures at the macrolevel** of the grain ensembles size, on average, $l_5 \in (1, 10)\, d$ and $l_6 \in (10, 100)\, d$ on which brittle macrofracture by transcrystalline or intergranular cleavage mechanisms at multicycle fatigue is possible. At a certain level of stresses with an increase of cycles number as a result of macrodefects growing, it is possible to form a leader macrocrack of a certain finite length L, its further growing is studied in fracture mechanics. At multicycle fatigue, the activity of the surface layer (heterogeneity of its chemical composition, residual stresses, material brittleness due to surface oxidation) in energy exchange with the environment determines the macrodefects evolution. V. E. Panin connects the moment of macrocrack occurrence with reaching of a critical molar volume at a positive value of Gibbs potential. The Paris equation is used for the crack growth rate. The fatigue macrocracks velocity is high is about the sound speed in metals.

Material macrofracture at multi- and gigacyclic fatigue is customarily characterized by the Weller curve, in general case it is described by different relations between the equivalent failure stress (determined by the stress state) and the number of N_f failure cycles.

In the region of multicycle fatigue of plastic materials, there are possible the inelastic deformation and growing of viscous failure processes. These processes are characterized by the appearance of plastic distortion at the critical curvature of the crystal lattice with the generation and growing of dislocations and a cellular substructure formation at the nanoscale level, its evolution on twinning and slip mechanisms, which leads to motion of grain ensembles and the appearance of microshear bands at the microlevel, to the formation of shear mesoscopic bands and structural phase decay of deformable material and porosity nucleation ending viscous macrocrack birth. At the macroscale level, intense glide of grains ensembles occur. At $N_f \in \left[5 * 10^3, 5 * 10^6\right]$ cycles the values of inelastic deformations does not exceed elastic deformations and inelastic deformation inhibits the brittle cracks growing. In ductile materials the process of viscous failure and brittle fatigue simultaneously occur. In fatigue fractographs of the surface chip it can be distinguished both the zone of the shear failure, i.e. the area of of inelastic deformations evolution and viscous cracks with characteristic pits, and the zone of brittle fracture by separation.

5.3 Modern Theoretical Approaches to Metal Fatigue Description

A brief overview of the main modern directions of theoretical research of metal failure processes at long-term and cyclic loadings allows to highlight the following scientific approaches.

In modern physical approaches the formulation of fracture criteria is based on the study of the mechanisms of structural interactions at loading on different scale levels, is characterized by the consideration of the metal atomic-crystalline structure, by the investigation of the evolution of defects, dislocations, slip bands, and crystalline lattice curvature and the establishment of relations of their interaction with elements of the metal internal structure. The energy analysis of the failure process is based on the principles of thermodynamics, synergetics, and structural—scaling transitions. Failure is considered as the last stage in the evolution of the metal internal structure, when the only way to dissipate the energy supplied to the material is to create defects of various scales after exhausting the possibilities for adjusting the structural elements to each other. In the framework of these investigations, the discrete and continuous structural dislocation theories of Zener-Stro-Petch, Smith, Barnby, Averbach, Louis, Cottrell, Ballough-Gilman, Akulov, McLean, Mugrabi, dilaton-frustron models of Kusov, Zhurkov, Petrov, structurally—energy theories of Panin, Botvina, Ivanova, Terentyev, Zakrzhevsky, Golsky, Heidzel, Weiner, Ito, Tomashi, as well as stochastic models of Ecobory, Freudenthal, Volkov, Ghonem-Provan, McKartney, Gal, Sobchik are developed.

In L. R. Botvina's works formulation of failure criteria is developed from the standpoint of the theory of similarity and the theory of phase transitions; it is noted the universality in the behavior of ensembles of defects in various physical processes: fatigue, creep, dynamic loading etc. The scale invariance (similarity) of the defect distribution curves at different scale levels is assumed, and the concentration criterion determining the macrocrack nucleation is formulated. V. I. Betekhtin and A. G. Kadomtsev propose a methodology of defect growing investigation, formulate the necessity to define the characteristic failure stages. Field equations for the free energy density are written based on the thermodynamic description of the microcrack nucleation and growing. Among the known theoretical works are the works of V. E. Panin on the physical mesomechanics of materials in which a solid in the fields of external influences is considered as a multilevel system in which all processes develop self-consistently at the nano-, micro-, mesoscale and macroscale levels, the surface layers and internal interfaces are functional subsystems. It is determined the main role of the nanoscale structural level. Hierarchical self-consistency of defect scales and transition to other levels are described on the basis of the dependence for the Gibbs thermodynamic potential from the material molar volume and wave equations for the flow of structural transformations and density of defects are formulated. A. A. Shanyavsky made an attempt to integrate in the uniform methodology the research of fatigue in metal physics, mesomechanics and synergetics. It is believed the metal reaches the limit state at each scale level due to the accumulation of the limiting energy for this level. A system of fatigue curves at the micro, meso, and macro levels is proposed.

According to multilevel fracture models, the nucleation of microcracks occurs at critical tensile microstresses. The constitutive relations of elastic-viscoplastic microdeformation with a power relation of flow and isotropic or anisotropic hardening are formulated. These relations contain parameters characterizing the evolution of slip bands and the density of statistically accumulated and geometrically necessary dislocations. Yield criterion is considered according to the anisotropic theory of plasticity. The probability of the mesocrack nucleation is defined as the ratio of the area of the boundaries of the pairs of grains with the microcracks to the total area of the boundaries of all grains. There are developed models in which the parameters of dislocation density near the grain boundaries and the density of dislocation substructures inside the grain are introduced into the defining relations for residual microstresses, and evolution equations for these parameters are formulated. In a number of models the stress-strain state of individual grains at the mesoscale is explored, the constitutive relations of elastic-viscomesoplasticity with isotropic and anisotropic hardening are assumed, the boundary conditions are determined in accordance with macroloading, and the distribution of mesostresses and mesostrains at grain level is found. According to other approaches, the formation of micropores (as a result of annihilation of dislocations in slip bands with the nucleation of vacancies and their coalescence) at the microlevel is determined by the maximum values of microstress intensities, and their growth is defined by the positive values of the first invariant of the microstress tensor. At the mesoscale, failure is to reach the limiting value of pore density. Evolution equations for pore density are formulated. To go to the macrolevel, the Voigt, Teiss or Krener models are used.

The most physical approaches and known multilevel models contain a significant number of not determined in macroexperiments material functions, structural parameters and physical mechanisms of failure. Difficulties are included also to obtain reliable statistical data for calculation on the models.

Phenomenological approaches to the research of failure processes at long-term and cyclic loading, based on a generalization of a large number of experimental data, are suggested that the macrofailure state of macrovolume is reached if some universal function of the stress tensor invariants reaches a certain value. There are determined a system of experiments to find the material constants of this function, and there are verified the theories for various types of loading. One of the most general phenomenological theories is the failure loading processes theory proposed by B. I. Zavoychinsky, by when the certain functional reaches its maximum values at the failure process. The main known strength theories are the special cases of this approach.

The most phenomenological approaches to failure estimation at long-term and cyclic loading suggest the existence of a damage function $\Omega = \Omega(t) : \Omega|_{t=0} = 0$, failure is determined by the equation: $\Omega|_{t=t_f} = 1$, t_f is the material life time and there are considered various kinetic equations for the damage function $\Omega = \Omega(t)$. One of the first approaches is the hypothesis of damage linear summation by D. Bally, A. Palmgren, M. Miner, formulated for uniaxial multi-frequency loads with an asymmetric cycle. The works of L. M. Kachanov, Yu. N. Rabotnov, V. Bolotin and V. Novozhilov are the most famous of them. These works gave the beginning to a whole direction of damage models creation. It is assumed that there is a damage tensor of the first or second ranks at a complex stress state. Failure is determined as to equal to unity of some measure of this tensor. For example, A. A. Ilyushin's damage tensor. The damage functions (tensors) are introduced in the structure of the constitutive relations. Evolution equations for damage tensor dependence from loading process and time are defined, and identification of their material functions is given. To determine the durability at irregular loads, linear and nonlinear criteria for damage summing are used. In the framework of this approaches there are made attempts to take into account the evolution of material damage, damage tensors are considered on the micro-, meso- and macrolevels. There are developed the stochastic phenomenological models, for example A. I. Radchenko's discrete-probabilistic model of damage accumulation, Monte-Carlo-based approaches. Seems the most perspective the stochastic models of damage accumulation based on Markov chains, for example, the models of J. Bogdanoff and F. Kozin.

The fatigue curve of a number of metallic materials in the transition zone between multi- and gigacyclic fatigue may have a bifurcation region characterized by a change of failure mechanisms. We can note the Reicher's hypothesis of damage spectral summation, according to which the damage accumulation is considered from the standpoint of self-organized transitions through bifurcation areas, when the principle of damage evolution changes after each transition. But the question is to sum of damage at different scale levels with different accumulation mechanisms. It was experimentally obtained that some samples after certain operating times have the durability longer than without the times, it takes place the hardening of the samples at the microlevel associating with the accumulation of defects before failure states (the transition to the subsequent level with an increase in the defect size). To describe these processes, the well-known hypotheses of damage are not applicable.

Many damage theories are not considered the failure process, hence, for example, the sequence of application of stresses of various levels is not taken into account.

One of the significant disadvantages of modern damage theories is also that, considering the constitutive relations of models within the same scale level, material functions are determined, as a rule, on total fracture.

To determine the effectiveness of the frequently used damage theories the author wrote down the basic equations in a general form, namely the left side of these equations includes the durability and the loading form, and the right side is equal to unity at failure states. According to the results of numerous experiments for a wide class of metals and alloys at various stress states, in particular, at irregular loads, the values of the left side of these relations were determined. It was established that its values are in the range from 0.1 to 10. A statistical analysis of this interval shows that the damage function is a random function with an average value of unity and its probability distribution density on the truncated normal (log-normal) law.

Conducted analysis, the results of numerous experimental studies and the known fatigue relations show the method of structure life evaluation based on the approaches of damage theories is not universal, the range of applicability of these theories is the loading processes that are close to the processes of the material functions and parameters determination.

These circumstances significantly narrow down the possibility to use the damage theories for evaluation of structure durability.

In the framework of experimental and theoretical works on fracture mechanics there are investigated the final fracture stages, namely the evolution of macrocracks in metals and alloys at long-term and cyclic loading at the macrolevel and nucleation of the leader cracks, there are determined the relations between the crack propagation and material structure, there are developed deformation, force and energy fracture criteria. Fundamental are the works of Griffiths, Dugdale, N. F. Morozov, E. M. Morozov, Bolotin, Cherepanov, Kornev, Rice, McKartney, Gale, Sih, Nott, Orovan, Taira, Tanaki, Hellan and others.

5.4 Scale-Structural Fatigue Theory at Complex Stress State

Experimental and theoretical studies on solid state physics, metal science and solid mechanics have shown at symmetric multicycle loading the defect structure evolution in the range of characteristic sizes, on average, $l \in \left(10^{-5}, 10\right)$ mm, is a hierarchical multilevel process. The proposed model is formulated in the framework of the physical and mechanical approach as a system of hypotheses about the probability of defect growing at various scale-structural levels.

Let us turn to the presentation of the model.

There are investigated the following multiaxial loading processes:

$$\begin{gathered}\sigma_{kk}(t) = \alpha_k \sigma_a f(t), \ f(t) = \alpha + \sin \omega t, \ k = 1, 2, 3, \ t \in [0, T] \\ |\sigma_{11}| \geq |\sigma_{22}| \geq |\sigma_{33}|, \alpha_k = \frac{\sigma_{kk}}{|\sigma_{11}|}, \alpha_0 = \frac{\sigma_0}{\sigma_{11}}, \sigma_0 = \frac{1}{3}\sum_{k=1}^{3} \sigma_{kk},\end{gathered} \tag{5.1}$$

where σ_a is the amplitude of maximum principal stresses, ω is the frequency of the stress variation, α_k is the ratio of the principal stresses, α is the cycle asymmetry parameter.

The stochastic process of fatigue failure at loading (5.1) is considered at six scale-structural levels. It is introduced the notion of the i-level defect, $i = 1, \ldots, 6$, defined the average size $l_i = l_i(t)$ and density $q_i = q_i(t)$ in some representative volume V_c (in which it is possible the nucleation of a leader macrocrack of a certain finite length L). The nanoscale structures considered above are related to I-level defects, microdefects are II-level defects, short non-propagating microcracks are III-level ones, mesoscale short propagating cracks are the IV-level ones, macrocracks are V-level and VI-level defects. The scale hierarchy mean the introduction of scale-structural level boundaries. The failure state of i-level defect is characterized by size $l_{f,i}$ and density $q_{f,i}$, $i = 1, \ldots, 6$.

It is proposed the nucleation of each level defect is a result of the consistent nucleation, growing and fusion of all previous level defects. At each level, a continuous increasing averaging function $l_i^* = l_i^*(t)$: $l_i^*(t) = l_i(t)\,(q_i(t)\,V_c)^{\gamma}$, $t \in [0, T]$, is introduced, γ is the material constant, $i = 1, \ldots, 6$. In the case of the i-level defect failure state this function reaches its limiting value, $l_{f,i}$, $i = 1, \ldots, 6$.

Due to the random distribution of defects on the volume V_c, the fatigue process is stochastic, and the values l_i^* for each time moment τ are random variables. Here is introduced function $F_i = F_i\left(l_i^*\right), 0 \le F_i \le 1$, of probability distribution of random value l_i^*, determining the probability with which a random variable l_i^* takes values less than its limit value $l_{f,i}$ at a time τ: $F_i = P\left(l_i^* < l_{f,i}^*\right)$, $P\left(l_i^* = l_{f,i}^*\right) = 0$, $i = 1, \ldots, 6$. It is proposed to consider truncated normal distributions with a density: $f_i = f_i\left(l_i^*\right)$: $F_i = \int_o^{l_i^*} f_i(x)\,dx$, of the following form:

$$f_i\left(l_i^*\right) = \frac{A_i}{\sqrt{2\pi D_i}} \exp\left(-\frac{\left(l_i^* - M_i\right)^2}{2D_i}\right), \quad c_i = \int_o^{\infty} f_i(x)\,dx, \; i = 1, \ldots, 6,$$

where $M_i = \int_{-\infty}^{\infty} x f_i(x)\,dx$ and $D_i = \int_{-\infty}^{\infty} (x - M_i)^2 f_i(x)\,dx$ are the mean and variance of a random value l_i^*, respectively.

The probability of brittle failure on i-level defects is determined by the function $Q_i = Q_i(t)$, $i = 1, \ldots, 6$, $t \in [0, T]$, in the form:

$$Q_i(t) = Q_{i,th} - F_i\left(l_i^*(t)\right), \quad 0 \le Q_{i,th} \le 1.$$

It is formulated a recurrence system of constitutive relations for function $Q_i = Q_i(t)$. These relations include times t_{i+1} when the i-level defect reaches the failure state with probability $Q_{i,th}$, the averaging function $l_i^* = l_i^*(t)$ is equal the value $l_{f,i}^*$, and it is beginning to form $(i+1)$-level defects. There are determined the series of fatigue curves on the failure levels:

$$Q_i(t_{i+1}) = Q_{i,th}, \quad i = 1, \ldots, 6.$$

At the macrolevel, the failure probability $Q = Q(t)$, $t \in [0, T]$, $0 \le Q \le 1$, is defined as:

$$Q(t) = \sum_{i=4}^{6} \frac{Q_i(t)}{1 - Q_i(t)} \prod_{j=4}^{6} \left(1 - Q_j(t)\right),$$

where $Q_i = Q_i(t)$ are the failure probabilities on i-level defects, $i = 4, 5, 6$. Fatigue curve on leader macrocrack of a certain finite length L is determined according to the equation:

$$Q(t_f) = 1,$$

t_f is the durability on the leader macrocracking.

For loading (5.1) assuming a uniform distribution of defects in volume V_c, the following system of relations for $Q_i = Q_i(\sigma_a, n)$ (value n is the number of loading cycles) and fatigue curves on i-level defects are proposed [10, 25–29]:

- at the microlevel $\sigma_a \ge \sigma_{i-1}$, $\lg n \ge \lg n_i(\sigma_a)$, $\sigma_0 = 0$, $n_1 = 1$, we have:

$$Q_i = F_i\left(\frac{\sigma_a - \sigma_{i-1}}{\sigma_i - \sigma_{i-1}}\right) R_i\left(\frac{\lg n - \lg n_i(\sigma_a)}{\lg N_i - \lg n_i(\sigma_i)}\right), \tag{5.2}$$

$$Q_i(n_{i+1}) = Q_{i,th};\ i = 1, 2, 3 \tag{5.3}$$

- at the mesoscale level $\sigma_3 \le \sigma_a$, $\lg n \ge \lg n_4(\sigma_a)$

$$Q_4 = F_4\left(\frac{\sigma_a - \sigma_3}{\sigma_4 - \sigma_3}\right) R_4\left(\frac{\lg n - \lg n_4(\sigma_a)}{\lg N_4 - \lg n_4(\sigma_4)}\right), \tag{5.4}$$

$$Q_4(n_5) = Q_{th}, \tag{5.5}$$

- at the macrolevel, if $\sigma_4 \le \sigma_a$, $\lg n \ge \lg n_5(\sigma_a)$

$$\begin{aligned} Q_5 &= F_5\left(\frac{\sigma_a - \sigma_4}{\sigma_5 - \sigma_4}\right) R_5\left(\frac{\lg n - \lg n_4(\sigma_a)}{\lg N_5 - \lg n_4(\sigma_5)}\right), \\ Q_4 &= G_4\left(\frac{\sigma_5 - \sigma_a}{\sigma_5 - \sigma_4}\right) R_4\left(\frac{\lg n - \lg n_4(\sigma_a)}{\lg N_4 - \lg n_4(\sigma_4)}\right), \quad Q_6 = 0, \end{aligned} \tag{5.6}$$

$$Q_5(n_6) = Q_{th}, \tag{5.7}$$

- at the macrolevel, if $\sigma_5 \le \sigma_a$, $\lg n \ge \lg n_6(\sigma_a)$

$$\begin{aligned} Q_6 &= F_6\left(\frac{\sigma_a - \sigma_5}{\sigma_6 - \sigma_5}\right) R_6\left(\frac{\lg n - \lg n_4(\sigma_a)}{\lg N_6 - \lg n_4(\sigma_6)}\right), \\ Q_5 &= G_5\left(\frac{\sigma_6 - \sigma_a}{\sigma_6 - \sigma_5}\right) R_5\left(\frac{\lg n - \lg n_4(\sigma_a)}{\lg N_5 - \lg n_4(\sigma_5)}\right), \quad Q_4 = 0, \end{aligned} \tag{5.8}$$

$$Q_6(n_7) = Q_{th}. \tag{5.9}$$

In the system (5.2), (5.4), (5.6), (5.8) the following equation is connected with the previous one through the number of cycles $n_{i+1} = n_{i+1}(\sigma_a)$ when the function $l_i^* = l_i^*(n)$ reaches the limit value $l_{f,i}^*$, the i-level defect level is in the failure state and it is beginning $(i+1)$-level defecting.

The system of material functions σ_i, $i = 1, \ldots, 6$, included in (5.2)–(5.9), is considered in the form:

$$\sigma_i = \sigma_i(N_i, \omega)\tilde{\sigma}_i(\alpha_2, \alpha_3, \tilde{\eta}_i, \hat{\eta}_i)\tilde{\sigma}_i^0(\alpha, \eta_i^0, \tilde{\eta}_i^0, \hat{\eta}_i^0), \tag{5.10}$$

where $\tilde{\eta}_i = \frac{\sigma_i(N_i,\omega)}{\tilde{\sigma}_i(N_i,\omega)}$, $\hat{\eta}_i = \frac{\sigma_i(N_i,\omega)}{\hat{\sigma}_i(N_i,\omega)}$, $\eta_i^o = \frac{\sigma_i^o(N_i,\omega)}{\sigma_i^o(N_i,\omega)}$, $\tilde{\eta}_i^o = \frac{\sigma_i^o(N_i,\omega)}{\tilde{\sigma}_i^o(N_i,\omega)}$, $\hat{\eta}_i^o = \frac{\sigma_i^o(N_i,\omega)}{\hat{\sigma}_i^o(N_i,\omega)}$.

Functions $\sigma_i = \sigma_i(N_i, \omega)$ are the system of material constants at the uniaxial loading model ($\alpha_1 = 1$, $\alpha = \alpha_2 = \alpha_3 = 0$), namely the number of cycles, N_i, $i = 1, \ldots, 6$, and the corresponding amplitudes σ_i when the i-level defect reaches the failure state and the function $l_i^* = l_i^*(n)$ is became equal the limit value $l_{f,i}^*$.

It is proposed the method for identifying of these functions. Analogically the systems of material constants $\tilde{\sigma}_i = \tilde{\sigma}_i(N_i, \omega)$ at shear ($\alpha_1 = 1, \alpha_2 = -1, \alpha = \alpha_3 = 0$) and $\hat{\sigma}_i = \hat{\sigma}_i(N_i, \omega)$ at biaxial uniform loading ($\alpha_1 = \alpha_2 = 1, \alpha = \alpha_3 = 0$) are determined. In the general case for loads (5.1) with an asymmetric cycle, it is necessary to additionally specify the systems of material constants $\sigma_i^0 = \sigma_i^0(N_i, \omega)$ for pulsating uniaxial and $\hat{\sigma}_i^0 = \hat{\sigma}_i^0(N_i, \omega)$ for pulsating biaxial loads and $\tilde{\sigma}_i^0 = \tilde{\sigma}_i^0(N_i, \omega)$ for pulsating shear. In the general case of symmetric loading (5.1), $\alpha = 0$, the functions $\tilde{\sigma}_i = \tilde{\sigma}_i(\alpha_2, \alpha_3, \tilde{\eta}_i, \hat{\eta}_i)$ in expression (5.10) are selected in the way:

for brittle materials

- at $-1 \le \alpha_2 \le 0 : \tilde{\sigma}_i = \left[6 - \tilde{\eta}_i - \alpha_2\left(2\tilde{\eta}_i - 6\right) + \alpha_0\left(3\tilde{\eta}_i - 15\right)\right]^{-1}$,
- at $0 \le \alpha_2 \le 1, \alpha_3 \ge 0 : \tilde{\sigma}_i = \left[1 + \alpha_2(\hat{\eta}_i - 1) + \alpha_3\left(\hat{\eta}_i - 1\right)\right]^{-1}$,
- at $0 \le \alpha_2 \le 1 \alpha_3 < 0 : \tilde{\sigma}_i = \left[6 - \tilde{\eta}_i - \alpha_3(2\tilde{\eta}_i - 6) + \alpha_0\left(3\tilde{\eta}_i - 15\right)\right]^{-1}, i = 1, \ldots, 6$

for plastic materials

- at $-1 \le \alpha_2 \le 0$: $\tilde{\sigma}_i = \left[3\alpha_0\left(1 + \alpha_2\right) + \frac{1}{2}\tilde{\eta}_i^2\left(1 - \alpha_2 - 3\alpha_0\right)\right]^{-1/2}$,
- at $0 \le \alpha_2 \le 1, \alpha_3 \ge 0$: $\tilde{\sigma}_i = \left[3\alpha_0\left(1 + \alpha_2\right) + \frac{1}{2}\hat{\eta}_i^2\left(1 - \alpha_2 - 3\alpha_0\right)\left(1 - \alpha_2 - 3\alpha_0\right)\right]^{-1/2}$
- at $0 \le \alpha_2 \le 1, \alpha_3 < 0$:$\tilde{\sigma}_i = \left[3\alpha_0\left(1 + \alpha_3\right) + \frac{1}{2}\tilde{\eta}_i^2\left(1 - \alpha_3 - 3\alpha_0\right)\right]^{-1/2}, i = 1, \ldots, 6$

In the author's works [10, 25–29], some particular cases of loading with asymmetric cycles are considered and there are written functions $\tilde{\sigma}_i^0 = \tilde{\sigma}_i^0(\alpha, \eta_i^0, \tilde{\eta}_i^0, \hat{\eta}_i^0)$ of expression (5.10) for various metals and alloys.

The proposed model allows to select the type of functions $F_i = F_i\left(\sigma_a\right)$ and $R_i = R_i\left(n\right), i = 1, \ldots, 6$, in expressions (5.2)–(5.8) for various materials. There are considered functions of the following form:

$$
\begin{aligned}
F_i &= \left(\frac{\sigma_a - \sigma_{i-1}}{\sigma_i - \sigma_{i-1}}\right)^{\beta_i}, R_i = \left(\frac{\lg n - \lg n_i\left(\sigma_a\right)}{\lg N_i - \lg n_i\left(\sigma_i\right)}\right)^{\phi_i}, \quad i = 1, \ldots, 4; \\
F_i &= \left(\frac{\sigma_a - \sigma_{i-1}}{\sigma_i - \sigma_{i-1}}\right)^{\chi_i}, R_i = \left(\frac{\lg n - \lg n_4\left(\sigma_a\right)}{\lg N_i - \lg n_4\left(\sigma_i\right)}\right)^{\phi_i}, \quad i = 5, 6;
\end{aligned}
\tag{5.11}
$$

$$
G_i = \left(\frac{\sigma_{i+1} - \sigma_a}{\sigma_{i+1} - \sigma_i}\right)^{\chi_i}, i = 4, 5 \tag{5.12}
$$

where for material functions $\beta_i = \beta_i(\alpha_2, \alpha_3, \omega)$, $\phi_i = \phi_i(\alpha_2, \alpha_3, \omega)$, $\chi_i = \chi_i(\alpha_2, \alpha_3, \omega)$ it is necessary to set additional basic characteristics of the model.

To determine the functions $\sigma_i = \sigma_i\left(N_i, \omega\right), i = 1, \ldots, 6$, in (5.10) it is necessary to conduct a sufficiently large number of macroexperiments at uniaxial symmetric loading with thin section processing by standard investigations of the microstructure. Well-known works, reference books and others contain a limited amount of these data. For materials with an endurance limit and a horizontal section at gigacycle fatigue, the method for determining material functions based on the known data on the metal fatigue was proposed and it was confirmed by comparing obtained results with experimental data for a representative series of metals and alloys.

Let us introduce some propositions it this method.

At first here is considered the metals and alloys with a horizontal section of the fatigue curve and endurance limit. The base amplitude $\sigma_4 = \sigma_4\,(N_4, \omega)$ is determined of the known material endurance limit σ_{-1} at symmetric axial loading and number of cycles N_4: $\sigma_4\,(N_4, \omega) = \sigma_{-1}$, when IV-scale level defects reach the fracture state, the function $l_4^* = l_4^*\,(N_4)$ equals the limit value $l_{f,4}^*$ and take place the macrofracture on short cracks with probability Q_{th}.

As a result of analysis of the known data on fatigue microfracture on the microlevel here can assume that the failure probability on defects I, II and III-levels are described with a sufficient degree of accuracy the relations (5.2), $\beta_i = \chi_i = 1,\ \phi_i = 1/2$. Noticeable changes of the metal micro- and macrostructure (the changes of average size and density of micro- and macrocracks) are observed when the number of loading cycles is changed by decimal orders. So for the base numbers of cycles N_i the following expressions can be selected on the number of cycles N_4:

$$\lg N_i = \lg N_4 + 4 - i,\ i = 1, \ldots, 4,\ \lg N_5 = \lg N_4 - 2,\ \lg N_6 = \lg N_4 - 3.$$

Besides that it is assumed at the amplitude σ_{-1} and number of cycles N_6 the I–level defect failure state is reached. The II–level defect failure takes place at the amplitude σ_{-1} and the number of cycles N_5 and III–level defects failure is at the amplitude σ_{-1} and the number of cycles $\lg N_7 = \lg N_4 - 1$. So at the number of cycles $N_4 = 2 * 10^7 cycles$ the basic amplitudes for microlevel defects are determined through the endurance limit σ_{-1} in the following form ($Q_{th} = 1$):

$$\sigma_1 = 0.4\sigma_{-1},\ \sigma_2 = 0.75\sigma_{-1},\ \sigma_3 = 0.86\sigma_{-1}.$$

According to the known results of the fatigue analysis for metals and alloys the base amplitudes for VI–level defects $\sigma_6 = \sigma_6(N_6, \omega)$ can be selected with a good degree of accuracy on the corresponding yield strengths. Note that it can be considered other values of material constants in the presence of experimental data.

For materials whose fatigue limits at gigacyclic fatigue decrease and the fatigue curve is described by different equations in the area of multicyclic and gigacyclic fatigue, based on the analysis of a numerous famous works, two characteristic fatigue limits are introduced into the model. It is assumed there is an experimentally determined the stress level with amplitude $\sigma_{-1} = \sigma_{-1}(N_{-1}, \omega)$ on IV-level defect macrofracture, at which there is a transition from the region of multicycle fatigue to gigacycle fatigue with a change in the fracture mechanisms. It is also believed there is an experimentally determined failure amplitude $\sigma_{-2}(N_{-2}, \omega) = \sigma_4, \sigma_{-2} < \sigma_{-1}$, at which IV-level defect macrofracture is observed at the number of cycles $N_{-2} = N_4$, $N_{-2} > N_{-1}$ and the fatigue curve has a horizontal section $\sigma_{-2} = const,\ n > N_{-2}$. In the field of gigacyclic fatigue at the I, II, III, IV-level defect failure can be possible. When choosing functions in (5.11) it is required that Eq. (5.5) pass through a point $\sigma_{-1} = \sigma_{-1}(N_{-1}, \omega)$. The basic numbers of cycles are chosen equal $N_i = N_4$, $i = 1, 2, 3$. Basic constants are determined so that the graph of the Eq. (5.3) at

$i = 1, 2, 3$ and went through points and (σ_{-2}, lgN_{-1}), $(\sigma_{-2}, (2lgN_{-1} + lgN_{-2})/3)$ and $(\sigma_{-2}, (lgN_{-1} + 2lgN_{-2})/3)$ respectively.

To justify the reliability of the proposed approach, the author's works are presented the results of the analysis of experimental data on the defects evolution of different scale-structural levels for a representative number of structural steels: carbon, austenitic, martensitic, alloyed, etc.; cast irons; metals: molybdenum, nickel, lead, titanium, etc.; nickel, magnesium, aluminum, titanium alloys at various processes of proportional loading.

5.5 Practical Implementation of the Scientific Results

One of the main fundamental directions of modern development of structure operation technogenic safety problems is the creation of algorithms for predicting of the residual resource of structures, taking into account the principles of multiscale failure of structural elements [23, 32–36]. A method has been developed for estimation of the long structure durability (gas and oil main pipelines, technological product pipelines), taking into account the mechanical-corrosion and stress-corrosion damage of structural elements during operational loading [10, 23, 30, 31, 37–39].

The nature of the changes in operational loads, the significant heterogeneity of the mechanical characteristics of materials, the variation of structural technological factors, and the need to take into account of technological and operational defects make it necessary to use probabilistic methods for assessing of the durability and crack resistance of structural elements. At the durability calculation the probabilistic parameters of material properties, such as characteristics of crack opening, Weller curve, Coffin-Manson equation, Paris relation, etc., must be introduced. It is necessary to considered the random stationary failure processes.

Sections of long structures such as oil and gas pipelines consist of a large number of structural elements at the internal pressure of the pumped product, the action of mass forces and the temperature field. The structural scheme is a serial connection of structural elements: the base metal, welded joints, tees, bends. The internal pressure loading drops from the compressor station along the length, it is occurred random pressure fluctuations with an amplitude of up to 10–15% of the maximum value. There is a planned (annual, monthly) increase (decrease) of pressure, depending on the volume of consumption. It is experimentally found these vibrations determine the fracture of structural elements, especially in areas of stress concentration. In the general case, operational loading is considered as a random function in the form of the sum of the constant component of the stress and the discrete spectrum with a set of stress amplitudes, frequencies, and phase shifts between the components, which are random variables. In the particular case, there are considered the various model loads of the q-element, $q = 1, \ldots, Q$, k-section, $k = 1, \ldots, K$, at multicycle fatigue in the time interval $t \in [0, T]$ by ring stress $\sigma_{\theta,k,q} = \sigma_1(t)$ and axial stress $\sigma_{z,k,q} = \sigma_2(t)$ of the form (5.1), $\alpha_3 = 0$. The inhomogeneous stress state arising in welded and tee joints, bends and bottoms, is taken into account by introducing stress concentration factors $K_{\theta,q}$ in the tangential and $K_{z,q}$ in the axial directions, respectively.

According to the proposed approach, the failure probability (structural risk) $Q_{k,q} = Q_{k,q}(\tau)$ of number $n_{k,q}$ q-elements of the k-sections is determined in the following form that is a generalization of the Poisson distribution function [10, 37–39] (for the first failure):

$$Q_{k,q}(t) = \varphi_{k,q}(t)e^{1-\varphi_{k,q}(t)},\ \varphi_{k,q}(t) = \lambda_q n_{k,q}\frac{\tilde{t}}{t_{f,k,q}}t \tag{5.13}$$

where $t_{f,k,q}$ is the durability of the q-element of the k-section, $k = 1, \ldots, K$, $q = 1, \ldots, Q$, which is determined on the fatigue curves of various defect levels on the proposed scale-structural fatigue theory, the parameter $\tilde{t}$ is the economically and socially acceptable structure design life, λ_q is the coefficient of the intensity of the element failure flow, determined by fracture statistics at the same operation of similar structures to achieve the appropriate defect levels.

The structure failure probability $Q = Q(t)$ is determined through the failure probability of its sections $Q_{k,q} = Q_{k,q}(t)$ as follows:

$$\begin{aligned} Q(t) &= \sum_{k=1}^{K}\left(\frac{Q_k(t)}{1 - Q_k(t)}\right)\prod_{k=1}^{K}[1 - Q_k(t)], \\ Q_k(t) &= \sum_{q=1}^{\mathrm{Q}}\left(\frac{Q_{k,q}(t)}{1 - Q_{k,q}(t)}\right)\prod_{q=1}^{\mathrm{Q}}\left[1 - Q_{k,q}(t)\right], \end{aligned} \tag{5.14}$$

or

$$Q(t) = 1 - \prod_{k=1}^{K}[1 - Q_k(t)],\quad Q_k(t) = 1 - \prod_{q=1}^{\mathrm{Q}}\left[1 - Q_{k,q}(t)\right], \tag{5.15}$$

where $Q_{k,q} = Q_{k,q}(t)$ are defined in (5.13). Expression (5.14) determines the sum of independent events—fractures of k-section in the absence of fractures of the rest, expression (5.15) is the sum of independent fractures of at least the k-section.

The criterion of structural reliability is formulated in the form ($t_{f,k,q} \geq \tilde{t}$):

$$Q(t) \leq \tilde{Q},\quad k = 1, \ldots, K,\ q = 1, \ldots, 7,$$

where $Q = Q(t)$ is the structure failure probability, $t \in [0, T]$, determined by (5.14) or (5.15) taking into account (5.13), $\tilde{Q}$ is its acceptable value.

According to the proposed method, the design durability of various linear sections between compressor stations, sections of gas recovery units, gas reduction units, gas metering stations, gas purification plants, compressor stations pipings, technological pipelines was evaluated [10, 23, 32, 37–39].

It was considered three main technogenic spheres that are victims of structural fracture, namely the people, objects and environment, and, accordingly, the concepts of the probability of people damage (social risk), of industrial object destruction (industrial risk) and flora and fauna failure (environmental risk), located in a potentially dangerous area near the structure at the potential fracture, are introduced. There were formulated the system of criteria of structure operation industrial, social and environmental safety in complex climatic conditions [10, 23, 32, 37–39]. On their basis the expression for the design durability of the structure operation safety is determined as a random function of the material properties, structural scheme, operational loading, distribution of the failure flow at a similar structural scheme system operation.

References

1. Tyutin, M.R., Botvina, L.R., Sinev, I.O.: Changes in the physical properties and the damage of low- and medium- carbon steels during tension. Rus. Metall. (Metally) **7**, 671–676 (2018)
2. Betekhtin, V.I., Kadomtsev, A.G., Narykova, M.V., Bannikov, M.V., Abaimov, S.G., Akhatov, ISh., Palin-Luc, T., Naimark, O.B.: Experimental and theoretical study of multiscale damage-failure transition in very high cycle fatigue. Phys. Mesomech. **20**(1), 82–93 (2017)
3. Terentyev, V.F., Korableva, S.A.: Fatigue of Metals. Nauka, Moscow (2015). (in Russian)
4. Shaniavsky, A.A., Soldatenkov, A.P.: Scale levels of the fatigue limit of metals. Phys. Mesomech. **22**(1), 44–53 (2019)
5. Furuya, Y., Hirukawa, H., Takeuchi, E.: Gigacycle fatigue in high strength steels. Sci. Technol. Adv. Mater. **20**(1), 643–656 (2019). https://doi.org/10.1080/14686996.2019.1610904
6. Volegov, P.S., Gribov, V.S., Trusov, P.V.: Damage and failure: a review of experimental work Phys. Mesomech. **18**(3), 11–24 (2015)
7. Makhutov, N.A.: Complex investigation of fracture of materials and structures. Factory laboratory. Diagn. Mater. **84**(11), 46–51 (2018)
8. Murakami, Y.: Metal Fatigue: Effects of Small Defects and Nonmetallic Inclusions. Elsevier, Oxford (2019)
9. Morel, F., Guerchais, R., Saintier, N.: Competition between microstructure and defect in multiaxial high cycle fatigue. Frattura ed Integrita Strutturale **33**, 404–414 (2015)
10. Zavoychinskaya, E.B.: Fatigue scale-structural failure and durability of structures with proportional loading processes. Ref. Doc. Dissert, Genesis LLC, Moscow (2018). (in Russian)
11. Botvina, L.R.: Failure. Kinetics, Mechanisms, General Relations. Science, Moscow (2008) (in Russian)
12. Ivanova, V.S., Terentyev, V.F.: The Nature of Metal Fatigue. Metallurgy, Moscow (1975). (in Russian)
13. Botvina, L.R., Soldatenkov, A.P.: On the concentration criterion of fracture. Metallofiz. Noveish. Tekhnol **39**, 477–490 (2017)
14. Shaniavsky, A.A., Soldatenkov A.P.: New paradigms in the description of metal fatigue. Bull. Perm Nat. Iss. Polytechnic un-that **1**, 196–207 (2019)
15. Mareau, C., Morel, F.: A Continuum damage mechanics – based approach for the high cycle fatigue behavior of metallic polycrystals. Int. J. Damage Mech. **XX**(X), 1–21 (2018). https://doi.org/10.1177/1056789518795204
16. Ai, Y., Zhu, S.P., Liao, D., Correia, J.A.F.O., Souto, C., De Jesus, A.M.P., Keshtegar, B.: Probabilistic modeling of fatigue life distribution and size effect of components with random defects. Int. J. Fatigue **126**, 165–173 (2019)

17. Amouzou, K., Charkaluk, E.A.: New high-cycle fatigue criterion based on a self-consistent scheme for hard metals under non-proportional loading. In: Proceedings of the 17th international conference on new trends in fatigue and fracture (2018). https://doi.org/10.1007/978-3-319-70365-7_4
18. Volegov, P.S., Gribov, V.S., Trusov, P.V.: Damage and failure: models based on physical theories of plasticity. Phys. Mesomech. **18**(6), 12–23 (2015)
19. Volegov, P.S., Gribov, V.S., Trusov, P.V.: Damage and failure: classical continuum theories. Phys. Mesomech. **18**(4), 68–86 (2015)
20. Nikitin, I.S., Burago, N.G., Nikitin, A.D., Yakushev, V.L.: Determination of the critical plane and assessment of fatigue life under various modes of cyclic loading. Bull. Perm Nat. Iss. Polytechnic un-that **4**, 238–252 (2017)
21. Stepanova, L.V., Igonin, S.A.: Description of deterioration processes: damage parameter of Y.N. Rabotnov: historical remarks, fundamental results and contemporary state. Bull. SamSU - Nat. Sci. Ser. 2014, **3**, (114), 97–114 (2014)
22. Volkov, I.A., Igumnov, L.A.: Introduction to the Continuum Mechanics of a Damaged Medium. Fizmatlit, Moscow (2017). (in Russian)
23. Zavoychinsky, B.I.: Durability of Main and Technological Pipelines (Theory, Calculation Methods, Design). Nedra, Moscow (1992)
24. Kiyko, I.A., Zavoychinskaya, E.B.: Introduction to the Theory of Failure Processes of Solids. MSU Pub. House, Moscow (2004)
25. Zavoychinskaya, E.B.: On the theory of scale structural fatigue of metals at the proportional loading. J. Phy. **1431**, 012024–012032 (2020). https://doi.org/10.1088/1742-6596/1431/1/012024
26. Zavoychinskaya, E.B.: On the stochastic theory of scale-structural metal fatigue. "Modern problems of mathematics and mechanics". Mater. International Conference dedicated to the 80th anniversary of acad. V. A. Sadovnichy, Llc MAX Press., Moscow, 694–697 (2019)
27. Zavoychinskaya, E.B.: Fatigue fracture of metals under complex stress state with consideration of structural changes. Moscow Univ. Mech. Bull. **74**(2), 36–40 (2019). https://doi.org/10.3103/S002713301902002X
28. Zavoychinskaya, E.B.: About the criterion of scale-structural metal fatigue at complex stress state. In: Proceedings of the XII All-Russian Congress on Fundamental Problems of Theoretical and Applied Mechanics. Ufa, vol. 3, pp. 607–609 (2019). https://doi.org/10.22226/2410-3535-2019-congress-v3
29. Zavoychinskaya, E.B.: On the theory of stage-by-stage metal fatigue at complex stress state. J. Mach. Manuf. Reliability, **47**(1), 72–80 (2018). https://doi.org/10.3103/S1052618818010156
30. Zavoychinskaya, E.B., Ovchinnikova, N.V.: To estimation of long structure durability at complex climatic conditions. In: Materials of XXV International Symposium "Dynamic and Technological Problems of Mechanics of Structures and Continuous Media" Named A.G. Gorshkov, TRP Llc, Moscow, vol. 2, pp. 163–171 (2019)
31. Zavoychinskaya, E.B.: Micro- and macromechanics of fracture of structural elements. Mech. Solids **3**, 304–323 (2012)
32. Engineering. Encyclopedia. T. IV-3. Machine Reliability. In: Klyuev, V.V. (ed.) Engineering, Moscow (2003)
33. Makhutov, N.A.: The Safety of Russia. Law, social-economical and scientific-technical aspects. MGF "Knowledge", Moscow (2018)
34. Makhutov, N.A., Matvienko, Yu.G., Romanov, A.N. (eds.): Problems of Strength, Technological Safety and Structural Materials Science. Lenand, Moscow (2018)
35. Bondar, V.S. (ed.): Resource of Materials and Structures: Monograph. Moscow Polytechnic, Moscow (2019)
36. Makhutov, N.A. (ed.) Strength, Resource, Survivability and Safety of Machines. Book house "Librocom", Moscow (2019)
37. Zavoychinsky, B.I., Zavoychinskaya, E.B., Volchanin, A.V.: Probabilistic assessment of the residual operational safety of long structures. Direct. Mag. **7**, 41–46; Eng. J. **12**, 33–36 (2012)

38. Zavoychinsky, B.I., Giller, G.P., Zavoychinskaya, E.B.: Recommendations for Assessing of the Safety and Durability of Gas Pipelines during Design. IRC Gazprom, Moscow (2002)
39. Zavoychinsky, B.I., Tutnov, I.A., Zavoychinskaya, E.B.: Recommendations for Assessing of the Safety and Durability of Main Gas Pipelines during Design. IRC Gazprom, Moscow (2000)

Chapter 6
On Tikhonov Regularization of Optimal Distributed Control Problem for an Ill-Posed Elliptic Equation with p-Laplace Operator and L^1-type of Non-linearity

Peter I. Kogut and Olha P. Kupenko

Abstract We discuss the existence of solutions to an optimal control problem for the Dirichlet boundary value problem for strongly non-linear p-Laplace equations with L_1-type of nonlinearity and $p \geq 2$. The control variable u is taken as a distributed control. The optimal control problem is to minimize the discrepancy between a given distribution $y_d \in L^p(\Omega)$ and the current system state. We deal with such case of nonlinearity when we cannot expect to have a solution of the original boundary value problem for each admissible control. Instead of this we make use of a variant of the classical Tikhonov regularization. We eliminate the differential constraints between control and state and allow such pairs run freely in their respective sets of feasibility by introducing some additional variable which plays the role of "defect". We show that this special residual function can be determined in a unique way. We introduce a special family of regularized optimization problems and show that each of these problems is consistent, well-posed, and their solutions allow to attain (in the limit) an optimal solution of the original problem as the parameter of regularization tends to zero. As a consequence, we establish sufficient conditions of the existence of optimal solutions to the given class of nonlinear Dirichlet BVP and propose the way for their approximation.

P. I. Kogut
Department of Differential Equations, Oles Honchar Dnipro National University, Gagarin av., 72, Dnipro 49010, Ukraine
e-mail: p.kogut@i.ua

O. P. Kupenko (✉)
Department of System Analysis and Control, National University of Technology "Dnipro Polytechnics", Yavornitsky av., 19, Dnipro 49005, Ukraine

Institute for Applied and System Analysis, National Technical University of Ukraine "Kiev Polytechnic Institute", Peremogy av., 37, building 35, Kiev 03056, Ukraine
e-mail: kupenko.olga@gmail.com

V. A. Sadovnichiy and M. Z. Zgurovsky (eds.), *Contemporary Approaches and Methods in Fundamental Mathematics and Mechanics*, Understanding Complex Systems,
https://doi.org/10.1007/978-3-030-50302-4_6

6.1 Introduction

Let Ω be a given bounded open subset of R^N $(N \geq 1)$ with a sufficiently smooth boundary $\partial\Omega$. We assume that Ω lies locally on one side of its boundary $\partial\Omega$.

We deal with the following optimal control problem (OCP) for nonlinear p-Laplace elliptic equation.

$$\text{Minimize } J(u, y) = \frac{1}{2}\int_\Omega |y - y_d|^p\, dx + \frac{1}{2}\int_\Omega |u|^2\, dx \tag{6.1}$$

$$\text{subject to constrains}$$

$$-\operatorname{div}\left(|\nabla y|^{p-2}\nabla y\right) = f(y) + u \quad \text{in } \Omega, \tag{6.2}$$

$$y = 0 \quad \text{on } \partial\Omega, \tag{6.3}$$

$$u \in U_\partial := L^2(\Omega), \tag{6.4}$$

where $y_d \in L^p(\Omega)$ is a given distribution, $p \geq 2$, $f : \mathbb{R} \to [0, +\infty)$ is a given function such that $\mathbb{R} \ni t \mapsto f(t) \in [0, +\infty)$ is a monotonically increasing mapping and $f \in C^1_{loc}(\mathbb{R})$.

Equations like (6.2) appear in a number of applications. In particular, it has been actively investigated in connection with combustion theory and in the study of stellar structures (see, for instance, [6, 11, 12, 14, 19] for the case $p = 2$). In that case it has been shown that the indicated BVP is ill-posed, in general. It means that there is no reason to assert the existence of weak solutions to (6.2)–(6.4) for a given control $u \in L^2(\Omega)$ and for $p \geq 2$, or to suppose that such solution, even if it exists, is unique.

Thus, in the context of the optimal control problem that we deal with in this paper, it is unknown whether this problem is consistent and admits at least one solution. It is worth to note here that pioneering works on optimal control problems governed by PDEs and the control of blowing-up nonlinear problems have been examined thoroughly by J. L. Lions (see [10, 13, 27–30], for instance). The optimal control problem (6.1)–(6.4) in the case of $f(y) = e^y$ and $p = 2$ was discussed in detail by Casas, Kavian, and Puel [4]. The problem of existence and uniqueness of the underlying boundary value problem and the corresponding optimal control problem was treated and an optimality system has been derived and analyzed. However, to the best knowledge of author, the existence of optimal controls for the problem (6.1)–(6.4) with $p \geq 2$ and more general assumptions on $f(y)$ remain arguably open questions for nowadays. Other important references, that also deal with the approximation issues for ill-posed optimal control problems and their asymptotic analysis without any attempt to be exhaustive, are [5, 8, 9, 16–19, 23, 25, 26, 35].

The novelty of this paper is that we discuss the existence of optimal pairs to OCP (6.1)–(6.4) using an indirect approach based on the classical Tikhonov regularization technique in its special implementation. The idea to involve the Tikhonov regularization is inspired by the following reason: the main characteristic feature of BVP (6.2)–(6.3) is the fact that because of the specificity of non-linearity $f(y)$ (in many particular implementations of the model (6.2)–(6.3), $F(u) = \lambda e^u$ and $p = 2$ [6, 11]), we have no a priori estimate for the weak solutions in the standard Sobolev space $W_0^{1,p}(\Omega)$. As a result, the consistency of OCP (6.1)–(6.4) and existence of optimal pairs can be established only if we impose rather strict assumptions on the initial data. In particular, it was shown in [31] that the set of optimal solutions of (6.1)–(6.4) is nonempty provided $p = 2$, $N > 2$, the domain Ω is star-shaped with respect to some interior point x_0, and the set of feasible pairs Ξ contains at least one pair (u, y) such that $f(y) \in L^2(\Omega)$.

Therefore, our main interest in this paper is to show that these assumptions can be essentially weakened or even eliminated. With that in mind, we introduce the additional variable z (the so-called "defect" in the state equation) into the regularized problem in order to let the pairs "control– statel" (u, y) run freely in the feasible space $L^2(\Omega) \times W_0^{1,p}(\Omega)$ so that there is no dependence of y on u. At the same time, there is a principle difference between the 'standard' implementation of the Tikhonov regularization of OCPs (see, for instance, [33, 34]) and the proposed scheme. This difference lies in the exploitation of the terms $\varepsilon\|f(y)\|^r_{L^r(\Omega)}$ and $\varepsilon^{-1}\|\nabla z\|_{L^{p'}(\Omega,\mathbb{R}^N)}$ in the perturbed cost functional $J_\varepsilon(u, y, z)$. We show that the boundedness of these terms on the set Ξ of feasible solutions to the original problem plays a crucial role in the study of asymptotic behaviour of global solutions to regularized OCPs. Having introduced a special family of optimization problems, we also show that there exists an optimal solution to the original OCP that can be attained by the sequence of optimal solutions for the regularized minimization problems (for benefit of this approach and its comparison with other ones, we refer to the recent papers [5, 20, 21, 24]).

In this paper we mainly focus on the case of p-Laplace operator with $p \geq 2$. Since, in the case $1 < p \leq 2$, we obviously have:

$$-\Delta_p := -\operatorname{div}\left(|\nabla \cdot|^{p-2}\nabla \cdot\right) : W_0^{1,p}(\Omega) \mapsto W^{-1,q}(\Omega) \subset H^{-1}(\Omega),$$

it means that we can utilize the approximation approach that has been recently proposed in [22].

The paper is organized as follows. In Sect. 6.2 we give some preliminaries and describe in details the characteristic features of OCP (6.1)–(6.4). The Tikhonov regularization of the original OCP is discussed in Sect. 6.3. The key result of this section is Theorem 6.3, where we announce the sufficient conditions of the existence of optimal solutions to the regularized problems. The details of the indirect approach to the study of the original optimal control problem are discussed in Sect. 6.4. The key points of such approach are summarized in Theorem 6.4.

6.2 Preliminaries

Following the standard notation, by $W_0^{1,p}(\Omega)$ we denote the Sobolev space which is defined as the closure of

$$C_0^\infty(\Omega) = \left\{\varphi \in C_0^\infty(\mathbb{R}^N) \ : \ \varphi = 0 \text{ on } \partial\Omega\right\}$$

with respect to the norm

$$\|y\|_{W_0^{1,p}(\Omega)} = \left(\int_\Omega |\nabla y|^p \, dx\right)^{1/p}.$$

So, we can suppose that each element of the space $W_0^{1,p}(\Omega)$ has zero trace at $\partial\Omega$. Let $W^{-1,p'}(\Omega) := \left(W_0^{1,p}(\Omega)\right)^*$ be the dual space to $W_0^{1,p}(\Omega)$.

In order to make a precise meaning of the weak solution to BVP (6.2)–(6.3) in the sense of distributions (or shortly, distributional solution), we begin with the following concept.

Definition 6.1 Let $u \in U_\partial$ be a given control function. We say that $y = y(u)$ is a weak solution to the boundary value problem (6.2)–(6.3) in the sense of distributions, if it belongs to the class of functions

$$H_f = \left\{y \in W_0^{1,p}(\Omega) \ : \ f(y) \in L^1(\Omega)\right\}$$

and the integral identity

$$\int_\Omega \left(|\nabla y|^{p-2}\nabla y, \nabla\varphi\right) dx = \int_\Omega f(y)\varphi \, dx + \int_\Omega u\varphi \, dx \tag{6.5}$$

holds for every test function $\varphi \in C_0^\infty(\Omega)$.

However, it is unknown whether the original BVP admits at least one weak solution in the sense of Definition 6.1 for each admissible control $u \in U_\partial$. Moreover, as follows from (6.5), the continuity of the mapping $\varphi \mapsto [y, \varphi]_f := \int_\Omega f(y)\varphi \, dx$ on the set $W_0^{1,p}(\Omega)$ is not evident. For the details related with this issue, we refer to the classical paper Casas, Kavian, and Puel [4] (see also the recent papers [7, 20–22, 24, 31]).

Before proceeding further, we make use of the following observation. Assume that for a given $u \in U_\partial \subset L^2(\Omega)$, we have: $y \in H_f$ and the pair (u, y) is related by integral identity (6.5). Then for each test function $\varphi \in C_0^\infty(\Omega)$, the following estimate

$$\begin{aligned}
\int_\Omega f(y)\varphi \, dx &\le \left|\int_\Omega \left(|\nabla y|^{p-2}\nabla y, \nabla\varphi\right) dx\right| + \left|\int_\Omega u\varphi \, dx\right| \\
&\le \|y\|^{p-1}_{W_0^{1,p}(\Omega)} \|\varphi\|_{W_0^{1,p}(\Omega)} + \|u\|_{L^{p'}(\Omega)} \|\varphi\|_{L^p(\Omega)} \\
&\overset{\text{by Friedrichs ineq.}}{\le} \left[\|y\|^{p-1}_{W_0^{1,p}(\Omega)} + C_\Omega |\Omega|^{\frac{2-p'}{2p'}} \|u\|_{L^2(\Omega)}\right] \|\varphi\|_{W_0^{1,p}(\Omega)}
\end{aligned} \tag{6.6}$$

holds true. Here, $C_\Omega = \operatorname{diam} \Omega$ and this constant comes from the Friedrich's inequality

$$\|y\|_{L^p(\Omega)} \le \operatorname{diam}\Omega\ \|\nabla y\|_{L^p(\Omega;\mathbb{R}^N)}, \quad \forall y \in W_0^{1,p}(\Omega). \tag{6.7}$$

Hence, the mapping $\varphi \mapsto [y,\varphi]_f$ can be extended by continuity onto the set of all $\varphi \in W_0^{1,p}(\Omega)$ using (6.6) and the standard rule

$$[y,\varphi]_f = \lim_{\varepsilon\to 0}[y,\varphi_\varepsilon]_f, \tag{6.8}$$

where $\{\varphi_\varepsilon\}_{\varepsilon>0} \subset C_0^\infty(\Omega)$ and $\varphi_\varepsilon \to \varphi$ strongly in $W_0^{1,p}(\Omega)$ as $\varepsilon \to 0$. In particular, if $y \in H_f$, then we can define the value $[y,y]_f$ and this one is finite for every $y \in H_f$. As a consequence, we deduce: if $y \in H_f$ is a weak solution to boundary value problem (6.2)–(6.3), then y satisfies the energy equality

$$\int_\Omega |\nabla y|^p\,dx = [y,y]_f + \langle u, y\rangle_{W^{-1,p'}(\Omega);W_0^{1,p}(\Omega)}, \tag{6.9}$$

where $\langle\cdot,\cdot\rangle_{W^{-1,p'}(\Omega);W_0^{1,p}(\Omega)} : W^{-1,p'}(\Omega)\times W_0^{1,p}(\Omega) \to \mathbb{R}$ denotes the duality pairing between $W^{-1,p'}(\Omega)$ and $W_0^{1,p}(\Omega)$. We notice that, by continuity of embeddings $L^2(\Omega) \hookrightarrow L^{p'}(\Omega)$ and $L^{p'}(\Omega) \hookrightarrow W^{-1,p'}(\Omega)$, the second term in the right-hand side is correctly defined.

However, it is unknown whether the value $[y,y]_f$ preserves a constant sign for all $y \in H_f$. Therefore, we cannot make use of the energy equality (6.9) in order to derive a priori estimate in $\|\cdot\|_{W_0^{1,p}(\Omega)}$-norm for the weak solutions.

To specify the term $[y,y]_f$, we have the following result (unlike similar results in [4, 20, 21], here we do not assume here that the function $f(y)$ is of the exponential type).

Lemma 6.1 *Let $u \in L^2(\Omega)$ be a given control, and let $y = y(u) \in H_f$ be a weak solution of BVP* (6.2)–(6.3) *in the sense of distribution. Then $f(y) \in W^{-1,p'}(\Omega)$ and*

$$[y,z]_f = \langle f(y), z\rangle_{W^{-1,p'}(\Omega);W_0^{1,p}(\Omega)} = \int_\Omega z\,f(y)\,dx, \quad \forall z \in W_0^{1,p}(\Omega), \tag{6.10}$$

i.e. $z\,f(y) \in L^1(\Omega)$ for every $z \in W_0^{1,p}(\Omega)$.

Proof Let $z \in W_0^{1,p}(\Omega)\cap L^\infty(\Omega)$ be an arbitrary element. Since $f(y) \in L^1(\Omega)$, it follows that the term $\int_\Omega z\,f(y)\,dx$ is well defined. Let $\{\varphi_\varepsilon\}_{\varepsilon>0} \subset C_0^\infty(\Omega)$ be a sequence such that $\varphi_\varepsilon \to z$ in $W_0^{1,p}(\Omega)$. Moreover, in this case we can suppose that

$$\sup_{\varepsilon>0}\|\varphi_\varepsilon\|_{L^\infty(\Omega)} < +\infty \quad \text{and} \quad \varphi_\varepsilon \overset{*}{\rightharpoonup} z \ \text{ in } L^\infty(\Omega).$$

Hence, due to the fact that $y \in H_f$, we get

$$\int_\Omega z\,f(y)\,dx = \lim_{\varepsilon\to 0}\int_\Omega \varphi_\varepsilon f(y)\,dx = \lim_{\varepsilon\to 0}[y,\varphi_\varepsilon]_f \overset{\text{by (6.13)}}{=} [y,z]_f. \tag{6.11}$$

Thus, we arrive at the relation (6.10) for each $z \in W_0^{1,p}(\Omega) \cap L^\infty(\Omega)$.

Let us take now $z \in W_0^{1,p}(\Omega)$ such that $z \geq 0$ almost everywhere in Ω. For every $\varepsilon > 0$, let $T_\varepsilon : \mathbb{R} \to \mathbb{R}$ be the truncation operator defined by

$$T_\varepsilon(s) = \max\left\{\min\left\{s, \varepsilon^{-1}\right\}, -\varepsilon^{-1}\right\}. \tag{6.12}$$

The following property of T_ε is well-known (see [15]): If $z \in W_0^{1,p}(\Omega)$ then

$$T_\varepsilon(z) \in L^\infty(\Omega) \cap W_0^{1,p}(\Omega)\ \forall\, \varepsilon > 0 \quad \text{and} \quad T_\varepsilon(z) \to z \ \text{in} W_0^{1,p}(\Omega) \text{as} \varepsilon \to 0.$$

Hence, we can suppose that $T_\varepsilon(z) \to z$ almost everywhere in Ω and the sequence $\{T_\varepsilon(z)\}_{\varepsilon>0}$ is monotonically increasing. Taking into account the initial assumptions on function $f(\cdot)$, we see that $\{T_\varepsilon(z) f(y)\}_{\varepsilon>0}$ is a pointwise non-decreasing sequence, and also $T_\varepsilon(z) f(y) \to z\, f(y)$ for almost all $x \in \Omega$. Therefore, by monotone convergence theorem, $z\, f(y)$ is a measurable function on Ω, and

$$\lim_{\varepsilon\to 0} \int_\Omega T_\varepsilon(z) f(y)\, dx = \int_\Omega z\, f(y)\, dx.$$

Thus, (6.10) holds true for each $z \in W_0^{1,p}(\Omega)$ such that $z \geq 0$.

As for a general case, i.e. $z \in W_0^{1,p}(\Omega)$, it is enough to note that $z = z^+ - z^-$ with $z^+, z^- \in W_0^{1,p}(\Omega)$ and $z^+, z^- \geq 0$ in Ω, where

$$z^+ := \max\{z, 0\}, \quad z^- := \max\{-z, 0\}.$$

To complete the proof, it remains to observe that, for any $z \in W_0^{1,p}(\Omega)$ and each sequence $\{\varphi_\varepsilon\}_{\varepsilon>0} \subset C_0^\infty(\Omega)$ such that $\varphi_\varepsilon \to z$ in $W_0^{1,p}(\Omega)$, we have:

$$\begin{aligned}
\int_\Omega z\, f(y)\, dx &= \lim_{\varepsilon\to 0} \int_\Omega \varphi_\varepsilon f(y)\, dx \\
&\overset{\text{by (6.6)}}{\leq} \lim_{\varepsilon\to 0} \left(\|y\|^{p-1}_{W_0^{1,p}(\Omega)} + C_\Omega |\Omega|^{\frac{2-p'}{2p'}} \|u\|_{L^2(\Omega)} \right) \|\varphi_\varepsilon\|_{W_0^{1,p}(\Omega)} \\
&= \left(\|y\|^{p-1}_{W_0^{1,p}(\Omega)} + C_\Omega |\Omega|^{\frac{2-p'}{2p'}} \|u\|_{L^2(\Omega)} \right) \|z\|_{W_0^{1,p}(\Omega)}.
\end{aligned}$$

Hence, $\quad f(y) \in W^{-1,p'}(\Omega), \quad \langle f(y), z\rangle_{W^{-1,p'}(\Omega); W_0^{1,p}(\Omega)} = \displaystyle\int_\Omega z\, f(y)\, dx, \quad \forall\, z \in W_0^{1,p}(\Omega)$, and

$$\|f(y)\|_{W^{-1,p'}(\Omega)} \leq \left(\|y\|^{p-1}_{W_0^{1,p}(\Omega)} + C_\Omega |\Omega|^{\frac{2-p'}{2p'}} \|u\|_{L^2(\Omega)} \right).$$

As a direct consequence of Lemma 6.1 and relation (6.9), we can specify the energy equality (6.9) as follows.

Corollary 6.1 *Let $u \in U_\partial$ be a given control and let $y = y(u) \in W_0^{1,p}(\Omega)$ be a weak solution to BVP (6.2)–(6.3) in the sense of Definition 6.1. Then the energy equality for y takes the form*

$$\int_\Omega |\nabla y|^p \, dx = \int_\Omega y\, f(y)\, dx + \int_\Omega u y \, dx. \tag{6.13}$$

Since it is unknown whether there exists a weak solution to BVP (6.2)–(6.3) for a given $u \in U_\partial$, or whether it is unique, it motivates us to introduce the following set.

Definition 6.2 We say that a pair (u, y) is a feasible solution for optimal control problem (6.1)–(6.4) if $u \in U_\partial$, $y \in H_f$, $f(y) \in L^1(\Omega)$, and the pair (u, y) is related by integral identity (6.5). By $\Xi \subset L^2(\Omega) \times W_0^{1,p}(\Omega)$ we denote the set of all feasible solutions.

As for the optimal control problem (6.1)–(6.4), it was mentioned in [4] that its study is a non-trivial matter even for distributed controls because of the specificity of non-linearity $f(y)$ (in [4] the authors set up $f(y) = e^y$). The main difficulties in this case are strongly related with the following circumstances:

- The set of feasible solutions can be empty, in general;
- Even if $\Xi \neq \emptyset$, we have no a priori estimate for the weak solutions of (6.2)–(6.3) with arbitrary $u \in U_\partial$;
- Some a priori estimates can be established if only $N > 2$, $p = 2$, the domain Ω has a sufficiently smooth boundary and it is star-shaped with respect to some interior point x_0, i.e.
$$(\sigma - x_0, \nu(\sigma)) \geq 0, \quad \text{for a.a.}\, \sigma \in \partial\Omega,$$
where $\nu(\sigma)$ denotes the outward unit normal vector to $\partial\Omega$ at the point σ, and the considered weak solutions $y(u)$ of (6.2)–(6.3) satisfies the extra property $f(y) \in L^2(\Omega)$ or (see [7, Definition 1.1])
$$\int_\Omega |\nabla y|^2 \, dx \leq \frac{2N}{N-2} \int_\Omega F(y)\, dx - \frac{2}{N-2} \int_\Omega u\, (x - x_0, \nabla y)\, dx, \tag{6.14}$$
$$\int_{\Gamma_N} F(y(\sigma))\, (\nu(\sigma), \sigma - x_0)\, d\mathcal{H}^{N-1} \geq -F(0) \int_{\Gamma_D} (\nu(\sigma), \sigma - x_0)\, d\mathcal{H}^{N-1}, \tag{6.15}$$
with some $x_0 \in \operatorname{int} \Omega$, where $F \in C^1_{loc}(\mathbb{R})$ is a nonlinear function such that $F'(t) = f(t)$.

- Since we have no estimates for the states (especially without the above mentioned extra property $f(y) \in L^2(\Omega)$), it follows that we cannot deduce the boundedness in $L^2(\Omega) \times W_0^{1,p}(\Omega)$ of a minimizing sequence to the problem (6.1)–(6.4);
- Even if a minimizing sequence $\{(u_k, y_k) \in \Xi\}_{n\in\mathbb{N}}$ is weakly compact in $L^2(\Omega) \times W_0^{1,p}(\Omega)$, it does not allow us to pass to the limit in integral identity (6.5) and inequalities (6.14)–(6.15) as $k \to \infty$ and, therefore, the existence of an optimal pair to the problem (6.1)–(6.4) can not be concluded by arguments of Direct Methods in the Calculus of Variations.

Although this list can be extended by many other options, we can summarize this issue by the following existence result (for the proof we refer to [31, Theorem 4.3]).

Theorem 6.1 *Let us assume that the following conditions hold true: $N > 2$, $p \geq 2$, the domain Ω has a $C^{0,1}$-boundary, this domain is star-shaped with respect to some interior point x_0, and for given $f \in C_{loc}(\mathbb{R})$ and U_∂, the set of feasible solutions Ξ with extra properties* (6.14)–(6.15) *is nonempty. Assume also that $F \in C^1_{loc}(\mathbb{R})$ is a nonlinear function such that $F'(t) = f(t)$ and*

$$C_F F'(z) \geq F(z), \quad \forall z \in \mathbb{R}, \quad \textit{and} \quad \left| \int_{-\infty}^{0} z F'(z)\, dz \right| < +\infty \tag{6.16}$$

for some constant $C_F > 0$. Then there exists a unique pair $(u^0, y^0) \in \Xi$ such that

$$(u^0, y^0) \in \Xi_0 \subset \Xi, \quad J(u^0, y^0) = \inf_{(u,y)\in\Xi_0} J(u, y), \tag{6.17}$$

where

$$\Xi_0 = \{(u, y) \in \Xi \mid \textit{inequalities (6.14)–(6.15) hold true}\}. \tag{6.18}$$

At the same time, because of the state constraints (6.14)–(6.15), it is clear that the constrained minimization problems

$$\left\langle \inf_{(u,y)\in\Xi} J(u, y) \right\rangle \quad \text{and} \quad \left\langle \inf_{(u,y)\in\Xi_0} J(u, y) \right\rangle$$

are drastically distinguished. Therefore, our main interest in this paper is to show that the main restrictions coming from Theorem 6.1 can be eliminated by introducing a new additional variable z into the problem which lets pairs (u, y) run freely in the feasible space $L^2(\Omega) \times W_0^{1,2}(\Omega)$ so that there is no dependence of y on u.

6.3 On Tikhonov Regularization of the Original OCP

Let us introduce the Tikhonov regularized optimal control problem associated to the original OCP (6.1)–(6.4). Let $\varepsilon > 0$ be a given small parameter. Then the regularized problem we are going to consider takes the form (for comparison, we refer to [22, 33, 34])

$$\begin{aligned}\text{Minimize } J_\varepsilon(u, y, z) = &\frac{1}{p}\int_\Omega |y - y_d|^p\,dx + \frac{1}{2}\int_\Omega |u|^2\,dx\\ &+ \frac{1}{\varepsilon p'}\int_\Omega |\nabla z|^{p'}\,dx + \frac{\varepsilon}{r}\int_\Omega |f(y)|^r\,dx,\end{aligned} \tag{6.19}$$

subject to constrains

$$u \in U_\partial := L^2(\Omega),\ \ y \in W_0^{1,p}(\Omega), \tag{6.20}$$

$$-\Delta z = \operatorname{div}\left(|\nabla y|^{p-2}\nabla y\right) + f(y) + u \quad \text{in } \Omega, \tag{6.21}$$

$$z = 0 \quad \text{on } \partial\Omega, \tag{6.22}$$

where

$$p' = \frac{p}{p-1} \quad \text{and} \quad N > r > \max\left\{1, \frac{Np'}{N+p'}\right\}. \tag{6.23}$$

To begin with, let us stress again that the main reason to introduce the additional variable z into the regularized problem is to let the pairs (u, y) run freely in the feasible space $U_\partial \times W_0^{1,p}(\Omega)$ so that there is no dependence of y on u. On the other hand, there is a principle difference between the standard scheme of the Tikhonov regularization of OCPs (see, for instance, [33, 34]) and the proposed regularization in the form (6.19)–(6.22). This difference lies in the exploitation of the terms $\|z\|^{p'}_{W_0^{1,p'}(\Omega)}/(\varepsilon p')$ and $\varepsilon\|f(y)\|^r_{L^r(\Omega)}/r$ in the perturbed cost functional $J_\varepsilon(u, y, z)$. As it will be shown later on, the boundedness of these terms on the set Ξ of feasible solutions to the original problem (see Lemma 6.1), plays a crucial role in the study of asymptotic behaviour of global solutions $\left\{(u_\varepsilon^0, y_\varepsilon^0, z_\varepsilon^0)\right\}_{0<\varepsilon\le\varepsilon_0}$ as ε tends to zero.

Our main assumptions in this section are:

(a) Ω is a bounded open domain in $\mathbb{R}^N$, $N \ge 1$, of class $\mathscr{D}^s$ with $s > p$. Here, following Meyers [32], we say that $\Omega \in \mathscr{D}^s$ if the equation

$$-\Delta u = -\operatorname{div} g$$

has a unique solution u in $W_0^{1,s}(\Omega)$ for every $g \in L^s(\Omega; \mathrm{R}^N)$ and $\|u\|_{W_0^{1,s}(\Omega)} \le C_s\|g\|_{L^s(\Omega;\mathrm{R}^N)}$ holds for some constant C_s independent of g. We notice that this property constitutes a condition of regularity on the boundary $\partial\Omega$, and inclusion $\Omega \in \mathscr{D}^s$ holds for any value of $s > p'$ if the boundary is sufficiently smooth.

The proof is based on the Calderon-Zygmund inequality for singular integrals (see [3, 32] and [1, Theorem 15.3]);

(b) $2 \leq p < +\infty$;

(c) $f : \mathbb{R} \to [0, +\infty)$ is a given monotonically increasing mapping such that $f \in C_{loc}(\mathbb{R})$.

Remark 6.1 If for a given $y \in W_0^{1,p}(\Omega)$, we have

$$f(y) \in L^r(\Omega) \quad \text{with} \quad r > \max\left\{1, \frac{Np'}{N+p'}\right\},$$

then $f(y) \in W^{-1,p'}(\Omega)$. Indeed, the Sobolev embedding theorem implies that

$$W_0^{1,p}(\Omega) \hookrightarrow L^{p^*}(\Omega) \quad \text{continuously for} p^* = \frac{Np}{N-p}, \text{provided} p \in [2, N),$$

and

$$W_0^{1,p}(\Omega) \hookrightarrow L^{p^*}(\Omega) \quad \text{for arbitrary} p^* \geq 1 \text{if} p \geq N.$$

Hence, by the duality arguments, we deduce that

$$L^r(\Omega) \hookrightarrow L^{(p^*)'}(\Omega) \hookrightarrow W^{-1,p'}(\Omega)$$

with continuous injections if only

$$r > (p^*)' := \frac{p^*}{p^*-1} = \frac{Np}{(N+1)p-N} = \frac{Np'}{N+p'} \quad \text{and} \quad 2 \leq p < N. \tag{6.24}$$

If $p \geq N$, then the embedding $L^r(\Omega) \hookrightarrow W^{-1,p'}(\Omega)$ is also valid for all exponents r satisfying condition

$$N > r > \max\left\{1, \frac{Np'}{N+p'}\right\}. \tag{6.25}$$

Thus, there exists a constant $C > 0$ independent of $f(y)$ such that

$$\|f(y)\|_{W^{-1,p'}(\Omega)} \leq C\|f(y)\|_{L^r(\Omega)}. \tag{6.26}$$

Since $2 \geq \max\left\{1, \frac{Np'}{N+p'}\right\}$ for all $p \geq 2$, it follows that, for a given $y \in W_0^{1,p}(\Omega)$, we also have

$$h := f(y) + u \in L^{(p^*)'}(\Omega) \hookrightarrow W^{-1,p'}(\Omega).$$

Definition 6.3 We say that a tuple $(u, y, z) \in L^2(\Omega) \times W_0^{1,p}(\Omega) \times W_0^{1,p'}(\Omega)$ is a feasible solution to regularized problem (6.19)–(6.22) (in symbols, $(u, y, z) \in \Lambda_\varepsilon$), if $f(y) \in L^r(\Omega)$ with some $r \in \left(\max\left\{1, \frac{Np'}{N+p'}\right\}, N\right)$, $J_\varepsilon(u, y, z) < +\infty$, and the following variational equality

$$a(z, \varphi) = \mathscr{F}(\varphi) \tag{6.27}$$

holds true for all $\varphi \in W_0^{1,p}(\Omega)$, where

$$\mathscr{F}(\varphi) := -\langle |\nabla y|^{p-1} \nabla y, \nabla \varphi \rangle_{L^{p'}(\Omega;\mathbb{R}^N);L^p(\Omega;\mathbb{R}^N)} + (u, \varphi)_{L^2(\Omega)} + \langle f(y), \varphi \rangle_{W^{-1,p'}(\Omega);W_0^{1,p}(\Omega)},$$

and $a : W_0^{1,p'}(\Omega) \times W_0^{1,p}(\Omega) \to \mathbb{R}$ denotes the bilinear form

$$a(z, \varphi) = \int_\Omega (\nabla z, \nabla \varphi)\, dx.$$

In order to show that, for each $\varepsilon > 0$, the set of feasible solutions Λ_ε to regularized problem (6.19)–(6.22) is nonempty, we make use of the following result (see Theorem 1 in [32]).

Theorem 6.2 *Let Ω be a bounded domain of class $\mathscr{D}^s$ for some $s > p \geq 2$. Let $f \in L^q(\Omega; \mathbb{R}^N)$ be a given vector field, where the exponent q is subjected to constrains*

$$s' < p' \leq q \leq p < s. \tag{6.28}$$

Let h be an element of $L^r(\Omega)$ with $r^* = \frac{Nr}{N-r} \geq q$ for some $r \in (1, N)$. Then there exists a unique function $z \in W_0^{1,q}(\Omega)$ such that

$$\|\nabla z\|_{L^q(\Omega;\mathbb{R}^N)} \leq C \left(\|f\|_{L^q(\Omega;\mathbb{R}^N))} + \|h\|_{L^r(\Omega)} \right) \quad \text{with } C = C(\Omega, q, r), \tag{6.29}$$

and z is a solution in the sense of distributions of the following Dirichlet boundary value problem

$$-\Delta z = -\operatorname{div} f + h \quad \text{in } \Omega,$$
$$z = 0 \quad \text{on} \quad \partial\Omega,$$

that is

$$\int_\Omega (\nabla z, \nabla \varphi)\, dx = \int_\Omega [(f, \nabla \varphi) + h\varphi]\, dx, \quad \forall \varphi \in C_0^\infty(\Omega). \tag{6.30}$$

Let (u, y) be an arbitrary pair in $L^2(\Omega) \times W_0^{1,p}(\Omega)$ such that

$$f(y) \in L^r(\Omega) \quad \text{with some} \quad r \in \left(\max\left\{1, \frac{Np'}{N + p'}\right\}, N \right).$$

Then $r^* = \frac{Nr}{N-r} \ge p'$ (see Remark 6.1) and $2^* := \frac{2N}{N+2} > p'$ for $p \ge 2$. Therefore (see [32, p.199]), there exists a function $v \in L^1(\Omega)$ such that

$$\Delta v = \text{div } (\nabla v) = f(y) + u \text{ in } \Omega$$

and

$$\|\nabla v\|_{L^{r^*}(\Omega;\mathbb{R}^N))} \le C \left(\|f(y)\|_{L^r(\Omega)} + \|u\|_{L^2(\Omega)} \right),$$

where $C > 0$ is a constant depending only on Ω, q, and r. Thus, we can rewrite equations (6.21)–(6.22) as follows

$$-\Delta z = \text{div } \left(|\nabla y|^{p-2} \nabla y + \nabla v \right) \quad \text{in } \Omega, \tag{6.31}$$

$$z = 0 \quad \text{on } \partial\Omega. \tag{6.32}$$

Since the condition $r \in \left(\max \left\{ 1, \frac{Np'}{N+p'} \right\}, N \right)$ implies that $r^* = \frac{Nr}{N-r} \ge p'$, it follows that the vector field $|\nabla y|^{p-2} \nabla y + \nabla v$ is in $L^{p'}(\Omega; \mathbb{R}^N)$. As a result, in view of Theorem 6.2, we deduce that boundary value problem (6.31)–(6.32) is solvable and its solution in the sense of distribution $z = z(u, y)$ is unique in $W_0^{1,p'}(\Omega)$. Hence, (see Definition 6.3) the tuple $(u, y, z(u, y)) \in L^2(\Omega) \times W_0^{1,p}(\Omega) \times W_0^{1,p'}(\Omega)$ is a feasible solution to regularized problem (6.19)–(6.22). Thus, $\Lambda_\varepsilon \neq \emptyset$ and this implies that regularized optimal control problem (6.19)–(6.22) is consistent for all $\varepsilon > 0$.

Our next intension is to discuss the issue related to the existence of optimal solutions of the regularized problems (6.19)–(6.22).

Theorem 6.3 *Assume that conditions* (6.23) *and (a)–(c) indicated before are valid. Then for each* $\varepsilon > 0$ *there is a triplet* $(u_\varepsilon^0, y_\varepsilon^0, z_\varepsilon^0) \in \Lambda_\varepsilon$ *such that*

$$J_\varepsilon(u_\varepsilon^0, y_\varepsilon^0, z_\varepsilon^0) = \inf_{(u,y,z) \in \Lambda_\varepsilon} J_\varepsilon(u, y, z). \tag{6.33}$$

If, in addition, the function $f : \mathbb{R} \to [0, +\infty)$ *is convex, then* $(u_\varepsilon^0, y_\varepsilon^0, z_\varepsilon^0)$ *is a unique triple in* Λ_ε *with property* (6.33).

Proof Let $\varepsilon > 0$ be a given value. Since the cost functional $J_\varepsilon : \Lambda_\varepsilon \to \mathbb{R}$ is non-negative on Λ_ε, we may suppose that there exist a $\mu_\varepsilon \ge 0$ and a sequence $\left\{ (u_{\varepsilon,k}, y_{\varepsilon,k}, z_{\varepsilon,k}) \right\}_{k \in \mathbb{N}} \subset \Lambda_\varepsilon$ such that

$$\mu_\varepsilon = \inf_{(u,y,z) \in \Lambda_\varepsilon} J_\varepsilon(u, y, z) = \lim_{k \to \infty} J_\varepsilon(u_{\varepsilon,k}, y_{\varepsilon,k}, z_{\varepsilon,k}), \tag{6.34}$$

$$\mu_\varepsilon \le J_\varepsilon(u_{\varepsilon,k+1}, y_{\varepsilon,k+1}, z_{\varepsilon,k+1}) \le J_\varepsilon(u_{\varepsilon,k}, y_{\varepsilon,k}, z_{\varepsilon,k}) \le \mu_\varepsilon + 1, \quad \forall k \in \mathbb{N}. \tag{6.35}$$

Then we can immediately deduce from (6.35) that the sequences

$$\left\{ y_{\varepsilon,k} \right\}_{k \in \mathbb{N}}, \quad \left\{ u_{\varepsilon,k} \right\}_{k \in \mathbb{N}}, \quad \left\{ z_{\varepsilon,k} \right\}_{k \in \mathbb{N}}, \quad \text{and} \quad \left\{ f\left(y_{\varepsilon,k} \right) \right\}_{k \in \mathbb{N}}$$

are uniformly bounded in $L^p(\Omega)$, $L^2(\Omega)$, $W_0^{1,p'}(\Omega)$, and $L^r(\Omega)$, respectively. In particular,

$$\sup_{k\in\mathbb{N}} \|u_{\varepsilon,k}\|^2_{L^2(\Omega)} \leq 2(\mu_\varepsilon + 1), \quad \sup_{k\in\mathbb{N}} \|z_{\varepsilon,k}\|^{p'}_{W_0^{1,p'}(\Omega)} \leq \varepsilon p'(\mu_\varepsilon + 1), \tag{6.36}$$

$$\sup_{k\in\mathbb{N}} \|y_{\varepsilon,k}\|^p_{L^p(\Omega)} \leq 2^{p-1}\left(p(\mu_\varepsilon + 1) + \|y_d\|^p_{L^p(\Omega)}\right), \tag{6.37}$$

$$\sup_{k\in\mathbb{N}} \|f\left(y_{\varepsilon,k}\right)\|^r_{L^r(\Omega)} \leq \frac{r(\mu_\varepsilon + 1)}{\varepsilon}. \tag{6.38}$$

Hence, without loss of generality, we can suppose that there exist elements $y_\varepsilon \in L^p(\Omega)$, $u_\varepsilon \in L^2(\Omega)$, $\xi_\varepsilon \in L^r(\Omega)$ and $z_\varepsilon \in W_0^{1,p'}(\Omega)$ such that

$$y_{\varepsilon,k} \rightharpoonup y_\varepsilon \quad \text{in } L^p(\Omega), \tag{6.39}$$

$$u_{\varepsilon,k} \rightharpoonup u_\varepsilon \quad \text{in } L^2(\Omega), \tag{6.40}$$

$$z_{\varepsilon,k} \rightharpoonup z_\varepsilon \quad \text{in } W_0^{1,p'}(\Omega), \tag{6.41}$$

$$f\left(y_{\varepsilon,k}\right) \rightharpoonup \xi_\varepsilon \quad \text{in } L^r(\Omega) \tag{6.42}$$

as $k \to \infty$.

Let us show that in fact $y_{\varepsilon,k} \rightharpoonup y_\varepsilon$ in $W_0^{1,p}(\Omega)$. Indeed, from (6.27), using $y_{\varepsilon,k} \in W_0^{1,p}(\Omega)$ as a test function, we find that

$$\int_\Omega \left[|\nabla y_{\varepsilon,k}|^p + \left(\nabla z_{\varepsilon,k}, \nabla y_{\varepsilon,k}\right) - u_{\varepsilon,k} y_{\varepsilon,k}\right] dx \overset{\text{by Remark 6.1}}{=} \left\langle f\left(y_{\varepsilon,k}\right), y_{\varepsilon,k}\right\rangle_{W^{-1,p'}(\Omega);\, W_0^{1,p}(\Omega)} \tag{6.43}$$

for all $k \in \mathbb{N}$. Then, utilizing the Friedrich's inequality (6.8), we obtain

$$\begin{aligned} \|\nabla y_{\varepsilon,k}\|^p_{L^p(\Omega;\mathbb{R}^N)} \leq &\left[\|z_{\varepsilon,k}\|_{W_0^{1,p'}(\Omega;\mathbb{R}^N)} + \|f\left(y_{\varepsilon,k}\right)\|_{W^{-1,p'}(\Omega)}\right] \|\nabla y_{\varepsilon,k}\|_{L^p(\Omega;\mathbb{R}^N)} \\ &+ \operatorname{diam}\Omega |\Omega|^{\frac{p-2}{2p}} \|u_{\varepsilon,k}\|_{L^2(\Omega)} \|\nabla y_{\varepsilon,k}\|_{L^p(\Omega;\mathbb{R}^N)}. \end{aligned} \tag{6.44}$$

From this and estimates (6.36)–(6.38), we deduce the existence of a constant $C^*_\varepsilon > 0$ such that $C^*_\varepsilon = C^*(\Omega, \mu_\varepsilon, \varepsilon)$ and

$$\sup_{k\in\mathbb{N}} \|\nabla y_{\varepsilon,k}\|_{L^p(\Omega;\mathbb{R}^N)} \leq C^*_\varepsilon < +\infty. \tag{6.45}$$

Thus, in view of reflexivity of $W_0^{1,p}(\Omega)$, we can suppose that (up to a subsequence)

$$y_{\varepsilon,k} \rightharpoonup y_\varepsilon \ \text{in } W_0^{1,p}(\Omega), \quad y_{\varepsilon,k} \to y_\varepsilon \ \text{in } L^p(\Omega), \quad y_{\varepsilon,k}(x) \to y_\varepsilon(x) \ \text{a.e. in } \Omega. \tag{6.46}$$

Utilizing the pointwise convergence $(6.46)_3$ and assumption (c), we see that $f(y_{\varepsilon,k}) \to f(y_\varepsilon)$ almost everywhere in Ω as $k \to \infty$. Let us show that this fact together with estimate (6.38) implies the strong convergence

$$f(y_{\varepsilon,k}) \to f(y_\varepsilon) \ \text{ in } W^{-1,p'}(\Omega) \ \text{ as } k \to \infty. \tag{6.47}$$

With that in mind, we recall that a sequence $\{g_k\}_{k\in\mathbb{N}}$ is called equi-integrable on Ω if for any $\delta > 0$, there is a $\tau = \tau(\delta)$ such that $\int_E |g_k|\,dx < \delta$ for every measurable subset $E \subset \Omega$ of Lebesgue measure $|E| < \tau$. In order to check whether the sequence $\left\{f(y_{\varepsilon,k})\right\}_{k\in\mathbb{N}}$ is equi-integrable on Ω, we note that

$$\begin{aligned}
&\left|\left\langle f\left(y_{\varepsilon,k}\right), y_{\varepsilon,k}\right\rangle_{W^{-1,p'}(\Omega);W_0^{1,p}(\Omega)}\right| \overset{\text{by (6.43)}}{\leq} \sup_{k\in\mathbb{N}} \|\nabla y_{\varepsilon,k}\|^p_{L^p(\Omega;\mathbb{R}^N)}\\
&\quad + \sup_{k\in\mathbb{N}}\left[\|z_{\varepsilon,k}\|_{W_0^{1,p'}(\Omega)} + \operatorname{diam}\Omega\,|\Omega|^{\frac{p-2}{2p}}\|u_{\varepsilon,k}\|_{L^2(\Omega)}\right]\|\nabla y_{\varepsilon,k}\|_{L^p(\Omega;\mathbb{R}^N)}\\
&\overset{\text{by (6.45)}}{\leq} \left[\left(\varepsilon p'(\mu_\varepsilon+1)\right)^{\frac{1}{p'}} + \operatorname{diam}|\Omega|^{\frac{p-2}{2p}}\sqrt{2(\mu_\varepsilon+1)}\right]C_\varepsilon^* + (C_\varepsilon^*)^p =: C_\varepsilon^{**}.
\end{aligned} \tag{6.48}$$

Let us single out $m > 0$ and $\tau > 0$ such that

$$m > \frac{2C_\varepsilon^{**}}{\delta}, \quad \tau = \frac{\delta}{2f(m)}.$$

Then for every measurable set $S \subset \Omega$ with $|S| < \tau$, we have

$$\begin{aligned}
\int_S f\left(y_{\varepsilon,k}\right)dx &\leq \int_{\{x\in S\,:\,|y_{\varepsilon,k}(x)|>m\}} f\left(y_{\varepsilon,k}\right)dx + \int_{\{x\in S\,:\,|y_{\varepsilon,k}(x)|\leq m\}} f\left(y_{\varepsilon,k}\right)dx\\
&\leq \frac{1}{m}\left|\int_{\{x\in S\,:\,|y_{\varepsilon,k}(x)|>m\}} y_k f\left(y_{\varepsilon,k}\right)dx\right| + \int_{\{x\in S\,:\,|y_{\varepsilon,k}(x)|\leq m\}} f(m)\,dx\\
&\leq \frac{1}{m}\left|\left\langle f\left(y_{\varepsilon,k}\right), y_{\varepsilon,k}\right\rangle_{W^{-1,p'}(\Omega);W_0^{1,p}(\Omega)}\right| + \int_{\{x\in S\,:\,|y_{\varepsilon,k}(x)|\leq m\}} f(m)\,dx\\
&\overset{\text{by (6.48)}}{\leq} \frac{C_\varepsilon^{**}}{m} + f(m)|S| \leq \frac{\delta}{2} + \frac{\delta}{2}.
\end{aligned}$$

Thus, $\left\{f(y_{\varepsilon,k})\right\}_{k\in\mathbb{N}}$ is an equi-integrable sequence on Ω for which $f(y_{\varepsilon,k}) \to f(y_\varepsilon)$ almost everywhere in Ω as $k \to \infty$. Hence, by Vitali's lemma, we deduce that $f(y_{\varepsilon,k}) \to f(y_\varepsilon)$ strongly in $L^1(\Omega)$. Since this sequence is bounded in $L^r(\Omega)$, it follows that

$$f(y_{\varepsilon,k}) \to f(y_\varepsilon) \quad \text{strongly in} L^r(\Omega) \text{as} k \to \infty. \tag{6.49}$$

To deduce the announced property (6.47), it remains to utilize the strong convergence (6.49) and continuity of the embedding $L^r(\Omega) \hookrightarrow W^{-1,p'}(\Omega)$ (see Remark 6.1).

Thus, in order to decide that $(u_\varepsilon, y_\varepsilon, z_\varepsilon)$ is a feasible solution to the regularized problems (6.19)–(6.22), we have to show that this tuple is related by the variational equality (6.27). With that in mind, we make use of the following integral identity

$$\int_\Omega \left[\left(|\nabla y_{\varepsilon,k}|^{p-1}\nabla y_{\varepsilon,k}, \nabla\varphi\right) + \left(\nabla z_{\varepsilon,k}, \nabla\varphi\right)\right] dx \\ = \int_\Omega u_{\varepsilon,k}\varphi\, dx + \left\langle f\left(y_{\varepsilon,k}\right), \varphi\right\rangle_{W^{-1,p'}(\Omega);W_0^{1,p}(\Omega)}, \quad (6.50)$$

which holds true for each $\varphi \in W_0^{1,p}(\Omega)$, $k \in \mathbb{N}$, and $\varepsilon > 0$. Taking into account properties (6.39)–(6.42), and (6.47), the limit passage in (6.50) as $k \to \infty$ becomes trivial. As a result, we arrive at the following relation

$$\int_\Omega \left[\left(|\nabla y_\varepsilon|^{p-1}\nabla y_\varepsilon, \nabla\varphi\right) + (\nabla z_\varepsilon, \nabla\varphi)\right] dx \int_\Omega u_\varepsilon\varphi\, dx + = \langle f(y_\varepsilon), \varphi\rangle_{W^{-1,p'}(\Omega);W_0^{1,p}(\Omega)}.$$

To conclude the proof, let us show that in fact the triplet $(u_\varepsilon, y_\varepsilon, z_\varepsilon)$ is optimal to the problem (6.19)–(6.22). Indeed, in view of the strong convergence (6.49) and lower semi-continuity of norms in reflexive Banach spaces $W_0^{1,p'}(\Omega)$ and $L^2(\Omega)$ with respect to the weak convergence, the limit passage in (6.34) immediately leads us to the relation

$$\mu_\varepsilon = \inf_{(u,y,z)\in\Lambda_\varepsilon} J_\varepsilon(u, y, z) = \lim_{k\to\infty} J_\varepsilon(u_{\varepsilon,k}, y_{\varepsilon,k}, z_{\varepsilon,k}) \geq \liminf_{k\to\infty} J_\varepsilon(u_{\varepsilon,k}, y_{\varepsilon,k}, z_{\varepsilon,k}) \\ \geq J_\varepsilon(u_\varepsilon, y_\varepsilon, z_\varepsilon) \geq \inf_{(u,y,z)\in\Lambda_\varepsilon} J_\varepsilon(u, y, z) = \mu_\varepsilon.$$

Thus, the equality (6.33) holds true with $(u_\varepsilon^0, y_\varepsilon^0, z_\varepsilon^0) = (u_\varepsilon, y_\varepsilon, z_\varepsilon)$ and, therefore, the tuple $(u_\varepsilon, y_\varepsilon, z_\varepsilon)$ is optimal for regularized problems (6.19)–(6.22).

As for uniqueness of this solution, in the case $f : \mathbb{R} \to [0, +\infty)$ is a convex function, it is a direct consequence of the uniform convexity of the space $L^r(\Omega)$, the inequality

$$\|f(\lambda y_1 + (1-\lambda)y_2)\|^r_{L^r(\Omega)} \leq \|\lambda f(y_1) + (1-\lambda)f(y_2)\|^r_{L^r(\Omega)} \\ < \lambda\|f(y_1)\|^r_{L^r(\Omega)} + (1-\lambda)\|f(y_2)\|^r_{L^r(\Omega)}, \quad \forall y_1, y_2 \in W_0^{1,p}(\Omega), \quad \forall \lambda \in (0,1),$$

and the strict convexity of the rest part of the cost functional (6.19).

6.4 Asymptotic Analysis of Regularized Optimal Control Problem

Our main aim in this section is to find out whether the original optimal control problem (6.1)–(6.4) is solvable under assumptions (a)–(c) and its optimal solutions can be attained (in some sense) by optimal solutions to the regularized problem (6.19)–(6.22).

The key point of our consideration is that, in contrast to the well-known approaches (see, for instance, [4, 7, 20, 21]), we do not assume here the fulfillment of the "standard" extra properties such that the domain Ω is an open subset of $\mathbb{R}^N$ with $N > 2$, this domain should be star-shaped, and the exists a weak solution $y \in W_0^{1,p}(\Omega)$ of Dirichlet problem (6.2)–(6.3) satisfying $f(y) \in L^2(\Omega)$ or inequalities (6.14)–(6.15). Because of this, the existence of at least one optimal pair to the problem (6.1)–(6.4) is an open question provided we restrict our consideration only by assumptions (a)–(c).

In what follows, in order to guarantee the consistency of the original problem (6.1)–(6.4), we notice that the set of feasible solutions Ξ to the problem (6.1)–(6.4) is nonempty. Indeed, let $\widetilde{y} \in C_0^\infty(\Omega)$ be an arbitrary function. Then $\widetilde{y} \in W_0^{1,p}(\Omega)$ and $f(\widetilde{y}) \in L^1(\Omega)$, and hence, $y \in H_f$. Defining the control $\widetilde{u} \in L^2(\Omega)$ as follows $\widetilde{u} = -\operatorname{div}\left(|\nabla \widetilde{y}|^{p-2}\nabla \widetilde{y}\right) - f(\widetilde{y})$, we see that $(\widetilde{u}, \widetilde{y})$ is a feasible pair to the problem (6.1)–(6.4).

The following result is crucial in this paper and it shows that solvability of the original OCP (6.1)–(6.4) can be established by an indirect approach based on the variant of Tikhonov regularization.

Theorem 6.4 *Assume that conditions* (6.23) *and (a)–(c) indicated in Sect. 6.3 hold true. Let* $\left\{(u_\varepsilon^0, y_\varepsilon^0, z_\varepsilon^0) \in \Lambda_\varepsilon\right\}_{\varepsilon\to 0}$ *be a sequence of optimal solutions to regularized problems* (6.19)–(6.22) *when the parameter* $\varepsilon > 0$ *varies in a strictly decreasing sequence of positive numbers converging to* 0. *Then there is a subsequence of* $\left\{(u_\varepsilon^0, y_\varepsilon^0, z_\varepsilon^0)\right\}_{\varepsilon\to 0}$, *still denoted by the suffix* ε, *such that*

$$
\begin{gathered}
u_\varepsilon^0 \rightharpoonup u^0 \ \ in L^2(\Omega), \quad y_\varepsilon^0 \rightharpoonup y^0 \ \ in W_0^{1,p}(\Omega), \quad z_\varepsilon^0 \to 0 \ \ in W_0^{1,p'}(\Omega), \\
-\operatorname{div}\left(|\nabla y^0|^{p-2}\nabla y^0\right) = f(y^0) + u^0 \ \ in \ \Omega, \\
y^0 = 0 \ \ on \ \partial\Omega, \\
(u^0, y^0) \in \Xi, \quad J(u^0, y^0) = \inf_{(u,y)\in\Xi} J(u, y).
\end{gathered}
$$

Proof Since $\Xi \neq \emptyset$, it follows that there exists a pair (u^*, y^*) such that (u^*, y^*) is a feasible solution to the original optimal control problem (6.1)–(6.4). Hence, in view of Definition 6.2 and Lemma 6.1, $u^* \in L^2(\Omega)$, $y^* \in W_0^{1,p}(\Omega)$, $f(y^*) \in L^1(\Omega) \cap W^{-1,p'}(\Omega)$, and the pair (u^*, y^*) is related by integral identity (6.5). Therefore, for each $\varepsilon > 0$, the triplet $(u^*, y^*, 0)$ is a feasible solution for regularized problem (6.19)–(6.22), i.e, $(u^*, y^*, 0) \in \Lambda_\varepsilon$ for all $\varepsilon > 0$. Taking this fact into account, we see that

$$
\begin{aligned}
J_\varepsilon(u_\varepsilon^0, y_\varepsilon^0, z_\varepsilon^0) &= \frac{1}{p}\int_\Omega |y_\varepsilon^0 - y_d|^p\,dx + \frac{1}{2}\int_\Omega |u_\varepsilon^0|^2\,dx \\
&+ \frac{1}{\varepsilon p'}\int_\Omega |\nabla z_\varepsilon^0|^{p'}\,dx + \frac{\varepsilon}{r}\|f(y_\varepsilon^0)\|_{L^r(\Omega)}^r \\
&= \inf_{(u,y,z)\in\Lambda_\varepsilon} J_\varepsilon(u,y,z) \le J_\varepsilon(u^*, y^*, 0) \\
&= \frac{1}{p}\int_\Omega |y^* - y_d|^p\,dx + \frac{1}{2}\int_\Omega |u^*|^2\,dx + \frac{\varepsilon}{r}\|f(y^*)\|_{L^r(\Omega)} \\
&- C_1 + \varepsilon C_2 < +\infty.
\end{aligned}\tag{6.51}
$$

Since this relation holds true for each $\varepsilon > 0$ varying in a given interval $(0, \varepsilon^0]$, it follows that

$$
\sup_{\varepsilon\in(0,\varepsilon^0]} \|u_\varepsilon^0\|_{L^2(\Omega)}^2 \le 2C_1, \quad \sup_{\varepsilon\in(0,\varepsilon^0]} \left[\frac{1}{\varepsilon p'}\|z_\varepsilon^0\|_{W_0^{1,p'}(\Omega)}^{p'}\right] \le C_1 + \varepsilon^0 C_2, \tag{6.52}
$$

$$
\sup_{\varepsilon\in(0,\varepsilon^0]} \|y_\varepsilon^0\|_{L^p(\Omega)}^p \le 2^{p-1}\left(p\,C_1 + \|y_d\|_{L^p(\Omega)}^p\right), \tag{6.53}
$$

$$
\sup_{\varepsilon\in(0,\varepsilon^0]} \|f\left(y_\varepsilon^0\right)\|_{L^r(\Omega)}^r \le rC_2. \tag{6.54}
$$

Hence, the sequences $\left\{u_\varepsilon^0\right\}_{\varepsilon\in(0,\varepsilon^0]}$, $\left\{y_\varepsilon^0\right\}_{\varepsilon\in(0,\varepsilon^0]}$, and $\left\{f(y_\varepsilon^0)\right\}_{\varepsilon\in(0,\varepsilon^0]}$ are weakly compact in $L^2(\Omega)$, $L^p(\Omega)$, and $L^r(\Omega)$ respectively, whereas estimate (6.52) implies that the sequence $\left\{z_\varepsilon^0\right\}_{\varepsilon\to 0}$ is strongly convergent to 0 in $W_0^{1,p'}(\Omega)$. So, we can suppose that there exist elements $u^0 \in L^2(\Omega)$, $y^0 \in L^p(\Omega)$, $\xi \in L^r(\Omega)$, and a sequence $\{\varepsilon_k\}_{k\in\mathbb{R}}$ monotonically converging to zero as $k \to \infty$ such that

$$
u_{\varepsilon_k}^0 \to u^0 \text{ weakly in } L^2(\Omega), \tag{6.55}
$$

$$
y_{\varepsilon_k}^0 \to y^0 \text{ weakly in } L^p(\Omega), \tag{6.56}
$$

$$
f(y_{\varepsilon_k}^0) \to \xi \text{ weakly in } L^r(\Omega) \text{ and } W^{-1,p'}(\Omega), \tag{6.57}
$$

$$
z_{\varepsilon_k}^0 \to 0 \text{ strongly in } W_0^{1,p'}(\Omega) \text{ as } k \to \infty. \tag{6.58}
$$

To begin with, let us show that, in fact, we have the weak convergence $y_{\varepsilon_k}^0 \rightharpoonup y^0$ in $W_0^{1,p}(\Omega)$. Indeed, arguing as in the proof of Theorem 6.3, we make use of the integral identity

$$
\int_\Omega \left[|\nabla y_{\varepsilon_k}^0|^p + \left(\nabla z_{\varepsilon_k}^0, \nabla y_{\varepsilon_k}^0\right) - u_{\varepsilon_k}^0 y_{\varepsilon_k}^0\right] dx = \left\langle f\left(y_{\varepsilon_k}^0\right), y_{\varepsilon_k}^0\right\rangle_{W^{-1,p'}(\Omega);W_0^{1,p}(\Omega)}, \tag{6.59}
$$

which holds true for each $\varepsilon \in (0, \varepsilon^0]$ and reflects the fact that the triplets

$$
\left\{(u_{\varepsilon_k}^0, y_{\varepsilon_k}^0, z_{\varepsilon_k}^0)\right\}_{k\in\mathbb{R}}
$$

are feasible to the problem (6.19)–(6.22) for each $k \in \mathbb{R}$. Then, by analogy with estimate (6.44), we deduce from (6.59) that

$$\|\nabla y^0_{\varepsilon_k}\|^p_{L^p(\Omega;\mathbb{R}^N)} \le \left[\|z^0_{\varepsilon_k}\|_{W_0^{1,p'}(\Omega)} + \|f\left(y^0_{\varepsilon_k}\right)\|_{W^{-1,p'}(\Omega)}\right] \|\nabla y^0_{\varepsilon_k}\|_{L^p(\Omega;\mathbb{R}^N)} + \operatorname{diam}\Omega\, |\Omega|^{\frac{p-2}{2p}} \|u^0_{\varepsilon_k}\|_{L^2(\Omega)} \|\nabla y^0_{\varepsilon_k}\|_{L^p(\Omega;\mathbb{R}^N)}.$$

Hence, estimates (6.52)–(6.54) imply that

$$\sup_{k\in\mathbb{N}} \|\nabla y^0_{\varepsilon_k}\|^{p-1}_{L^p(\Omega;\mathbb{R}^N)} \le \left[\left(\varepsilon^0 p'(C_1 + \varepsilon^0 C_2)\right)^{\frac{1}{p'}} + (rC_2)^{\frac{1}{r}} + \operatorname{diam}\Omega\, |\Omega|^{\frac{p-2}{2p}} \sqrt{2C_1}\right].$$

Thus, without loss of generality, we can suppose that (up to a subsequence)

$$y^0_{\varepsilon_k} \rightharpoonup y^0 \text{ in } W_0^{1,p}(\Omega), \quad y^0_{\varepsilon_k} \to y^0 \text{ in } L^p(\Omega), \quad y^0_{\varepsilon_k}(x) \to y^0(x) \text{ a.e. in } \Omega. \tag{6.60}$$

Utilizing the pointwise convergence $(6.60)_3$ and (c)-property, we see that $f(y^0_{\varepsilon_k}) \to f(y^0)$ almost everywhere in Ω as $k \to \infty$. Then, following the arguments of the proof Theorem 6.3, it can be shown that the sequence $\left\{f(y^0_{\varepsilon_k})\right\}_{k\in\mathbb{N}}$ is equi-integrable. As a result, we deduce from this the strong convergence

$$f(y^0_{\varepsilon_k}) \to f(y^0) \text{ in } L^r(\Omega) \text{ and} W^{-1,p'}(\Omega) \text{as } k \to \infty \tag{6.61}$$

and the fact that $\xi = f(y^0)$.

We are now in a position to show that (u^0, y^0) is a feasible solution to the original OCP (6.1)–(6.4). Indeed, in view of the initial assumptions (for the details we refer to the proof of Theorem 6.3), it remains to show that the pair (u^0, y^0) is related by the integral identity (6.5). To this end, we note that $(u^0_{\varepsilon_k}, y^0_{\varepsilon_k}, z^0_{\varepsilon_k}) \in \Lambda_{\varepsilon_k}$ for all $k \in \mathbb{R}$. Hence, the equality

$$\int_\Omega \left[\left(|\nabla y^0_{\varepsilon_k}|^{p-1} \nabla y^0_{\varepsilon_k}, \nabla\varphi\right) + \left(\nabla z^0_{\varepsilon_k}, \nabla\varphi\right) - u^0_{\varepsilon_k}\varphi\right] dx = \left\langle f\left(y^0_{\varepsilon_k}\right), \varphi\right\rangle_{W^{-1,p'}(\Omega); W_0^{1,p}(\Omega)} \tag{6.62}$$

holds true for each test function $\varphi \in C_0^\infty(\Omega)$. As a result, the limit passage in (6.62) as $k \to \infty$ becomes trivial, and it immediately leads us to the expected integral identity (6.5). Thus, combining all obtained properties of the pair (u^0, y^0), we finally deduce that $(u^0, y^0) \in \Xi$.

To conclude the proof, we have to show that $(u^0, y^0) \in \Xi$ in an optimal pair to the problem (6.1)–(6.4). To do so, we assume the converse, namely, there is a pair $(\widehat{u}, \widehat{y}) \in \Xi$ such that $J(\widehat{u}, \widehat{y}) < J(u^0, y^0)$. Then the triplet $(\widehat{u}, \widehat{y}, 0)$ is feasible to the regularized problem (i.e., $(\widehat{u}, \widehat{y}, 0) \in \Lambda_{\varepsilon_k}$ for each $k \in \mathbb{N}$). Hence,

$$J(\widehat{u},\widehat{y}) + \frac{\varepsilon_k}{r}\|f(\widehat{y})\|_{L^r(\Omega)} = J_{\varepsilon_k}(\widehat{u},\widehat{y},0) \\ \geq \inf_{(u,y,z)\in\Lambda_{\varepsilon_k}} J_{\varepsilon_k}(u,y,z) = J_{\varepsilon_k}\left(u^0_{\varepsilon_k}, y^0_{\varepsilon_k}, z^0_{\varepsilon_k}\right), \quad \forall k \in \mathbb{N}. \tag{6.63}$$

Therefore, passing in (6.63) to the limit as $k \to \infty$ and using the properties

$$\liminf_{k\to\infty} \int_\Omega |y^0_{\varepsilon_k} - y_d|^p\,dx \overset{\text{by (6.60)}}{=} \int_\Omega |y^0 - y_d|^p\,dx,$$

$$\liminf_{k\to\infty} \int_\Omega |u^0_{\varepsilon_k}|^2\,dx \overset{\text{by (6.55)}}{\geq} \int_\Omega |u^0|^2\,dx,$$

$$\lim_{k\to\infty}\left[\frac{1}{\varepsilon_k}\|z^0_{\varepsilon_k}\|^{p'}_{W_0^{1,p'}(\Omega)}\right] \overset{\text{by (6.52)}}{=} \text{const} > 0,$$

$$\lim_{k\to\infty}\left[\frac{\varepsilon_k}{2}\|f(y^0_{\varepsilon_k})\|_{L^r(\Omega)}\right] \overset{\text{by (6.61)}}{=} 0,$$

we obtain

$$J(\widehat{u},\widehat{y}) \geq \liminf_{k\to\infty} J_{\varepsilon_k}\left(u^0_{\varepsilon_k}, y^0_{\varepsilon_k}, z^0_{\varepsilon_k}\right) \\ \geq J(u^0,y^0) + \lim_{k\to\infty}\left[\frac{1}{\varepsilon_k}\|z^0_{\varepsilon_k}\|^{p'}_{W_0^{1,p'}(\Omega)}\right] \geq J(u^0,y^0).$$

As a result, this leads us to a contradiction. Thus, $(u^0, y^0) \in \Xi$ in an optimal pair to the problem (6.1)–(6.4).

To the end of proof, we note: if the original OCP admits a unique solution, then the given asymptotic analysis remains valid for each subsequence $\left\{(u^0_{\varepsilon_k}, y^0_{\varepsilon_k}, z^0_{\varepsilon_k})\right\}_{k\in\mathbb{R}}$ of the sequence of optimal solutions $\left\{(u^0_\varepsilon, y^0_\varepsilon, z^0_\varepsilon) \in \Lambda_\varepsilon\right\}_{\varepsilon\to 0}$. Therefore, the limits in (6.55)–(6.58) do not depend on the choice of a subsequence and, hence, $(u^0, y^0) \in \Xi$ is a unique limit pair for the entire sequence of optimal pairs $\left\{(u^0_\varepsilon, y^0_\varepsilon)\right\}_{\varepsilon>o}$.

References

1. Agmon, S., Douglis, A., Nirenberg, L.: Estimates near the boundary for solutions of elliptic partial differential equations satisfying general boundary conditions I. Com. Pure Appl. Math. **12**, 623–727 (1959)
2. Al'shin, A.B., Korpusov, M.O., Sveshnikov, A.G.: Blow-up in Nonlinear Sobolev Type Equations. De Gruyter series in Nonliniear Analysis and Applications, vol. 15. Walter de Gruyter GmbH, Berlin (2011)
3. Calderon, A.P., Zygmund, A.: On the existence of certain singular integrals. Acta. Math. **88**, 85–139 (1952)
4. Casas, E., Kavian, O., Puel, J.P.: Optimal control of an ill-posed elliptic semilinear equation with an exponential nonlinearity. ESAIM: Control Opt. Calculus Var. **3**, 361–380 (1998)

5. Casas, E., Kogut, P.I., Leugering, G.: Approximation of optimal control problems in the coefficient for the p-Laplace equation. I. Convergence result. SIAM J. Control Opt. **54**(3), 1406–1422 (2016)
6. Chandrasekhar, S.: An Introduction to the Study of Stellar Structures. Dover Publishing Inc., New York (1985)
7. D'Apice, C., De Maio, U., Kogut, P.I.: Optimal boundary control problem for ill-posed elliptic equation in domains with rugous boundary. I. Existence result and optimality conditions. to appear
8. D'Apice, C., Kogut, P.I., Manzo, R.: On approximation of entropy solutions for one system of nonlinear hyperbolic conservation laws with impulse source terms. J. Control Sci. Eng. **2019** (2010), Article number 982369
9. D'Apice, C., Kogut, P.I., Manzo, R.: On relaxation of state constrained optimal control problem for a PDE-ODE model of supply chains. Netw. Heterog. Media **9**(3), 501–518 (2014)
10. Daz, J.I., Lions, J.-L.: On the approximate controllability for some explosive parabolic problems. International Series of Numerical Mathematics. vol. 133, pp. 115–132. Birkhuser Verlag, Basel (1999)
11. Franck-Kamenetskii, D.A.: Diffusion and Heat Transfer in Chemical Kinetics, 2nd edn. Plenum Press, New York (1969)
12. Gelfand, I.M.: Some problems in the theory of quasi-linear equations. Amer. Math. Soc. Transl., Ser. 2, **29**, 289–292 (1963)
13. Glowinski, R., Lions, J.-L., He, J.: Exact and Approximate Controllability for Distributed Parameter Systems. A Numerical Approach. Cambridge University Press, Cambridge (2008)
14. Joseph, D.D., Lundgren, T.S.: Quasilinear Dirichlet problems driven by positive sources. Arch. Rat. Mech. Anal. **49**, 241–269 (1973)
15. Kinderlehrer, D., Stampacchia, G.: An Introduction to Variational Inequalities and Their Applications. Academic, New York (1980)
16. Kogut, P.I., Leugering, G.: Optimal Control Problems for Partial Differential Equations on Reticulated Domains. Systems and Control. Approximation and Asymptotic Analysis, Series. Birkhäuser, Boston (2011)
17. Kogut, P.I.: On approximation of an optimal boundary control problem for linear elliptic equation with unbounded coefficients. Discrete Contin. Dyn. Sys. Ser. A **34**(5), 2105–2133 (2014)
18. Kogut, P.I.: Variational S-convergence of minimization problems. Part I. Definitions and basic properties. Problemy Upravleniya I Informatiki (Avtomatika), Series A, **5**, 29–42 (1996)
19. Kogut, P.I., Kupenko, O.P.: Approximation Methods in Optimization of Nonlinear Systems, De Gruyter Series in Nonlinear Analysis and Applications 32. Walter de Gruyter GmbH, Berlin (2019)
20. Kogut, P.I., Kupenko, O.P.: On optimal control problem for an ill-posed strongly nonlinear elliptic equation with p-Laplace operator and L^1-type of nonlinearity. Discrete Contin. Dyn. Syst. Ser. B **24**(3), 1273–1295 (2019)
21. Kogut, P.I., Kupenko, O.P.: On Approximation of an optimal control problem for ill-posed strongly nonlinear elliptic equation with p-Laplace operator. In: Modern Mathematics and Mechanics. Fundamentals, Problems and Challenges (Chapter 23), pp. 445–480. Sadovnichiy, V.A., Zgurovsky, M. (Eds.). Springer, Berlin (2019)
22. Kogut, P.I., Manzo, R.: Tikhonov regularization of optimal control problem for an ill-posed strongly nonlinear elliptic equation with an exponential type of non-linearity. In press
23. Kogut, P.I., Manzo, R.: On vector-valued approximation of state constrained optimal control problems for nonlinear hyperbolic conservation laws. J. Dyn. Control Syst. **19**(3), 381–404 (2013)
24. Kogut, P.I., Manzo, R., Putchenko, A.O.: On approximate solutions to the Neumann elliptic boundary value problem with non-linearity of exponential type. Bound. Value Probl. **2016**(1), 1–32 (2016)
25. Lasiecka, I., Triggiani, R.: Control Theory for Partial Differential Equations: Continuous and Approximation Theories. I. Abstract Parabolic Systems. Encyclopedia of Mathematics and its Applications, Vol. 74, Cambridge University Press, Cambridge (2000)

26. Lasiecka, I., Triggiani, R.: Control Theory for Partial Differential Equations: Continuous and Approximation Theories. II. Abstract Hyperbolic-like Systems over a Finite Time Horizon. Encyclopedia of Mathematics and its Applications, vol. 75, Cambridge University Press, Cambridge (2000)
27. Lions, J.-L.: Optimal control of non well posed distributed systems and related non linear partial differential equation. North-Holland Math. Stud. **61**(C), 3–16 (1982)
28. Lions, J.-L.: Control of Distributed Singular Systems. Gauthier-Villars, Paris (1985)
29. Lions, J.-L.: Optimal Control of Systems Governed by Partial Differential Equations. Springer, Berlin (1971)
30. Lions, J.-L.: Some Aspects of the Optimal Control of Distributed Parameter Systems. Regional Conference Series in Applied Mathematics. SIAM, Philadelphia (1972)
31. Manzo, R.: On Neumann boundary control problem for ill-posed strongly nonlinear elliptic equation with p-Laplace operator and L^1-type of nonlinearity. Ricerche di Matematica **68**(2), 769–802 (2019)
32. Meyers, N.G.: An L^p-estimate for the gradient of solutions of second order elliptic divergence equations. Annali della Scuola Normale Superiore di Pisa **17**(3), 189–206 (1963)
33. Pedregal, P.: On variant of Tikhonov regularization in optimal control under PDEs, pp. 1–23. arXiv.org. math. arXiv:1803.09096 (2019)
34. Pörnel, F., Wachsmuth, D.: Tikhonov regularization of optimal control problems governed by semi-linear partial differential equations. Math. Control Related Fields **8**(1), 315–335 (2018)
35. Tröltzsch, F.: Optimal Control of Partial Differential Equations. Theory, Methods and Applications, Translated from the 2005 German original by Jürgen Sprekels, Graduate Studies in Mathematics, vol. 112. American Mathematical Society, Providence, RI (2010)

Chapter 7
Symmetries and Conservation Laws of the Equations of Two-Dimensional Shallow Water Over Uneven Bottom

A. V. Aksenov and K. P. Druzhkov

Abstract A system of equations of two-dimensional shallow water over uneven bottom is considered. Overdetermined systems of equations for determining the symmetries and the conservation laws are obtained. The compatibilities of this overdetermined systems of equations are investigated. The general forms of the solutions of the overdetermined systems are found. The kernels of the symmetry operators and conservation laws are found. Cases of kernels extensions of symmetry operators and conservation laws are presented. The corresponding classifying equations are given. The results of the group classification have indicated that the system of equations of two-dimensional shallow water over uneven bottom cannot be linearized by point transformation in contrast to the system of equations of one-dimensional shallow water in the cases of horizontal and inclined bottom profiles.

7.1 Introduction and Main Result

The system of equations of one-dimensional shallow water above uneven bottom was considered in [1]. The group classification problem was solved, and all hydrodynamic conservation laws were found there. It was also shown that the system of equations of one-dimensional shallow water is linearizable by a point transformation of variables only in cases of horizontal and inclined bottom profiles.

In dimensionless variables, the system of two-dimensional shallow-water equations over uneven bottom has the following form [2]

A. V. Aksenov (✉) · K. P. Druzhkov
Lomonosov Moscow State University, GSP-1 Leninskie Gory,
119991 Moscow, Russian Federation
e-mail: aksenov@mech.math.msu.su

K. P. Druzhkov
e-mail: Konstantin.Druzhkov@gmail.com

V. A. Sadovnichiy and M. Z. Zgurovsky (eds.), *Contemporary Approaches and Methods in Fundamental Mathematics and Mechanics*, Understanding Complex Systems,
https://doi.org/10.1007/978-3-030-50302-4_7

$$
\begin{aligned}
& u_t + uu_x + vu_y + \eta_x = 0\,, \\
& v_t + uv_x + vv_y + \eta_y = 0\,, \\
& \eta_t + \big((\eta + h)u\big)_x + \big((\eta + h)v\big)_y = 0\,.
\end{aligned} \tag{7.1}
$$

Here $u = u(x, y, t)$, $v = v(x, y, t)$ are components of a depth-averaged horizontal velocity; $\eta = \eta(x, y, t)$ is a deviation of a free surface; $h = h(x, y)$; $z = -h$ is a bottom profile; $\eta + h \geqslant 0$.

In this paper, the problem of group classification is solved for the system of Eq. (7.1) and all hydrodynamic conservation laws are found for any profile of the bottom. This work is a further development of the works [1, 3].

7.2 Symmetries of the System of Equations of Two-Dimensional Shallow Water Over Uneven Bottom

7.2.1 System of Determining Equations

We seek symmetry operators of the system (7.1) as

$$
\begin{aligned}
X = {} & \xi^1(x, y, t, u, v, \eta)\partial_x + \xi^2(x, y, t, u, v, \eta)\partial_y + \xi^3(x, y, t, u, v, \eta)\partial_t + \\
& + \eta^1(x, y, t, u, v, \eta)\partial_u + \eta^2(x, y, t, u, v, \eta)\partial_v + \eta^3(x, y, t, u, v, \eta)\partial_\eta\,.
\end{aligned}
$$

Applying the criterion of invariance [4] we obtain an overdetermined linear homogeneous system of determining equations

$$
\begin{aligned}
& \xi^1_u = 0,\quad \xi^1_v = 0,\quad \xi^1_\eta = 0,\quad \xi^2_u = 0,\quad \xi^2_v = 0,\quad \xi^2_\eta = 0, \\
& \xi^3_u = 0,\quad \xi^3_v = 0,\quad \xi^3_\eta = 0,\quad \eta^1_\eta = \xi^3_x,\quad \eta^2_\eta = \xi^3_y,\quad \eta^3_u = 0,\qquad \eta^3_v = 0, \\
& -u\xi^1_x - v\xi^1_y - \xi^1_t + (u^2 + \eta + h)\xi^3_x + uv\xi^3_y + u\xi^3_t + \eta^1 - (\eta + h)\eta^1_\eta = 0\,, \\
& -u\xi^2_x - v\xi^2_y - \xi^2_t + uv\xi^3_x + v^2\xi^3_y + v\xi^3_t + \eta^2 = 0\,,\qquad -\xi^2_x + v\xi^3_x - \eta^1_v = 0\,, \\
& -\xi^1_x + 2u\xi^3_x + v\xi^3_y + \xi^3_t - \eta^1_u + \eta^3_\eta = 0\,,\qquad u\eta^1_x + v\eta^1_y + \eta^1_t + \eta^3_x = 0\,, \\
& (uh_x + vh_y)\xi^3_y + u\eta^2_x + v\eta^2_y + \eta^2_t - (uh_x + vh_y)\eta^2_\eta + \eta^3_y = 0\,,\qquad \eta^1_v + \eta^2_u = 0\,, \\
& \xi^1_y - u\xi^3_y + \eta^2_u = 0\,,\qquad -\xi^2_y + u\xi^3_x + 2v\xi^3_y + \xi^3_t - \eta^2_v + \eta^3_\eta = 0\,, \\
& (\eta + h)(-\xi^1_x + 2u\xi^3_x + v\xi^3_y + \xi^3_t + \eta^1_u - \eta^3_\eta) + h_x\xi^1 + h_y\xi^2 + \eta^3 = 0\,, \\
& (\eta + h)(-\xi^2_y + u\xi^3_x + 2v\xi^3_y + \xi^3_t + \eta^2_v - \eta^3_\eta) + h_x\xi^1 + h_y\xi^2 + \eta^3 = 0\,, \\
& -u\xi^1_x - v\xi^1_y - \xi^1_t + (u^2 + \eta + h)\xi^3_x + uv\xi^3_y + u\xi^3_t + (\eta + h)\eta^1_\eta + \eta^1 - \eta^3_u = 0\,, \\
& -u\xi^2_x - v\xi^2_y - \xi^2_t + uv\xi^3_x + (v^2 + \eta + h)\xi^3_y + v\xi^3_t + (\eta + h)\eta^2_\eta - \eta^3_v + \eta^2 = 0\,, \\
& u(uh_x + vh_y)\xi^3_x + v(uh_x + vh_y)\xi^3_y + (uh_x + vh_y)\xi^3_t + (\eta + h)(\eta^1_x + \eta^2_y) + u\eta^3_x + \\
& + v\eta^3_y + \eta^3_t - (uh_x + vh_y)\eta^3_\eta + (uh_{xx} + vh_{xy})\xi^1 + (uh_{xy} + vh_{yy})\xi^2 + h_x\eta^1 + h_y\eta^2 = 0.
\end{aligned} \tag{7.2}
$$

7.2.2 Analysis of the Classifying Equation

From the system of Eq. (7.2) we find

$$\xi^1 = -\dot{a}x - 2By + 2k\,, \qquad \eta^1 = (\dot{a} + 2C)u - 2Bv - \ddot{a}x + 2\dot{k}\,,$$
$$\xi^2 = -\dot{a}y + 2Bx + 2l\,, \qquad \eta^2 = (\dot{a} + 2C)v + 2Bu - \ddot{a}y + 2\dot{l}\,,$$
$$\xi^3 = -2a - 2Ct\,, \qquad \eta^3 = (2\dot{a} + 4C)\eta + \frac{x^2 + y^2}{2}\dddot{a} - 2\ddot{k}x - 2\ddot{l}y + \dot{f}\,.$$

Functions $a = a(t), k = k(t), l = l(t), f = f(t)$ and constants B, C are determined from the classifying equation

$$\big(-\dot{a}x - 2By + 2k\big)h_x + \big(-\dot{a}y + 2Bx + 2l\big)h_y - (2\dot{a} + 4C)h = \\ = -\frac{x^2 + y^2}{2}\dddot{a} + 2x\ddot{k} + 2y\ddot{l} - \dot{f}\,. \tag{7.3}$$

The kernel of the symmetry operators of the system of Eq. (7.1) is found from condition that the classifying Eq. (7.3) is satisfied for an arbitrary function $h(x, y)$. Whence it follows that the kernel of the symmetry operators is determined by the symmetry operator

$$X_1 = \partial_t\,. \tag{7.4}$$

Next we consider the cases in which the kernel of the symmetry operators can be extended.

The differentiation of the classifying Eq. (7.3) with respect to variable t gives the following equation

$$\big(-\ddot{a}x + 2\dot{k}\big)h_x + \big(-\ddot{a}y + 2\dot{l}\big)h_y - 2\ddot{a}h = -\frac{x^2 + y^2}{2}\ddddot{a} + 2x\dddot{k} + 2y\dddot{l} - \ddot{f}\,. \tag{7.5}$$

Assuming that all coefficients of the Eq. (7.5), depending on the variable t are constants, we write it in the form

$$\big(-a_2x + 2k_1\big)h_x + \big(-a_2y + 2l_1\big)h_y - 2a_2h = -\frac{x^2 + y^2}{2}a_4 + 2xk_3 + 2yl_3 - f_2\,. \tag{7.6}$$

In the case of $a_2 \neq 0$ from the general solution of the Eq. (7.6) we find the form of the profile h, in which the kernel of the symmetry operators can be extended under the condition $\ddot{a} \neq 0$

$$h = \frac{1}{(x + b_1)^2}\,H\Big(\frac{x + b_1}{y + b_2}\Big) + b_3(x^2 + y^2) + b_4x + b_5y + b_6\,, \tag{7.7}$$

where b_i are some constant values.

In the case $a_2 = 0,\ k_1^2 + l_1^2 \neq 0$ from the general solution of the Eq. (7.6) we find the types of the profile h for which extension of the kernel of the symmetry operators is possible under condition $\ddot{a} = 0, \dot{k}^2 + \dot{l}^2 \neq 0$

$$h = H(y + b_0 x) + b_1 xy + b_2 x^2 + b_3 x\,, \tag{7.8}$$

$$h = H(x) + b_1 y^2 + b_2 xy + b_3 y\,. \tag{7.9}$$

In the case of $\ddot{a} = 0, \dot{k} = 0, \dot{l} = 0$ from the analysis of the general solution of the classifying Eq. (7.3) implies that complementary profiles to the profiles (7.7)–(7.9) are the following ones

$$h = H\Big(\ln((x + b_1)^2 + (y + b_2)^2) + b_3 \,\mathrm{arctg}\Big(\frac{y + b_2}{x + b_1}\Big)\Big) \exp\Big(b_4 \,\mathrm{arctg}\Big(\frac{y + b_2}{x + b_1}\Big)\Big) + b_5,$$

$$h = H\Big(\ln((x + b_1)^2 + (y + b_2)^2) + b_3 \,\mathrm{arctg}\Big(\frac{y + b_2}{x + b_1}\Big)\Big) + b_4 \,\mathrm{arctg}\Big(\frac{y + b_2}{x + b_1}\Big),$$

$$h = H(x) \exp(b_1 y) + b_2, \qquad h = H(y + b_1 x) \exp(b_2 x) + b_3,$$

$$h = H\Big(\frac{x + b_1}{y + b_2}\Big)(x + b_1)^{b_3} + b_4, \qquad h = H\Big(\frac{x + b_1}{y + b_2}\Big) + b_3 \ln(x + b_1)\,.$$

Remark 7.1 The cases (7.8) and (7.9) are related by the equivalence transformation equivalence of the classifying Eq. (7.3). The transformation

$$\widetilde{x} = y + b_0 x\,, \quad \widetilde{y} = -x + b_0 y\,, \quad \widetilde{h} = (b_0^2 + 1)h$$

converts the profiles h of the case (7.9) to profiles h of the case (7.8). In this case, the desired functions are transformed in the following way

$$\widetilde{B} = B\,, \quad \widetilde{C} = C, \quad \dot{\widetilde{a}} = \dot{a}, \quad \dot{\widetilde{k}} = \dot{l} + b_0 \dot{k}, \quad \dot{\widetilde{l}} = -\dot{k} + b_0 \dot{l}, \quad \dot{\widetilde{f}} = (b_0^2 + 1)\dot{f}.$$

7.2.3 *Cases of Extension of the Kernel of Symmetry Operators*

Next we obtain a solution of the classifying Eq. (7.3) for each function h, considered in the previous subsection. For such functions h corresponding solutions allow to find all symmetry operators, additional to the symmetry operator (7.4). The following cases arise.

1. $\mathbf{h = H(x) + b_1 y^2 + b_2 xy + b_3 y}$.

1.1. $\mathbf{h = a_1 x + a_2 y + a_3}$.

In this case we obtain the following solution

$$a = C_1t^2 + C_2t + C_3\,, \quad k = -\frac{a_1C_1}{2}t^3 + \Big(\frac{a_2B}{2} - a_1C - \frac{3a_1C_2}{4}\Big)t^2 + C_4t + C_5\,,$$
$$l = -\frac{a_2C_1}{2}t^3 + \Big(-\frac{a_1B}{2} - a_2C - \frac{3a_2C_2}{4}\Big)t^2 + C_6t + C_7\,,$$
$$f = \frac{C_1(a_1^2+a_2^2)}{4}t^4 + \frac{(4C+3C_2)(a_1^2+a_2^2)}{6}t^3 + C_8 + (2a_3C_1 - a_1C_4 - a_2C_6)t^2 + \\ + (4a_3C + 2a_3C_2 - 2a_1C_5 - 2a_2C_7)t\,.$$

Thus, the kernel of symmetry operators is extended by the operators

$$X_2 = -\Big(x + \frac{3a_1}{2}t^2\Big)\partial_x - \Big(y + \frac{3a_2}{2}t^2\Big)\partial_y - 2t\partial_t + (u - 3a_1t)\partial_u + \\ + (v - 3a_2t)\partial_v + \Big(2\eta + 3a_1x + 3a_2y + \frac{3(a_1^2+a_2^2)}{2}t^2 + 2a_3\Big)\partial_\eta\,,$$
$$X_3 = -(2tx + a_1t^3)\partial_x - (2ty + a_2t^3)\partial_y - 2t^2\partial_t + (2tu - 2x - 3a_1t^2)\partial_u + \\ + (2tv - 2y - 3a_2t^2)\partial_v + (4t\eta + 6a_1tx + 6a_2ty + (a_1^2+a_2^2)t^3 + 4a_3t)\partial_\eta\,,$$
$$X_4 = t\partial_x + \partial_u - a_1t\partial_\eta\,, \qquad X_5 = t\partial_y + \partial_v - a_2t\partial_\eta\,,$$
$$X_6 = \partial_x - a_1\partial_\eta\,, \qquad X_7 = \partial_y - a_2\partial_\eta\,,$$
$$X_8 = (a_2t^2 - 2y)\partial_x + (2x - a_1t^2)\partial_y + (2a_2t - 2v)\partial_u + (2u - 2a_1t)\partial_v + (2a_1y - 2a_2x)\partial_\eta,$$
$$X_9 = -a_1t^2\partial_x - a_2t^2\partial_y - t\partial_t + (u - 2a_1t)\partial_u + (v - 2a_2t)\partial_v + \\ + (2\eta + 2a_1x + 2a_2y + (a_1^2+a_2^2)t^2 + 2a_3)\partial_\eta\,.$$

1.2. $\mathbf{h = a_1x^2 + a_2y^2 + a_3xy + a_4x + a_5y + a_6,\ a_1^2 + a_2^2 + a_3^2 \neq 0.}$
1.2.1. $\mathbf{h = a_1(x^2 + y^2) + a_4x + a_5y + a_6,\ a_1 > 0.}$
In this case we obtain the following solution

$$a = C_1e^{t\sqrt{8a_1}} + C_2e^{-t\sqrt{8a_1}} - Ct + C_3\,,$$
$$k = \frac{a_4C - a_4\sqrt{8a_1}(C_1e^{t\sqrt{8a_1}} - C_2e^{-t\sqrt{8a_1}}) - 2a_5B}{4a_1} + C_4e^{t\sqrt{2a_1}} + C_5e^{-t\sqrt{2a_1}}\,,$$
$$l = \frac{a_5C - a_5\sqrt{8a_1}(C_1e^{t\sqrt{8a_1}} - C_2e^{-t\sqrt{8a_1}}) + 2a_4B}{4a_1} + C_6e^{t\sqrt{2a_1}} + C_7e^{-t\sqrt{2a_1}}\,,$$
$$f = C_1\Big(2a_6 + \frac{a_4^2+a_5^2}{2a_1}\Big)e^{t\sqrt{8a_1}} + C_2\Big(2a_6 + \frac{a_4^2+a_5^2}{2a_1}\Big)e^{-t\sqrt{8a_1}} + Ct\Big(2a_6 - \frac{a_4^2+a_5^2}{2a_1}\Big) + \\ + \frac{\sqrt{2}}{\sqrt{a_1}}(a_4C_5 + a_5C_7)e^{-t\sqrt{2a_1}} - \frac{\sqrt{2}}{\sqrt{a_1}}(a_4C_4 + a_5C_6)e^{t\sqrt{2a_1}} + C_8\,.$$

Thus, the kernel of symmetry operators is extended by the operators

$$X_2 = e^{t\sqrt{8a_1}}\Big(-\Big(\sqrt{8a_1}x + \frac{2a_4}{\sqrt{2a_1}}\Big)\partial_x - \Big(\sqrt{8a_1}y + \frac{2a_5}{\sqrt{2a_1}}\Big)\partial_y - 2\partial_t +$$
$$+ (\sqrt{8a_1}u - 8a_1x - 4a_4)\partial_u + (\sqrt{8a_1}v - 8a_1y - 4a_5)\partial_v +$$
$$+ \Big(2\sqrt{8a_1}\eta + 4a_1(x^2+y^2)\sqrt{8a_1} + 4a_4\sqrt{8a_1}x + 4a_5\sqrt{8a_1}y +$$
$$+ \sqrt{8a_1}\Big(2a_6 + \frac{a_4^2+a_5^2}{2a_1}\Big)\Big)\partial_\eta\Big),$$
$$X_3 = e^{-t\sqrt{8a_1}}\Big(\Big(\sqrt{8a_1}x + \frac{2a_4}{\sqrt{2a_1}}\Big)\partial_x + \Big(\sqrt{8a_1}y + \frac{2a_5}{\sqrt{2a_1}}\Big)\partial_y - 2\partial_t +$$
$$+ (-\sqrt{8a_1}u - 8a_1x - 4a_4)\partial_u + (-\sqrt{8a_1}v - 8a_1y - 4a_5)\partial_v +$$
$$- \Big(2\sqrt{8a_1}\eta + 4a_1(x^2+y^2)\sqrt{8a_1} + 4a_4\sqrt{8a_1}x + 4a_5\sqrt{8a_1}y +$$
$$+ \sqrt{8a_1}\Big(2a_6 + \frac{a_4^2+a_5^2}{2a_1}\Big)\Big)\partial_\eta\Big),$$
$$X_4 = e^{t\sqrt{2a_1}}\Big(\partial_x + \sqrt{2a_1}\partial_u - (2a_1x + a_4)\partial_\eta\Big),$$
$$X_5 = e^{t\sqrt{2a_1}}\Big(\partial_y + \sqrt{2a_1}\partial_v - (2a_1y + a_5)\partial_\eta\Big),$$
$$X_6 = e^{-t\sqrt{2a_1}}\Big(\partial_x - \sqrt{2a_1}\partial_u - (2a_1x + a_4)\partial_\eta\Big),$$
$$X_7 = e^{-t\sqrt{2a_1}}\Big(\partial_y - \sqrt{2a_1}\partial_v - (2a_1y + a_5)\partial_\eta\Big),$$
$$X_8 = -\Big(2y + \frac{a_5}{a_1}\Big)\partial_x + \Big(2x + \frac{a_4}{a_1}\Big)\partial_y - 2v\partial_u + 2u\partial_v\,,$$
$$X_9 = \Big(x + \frac{a_4}{2a_1}\Big)\partial_x + \Big(y + \frac{a_5}{2a_1}\Big)\partial_y + u\partial_u + v\partial_v + \Big(2\eta + 2a_6 - \frac{a_4^2+a_5^2}{2a_1}\Big)\partial_\eta\,.$$

1.2.2. $\mathbf{h = a_1(x^2+y^2) + a_4x + a_5y + a_6,\ a_1 < 0}$.
In this case we obtain the following solution

$$a = C_1\sin(t\sqrt{-8a_1}) + C_2\cos(t\sqrt{-8a_1}) - Ct + C_3\,,$$
$$k = -\frac{C_1a_4\sqrt{-8a_1}\,\cos(t\sqrt{-8a_1}) - C_2a_4\sqrt{-8a_1}\,\sin(t\sqrt{-8a_1})}{4a_1} +$$
$$+\frac{a_4C - 2a_5B}{4a_1} + C_4\sin(t\sqrt{-2a_1}) + C_5\cos(t\sqrt{-2a_1})\,,$$
$$l = -\frac{C_1a_5\sqrt{-8a_1}\,\cos(t\sqrt{-8a_1}) - C_2a_5\sqrt{-8a_1}\,\sin(t\sqrt{-8a_1})}{4a_1} +$$
$$+\frac{2a_4B + a_5C}{4a_1} + C_6\sin(t\sqrt{-2a_1}) + C_7\cos(t\sqrt{-2a_1})\,,$$

$$f = C_1\Big(2a_6 + \frac{a_4^2+a_5^2}{2a_1}\Big)\sin(t\sqrt{-8a_1}) + C_2\Big(2a_6 + \frac{a_4^2+a_5^2}{2a_1}\Big)\cos(t\sqrt{-8a_1}) +$$
$$\frac{\sqrt{2}}{\sqrt{-a_1}}\big((a_4C_4 + a_5C_6)\cos(t\sqrt{-2a_1}) - (a_4C_5 + a_5C_7)\sin(t\sqrt{-2a_1})\big) +$$
$$+ Ct\Big(2a_6 - \frac{a_4^2+a_5^2}{2a_1}\Big) + C_8\,.$$

Thus, the kernel of symmetry operators is extended by the operators

$$X_2 = \cos(t\sqrt{-8a_1})\Big(\Big(\frac{2a_4}{\sqrt{-2a_1}} - \sqrt{-8a_1}x\Big)\partial_x + \Big(\frac{2a_5}{\sqrt{-2a_1}} - \sqrt{-8a_1}y\Big)\partial_y\Big) -$$
$$- 2\sin(t\sqrt{-8a_1})\partial_t + (\sqrt{-8a_1}\cos(t\sqrt{-8a_1})u - \sin(t\sqrt{-8a_1})(8a_1x + 4a_4))\partial_u +$$
$$+ (\sqrt{-8a_1}\cos(t\sqrt{-8a_1})v - (8a_1y + 4a_5)\sin(t\sqrt{-8a_1}))\partial_v +$$
$$+ \sqrt{-8a_1}\cos(t\sqrt{-8a_1})\Big(2\eta + 4a_1(x^2+y^2) + 4a_4x + 4a_5y + 2a_6 + \frac{a_4^2+a_5^2}{2a_1}\Big)\partial_\eta\,,$$
$$X_3 = \sin(t\sqrt{-8a_1})\Big(\Big(\sqrt{-8a_1}x - \frac{2a_4}{\sqrt{-2a_1}}\Big)\partial_x + \Big(\sqrt{-8a_1}y - \frac{2a_5}{\sqrt{-2a_1}}\Big)\partial_y\Big) -$$
$$- 2\cos(t\sqrt{-8a_1})\partial_t - (\sqrt{-8a_1}\sin(t\sqrt{-8a_1})u + \cos(t\sqrt{-8a_1})(8a_1x + 4a_4))\partial_u +$$
$$+ (-\sqrt{-8a_1}\sin(t\sqrt{-8a_1})v - (8a_1y + 4a_5)\cos(t\sqrt{-8a_1}))\partial_v -$$
$$- \sqrt{-8a_1}\sin(t\sqrt{-8a_1})\Big(2\eta + 4a_1(x^2+y^2) + 4a_4x + 4a_5y + 2a_6 + \frac{a_4^2+a_5^2}{2a_1}\Big)\partial_\eta\,,$$
$$X_4 = \sin(t\sqrt{-2a_1})\partial_x + \sqrt{-2a_1}\cos(t\sqrt{-2a_1})\partial_u - \sin(t\sqrt{-2a_1})(2a_1x + a_4)\partial_\eta\,,$$
$$X_5 = \sin(t\sqrt{-2a_1})\partial_y + \sqrt{-2a_1}\cos(t\sqrt{-2a_1})\partial_v - \sin(t\sqrt{-2a_1})(2a_1y + a_5)\partial_\eta\,,$$
$$X_6 = \cos(t\sqrt{-2a_1})\partial_x - \sqrt{-2a_1}\sin(t\sqrt{-2a_1})\partial_u - \cos(t\sqrt{-2a_1})(2a_1x + a_4)\partial_\eta\,,$$
$$X_7 = \cos(t\sqrt{-2a_1})\partial_y - \sqrt{-2a_1}\sin(t\sqrt{-2a_1})\partial_v - \cos(t\sqrt{-2a_1})(2a_1y + a_5)\partial_\eta\,,$$
$$X_8 = -\Big(2y + \frac{a_5}{a_1}\Big)\partial_x + \Big(2x + \frac{a_4}{a_1}\Big)\partial_y - 2v\partial_u + 2u\partial_v\,,$$
$$X_9 = \Big(x + \frac{a_4}{2a_1}\Big)\partial_x + \Big(y + \frac{a_5}{2a_1}\Big)\partial_x + u\partial_u + v\partial_v + \Big(2\eta + 2a_6 - \frac{a_4^2+a_5^2}{2a_1}\Big)\partial_\eta\,.$$

1.2.3. $\mathbf{h = a_1x^2 + a_2y^2 + a_4x + a_5y + a_6,\ a_1 \neq a_2,\ a_1 > 0,\ a_2 > 0.}$
In this case we obtain the following solution

$$B = 0\,, \qquad a = -Ct + C_1\,, \qquad k = C_2e^{t\sqrt{2a_1}} + C_3e^{-t\sqrt{2a_1}} + \frac{a_4C}{4a_1}\,,$$
$$l = C_4e^{t\sqrt{2a_2}} + C_5e^{-t\sqrt{2a_2}} + \frac{a_5C}{4a_2}\,,$$

$$f = -\frac{a_4 C_2\sqrt{2}}{\sqrt{a_1}} e^{t\sqrt{2a_1}} + \frac{a_4 C_3\sqrt{2}}{\sqrt{a_1}} e^{-t\sqrt{2a_1}} - \frac{a_5 C_4\sqrt{2}}{\sqrt{a_2}} e^{t\sqrt{2a_2}} + \\ + \frac{a_5 C_5\sqrt{2}}{\sqrt{a_2}} e^{-t\sqrt{2a_2}} + Ct\Big(2a_6 - \frac{a_4^2}{2a_1} - \frac{a_5^2}{2a_2}\Big) + C_6\,.$$

Thus, the kernel of symmetry operators is extended by the operators

$$X_2 = e^{t\sqrt{2a_1}}\Big(\partial_x + \sqrt{2a_1}\,\partial_u - (2a_1 x + a_4)\partial_\eta\Big),$$
$$X_3 = e^{t\sqrt{2a_2}}\Big(\partial_y + \sqrt{2a_2}\,\partial_v - (2a_2 y + a_5)\partial_\eta\Big),$$
$$X_4 = e^{-t\sqrt{2a_1}}\Big(\partial_x - \sqrt{2a_1}\,\partial_u - (2a_1 x + a_4)\partial_\eta\Big),$$
$$X_5 = e^{-t\sqrt{2a_2}}\Big(\partial_y - \sqrt{2a_2}\,\partial_v - (2a_2 y + a_5)\partial_\eta\Big),$$
$$X_6 = \Big(x + \frac{a_4}{2a_1}\Big)\partial_x + \Big(y + \frac{a_5}{2a_2}\Big)\partial_y + u\partial_u + v\partial_v + \Big(2\eta + 2a_6 - \frac{a_4^2}{2a_1} - \frac{a_5^2}{2a_2}\Big)\partial_\eta\,.$$

1.2.4. $\mathbf{h = a_1 x^2 + a_4 x + a_5 y + a_6,\ a_1 > 0.}$
In this case we obtain the following solution

$$B = 0\,,\qquad a = -Ct + C_1\,,\qquad k = C_2 e^{t\sqrt{2a_1}} + C_3 e^{-t\sqrt{2a_1}} + \frac{a_4 C}{4a_1}\,,$$
$$l = -\frac{a_5 C}{4} t^2 + C_4 t + C_5\,,$$
$$f = -\frac{a_4 C_2\sqrt{2}}{\sqrt{a_1}} e^{t\sqrt{2a_1}} + \frac{a_4 C_3\sqrt{2}}{\sqrt{a_1}} e^{-t\sqrt{2a_1}} + \frac{a_5^2 C}{6} t^3 - a_5 C_4 t^2 + \\ + \Big(2a_6 C - \frac{a_4^2 C}{2a_1} - 2a_5 C_5\Big)t + C_6\,.$$

Thus, the kernel of symmetry operators is extended by the operators

$$X_2 = e^{t\sqrt{2a_1}}\Big(\partial_x + \sqrt{2a_1}\,\partial_u - (2a_1 x + a_4)\partial_\eta\Big),\qquad X_3 = t\partial_y + \partial_v - a_5 t\partial_\eta\,,$$
$$X_4 = e^{-t\sqrt{2a_1}}\Big(\partial_x - \sqrt{2a_1}\,\partial_u - (2a_1 x + a_4)\partial_\eta\Big),\qquad X_5 = \partial_y - a_5\partial_\eta\,,$$
$$X_6 = \Big(x + \frac{a_4}{2a_1}\Big)\partial_x + \Big(y - \frac{a_5}{2} t^2\Big)\partial_y + u\partial_u + (v - a_5 t)\partial_v + \\ + \Big(2\eta + a_5 y + \frac{a_5^2}{2} t^2 + 2a_6 - \frac{a_4^2}{2a_1}\Big)\partial_\eta\,.$$

1.2.5. $\mathbf{h = a_1 x^2 + a_2 y^2 + a_4 x + a_5 y + a_6,\ a_1 > 0,\ a_2 < 0.}$
In this case we obtain the following solution

$$B=0\,,\qquad a=-Ct+C_1\,,\qquad k=C_2e^{t\sqrt{2a_1}}+C_3e^{-t\sqrt{2a_1}}+\frac{a_4C}{4a_1}\,,$$
$$l=C_4\sin(t\sqrt{-2a_2})+C_5\cos(t\sqrt{-2a_2})+\frac{a_5C}{4a_2}\,,$$
$$f=-\frac{a_4C_2\sqrt{2}}{\sqrt{a_1}}e^{t\sqrt{2a_1}}+\frac{a_4C_3\sqrt{2}}{\sqrt{a_1}}e^{-t\sqrt{2a_1}}+\frac{a_5C_4\sqrt{2}}{\sqrt{-a_2}}\cos(t\sqrt{-2a_2})-$$
$$-\frac{a_5C_5\sqrt{2}}{\sqrt{-a_2}}\sin(t\sqrt{-2a_2})+Ct\Big(2a_6-\frac{a_4^2}{2a_1}-\frac{a_5^2}{2a_2}\Big)+C_6\,.$$

Thus, the kernel of symmetry operators is extended by the operators

$$X_2=e^{t\sqrt{2a_1}}\Big(\partial_x+\sqrt{2a_1}\,\partial_u-(2a_1x+a_4)\partial_\eta\Big),$$
$$X_3=\sin(t\sqrt{-2a_2})\partial_y+\sqrt{-2a_2}\cos(t\sqrt{-2a_2})\partial_v-(2a_2y+a_5)\sin(t\sqrt{-2a_2})\partial_\eta\,,$$
$$X_4=e^{-t\sqrt{2a_1}}\Big(\partial_x-\sqrt{2a_1}\,\partial_u-(2a_1x+a_4)\partial_\eta\Big),$$
$$X_5=\cos(t\sqrt{-2a_2})\partial_y-\sqrt{-2a_2}\sin(t\sqrt{-2a_2})\partial_v-(2a_2y+a_5)\cos(t\sqrt{-2a_2})\partial_\eta\,,$$
$$X_6=\Big(x+\frac{a_4}{2a_1}\Big)\partial_x+\Big(y+\frac{a_5}{2a_2}\Big)\partial_y+u\partial_u+v\partial_v+\Big(2\eta+2a_6-\frac{a_4^2}{2a_1}-\frac{a_5^2}{2a_2}\Big)\partial_\eta\,.$$

1.2.6. $\mathbf{h=a_2y^2+a_4x+a_5y+a_6,\ a_2<0.}$
In this case we obtain the following solution

$$B=0\,,\qquad a=-Ct+C_1\,,\qquad k=-\frac{a_4C}{4}t^2+C_2t+C_3\,,$$
$$l=C_4\sin(t\sqrt{-2a_2})+C_5\cos(t\sqrt{-2a_2})+\frac{a_5C}{4a_2}\,,$$
$$f=\frac{a_5C_4\sqrt{2}}{\sqrt{-a_2}}\cos(t\sqrt{-2a_2})-\frac{a_5C_5\sqrt{2}}{\sqrt{-a_2}}\sin(t\sqrt{-2a_2})+\frac{a_4^2C}{6}t^3-a_4C_2t^2+$$
$$+\Big(2a_6C-\frac{a_5^2C}{2a_2}-2a_4C_3\Big)t+C_6\,.$$

Thus, the kernel of symmetry operators is extended by the operators

$$X_2=t\partial_x+\partial_u-a_4t\partial_\eta\,,$$
$$X_3=\sin(t\sqrt{-2a_2})\partial_y+\sqrt{-2a_2}\cos(t\sqrt{-2a_2})\partial_v-(2a_2y+a_5)\sin(t\sqrt{-2a_2})\partial_\eta\,,$$
$$X_4=\partial_x-a_4\partial_\eta\,,$$
$$X_5=\cos(t\sqrt{-2a_2})\partial_y-\sqrt{-2a_2}\sin(t\sqrt{-2a_2})\partial_v-(2a_2y+a_5)\cos(t\sqrt{-2a_2})\partial_\eta\,,$$

$$X_6 = \Big(x - \frac{a_4}{2}t^2\Big)\partial_x + \Big(y + \frac{a_5}{2a_2}\Big)\partial_y + (u - a_4 t)\partial_u + v\partial_v +$$
$$+ \Big(2\eta + a_4 x + \frac{a_4^2}{2}t^2 + 2a_6 - \frac{a_5^2}{2a_2}\Big)\partial_\eta\,.$$

1.2.7. $\mathbf{h = a_1x^2 + a_2y^2 + a_4x + a_5y + a_6,\ a_1 < 0,\ a_2 < 0.}$
In this case we obtain the following solution

$$B = 0\,, \qquad a = -Ct + C_1\,, \qquad k = C_2\sin(t\sqrt{-2a_1}) + C_3\cos(t\sqrt{-2a_1}) + \frac{a_4C}{4a_1}\,,$$
$$l = C_4\sin(t\sqrt{-2a_2}) + C_5\cos(t\sqrt{-2a_2}) + \frac{a_5C}{4a_2}\,,$$
$$f = \frac{a_4C_2\sqrt{2}}{\sqrt{-a_1}}\cos(t\sqrt{-2a_1}) - \frac{a_4C_3\sqrt{2}}{\sqrt{-a_1}}\sin(t\sqrt{-2a_1}) + \frac{a_5C_4\sqrt{2}}{\sqrt{-a_2}}\cos(t\sqrt{-2a_2}) -$$
$$- \frac{a_5C_5\sqrt{2}}{\sqrt{-a_2}}\sin(t\sqrt{-2a_2}) + Ct\Big(2a_6 - \frac{a_5^2}{2a_2} - \frac{a_4^2}{2a_1}\Big) + C_6\,.$$

Thus, the kernel of symmetry operators is extended by the operators

$$X_2 = \sin(t\sqrt{-2a_1})\partial_x + \sqrt{-2a_1}\cos(t\sqrt{-2a_1})\partial_u - (2a_1x + a_4)\sin(t\sqrt{-2a_1})\partial_\eta\,,$$
$$X_3 = \sin(t\sqrt{-2a_2})\partial_y + \sqrt{-2a_2}\cos(t\sqrt{-2a_2})\partial_v - (2a_2y + a_5)\sin(t\sqrt{-2a_2})\partial_\eta\,,$$
$$X_4 = \cos(t\sqrt{-2a_1})\partial_x - \sqrt{-2a_1}\sin(t\sqrt{-2a_1})\partial_u - (2a_1x + a_4)\cos(t\sqrt{-2a_1})\partial_\eta\,,$$
$$X_5 = \cos(t\sqrt{-2a_2})\partial_y - \sqrt{-2a_2}\sin(t\sqrt{-2a_2})\partial_v - (2a_2y + a_5)\cos(t\sqrt{-2a_2})\partial_\eta\,,$$
$$X_6 = \Big(x + \frac{a_4}{2a_1}\Big)\partial_x + \Big(y + \frac{a_5}{2a_2}\Big)\partial_y + u\partial_u + v\partial_v + \Big(2\eta + 2a_6 - \frac{a_4^2}{2a_1} - \frac{a_5^2}{2a_2}\Big)\partial_\eta\,.$$

1.2.8. $\mathbf{h = a_1x^2 + a_2y^2 + a_3xy + a_4x + a_5y + a_6,\ a_3 \neq 0,}$
$\mathbf{a_1 + a_2 > \sqrt{(a_1 - a_2)^2 + a_3^2}.}$

Denote $s = \sqrt{(a_1 - a_2)^2 + a_3^2}$, $S_1 = \sqrt{a_1 + a_2 + s}$, $S_2 = \sqrt{a_1 + a_2 - s}$. In this case we obtain the following solution

$$B = 0\,, \qquad a = -Ct + C_1\,,$$
$$k = \frac{C(a_3a_5 - 2a_2a_4)}{2(a_3^2 - 4a_1a_2)} + C_2e^{tS_2} + C_3e^{-tS_2} + C_4e^{tS_1} + C_5e^{-tS_1}\,,$$
$$l = \frac{a_4C - 4a_1k + 2\ddot{k}}{2a_3}\,,$$

$$f=\Big(-2a_4+\frac{4a_1a_5}{a_3}\Big)\Big(\frac{C(a_3a_5-2a_2a_4)}{2(a_3^2-4a_1a_2)}t+\frac{C_2}{S_2}e^{tS_2}-\frac{C_3}{S_2}e^{-tS_2}+ \\ +\frac{C_4}{S_1}e^{tS_1}-\frac{C_5}{S_1}e^{-tS_1}\Big)+Ct\Big(2a_6-\frac{a_4a_5}{a_3}\Big)-\frac{2a_5}{a_3}\dot{k}+C_6\,.$$

Thus, the kernel of symmetry operators is extended by the operators

$$X_2=e^{tS_2}\Big(\partial_x+\frac{S_2^2-2a_1}{a_3}\partial_y+S_2\partial_u+\frac{S_2^3-2a_1S_2}{a_3}\partial_v- \\ -\Big(S_2^2x+\frac{S_2^4-2a_1S_2^2}{a_3}y+a_4+\frac{a_5(S_2^2-2a_1)}{a_3}\Big)\partial_\eta\Big),$$

$$X_3=e^{-tS_2}\Big(\partial_x+\frac{S_2^2-2a_1}{a_3}\partial_y-S_2\partial_u-\frac{S_2^3-2a_1S_2}{a_3}\partial_v- \\ -\Big(S_2^2x+\frac{S_2^4-2a_1S_2^2}{a_3}y+a_4+\frac{a_5(S_2^2-2a_1)}{a_3}\Big)\partial_\eta\Big),$$

$$X_4=e^{tS_1}\Big(\partial_x+\frac{S_1^2-2a_1}{a_3}\partial_y+S_1\partial_u+\frac{S_1^3-2a_1S_1}{a_3}\partial_v- \\ -\Big(S_1^2x+\frac{S_1^4-2a_1S_1^2}{a_3}y+a_4+\frac{a_5(S_1^2-2a_1)}{a_3}\Big)\partial_\eta\Big),$$

$$X_5=e^{-tS_1}\Big(\partial_x+\frac{S_1^2-2a_1}{a_3}\partial_y-S_1\partial_u-\frac{S_1^3-2a_1S_1}{a_3}\partial_v- \\ -\Big(S_1^2x+\frac{S_1^4-2a_1S_1^2}{a_3}y+a_4+\frac{a_5(S_1^2-2a_1)}{a_3}\Big)\partial_\eta\Big),$$

$$X_6=\Big(x+\frac{a_3a_5-2a_2a_4}{a_3^2-4a_1a_2}\Big)\partial_x+\Big(y+\frac{a_3a_4-2a_1a_5}{a_3^2-4a_1a_2}\Big)\partial_y+u\partial_u+ \\ +v\partial_v+2\Big(\eta+a_6+\frac{a_1a_5^2+a_2a_4^2-a_3a_4a_5}{a_3^2-4a_1a_2}\Big)\partial_\eta\,.$$

1.2.9.

$$\mathbf{h}=\Big(\sqrt{\mathbf{a_1}}\mathbf{x}+\frac{\mathbf{a_3}}{2\sqrt{\mathbf{a_1}}}\mathbf{y}\Big)^2+\mathbf{a_4x}+\mathbf{a_5y}+\mathbf{a_6},\ \mathbf{a_1}>\mathbf{0},$$

$$\mathbf{a_3}\neq\mathbf{0},\qquad \mathbf{a_1}+\mathbf{a_2}=\sqrt{(\mathbf{a_1}-\mathbf{a_2})^2+\mathbf{a_3^2}}.$$

Denote

$$S=\sqrt{2a_1+\frac{a_3^2}{2a_1}},\ A_1=\frac{2a_1a_5-a_3a_4}{4a_1^2+a_3^2},\ A_2=\frac{2a_1a_4+a_3a_5}{4a_1^2+a_3^2}\,.$$

In this case we obtain the following solution

$$B=0\,,\qquad a=-Ct+C_1\,,\qquad k=C_2e^{tS}+C_3e^{-tS}+\frac{a_3A_1C}{4}\,t^2+C_4t+C_5\,,$$

$$l=\frac{a_4C-4a_1k+2\ddot{k}}{2a_3}\,,$$

$$f=\Big(-2a_4+\frac{4a_1a_5}{a_3}\Big)\Big(\frac{C_2}{S}\,e^{tS}-\frac{C_3}{S}\,e^{-tS}\Big)+$$
$$+\Big(-2a_4+\frac{4a_1a_5}{a_3}\Big)\Big(\frac{a_3A_1C}{12}\,t^3+\frac{C_4}{2}\,t^2+C_5t\Big)+Ct\Big(2a_6-\frac{a_4a_5}{a_3}\Big)-\frac{2a_5}{a_3}\,\dot{k}+C_6\,.$$

Thus, the kernel of symmetry operators is extended by the operators

$$X_2=e^{tS}\Big(\partial_x+\frac{S^2-2a_1}{a_3}\partial_y+S\partial_u+\frac{S^3-2a_1S}{a_3}\partial_v-$$
$$-\Big(S^2x+\frac{S^4-2a_1S^2}{a_3}\,y+a_4+\frac{a_5(S^2-2a_1)}{a_3}\Big)\partial_\eta\Big)\,,$$

$$X_3=e^{-tS}\Big(\partial_x+\frac{S^2-2a_1}{a_3}\partial_y-S\partial_u-\frac{S^3-2a_1S}{a_3}\partial_v-$$
$$-\Big(S^2x+\frac{S^4-2a_1S^2}{a_3}\,y+a_4+\frac{a_5(S^2-2a_1)}{a_3}\Big)\partial_\eta\Big)\,,$$

$$X_4=t\partial_x-\frac{2a_1}{a_3}\,t\partial_y+\partial_u-\frac{2a_1}{a_3}\partial_v+\Big(-a_4+\frac{2a_1a_5}{a_3}\Big)t\partial_\eta\,,$$

$$X_5=\partial_x-\frac{2a_1}{a_3}\partial_y+\Big(\frac{2a_1a_5}{a_3}-a_4\Big)\partial_\eta\,,$$

$$X_6=\Big(x+\frac{a_3A_1}{2}\,t^2\Big)\partial_x+\Big(y+\frac{2a_1A_2}{a_3}-a_1A_1t^2\Big)\partial_y+(u+a_3A_1t)\partial_u+(v-2a_1A_1t)\partial_v+$$
$$+\Big(2\eta-a_3A_1x+2a_1A_1y+\frac{(2a_1a_5-a_3a_4)A_1}{2}\,t^2+2a_6-\frac{2a_1a_5A_2}{a_3}\Big)\partial_\eta\,.$$

1.2.10. $\mathbf{h=a_1x^2+a_2y^2+a_3xy+a_4x+a_5y+a_6,\ a_3\neq 0,}$
$\mathbf{-\sqrt{(a_1-a_2)^2+a_3^2}<a_1+a_2<\sqrt{(a_1-a_2)^2+a_3^2}.}$

Denote $s=\sqrt{(a_1-a_2)^2+a_3^2}$, $S_1=\sqrt{a_1+a_2+s}$, $S_3=\sqrt{s-a_1-a_2}$. In this case we obtain the following solution

$$B=0\,,\qquad a=-Ct+C_1\,,$$
$$k=\frac{C(a_3a_5-2a_2a_4)}{2(a_3^2-4a_1a_2)}+C_2\sin(tS_3)+C_3\cos(tS_3)+C_4e^{tS_1}+C_5e^{-tS_1}\,,$$
$$l=\frac{a_4C-4a_1k+2\ddot{k}}{2a_3}\,,$$

$$f = \Big(-2a_4 + \frac{4a_1a_5}{a_3}\Big)\Big(\frac{C(a_3a_5 - 2a_2a_4)}{2(a_3^2 - 4a_1a_2)}t - \frac{C_2}{S_3}\cos(tS_3) + \frac{C_3}{S_3}\sin(tS_3) +$$
$$+ \frac{C_4}{S_1}e^{tS_1} - \frac{C_5}{S_1}e^{-tS_1}\Big) + Ct\Big(2a_6 - \frac{a_4a_5}{a_3}\Big) - \frac{2a_5}{a_3}\dot{k} + C_6\,.$$

Thus, the kernel of symmetry operators is extended by the operators

$$X_2 = \sin(tS_3)\partial_x - \frac{S_3^2 + 2a_1}{a_3}\sin(tS_3)\partial_y + S_3\cos(tS_3)\partial_u - \frac{S_3^3 + 2a_1S_3}{a_3}\cos(tS_3)\partial_v +$$
$$+ \sin(tS_3)\Big(S_3^2x - \frac{S_3^4 + 2a_1S_3^2}{a_3}y - a_4 + \frac{a_5(2a_1 + S_3^2)}{a_3}\Big)\partial_\eta\Big),$$
$$X_3 = \cos(tS_3)\partial_x - \frac{S_3^2 + 2a_1}{a_3}\cos(tS_3)\partial_y - S_3\sin(tS_3)\partial_u + \frac{S_3^3 + 2a_1S_3}{a_3}\sin(tS_3)\partial_v +$$
$$+ \cos(tS_3)\Big(S_3^2x - \frac{S_3^4 + 2a_1S_3^2}{a_3}y - a_4 + \frac{a_5(2a_1 + S_3^2)}{a_3}\Big)\partial_\eta\Big),$$
$$X_4 = e^{tS_1}\Big(\partial_x + \frac{S_1^2 - 2a_1}{a_3}\partial_y + S_1\partial_u + \frac{S_1^3 - 2a_1S_1}{a_3}\partial_v -$$
$$-\Big(S_1^2x + \frac{S_1^4 - 2a_1S_1^2}{a_3}y + a_4 + \frac{a_5(S_1^2 - 2a_1)}{a_3}\Big)\partial_\eta\Big),$$
$$X_5 = e^{-tS_1}\Big(\partial_x + \frac{S_1^2 - 2a_1}{a_3}\partial_y - S_1\partial_u - \frac{S_1^3 - 2a_1S_1}{a_3}\partial_v -$$
$$-\Big(S_1^2x + \frac{S_1^4 - 2a_1S_1^2}{a_3}y + a_4 + \frac{a_5(S_1^2 - 2a_1)}{a_3}\Big)\partial_\eta\Big),$$
$$X_6 = \Big(x + \frac{a_3a_5 - 2a_2a_4}{a_3^2 - 4a_1a_2}\Big)\partial_x + \Big(y + \frac{a_3a_4 - 2a_1a_5}{a_3^2 - 4a_1a_2}\Big)\partial_y + u\partial_u +$$
$$+ v\partial_v + 2\Big(\eta + a_6 + \frac{a_1a_5^2 + a_2a_4^2 - a_3a_4a_5}{a_3^2 - 4a_1a_2}\Big)\partial_\eta\,.$$

1.2.11.

$$\mathbf{h} = -\Big(\sqrt{-\mathbf{a}_1}\mathbf{x} - \frac{\mathbf{a}_3}{2\sqrt{-\mathbf{a}_1}}\mathbf{y}\Big)^2 + \mathbf{a}_4\mathbf{x} + \mathbf{a}_5\mathbf{y} + \mathbf{a}_6,\ \mathbf{a}_1 < \mathbf{0},$$
$$\mathbf{a}_3 \neq \mathbf{0}, \qquad \mathbf{a}_1 + \mathbf{a}_2 = -\sqrt{(\mathbf{a}_1 - \mathbf{a}_2)^2 + \mathbf{a}_3^2}.$$

Denote

$$S = \sqrt{-2a_1 - \frac{a_3^2}{2a_1}},\ A_1 = \frac{2a_1a_5 - a_3a_4}{4a_1^2 + a_3^2},\ A_2 = \frac{2a_1a_4 + a_3a_5}{4a_1^2 + a_3^2}.$$

In this case we obtain the following solution

$$B=0\,,\qquad a=-Ct+C_1\,,\qquad k=C_2\sin(tS)+C_3\cos(tS)+\frac{a_3A_1C}{4}t^2+C_4t+C_5\,,$$
$$l=\frac{a_4C-4a_1k+2\ddot{k}}{2a_3}\,,$$
$$f=\Big(-2a_4+\frac{4a_1a_5}{a_3}\Big)\Big(-\frac{C_2}{S}\cos(tS)+\frac{C_3}{S}\sin(tS)\Big)+$$
$$+\Big(-2a_4+\frac{4a_1a_5}{a_3}\Big)\Big(\frac{a_3A_1C}{12}t^3+\frac{C_4}{2}t^2+C_5t\Big)+Ct\Big(2a_6-\frac{a_4a_5}{a_3}\Big)-\frac{2a_5}{a_3}\dot{k}+C_6\,.$$

Thus, the kernel of symmetry operators is extended by the operators

$$X_2=\sin(tS)\partial_x-\frac{S^2+2a_1}{a_3}\sin(tS)\partial_y+S\cos(tS)\partial_u-\frac{S^3+2a_1S}{a_3}\cos(tS)\partial_v+$$
$$+\sin(tS)\Big(S^2x-\frac{S^4+2a_1S^2}{a_3}y-a_4+\frac{a_5(2a_1+S^2)}{a_3}\Big)\partial_\eta\Big),$$
$$X_3=\cos(tS)\partial_x-\frac{S^2+2a_1}{a_3}\cos(tS)\partial_y-S\sin(tS)\partial_u+\frac{S^3+2a_1S}{a_3}\sin(tS)\partial_v+$$
$$+\cos(tS)\Big(S^2x-\frac{S^4+2a_1S^2}{a_3}y-a_4+\frac{a_5(2a_1+S^2)}{a_3}\Big)\partial_\eta\Big),$$
$$X_4=t\partial_x-\frac{2a_1}{a_3}t\partial_y+\partial_u-\frac{2a_1}{a_3}\partial_v+\Big(-a_4+\frac{2a_1a_5}{a_3}\Big)t\partial_\eta\,,$$
$$X_5=\partial_x-\frac{2a_1}{a_3}\partial_y+\Big(\frac{2a_1a_5}{a_3}-a_4\Big)\partial_\eta\,,$$
$$X_6=\Big(x+\frac{a_3A_1}{2}t^2\Big)\partial_x+\Big(y+\frac{2a_1A_2}{a_3}-a_1A_1t^2\Big)\partial_y+(u+a_3A_1t)\partial_u+(v-2a_1A_1t)\partial_v+$$
$$+\Big(2\eta-a_3A_1x+2a_1A_1y+\frac{(2a_1a_5-a_3a_4)A_1}{2}t^2+2a_6-\frac{2a_1a_5A_2}{a_3}\Big)\partial_\eta\,.$$

1.2.12. $\mathbf{h=a_1x^2+a_2y^2+a_3xy+a_4x+a_5y+a_6}$, $\mathbf{a_3\neq 0}$, $\mathbf{a_1+a_2<-\sqrt{(a_1-a_2)^2+a_3^2}}$.

Denote $s=\sqrt{(a_1-a_2)^2+a_3^2}$, $S_3=\sqrt{s-a_1-a_2}$, $S_4=\sqrt{-s-a_1-a_2}$. In this case we obtain the following solution

$$B=0\,,\qquad a=-Ct+C_1\,,$$
$$k=\frac{C(a_3a_5-2a_2a_4)}{2(a_3^2-4a_1a_2)}+C_2\sin(tS_4)+C_3\cos(tS_4)+C_4\sin(tS_3)+C_5\cos(tS_3)\,,$$
$$l=\frac{a_4C-4a_1k+2\ddot{k}}{2a_3}\,,$$

$$f = \Big(-2a_4 + \frac{4a_1a_5}{a_3}\Big)\Big(\frac{C(a_3a_5 - 2a_2a_4)}{2(a_3^2 - 4a_1a_2)}t - \frac{C_2}{S_4}\cos(tS_4) + \frac{C_3}{S_4}\sin(tS_4) -$$
$$-\frac{C_4}{S_3}\cos(tS_3) + \frac{C_5}{S_3}\sin(tS_3)\Big) + Ct\Big(2a_6 - \frac{a_4a_5}{a_3}\Big) - \frac{2a_5}{a_3}\dot{k} + C_6 .$$

Thus, the kernel of symmetry operators is extended by the operators

$$X_2 = \sin(tS_3)\partial_x - \frac{S_3^2 + 2a_1}{a_3}\sin(tS_3)\partial_y + S_3\cos(tS_3)\partial_u - \frac{S_3^3 + 2a_1S_3}{a_3}\cos(tS_3)\partial_v +$$
$$+ \sin(tS_3)\Big(S_3^2x - \frac{S_3^4 + 2a_1S_3^2}{a_3}y - a_4 + \frac{a_5(2a_1 + S_3^2)}{a_3}\Big)\partial_\eta\Big),$$
$$X_3 = \cos(tS_3)\partial_x - \frac{S_3^2 + 2a_1}{a_3}\cos(tS_3)\partial_y - S_3\sin(tS_3)\partial_u + \frac{S_3^3 + 2a_1S_3}{a_3}\sin(tS_3)\partial_v +$$
$$+ \cos(tS_3)\Big(S_3^2x - \frac{S_3^4 + 2a_1S_3^2}{a_3}y - a_4 + \frac{a_5(2a_1 + S_3^2)}{a_3}\Big)\partial_\eta\Big),$$
$$X_4 = \sin(tS_4)\partial_x - \frac{S_4^2 + 2a_1}{a_3}\sin(tS_4)\partial_y + S_4\cos(tS_4)\partial_u - \frac{S_4^3 + 2a_1S_4}{a_3}\cos(tS_4)\partial_v +$$
$$+ \sin(tS_4)\Big(S_4^2x - \frac{S_4^4 + 2a_1S_4^2}{a_3}y - a_4 + \frac{a_5(2a_1 + S_4^2)}{a_3}\Big)\partial_\eta\Big),$$
$$X_5 = \cos(tS_4)\partial_x - \frac{S_4^2 + 2a_1}{a_3}\cos(tS_4)\partial_y - S_4\sin(tS_4)\partial_u + \frac{S_4^3 + 2a_1S_4}{a_3}\sin(tS_4)\partial_v +$$
$$+ \cos(tS_4)\Big(S_4^2x - \frac{S_4^4 + 2a_1S_4^2}{a_3}y - a_4 + \frac{a_5(2a_1 + S_4^2)}{a_3}\Big)\partial_\eta\Big),$$
$$X_6 = \Big(x + \frac{a_3a_5 - 2a_2a_4}{a_3^2 - 4a_1a_2}\Big)\partial_x + \Big(y + \frac{a_3a_4 - 2a_1a_5}{a_3^2 - 4a_1a_2}\Big)\partial_y + u\partial_u +$$
$$+ v\partial_v + 2\Big(\eta + a_6 + \frac{a_1a_5^2 + a_2a_4^2 - a_3a_4a_5}{a_3^2 - 4a_1a_2}\Big)\partial_\eta .$$

1.3. $\mathbf{h = H(x) + b_1y^2 + b_2y,\ H''' \neq 0}$.
1.3.1. $\mathbf{h = H(x) + b_1y^2 + b_2y,\ H''' \neq 0,\ b_1 > 0}$.
In this case we obtain the following solution

$$B = 0, \quad C = 0, \quad a = C_1, \quad k = 0, \quad l = C_2e^{t\sqrt{2b_1}} + C_3e^{-t\sqrt{2b_1}},$$
$$f = -2b_2\Big(\frac{C_2}{\sqrt{2b_1}}e^{t\sqrt{2b_1}} - \frac{C_3}{\sqrt{2b_1}}e^{-t\sqrt{2b_1}}\Big) + C_4 .$$

Thus, the kernel of symmetry operators is extended by the operators

$$X_2 = e^{t\sqrt{2b_1}}(\partial_y + \sqrt{2b_1}\partial_v - (2b_1y + b_2)\partial_\eta),$$
$$X_3 = e^{-t\sqrt{2b_1}}(\partial_y - \sqrt{2b_1}\partial_v - (2b_1y + b_2)\partial_\eta).$$

1.3.1.1.

$$\mathbf{h} = \frac{\mathbf{a_1}}{(\mathbf{x}+\mathbf{a_2})^2} + \mathbf{b_1}((\mathbf{x}+\mathbf{a_2})^2 + \mathbf{y}^2) + \mathbf{b_2 y} + \mathbf{b_3},\ \mathbf{b_1} > \mathbf{0},\ \mathbf{a_1} \neq \mathbf{0}.$$

In this case we obtain additional solution to the solution from the case 1.3.1

$$B = 0\,, \quad C = 0\,, \quad a = C_5 e^{t\sqrt{8b_1}} + C_6 e^{-t\sqrt{8b_1}}\,,$$
$$k = -a_2\sqrt{2b_1}(C_5 e^{t\sqrt{8b_1}} - C_6 e^{-t\sqrt{8b_1}})\,, \qquad l = -\frac{b_2}{\sqrt{2b_1}}(C_5 e^{t\sqrt{8b_1}} - C_6 e^{-t\sqrt{8b_1}})\,,$$
$$f = \frac{4b_1(b_3 + 2a_2^2 b_1) + b_2^2}{2b_1}(C_5 e^{t\sqrt{8b_1}} + C_6 e^{-t\sqrt{8b_1}})\,.$$

Then the kernel of symmetry operators is extended by the operators X_2, X_3 from the case 1.3.1 and by the additional symmetry operators

$$X_4 = e^{t\sqrt{8b_1}}\Big(-\sqrt{8b_1}(x+a_2)\partial_x - \Big(\sqrt{8b_1}y + \frac{2b_2}{\sqrt{2b_1}}\Big)\partial_y - 2\partial_t +$$
$$+ (\sqrt{8b_1}u - 8b_1(x+a_2))\partial_u + (\sqrt{8b_1}v - 8b_1 y - 4b_2)\partial_v +$$
$$+ \sqrt{8b_1}\Big(2\eta + 4b_1((x+a_2)^2 + y^2) + 4b_2 y + 2b_3 + \frac{b_2^2}{2b_1}\Big)\partial_\eta\Big),$$
$$X_5 = e^{-t\sqrt{8b_1}}\Big(\sqrt{8b_1}(x+a_2)\partial_x + \Big(\sqrt{8b_1}y + \frac{2b_2}{\sqrt{2b_1}}\Big)\partial_y - 2\partial_t -$$
$$- (\sqrt{8b_1}u + 8b_1(x+a_2))\partial_u - (\sqrt{8b_1}v + 8b_1 y + 4b_2)\partial_v -$$
$$- \sqrt{8b_1}\Big(2\eta + 4b_1((x+a_2)^2 + y^2) + 4b_2 y + 2b_3 + \frac{b_2^2}{2b_1}\Big)\partial_\eta\Big).$$

1.3.2. $\mathbf{h} = \mathbf{H(x)} + \mathbf{b_2 y},\ \mathbf{H'''} \neq \mathbf{0}.$
In this case we obtain the following solution

$$B = 0\,, \quad C = 0\,, \quad a = C_1\,, \quad k = 0\,, \quad l = C_2 t + C_3\,,$$
$$f = -b_2 C_2 t^2 - 2b_2 C_3 t + C_4\,.$$

Thus, the kernel of symmetry operators is extended by the operators

$$X_2 = t\partial_y + \partial_v - b_2 t\partial_\eta\,, \qquad X_3 = \partial_y - b_2\partial_\eta\,.$$

1.3.2.1.

$$\mathbf{h} = \frac{\mathbf{a_1}}{(\mathbf{x}+\mathbf{a_2})^2} + \mathbf{b_2 y} + \mathbf{b_3},\ \mathbf{a_1} \neq \mathbf{0}.$$

In this case we obtain additional solution to the solution from the case 1.3.2

$$B=0\,,\qquad C=0\,,\qquad a=C_5t^2+C_6t\,,\qquad k=-\frac{a_2}{2}\,(2C_5t+C_6)\,,$$
$$l=-\frac{3b_2}{2}\Big(\frac{C_5}{3}\,t^3+\frac{C_6}{2}\,t^2\Big),\qquad f=\frac{b_2^2C_5}{4}\,t^4+\frac{b_2^2C_6}{2}\,t^3+2b_3C_5t^2+2b_3C_6t\,.$$

Then the kernel of symmetry operators is extended by the operators X_2, X_3 from the case 1.3.2 and by the additional symmetry operators

$$X_4=-t\,(x+a_2)\partial_x-t\Big(y+\frac{b_2}{2}\,t^2\Big)\partial_y-t^2\partial_t+(tu-(x+a_2))\partial_u+$$
$$+\Big(tv-y-\frac{3b_2}{2}\,t^2\Big)\partial_v+t\Big(2\eta+3b_2y+\frac{b_2^2}{2}\,t^2+2b_3\Big)\partial_\eta\,,$$
$$X_5=-(x+a_2)\partial_x-\Big(y+\frac{3b_2}{2}\,t^2\Big)\partial_y-2t\,\partial_t+u\,\partial_u+$$
$$+(v-3b_2t)\partial_v+\Big(2\eta+3b_2y+\frac{3b_2^2}{2}\,t^2+2b_3\Big)\partial_\eta\,.$$

1.3.2.2. $\mathbf{h=a_1(x+a_2)^{a_3}+b_2y+b_3,\ a_1\neq 0,\ a_3\in\mathbb{Z}\setminus\{-2,0,1,2\}}$ or $\mathbf{h=a_1|x+a_2|^{a_3}+b_2y+b_3,\ a_1\neq 0,\ a_3\neq -2,0,1,2.}$
In this case we obtain additional solution to the solution from the case 1.3.2

$$B=0\,,\qquad a=-\frac{4C}{a_3+2}\,t\,,\qquad k=\frac{2a_2C}{a_3+2}\,,$$
$$l=\frac{(1-a_3)b_2C}{a_3+2}\,t^2\,,\qquad f=-\frac{2(1-a_3)b_2^2C}{3(a_3+2)}\,t^3+\frac{4a_3b_3C}{a_3+2}\,t\,.$$

Then the kernel of symmetry operators is extended by the operators X_2, X_3 from the case 1.3.2 and by the additional symmetry operator

$$X_4=2(x+a_2)\partial_x+(2y+(1-a_3)b_2t^2)\partial_y-(a_3-2)t\,\partial_t+a_3u\,\partial_u+$$
$$+(a_3v+2(1-a_3)b_2t)\partial_v+(2a_3\eta-(1-a_3)(2b_2y+b_2^2t^2)+2a_3b_3)\partial_\eta\,.$$

1.3.2.3. $\mathbf{h=a_1e^{a_2x}+b_2y+b_3,\ a_1\neq 0,\ a_2\neq 0.}$
In this case we obtain additional solution to the solution from the case 1.3.2

$$B=0\,,\qquad a=0\,,\qquad k=\frac{2C}{a_2}\,,\qquad l=-b_2Ct^2\,,\qquad f=\frac{2b_2^2C}{3}\,t^3+4b_3Ct\,.$$

Then the kernel of symmetry operators is extended by the operators X_2, X_3 from the case 1.3.2 and by the additional symmetry operator

$$X_4=\frac{2}{a_2}\,\partial_x-b_2t^2\partial_y-t\partial_t+u\,\partial_u+(v-2b_2t)\partial_v+(2\eta+2b_2y+b_2^2t^2+2b_3)\partial_\eta\,.$$

1.3.2.4. $\mathbf{h = a_1 \ln |x + a_2| + b_2 y + b_3,\ a_1 \neq 0.}$
In this case we obtain additional solution to the solution from the case 1.3.2

$$B = 0\,, \qquad a = -2Ct\,, \qquad k = a_2 C\,, \qquad l = \frac{b_2 C}{2}\, t^2\,, \qquad f = -\frac{b_2^2 C}{3}\, t^3 - 2a_1 Ct\,.$$

Then the kernel of symmetry operators is extended by the operators X_2, X_3 from the case 1.3.2 and by the additional symmetry operator

$$X_4 = (x + a_2)\partial_x + \Big(y + \frac{b_2}{2}\, t^2\Big)\partial_y + t\partial_t + b_2 t\partial_v - \Big(b_2 y + \frac{b_2^2}{2}\, t^2 + a_1\Big)\partial_\eta\,.$$

1.3.3. $\mathbf{h = H(x) + b_1 y^2 + b_2 y,\ H''' \neq 0,\ b_1 < 0.}$
In this case we obtain the following solution

$$\begin{aligned}
&B = 0\,, \qquad C = 0\,, \qquad a = C_1\,, \qquad k = 0\,,\\
&l = C_2 \sin(t\sqrt{-2b_1}) + C_3 \cos(t\sqrt{-2b_1})\,,\\
&f = -2b_2\Big(-\frac{C_2}{\sqrt{-2b_1}}\, \cos(t\sqrt{-2b_1}) + \frac{C_3}{\sqrt{-2b_1}}\, \sin(t\sqrt{-2b_1})\Big) + C_4\,.
\end{aligned}$$

Thus, the kernel of symmetry operators is extended by the operators

$$\begin{aligned}
X_2 &= \sin(t\sqrt{-2b_1})\partial_y + \sqrt{-2b_1}\cos(t\sqrt{-2b_1})\partial_v - (2b_1 y + b_2)\sin(t\sqrt{-2b_1})\partial_\eta\,,\\
X_3 &= \cos(t\sqrt{-2b_1})\partial_y - \sqrt{-2b_1}\sin(t\sqrt{-2b_1})\partial_v - (2b_1 y + b_2)\cos(t\sqrt{-2b_1})\partial_\eta\,.
\end{aligned}$$

1.3.3.1.

$$\mathbf{h = \frac{a_1}{(x + a_2)^2} + b_1((x + a_2)^2 + y^2) + b_2 y + b_3,\ b_1 < 0,\ a_1 \neq 0.}$$

In this case we obtain additional solution to the solution from the case 1.3.3

$$\begin{aligned}
&B = 0\,, \qquad C = 0\,, \qquad a = C_5 \sin(t\sqrt{-8b_1}) + C_6 \cos(t\sqrt{-8b_1})\,,\\
&k = -a_2\sqrt{-2b_1}(C_5 \cos(t\sqrt{-8b_1}) - C_6 \sin(t\sqrt{-8b_1}))\,,\\
&l = \frac{b_2}{\sqrt{-2b_1}}(C_5 \cos(t\sqrt{-8b_1}) - C_6 \sin(t\sqrt{-8b_1}))\,,\\
&f = \frac{4b_1(b_3 + 2a_2^2 b_1) + b_2^2}{2b_1}(C_5 \sin(t\sqrt{-8b_1}) + C_6 \cos(t\sqrt{-8b_1}))\,.
\end{aligned}$$

Then the kernel of symmetry operators is extended by the operators X_2, X_3 from the case 1.3.3 and by the additional symmetry operators

$$X_4 = -\cos(t\sqrt{-8b_1})\Big(\sqrt{-8b_1}(x+a_2)\partial_x + \Big(\sqrt{-8b_1}y - \frac{2b_2}{\sqrt{-2b_1}}\Big)\partial_y\Big) -$$
$$-2\sin(t\sqrt{-8b_1})\partial_t + (\sqrt{-8b_1}\cos(t\sqrt{-8b_1})u - 8b_1\sin(t\sqrt{-8b_1})(x+a_2))\partial_u +$$
$$+(\sqrt{-8b_1}\cos(t\sqrt{-8b_1})v - 4\sin(t\sqrt{-8b_1})(2b_1y+b_2))\partial_v +$$
$$+\sqrt{-8b_1}\cos(t\sqrt{-8b_1})\Big(2\eta + 4b_1((x+a_2)^2+y^2) + 4b_2y + 2b_3 + \frac{b_2^2}{2b_1}\Big)\partial_\eta\,,$$
$$X_5 = \sqrt{-8b_1}\sin(t\sqrt{-8b_1})(x+a_2)\partial_x + \sin(t\sqrt{-8b_1})\Big(\sqrt{-8b_1}y - \frac{2b_2}{\sqrt{-2b_1}}\Big)\partial_y -$$
$$-2\cos(t\sqrt{-8b_1})\partial_t - (\sqrt{-8b_1}\sin(t\sqrt{-8b_1})u + 8b_1\cos(t\sqrt{-8b_1})(x+a_2))\partial_u -$$
$$-(\sqrt{-8b_1}\sin(t\sqrt{-8b_1})v + 4\cos(t\sqrt{-8b_1})(2b_1y+b_2))\partial_v -$$
$$-\sqrt{-8b_1}\sin(t\sqrt{-8b_1})\Big(2\eta + 4b_1((x+a_2)^2+y^2) + 4b_2y + 2b_3 + \frac{b_2^2}{2b_1}\Big)\partial_\eta\,.$$

2.

$$\mathbf{h = H(y + b_0x) + \frac{b_1}{b_0^2+1}(-x + b_0y)^2 + \frac{b_2}{b_0^2+1}(-x + b_0y),\ H''' \neq 0.}$$

2.1.

$$\mathbf{h = H(y + b_0x) + \frac{b_1}{b_0^2+1}(-x + b_0y)^2 + \frac{b_2}{b_0^2+1}(-x + b_0y),\ H''' \neq 0,\ b_1 > 0.}$$

In this case we obtain the following solution

$$B = 0\,, \qquad C = 0\,, \qquad a = C_1\,, \qquad k = -\frac{1}{b_0^2+1}(C_2e^{t\sqrt{2b_1}} + C_3e^{-t\sqrt{2b_1}})\,,$$
$$l = \frac{b_0}{b_0^2+1}(C_2e^{t\sqrt{2b_1}} + C_3e^{-t\sqrt{2b_1}})\,,$$
$$f = -\frac{2b_2}{b_0^2+1}\Big(\frac{C_2}{\sqrt{2b_1}}e^{t\sqrt{2b_1}} - \frac{C_3}{\sqrt{2b_1}}e^{-t\sqrt{2b_1}}\Big) + C_4\,.$$

Thus, the kernel of symmetry operators is extended by the operators

$$X_2 = e^{t\sqrt{2b_1}}(-\partial_x + b_0\partial_y - \sqrt{2b_1}\partial_u + b_0\sqrt{2b_1}\partial_v + (2b_1x - 2b_0b_1y - b_2)\partial_\eta)\,,$$
$$X_3 = e^{-t\sqrt{2b_1}}(-\partial_x + b_0\partial_y + \sqrt{2b_1}\partial_u - b_0\sqrt{2b_1}\partial_v + (2b_1x - 2b_0b_1y - b_2)\partial_\eta)\,.$$

2.1.1.

$$\mathbf{h} = \frac{1}{\mathbf{b_0^2+1}}\Big(\frac{\mathbf{a_1}}{(\mathbf{y}+\mathbf{b_0x}+\mathbf{a_2})^2}+\mathbf{b_1}((\mathbf{y}+\mathbf{b_0x}+\mathbf{a_2})^2+(\mathbf{b_0y}-\mathbf{x})^2)+\mathbf{b_2}(\mathbf{b_0y}-\mathbf{x})+\mathbf{b_3}\Big),$$

$\mathbf{b_1 > 0,\ a_1 \neq 0.}$

Denote

$$S_1 = \frac{1}{b_0^2+1}\Big(\frac{b_2}{2b_1} - b_0a_2\Big), \qquad S_2 = \frac{1}{b_0^2+1}\Big(-\frac{b_0b_2}{2b_1} - a_2\Big).$$

In this case we obtain additional solution to the solution from the case 2.1

$$B = 0, \qquad C = 0, \qquad a = C_5e^{t\sqrt{8b_1}} + C_6e^{-t\sqrt{8b_1}},$$
$$k = S_1\sqrt{2b_1}(C_5e^{t\sqrt{8b_1}} - C_6e^{-t\sqrt{8b_1}}), \qquad l = S_2\sqrt{2b_1}(C_5e^{t\sqrt{8b_1}} - C_6e^{-t\sqrt{8b_1}}),$$
$$f = \frac{4b_1(b_3+2a_2^2b_1)+b_2^2}{2b_1(b_0^2+1)}(C_5e^{t\sqrt{8b_1}} + C_6e^{-t\sqrt{8b_1}}).$$

Then the kernel of symmetry operators is extended by the operators X_2, X_3 from the case 2.1 and by the additional symmetry operators

$$X_4 = e^{t\sqrt{8b_1}}\Big(\sqrt{2b_1}(S_1-x)\partial_x - \sqrt{2b_1}(y-S_2)\partial_y - \partial_t + (\sqrt{2b_1}u - 4b_1(x-S_1))\partial_u +$$
$$+ (\sqrt{2b_1}v - 4b_1(y-S_2))\partial_v + \frac{\sqrt{2b_1}}{b_0^2+1}\Big(2(b_0^2+1)\eta +$$
$$+ 4b_1((y+b_0x+a_2)^2+(b_0y-x)^2) + 4b_2(b_0y-x) + 2b_3 + \frac{b_2^2}{2b_1}\Big)\partial_\eta\Big),$$
$$X_5 = e^{-t\sqrt{8b_1}}\Big(\sqrt{2b_1}(x-S_1)\partial_x + \sqrt{2b_1}(y-S_2)\partial_y - \partial_t - (\sqrt{2b_1}u + 4b_1(x-S_1))\partial_u -$$
$$- (\sqrt{2b_1}v + 4b_1(y-S_2))\partial_v - \frac{\sqrt{2b_1}}{b_0^2+1}\Big(2(b_0^2+1)\eta +$$
$$+ 4b_1((y+b_0x+a_2)^2+(b_0y-x)^2)+4b_2(b_0y-x) + 2b_3+\frac{b_2^2}{2b_1}\Big)\partial_\eta\Big).$$

2.2.

$$\mathbf{h} = \mathbf{H}(\mathbf{y}+\mathbf{b_0x}) + \frac{\mathbf{b_2}}{\mathbf{b_0^2+1}}(-\mathbf{x}+\mathbf{b_0y}),\ \mathbf{H'''} \neq \mathbf{0}.$$

In this case we obtain the following solution

$$B = 0, \qquad C = 0, \qquad a = C_1, \qquad k = -\frac{1}{b_0^2+1}(C_2t + C_3),$$

$$l = \frac{b_0}{b_0^2+1}(C_2 t + C_3)\,, \qquad f = \frac{1}{b_0^2+1}(-b_2 C_2 t^2 - 2b_2 C_3 t + C_4)\,.$$

Thus, the kernel of symmetry operators is extended by the operators

$$X_2 = -t\partial_x + b_0 t\partial_y - \partial_u + b_0\partial_v - b_2 t\partial_\eta\,, \qquad X_3 = -\partial_x + b_0\partial_y - b_2\partial_\eta\,.$$

2.2.1.

$$\mathbf{h} = \frac{\mathbf{1}}{\mathbf{b_0^2+1}}\Big(\frac{\mathbf{a_1}}{(\mathbf{y}+\mathbf{b_0 x}+\mathbf{a_2})^2} + \mathbf{b_2}(-\mathbf{x}+\mathbf{b_0 y}) + \mathbf{b_3}\Big),\ \mathbf{a_1} \neq \mathbf{0}.$$

In this case we obtain additional solution to the solution from the case 2.2

$$B = 0\,, \qquad C = 0\,, \qquad a = C_5 t^2 + C_6 t\,,$$

$$k = \frac{1}{b_0^2+1}\Big(-\frac{a_2 b_0}{2}(2C_5 t + C_6) + \frac{3b_2}{2}\Big(\frac{C_5}{3}t^3 + \frac{C_6}{2}t^2\Big)\Big)\,,$$

$$l = \frac{1}{b_0^2+1}\Big(-\frac{a_2}{2}(2C_5 t + C_6) - \frac{3b_0 b_2}{2}\Big(\frac{C_5}{3}t^3 + \frac{C_6}{2}t^2\Big)\Big)\,,$$

$$f = \frac{1}{b_0^2+1}\Big(\frac{b_2^2 C_5}{4}t^4 + \frac{b_2^2 C_6}{2}t^3 + 2b_3 C_5 t^2 + 2b_3 C_6 t\Big)\,.$$

Then the kernel of symmetry operators is extended by the operators X_2, X_3 from the case 2.2 and by the additional symmetry operators

$$X_4 = t\Big(\frac{1}{b_0^2+1}(b_2 t^2 - 2b_0 a_2) - 2x\Big)\partial_x - t\Big(2y + \frac{1}{b_0^2+1}(b_0 b_2 t^2 + 2a_2)\Big)\partial_y - 2t^2\partial_t +$$
$$+\Big(2tu - 2x + \frac{1}{b_0^2+1}(3b_2 t^2 - 2b_0 a_2)\Big)\partial_u + \Big(2tv - 2y - \frac{1}{b_0^2+1}(3b_0 b_2 t^2 + 2a_2)\Big)\partial_v +$$
$$+t\Big(4\eta + \frac{6b_2}{b_0^2+1}(-x + b_0 y) + \frac{b_2^2}{b_0^2+1}t^2 + \frac{4b_3}{b_0^2+1}\Big)\partial_\eta\,,$$

$$X_5 = \Big(\frac{1}{b_0^2+1}(3b_2 t^2 - 2b_0 a_2) - 2x\Big)\partial_x - \Big(2y + \frac{1}{b_0^2+1}(3b_0 b_2 t^2 + 2a_2)\Big)\partial_y - 4t\partial_t +$$
$$+\Big(2u + \frac{6b_2}{b_0^2+1}t\Big)\partial_u + \Big(2v - \frac{6b_0 b_2}{b_0^2+1}t\Big)\partial_v +$$
$$+\Big(4\eta + \frac{6b_2}{b_0^2+1}(-x + b_0 y) + \frac{3b_2^2}{b_0^2+1}t^2 + \frac{4b_3}{b_0^2+1}\Big)\partial_\eta\,.$$

2.2.2.

$$\mathbf{h} = \frac{\mathbf{1}}{\mathbf{b_0^2+1}}(\mathbf{a_1(y+b_0x+a_2)^{a_3}+b_2(b_0y-x)+b_3}),\ \mathbf{a_1 \neq 0},\ \mathbf{a_3} \in \mathbb{Z} \setminus \{\mathbf{-2, 0, 1, 2}\}$$

$$\text{or } \mathbf{h} = \frac{\mathbf{1}}{\mathbf{b_0^2+1}}(\mathbf{a_1|y+b_0x+a_2|^{a_3}+b_2(-x+b_0y)+b_3}),\ \mathbf{a_1 \neq 0},\ \mathbf{a_3 \neq -2, 0, 1, 2}.$$

In this case we obtain additional solution to the solution from the case 2.2

$$B = 0\,, \qquad a = -\frac{4C}{a_3+2}\,t\,, \qquad k = \frac{1}{b_0^2+1}\Big(\frac{2b_0a_2C}{a_3+2} - \frac{(1-a_3)b_2C}{a_3+2}\,t^2\Big),$$

$$l = \frac{1}{b_0^2+1}\Big(\frac{2a_2C}{a_3+2} + \frac{(1-a_3)b_0b_2C}{a_3+2}\,t^2\Big),$$

$$f = -\frac{2(1-a_3)b_2^2C}{3(a_3+2)(b_0^2+1)}\,t^3 + \frac{4a_3b_3C}{(a_3+2)(b_0^2+1)}\,t\,.$$

Then the kernel of symmetry operators is extended by the operators X_2, X_3 from the case 2.2 and by the additional symmetry operator

$$X_4 = \Big(2x + \frac{2b_0a_2 - (1-a_3)b_2t^2}{b_0^2+1}\Big)\partial_x + \Big(2y + \frac{2a_2 + (1-a_3)b_0b_2t^2}{b_0^2+1}\Big)\partial_y -$$
$$- (a_3-2)t\,\partial_t + \Big(a_3u + \frac{2(a_3-1)b_2}{b_0^2+1}\,t\Big)\partial_u + \Big(a_3v - \frac{2(a_3-1)b_0b_2}{b_0^2+1}\,t\Big)\partial_v +$$
$$+ \Big(2a_3\eta - \frac{(1-a_3)}{b_0^2+1}(2b_2(-x+b_0y) + b_2^2t^2) + \frac{2a_3b_3}{b_0^2+1}\Big)\partial_\eta\,.$$

2.2.3.

$$\mathbf{h} = \frac{\mathbf{1}}{\mathbf{b_0^2+1}}(\mathbf{a_1e^{a_2(y+b_0x)}+b_2(-x+b_0y)+b_3}),\ \mathbf{a_1 \neq 0},\ \mathbf{a_2 \neq 0}.$$

In this case we obtain additional solution to the solution from the case 2.2

$$B = 0\,, \quad a = 0\,, \quad k = \frac{1}{b_0^2+1}\Big(\frac{2b_0C}{a_2} + b_2Ct^2\Big), \quad l = \frac{1}{b_0^2+1}\Big(\frac{2C}{a_2} - b_0b_2Ct^2\Big),$$

$$f = \frac{1}{b_0^2+1}\Big(\frac{2b_2^2C}{3}\,t^3 + 4b_3Ct\Big).$$

Then the kernel of symmetry operators is extended by the operators X_2, X_3 from the case 2.2 and by the additional symmetry operator

$$X_4 = \frac{1}{b_0^2+1}\Big(\frac{2b_0}{a_2} + b_2 t^2\Big)\,\partial_x + \frac{1}{b_0^2+1}\Big(\frac{2}{a_2} - b_0 b_2 t^2\Big)\partial_y - t\partial_t + \Big(u + \frac{2b_2}{b_0^2+1}\,t\Big)\partial_u +$$
$$+\Big(v - \frac{2b_0 b_2}{b_0^2+1}\,t\Big)\partial_v + \Big(2\eta + \frac{2b_2}{b_0^2+1}(-x+b_0 y) + \frac{b_2^2}{b_0^2+1}\,t^2 + \frac{2b_3}{b_0^2+1}\Big)\partial_\eta\,.$$

2.2.4.

$$\mathbf{h} = \frac{\mathbf{1}}{\mathbf{b_0^2+1}}(\mathbf{a_1}\ln|\mathbf{y}+\mathbf{b_0 x}+\mathbf{a_2}| + \mathbf{b_2}(-\mathbf{x}+\mathbf{b_0 y}) + \mathbf{b_3}),\ \mathbf{a_1} \neq \mathbf{0}.$$

In this case we obtain additional solution to the solution from the case 2.2

$$B = 0\,, \qquad a = -2Ct\,, \qquad k = \frac{1}{b_0^2+1}\Big(b_0 a_2 C - \frac{b_2 C}{2}\,t^2\Big),$$
$$l = \frac{1}{b_0^2+1}\Big(a_2 C + \frac{b_0 b_2 C}{2}\,t^2\Big), \qquad f = \frac{1}{b_0^2+1}\Big(-\frac{b_2^2 C}{3}\,t^3 - 2a_1 C t\Big).$$

Then the kernel of symmetry operators is extended by the operators X_2, X_3 from the case 2.2 and by the additional symmetry operator

$$X_4 = \Big(x + \frac{1}{b_0^2+1}\Big(b_0 a_2 - \frac{b_2}{2}\,t^2\Big)\Big)\partial_x + \Big(y + \frac{1}{b_0^2+1}\Big(a_2 + \frac{b_0 b_2}{2}\,t^2\Big)\Big)\partial_y + t\partial_t -$$
$$-\frac{b_2}{b_0^2+1}\,t\partial_u + \frac{b_0 b_2}{b_0^2+1}\,t\partial_v - \frac{1}{b_0^2+1}\Big(b_2(-x+b_0 x) + \frac{b_2^2}{2}\,t^2 + a_1\Big)\partial_\eta\,.$$

2.3.

$$\mathbf{h} = \mathbf{H}(\mathbf{y}+\mathbf{b_0 x}) + \frac{\mathbf{b_1}}{\mathbf{b_0^2+1}}(-\mathbf{x}+\mathbf{b_0 y})^2 + \frac{\mathbf{b_2}}{\mathbf{b_0^2+1}}(-\mathbf{x}+\mathbf{b_0 y}),\ \mathbf{H}''' \neq \mathbf{0},\ \mathbf{b_1} < \mathbf{0}.$$

In this case we obtain the following solution

$$B = 0\,, \qquad C = 0\,, \qquad a = C_1\,,$$
$$k = -\frac{1}{b_0^2+1}(C_2\sin(t\sqrt{-2b_1}) + C_3\cos(t\sqrt{-2b_1}))\,,$$
$$l = \frac{b_0}{b_0^2+1}(C_2\sin(t\sqrt{-2b_1}) + C_3\cos(t\sqrt{-2b_1}))\,,$$
$$f = -\frac{2b_2}{b_0^2+1}\Big(-\frac{C_2}{\sqrt{-2b_1}}\cos(t\sqrt{-2b_1}) + \frac{C_3}{\sqrt{-2b_1}}\sin(t\sqrt{-2b_1})\Big) + C_4\,.$$

Thus, the kernel of symmetry operators is extended by the operators

$$X_2 = -\sin(t\sqrt{-2b_1})\partial_x + b_0\sin(t\sqrt{-2b_1})\partial_y - \sqrt{-2b_1}\cos(t\sqrt{-2b_1})\partial_u + \\ + b_0\sqrt{-2b_1}\cos(t\sqrt{-2b_1})\partial_v - (2b_1(-x+b_0y)+b_2)\sin(t\sqrt{-2b_1})\partial_\eta\,,$$

$$X_3 = -\cos(t\sqrt{-2b_1})\partial_x + b_0\cos(t\sqrt{-2b_1})\partial_y + \sqrt{-2b_1}\sin(t\sqrt{-2b_1})\partial_u - \\ - b_0\sqrt{-2b_1}\sin(t\sqrt{-2b_1})\partial_v - (2b_1(-x+b_0y)+b_2)\cos(t\sqrt{-2b_1})\partial_\eta\,.$$

2.3.1.

$$\mathbf{h} = \frac{1}{\mathbf{b_0^2+1}}\Big(\frac{\mathbf{a_1}}{(\mathbf{y+b_0x+a_2})^2} + \mathbf{b_1}((\mathbf{y+b_0x+a_2})^2 + (\mathbf{x-b_0y})^2) + \mathbf{b_2}(\mathbf{b_0y-x}) + \mathbf{b_3}\Big),$$

$\mathbf{b_1 < 0,\ a_1 \neq 0.}$

Denote

$$S_1 = \frac{1}{b_0^2+1}\Big(\frac{b_2}{2b_1} - b_0a_2\Big), \qquad S_2 = \frac{1}{b_0^2+1}\Big(-\frac{b_0b_2}{2b_1} - a_2\Big).$$

In this case we obtain additional solution to the solution from the case 2.3

$$B = 0, \qquad C = 0, \qquad a = C_5\sin(t\sqrt{-8b_1}) + C_6\cos(t\sqrt{-8b_1}),$$
$$k = S_1\sqrt{-2b_1}(C_5\cos(t\sqrt{-8b_1}) - C_6\sin(t\sqrt{-8b_1}))\,,$$
$$l = S_2\sqrt{-2b_1}(C_5\cos(t\sqrt{-8b_1}) - C_6\sin(t\sqrt{-8b_1}))\,,$$
$$f = \frac{4b_1(b_3+2a_2^2b_1)+b_2^2}{2b_1(b_0^2+1)}(C_5\sin(t\sqrt{-8b_1}) + C_6\cos(t\sqrt{-8b_1}))\,.$$

Then the kernel of symmetry operators is extended by the operators X_2, X_3 from the case 2.3 and by the additional symmetry operators

$$X_4 = -\sqrt{-2b_1}\cos(t\sqrt{-8b_1})(x-S_1)\partial_x - \sqrt{-2b_1}\cos(t\sqrt{-8b_1})(y-S_2)\partial_y - \\ - \sin(t\sqrt{-8b_1})\partial_t + (\sqrt{-2b_1}\cos(t\sqrt{-8b_1})u - 4b_1\sin(t\sqrt{-8b_1})(x-S_1))\partial_u + \\ + (\sqrt{-2b_1}\cos(t\sqrt{-8b_1})v - 4b_1\sin(t\sqrt{-8b_1})(y-S_2))\partial_v + \\ + \frac{\sqrt{-2b_1}}{b_0^2+1}\cos(t\sqrt{-8b_1})\Big(2(b_0^2+1)\eta + 4b_1((y+b_0x+a_2)^2 + (b_0y-x)^2) + \\ + 4b_2(b_0y-x) + 2b_3 + \frac{b_2^2}{2b_1}\Big)\partial_\eta\,,$$

$$X_5 = \sqrt{-2b_1}\sin(t\sqrt{-8b_1})(x - S_1)\partial_x + \sqrt{-2b_1}\sin(t\sqrt{-8b_1})(y - S_2)\partial_y -$$
$$- \cos(t\sqrt{-8b_1})\partial_t - (\sqrt{-2b_1}\sin(t\sqrt{-8b_1})u + 4b_1\cos(t\sqrt{-8b_1})(x - S_1))\partial_u -$$
$$- (\sqrt{-2b_1}\sin(t\sqrt{-8b_1})v + 4b_1\cos(t\sqrt{-8b_1})(y - S_2))\partial_v -$$
$$- \frac{\sqrt{-2b_1}}{b_0^2+1}\sin(t\sqrt{-8b_1})\Big(2(b_0^2+1)\eta + 4b_1((y + b_0x + a_2)^2 + (b_0y - x)^2) +$$
$$+ 4b_2(b_0y - x) + 2b_3 + \frac{b_2^2}{2b_1}\Big)\partial_\eta .$$

3.

$$\mathbf{h = \frac{1}{(x+a_1)^2} H\Big(\frac{x+a_1}{y+a_2}\Big) + a_3((x+a_1)^2 + (y+a_2)^2) + a_4,\ H \neq 0.}$$

3.1.

$$\mathbf{h = \frac{1}{(x+a_1)^2} H\Big(\frac{x+a_1}{y+a_2}\Big) + a_3((x+a_1)^2 + (y+a_2)^2) + a_4,\ H \neq 0,\ a_3 > 0.}$$

In this case we obtain the following solution

$$B = 0, \quad C = 0, \quad a = C_1e^{t\sqrt{8a_3}} + C_2e^{-t\sqrt{8a_3}} + C_3,$$
$$k = -\frac{a_1\sqrt{8a_3}}{2}(C_1e^{t\sqrt{8a_3}} - C_2e^{-t\sqrt{8a_3}}), \qquad l = -\frac{a_2\sqrt{8a_3}}{2}(C_1e^{t\sqrt{8a_3}} - C_2e^{-t\sqrt{8a_3}}),$$
$$f = 2(2a_3(a_1^2 + a_2^2) + a_4)(C_1e^{t\sqrt{8a_3}} + C_2e^{-t\sqrt{8a_3}}) + C_4 .$$

Thus, the kernel of symmetry operators is extended by the operators

$$X_2 = e^{t\sqrt{8a_3}}\Big(-\sqrt{8a_3}(x+a_1)\partial_x - \sqrt{8a_3}(y+a_2)\partial_y - 2\partial_t + (\sqrt{8a_3}u - 8a_3(x+a_1))\partial_u +$$
$$+ (\sqrt{8a_3}v - 8a_3(y+a_2))\partial_v + 2\sqrt{8a_3}(\eta + 2a_3((x+a_1)^2 + (y+a_2)^2) + a_4)\partial_\eta\Big),$$
$$X_3 = e^{-t\sqrt{8a_3}}\Big(\sqrt{8a_3}(x+a_1)\partial_x + \sqrt{8a_3}(y+a_2)\partial_y - 2\partial_t - (\sqrt{8a_3}u + 8a_3(x+a_1))\partial_u -$$
$$- (\sqrt{8a_3}v + 8a_3(y+a_2))\partial_v - 2\sqrt{8a_3}(\eta + 2a_3((x+a_1)^2 + (y+a_2)^2) + a_4)\partial_\eta\Big).$$

3.1.1.

$$\mathbf{h = \frac{b_1\exp\Big(b_2\ \mathrm{arctg}\Big(\frac{x+a_1}{y+a_2}\Big)\Big)}{(x+a_1)^2 + (y+a_2)^2} + a_3((x+a_1)^2 + (y+a_2)^2) + a_4,\ a_3 > 0,\ b_1 \neq 0.}$$

In this case we obtain additional solution to the solution from the case 3.1

$$C = -\frac{b_2}{2} B\,, \qquad a = \frac{b_2 B}{2} t\,, \qquad k = -\Big(a_2 + \frac{a_1 b_2}{4}\Big) B\,, \qquad l = \Big(a_1 - \frac{a_2 b_2}{4}\Big) B\,,$$
$$f = -a_4 b_2 B t\,.$$

Then the kernel of symmetry operators is extended by the operators X_2, X_3 from the case 3.1 and by the additional symmetry operator

$$X_4 = -2\Big(y + a_2 + \frac{(x+a_1)b_2}{4}\Big)\partial_x + 2\Big(x + a_1 - \frac{(y+a_2)b_2}{4}\Big)\partial_y - \\ - \Big(\frac{b_2}{2} u + 2v\Big)\partial_u + \Big(-\frac{b_2}{2} v + 2u\Big)\partial_v - b_2(\eta + a_4)\partial_\eta\,.$$

3.2.

$$\mathbf{h} = \frac{\mathbf{1}}{(\mathbf{x}+\mathbf{a_1})^2}\,\mathbf{H}\Big(\frac{\mathbf{x}+\mathbf{a_1}}{\mathbf{y}+\mathbf{a_2}}\Big) + \mathbf{a_4},\ \mathbf{H} \neq \mathbf{0}.$$

In this case we obtain the following solution

$$B = 0\,, \qquad C = 0\,, \qquad a = C_1 t^2 + C_2 t + C_3\,, \qquad k = -\frac{a_1}{2}(2C_1 t + C_2)\,,$$
$$l = -\frac{a_2}{2}(2C_1 t + C_2)\,, \qquad f = 2a_4(C_1 t^2 + C_2 t) + C_4\,.$$

Thus, the kernel of symmetry operators is extended by the operators

$$X_2 = -t(x+a_1)\partial_x - t(y+a_2)\partial_y - t^2\partial_t + (tu - (x+a_1))\partial_u + \\ + (tv - (y+a_2))\partial_v + 2t(\eta + a_4)\partial_\eta\,,$$
$$X_3 = -(x+a_1)\partial_x - (y+a_2)\partial_y - 2t\partial_t + u\partial_u + v\partial_v + 2(\eta + a_4)\partial_\eta\,.$$

3.2.1.

$$\mathbf{h} = \frac{\mathbf{b_1}\exp\Big(\mathbf{b_2}\ \mathrm{arctg}\Big(\frac{\mathbf{x}+\mathbf{a_1}}{\mathbf{y}+\mathbf{a_2}}\Big)\Big)}{(\mathbf{x}+\mathbf{a_1})^2 + (\mathbf{y}+\mathbf{a_2})^2} + \mathbf{a_4},\ \mathbf{b_1} \neq \mathbf{0}.$$

In this case we obtain additional solution to the solution from the case 3.2

$$C = -\frac{b_2}{2} B\,, \qquad a = 0\,, \qquad k = -a_2 B\,, \qquad l = a_1 B\,, \qquad f = -2a_4 b_2 B t\,.$$

Then the kernel of symmetry operators is extended by the operators X_2, X_3 from the case 3.2 and by the additional symmetry operator

$$X_4 = -2(y + a_2)\partial_x + 2(x + a_1)\partial_y + b_2 t\partial_t - (b_2 u + 2v)\partial_u + (2u - b_2 v)\partial_v - \\ - 2b_2(\eta + a_4)\partial_\eta .$$

3.3.

$$\mathbf{h} = \frac{\mathbf{1}}{(\mathbf{x}+\mathbf{a_1})^2}\mathbf{H}\Big(\frac{\mathbf{x}+\mathbf{a_1}}{\mathbf{y}+\mathbf{a_2}}\Big) + \mathbf{a_3}((\mathbf{x}+\mathbf{a_1})^2 + (\mathbf{y}+\mathbf{a_2})^2) + \mathbf{a_4},\ \mathbf{H} \neq \mathbf{0},\ \mathbf{a_3} < \mathbf{0}.$$

In this case we obtain the following solution

$$B = 0, \quad C = 0, \quad a = C_1 \sin(t\sqrt{-8a_3}) + C_2 \cos(t\sqrt{-8a_3}) + C_3 ,$$
$$k = -\frac{a_1\sqrt{-8a_3}}{2}(C_1 \cos(t\sqrt{-8a_3}) - C_2 \sin(t\sqrt{-8a_3})) ,$$
$$l = -\frac{a_2\sqrt{-8a_3}}{2}(C_1 \cos(t\sqrt{-8a_3}) - C_2 \sin(t\sqrt{-8a_3})) ,$$
$$f = 2(2a_3(a_1^2 + a_2^2) + a_4)(C_1 \sin(t\sqrt{-8a_3}) + C_2 \cos(t\sqrt{-8a_3})) + C_4 .$$

Thus, the kernel of symmetry operators is extended by the operators

$$\begin{aligned} X_2 = &-\sqrt{-8a_3}\cos(t\sqrt{-8a_3})(x + a_1)\partial_x - \sqrt{-8a_3}\cos(t\sqrt{-8a_3})(y + a_2)\partial_y - \\ &- 2\sin(t\sqrt{-8a_3})\partial_t + (\sqrt{-8a_3}\cos(t\sqrt{-8a_3})u - 8a_3 \sin(t\sqrt{-8a_3})(x + a_1))\partial_u + \\ &+ (\sqrt{-8a_3}\cos(t\sqrt{-8a_3})v - 8a_3 \sin(t\sqrt{-8a_3})(y + a_2))\partial_v + \\ &+ 2\sqrt{-8a_3}\cos(t\sqrt{-8a_3})(\eta + 2a_3((x + a_1)^2 + (y + a_2)^2) + a_4)\partial_\eta , \\ X_3 = &\sqrt{-8a_3}\sin(t\sqrt{-8a_3})(x + a_1)\partial_x + \sqrt{-8a_3}\sin(t\sqrt{-8a_3})(y + a_2)\partial_y - \\ &- 2\cos(t\sqrt{-8a_3})\partial_t - (\sqrt{-8a_3}\sin(t\sqrt{-8a_3})u + 8a_3 \cos(t\sqrt{-8a_3})(x + a_1))\partial_u - \\ &- (\sqrt{-8a_3}\sin(t\sqrt{-8a_3})v + 8a_3 \cos(t\sqrt{-8a_3})(y + a_2))\partial_v - \\ &- 2\sqrt{-8a_3}\sin(t\sqrt{-8a_3})(\eta + 2a_3((x + a_1)^2 + (y + a_2)^2) + a_4)\partial_\eta . \end{aligned}$$

3.3.1.

$$\mathbf{h} = \frac{\mathbf{b_1}\exp\Big(\mathbf{b_2}\ \mathrm{arctg}\Big(\frac{\mathbf{x}+\mathbf{a_1}}{\mathbf{y}+\mathbf{a_2}}\Big)\Big)}{(\mathbf{x}+\mathbf{a_1})^2 + (\mathbf{y}+\mathbf{a_2})^2} + \mathbf{a_3}((\mathbf{x}+\mathbf{a_1})^2 + (\mathbf{y}+\mathbf{a_2})^2) + \mathbf{a_4},\ \mathbf{a_3} < \mathbf{0},\ \mathbf{b_1} \neq \mathbf{0}.$$

In this case we obtain additional solution to the solution from the case 3.3

$$C = -\frac{b_2}{2}\,B\,, \qquad a = \frac{b_2 B}{2}\,t\,, \qquad k = -\Big(a_2 + \frac{a_1 b_2}{4}\Big)B\,,$$
$$l = \Big(a_1 - \frac{a_2 b_2}{4}\Big)B\,, \qquad f = -a_4 b_2 B t\,.$$

Then the kernel of symmetry operators is extended by the operators X_2, X_3 from the case 3.3 and by the additional symmetry operator

$$X_4 = -2\Big(y + a_2 + \frac{(x+a_1)b_2}{4}\Big)\partial_x + 2\Big(x + a_1 - \frac{(y+a_2)b_2}{4}\Big)\partial_y - \\ -\Big(\frac{b_2}{2}\,u + 2v\Big)\partial_u + \Big(-\frac{b_2}{2}\,v + 2u\Big)\partial_v - b_2(\eta + a_4)\partial_\eta\,.$$

4.

$$\mathbf{h} = \mathbf{H}\Big(\ln\big((\mathbf{x}+\mathbf{b_1})^2 + (\mathbf{y}+\mathbf{b_2})^2\big) + \mathbf{b_3}\,\mathrm{arctg}\Big(\frac{\mathbf{y}+\mathbf{b_2}}{\mathbf{x}+\mathbf{b_1}}\Big)\Big)\exp\Big(\mathbf{b_4}\,\mathrm{arctg}\Big(\frac{\mathbf{y}+\mathbf{b_2}}{\mathbf{x}+\mathbf{b_1}}\Big)\Big) + \mathbf{b_5}\,,$$

$\mathbf{b_3} \neq \mathbf{0}$, $\mathbf{b_4} \neq \mathbf{0}$, except cases 1–3.

In this case we obtain the following solution

$$C = \frac{b_4 - b_3}{2}\,B\,, \qquad a = b_3 B t + C_1\,, \qquad k = -\frac{2b_2 + b_1 b_3}{2}\,B\,,$$
$$l = \frac{2b_1 - b_2 b_3}{2}\,B\,, \qquad f = 2b_4 b_5 B t + C_2\,.$$

Thus the kernel of symmetry operators is extended by the operator

$$X_2 = -\,(b_3(x+b_1) + 2(y+b_2))\partial_x - (b_3(y+b_2) - 2(x+b_1))\partial_y - \\ -\,(b_3 + b_4)t\,\partial_t + (b_4 u - 2v)\partial_u + (b_4 v + 2u)\partial_v + 2b_4(\eta + b_5)\partial_\eta\,.$$

4.1.

$$\mathbf{h} = \mathbf{a_1}\big((\mathbf{x}+\mathbf{b_1})^2 + (\mathbf{y}+\mathbf{b_2})^2\big)^{\mathbf{a_2}}\exp\Big(\mathbf{a_3}\,\mathrm{arctg}\Big(\frac{\mathbf{y}+\mathbf{b_2}}{\mathbf{x}+\mathbf{b_1}}\Big)\Big) + \mathbf{b_5}\,,$$
$$\mathbf{a_1} \neq \mathbf{0}, \qquad \mathbf{a_2} \neq -\mathbf{1}, \qquad \mathbf{a_2^2} + \mathbf{a_3^2} > \mathbf{0}\,.$$

In this case we obtain additional solution to the solution from the case 4 at $b_4 = a_3 - a_2 b_3$

$$B = 0\,, \qquad C = -\frac{1}{2}\,(1 + a_2)C_3\,, \qquad a = C_3 t\,, \qquad k = -\frac{b_1}{2}\,C_3\,,$$
$$l = -\frac{b_2}{2}\,C_3\,, \qquad f = -2a_2 b_5 C_3 t\,.$$

Then the kernel of symmetry operators is extended by the operators X_2, X_3 from the case 4 at $b_4 = a_3 - a_2 b_3$ and by the additional symmetry operator

$$X_3 = (x + b_1)\partial_x + (y + b_2)\partial_y + (1 - a_2)t\partial_t + a_2 u\partial_u + a_2 v\partial_v + 2a_2(\eta + b_5)\partial_\eta\,.$$

5.

$$\mathbf{h} = \mathbf{H}\Big(\ln\big((\mathbf{x} + \mathbf{b}_1)^2 + (\mathbf{y} + \mathbf{b}_2)^2\big) + \mathbf{b}_3 \operatorname{arctg}\Big(\frac{\mathbf{y} + \mathbf{b}_2}{\mathbf{x} + \mathbf{b}_1}\Big)\Big) + \mathbf{b}_4 \operatorname{arctg}\Big(\frac{\mathbf{y} + \mathbf{b}_2}{\mathbf{x} + \mathbf{b}_1}\Big),$$

$$\mathbf{b}_3 \neq \mathbf{0},\ \mathbf{b}_4 \neq \mathbf{0},$$

except cases 1–4.
In this case we obtain the following solution

$$C = -\frac{b_3}{2}B\,, \qquad a = b_3 Bt + C_1\,, \qquad k = -\Big(\frac{b_1 b_3}{2} + b_2\Big)B\,,$$
$$l = \Big(b_1 - \frac{b_2 b_3}{2}\Big)B\,, \qquad f = -2b_4 Bt + C_2\,.$$

Thus, the kernel of symmetry operators is extended by the operator

$$X_2 = (b_3(x + b_1) + 2(y + b_2))\partial_x + (b_3(y + b_2) - 2(x + b_1))\partial_y + b_3 t\partial_t + \\ + 2v\partial_u - 2u\partial_v + 2b_4\partial_\eta\,.$$

5.1.

$$\mathbf{h} = \mathbf{a}_1 \ln\big((\mathbf{x} + \mathbf{b}_1)^2 + (\mathbf{y} + \mathbf{b}_2)^2\big) + \mathbf{a}_3 \operatorname{arctg}\Big(\frac{\mathbf{y} + \mathbf{b}_2}{\mathbf{x} + \mathbf{b}_1}\Big) + \mathbf{a}_2, \qquad \mathbf{a}_1^2 + \mathbf{a}_3^2 > \mathbf{0}\,.$$

In this case we obtain additional solution to the solution from the case 5 at $b_4 = a_3 - a_1 b_3$

$$B = 0\,, \quad C = -\frac{1}{2}C_3\,, \quad a = C_3 t\,, \quad k = -\frac{b_1}{2}C_3\,, \quad l = -\frac{b_2}{2}C_3\,, \quad f = 2a_1 C_3 t\,.$$

Then the kernel of symmetry operators is extended by the operators X_2, X_3 from the case 5 at $b_4 = a_3 - a_1 b_3$ and by the additional symmetry operator

$$X_3 = (x + b_1)\partial_x + (y + b_2)\partial_y + t\partial_t - 2a_1\partial_\eta\,.$$

6. $\mathbf{h} = \mathbf{H}(\mathbf{x}) \exp(\mathbf{b}_1 \mathbf{y}) + \mathbf{b}_2,\ \mathbf{b}_1 \neq \mathbf{0}$, except cases 1–5.
In this case we obtain the following solution

$$B=0\,,\quad C=\frac{b_1}{2}\,C_3\,,\quad a=C_1\,,\quad k=0\,,\quad l=C_3\,,\quad f=2b_1b_2C_3t+C_2\,.$$

Thus, the kernel of symmetry operators is extended by the operator

$$X_2=2\partial_y-b_1t\partial_t+b_1u\partial_u+b_1v\partial_v+2b_1(\eta+b_2)\partial_\eta\,.$$

7. $\mathbf{h=H(y+b_1x)}\exp(\mathbf{b_2x})+\mathbf{b_3},\ \mathbf{b_2\neq 0}$, except cases 1–6.
In this case we obtain the following solution

$$B=0\,,\quad C=\frac{b_2}{2}\,C_3\,,\quad a=C_1\,,\quad k=C_3\,,\quad l=-b_1C_3\,,\quad f=2b_2b_3C_3t+C_2\,.$$

Thus, the kernel of symmetry operators is extended by the operator

$$X_2=2\partial_x-2b_1\partial_y-b_2t\partial_t+b_2u\partial_u+b_2v\partial_v+2b_2(\eta+b_3)\partial_\eta\,.$$

8.

$$\mathbf{h=H\Big(\frac{x+b_1}{y+b_2}\Big)(x+b_1)^{b_3}+b_4\,,\ b_3\in\mathbb{Z}\setminus\{-2,0\}}\ \text{ or}$$
$$\mathbf{h=H\Big(\frac{x+b_1}{y+b_2}\Big)|x+b_1|^{b_3}+b_4\,,\ b_3\neq-2,\ 0,}\ \text{ except cases 1–7.}$$

In this case we obtain the following solution

$$B=0\,,\qquad C=-\frac{2+b_3}{4}\,C_3\,,\qquad a=C_3t+C_1\,,\qquad k=-\frac{b_1}{2}\,C_3\,,$$
$$l=-\frac{b_2}{2}\,C_3\,,\qquad f=-b_3b_4C_3t+C_2\,.$$

Thus, the kernel of symmetry operators is extended by the operator

$$X_2=2(x+b_1)\partial_x+2(y+b_2)\partial_y+(2-b_3)t\partial_t+b_3u\partial_u+b_3v\partial_v+2b_3(\eta+b_4)\partial_\eta\,.$$

9.

$$\mathbf{h=H\Big(\frac{x+b_1}{y+b_2}\Big)+b_3\ln|x+b_1|\,,\ b_3\neq 0,}\ \text{ except cases 1–8.}$$

In this case we obtain the following solution

$$B=0\,,\qquad C=-\frac{1}{2}\,C_3\,,\qquad a=C_3t+C_1\,,\qquad k=-\frac{b_1}{2}\,C_3\,,$$
$$l=-\frac{b_2}{2}\,C_3\,,\qquad f=b_3C_3t+C_2\,.$$

Thus, the kernel of symmetry operators is extended by the operator

$$X_2 = (x + b_1)\partial_x + (y + b_2)\partial_y + t\partial_t - b_3\partial_\eta\,.$$

Remark 7.2 Analysis of group classification results shows the impossibility of linearizing of the two-dimensional shallow water system over uneven bottom by a point transformation in contrast to the system of equations of one-dimensional shallow water in cases of horizontal and inclined bottom profiles [1].

7.2.4 Nonlinearity of the System of Equations of Two-Dimensional Shallow Water Over Uneven Bottom

Analysis of group classification results shows the impossibility of linearizing of the two-dimensional shallow water system over uneven bottom by a point transformation in contrast to the system of equations of one-dimensional shallow water in cases of horizontal and inclined bottom profiles [1].

7.3 Conservation Laws of the Equations of Two-Dimensional Shallow Water Over Uneven Bottom

7.3.1 Determining System of Equations

By hydrodynamic conservation laws of the system of Eq. (7.1) we shall mean divergent expressions of the form

$$D_x(P) + D_y(Q) + D_t(R)\,,$$

vanishing on the system of Eq. (7.1):

$$\big(D_x(P) + D_y(Q) + D_t(R)\big)|_{(7.1)} = 0\,.$$

Here $P = P(x, y, t, u, v, \eta)$, $Q = Q(x, y, t, u, v, \eta)$, $R = R(x, y, t, u, v, \eta)$.

The determining system of equations for conservation laws is the following overdetermined system of linear equations

$$\begin{aligned} &P_u - uR_u - (\eta + h)R_\eta = 0\,, \qquad Q_u - vR_u = 0\,, \qquad P_v - uR_v = 0\,,\\ &Q_v - vR_v - (\eta + h)R_\eta = 0\,, \qquad P_\eta - R_u - uR_\eta = 0\,,\\ &Q_\eta - R_v - vR_\eta = 0\,, \qquad P_x + Q_y + R_t - (uh_x + vh_y)R_\eta = 0\,. \end{aligned} \tag{7.10}$$

7.3.2 Analysis of the Classifying Equation

One can express functions P and Q in terms of R and constants of integration from the second and third equations of the system (7.10). Next, after examining the system of Eq. (7.10) for compatibility, we obtain

$$P = (\eta + h)^2\Big(au - \frac{\dot a}{2}x - By + k\Big) + uR\,,$$
$$Q = (\eta + h)^2\Big(av - \frac{\dot a}{2}y + Bx + l\Big) + vR\,,$$
$$R = (\eta + h)\Big(\frac{x^2 + y^2}{2}\ddot a - 2\dot k x - 2\dot l y + 2ku + 2lv +$$
$$+ 2B(xv - yu) - \dot a(xu + yv) + a(u^2 + v^2 + \eta - h) + f\Big)\,,$$

where functions $a = a(t), k = k(t), l = l(t), f = f(t)$ and a constant B are determined from the classifying equation

$$(-\dot a x - 2By + 2k)h_x + (-\dot a y + 2Bx + 2l)h_y - 2\dot a h =$$
$$= -\frac{x^2 + y^2}{2}\dddot a + 2x\ddot k + 2y\ddot l - \dot f\,. \tag{7.11}$$

Remark 7.3 The classifying Eq. (7.11) coincides with the classifying Eq. (7.3) at $C = 0$. Thus analysis of the classifying Eq. (7.11) coincides with the analysis of the classifying Eq. (7.3) and cases of additional conservation laws can be obtained as in the Sect. 7.2.

The kernel of the conservation laws of the system of Eq. (7.1) consists of the following conservation laws

$$\begin{aligned} &P_0 = u(\eta + h)\,, && P_1 = u(\eta + h)(u^2 + v^2 + 2\eta)\,,\\ &Q_0 = v(\eta + h)\,, && Q_1 = v(\eta + h)(u^2 + v^2 + 2\eta)\,,\\ &R_0 = \eta + h\,, && R_1 = (\eta + h)(u^2 + v^2 + \eta - h)\,. \end{aligned} \tag{7.12}$$

7.3.3 Cases of Additional Conservation Laws

Find conservation laws, additional to the conservation laws (7.12). The following cases arise.

1. $\mathbf{h = H(x) + b_1 y^2 + b_2 xy + b_3 y}$

1.1. $\mathbf{h = a_1 x + a_2 y + a_3}$

In this case additional conservation laws correspond to functions

$$P_2 = (\eta+h)^2\Big(tu - \frac{x}{2} - \frac{3a_1}{4}t^2\Big) + uR_2, \quad Q_2 = (\eta+h)^2\Big(tv - \frac{y}{2} - \frac{3a_2}{4}t^2\Big) + vR_2,$$
$$R_2 = (\eta+h)\Big(3a_1tx + 3a_2ty - \frac{3a_1}{2}t^2u - \frac{3a_2}{2}t^2v - xu - yv +$$
$$+ t(u^2+v^2+\eta-h) + \frac{a_1^2+a_2^2}{2}t^3 + 2a_3t\Big),$$
$$P_3 = (\eta+h)^2\Big(t^2u - tx - \frac{a_1}{2}t^3\Big) + uR_3, \quad Q_3 = (\eta+h)^2\Big(t^2v - ty - \frac{a_2}{2}t^3\Big) + vR_3,$$
$$R_3 = (\eta+h)\Big(x^2+y^2+3a_1t^2x+3a_2t^2y - a_1t^3u - a_2t^3v - 2t(xu+yv) +$$
$$+ t^2(u^2+v^2+\eta-h) + \frac{a_1^2+a_2^2}{4}t^4 + 2a_3t^2\Big),$$
$$P_4 = \frac{t(\eta+h)^2}{2} + uR_4, \quad Q_4 = vR_4, \quad R_4 = (\eta+h)\Big(-x+tu-\frac{a_1}{2}t^2\Big),$$
$$P_5 = uR_5, \quad Q_5 = \frac{t(\eta+h)^2}{2} + vR_5, \quad R_5 = (\eta+h)\Big(-y+tv-\frac{a_2}{2}t^2\Big),$$
$$P_6 = \frac{(\eta+h)^2}{2} + uR_6, \quad Q_6 = vR_6, \quad R_6 = (\eta+h)(u-a_1t),$$
$$P_7 = uR_7, \quad Q_7 = \frac{(\eta+h)^2}{2} + vR_7, \quad R_7 = (\eta+h)(v-a_2t),$$
$$P_8 = (\eta+h)^2\Big(-y+\frac{a_2}{2}t^2\Big) + uR_8, \quad Q_8 = (\eta+h)^2\Big(x-\frac{a_1}{2}t^2\Big) + vR_8,$$
$$R_8 = (\eta+h)(-2a_2tx + 2a_1ty + a_2t^2u - a_1t^2v + 2xv - 2yu).$$

1.2. $\mathbf{h = a_1x^2 + a_2y^2 + a_3xy + a_4x + a_5y + a_6,\ a_1^2 + a_2^2 + a_3^2 > 0.}$
1.2.1. $\mathbf{h = a_1(x^2+y^2) + a_4x + a_5y + a_6,\ a_1 > 0.}$
In this case additional conservation laws correspond to functions

$$P_2 = (\eta+h)^2e^{t\sqrt{8a_1}}\Big(u - \sqrt{2a_1}x - \frac{a_4\sqrt{2a_1}}{2a_1}\Big) + uR_2,$$
$$Q_2 = (\eta+h)^2e^{t\sqrt{8a_1}}\Big(v - \sqrt{2a_1}y - \frac{a_5\sqrt{2a_1}}{2a_1}\Big) + vR_2,$$
$$R_2 = (\eta+h)e^{t\sqrt{8a_1}}\Big(3h - \frac{a_4\sqrt{2a_1}}{a_1}u - \frac{a_5\sqrt{2a_1}}{a_1}v - \sqrt{8a_1}(xu+yv) +$$
$$+ u^2+v^2+\eta-2a_6 + \frac{a_4^2+a_5^2}{2a_1}\Big),$$

$$P_3 = (\eta + h)^2 e^{-t\sqrt{8a_1}}\Big(u + \sqrt{2a_1}x + \frac{a_4\sqrt{2a_1}}{2a_1}\Big) + uR_3,$$

$$Q_3 = (\eta + h)^2 e^{-t\sqrt{8a_1}}\Big(v + \sqrt{2a_1}y + \frac{a_5\sqrt{2a_1}}{2a_1}\Big) + vR_3,$$

$$R_3 = (\eta + h)e^{-t\sqrt{8a_1}}\Big(3h + \frac{a_4\sqrt{2a_1}}{a_1}u + \frac{a_5\sqrt{2a_1}}{a_1}v + \sqrt{8a_1}(xu + yv) +$$

$$+ u^2 + v^2 + \eta - 2a_6 + \frac{a_4^2 + a_5^2}{2a_1}\Big),$$

$$P_4 = \frac{(\eta+h)^2 e^{t\sqrt{2a_1}}}{2} + uR_4, \quad Q_4 = vR_4, \quad R_4 = (\eta+h)e^{t\sqrt{2a_1}}\Big(u - \sqrt{2a_1}x - \frac{a_4}{\sqrt{2a_1}}\Big),$$

$$P_5 = uR_5, \quad Q_5 = \frac{(\eta+h)^2 e^{t\sqrt{2a_1}}}{2} + vR_5, \quad R_5 = (\eta+h)e^{t\sqrt{2a_1}}\Big(v - \sqrt{2a_1}y - \frac{a_5}{\sqrt{2a_1}}\Big),$$

$$P_6 = \frac{(\eta + h)^2 e^{-t\sqrt{2a_1}}}{2} + uR_6, \qquad Q_6 = vR_6,$$

$$R_6 = (\eta + h)e^{-t\sqrt{2a_1}}\Big(\sqrt{2a_1}x + u + \frac{a_4}{\sqrt{2a_1}}\Big),$$

$$P_7 = uR_7, \qquad Q_7 = \frac{(\eta + h)^2 e^{-t\sqrt{2a_1}}}{2} + vR_7,$$

$$R_7 = (\eta + h)e^{-t\sqrt{2a_1}}\Big(\sqrt{2a_1}y + v + \frac{a_5}{\sqrt{2a_1}}\Big),$$

$$P_8 = (\eta + h)^2\Big(-y - \frac{a_5}{2a_1}\Big) + uR_8, \qquad Q_8 = (\eta + h)^2\Big(x + \frac{a_4}{2a_1}\Big) + vR_8,$$

$$R_8 = (\eta + h)\Big(-\frac{a_5}{a_1}u + \frac{a_4}{a_1}v + 2xv - 2yu\Big).$$

1.2.2. $\mathbf{h = a_1(x^2 + y^2) + a_4x + a_5y + a_6,\ a_1 < 0.}$
In this case additional conservation laws correspond to functions

$$P_2 = (\eta + h)^2\Big(\sin(t\sqrt{-8a_1})u - \sqrt{-2a_1}\Big(x + \frac{a_4}{2a_1}\Big)\cos(t\sqrt{-8a_1})\Big) + uR_2,$$

$$Q_2 = (\eta + h)^2\Big(\sin(t\sqrt{-8a_1})v - \sqrt{-2a_1}\Big(y + \frac{a_5}{2a_1}\Big)\cos(t\sqrt{-8a_1})\Big) + vR_2,$$

$$R_2 = (\eta + h)\Big(3h\sin(t\sqrt{-8a_1}) - \frac{a_4\sqrt{-2a_1}}{a_1}\cos(t\sqrt{-8a_1})u -$$

$$- \frac{a_5\sqrt{-2a_1}}{a_1}\cos(t\sqrt{-8a_1})v - \sqrt{-8a_1}\cos(t\sqrt{-8a_1})(xu + yv) +$$

$$+ \sin(t\sqrt{-8a_1})\Big(u^2 + v^2 + \eta - 2a_6 + \frac{a_4^2 + a_5^2}{2a_1}\Big)\Big),$$

$$P_3 = (\eta + h)^2\Big(\cos(t\sqrt{-8a_1})u + \sqrt{-2a_1}\Big(x + \frac{a_4}{2a_1}\Big)\sin(t\sqrt{-8a_1})\Big) + uR_3,$$
$$Q_3 = (\eta + h)^2\Big(\cos(t\sqrt{-8a_1})v + \sqrt{-2a_1}\Big(y + \frac{a_5}{2a_1}\Big)\sin(t\sqrt{-8a_1})\Big) + vR_3,$$
$$R_3 = (\eta + h)\Big(3h\cos(t\sqrt{-8a_1}) + \frac{a_4\sqrt{-2a_1}}{a_1}\sin(t\sqrt{-8a_1})u +$$
$$+\frac{a_5\sqrt{-2a_1}}{a_1}\sin(t\sqrt{-8a_1})v + \sqrt{-8a_1}\sin(t\sqrt{-8a_1})(xu + yv) +$$
$$+\cos(t\sqrt{-8a_1})\Big(u^2 + v^2 + \eta - 2a_6 + \frac{a_4^2 + a_5^2}{2a_1}\Big)\Big),$$
$$P_4 = \frac{(\eta + h)^2\sin(t\sqrt{-2a_1})}{2} + uR_4, \quad Q_4 = vR_4,$$
$$R_4 = (\eta + h)\Big(-\sqrt{-2a_1}\cos(t\sqrt{-2a_1})\Big(x + \frac{a_4}{2a_1}\Big) + \sin(t\sqrt{-2a_1})u\Big),$$
$$P_5 = uR_5, \quad Q_5 = \frac{(\eta + h)^2\sin(t\sqrt{-2a_1})}{2} + vR_5,$$
$$R_5 = (\eta + h)\Big(-\sqrt{-2a_1}\cos(t\sqrt{-2a_1})\Big(y + \frac{a_5}{2a_1}\Big) + \sin(t\sqrt{-2a_1})v\Big),$$
$$P_6 = \frac{(\eta + h)^2\cos(t\sqrt{-2a_1})}{2} + uR_6, \quad Q_6 = vR_6,$$
$$R_6 = (\eta + h)\Big(\sqrt{-2a_1}\sin(t\sqrt{-2a_1})\Big(x + \frac{a_4}{2a_1}\Big) + \cos(t\sqrt{-2a_1})u\Big),$$
$$P_7 = uR_7, \quad Q_7 = \frac{(\eta + h)^2\cos(t\sqrt{-2a_1})}{2} + vR_7,$$
$$R_7 = (\eta + h)\Big(\sqrt{-2a_1}\sin(t\sqrt{-2a_1})\Big(y + \frac{a_5}{2a_1}\Big) + \cos(t\sqrt{-2a_1})v\Big),$$
$$P_8 = (\eta + h)^2\Big(-y - \frac{a_5}{2a_1}\Big) + uR_8, \quad Q_8 = (\eta + h)^2\Big(x + \frac{a_4}{2a_1}\Big) + vR_8,$$
$$R_8 = (\eta + h)\Big(-\frac{a_5}{a_1}u + \frac{a_4}{a_1}v + 2xv - 2yu\Big).$$

1.2.3. $\mathbf{h = a_1x^2 + a_2y^2 + a_4x + a_5y + a_6,\ a_1 \neq a_2,\ a_1 > 0,\ a_2 > 0.}$
In this case additional conservation laws correspond to functions

$$P_2 = \frac{(\eta+h)^2e^{t\sqrt{2a_1}}}{2} + uR_2, \quad Q_2 = vR_2, \quad R_2 = (\eta+h)e^{t\sqrt{2a_1}}\Big(u - \sqrt{2a_1}x - \frac{a_4}{\sqrt{2a_1}}\Big),$$
$$P_3 = uR_3, \quad Q_3 = \frac{(\eta+h)^2e^{t\sqrt{2a_2}}}{2} + vR_3, \quad R_3 = (\eta+h)e^{t\sqrt{2a_2}}\Big(v - \sqrt{2a_2}y - \frac{a_5}{\sqrt{2a_2}}\Big),$$
$$P_4 = \frac{(\eta + h)^2e^{-t\sqrt{2a_1}}}{2} + uR_4, \quad Q_4 = vR_4,$$
$$R_4 = (\eta + h)e^{-t\sqrt{2a_1}}\Big(\sqrt{2a_1}x + u + \frac{a_4}{\sqrt{2a_1}}\Big),$$

$$P_5 = uR_5, \quad Q_5 = \frac{(\eta+h)^2 e^{-t\sqrt{2a_2}}}{2} + vR_5,$$
$$R_5 = (\eta+h)e^{-t\sqrt{2a_2}}\Big(\sqrt{2a_2}y + v + \frac{a_5}{\sqrt{2a_2}}\Big).$$

1.2.4. $\mathbf{h = a_1x^2 + a_4x + a_5y + a_6,\ a_1 > 0.}$
In this case additional conservation laws correspond to functions

$$P_2 = \frac{(\eta+h)^2 e^{t\sqrt{2a_1}}}{2} + uR_2, \quad Q_2 = vR_2, \quad R_2 = (\eta+h)e^{t\sqrt{2a_1}}\Big(u - \sqrt{2a_1}x - \frac{a_4}{\sqrt{2a_1}}\Big),$$
$$P_3 = uR_3, \quad Q_3 = \frac{(\eta+h)^2 t}{2} + vR_3, \quad R_3 = (\eta+h)\Big(-y + tv - \frac{a_5t^2}{2}\Big),$$
$$P_4 = \frac{(\eta+h)^2 e^{-t\sqrt{2a_1}}}{2} + uR_4, \quad Q_4 = vR_4,$$
$$R_4 = (\eta+h)e^{-t\sqrt{2a_1}}\Big(\sqrt{2a_1}x + u + \frac{a_4}{\sqrt{2a_1}}\Big),$$
$$P_5 = uR_5, \quad Q_5 = \frac{(\eta+h)^2}{2} + vR_5, \quad R_5 = (\eta+h)(v - a_5t)\,.$$

1.2.5. $\mathbf{h = a_1x^2 + a_2y^2 + a_4x + a_5y + a_6,\ a_1 > 0,\ a_2 < 0.}$
In this case additional conservation laws correspond to functions

$$P_2 = \frac{(\eta+h)^2 e^{t\sqrt{2a_1}}}{2} + uR_2, \quad Q_2 = vR_2, \quad R_2 = (\eta+h)e^{t\sqrt{2a_1}}\Big(u - \sqrt{2a_1}x - \frac{a_4}{\sqrt{2a_1}}\Big),$$
$$P_3 = uR_3, \qquad Q_3 = \frac{(\eta+h)^2 \sin(t\sqrt{-2a_2})}{2} + vR_3,$$
$$R_3 = (\eta+h)\Big(-\sqrt{-2a_2}\cos(t\sqrt{-2a_2})\Big(y + \frac{a_5}{2a_2}\Big) + \sin(t\sqrt{-2a_2})v\Big),$$
$$P_4 = \frac{(\eta+h)^2 e^{-t\sqrt{2a_1}}}{2} + uR_4, \qquad Q_4 = vR_4,$$
$$R_4 = (\eta+h)e^{-t\sqrt{2a_1}}\Big(\sqrt{2a_1}x + u + \frac{a_4}{\sqrt{2a_1}}\Big),$$
$$P_5 = uR_5, \qquad Q_5 = \frac{(\eta+h)^2 \cos(t\sqrt{-2a_2})}{2} + vR_5,$$
$$R_5 = (\eta+h)\Big(\sqrt{-2a_2}\sin(t\sqrt{-2a_2})\Big(y + \frac{a_5}{2a_2}\Big) + \cos(t\sqrt{-2a_2})v\Big).$$

1.2.6. $\mathbf{h = a_2y^2 + a_4x + a_5y + a_6,\ a_2 < 0.}$
In this case additional conservation laws correspond to functions

$$P_2 = \frac{(\eta+h)^2 t}{2} + uR_2, \qquad Q_2 = vR_2, \qquad R_2 = (\eta+h)\Big(-x + tu - \frac{a_4}{2}t^2\Big),$$

$$P_3 = uR_3, \qquad Q_3 = \frac{(\eta+h)^2 \sin(t\sqrt{-2a_2})}{2} + vR_3,$$

$$R_3 = (\eta+h)\Big(-\sqrt{-2a_2}\cos(t\sqrt{-2a_2})\Big(y + \frac{a_5}{2a_2}\Big) + \sin(t\sqrt{-2a_2})v\Big),$$

$$P_4 = \frac{(\eta+h)^2}{2} + uR_4, \qquad Q_4 = vR_4, \qquad R_4 = (\eta+h)(u - a_4 t),$$

$$P_5 = uR_5, \qquad Q_5 = \frac{(\eta+h)^2 \cos(t\sqrt{-2a_2})}{2} + vR_5,$$

$$R_5 = (\eta+h)\Big(\sqrt{-2a_2}\sin(t\sqrt{-2a_2})\Big(y + \frac{a_5}{2a_2}\Big) + \cos(t\sqrt{-2a_2})v\Big).$$

1.2.7. $\mathbf{h = a_1x^2 + a_2y^2 + a_4x + a_5y + a_6,\ a_1 < 0,\ a_2 < 0.}$
In this case additional conservation laws correspond to functions

$$P_2 = \frac{(\eta+h)^2 \sin(t\sqrt{-2a_1})}{2} + uR_2, \quad Q_2 = vR_2,$$

$$R_2 = (\eta+h)\Big(-\sqrt{-2a_1}\cos(t\sqrt{-2a_1})\Big(x + \frac{a_4}{2a_1}\Big) + \sin(t\sqrt{-2a_1})u\Big),$$

$$P_3 = uR_3, \quad Q_3 = \frac{(\eta+h)^2 \sin(t\sqrt{-2a_2})}{2} + vR_3,$$

$$R_3 = (\eta+h)\Big(-\sqrt{-2a_2}\cos(t\sqrt{-2a_2})\Big(y + \frac{a_5}{2a_2}\Big) + \sin(t\sqrt{-2a_2})v\Big),$$

$$P_4 = \frac{(\eta+h)^2 \cos(t\sqrt{-2a_1})}{2} + uR_4, \quad Q_4 = vR_4,$$

$$R_4 = (\eta+h)\Big(\sqrt{-2a_1}\sin(t\sqrt{-2a_1})\Big(x + \frac{a_4}{2a_1}\Big) + \cos(t\sqrt{-2a_1})u\Big),$$

$$P_5 = uR_5, \quad Q_5 = \frac{(\eta+h)^2 \cos(t\sqrt{-2a_2})}{2} + vR_5,$$

$$R_5 = (\eta+h)\Big(\sqrt{-2a_2}\sin(t\sqrt{-2a_2})\Big(y + \frac{a_5}{2a_2}\Big) + \cos(t\sqrt{-2a_2})v\Big).$$

1.2.8. $\mathbf{h = a_1x^2 + a_2y^2 + a_3xy + a_4x + a_5y + a_6,\ a_3 \neq 0,}$
$\mathbf{a_1 + a_2 > \sqrt{(a_1 - a_2)^2 + a_3^2}.}$
Denote $s = \sqrt{(a_1 - a_2)^2 + a_3^2}$, $S_1 = \sqrt{a_1 + a_2 + s}$, $S_2 = \sqrt{a_1 + a_2 - s}$. In this case additional conservation laws correspond to functions

$$P_2 = \frac{(\eta+h)^2 e^{tS_2}}{2} + uR_2, \quad Q_2 = \frac{(\eta+h)^2 (S_2^2 - 2a_1) e^{tS_2}}{2a_3} + vR_2,$$

$$R_2 = (\eta+h)e^{tS_2}\Big(u - S_2 x - \frac{S_2(S_2^2 - 2a_1)}{a_3} y + \frac{S_2^2 - 2a_1}{a_3} v - \frac{1}{S_2}\Big(a_4 + \frac{(S_2^2 - 2a_1)a_5}{a_3}\Big)\Big),$$

$$P_3 = \frac{(\eta+h)^2 e^{-tS_2}}{2} + uR_3, \quad Q_3 = \frac{(\eta+h)^2(S_2^2-2a_1)e^{-tS_2}}{2a_3} + vR_3,$$

$$R_3 = (\eta+h)e^{-tS_2}\Big(u + S_2 x + \frac{S_2(S_2^2-2a_1)}{a_3}y + \frac{S_2^2-2a_1}{a_3}v + \frac{1}{S_2}\Big(a_4 + \frac{(S_2^2-2a_1)a_5}{a_3}\Big)\Big),$$

$$P_4 = \frac{(\eta+h)^2 e^{tS_1}}{2} + uR_4, \quad Q_4 = \frac{(\eta+h)^2(S_1^2-2a_1)e^{tS_1}}{2a_3} + vR_4,$$

$$R_4 = (\eta+h)e^{tS_1}\Big(u - S_1 x - \frac{S_1(S_1^2-2a_1)}{a_3}y + \frac{S_1^2-2a_1}{a_3}v - \frac{1}{S_1}\Big(a_4 + \frac{(S_1^2-2a_1)a_5}{a_3}\Big)\Big),$$

$$P_5 = \frac{(\eta+h)^2 e^{-tS_1}}{2} + uR_5, \quad Q_5 = \frac{(\eta+h)^2(S_1^2-2a_1)e^{-tS_1}}{2a_3} + vR_5,$$

$$R_5 = (\eta+h)e^{-tS_1}\Big(u + S_1 x + \frac{S_1(S_1^2-2a_1)}{a_3}y + \frac{S_1^2-2a_1}{a_3}v + \frac{1}{S_1}\Big(a_4 + \frac{(S_1^2-2a_1)a_5}{a_3}\Big)\Big).$$

1.2.9.

$$\mathbf{h} = \Big(\sqrt{\mathbf{a_1}}\mathbf{x} + \frac{\mathbf{a_3}}{2\sqrt{\mathbf{a_1}}}\mathbf{y}\Big)^2 + \mathbf{a_4 x} + \mathbf{a_5 y} + \mathbf{a_6}, \ \mathbf{a_1} > \mathbf{0},$$

$$\mathbf{a_3} \neq \mathbf{0}, \quad \mathbf{a_1} + \mathbf{a_2} = \sqrt{(\mathbf{a_1} - \mathbf{a_2})^2 + \mathbf{a_3^2}}.$$

Denote

$$S = \sqrt{2a_1 + \frac{a_3^2}{2a_1}}.$$

In this case additional conservation laws correspond to functions

$$P_2 = \frac{(\eta+h)^2 e^{tS}}{2} + uR_2, \quad Q_2 = \frac{(\eta+h)^2(S^2-2a_1)e^{tS}}{2a_3} + vR_2,$$

$$R_2 = (\eta+h)e^{tS}\Big(u - Sx - \frac{S(S^2-2a_1)}{a_3}y + \frac{S^2-2a_1}{a_3}v - \frac{1}{S}\Big(a_4 + \frac{(S^2-2a_1)a_5}{a_3}\Big)\Big),$$

$$P_3 = \frac{(\eta+h)^2 e^{-tS}}{2} + uR_3, \quad Q_3 = \frac{(\eta+h)^2(S^2-2a_1)e^{-tS}}{2a_3} + vR_3,$$

$$R_3 = (\eta+h)e^{-tS}\Big(u + Sx + \frac{S(S^2-2a_1)}{a_3}y + \frac{S^2-2a_1}{a_3}v + \frac{1}{S}\Big(a_4 + \frac{(S^2-2a_1)a_5}{a_3}\Big)\Big),$$

$$P_4 = \frac{(\eta+h)^2 t}{2} + uR_4, \quad Q_4 = -\frac{a_1(\eta+h)^2 t}{a_3} + vR_4,$$

$$R_4 = (\eta+h)\Big(-x + \frac{2a_1}{a_3}y + tu - \frac{2a_1}{a_3}tv + \Big(-a_4 + \frac{2a_1a_5}{a_3}\Big)\frac{t^2}{2} - \frac{a_5}{a_3}\Big),$$

$$P_5 = \frac{(\eta+h)^2}{2} + uR_5, \quad Q_5 = -\frac{a_1(\eta+h)^2}{a_3} + vR_5,$$
$$R_5 = (\eta+h)\Big(u - \frac{2a_1}{a_3}v + \Big(-a_4 + \frac{2a_1a_5}{a_3}\Big)t\Big).$$

1.2.10. $\mathbf{h = a_1x^2 + a_2y^2 + a_3xy + a_4x + a_5y + a_6,\ a_3 \neq 0,}$
$\mathbf{-\sqrt{(a_1-a_2)^2 + a_3^2} < a_1 + a_2 < \sqrt{(a_1-a_2)^2 + a_3^2}.}$

Denote $s = \sqrt{(a_1-a_2)^2 + a_3^2}$, $S_1 = \sqrt{a_1+a_2+s}$, $S_3 = \sqrt{s-a_1-a_2}$. In this case additional conservation laws correspond to functions

$$P_2 = \frac{(\eta+h)^2\sin(tS_3)}{2} + uR_2, \quad Q_2 = -\frac{(\eta+h)^2(S_3^2+2a_1)\sin(tS_3)}{2a_3} + vR_2,$$
$$R_2 = (\eta+h)\Big(-S_3\cos(tS_3)x + \frac{S_3(S_3^2+2a_1)}{a_3}\cos(tS_3)y + \sin(tS_3)u -$$
$$-\frac{S_3^2+2a_1}{a_3}\sin(tS_3)v - \frac{\cos(tS_3)}{S_3}\Big(-a_4 + \frac{(S_3^2+2a_1)a_5}{a_3}\Big)\Big),$$
$$P_3 = \frac{(\eta+h)^2\cos(tS_3)}{2} + uR_3, \quad Q_3 = -\frac{(\eta+h)^2(S_3^2+2a_1)\cos(tS_3)}{2a_3} + vR_3,$$
$$R_3 = (\eta+h)\Big(S_3\sin(tS_3)x - \frac{S_3(S_3^2+2a_1)}{a_3}\sin(tS_3)y + \cos(tS_3)u -$$
$$-\frac{S_3^2+2a_1}{a_3}\cos(tS_3)v + \frac{\sin(tS_3)}{S_3}\Big(-a_4 + \frac{(S_3^2+2a_1)a_5}{a_3}\Big)\Big),$$
$$P_4 = \frac{(\eta+h)^2e^{tS_1}}{2} + uR_4, \quad Q_4 = \frac{(\eta+h)^2(S_1^2-2a_1)e^{tS_1}}{2a_3} + vR_4,$$
$$R_4 = (\eta+h)e^{tS_1}\Big(u - S_1x - \frac{S_1(S_1^2-2a_1)}{a_3}y + \frac{S_1^2-2a_1}{a_3}v - \frac{1}{S_1}\Big(a_4 + \frac{(S_1^2-2a_1)a_5}{a_3}\Big)\Big),$$
$$P_5 = \frac{(\eta+h)^2e^{-tS_1}}{2} + uR_5, \quad Q_5 = \frac{(\eta+h)^2(S_1^2-2a_1)e^{-tS_1}}{2a_3} + vR_5,$$
$$R_5 = (\eta+h)e^{-tS_1}\Big(u + S_1x + \frac{S_1(S_1^2-2a_1)}{a_3}y + \frac{S_1^2-2a_1}{a_3}v + \frac{1}{S_1}\Big(a_4 + \frac{(S_1^2-2a_1)a_5}{a_3}\Big)\Big).$$

1.2.11.

$$\mathbf{h = -\Big(\sqrt{-a_1}x - \frac{a_3}{2\sqrt{-a_1}}y\Big)^2 + a_4x + a_5y + a_6,\ a_1 < 0,}$$
$$\mathbf{a_3 \neq 0, \quad a_1 + a_2 = -\sqrt{(a_1-a_2)^2 + a_3^2}.}$$

Denote

$$S = \sqrt{-2a_1 - \frac{a_3^2}{2a_1}}.$$

In this case additional conservation laws correspond to functions

$$P_2 = \frac{(\eta + h)^2 \sin(tS)}{2} + uR_2, \quad Q_2 = -\frac{(\eta + h)^2 (S^2 + 2a_1) \sin(tS)}{2a_3} + vR_2,$$

$$R_2 = (\eta + h)\Big(-S\cos(tS)x + \frac{S(S^2 + 2a_1)}{a_3}\cos(tS)y + \sin(tS)u -$$
$$- \frac{S^2 + 2a_1}{a_3}\sin(tS)v - \frac{\cos(tS)}{S}\Big(-a_4 + \frac{(S^2 + 2a_1)a_5}{a_3}\Big)\Big),$$

$$P_3 = \frac{(\eta + h)^2 \cos(tS)}{2} + uR_3, \quad Q_3 = -\frac{(\eta + h)^2 (S^2 + 2a_1) \cos(tS)}{2a_3} + vR_3,$$

$$R_3 = (\eta + h)\Big(S\sin(tS)x - \frac{S(S^2 + 2a_1)}{a_3}\sin(tS)y + \cos(tS)u -$$
$$- \frac{S^2 + 2a_1}{a_3}\cos(tS)v + \frac{\sin(tS)}{S}\Big(-a_4 + \frac{(S^2 + 2a_1)a_5}{a_3}\Big)\Big),$$

$$P_4 = \frac{(\eta + h)^2 t}{2} + uR_4, \quad Q_4 = -\frac{a_1(\eta + h)^2 t}{a_3} + vR_4,$$

$$R_4 = (\eta + h)\Big(-x + \frac{2a_1}{a_3}y + tu - \frac{2a_1}{a_3}tv + \Big(-a_4 + \frac{2a_1 a_5}{a_3}\Big)\frac{t^2}{2} - \frac{a_5}{a_3}\Big),$$

$$P_5 = \frac{(\eta + h)^2}{2} + uR_5, \quad Q_5 = -\frac{a_1(\eta + h)^2}{a_3} + vR_5,$$

$$R_5 = (\eta + h)\Big(u - \frac{2a_1}{a_3}v + \Big(-a_4 + \frac{2a_1 a_5}{a_3}\Big)t\Big).$$

1.2.12. $\mathbf{h = a_1x^2 + a_2y^2 + a_3xy + a_4x + a_5y + a_6,\ a_3 \neq 0,}$
$\mathbf{a_1 + a_2 < -\sqrt{(a_1 - a_2)^2 + a_3^2}.}$

Denote $s = \sqrt{(a_1 - a_2)^2 + a_3^2}$, $S_3 = \sqrt{s - a_1 - a_2}$, $S_4 = \sqrt{-s - a_1 - a_2}$. In this case additional conservation laws correspond to functions

$$P_2 = \frac{(\eta + h)^2 \sin(tS_3)}{2} + uR_2, \quad Q_2 = -\frac{(\eta + h)^2 (S_3^2 + 2a_1) \sin(tS_3)}{2a_3} + vR_2,$$

$$R_2 = (\eta + h)\Big(-S_3\cos(tS_3)x + \frac{S_3(S_3^2 + 2a_1)}{a_3}\cos(tS_3)y + \sin(tS_3)u -$$
$$- \frac{S_3^2 + 2a_1}{a_3}\sin(tS_3)v - \frac{\cos(tS_3)}{S_3}\Big(-a_4 + \frac{(S_3^2 + 2a_1)a_5}{a_3}\Big)\Big),$$

$$P_3 = \frac{(\eta+h)^2\cos(tS_3)}{2} + uR_3, \quad Q_3 = -\frac{(\eta+h)^2(S_3^2+2a_1)\cos(tS_3)}{2a_3} + vR_3,$$

$$R_3 = (\eta+h)\Big(S_3\sin(tS_3)x - \frac{S_3(S_3^2+2a_1)}{a_3}\sin(tS_3)y + \cos(tS_3)u -$$

$$-\frac{S_3^2+2a_1}{a_3}\cos(tS_3)v + \frac{\sin(tS_3)}{S_3}\Big(-a_4 + \frac{(S_3^2+2a_1)a_5}{a_3}\Big)\Big),$$

$$P_4 = \frac{(\eta+h)^2\sin(tS_4)}{2} + uR_4, \quad Q_4 = -\frac{(\eta+h)^2(S_4^2+2a_1)\sin(tS_4)}{2a_3} + vR_4,$$

$$R_4 = (\eta+h)\Big(-S_4\cos(tS_4)x + \frac{S_4(S_4^2+2a_1)}{a_3}\cos(tS_4)y + \sin(tS_4)u -$$

$$-\frac{S_4^2+2a_1}{a_3}\sin(tS_4)v - \frac{\cos(tS_4)}{S_4}\Big(-a_4 + \frac{(S_4^2+2a_1)a_5}{a_3}\Big)\Big),$$

$$P_5 = \frac{(\eta+h)^2\cos(tS_4)}{2} + uR_5, \quad Q_5 = -\frac{(\eta+h)^2(S_4^2+2a_1)\cos(tS_4)}{2a_3} + vR_5,$$

$$R_5 = (\eta+h)\Big(S_4\sin(tS_4)x - \frac{S_4(S_4^2+2a_1)}{a_3}\sin(tS_4)y + \cos(tS_4)u -$$

$$-\frac{S_4^2+2a_1}{a_3}\cos(tS_4)v + \frac{\sin(tS_4)}{S_4}\Big(-a_4 + \frac{(S_4^2+2a_1)a_5}{a_3}\Big)\Big).$$

1.3. $\mathbf{h = H(x) + b_1y^2 + b_2y, H''' \neq 0}$.

1.3.1. $\mathbf{h = H(x) + b_1y^2 + b_2y, H''' \neq 0, b_1 > 0}$.

In this case additional conservation laws correspond to functions

$$P_2 = uR_2, \quad Q_2 = \frac{(\eta+h)^2e^{t\sqrt{2b_1}}}{2} + vR_2, \quad R_2 = (\eta+h)e^{t\sqrt{2b_1}}\Big(v - \sqrt{2b_1}\,y - \frac{b_2}{\sqrt{2b_1}}\Big),$$

$$P_3 = uR_3, \quad Q_3 = \frac{(\eta+h)^2e^{-t\sqrt{2b_1}}}{2} + vR_3, \quad R_3 = (\eta+h)e^{-t\sqrt{2b_1}}\Big(v + \sqrt{2b_1}\,y + \frac{b_2}{\sqrt{2b_1}}\Big).$$

1.3.1.1.

$$\mathbf{h} = \frac{\mathbf{a_1}}{(\mathbf{x}+\mathbf{a_2})^2} + \mathbf{b_1}((\mathbf{x}+\mathbf{a_2})^2 + \mathbf{y}^2) + \mathbf{b_2 y} + \mathbf{b_3}, \ \mathbf{b_1} > \mathbf{0}, \ \mathbf{a_1} \neq \mathbf{0}.$$

In this case additional conservation laws correspond to functions (P_2, Q_2, R_2) and (P_3, Q_3, R_3) from the case 1.3.1 and to functions

$$P_4 = (\eta + h)^2 e^{t\sqrt{8b_1}}(u - \sqrt{2b_1}(x + a_2)) + uR_4,$$
$$Q_4 = (\eta + h)^2 e^{t\sqrt{8b_1}}\Big(v - \sqrt{2b_1}\Big(y + \frac{b_2}{2b_1}\Big)\Big) + vR_4,$$
$$R_4 = (\eta + h)e^{t\sqrt{8b_1}}\Big(4b_1((x + a_2)^2 + y^2) + 4b_2 y - 2a_2\sqrt{2b_1}u - \\ - \frac{2b_2}{\sqrt{2b_1}}v - \sqrt{8b_1}(xu + yv) + u^2 + v^2 + \eta - h + \frac{4b_1b_3 + b_2^2}{2b_1}\Big),$$
$$P_5 = (\eta + h)^2 e^{-t\sqrt{8b_1}}(u + \sqrt{2b_1}(x + a_2)) + uR_5,$$
$$Q_5 = (\eta + h)^2 e^{-t\sqrt{8b_1}}\Big(v + \sqrt{2b_1}\Big(y + \frac{b_2}{2b_1}\Big)\Big) + vR_5,$$
$$R_5 = (\eta + h)e^{-t\sqrt{8b_1}}\Big(4b_1((x + a_2)^2 + y^2) + 4b_2 y + 2a_2\sqrt{2b_1}u + \\ + \frac{2b_2}{\sqrt{2b_1}}v + \sqrt{8b_1}(xu + yv) + u^2 + v^2 + \eta - h + \frac{4b_1b_3 + b_2^2}{2b_1}\Big).$$

1.3.2. $\mathbf{h = H(x) + b_2 y,\ H''' \neq 0}$.

In this case additional conservation laws correspond to functions

$$P_2 = uR_2,\quad Q_2 = \frac{(\eta + h)^2 t}{2} + vR_2,\quad R_2 = (\eta + h)\Big(-y + tv - \frac{b_2}{2}t^2\Big),$$
$$P_3 = uR_3,\quad Q_3 = \frac{(\eta + h)^2}{2} + vR_3,\quad R_3 = (\eta + h)(v - b_2 t).$$

1.3.2.1.

$$\mathbf{h = \frac{a_1}{(x + a_2)^2} + b_2 y + b_3,\ a_1 \neq 0}.$$

In this case additional conservation laws correspond to functions (P_2, Q_2, R_2) and (P_3, Q_3, R_3) from the case 1.3.2 and to functions

$$P_4 = (\eta + h)^2\Big(\frac{t^2}{2}u - \frac{t}{2}(x + a_2)\Big) + uR_4,\quad Q_4 = (\eta + h)^2\Big(\frac{t^2}{2}v - \frac{t}{2}y - \frac{b_2}{4}t^3\Big) + vR_4,$$
$$R_4 = (\eta + h)\Big(\frac{x^2 + y^2}{2} + a_2 x + \frac{3b_2}{2}t^2 y - a_2 tu - \frac{b_2}{2}t^3 v - t(xu + yv) + \\ + \frac{t^2}{2}(u^2 + v^2 + \eta - h) + \frac{b_2^2}{8}t^4 + b_3 t^2\Big),$$
$$P_5 = (\eta + h)^2\Big(tu - \frac{1}{2}(x + a_2)\Big) + uR_5,\quad Q_5 = (\eta + h)^2\Big(tv - \frac{1}{2}y - \frac{3b_2}{4}t^2\Big) + vR_5,$$
$$R_5 = (\eta + h)\Big(3b_2 ty - a_2 u - \frac{3b_2}{2}t^2 v - (xu + yv) + t(u^2 + v^2 + \eta - h) + \frac{b_2^2}{2}t^3 + 2b_3 t\Big).$$

1.3.3. $\mathbf{h = H(x) + b_1y^2 + b_2y,\ H''' \neq 0,\ b_1 < 0.}$
In this case additional conservation laws correspond to functions

$$P_2 = uR_2, \quad Q_2 = \frac{(\eta+h)^2 \sin(t\sqrt{-2b_1})}{2} + vR_2,$$
$$R_2 = (\eta+h)\Big(-\sqrt{-2b_1}\cos(t\sqrt{-2b_1})\Big(y+\frac{b_2}{2b_1}\Big) + \sin(t\sqrt{-2b_1})v\Big),$$
$$P_3 = uR_3, \quad Q_3 = \frac{(\eta+h)^2 \cos(t\sqrt{-2b_1})}{2} + vR_3,$$
$$R_3 = (\eta+h)\Big(\sqrt{-2b_1}\sin(t\sqrt{-2b_1})\Big(y+\frac{b_2}{2b_1}\Big) + \cos(t\sqrt{-2b_1})v\Big).$$

1.3.3.1.

$$\mathbf{h = \frac{a_1}{(x+a_2)^2} + b_1((x+a_2)^2 + y^2) + b_2y + b_3,\ b_1 < 0,\ a_1 \neq 0\,.}$$

In this case additional conservation laws correspond to functions (P_2, Q_2, R_2) and (P_3, Q_3, R_3) from the case 1.3.3 and to functions

$$P_4 = (\eta+h)^2(\sin(t\sqrt{-8b_1})u - \sqrt{-2b_1}\cos(t\sqrt{-8b_1})(x+a_2)) + uR_4,$$
$$Q_4 = (\eta+h)^2\Big(\sin(t\sqrt{-8b_1})v - \sqrt{-2b_1}\cos(t\sqrt{-8b_1})\Big(y+\frac{b_2}{2b_1}\Big)\Big) + vR_4,$$
$$\begin{aligned}R_4 = (\eta+h)\Big(&4\big(b_1((x+a_2)^2+y^2)+b_2y\big)\sin(t\sqrt{-8b_1}) - \\ &- 2a_2\sqrt{-2b_1}\cos(t\sqrt{-8b_1})u + \frac{2b_2}{\sqrt{-2b_1}}\cos(t\sqrt{-8b_1})v - \\ &- \sqrt{-8b_1}\cos(t\sqrt{-8b_1})(xu+yv) + \sin(t\sqrt{-8b_1})(u^2+v^2+\eta-h) + \\ &+ \frac{4b_1b_3+b_2^2}{2b_1}\sin(t\sqrt{-8b_1})\Big),\end{aligned}$$
$$P_5 = (\eta+h)^2(\cos(t\sqrt{-8b_1})u + \sqrt{-2b_1}\sin(t\sqrt{-8b_1})(x+a_2)) + uR_5,$$
$$Q_5 = (\eta+h)^2\Big(\cos(t\sqrt{-8b_1})v + \sqrt{-2b_1}\sin(t\sqrt{-8b_1})\Big(y+\frac{b_2}{2b_1}\Big)\Big) + vR_5,$$
$$\begin{aligned}R_5 = (\eta+h)\Big(&4\big(b_1((x+a_2)^2+y^2)+b_2y\big)\cos(t\sqrt{-8b_1}) + \\ &+ 2a_2\sqrt{-2b_1}\sin(t\sqrt{-8b_1})u - \frac{2b_2}{\sqrt{-2b_1}}\sin(t\sqrt{-8b_1})v + \\ &+ \sqrt{-8b_1}\sin(t\sqrt{-8b_1})(xu+yv) + \cos(t\sqrt{-8b_1})(u^2+v^2+\eta-h) + \\ &+ \frac{4b_1b_3+b_2^2}{2b_1}\cos(t\sqrt{-8b_1})\Big).\end{aligned}$$

2.

$$\mathbf{h} = \mathbf{H}(\mathbf{y} + \mathbf{b_0 x}) + \frac{\mathbf{b_1}}{\mathbf{b_0^2 + 1}}(-\mathbf{x} + \mathbf{b_0 y})^2 + \frac{\mathbf{b_2}}{\mathbf{b_0^2 + 1}}(-\mathbf{x} + \mathbf{b_0 y}),\ \mathbf{H}''' \neq \mathbf{0}.$$

2.1.

$$\mathbf{h} = \mathbf{H}(\mathbf{y} + \mathbf{b_0 x}) + \frac{\mathbf{b_1}}{\mathbf{b_0^2 + 1}}(-\mathbf{x} + \mathbf{b_0 y})^2 + \frac{\mathbf{b_2}}{\mathbf{b_0^2 + 1}}(-\mathbf{x} + \mathbf{b_0 y}),\ \mathbf{H}''' \neq \mathbf{0},\ \mathbf{b_1} > \mathbf{0}.$$

In this case additional conservation laws correspond to functions

$$P_2 = -(\eta + h)^2 \frac{e^{t\sqrt{2b_1}}}{b_0^2 + 1} + uR_2, \quad Q_2 = b_0(\eta + h)^2 \frac{e^{t\sqrt{2b_1}}}{b_0^2 + 1} + vR_2,$$

$$R_2 = (\eta + h)e^{t\sqrt{2b_1}}\Big(\frac{2\sqrt{2b_1}}{b_0^2 + 1}x - \frac{2b_0\sqrt{2b_1}}{b_0^2 + 1}y - \frac{2}{b_0^2 + 1}u + \frac{2b_0}{b_0^2 + 1}v - \frac{2b_2}{(b_0^2 + 1)\sqrt{2b_1}}\Big),$$

$$P_3 = -(\eta + h)^2 \frac{e^{-t\sqrt{2b_1}}}{b_0^2 + 1} + uR_3, \quad Q_3 = b_0(\eta + h)^2 \frac{e^{-t\sqrt{2b_1}}}{b_0^2 + 1} + vR_3,$$

$$R_3 = (\eta + h)e^{-t\sqrt{2b_1}}\Big(\frac{2b_0\sqrt{2b_1}}{b_0^2 + 1}y - \frac{2\sqrt{2b_1}}{b_0^2 + 1}x - \frac{2}{b_0^2 + 1}u + \frac{2b_0}{b_0^2 + 1}v + \frac{2b_2}{(b_0^2 + 1)\sqrt{2b_1}}\Big).$$

2.1.1.

$$\mathbf{h} = \frac{\mathbf{1}}{\mathbf{b_0^2 + 1}}\Big(\frac{\mathbf{a_1}}{(\mathbf{y} + \mathbf{b_0 x} + \mathbf{a_2})^2} + \mathbf{b_1}((\mathbf{y} + \mathbf{b_0 x} + \mathbf{a_2})^2 + (\mathbf{x} - \mathbf{b_0 y})^2) + \mathbf{b_2}(\mathbf{b_0 y} - \mathbf{x}) + \mathbf{b_3}\Big),$$

$\mathbf{b_1} > \mathbf{0},\ \mathbf{a_1} \neq \mathbf{0}.$

Denote

$$S_1 = \frac{1}{b_0^2 + 1}\Big(\frac{b_2}{2b_1} - b_0 a_2\Big), \qquad S_2 = \frac{1}{b_0^2 + 1}\Big(-\frac{b_0 b_2}{2b_1} - a_2\Big).$$

In this case additional conservation laws correspond to functions (P_2, Q_2, R_2) and (P_3, Q_3, R_3) from the case 2.1 and to functions

$$P_4 = (\eta + h)^2 e^{t\sqrt{8b_1}}(u - \sqrt{2b_1}(x - S_1)) + uR_4,$$
$$Q_4 = (\eta + h)^2 e^{t\sqrt{8b_1}}(v - \sqrt{2b_1}(y - S_2)) + vR_4,$$
$$R_4 = (\eta + h)e^{t\sqrt{8b_1}}\Big(4b_1(x^2 + y^2) - 8b_1(S_1x + S_2y) - \sqrt{8b_1}((x - S_1)u + (y - S_2)v) +$$
$$+ u^2 + v^2 + \eta - h + \frac{4b_1(b_3 + 2a_2^2 b_1) + b_2^2}{2b_1(b_0^2 + 1)}\Big),$$
$$P_5 = (\eta + h)^2 e^{-t\sqrt{8b_1}}(u + \sqrt{2b_1}(x - S_1)) + uR_5,$$
$$Q_5 = (\eta + h)^2 e^{-t\sqrt{8b_1}}(v + \sqrt{2b_1}(y - S_2)) + vR_5,$$
$$R_5 = (\eta + h)e^{-t\sqrt{8b_1}}\Big(4b_1(x^2 + y^2) - 8b_1(S_1x + S_2y) + \sqrt{8b_1}((x - S_1)u + (y - S_2)v) +$$
$$+ u^2 + v^2 + \eta - h + \frac{4b_1(b_3 + 2a_2^2 b_1) + b_2^2}{2b_1(b_0^2 + 1)}\Big).$$

2.2.

$$\mathbf{h = H(y + b_0x) + \frac{b_2}{b_0^2 + 1}(-x + b_0y),\ H''' \neq 0.}$$

In this case additional conservation laws correspond to functions

$$P_2 = -(\eta + h)^2 \frac{t}{b_0^2 + 1} + uR_2,\quad Q_2 = b_0(\eta + h)^2 \frac{t}{b_0^2 + 1} + vR_2,$$
$$R_2 = (\eta + h)\Big(\frac{2}{b_0^2 + 1}x - \frac{2b_0}{b_0^2 + 1}y - \frac{2t}{b_0^2 + 1}u + \frac{2b_0t}{b_0^2 + 1}v - \frac{b_2t^2}{b_0^2 + 1}\Big),$$
$$P_3 = -(\eta + h)^2 \frac{1}{b_0^2 + 1} + uR_3,\quad Q_3 = b_0(\eta + h)^2 \frac{1}{b_0^2 + 1} + vR_3,$$
$$R_3 = (\eta + h)\Big(-\frac{2}{b_0^2 + 1}u + \frac{2b_0}{b_0^2 + 1}v - \frac{2b_2t}{b_0^2 + 1}\Big).$$

2.2.1.

$$\mathbf{h = \frac{1}{b_0^2 + 1}\Big(\frac{a_1}{(y + b_0x + a_2)^2} + b_2(-x + b_0y) + b_3\Big),\ a_1 \neq 0.}$$

In this case additional conservation laws correspond to functions (P_2, Q_2, R_2) and (P_3, Q_3, R_3) from the case 2.2 and to functions

$$P_4 = (\eta + h)^2\Big(t^2 u - tx + \frac{1}{b_0^2+1}\Big(\frac{b_2}{2}\,t^3 - b_0 a_2 t\Big)\Big) + uR_4\,,$$

$$Q_4 = (\eta + h)^2\Big(t^2 v - ty + \frac{1}{b_0^2+1}\Big(-\frac{b_0 b_2}{2}\,t^3 - a_2 t\Big)\Big) + vR_4\,,$$

$$R_4 = (\eta + h)\Big(x^2 + y^2 - \frac{2}{b_0^2+1}\Big(\frac{3b_2}{2}\,t^2 - b_0 a_2\Big)x - \frac{2}{b_0^2+1}\Big(-\frac{3b_0 b_2}{2}\,t^2 - a_2\Big)y +$$
$$+ \frac{2}{b_0^2+1}\Big(\frac{b_2}{2}\,t^3 - b_0 a_2 t\Big)u + \frac{2}{b_0^2+1}\Big(-\frac{b_0 b_2}{2}\,t^3 - a_2 t\Big)v - 2t(xu + yv) +$$
$$+ t^2(u^2 + v^2 + \eta - h) + \frac{1}{b_0^2+1}\Big(\frac{b_2^2}{4}\,t^4 + 2b_3 t^2\Big)\Big),$$

$$P_5 = (\eta + h)^2\Big(2tu - x + \frac{1}{b_0^2+1}\Big(\frac{3b_2}{2}\,t^2 - b_0 a_2\Big)\Big) + uR_5\,,$$

$$Q_5 = (\eta + h)^2\Big(2tv - y + \frac{1}{b_0^2+1}\Big(-\frac{3b_0 b_2}{2}\,t^2 - a_2\Big)\Big) + vR_5\,,$$

$$R_5 = (\eta + h)\Big(-\frac{6b_2}{b_0^2+1}\,tx + \frac{6b_0 b_2}{b_0^2+1}\,ty + \frac{2}{b_0^2+1}\Big(\frac{3b_2}{2}\,t^2 - b_0 a_2\Big)u +$$
$$+ \frac{2}{b_0^2+1}\Big(-\frac{3b_0 b_2}{2}\,t^2 - a_2\Big)v - 2(xu + yv) +$$
$$+ 2t(u^2 + v^2 + \eta - h) + \frac{1}{b_0^2+1}(b_2^2 t^3 + 4b_3 t)\Big).$$

2.3.

$$\mathbf{h} = \mathbf{H}(\mathbf{y} + \mathbf{b_0 x}) + \frac{\mathbf{b_1}}{\mathbf{b_0^2 + 1}}(-\mathbf{x} + \mathbf{b_0 y})^2 + \frac{\mathbf{b_2}}{\mathbf{b_0^2 + 1}}(-\mathbf{x} + \mathbf{b_0 y}),\ \mathbf{H'''} \neq \mathbf{0},\ \mathbf{b_1} < \mathbf{0}.$$

In this case additional conservation laws correspond to functions

$$P_2 = -(\eta + h)^2\,\frac{\sin(t\sqrt{-2b_1})}{b_0^2+1} + uR_2\,,\quad Q_2 = b_0(\eta + h)^2\,\frac{\sin(t\sqrt{-2b_1})}{b_0^2+1} + vR_2\,,$$

$$R_2 = (\eta + h)\Big(\frac{2\sqrt{-2b_1}}{b_0^2+1}\,\cos(t\sqrt{-2b_1})x - \frac{2b_0\sqrt{-2b_1}}{b_0^2+1}\,\cos(t\sqrt{-2b_1})y -$$
$$- \frac{2}{b_0^2+1}\,\sin(t\sqrt{-2b_1})u + \frac{2b_0}{b_0^2+1}\,\sin(t\sqrt{-2b_1})v + \frac{2b_2}{(b_0^2+1)\sqrt{-2b_1}}\,\cos(t\sqrt{-2b_1})\Big),$$

$$P_3 = -(\eta + h)^2\,\frac{\cos(t\sqrt{-2b_1})}{b_0^2+1} + uR_3\,,\quad Q_3 = b_0(\eta + h)^2\,\frac{\cos(t\sqrt{-2b_1})}{b_0^2+1} + vR_3\,,$$

$$R_3 = (\eta + h)\Big(-\frac{2\sqrt{-2b_1}}{b_0^2+1}\,\sin(t\sqrt{-2b_1})x + \frac{2b_0\sqrt{-2b_1}}{b_0^2+1}\,\sin(t\sqrt{-2b_1})y -$$
$$- \frac{2}{b_0^2+1}\,\cos(t\sqrt{-2b_1})u + \frac{2b_0}{b_0^2+1}\,\cos(t\sqrt{-2b_1})v - \frac{2b_2}{(b_0^2+1)\sqrt{-2b_1}}\,\sin(t\sqrt{-2b_1})\Big).$$

2.3.1.

$$\mathbf{h} = \frac{1}{\mathbf{b_0^2+1}}\Big(\frac{\mathbf{a_1}}{(\mathbf{y}+\mathbf{b_0x}+\mathbf{a_2})^2} + \mathbf{b_1}((\mathbf{y}+\mathbf{b_0x}+\mathbf{a_2})^2 + (\mathbf{x}-\mathbf{b_0y})^2) + \mathbf{b_2}(\mathbf{b_0y}-\mathbf{x}) + \mathbf{b_3}\Big),$$
$$\mathbf{b_1} < \mathbf{0},\ \mathbf{a_1} \neq \mathbf{0}.$$

Denote

$$S_1 = \frac{1}{b_0^2+1}\Big(\frac{b_2}{2b_1} - b_0 a_2\Big), \qquad S_2 = \frac{1}{b_0^2+1}\Big(-\frac{b_0 b_2}{2b_1} - a_2\Big).$$

In this case additional conservation laws correspond to functions (P_2, Q_2, R_2) and (P_3, Q_3, R_3) from the case 2.3 and to functions

$$\begin{aligned}
P_4 &= (\eta+h)^2(\sin(t\sqrt{-8b_1})u - \sqrt{-2b_1}\cos(t\sqrt{-8b_1})(x-S_1)) + uR_4,\\
Q_4 &= (\eta+h)^2(\sin(t\sqrt{-8b_1})v - \sqrt{-2b_1}\cos(t\sqrt{-8b_1})(y-S_2)) + vR_4,\\
R_4 &= (\eta+h)\Big(4b_1\sin(t\sqrt{-8b_1})(x^2+y^2) - 8b_1\sin(t\sqrt{-8b_1})(S_1x+S_2y) -\\
&\quad - \sqrt{-8b_1}\cos(t\sqrt{-8b_1})((x-S_1)u + (y-S_2)v) +\\
&\quad + \sin(t\sqrt{-8b_1})(u^2+v^2+\eta-h) + \frac{4b_1(b_3+2a_2^2b_1)+b_2^2}{2b_1(b_0^2+1)}\sin(t\sqrt{-8b_1})\Big),\\
P_5 &= (\eta+h)^2(\cos(t\sqrt{-8b_1})u + \sqrt{-2b_1}\sin(t\sqrt{-8b_1})(x-S_1)) + uR_5,\\
Q_5 &= (\eta+h)^2(\cos(t\sqrt{-8b_1})v + \sqrt{-2b_1}\sin(t\sqrt{-8b_1})(y-S_2)) + vR_5,\\
R_5 &= (\eta+h)\Big(4b_1\cos(t\sqrt{-8b_1})(x^2+y^2) - 8b_1\cos(t\sqrt{-8b_1})(S_1x+S_2y) +\\
&\quad + \sqrt{-8b_1}\sin(t\sqrt{-8b_1})((x-S_1)u + (y-S_2)v) +\\
&\quad + \cos(t\sqrt{-8b_1})(u^2+v^2+\eta-h) + \frac{4b_1(b_3+2a_2^2b_1)+b_2^2}{2b_1(b_0^2+1)}\cos(t\sqrt{-8b_1})\Big).
\end{aligned}$$

3.

$$\mathbf{h} = \frac{1}{(\mathbf{x}+\mathbf{a_1})^2}\mathbf{H}\Big(\frac{\mathbf{x}+\mathbf{a_1}}{\mathbf{y}+\mathbf{a_2}}\Big) + \mathbf{a_3}((\mathbf{x}+\mathbf{a_1})^2 + (\mathbf{y}+\mathbf{a_2})^2) + \mathbf{a_4},\ \mathbf{H} \neq \mathbf{0}.$$

3.1.

$$\mathbf{h} = \frac{1}{(\mathbf{x}+\mathbf{a_1})^2}\mathbf{H}\Big(\frac{\mathbf{x}+\mathbf{a_1}}{\mathbf{y}+\mathbf{a_2}}\Big) + \mathbf{a_3}((\mathbf{x}+\mathbf{a_1})^2 + (\mathbf{y}+\mathbf{a_2})^2) + \mathbf{a_4},\ \mathbf{H} \neq \mathbf{0},\ \mathbf{a_3} > \mathbf{0}.$$

In this case additional conservation laws correspond to functions

$$
\begin{aligned}
P_2 &= (\eta + h)^2 e^{t\sqrt{8a_3}}(u - \sqrt{2a_3}(x + a_1)) + uR_2,\\
Q_2 &= (\eta + h)^2 e^{t\sqrt{8a_3}}(v - \sqrt{2a_3}(y + a_2)) + vR_2,\\
R_2 &= (\eta + h) e^{t\sqrt{8a_3}}\big(4a_3((x + a_1)^2 + (y + a_2)^2) -\\
&\quad - \sqrt{8a_3}((x + a_1)u + (y + a_2)v) + u^2 + v^2 + \eta - h + 2a_4\big),\\
P_3 &= (\eta + h)^2 e^{-t\sqrt{8a_3}}(u + \sqrt{2a_3}(x + a_1)) + uR_3,\\
Q_3 &= (\eta + h)^2 e^{-t\sqrt{8a_3}}(v + \sqrt{2a_3}(y + a_2)) + vR_3,\\
R_3 &= (\eta + h) e^{-t\sqrt{8a_3}}\big(4a_3((x + a_1)^2 + (y + a_2)^2) +\\
&\quad + \sqrt{8a_3}((x + a_1)u + (y + a_2)v) + u^2 + v^2 + \eta - h + 2a_4\big).
\end{aligned}
$$

3.1.1.

$$
\mathbf{h = \frac{b_1}{(x + a_1)^2 + (y + a_2)^2} + a_3((x + a_1)^2 + (y + a_2)^2) + a_4,\ a_3 > 0,\ b_1 \neq 0.}
$$

In this case additional conservation laws correspond to functions (P_2, Q_2, R_2) and (P_3, Q_3, R_3) from the case 3.1 and to functions

$$
\begin{aligned}
P_4 &= -(\eta + h)^2(y + a_2) + uR_4, \quad Q_4 = (\eta + h)^2(x + a_1) + vR_4,\\
R_4 &= 2(\eta + h)((x + a_1)v - (y + a_2)u).
\end{aligned}
$$

3.2.

$$
\mathbf{h = \frac{1}{(x + a_1)^2} H\Big(\frac{x + a_1}{y + a_2}\Big) + a_4,\ H \neq 0.}
$$

In this case additional conservation laws correspond to functions

$$
\begin{aligned}
P_2 &= (\eta + h)^2\Big(\frac{t^2}{2}u - \frac{t}{2}(x + a_1)\Big) + uR_2, \quad Q_2 = (\eta + h)^2\Big(\frac{t^2}{2}v - \frac{t}{2}(y + a_2)\Big) + vR_2,\\
R_2 &= (\eta + h)\Big(\frac{x^2 + y^2}{2} + a_1x + a_2y - t(x + a_1)u - t(y + a_2)v +\\
&\quad + \frac{t^2}{2}(u^2 + v^2 + \eta - h) + a_4t^2\Big),\\
P_3 &= (\eta + h)^2\Big(tu - \frac{1}{2}(x + a_1)\Big) + uR_3, \quad Q_3 = (\eta + h)^2\Big(tv - \frac{1}{2}(y + a_2)\Big) + vR_3,\\
R_3 &= (\eta + h)(-(x + a_1)u - (y + a_2)v + t(u^2 + v^2 + \eta - h) + 2a_4t).
\end{aligned}
$$

3.2.1.

$$\mathbf{h} = \frac{\mathbf{b_1}}{(\mathbf{x}+\mathbf{a_1})^2 + (\mathbf{y}+\mathbf{a_2})^2} + \mathbf{a_4}, \ \mathbf{b_1} \neq \mathbf{0}.$$

In this case additional conservation laws correspond to functions (P_2, Q_2, R_2) and (P_3, Q_3, R_3) from the case 3.2 and to functions

$$P_4 = -(\eta+h)^2(y+a_2) + uR_4, \quad Q_4 = (\eta+h)^2(x+a_1) + vR_4,$$
$$R_4 = 2(\eta+h)((x+a_1)v - (y+a_2)u).$$

3.3.

$$\mathbf{h} = \frac{\mathbf{1}}{(\mathbf{x}+\mathbf{a_1})^2}\mathbf{H}\Big(\frac{\mathbf{x}+\mathbf{a_1}}{\mathbf{y}+\mathbf{a_2}}\Big) + \mathbf{a_3}((\mathbf{x}+\mathbf{a_1})^2 + (\mathbf{y}+\mathbf{a_2})^2) + \mathbf{a_4}, \ \mathbf{H} \neq \mathbf{0}, \ \mathbf{a_3} < \mathbf{0}.$$

In this case additional conservation laws correspond to functions

$$\begin{aligned}
P_2 &= (\eta+h)^2(\sin(t\sqrt{-8a_3})u - \sqrt{-2a_3}\cos(t\sqrt{-8a_3})(x+a_1)) + uR_2,\\
Q_2 &= (\eta+h)^2(\sin(t\sqrt{-8a_3})v - \sqrt{-2a_3}\cos(t\sqrt{-8a_3})(y+a_2)) + vR_2,\\
R_2 &= (\eta+h)\big(4a_3\sin(t\sqrt{-8a_3})((x+a_1)^2 + (y+a_2)^2) -\\
&\quad - \sqrt{-8a_3}\cos(t\sqrt{-8a_3})((x+a_1)u + (y+a_2)v) +\\
&\quad + \sin(t\sqrt{-8a_3})(u^2 + v^2 + \eta - h + 2u_4)\big),\\
P_3 &= (\eta+h)^2(\cos(t\sqrt{-8a_3})u + \sqrt{-2a_3}\sin(t\sqrt{-8a_3})(x+a_1)) + uR_3,\\
Q_3 &= (\eta+h)^2(\cos(t\sqrt{-8a_3})v + \sqrt{-2a_3}\sin(t\sqrt{-8a_3})(y+a_2)) + vR_3,\\
R_3 &= (\eta+h)\big(4a_3\cos(t\sqrt{-8a_3})((x+a_1)^2 + (y+a_2)^2) +\\
&\quad + \sqrt{-8a_3}\sin(t\sqrt{-8a_3})((x+a_1)u + (y+a_2)v) +\\
&\quad + \cos(t\sqrt{-8a_3})(u^2 + v^2 + \eta - h + 2a_4)\big).
\end{aligned}$$

3.3.1.

$$\mathbf{h} = \frac{\mathbf{b_1}}{(\mathbf{x}+\mathbf{a_1})^2 + (\mathbf{y}+\mathbf{a_2})^2} + \mathbf{a_3}((\mathbf{x}+\mathbf{a_1})^2 + (\mathbf{y}+\mathbf{a_2})^2) + \mathbf{a_4}, \ \mathbf{a_3} < \mathbf{0}, \ \mathbf{b_1} \neq \mathbf{0}.$$

In this case additional conservation laws correspond to functions (P_2, Q_2, R_2) and (P_3, Q_3, R_3) from the case 3.3 and to functions

$$P_4 = -(\eta+h)^2(y+a_2) + uR_4, \quad Q_4 = (\eta+h)^2(x+a_1) + vR_4,$$
$$R_4 = 2(\eta+h)((x+a_1)v - (y+a_2)u).$$

4.

$$\mathbf{h} = \mathbf{H}\Big(\ln\big((\mathbf{x}+\mathbf{b_1})^2+(\mathbf{y}+\mathbf{b_2})^2\big)+\mathbf{b_3}\,\mathrm{arctg}\Big(\frac{\mathbf{y}+\mathbf{b_2}}{\mathbf{x}+\mathbf{b_1}}\Big)\Big)\exp\Big(\mathbf{b_3}\,\mathrm{arctg}\Big(\frac{\mathbf{y}+\mathbf{b_2}}{\mathbf{x}+\mathbf{b_1}}\Big)\Big)+\mathbf{a_4},$$

$\mathbf{b_3} \neq \mathbf{0}$, except cases 1–3.

In this case an additional conservation law of the system of Eq. (7.1) corresponds to functions

$$P_2 = (\eta+h)^2\Big(b_3tu - \frac{b_3}{2}x - y - \frac{2b_2+b_1b_3}{2}\Big) + uR_2\,,$$
$$Q_2 = (\eta+h)^2\Big(b_3tv - \frac{b_3}{2}y + x + \frac{2b_1-b_2b_3}{2}\Big) + vR_2\,,$$
$$R_2 = (\eta+h)\Big(-(2b_2+b_1b_3)u + (2b_1-b_2b_3)v +$$
$$+\,2(xv-yu) - b_3(xu+yv) + b_3t(u^2+v^2+\eta-h) + 2b_3a_4t\Big).$$

5.

$$\mathbf{h} = \mathbf{H}((\mathbf{x}+\mathbf{b_1})^2+(\mathbf{y}+\mathbf{b_2})^2)) + \mathbf{a_3}\,\mathrm{arctg}\Big(\frac{\mathbf{y}+\mathbf{b_2}}{\mathbf{x}+\mathbf{b_1}}\Big),\quad \mathbf{a_3}\neq\mathbf{0},\quad \text{except cases 1–4.}$$

In this case an additional conservation law of the system of Eq. (7.1) corresponds to functions

$$P_2 = (\eta+h)^2(-y-b_2) + uR_2\,,\quad Q_2 = (\eta+h)^2(x+b_1) + vR_2\,,$$
$$R_2 = (\eta+h)(-2b_2u + 2b_1v + 2(xv-yu) - 2a_3t)\,.$$

7.4 Conclusion

Results of the group classification can be used for obtaining new exact solutions. Results of the classification of conservation laws can be used in numerical modeling of shallow water movement over uneven bottom.

Acknowledgements The research was supported by the Russian Science Foundation (grant no. 18-11-00238).

References

1. Aksenov, A.V., Druzhkov, K.P.: Conservation laws and symmetries of the shallow water system above rough bottom. J. Phys.: Conf. Ser. **722**(012001), 1–7 (2016)
2. Stoker, J.J.: Water Waves. The Mathematical Theory With Applications. Interscience Publishers, New York (1957)
3. Aksenov, A.V., Druzhkov, K.P.: Symmetries of the equations of two-dimensional shallow water over a rough bottom. J. Phys.: Conf. Ser. **1205**(012002), 1–7 (2019)
4. Ovsiannikov, L.V.: Group Analysis of Differential Equations. Academic, New York (1982)

Chapter 8
Existence and Stability Analysis of Solutions for an Ultradian Glucocorticoid Rhythmicity and Acute Stress Model

Casey Johnson, Roman M. Taranets, Nataliya Vasylyeva, and Marina Chugunova

Abstract The hypothalamic pituitary adrenal (HPA) axis responds to physical and mental challenge to maintain homeostasis in part by controlling the body's cortisol (CORT) level. Dysregulation of the HPA axis is implicated in numerous stress-related diseases. For a structured model of the HPA axis that includes the glucocorticoid receptor (GR) but does not take into account the system response delay, we analyze linear and non-linear stability of stationary solutions. For a second mathematical model that describes the mechanism of the HPA axis self-regulatory activities and takes into account a delay of system response, we prove existence of periodic solutions under certain assumptions on ranges of parameter values and analyze stability of these solutions with respect to the time delay value.

Keywords Nonlinear differential equations · HPA · Time delay · Stability · Periodic solutions

8.1 Introduction

Hormones control a vast array of bodily functions, including sexual reproduction and sexual development, whole-body metabolism, blood glucose levels and so on (see [1]). Hormones are produced, in main, and released from diverse places including the hypothalamus, pituitary, and the adrenal gland. Hormones are capable of a diffusion whole-body effect, as well as a localized effect, depending on the distance between the production site and the site of action. In many ways, the endocrine system is similar to the nervous system, in that it is an intercellular signaling system in which

C. Johnson (✉) · M. Chugunova
Institute of Mathematical Sciences, Claremont Graduate University, Claremont, CA, USA
e-mail: casey.johnson@cgu.edu

R. M. Taranets · N. Vasylyeva
Institute of Applied Mathematics and Mechanics of the NASU,
Dobrovol'skogo Str. 1, Sloviansk 84100, Ukraine
e-mail: taranets_r@yahoo.com

V. A. Sadovnichiy and M. Z. Zgurovsky (eds.), *Contemporary Approaches and Methods in Fundamental Mathematics and Mechanics*, Understanding Complex Systems,
https://doi.org/10.1007/978-3-030-50302-4_8

cells communicate via cellular secretions. Further, the distance between the sites of hormone production and action, and the complexities inherent in the mode of transport, make it extraordinarily difficult to construct quantitative models of hormonal control.

The hypothalamus pituitary adrenal (HPA) axis is a central neuroendocrine system, which consists of the hypothalamus, pituitary, and adrenal glands. The paraventricular nucleus of the hypothalamus secrets corticotropin releasing hormone (CRH), which is transferred to the pituitary and stimulates the synthesis and release of adrenocorticotropic hormone (ACTH). ACTH moves through the bloodstream and reaches the adrenal gland in which it stimulates the secretion of cortisol. In response to stress, the concentrations of the HPA axis hormones are increased. A brief review of the HPA axis and the various factors that regulate its functions are described in [2].

Disruption of HPA axis regulation is known to contribute to a number of stress-related disorders. For example, increased CORT has been shown in patients with major depressive disorder and is believed to maximize therapeutic antidepressant effects (see [3, 4]), and decreased CORT has been observed in people with post-traumatic stress disorder (see [5]).

Multiple models of the HPA axis have been developed to characterize the oscillations seen in the hormone concentrations and to examine HPA axis dysfunction. Most of these models have been constructed using deterministic coupled ordinary differential equations (see [6]). A major inconsistency among different existing HPA models that was mentioned in [7] is related to their treatment of the circadian and ultradian oscillations. For example, the authors of [8, 9] assumed that both oscillations can be generated inside the HPA axis system by interaction of its elements; the authors of [10–12] treat the circadian and ultradian oscillations differently assuming that only ultradian oscillations are HPA axis based but at the same time circadian rhythms are due to external input. Only one model made no explicit assumption about the origin of the oscillations and was developed to replicate the HPA axis response to CRH injection (see [13]). It has also been suggested that the ultradian rhythm arises from the introduction of a time delay (see [9]). Other models based on delay-differential equations include [14, 15]. The results of these models have been used to develop treatment strategies by maximally exploiting system dynamics to stimulate recovery (see [16]).

To determine if delay-differential equations could predict the general features of CORT production, the experimental data was compared to a simulated CORT curve in [15]. It was not possible to obtain any experimental fitting of ACTH for the model since hypothalamic derived CRH cannot be measured.

Inclusion of the glucocorticoid receptor (GR) in a HPA axis model reveals 'bi-stability' (see [15]). To be more concrete, there arises a nonlinear Gauss type function with compact support, which is characterized by the parameter p_4. This Hill function arises as a result of 'inner' nonlinearity in the physiological system which is produced by the stress impulse, which is activated by the outer impulse that is called by an acute stress. This situation is provided formally by the two parameters p_4 and CRH. The amplitude of the Hill function determines GR density in the pituitary, which is coupled nonlinearly in reaction with regulated levels of CORT which in turn mediate a wide range of physiological processes, including metabolic, immunological and cognitive function (see [17, 20]).

The stress response is subserved by the stress system which is located both in the central nervous system and the periphery. The principal effects to the stress system include the CRH. The secretion of CRH causes the anterior pituitary to synthesize ACTH which then stimulates the adrenal glands to release CORT that regulate the blood concentration of CRH and ACTH via different negative feedback mechanisms. A model has been developed that links the HPA axis and the memory system in the stress reaction (see [18]). The HPA axis is the subject of intensive research in endocrinology. A study has shown that CRH may have a positive very short loop feedback action that enhances stress-induced ACTH released (see [19]). This model is based on the feed-forward and feedback interactions between the anterior pituitary and adrenal glands. Because responsiveness of the stress system to stressors is crucial for life, it is important to consider the simpler case when distributions of hormones in the system become unstable by action on stress, and further to consider influence on the delay time as response of the physiological system on action on stress.

Mathematically, it means that we can consider two mathematical models: the first one is described by a system of ordinary differential equations with initial distributions of hormones at a point $t = 0$, and the second one is based upon a system of differential equations with initial distributions of hormones on the interval $[-\tau, 0)$, where τ is a time delay. It turns out that bi-stability is present in both models, i.e. limit distributions of hormones may be stable or unstable depending on parameter values. In the model with a time delay, periodic solutions arise for special distributions of hormones when there is a connection between the concentrations of hormones at end points. Thus, the initial distributions of hormones must be coupled in a special manner. This condition may be considered as a 'normal' reaction of the organism on action of stress.

In this paper, we study a system of delay differential equations (see [15, 20]):

$$\frac{da}{dt} = \frac{CRH}{1+p_2 or} - p_3 a =: f_1, \tag{8.1}$$

$$\frac{dr}{dt} = \frac{(or)^2}{p_4+(or)^2} + p_5 - p_6 r =: f_2, \tag{8.2}$$

$$\frac{do}{dt} = a(t-\tau) - o =: f_3, \tag{8.3}$$

with initial conditions

$$a(t) = a_\tau(t)\ \forall t \in [-\tau, 0],\ r(0) = r_0,\ o(0) = o_0, \tag{8.4}$$

where

$$0 \leqslant a_\tau(t) \in C^1[-\tau, 0],\ r_0 > 0,\ o_0 > 0. \tag{8.5}$$

Based on the principles of mass action kinetics, these equations describe the production and degradation of the hormones ACTH (a), and CORT (o), as well as GR density (r) in the pituitary. Here, the parameters p_{2-6} represent dimensionless forms of rate constants of the system, and the dimensionless parameter τ represents a discrete delay, which accounts for the delayed response of the adrenal gland to ACTH. The dimensionless time $t = 0$ corresponds to the maximal value of an ACTH pulse.

The paper is organized as follows. In Sect. 8.2 we study the stability of the system (8.1)–(8.3) without a delay for the given initial distributions of hormones. In Sect. 8.3 we analyze the system (8.1)–(8.3) with delay in terms of stability, solvability, and existence of periodic solutions. Moreover, the existence of periodic solutions is proved. The system without delay is always asymptotically stable for strictly positive parameter values. When delay is introduced, the system always becomes unstable after a certain threshold delay value is reached. The theoretical results are then compared with numerical simulations.

8.2 Stability Analysis of the Model Without a Time Delay

We consider the following nonlinear ODEs without a delay:

$$\frac{da}{dt} = \frac{A}{1+p_2or} - p_3a, \tag{8.6}$$

$$\frac{dr}{dt} = \frac{(or)^2}{p_4+(or)^2} + p_5 - p_6r, \tag{8.7}$$

$$\frac{do}{dt} = a - o, \tag{8.8}$$

with initial conditions

$$a(0) = a_0 \geqslant 0, \; r(0) = r_0 > 0, \; o(0) = o_0 > 0, \tag{8.9}$$

where $A := \text{CRH} \geqslant 0$, and $p_i \geqslant 0$. Using Picard's iteration method, we can show existence of a unique global in time non-negative solution.

Lemma 8.1 *Assume that $A > 0$ and $p_i > 0$. Then the system (8.6)–(8.8) has a unique fixed point and this point is asymptotically stable if the following conditions are satisfied:*

$$\frac{1}{8} < p_5 < \frac{p_3 - p_6(p_3+1)}{4p_6(p_3+1)} \; \textit{if}\, 0 < p_6 < \frac{p_3}{p_3+1},$$
$$p_5 > \max\{\tfrac{1}{8}, \tfrac{(p_6(p_3+1)-p_3)^2}{p_6(p_3+1)+p_3}\} \; \textit{if}\, \frac{p_3}{p_3+1} \leqslant p_6 < p_3+1,$$
$$p_5 > \max\{\tfrac{1}{8}, \tfrac{(p_6(p_3+1)-p_3)^2}{p_6(p_3+1)+p_3}, \tfrac{(p_3-p_6+1)^2}{2p_6(p_3+p_6+1)}\} \; \textit{if}\, p_6 \geqslant p_3+1.$$

Proof (*Proof of Lemma* 8.1) By (8.6)–(8.8) we obtain the following equations for the nullclines:

$$o = a, \quad \frac{A}{1+p_2 or} - p_3 a = 0, \quad \frac{(or)^2}{p_4+(or)^2} + p_5 - p_6 r = 0. \tag{8.10}$$

The algebraic system (8.10) has a nonnegative solution in the following domain:

$$D := \{(a, r, o) \in R_+^3 : a = o,\ 0 \leqslant a \leqslant \frac{A}{p_3},\ \frac{p_5}{p_6} \leqslant r \leqslant \frac{p_5+1}{p_6}\}.$$

From (8.10) we have

$$o = a = \frac{1}{2p_2 r}\left[\sqrt{1 + \frac{4p_2 A}{p_3} r} - 1\right],$$

$$\frac{1}{4p_2^2}\left[\sqrt{1+\frac{4p_2 A}{p_3} r} - 1\right]^2 = \frac{p_4(p_6 r - p_5)}{1+p_5-p_6 r} \Rightarrow A = \frac{p_3}{r}\sqrt{\frac{p_4(p_6 r-p_5)}{1+p_5-p_6 r}}\left(1 + p_2\sqrt{\frac{p_4(p_6 r-p_5)}{1+p_5-p_6 r}}\right). \tag{8.11}$$

Note that the Eq. (8.11) has a unique solution $r^* \in (\frac{p_5}{p_6}, \frac{p_5+1}{p_6})$ in D. Really, the function $f_1(r) := \frac{1}{4p_2^2}[\sqrt{1+\frac{4p_2 A}{p_3} r} - 1]^2$ is nonnegative and monotone increasing on $[0, +\infty)$ such that $f_1(0) = 0$, $f_1(+\infty) = +\infty$, but the function $f_2(r) := \frac{p_4(p_6 r - p_5)}{1+p_5-p_6 r}$ is nonnegative and monotone increasing on $[\frac{p_5}{p_6}, \frac{p_5+1}{p_6}]$ such that $f_2''(r) > 0$, $f_2(\frac{p_5}{p_6}) = 0$, $f_2(\frac{p_5+1}{p_6}) = +\infty$. Therefore, there is only one intersection of $f_1(r)$ and $f_2(r)$ on the interval $[\frac{p_5}{p_6}, \frac{p_5+1}{p_6}]$. On the other hand, let us denote by

$$z := \sqrt{1+\frac{4p_2 A}{p_3} r} - 1 \geqslant 0 \Rightarrow r = \frac{p_3}{4p_2 A} z(z+2).$$

Then (8.11) can be rewritten in the following form

$$z^4 + 2z^3 + C_1 z^2 + C_2 z - C_3 = 0, \tag{8.12}$$

where

$$C_1 := \frac{4p_2(p_2 p_3 p_4 p_6 - A(p_5+1))}{p_3 p_6},\ C_2 := 8p_2^2 p_4 > 0,\ C_3 := \frac{16Ap_2^3 p_4 p_5}{p_3 p_6} > 0.$$

So, we can find the explicit value of $r^* \in (\frac{p_5}{p_6}, \frac{p_5+1}{p_6})$ as a solution of (8.12).

As a result, the system (8.6)–(8.8) has only one fixed point (a^*, r^*, o^*) in D. Here,

$$a^* = o^* = \frac{1}{2p_2 r^*}\left[\sqrt{1+\frac{4p_2 A}{p_3} r^*} - 1\right] = \frac{1}{r^*}\sqrt{\frac{p_4(p_6 r^* - p_5)}{1+p_5-p_6 r^*}},$$

and r^* is the solution of (8.11) or (8.12).

Next, we find the Jacobian matrix J^* for (8.6)–(8.8) at the fixed point (a^*, r^*, o^*).

$$J^* = \begin{pmatrix} -p_3 & -K_1 & -K_3 \\ 0 & -p_6 + K_2 & K_4 \\ 1 & 0 & -1 \end{pmatrix},$$

where

$$K_1 = \frac{Ap_2a^*}{(1+p_2a^*r^*)^2} = \frac{2A(\sqrt{1+\frac{4p_2A}{p_3}r^*}-1)}{r^*(1+\sqrt{1+\frac{4p_2A}{p_3}r^*})^2} = \frac{Ap_2\sqrt{p_4(p_6r^*-p_5)(1+p_5-p_6r^*)}}{r^*(p_2\sqrt{p_4(p_6r^*-p_5)}+\sqrt{1+p_5-p_6r^*})^2} =$$

$$\frac{p_2p_3p_4(p_6r^*-p_5)}{(r^*)^2(p_2\sqrt{p_4(p_6r^*-p_5)}+\sqrt{1+p_5-p_6r^*})^2}\left(1 + p_2\sqrt{\frac{p_4(p_6r^*-p_5)}{1+p_5-p_6r^*}}\right) \geqslant 0,$$

$$K_2 = \frac{2p_4r^*(a^*)^2}{(p_4+(a^*r^*)^2)^2} = \frac{1}{2p_2^2p_4r^*}\left[\sqrt{1+\frac{4p_2A}{p_3}r^*} - 1\right]^2(1+p_5-p_6r^*)^2 =$$

$$\frac{2}{r^*}(p_6r^* - p_5)(1 + p_5 - p_6r^*) \geqslant 0 \text{ and } 0 \leqslant K_2 \leqslant 2p_6(\sqrt{1+p_5} - \sqrt{p_5})^2,$$

$$K_3 = \frac{r^*}{a^*}K_1 = \frac{p_2p_3\sqrt{p_4(p_6r^*-p_5)}}{\sqrt{1+p_5-p_6r^*}+p_2\sqrt{p_4(p_6r^*-p_5)}} \text{ and } 0 \leqslant K_3 \leqslant p_3,$$

$$K_4 = \frac{r^*}{a^*}K_2 = \frac{2r^*}{\sqrt{p_4}}(1 + p_5 - p_6r^*)^{\frac{3}{2}}\sqrt{(p_6r^* - p_5)} \geqslant 0.$$

Next, we will analyze the stability of the fixed point. First, we look for eigenvalues for J^*. So,

$$|J^* - \lambda I| = \begin{vmatrix} -p_3 - \lambda & -K_1 & -K_3 \\ 0 & -p_6 + K_2 - \lambda & K_4 \\ 1 & 0 & -1 - \lambda \end{vmatrix} = 0,$$

whence we obtain the characteristic equation:

$$(\lambda + 1)(\lambda + p_3)(\lambda + p_6 - K_2) + K_3(\lambda + p_6) = 0,$$

i.e.

$$\lambda^3 + \alpha_1\lambda^2 + \alpha_2\lambda + \alpha_3 = 0, \tag{8.13}$$

where

$$\alpha_1 = p_3 + p_6 - K_2 + 1,\ \alpha_2 = p_3 + p_6 - K_2 + p_3(p_6 - K_2) + K_3,\ \alpha_3 = p_3(p_6 - K_2) + p_6K_3.$$

Let us denote by

$$\Delta := 18\alpha_1\alpha_2\alpha_3 - 4\alpha_1^3\alpha_3 + \alpha_1^2\alpha_2^2 - 4\alpha_2^3 - 27\alpha_3^2.$$

If $\Delta > 0$, then (8.13) has three distinct real roots. If $\Delta = 0$, then (8.13) has a multiple root and all of its roots are real. If $\Delta < 0$, then (8.13) has one real root and two complex roots.

To analyze stability, we will use Lemma 8.4 from the Appendix. Let $x_1 := \frac{p_5}{p_6} \leqslant x := r^* \leqslant x_2 := \frac{p_5+1}{p_6}$. Then in our case,

$$\alpha_1 > 0 \Leftrightarrow 0 \leqslant K_2 < p_6 + p_3 + 1 \Leftrightarrow (x - x_1)(x_2 - x) < \frac{p_6+p_3+1}{2p_6^2}x,$$

is true provided

$$\frac{(\sqrt{p_5+1}-\sqrt{p_5})^2}{p_6} < \frac{p_6+p_3+1}{2p_6^2} \Rightarrow p_5 > 0 \text{ if } p_6 < p_3 + 1,$$

$$p_5 > \frac{(p_3-p_6+1)^2}{2p_6(p_3+p_6+1)} \text{ if } p_6 \geqslant p_3 + 1,$$

and

$$\alpha_3 > 0 \Leftrightarrow 0 \leqslant K_2 < p_6(1 + \tfrac{K_3}{p_3}) \Leftrightarrow (x - x_1)(x_2 - x) < \frac{x}{2p_6}\left[1 + \frac{p_2\sqrt{p_4(x-x_1)}}{\sqrt{x_2-x}+p_2\sqrt{p_4(x-x_1)}}\right].$$

Which is true for

$$\frac{(\sqrt{p_5+1}-\sqrt{p_5})^2}{p_6} < \frac{1}{2p_6} \Rightarrow p_5 > \frac{1}{8}.$$

As $\alpha_1 > 0$ and $\alpha_3 > 0$ then

$$\alpha_2 > 0 \Rightarrow 0 \leqslant K_2 \leqslant p_6 + \frac{K_3+p_3}{p_3+1}, \tag{8.14}$$

$$\alpha_1\alpha_2 > \alpha_3 \Leftrightarrow K_2^2 - K_2(p_3 + 1 + 2p_6 + \tfrac{K_3}{p_3+1}) + (p_3 + p_6)(p_6 + 1) + K_3 > 0. \tag{8.15}$$

From (8.14) and (8.15) it follows that

$$0 \leqslant K_2 \leqslant p_6 + \frac{K_3+p_3}{p_3+1} \Leftrightarrow (x - x_1)(x_2 - x) \leqslant \frac{x}{2p_6^2}\left[p_6 + \frac{p_3}{p_3+1} + \frac{1}{p_3+1}\frac{p_2p_3\sqrt{p_4(x-x_1)}}{\sqrt{x_2-x}+p_2\sqrt{p_4(x-x_1)}}\right].$$

This is true provided

$$\frac{(\sqrt{p_5+1}-\sqrt{p_5})^2}{p_6} < \frac{1}{2p_6^2}(p_6 + \tfrac{p_3}{p_3+1}), \quad \Leftrightarrow 2p_5 + \tfrac{1}{2} - \frac{p_3}{2p_6(p_3+1)} < 2\sqrt{p_5(p_5 + 1)},$$
$$\Rightarrow 0 < p_5 < \frac{p_3-p_6(p_3+1)}{4p_6(p_3+1)} \text{ if } 0 < p_6 < \frac{p_3}{p_3+1},$$

and

$$p_5 \geqslant \frac{(p_6(p_3+1)-p_3)^2}{p_6(p_3+1)+p_3} \text{ if } p_6 \geqslant \frac{p_3}{p_3+1}.$$

Therefore, the fixed point is asymptotically stable if:

$$\frac{1}{8} < p_5 < \frac{p_3 - p_6(p_3+1)}{4p_6(p_3+1)} \text{ if } 0 < p_6 < \frac{p_3}{p_3+1},$$
$$p_5 > \max\{\frac{1}{8}, \frac{(p_6(p_3+1)-p_3)^2}{p_6(p_3+1)+p_3}\} \text{ if } \frac{p_3}{p_3+1} \leqslant p_6 < p_3+1,$$
$$p_5 > \max\{\frac{1}{8}, \frac{(p_6(p_3+1)-p_3)^2}{p_6(p_3+1)+p_3}, \frac{(p_3-p_6+1)^2}{2p_6(p_3+p_6+1)}\} \text{ if } p_6 \geqslant p_3+1.$$

8.2.1 Stability for Different Parameter Values

Case 1: If $A = 0$ then the system (8.6)–(8.8) has the fixed point $(0, \frac{p_5}{p_6}, 0)$. The corresponding characteristic equation is

$$(\lambda + 1)(\lambda + p_3)(\lambda + p_6) = 0,$$

whence $\lambda_i = -1, \ -p_3, \ -p_6 < 0$. As a result, $(0, \frac{p_5}{p_6}, 0)$ is stable node.
Case 2: If $p_2 = 0$ then the system (8.6)–(8.8) has the fixed point $(\frac{A}{p_3}, r^*, \frac{A}{p_3})$, where r^* is a solution of the equation:

$$\frac{p_4}{p_4 + (\frac{A}{p_3})^2 r^2} = 1 + p_5 - p_6 r.$$

This equation can also be written as

$$r\left[\frac{p_6}{p_4 p_5}\left(\frac{A}{p_3}\right)^2 r^2 - \frac{1+p_5}{p_4 p_5}\left(\frac{A}{p_3}\right)^2 r + \frac{p_6}{p_5}\right] = 1$$

which yields the following cases:

- if $p_4 > [\frac{A(1+p_5)}{p_3 p_6}]^2$ then there is one real root;
- if $p_4 = [\frac{A(1+p_5)}{p_3 p_6}]^2$ then $r(r - \frac{1+p_5}{2p_6})^2 = \frac{p_4 p_5}{p_6}$, whence

  if $\frac{p_5 A^2}{p_3^2 p_6} < \frac{1}{54}$ then we have three real roots,
  if $\frac{p_5 A^2}{p_3^2 p_6} = \frac{1}{54}$ then we have two real roots,
  if $\frac{p_5 A^2}{p_3^2 p_6} > \frac{1}{54}$ then we have one real root;

- if $p_4 < [\frac{A(1+p_5)}{p_3 p_6}]^2$ then $r(r - r_1)(r - r_2) = \frac{p_4 p_5}{p_6}(\frac{p_3}{A})^2$, where
  $r_{1,2} = \frac{1}{2p_6}[1 + p_5 \pm \sqrt{(p_5+1)^2 - p_4 p_6^2 (\frac{p_3}{A})^2}]$, whence

  if $\frac{p_4 p_5}{p_6}(\frac{p_3}{A})^2 < \frac{(r_1+r_2-K)(r_2-2r_1-K)(r_1-2r_2-K)}{27}$ then we have three real roots,
  if $\frac{p_4 p_5}{p_6}(\frac{p_3}{A})^2 = \frac{(r_1+r_2-K)(r_2-2r_1-K)(r_1-2r_2-K)}{27}$ then we have two real roots,
  if $\frac{p_4 p_5}{p_6}(\frac{p_3}{A})^2 > \frac{(r_1+r_2-K)(r_2-2r_1-K)(r_1-2r_2-K)}{27}$ then we have one real root,

where $K^2 = r_1^2 - r_1 r_2 + r_2^2$.

The corresponding characteristic equation is

$$(\lambda + 1)(\lambda + p_3)(\lambda + p_6 - K_2) = 0,$$

whence $\lambda_i = -1, -p_3, -p_6 + K_2$. If $K_2 < p_6$ then $(\frac{A}{p_3}, r^*, \frac{A}{p_3})$ is stable node. If $K_2 > p_6$ then $(\frac{A}{p_3}, r^*, \frac{A}{p_3})$ is saddle. If $K_2 = p_6$ then it is a non-hyperbolic fixed point.

Case 3: If $p_3 = 0$ then the system (8.6)–(8.8) has the fixed point $(+\infty, \frac{p_5+1}{p_6}, +\infty)$. The corresponding characteristic equation is

$$\lambda(\lambda + 1)(\lambda + p_6) = 0,$$

whence $\lambda_i = -1, 0, -p_6$. As a result, $(+\infty, \frac{p_5+1}{p_6}, +\infty)$ is non-hyperbolic fixed point.

Case 4: If $p_4 = 0$ then the system (8.6)–(8.8) has the fixed point $(a^*, \frac{p_5+1}{p_6}, a^*)$, where

$$a^* = \frac{p_6}{2p_2(p_5+1)}\left[\sqrt{1 + \frac{4Ap_2(p_5+1)}{p_3 p_6}} - 1\right].$$

In this case, we have that $K_2 = K_3 = 0$ and $(\lambda + 1)(\lambda + p_3)(\lambda + p_6) = 0$, whence $\lambda_i = -1, -p_3, -p_6$. Hence, the fixed point is a stable node.

Case 5: If $p_2 = p_4 = 0$ then we obtain the explicit solution

$$\begin{aligned}
a(t) &= (a_0 - \tfrac{A}{p_3})e^{-p_3 t} + \tfrac{A}{p_3} \to \tfrac{A}{p_3} \text{ as } t \to +\infty,\\
r(t) &= (r_0 - \tfrac{p_5+1}{p_6})e^{-p_6 t} + \tfrac{1+p_5}{p_6} \to \tfrac{1+p_5}{p_6} \text{ as } t \to +\infty,\\
o(t) &= (o_0 - \tfrac{A}{p_3})e^{-t} + (a_0 - \tfrac{A}{p_3})e^{-t}\int_0^t e^{(1-p_3)s}ds + \tfrac{A}{p_3} \to \tfrac{A}{p_3} \text{ as } t \to +\infty.
\end{aligned}$$

Case 6: If $p_5 = 0$ then the one of fixed points is $(\frac{A}{p_3}, 0, \frac{A}{p_3})$ and

$$(\lambda + 1)(\lambda + p_3)(\lambda + p_6) = 0,$$

whence $\lambda_i = -1, -p_3, -p_6$. Hence, this fixed point is stable node. In this case, by (8.10) we obtain that

$$\frac{A}{1+p_2 ar} = p_3 a \Leftrightarrow r = \frac{1}{p_2 a}\left(\frac{A}{p_3 a} - 1\right) \text{ provided } 0 < a < \frac{A}{p_3},$$

whence

$$\frac{(ar)^2}{p_4+(ar)^2} = p_6 r$$

implies that

$$r = 0 \text{ or } \frac{a^2 r}{p_4 + (ar)^2} = p_6.$$

Hence, we find that

$$a^3 + \left(\frac{p_6(1+p_2^2 p_4)}{p_2} - \frac{A}{p_3}\right)a^2 - 2\frac{Ap_6}{p_2 p_3}a = -\frac{p_6}{p_2}(\frac{A}{p_3})^2 \Leftrightarrow f(a) := a(a-a_1)(a-a_2) = -\frac{p_6}{p_2}(\frac{A}{p_3})^2,$$

where

$$a_{1,2} = \frac{1}{2}\left[-(\frac{p_6(1+p_2^2 p_4)}{p_2} - \frac{A}{p_3}) \pm \sqrt{(\frac{p_6(1+p_2^2 p_4)}{p_2} - \frac{A}{p_3})^2 + 8\frac{Ap_6}{p_2 p_3}}\right]$$

and $a_1 < 0 < a_2$. Note that $a_2 \leqslant \frac{A}{p_3}$ provided $p_2^2 p_4 \geqslant 1$. As a result,

- if $p_2^2 p_4 \geqslant 1$ then

  if $f_{min} > -\frac{p_6}{p_2}(\frac{A}{p_3})^2$ then no real roots;
  if $f_{min} = -\frac{p_6}{p_2}(\frac{A}{p_3})^2$ then one positive real root;
  if $f_{min} < -\frac{p_6}{p_2}(\frac{A}{p_3})^2$ then two positive real roots;

- if $p_2^2 p_4 < 1$ then

  if $f_{min} \geqslant -\frac{p_6}{p_2}(\frac{A}{p_3})^2$ then no real roots;
  if $f_{min} < -\frac{p_6}{p_2}(\frac{A}{p_3})^2$ then one positive real root.

Case 7: If $p_6 = p_5 = 0$ then we have the following system

$$a'(t) = \frac{A}{1+p_2 or} - p_3 a, \quad r'(t) = \frac{(or)^2}{p_4 + (or)^2}, \quad o'(t) = a - o.$$

If $r_0 = 0$ then we find the explicit solution

$$a(t) = (a_0 - \frac{A}{p_3})e^{-p_3 t} + \frac{A}{p_3}, \quad r(t) = 0,$$

$$o(t) = (o_0 - \frac{A}{p_3})e^{-t} + (o_0 - \frac{A}{p_3})e^{-t}\int_0^t e^{(1-p_3)s}ds + \frac{A}{p_3}.$$

If $r_0 \neq 0$ then we approximately have

$$a'(t) \approx -p_3(a - \frac{A}{p_3}) - \frac{A^2 p_2}{p_3}r, \quad r'(t) \approx \frac{1}{p_4}(\frac{A}{p_3})^2 r^2, \quad o'(t) = a - o,$$

whence

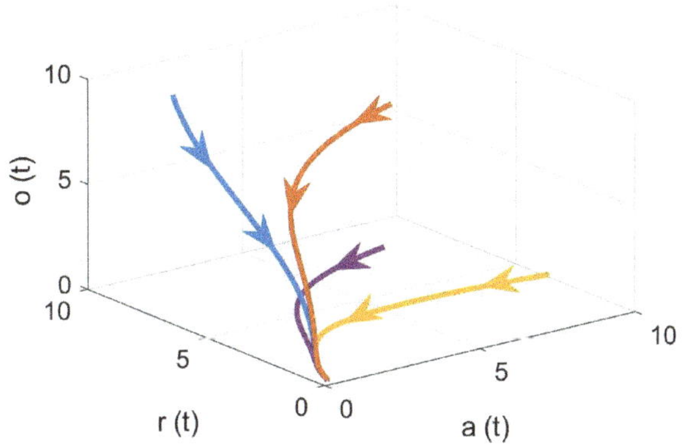

Fig. 8.1 A plot of different trajectories illustrating the stable node associated with parameter values given in Example 8.1

$$r(t) \approx \frac{r_0}{1-\frac{r_0}{p_4}(\frac{A}{p_3})^2 t} \to +\infty \text{ as } t \to T^* := \frac{p_4}{r_0}(\frac{p_3}{A})^2,$$

$$a(t) \approx a_0 e^{-p_3 t} + \frac{A}{p_3}(1 - e^{-p_3 t}) - \frac{A^2 p_2}{p_3} e^{-p_3 t} \int_0^t r(s) e^{p_3 s} ds,$$

$$o(t) \approx o_0 e^{-t} + e^{-t} \int_0^t a(s) e^s ds.$$

If $p_6 = 0$ but $p_5 \neq 0$ then $r(t)$ blows up in a finite time too.

Also, note that if $p_6 = 0$ and $p_5 = 0$ then the system (8.6)–(8.8) has the fixed point $(\frac{A}{p_3}, 0, \frac{A}{p_3})$. The corresponding characteristic equation is

$$\lambda(\lambda + 1)(\lambda + p_3) = 0,$$

whence $\lambda_i = -1, -p_3, 0$. As a result, $(\frac{A}{p_3}, 0, \frac{A}{p_3})$ is non-hyperbolic fixed point.

Example 8.1 Let $A = 1$, $p_2 = 15$, $p_3 = 7.2$, $p_4 = 0.05$, $p_5 = 0.11$, and $p_6 = 2.9$. Then $r^* \approx 0.03$, $a^* = o^* \approx 0.12$, $\alpha_1 = 11.1 - K_2 \approx 11.07$, $\alpha_2 = 30.98 - 8.2K_2 + K_3 \approx 30.78$, $\alpha_3 = 20.88 - 7.2K_2 + 2.9K_3 \approx 20.75$, whence we find that $\Delta \approx 2509.05 > 0$, $\alpha_1\alpha_2 > \alpha_3$. As a result, all characteristic roots are negative real numbers and the fixed point is stable node.

A visual representation of this stable node can be found in Fig. 8.1. This plot was created using the Matlab ode45 solver [21] using various starting values and the parameter values given above. The starting values were selected so that $a_0 > 0$, $r_0 > 0$ and $o_0 > 0$ to imitate real initial hormone levels.

Example 8.2 Let $A = 0.106$, $p_2 = 0$, $p_3 = 0.222$, $p_4 = 0.464$, $p_5 = 0.094$, and $p_6 = 0.418$. Then $r^* \approx 0.39, 0.83, 1.38$ and $a^* = o^* \approx 0.47$. Using similar calculations as above according to the defined values. If $r^* \approx 0.39$ then $\alpha_1 \approx 1.30$, $\alpha_2 \approx 0.32$, $\alpha_3 \approx 0.01$, and $K_2 \approx 0.33 < p_6$ which means this is a stable node. If $r^* \approx 0.83$ then $\alpha_1 \approx 1.18$, $\alpha_2 \approx 0.17$, and $\alpha_3 \approx -0.008$, and $K_2 \approx 0.45 > p_6$

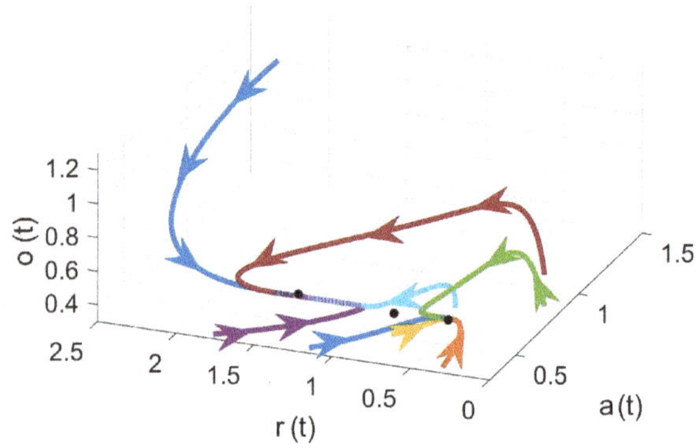

Fig. 8.2 A plot of different trajectories illustrating the unstable saddle-node with only realistic initial conditions and the above parameter values stated in Example 8.2

which means this is a saddle. If $r^* \approx 1.38$ then $\alpha_1 \approx 1.28$, $\alpha_2 \approx 0.29$, $\alpha_3 \approx 0.01$, and $K_2 \approx 0.35 < p_6$ which means this is a stable node.

This is illustrated in Fig. 8.2 using the stated above parameter values. The starting values were selected so that $a_0 > 0$, $r_0 > 0$ and $o_0 > 0$ to imitate real initial hormone levels.

8.2.2 *Lyapunov Stability Analysis*

In this subsection, we show the stability of the fixed point by using the Lyapunov function approach. We consider the system (8.6)-(8.8) and denote

$$W(t) := \tfrac{1}{2}[(a(t)-a^*)^2 + (r(t)-r^*)^2 + (o(t)-o^*)^2 + (o(t)r(t)-o^*r^*)^2],$$

where (a^*, r^*, o^*) is the fixed point $(a^* = o^*)$.

Lemma 8.2 (Stability) *Assume that*

$$A \geqslant 0,\ \ p_2 \geqslant 0,\ \ p_3 > \tfrac{1}{2},\ \ p_4 \geqslant 0,\ \ p_6 \min\{p_3 - \tfrac{1}{2}, p_6, 1\} > 0,$$

and

$$0 \leqslant p_5 < p_6 \min\{p_3 - \tfrac{1}{2}, p_6, 1\} - 1.$$

Then there exist $W^* > 0$, $A_0 > 0$ and $p_4^* > 0$ such that

$$W(t) \to 0 \text{ as } t \to +\infty \tag{8.16}$$

provided $W(0) < W^*$, $0 \leqslant A < A_0$, $p_4 > p_4^*$, hence, the fixed point (a^*, r^*, o^*) is stable. If $p_4 \leqslant p_4^*$ then there exist $A_0 \leqslant A_1 < A_2$ such that (8.16) holds provided $W(0) < W^*$, $A_1 < A < A_2$.

Proof (*Proof of Lemma* 8.2) Using the system (8.6)–(8.8), we have

$$\frac{d}{dt}W(t)-(or-o^*r^*)(o(t)r(t))'=-p_3(a-a^*)^2-p_6(r-r^*)^2-(o-o^*)^2$$
$$+(a-a^*)\Big[\frac{A}{1+p_2or}-\frac{A}{1+p_2o^*r^*}\Big]+(r-r^*)\Big[\frac{p_4}{p_4+(o^*r^*)^2}-\frac{p_4}{p_4+(or)^2}\Big]+(a-o^*)(o-o^*).$$

As

$$2(a-o^*)(o-o^*)\leqslant(a-a^*)^2+(o-o^*)^2,\quad \Big|\frac{1}{1+p_2or}-\frac{1}{1+p_2o^*r^*}\Big|\leqslant p_2|or-o^*r^*|,$$

$$\begin{aligned}(o(t)r(t))'&=r(a-o)+o[-\tfrac{p_4}{p_4+(or)^2}+1+p_5-p_6r]\\&=(r-r^*)(a-a^*)+r^*(a-a^*)+a^*(r-r^*)-(or-o^*r^*)-\\&\quad p_6(r-r^*)(o-o^*)-p_6o^*(r-r^*)+(o-o^*)\Big[\frac{p_4}{p_4+(o^*r^*)^2}-\frac{p_4}{p_4+(or)^2}\Big]\\&\quad+o^*\Big[\frac{p_4}{p_4+(o^*r^*)^2}-\frac{p_4}{p_4+(or)^2}\Big],\end{aligned}$$

$$\Big|\frac{p_4}{p_4+(o^*r^*)^2}-\frac{p_4}{p_4+(or)^2}\Big|\leqslant\frac{|or-o^*r^*|\cdot|or+o^*r^*|}{p_4+(o^*r^*)^2},\quad |or+o^*r^*|\leqslant|or-o^*r^*|+2o^*r^*,$$

then

$$\frac{d}{dt}W(t)\leqslant-\alpha W(t)+\beta W^{\frac{3}{2}}(t)+\gamma W^2(t),$$

i.e.

$$\frac{d}{dt}W(t)\leqslant\gamma W(t)\Big[W^{\frac{1}{2}}(t)+\frac{\beta-\sqrt{\beta^2+4\alpha\gamma}}{2\gamma}\Big]\Big[W^{\frac{1}{2}}(t)+\frac{\beta+\sqrt{\beta^2+4\alpha\gamma}}{2\gamma}\Big],\tag{8.17}$$

where

$$\alpha=2\Big[\min\{p_3-\tfrac{1}{2},p_6,1\}-Ap_2-r^*-(p_6+1)a^*-\frac{4o^*r^*}{p_4+(o^*r^*)^2}\Big]>0,$$

$$\beta=2^{\frac{3}{2}}\Big[p_6+1+\frac{3o^*r^*}{p_4+(o^*r^*)^2}\Big]\leqslant2^{\frac{3}{2}}\Big[p_6+1+3\min\Big\{\frac{1}{2p_4^{\frac{1}{2}}},\frac{2p_2}{\sqrt{1+\frac{4p_2p_5A}{p_6}}-1}\Big\}\Big],$$

$$\gamma=\frac{4o^*r^*}{p_4+(o^*r^*)^2}\leqslant4\min\Big\{\frac{1}{2p_4^{\frac{1}{2}}},\frac{2p_2}{\sqrt{1+\frac{4p_2p_5A}{p_6}}-1}\Big\},$$

provided

$$0<r^*+(p_6+1)a^*+\frac{4o^*r^*}{p_4+(o^*r^*)^2}<\min\{p_3-\tfrac{1}{2},p_6,1\}-Ap_2.\tag{8.18}$$

As $0 \leqslant a^* = o^* \leqslant \frac{A}{p_3}, \frac{p_5}{p_6} \leqslant r^* \leqslant \frac{p_5+1}{p_6}$ and $o^* r^* \geqslant \frac{1}{2p_2}[\sqrt{1+\frac{4p_2p_5A}{p_6}} - 1]$ then by (8.18) we get

$$\frac{p_6+1+p_2p_3}{p_3} A + \min\left\{\frac{2}{p_4^{\frac{1}{2}}}, \frac{8p_2}{\sqrt{1+\frac{4p_2p_5A}{p_6}}-1}\right\} < B := \min\{p_3 - \tfrac{1}{2}, p_6, 1\} - \frac{p_5+1}{p_6}.$$

Hence,

$$\frac{p_6+1+p_2p_3}{p_3} A + \frac{2}{p_4^{\frac{1}{2}}} < B \text{ and } A \leqslant \frac{2p_6p_4^{\frac{1}{2}}}{p_5}(1+2p_2p_4^{\frac{1}{2}}),$$

whence

$$0 \leqslant A < A_0 := \min\left\{\frac{2p_6p_4^{\frac{1}{2}}}{p_5}(1+2p_2p_4^{\frac{1}{2}}), \frac{p_3}{p_6+1+p_2p_3}\left(B - \frac{2}{p_4^{\frac{1}{2}}}\right)\right\},$$

or

$$F(A) := \frac{p_6+1+p_2p_3}{p_3} A + \frac{8p_2}{\sqrt{1+\frac{4p_2p_5A}{p_6}}-1} < B \text{ and } A > \frac{2p_6p_4^{\frac{1}{2}}}{p_5}(1+2p_2p_4^{\frac{1}{2}}).$$

As the function $F(A)$ has a unique minimum for positive A, denote by $A_{\min}$, then there exist $0 < A_1 < A_{\min} < A_2$ such that $F(A) < B$ provided $F(A_{\min}) < B$.

So, if $W(0) < [\frac{\sqrt{\beta^2+4\alpha\gamma}-\beta}{2\gamma}]^2$ then by (8.17) we deduce that

$$W(t) \to 0 \text{ as } t \to +\infty.$$

8.3 Analysis of the Model with a Time Delay

8.3.1 Stability Analysis with Respect to a Time Delay

Note the fixed point (a^*, r^*, o^*) for (8.6)–(8.8) coincides with the one for (8.1)–(8.3). Let us denote by

$$J_\tau := \begin{pmatrix} \frac{\partial f_1}{\partial a_\tau} & \frac{\partial f_1}{\partial r_\tau} & \frac{\partial f_1}{\partial o_\tau} \\ \frac{\partial f_2}{\partial a_\tau} & \frac{\partial f_2}{\partial r_\tau} & \frac{\partial f_2}{\partial o_\tau} \\ \frac{\partial f_3}{\partial a_\tau} & \frac{\partial f_3}{\partial r_\tau} & \frac{\partial f_3}{\partial o_\tau} \end{pmatrix},$$

where $a_\tau = a(t-\tau)$, $r_\tau = r(t-\tau)$, and $o_\tau = o(t-\tau)$. Then J_τ at the point (a^*, r^*, a^*) is equal

$$J_\tau^* := \begin{pmatrix} 0 & 0 & 0 \\ 0 & 0 & 0 \\ 1 & 0 & 0 \end{pmatrix}.$$

Now we will look for eigenvalues for the matrix $J^* + e^{-\lambda\tau} J^*_\tau$. So,

$$|J^* + e^{-\lambda\tau} J^*_\tau - \lambda I| = \begin{vmatrix} -p_3 - \lambda & -K_1 & -K_3 \\ 0 & -p_6 + K_2 - \lambda & K_4 \\ 1 + e^{-\lambda\tau} & 0 & -1 - \lambda \end{vmatrix} = 0,$$

whence we obtain the characteristic equation:

$$(\lambda + 1)(\lambda + p_3)(\lambda + p_6 - K_2) = -(1 + e^{-\lambda\tau})K_3(\lambda + p_6).$$

Time delays are known to affect the stability of a fixed point. They can induce stability switches in which the zeros of the characteristic equation may cross the imaginary axis as the delay, τ, increases. Looking at the characteristic equation as a function of τ, and analyzing the location of the roots and the direction of motion as they cross the imaginary axis (see [22]). Destabilization will happen at critical values τ_c which is when there is a pair of purely imaginary characteristic values. Following the ideas of papers [22, 23], let's rewrite the characteristic equation as

$$\begin{aligned} C(\lambda) &:= (\lambda + 1)(\lambda + p_3)(\lambda + p_6 - K_2) + (1 + e^{-\lambda\tau})K_3(\lambda + p_6) \\ &= ((\lambda + 1)(\lambda + p_3)(\lambda + p_6 - K_2) + K_3(\lambda + p_6)) + e^{-\lambda\tau} K_3(\lambda + p_6) \\ &= P(\lambda) + Q(\lambda)e^{-\lambda\tau} = 0. \end{aligned} \tag{8.19}$$

Then define

$$\begin{aligned} F(y) &= |P(iy)|^2 - |Q(iy)|^2 \\ &= y^6 + (p_6^2 - 2K_2p_6 + p_3^2 - 2K_3 + K_2^2 + 1)y^4 + (K_2^2 - 2K_2K_3 + 2K_3p_3 \\ &\quad -2K_2K_3p_3 + p_3^2 + K_2^2p_3^2 - 2K_2p_6 + 2K_2K_3p_6 - 2K_2p_3^2p_6 + p_6^2 - 2K_3p_6^2 \\ &\quad +p_3^2p_6^2)y^2 + (K_2^2p_3^2 - 2K_2K_3p_3p_6 - 2K_2p_3^2p_6 + 2K_3p_3p_6^2 + p_3^2p_6^2). \end{aligned} \tag{8.20}$$

We want to use Theorem 8.3 from the Appendix, so we need to check the following conditions:

1. $P(\lambda) = (\lambda + 1)(\lambda + p_3)(\lambda + p_6 - K_2) + K_3(\lambda + p_6)$ and $Q(\lambda) = K_3(\lambda + p_6)$ have no common imaginary zeros since each p_i are real values.
2. It is quick to see that $\overline{P(i\lambda)} = P(i\lambda)$ and $\overline{Q(i\lambda)} = Q(i\lambda)$ for real λ.
3. $P(0) + Q(0) = p_3(p_6 - K_2) + 2K_3p_6 \neq 0$ so this is an important restriction in order to use the Theorem 8.3.
4. Referring back to (8.13), we see that there are at most 3 roots of (8.19) if $\tau = 0$.
5. (8.20) has at most 6 real zeros for real y.

Therefore, by Theorem 8.3 from the Appendix, if $F(y)$ has no positive roots, the system is stable for all $\tau \geqslant 0$. If $F(y)$ has a simple positive root y_0, then there exists a pair of purely imaginary roots $\pm i v_0$ such that $v_0 = \sqrt{y_0}$. For this v_0, there is a countable sequence of $\{\tau_0^n\}$ of delays for which stability switches can occur. Also, there exists a positive τ_c such that the system is unstable for all $\tau > \tau_c$. Investigating this further, let $x = y^2$

$$F(x) = x^3 + b_1x^2 + b_2x + b_3, \tag{8.21}$$

where $b_1 = p_6^2 - 2K_2p_6 + p_3^2 - 2K_3 + K_2^2 + 1$, $b_2 = K_2^2 + -2K_2K_3 + 2K_3p_3 - 2K_2K_3p_3 + p_3^2 + K_2^2p_3^2 - 2K_2p_6 + 2K_2K_3p_6 - 2K_2p_3^2p_6 + p_6^2 - 2K_3p_6^2 + p_3^2p_6^2$, and $b_3 = K_2^2p_3^2 - 2K_2K_3p_3p_6 - 2K_2p_3^2p_6 + 2K_3p_3p_6^2 + p_3^2p_6^2$. Note that

$$F'(x) = 3x^2 + 2b_1x + b_2$$

and

$$\Delta_0 = b_1^2 - 3b_2. \tag{8.22}$$

Now analyzing the roots of (8.21),

- If $\Delta_0 \leqslant 0$, then $F'(0) \geqslant 0$ and $F(x)$ is monotonically non-decreasing. Further,
  - if $F(0) > 0$, then F has no positive roots and all the roots of the characteristic will remain to the left of the imaginary axis for all $\tau > 0$.
  - if $F(0) < 0$, then since $\lim_{x\to\infty} F(x) = \infty$, there is at least one positive root of F and thus the roots of the characteristic equation can cross the imaginary axis.
- If $\Delta_0 > 0$ then F has critical points

$$x_{c_1} = \frac{-b_1+\sqrt{\Delta_0}}{3}, \quad x_{c_2} = \frac{-b_1-\sqrt{\Delta_0}}{3}$$

  and if $x_{c_1} > 0$ and $F(x_{c_1}) < 0$, then F has positive roots (see [22]).

Stability switches are possible for each positive simple root x_j of (8.21) and the cross is from left to right if $F'(v_0) > 0$, and from right to left if $F'(v_0) < 0$ according to Theorem 1 (see [22]). Now let's analyze the characteristic quasi-polynomial (8.19) for $\lambda = iv$:

$$C(iv) = A_1 - A_2\cos(v\tau) - A_3\sin(v\tau) + i[A_4 - A_3\cos(v\tau) + A_2\sin(v\tau)] = 0,$$

where

$$A_1(v) = p_3p_6 - K_2p_3 + K_3p_6 - v^2(p_6 + p_3 - K_2 + 1), \quad A_2 = -K_3p_6,$$
$$A_4(v) = v(p_3 - K_2 - K_2p_3 + p_6 + p_3p_6 + K_3) - v^3, \quad A_3(v) = -K_3v.$$

So x_j $(j = 1, 2, 3)$ is a positive root of $F(x) = 0$ and $v_j = \sqrt{x_j}$. Then v_j satisfies (8.23) if its a solution to the system

$$\begin{cases} A_1(v) - A_2\cos(v\tau) - A_3(v)\sin(v\tau) = 0, \\ A_4(v) - A_3(v)\cos(v\tau) + A_2\sin(v\tau) = 0. \end{cases}$$

This yields

$$\sin(\nu\tau) = \frac{A_1(\nu)A_3(\nu)-A_2A_4(\nu)}{A_2^2+A_3^2(\nu)}, \quad \cos(\nu\tau) = \frac{A_1(\nu)A_2+A_3(\nu)A_4(\nu)}{A_2^2+A_3^2(\nu)},$$

provided $\max\{|A_1(\nu)A_3(\nu) - A_2A_4(\nu)|, |A_1(\nu)A_2 - A_3(\nu)A_4(\nu)|\} \leqslant A_2^2 + A_3^2(\nu)$,

Therefore, for every positive root ν_j, it yields the following sequence of delays $\{\tau_j^n\}$ for which there are pure imaginary roots (8.19):

$$\tau_j^n = \frac{1}{\nu_j}\left\{\arctan\left(\frac{A_1(\nu_j)A_3(\nu_j)-A_2A_4(\nu_j)}{A_1(\nu_j)A_2+A_3(\nu_j)A_4(\nu_j)}+\pi\, n\right)\right\} \quad \text{for } n = 0, 1, 2, \ldots \tag{8.23}$$

As a result the following statement holds.

Lemma 8.3 *The system (8.6)–(8.8) with delay and $p_3(p_6 - K_2) + 2K_3p_6 \neq 0$ is stable for all $\tau \geqslant 0$ if $F(0) > 0$ and $\Delta_0 \leqslant 0$ (where Δ_0 is from (8.22)). The system has stability switches at some $\{\tau_j^n\}$ for every positive root ν_j of (8.19). Furthermore, if $A = 0$, $p_2 = 0$ or $p_5 = p_6 = 0$ then the delay has no affect on the stability of the system.*

Example 8.3 This example illustrates the dynamics of eigenvalues with respect to the time delay for the following set of parameters $p_3 = 0.41$, $p_6 = 0.91$, $K_2 = 0.81$, and $K_3 = 0.41$ in the Eq. (8.19). Taking the real and imaginary parts, we rewrite the equation as a system

$$\begin{cases} -K_2p_3 + K_3p_6 + p_3p_6 - K_2x + K_3x + p_3x - K_2p_3x + p_6x + p_3p_6x + x^2 \\ \quad -K_2x^2 + p_3x^2 + p_6x^2 + x^3 - y^2 + K_2y^2 - p_3y^2 - p_6y^2 - 3xy^2 \\ \quad +e^{-\tau x}K_3p_6\cos(\tau y) + e^{-\tau x}K_3x\cos(\tau y) + e^{-\tau x}K_3y\sin(\tau y) = 0, \\ -K_2y + K_3y + p_3y - K_2p_3y + p_6y + p_3p_6y + 2xy - 2K_2xy + 2p_3xy \\ \quad +2p_6xy + 3x^2y - y^3 + e^{-\tau x}K_3y\cos(\tau y) - e^{-\tau x}K_3p_6\sin(\tau y) \\ \quad -e^{-\tau x}K_3x\sin(\tau y) = 0. \end{cases}$$

The red lines in Fig. 8.3 represent the solution curves for the first equation and the blue lines in Fig. 8.3 represent the solution curves for the second equation for different values of delay τ. The eigenvalues, which are roots of (8.19), correspond to intersections between the red and blue lines.

When there is no delay, i.e. $\tau = 0$, we only have three eigenvalues $\lambda \approx -0.9, -0.2 \pm 0.8i$ (see Fig. 8.3). When delay is non-zero, countably many eigenvalues originate from $-\infty$ and move toward the imaginary axis as τ increases (see Fig. 8.3). The eigenvalues can cross the imaginary axis only at the points $y_1 \approx \pm 0.7$ and $y_2 \approx \pm 0.25$ which are real roots of the Eq. (8.20) (see Fig. 8.4). The density of complex eigenvalues around these crossing points y_1, y_2 is increasing as the τ gets larger (see Fig. 8.3).

When the delay $\tau < \tau^* \approx 2$ (where τ^* is a critical value found as a solution of (8.23) with $\nu_1 = \sqrt{|y_1|}$) all eigenvalues are stable. The first stability switch happens at $\tau^* \approx 2$ when two complex conjugate eigenvalues cross the imaginary axis at $y_1 \approx \pm 0.7$ changing the sign of the real part from negative to positive. At a later time $\tau^* \approx 11$ (where this τ^* is a critical value found as a solution of (8.23) with

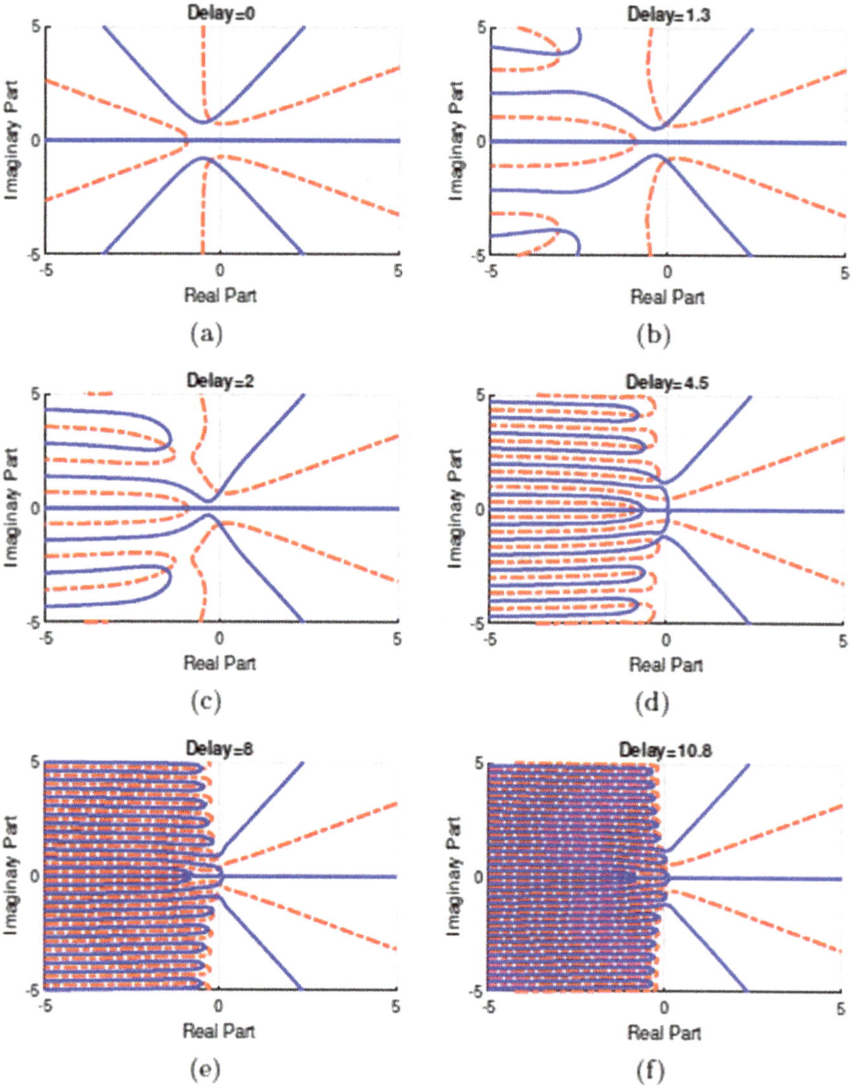

Fig. 8.3 Contour plots for different values of delay τ showing stability switches

$\nu_2 = \sqrt{|y_2|}$) this complex pair will cross the imaginary axis back changing the sign of the real part from positive to negative (see Fig. 8.4).

Solving (8.23) and taking into account the periodicity of the arctangent function one can obtained the infinite sequences of delays associated with ν_1 and another infinite sequence associated with ν_2 at which stability switches may happen. At time delays associated with ν_1 a complex conjugate pair of eigenvalues may cross the imaginary axis from left to right and for time delays associated with ν_2 the pair may

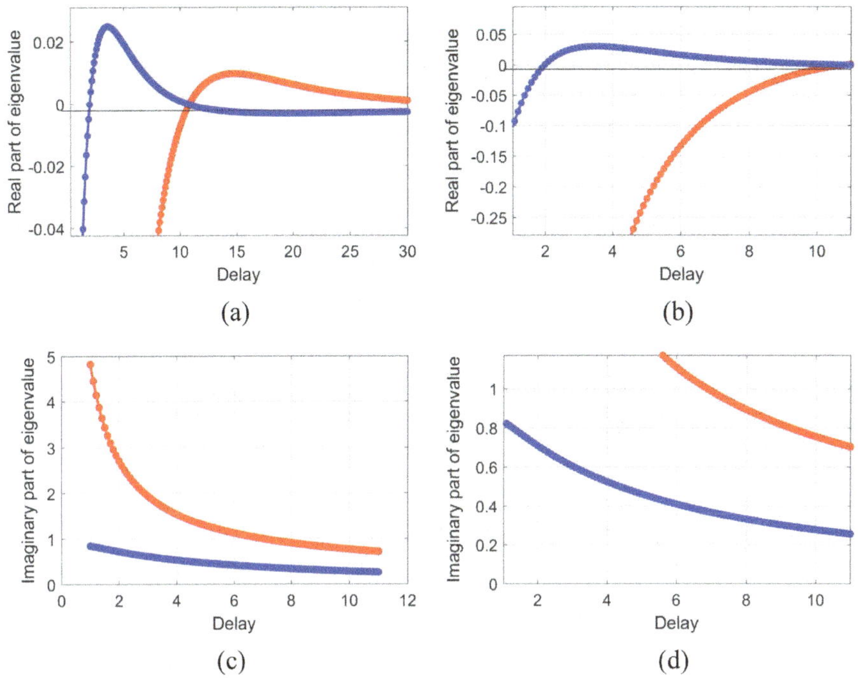

Fig. 8.4 Tracking two complex eigenvalues to see how the value of their real part changes and how the value of their imaginary part changes

cross the imaginary axis from right to left. If the derivative of $F(y)$ (see (8.19)) does not change sign at the corresponding τ^* from either of the two sequences above, then the crossing of the imaginary axis does not happen.

8.3.2 Global in Time Existence of Solutions

In this section, by Picard's method we prove the existence of solutions to problem (8.1)–(8.4).

Theorem 8.1 *Let* $A > 0$, $p_i > 0$, *and assumption* (8.5) *hold. Let*

$$a'_\tau(0) + p_3 a_\tau(0) = \frac{A}{1+p_2 o_0 r_0}, \tag{8.24}$$

then problem (8.1)–(8.4) has a unique non-negative solution $(a(t), r(t), o(t))$ *in* C^2 *for all* $t \geqslant 0$. *Moreover, for all time* $t \geqslant 0$ *and* $\tau > 0$ *there are estimates*

$$\frac{Ap_6}{p_3p_6+12Ap_2(p_5+1)}[1-e^{-p_3\tau}][1+(p_3\tau)^{-1}] \leqslant a(t) \leqslant \frac{3A}{p_3} + a_\tau(0),$$
$$\frac{Ap_6}{p_3p_6+12Ap_2(p_5+1)}[1-e^{-2\tau}][1-e^{-p_3\tau}][1+(p_3\tau)^{-1}] \leqslant o(t) \leqslant \frac{3A}{p_3} + a_\tau(0) + o_0 + \max_{t\in[-\tau,0]} a_\tau(t),$$
$$\frac{p_5}{p_6}[1-e^{-p_6\tau}][1+(p_6\tau)^{-1}] \leqslant r(t) \leqslant \frac{p_5+1}{p_6} + r_0.$$

Remark 8.1 It is apparent that for $a_\tau(t) = a_0 + \Lambda t^2 e^{-t}$, where Λ is a positive number, fitting condition (8.24) of this theorem is rewritten as

$$a_0 = \frac{A}{p_3(1+p_2 o_0 r_0)}, \quad o_0 > 0, \; r_0 > 0.$$

Proof (*Proof of Theorem* 8.1) We will construct a solution to (8.1)–(8.4) by the iterative process. First of all, we will look for a solution on the interval $[0, \tau]$. From (8.3) we obtain that

$$o(t) = e^{-t}o_0 + e^{-t}\int_0^t a_\tau(s-\tau)e^s\,ds = e^{-t}o_0 + e^{-(t-\tau)}\int_{-\tau}^{t-\tau} a_\tau(s)e^s\,ds =: o_1(t)\ \forall t \in [0,\tau]. \tag{8.25}$$

So, by (8.5) and (8.25) we have $o(t) \in C^2[0, \tau]$, and for all $t \in [0, \tau]$ we have

$$\underline{o}_1(t) \leqslant o(t) \leqslant \overline{o}_1(t) \tag{8.26}$$

where we put

$$\underline{o}_1(t) = e^{-t}o_0 + (1-e^{-t}) \min_{[-\tau,0]} a_\tau(t),$$
$$\overline{o}_1(t) = e^{-t}o_0 + (1-e^{-t}) \max_{[-\tau,0]} a_\tau(t).$$

Integrating (8.1) and (8.2) on the interval $[0, \tau]$, taking into account (8.25), we arrive at

$$a(t) = e^{-p_3t}a_\tau(0) + A\,e^{-p_3t}\int_0^t \frac{e^{p_3s}\,ds}{1+p_2r(s)o_1(s)}, \tag{8.27}$$

$$r(t) = e^{-p_6t}r_0 - p_4\,e^{-p_6t}\int_0^t \frac{e^{p_6s}\,ds}{p_4+r^2(s)o_1^2(s)} + \frac{p_5+1}{p_6}(1-e^{-p_6t}) \tag{8.28}$$

for all $t \in [0, \tau]$.

Introducing the functions

$$\underline{r}_1(t) = e^{-p_6t}r_0 + \frac{p_5}{p_6}(1-e^{-p_6t}),$$
$$\overline{r}_1(t) = e^{-p_6t}r_0 + \frac{p_5+1}{p_6}(1-e^{-p_6t}),$$

$$\underline{a}_1(t) = e^{-p_3 t} a_\tau(0) + \frac{A}{p_3[1+p_2 \max\limits_{[0,\tau]} \overline{o}_1(t) \max\limits_{[0,\tau]} \overline{r}_1(t)]}(1 - e^{-p_3 t}),$$

$$\overline{a}_1(t) = e^{-p_3 t} a_\tau(0) + \frac{A}{p_3}(1 - e^{-p_3 t}),$$

and taking advantage (8.27), (8.28), we find that

$$\underline{r}_1(t) \leqslant r(t) \leqslant \overline{r}_1(t), \quad \underline{a}_1(t) \leqslant a(t) \leqslant \overline{a}_1(t) \tag{8.29}$$

for all $t \in [0, \tau]$.

As a result, estimates (8.26), (8.29) imply positivity of $o(t),\ a(t),\ r(t)$ on $[0, \tau]$ provided $o_0 > 0$, $r_0 > 0$, and $a_\tau(0) > 0$.

As the right-hand side of (8.28) is Lipschitz continuous on r then there exists a unique solution of (8.28) on whole interval $[0, \tau]$ and, as a result, the one of (8.27). Moreover, obviously the solution $o(t),\ a(t),\ r(t) \in C^2[0, \tau]$ if the fitting condition (8.24) holds. Let us denote the corresponding solution to (8.25), (8.27), (8.28) on $[0, \tau]$ by $(o_1(t), a_1(t), r_1(t))$.

At this point we extend the constructed solution on the interval $[\tau, 2\tau]$. By (8.1)–(8.3) we get

$$\begin{aligned} o(t) &= e^{-(t-\tau)} o_1(\tau) + e^{-t} \int_\tau^t a(s-\tau) e^s\, ds \\ &= e^{-(t-\tau)} o_1(\tau) + e^{-(t-\tau)} \int_0^{t-\tau} a_1(s) e^s\, ds \\ &= e^{-t} o_0 + e^{-(t-\tau)} \left[\int_{-\tau}^0 a_\tau(s) e^s\, ds + \int_0^{t-\tau} a_1(s) e^s\, ds \right] \\ &=: o_2(t), \end{aligned} \tag{8.30}$$

$$r(t) = e^{-p_6(t-\tau)} r_1(\tau) - p_4 e^{-p_6 t} \int_\tau^t \frac{e^{p_6 s}\, ds}{p_4 + r^2(s) o_2^2(s)} + \frac{p_5+1}{p_6}(1 - e^{-p_6(t-\tau)}), \tag{8.31}$$

$$a(t) = e^{-p_3(t-\tau)} a_1(\tau) + A\, e^{-p_3 t} \int_\tau^t \frac{e^{p_3 s}\, ds}{1 + p_2 r(s) o_2(s)} \tag{8.32}$$

for all $t \in [\tau, 2\tau]$.

Recasting the arguments above, we obtain the one-valued solvability of system (8.30)–(8.32) such that $o_2(t),\ a_2(t),\ r_2(t) \in C^2[\tau, 2\tau]$. Besides, for all $t \in [\tau, 2\tau]$, the following estimates hold

$$\underline{o}_2(t) \leqslant o_2(t) \leqslant \overline{o}_2(t), \quad \underline{r}_2(t) \leqslant r_2(t) \leqslant \overline{r}_2(t), \quad \underline{a}_2(t) \leqslant a_2(t) \leq \overline{a}_2(t),$$

where we put

$$\begin{aligned}
\underline{o}_2(t) &= e^{-t} o_0 + e^{-t}(e^\tau - 1) \min_{[-\tau,0]} a_\tau(t) + (1 - e^{-(t-\tau)}) \min_{[0,\tau]} a_1(t),\\
\overline{o}_2(t) &= e^{-t} o_0 + e^{-t}(e^\tau - 1) \max_{[-\tau,0]} a_\tau(t) + (1 - e^{-(t-\tau)}) \max_{[0,\tau]} a_1(t),\\
\underline{r}_2(t) &= e^{-p_6(t-\tau)} r_1(\tau) + \tfrac{p_5}{p_6}(1 - e^{-p_6(t-\tau)}),\\
\overline{r}_2(t) &= e^{-p_6(t-\tau)} r_1(\tau) + \tfrac{p_5+1}{p_6}(1 - e^{-p_6(t-\tau)}),\\
\underline{a}_2(t) &= e^{-p_3(t-\tau)} a_1(\tau) + \frac{A}{p_3[1+p_2 \max\limits_{[\tau,2\tau]} \overline{o}_2(t) \max\limits_{[\tau,2\tau]} \overline{r}_2(t)]}(1 - e^{-p_3(t-\tau)}),\\
\overline{a}_2(t) &= e^{-p_3(t-\tau)} a_1(\tau) + \tfrac{A}{p_3}(1 - e^{-p_3(t-\tau)}).
\end{aligned}$$

Continuing this iteration procedure, we derive

$$o(t) = e^{-(t-(k-1)\tau)} o_{k-1}((k-1)\tau) + e^{-(t-\tau)} \int\limits_{(k-2)\tau}^{t-\tau} a_{k-1}(s) e^s \, ds =: o_k(t),$$

$$r(t) = e^{-p_6(t-(k-1)\tau)} r_{k-1}((k-1)\tau) - p_4 \, e^{-p_6 t} \int\limits_{(k-1)\tau}^{t} \frac{e^{p_6 s} \, ds}{p_4 + r^2(s) o_k^2(s)} + \tfrac{p_5+1}{p_6}(1 - e^{-p_6(t-(k-1)\tau)}),$$

$$a(t) = e^{-p_3(t-(k-1)\tau)} a_{k-1}((k-1)\tau) + A \, e^{-p_3 t} \int\limits_{(k-1)\tau}^{t} \frac{e^{p_3 s} \, ds}{1 + p_2 r(s) o_k(s)}$$

for all $t \in [(k-1)\tau, k\,\tau]$, $k \in \mathbb{N}$, where $a_0(t) = a_\tau(t)$.

This system has a unique solution $o_k(t), a_k(t), r_k(t) \in C^{k+1}[(k-1)\tau, k\,\tau]$. Moreover, setting

$$\begin{aligned}
\underline{o}_k(t) &= e^{-(t-(k-1)\tau)} o_{k-1}((k-1)\tau) + (1 - e^{-(t-(k-1)\tau)}) \min_{[(k-2)\tau,(k-1)\tau]} a_{k-1}(t),\\
\overline{o}_k(t) &= e^{-(t-(k-1)\tau)} o_{k-1}((k-1)\tau) + (1 - e^{-(t-(k-1)\tau)}) \max_{[(k-2)\tau,(k-1)\tau]} a_{k-1}(t),\\
\underline{r}_k(t) &= e^{-p_6(t-(k-1)\tau)} r_{k-1}((k-1)\tau) + \tfrac{p_5}{p_6}(1 - e^{-p_6(t-(k-1)\tau)}),\\
\overline{r}_k(t) &= e^{-p_6(t-(k-1)\tau)} r_{k-1}((k-1)\tau) + \tfrac{p_5+1}{p_6}(1 - e^{-p_6(t-(k-1)\tau)}),\\
\underline{a}_k(t) &= e^{-p_3(t-(k-1)\tau)} a_{k-1}((k-1)\tau) + \frac{A}{p_3[1+p_2 \max\limits_{[(k-1)\tau,k\tau]} \overline{o}_k(t) \max\limits_{[(k-1)\tau,k\tau]} \overline{r}_k(t)]}(1 - e^{-p_3(t-(k-1)\tau)}),\\
\overline{a}_k(t) &= e^{-p_3(t-(k-1)\tau)} a_{k-1}((k-1)\tau) + \tfrac{A}{p_3}(1 - e^{-p_3(t-(k-1)\tau)}),
\end{aligned}$$

we conclude

$$\underline{o}_k(t) \leqslant o_k(t) \leqslant \overline{o}_k(t), \quad \underline{r}_k(t) \leqslant r_k(t) \leqslant \overline{r}_k(t), \quad \underline{a}_k(t) \leqslant a_k(t) \leqslant \overline{a}_k(t),$$

for all $t \in [(k-1)\tau, k\,\tau]$.

Then straightforward calculations arrive at the estimates for all $t \in [(k-1)\tau, k\tau]$ and $k \in \mathbb{N}$

$$\overline{a}_k(t) \leqslant e^{-p_3 t} a_1(\tau) + \tfrac{A}{p_3}[1 - e^{-p_3(t-(k-1)\tau)}] + \tfrac{A}{p_3}[1 - e^{-p_3\tau}]e^{-p_3(t-(k-1)\tau)} \sum_{l=0}^{k-2} e^{-lp_3\tau},$$

$$\overline{r}_k(t) \leqslant e^{-p_6 t} r_1(\tau) + \tfrac{p_5+1}{p_6}[1 - e^{-p_6(t-(k-1)\tau)}] + \tfrac{p_5+1}{p_6}[1 - e^{-p_6\tau}] \sum_{l=0}^{k-2} e^{-lp_6\tau},$$

$$\overline{o}_k(t) \leqslant e^{-(k-2)\tau} o_1(\tau) + (1 - e^{-2\tau}) \sum_{l=0}^{k-3} e^{-l} \max_{[(k-l-3)\tau,(k-l-2)\tau]} a_{k-l-2}(t),$$

$$\begin{aligned} \underline{a}_k(t) \geqslant{}& \tfrac{Ap_6}{p_3p_6+12p_2A(p_5+1)}[1 - e^{-p_3\tau})] + e^{-p_3(t-(k-1)\tau)-(k-2)\tau} a_1(\tau) \\ &+ \tfrac{Ap_6}{p_3p_6+12p_2A(p_5+1)}[1 - e^{-p_3\tau}]e^{-p_3(t-(k-1)\tau)} \sum_{l=0}^{k-3} e^{-lp_3\tau}, \end{aligned}$$

$$\underline{r}_k(t) \geqslant \tfrac{p_5}{p_6}[1 - e^{-p_6(t-(k-1)\tau)}] + e^{-p_6 t} r_1(\tau) + \tfrac{p_5}{p_6}[1 - e^{-p_6\tau}] \sum_{l=0}^{k-2} e^{-lp_6\tau},$$

$$\overline{o}_k(t) \leqslant e^{-(k-2)\tau} o_1(\tau) + (1 - e^{-p_2\tau}) \sum_{l=0}^{k-3} e^{-l} \min_{[(k-l-3)\tau,(k-l-2)\tau]} a_{k-l-2}(t).$$

After that, taking into account relations (8.26), (8.29) and the straightforward inequality

$$\sum_{l=0}^{N} e^{-\gamma l} < \int_0^{+\infty} e^{-\gamma x} dx = \gamma^{-1} \quad \text{for} \quad \gamma > 0,$$

we deduce

$$\begin{aligned} \overline{a}_k(t) &\leqslant \tfrac{A}{p_3}(1 - e^{-p_3\tau})[1 + (p_3\tau)^{-1} + e^{-p_3(k-1)\tau}] + e^{-p_3k\tau} a_\tau(0), \\ \overline{r}_k(t) &\leqslant \tfrac{p_5+1}{p_6}(1 - e^{-p_6\tau})[1 + (p_6\tau)^{-1} + e^{-p_6(k-1)\tau}] + e^{-p_6k\tau} r_0, \\ \overline{o}_k(t) &\leqslant o_0 e^{-k\tau} + e^{-(k-1)\tau}(1 - e^{-\tau}) \max_{[-\tau,0]} a_\tau(t) + (1 - e^{-2\tau}) a_\tau(0) \\ &\quad + \tfrac{A}{p_3}(1 - e^{-2\tau})(1 - e^{-p_3\tau})[2 + (p_3\tau)^{-1}], \end{aligned}$$

$$\begin{aligned} \underline{a}_k(t) &\geqslant \tfrac{Ap_6}{p_3p_6+12p_2A(p_5+1)}[1 - e^{-p_3\tau}][1 + (p_3\tau)^{-1} + e^{-(k-2)\tau}] + e^{-(k-1)\tau} a_\tau(0), \\ \underline{r}_k(t) &\geqslant e^{-p_6k\tau} r_0 + \tfrac{p_5}{p_6}[1 - e^{-p_6\tau}][1 + (p_6\tau)^{-1} + e^{-p_6(k-2)\tau}], \end{aligned}$$

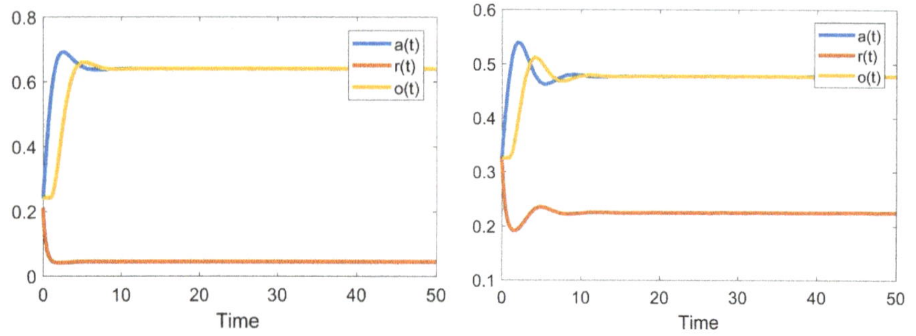

(a) (b)

Fig. 8.5 Plot of the solutions $a(t)$, $r(t)$, and $o(t)$ with the parameter values from Example 8.4

$$\begin{aligned}\underline{o}_k(t) &\geqslant e^{-k\tau}o_0 + +e^{-(k-2)\tau}(1-e^{-\tau})\min_{[-\tau,0]} a_\tau(t)\\&+\frac{Ap_6}{p_3p_6+12p_2A(p_5+1)}(1-e^{-2\tau})(1-e^{-p_3\tau})[1+(p_3\tau)^{-1}].\end{aligned}$$

Finally, letting $k \to +\infty$, we get

$$\begin{aligned}&\frac{Ap_6}{p_3p_6+12Ap_2(p_5+1)}[1-e^{-p_3\tau}][1+(p_3\tau)^{-1}] \leqslant \lim_{k\to+\infty} a_k(t) \leqslant \frac{A}{p_3}[1-e^{-p_3\tau}][1+(p_3\tau)^{-1}],\\&\frac{Ap_6}{p_3p_6+12Ap_2(p_5+1)}[1-e^{-p_3\tau}][1-e^{-2\tau}][1+(p_3\tau)^{-1}] \leqslant \lim_{k\to+\infty} o_k(t)\\&\qquad\qquad \leqslant [1-e^{-2\tau}]\Big[\frac{A}{p_3}[1-e^{-p_3\tau}][2+(p_3\tau)^{-1}]+a_\tau(0)\Big],\\&\frac{p_5}{p_6}[1-e^{-p_6\tau}][1+(p_6\tau)^{-1}] \leqslant \lim_{k\to+\infty} r_k(t) \leqslant \frac{p_5+1}{p_6}[1-e^{-p_6\tau}][1+(p_6\tau)^{-1}].\end{aligned}$$

As a result, problem (8.1)–(8.4) has a unique global in time solution, and the following estimates hold for any $t \geqslant 0$

$$\begin{aligned}\frac{Ap_6}{p_3p_6+12Ap_2(p_5+1)}[1-e^{-p_3\tau}][1+(p_3\tau)^{-1}] &< a(t) < \frac{3A}{p_3}+a_\tau(0),\\\frac{Ap_6}{p_3p_6+12Ap_2(p_5+1)}[1-e^{-p_3\tau}][1-e^{-2\tau}][1+(p_3\tau)^{-1}] &< o(t) < o_0+a_\tau(0)+\frac{3A}{p_3}+\max_{[-\tau,0]} a_\tau(t),\\\frac{p_5}{p_6}[1-e^{-p_6\tau}][1+(p_6\tau)^{-1}] &< r(t) < r_0+\frac{3(p_5+1)}{p_6}.\end{aligned}$$

That completes the proof of this theorem.

Example 8.4 Let $A = 1$, $p_2 = 11$, $p_3 = 1.2$, $p_4 = 0.05$, $p_5 = 0.11$, and $p_6 = 2.9$. Hence the initial conditions are $a_0 = \frac{A}{p_3}(1 + p_2\sqrt{\frac{p_4(p_6r_0-p_5)}{p_5+1-p_6r_0}})^{-1} \approx 0.2409$, $r_0 = \frac{1}{2}(\frac{p_5}{p_6} + \frac{p_5+1}{p_6}) \approx 0.2103$, and $o_0 = a_0 \approx 0.2409$. With these parameter values and setting delay $\tau = 1$ we solve (8.1)–(8.4) numerically using the Matlab solver dde23 [21]. Convergence of solutions toward the steady state, $a_s \approx 0.6395$, $r_s \approx 0.0431$, $o_s \approx 0.6395$, can be seen in Fig. 8.5a.

If we perturb the parameters by a bit and use $A = 1,\ p_2 = 7,\ p_3 = 1.2,\ p_4 = 0.05,\ p_5 = 0.51$, and $p_6 = 3.1$, then the initial conditions are $a_0 = \frac{A}{p_3}(1 + p_2 \sqrt{\frac{p_4(p_6 r_0 - p_5)}{p_5 + 1 - p_6 r_0}})^{-1} \approx 0.3249$, $r_0 = \frac{1}{2}(\frac{p_5}{p_6} + \frac{p_5+1}{p_6}) \approx 0.3258$, and $o_0 = a_0 \approx 0.3249$. Now initial levels of adrenocorticotropin and cortisol are below the initial level of glucocorticoid density that was opposite in the case above. But after a short time the order of all levels becomes the same as in the case (a) that can be seen in Fig. 8.5b. The steady state for these parameter values is $a_s \approx 0.4764, r_s \approx 0.2246$, and $o_s \approx 0.4764$.

8.3.3 Existence of Periodic Solutions

Theorem 8.2 *Under the conditions of Theorem 8.1, the system (8.1)–(8.5) has at least one C^2-smooth T-periodic solution.*

Proof (*Proof of Theorem* 8.2) The main line of proof follows (see [24, pp. 278–280]) (see also [25, Theorem 5]). Rewrite the system of (8.1)–(8.3) in the following form

$$\mathbf{x}'(t) = M\,\mathbf{x}(t) + B\,\mathbf{x}(t - \tau) + \mathbf{f}(\mathbf{x}(t)), \tag{8.33}$$

where

$$\mathbf{x}(t) = \begin{pmatrix} a(t) \\ r(t) \\ o(t) \end{pmatrix}, \quad \mathbf{f}(\mathbf{x}(t)) = \begin{pmatrix} \frac{A}{1+p_2 or} \\ p_5 + \frac{(or)^2}{p_4+(or)^2} \\ 0 \end{pmatrix},$$

$$M = \begin{pmatrix} -p_3 & 0 & 0 \\ 0 & -p_6 & 0 \\ 0 & 0 & -1 \end{pmatrix}, \quad B = \begin{pmatrix} 0\,0\,0 \\ 0\,0\,0 \\ 1\,0\,0 \end{pmatrix}.$$

Obviously, the right-hand side of (8.33) is T-periodic with respect to t as it does not depend on time explicitly. Without loss of generality, we may assume that

$$0 \leqslant \tau < T.$$

This is true because otherwise we could represent the τ in the form

$$0 < \tau = nT + \tau_1, \text{ where } n \in \mathbb{Z}^+,\ \tau_1 \in [0, T).$$

Then shift to the auxiliary equation

$$\mathbf{x}'(t) = M\,\mathbf{x}(t) + B\,\mathbf{x}(t - \tau_1) + \mathbf{f}(\mathbf{x}(t)),$$

the T-periodic solutions of which coincide with the T-periodic solutions of (8.33).

On the set of all vector-valued functions $\mathbf{x}(t)$ defined on $[0, T]$, let us define an operator S_τ by

$$S_\tau \mathbf{x}(t) := \begin{cases} \mathbf{x}(t-\tau) \text{ if } \tau \leqslant t \leqslant T, \\ \mathbf{x}(t-\tau+T) \text{ if } 0 \leqslant t < \tau. \end{cases}$$

Note that the T-periodic solutions of (8.33) coincides with the solutions of the following integral equations:

$$\mathbf{x}(t) = T(\tau, \mathbf{x}) := \mathbf{x}(0) + \int_0^t (M\,\mathbf{x}(s) + B\,S_\tau \mathbf{x}(s) + \mathbf{f}(\mathbf{x}(s)))\,ds.$$

The operator $T(\tau, \mathbf{x})$ maps every continuous vector-valued function $\mathbf{x}(t)$ into a continuous vector-valued function for $0 \leqslant t \leqslant T$, therefore $T(\tau, \mathbf{x})$ is compact in C. Next, we will show that for all T-periodic solutions $\mathbf{x}_p(t)$ there exists $R > 0$ such that

$$|\mathbf{x}_p(t)| \leqslant R < \infty. \tag{8.34}$$

In summary, we deduce that

$$|\mathbf{x}_p(t)| \leqslant |\mathbf{x}_p(0)| + \int_0^t [|M|\,|\mathbf{x}_p(s)| + |B|\,|S_\tau \mathbf{x}_p(s)| + |\mathbf{f}(\mathbf{x}_p(s))|]\,ds \leqslant$$

$$|\mathbf{x}_p(0)| + [(p_3^2 + p_6^2 + 1)^{\frac{1}{2}} + 1] \int_0^t |\mathbf{x}_p(s)|\,ds + (A^2 + (p_5 + 1)^2)^{\frac{1}{2}} t.$$

From here, using Grönwall's lemma, we arrive at

$$|\mathbf{x}_p(t)| \leqslant (|\mathbf{x}_p(0)| + a)e^{bT} - a,$$

where $a = \frac{(A^2+(p_5+1)^2)^{\frac{1}{2}}}{(p_3^2+p_6^2+1)^{\frac{1}{2}}+1}$, $b = (p_3^2 + p_6^2 + 1)^{\frac{1}{2}} + 1$. Hence, (8.34) holds with $R = (|\mathbf{x}_p(0)| + a)e^{bT} - a$. As a result, by the fixed point theorem the integral equation has at least one solution, and consequently the Eq. (8.33) has at least one T-periodic solution.

Example 8.5 Let $A = 1.5$, $p_2 = 1.8$, $p_3 = 0.2$, $p_4 = 5$, $p_5 = 0.11$, and $p_6 = 0.9$. Hence the initial conditions are $a_0 = \frac{A}{p_3}(1 + p_2\sqrt{\frac{p_4(p_6 r_0 - p_5)}{p_5+1-p_6 r_0}})^{-1} \approx 1.4926$, $r_0 = \frac{1}{2}(\frac{p_5}{p_6} + \frac{p_5+1}{p_6}) \approx 0.6778$, and $o_0 = a_0 \approx 1.4926$. With these parameter values and setting delay $\tau = 4$ we solve (8.1)–(8.4) numerically using the Matlab solver dde23 [21]. Periodicity of the solutions can be seen in Fig. 8.6a.

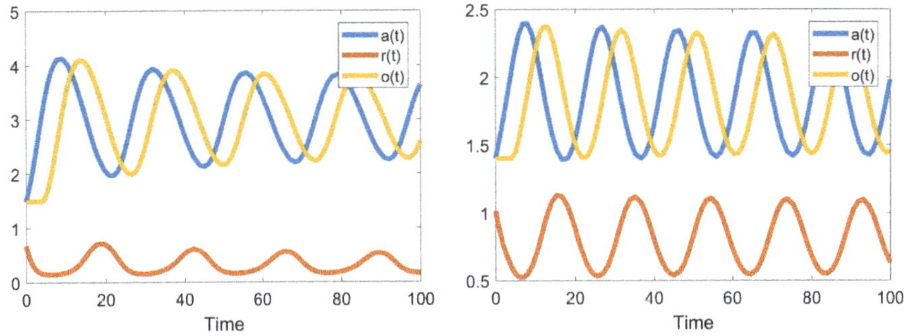

(a) (b)

Fig. 8.6 Plot of the solutions $a(t)$, $r(t)$, and $o(t)$ with the parameter values from Example 8.5

If we perturb the parameters by a bit and use $A = 2.1$, $p_2 = 2$, $p_3 = 0.3$, $p_4 = 4$, $p_5{=}0.21$, and $p_6{=}0.7$, then the initial conditions are $a_0{=}\frac{A}{p_3}(1 + p_2\sqrt{\frac{p_4(p_6r_0-p_5)}{p_5+1-p_6r_0}})^{-1} \approx 1.4$, $r_0 = \frac{1}{2}(\frac{p_5}{p_6} + \frac{p_5+1}{p_6}) \approx 1.0143$, and $o_0 = a_0 \approx 1.4$. This small perturbation of the parameters changes the amplitudes a lot as can be seen in Fig. 8.6b. Periodicity of solutions can be illustrated by plotting delayed function versus no delay function or function versus derivative as seen in Fig. 8.7.

8.4 Discussion

Existence of non-negative solutions, uniqueness of a steady state and its stability were analyzed for the minimal model of the HPA in [10]. The analytical difference between the model without delay that we studied and the minimal model is the type of the nonlinearities in the equation that describes production and degradation of the adrenocorticotropic hormone $a(t)$ and in the equation for the density $r(t)$ of glucocorticoid receptor. The nonlinear functions in the model that we analyze depend on the product $o(t)r(t)$ but in the minimal model of the HPA they depend on $o(t)$ only. We obtained similar results but for more complicated non-linearity terms. We also analyzed stability for all possible cases of parameter ranges. We added Lyapunov stability analysis to show the non-linear stability result and we believe that non-linear stability analysis has never done before for this type of model.

For the model with delay we obtained a critical time delay value when the originally stable system becomes unstable as a pair of complex eigenvalues crosses the imaginary axis. To illustrate the dynamics of complex eigenvalues with respect to time delay we used intersections of zero level sets between real and imaginary parts of the characteristic equation as a first approximation of a complex eigenvalue and after that we used Newton iterations to improve the accuracy. We believe that stabil-

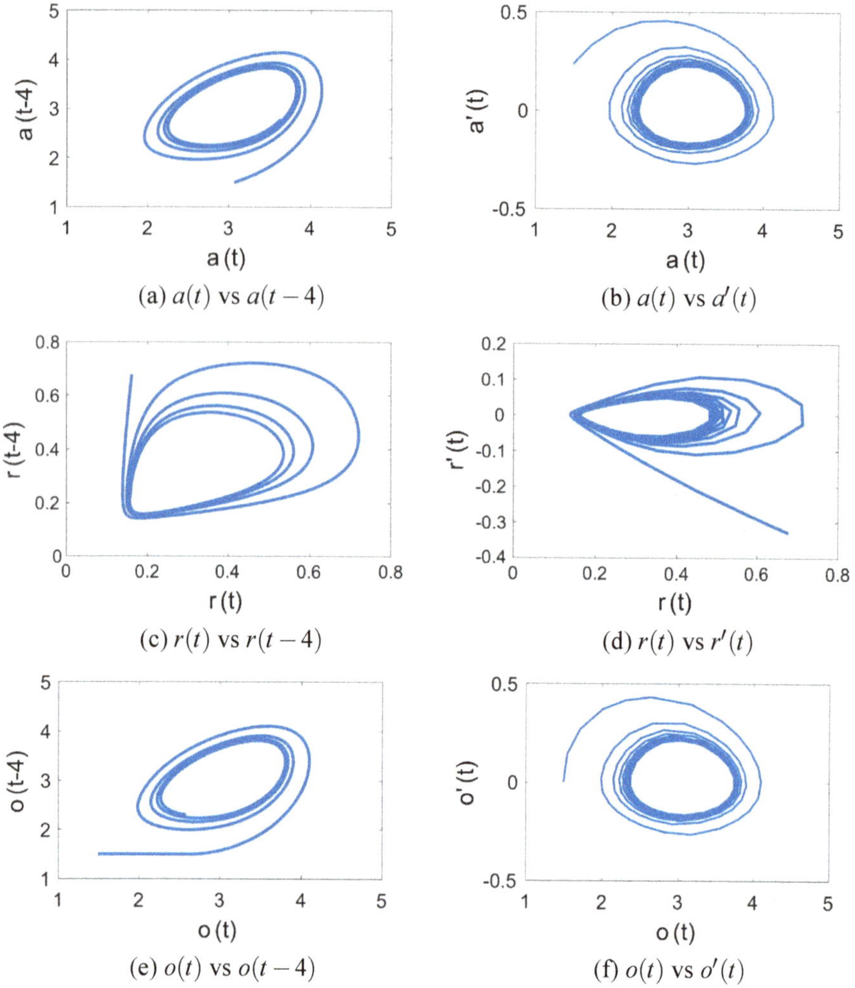

Fig. 8.7 Plots using parameter values $A = 1.5$, $p_2 = 1.8$, $p_3 = 0.2$, $p_4 = 5$, $p_5 = 0.11$, $p_6 = 0.9$

ity analysis for HPA model with respect to time delay was not done before and our results are new in this area.

For certain parameter values we rigorously proved existence of non-negative periodic solutions for a given period T (under an assumption that a given period does not coincide with the value of time delay) and illustrated different periodic solutions numerically. Our numerical simulations revealed that the period of the solution is very sensitive to small perturbations of parameter values.

Declaration of Interest Statement

We have nothing to declare.

8.5 Appendix

Lemma 8.4 (Routh Hurwitz Criteria for a Nonlinear System) *Suppose*

$$\dot{\boldsymbol{x}} = \boldsymbol{f}(\boldsymbol{x}),\ \boldsymbol{f} : \mathbb{R}^3 \to \mathbb{R}^3,\ \boldsymbol{x}(t_0) = \boldsymbol{x}_0. \tag{8.35}$$

Suppose $\boldsymbol{x}_s$ *is a fixed point of (8.35) and the characteristic polynomial at the fixed point is*

$$\lambda^3 + \alpha_1\lambda^2 + \alpha_2\lambda + \alpha_3 = 0, \quad \alpha_i \in \mathbb{R}^1.$$

If $\alpha_1 > 0$, $\alpha_3 > 0$ *and* $\alpha_1\alpha_2 > \alpha_3$, *then the fixed point is asymptotically stable. If* $\alpha_1 < 0$, $\alpha_3 < 0$ *or* $\alpha_1\alpha_2 < \alpha_3$, *then the fixed point is unstable.*

Theorem 8.3 (see [22]) *Consider Eq. (8.19), where* $P(z)$ *and* $Q(z)$ *are analytic functions in a right half-plane* $Re(z) > -\delta$, $\delta > 0$, *which satisfy the following conditions:*

(i) $P(z)$ *and* $Q(z)$ *have no common imaginary zeros;*
(ii) $\overline{P(-iy)} = P(iy)$ *and* $\overline{Q(-iy)} = Q(iy)$ *for real* y*;*
(iii) $P(0) + Q(0) \neq 0$*;*
(iv) There are at most a finite number of roots of (8.19) in the right half-plane when $\tau = 0$*;*
(v) $F(y) \equiv |P(iy)|^2 - |Q(iy)|^2$ *for real* y *has at most a finite number of real zeros.*

Under these conditions, the following statements are true.

(a) Suppose that the equation $F(y) = 0$ *has no positive roots. Then if (8.19) is stable at* $\tau = 0$ *it remains stable for all* $\tau \geqslant 0$, *whereas if it is unstable at* $\tau = 0$ *it remains unstable for all* $\tau \geqslant 0$.
(b) Suppose that the equation $F(y) = 0$ *has at least one positive root and that each positive root is simple. As* τ *increases, stability switches may occur. There exists a positive number* τ_c *such that the Eq. (8.19) is unstable for all* $\tau > \tau_c$. *As* τ *varies from* 0 *to* τ_c, *at most a finite number of stability switches may occur.*

References

1. Walker, J.J., Terry, J.R., Lightman, S.L.: Origin of ultradian pulsatility in the hypothalamicpituitary-adrenal axis. Proc. R. Soc. B. **277**, 1627–1633 (2010)
2. Papadimitriou, A., Priftis, K.: Regulation of the hypothalamicpituitary-adrenal axis. Neuroimmunomodulation **16**(5), 265–71 (2009)
3. Gold, P.W., Chrousos, G.P.: Organization of the stress system and its dysregulation in melancholic and atypical depression: high vs low CRH/NE states. Mol. Psychiatry **7**, 254–75 (2002)
4. Juruena, M.F., Cleare, A.J., Pariante, C.M.: The hypothalamic pituitary adrenal axis, glucocorticoid receptor function and relevance to depression. Rev. Bras. Psiquiatr. **26**, 189–201 (2004)
5. Rohleder, N., Joksimovic, L., Wolf, J.M., Kirschbaum, C.: Hypocortisolism and increased glucocorticoid sensitivity of pro-inflammatory cytokine production in Bosnian war refugees with posttraumatic stress disorder. Biol. Psychiatry **55**, 745–51 (2004)

6. Gonzalez-Heydrich, J., Steingard, R.J., Kohane, I.: A computer simulation of the hypothalamic-pituitary-adrenal axis. Proc. Ann. Symp. Comput. Appl. Med. Care **1010** (1994)
7. Hosseinichimeh, N., Rahmandad, H., Wittenborn, A.: Modeling the hypothalamus-pituitary-adrenal axis: A review and extension. Math. Biosci. **268**, 52–65 (2015)
8. Sriram, K., Rodriguez-Fernandez, M., Doyle, F.J.: III Modeling cortisol dynamics in the neuroendocrine axis distinguishes normal, depression, and post-traumatic stress disorder (PTSD) in humans. PLoS Comput. Biol. **8**(2), 1–15 (2012)
9. Bairagi, N., Chatterjee, S., Chattopadhyay, J.: Variability in the secretion of corticotropin-releasing hormone, adrenocorticotropic hormone and cortisol and understandability of the hypothalamic-pituitary-adrenal axis dynamics - a mathematical study based on clinical evidence. Math. Med. Biol. **25**(1), 37–63 (2008)
10. Vinther, F., Andersen, M., Ottesen, J.: The minimal model of the hypothalamic-pituitary-adrenal axis. J. Math. Biol. **63**(4), 663–90 (2011)
11. Andersen, M., Vinther, F., Ottesen, J.: Mathematical modeling of the hypothalamic-pituitary-adrenal gland (HPA) axis, including hippocampal mechanisms. Math. Biosci. **246**(1), 122–38 (2013)
12. Jelić, S., Čupić, Ž., Kolar-Anić, L.: Mathematical modeling of the hypothalamic-pituitary-adrenal system activity. Math. Biosci. **197**(2),173–87 (2005)
13. Conrad, M., Hubold, C., Fischer, B., Peters, A.: Modeling the hypothalamus-pituitary-adrenal system: Homeostasis by interacting positive and negative feedback. J. Biol. Phys. **35**(2), 149–62 (2009)
14. Lenbury, Y., Pornsawad, P.: A delay-differential equation model of the feedback-controlled hypothalamus-pituitary-adrenal axis in humans. Math. Med. Biol. **22**, 15–33 (2005)
15. Gupta, S., Aslakson, E., Gurbaxani, B.M., Vernon, S.D.: Inclusion of the glucocorticoid receptor in a hypothalamic pituitary adrenal axis model reveals bistability. Theor. Biol. Med. Model **4**, 8 (2007)
16. Ben-Zvi, A., Vernon, S.D., Broderick, G.: Model-based therapeutic correction of hypothalamic-pituitary-adrenal axis dysfunction. PLoS Comput. Biol. **5**(1), 1000273 (2009)
17. McEwen, B.S.: Physiology and neurobiology of stress and adaptation: central role of the brain. Physiol. Rev. **87**, 873–904 (2007)
18. Savić, D., Knežević, G., Opačić, G.: A mathematical model of stress reaction: Individual differences in threshold and duration. Psychobiology **28**(4),581–92 (2000)
19. Ono, N., Castro, J.D., McCann, S.: Ultrashort-loop positive feedback of corticotropin (ACTH)-releasing factor to enhance ACTH release in stress. Proc. Natl. Acad. Sci. **82**(10), 3528–31 (1985)
20. Rankin, J., Walker, J.J., Windle, R., Lightman, S.L., Terry, J.R.: Characterizing dynamic interactions between ultradian glucocorticoid rhythmicity and acute stress using the phase response curve. PLoS ONE **7**(2), e30978 (2012). https://doi.org/10.1371/journal.pone.0030978
21. MATLAB version 7.10.0. Natick, Massachusetts: The MathWorks Inc. (2017)
22. Cooke, K.L., van den Driessche, P.: On zeros of some transcendental equations. Funkcial. Ekvac **29**, 77–90 (1986)
23. Barresi, R., Lombardo, M.C., Sammartino, M.: Hopf bifurcation analysis of the generalized Lorenz system with time delayed feedback control (2015). arXiv:1406.4694
24. Krasnosel'skii, M.A.: The Operator of Translation along the Trajectories of Differential Equations. Translation of Mathematical Monographs, vol. 19, p. 294. American Mathematical Society, Providence (1968)
25. Krasnosel'skii, M.A.: An alternative principle for the existence of periodic solutions for differential equations with retarded argument. Dokl. Akad. Nauk SSSR **52**(4), 801–4 (1963)
26. Shampine, L.F., Thompson, S.: Solving delay differential equations with dde23. (Lecture notes). (2000)

Chapter 9
Mixed Dirichlet-Transmission Problems in Non-smooth Domains

Nataliya Vasylyeva

Abstract In this paper, we consider the Poisson equations in two dimensional non-smooth domains on the boundaries of which inhomogeneous Dirichlet and transmission conditions are imposed. Under certain assumptions on the parameter in the model, we prove the existence of a unique classical solution which belongs to the weighted Hölder classes.

9.1 Introduction

Transmission problems (also called interface problems) emerge in wide applications such as multiphase flows with and without phase change, in heat transfer, in electrokinetics or in the modeling of biomolecules' electrostatics. In particular, electromagnetic processes in ferromagnetic media with different dielectric constants are modeled with transmission problems, see e.g. [2, 3, 11] and references therein. The similar problems appear in solid mechanics and deal with study of composite materials [21]. We also quote works [10, 15, 23, 24], where authors study vibrating folded membranes, composite and folded plates, junctions in elastic multi-structures. Finally, we refer the paper [17], where transmission problems for the Laplace operator describe the initial concentration of media in tumors.

In the present paper, we focus on the interface problems for Poisson equations in non-smooth domains. Let $\Omega \subset \mathbb{R}^2$ be a bounded domain with a smooth boundary $\partial\Omega \in \mathbb{C}^{2+\beta}$, $\beta \in (0, 1)$. A simple curve $\Gamma \subseteq \bar{\Omega}$ splits Ω onto two sub-domains Ω_1 and Ω_2, such that

$$\Omega = \Omega_1 \cup \Omega_2 \cup \Gamma, \quad \Omega_1 \cap \Omega_2 = \oslash \quad \partial\Omega_1 \cap \partial\Omega_2 = \Gamma,$$

N. Vasylyeva (✉)
Institute of Applied Mathematics and Mechanics of NAS of Ukraine,
G.Batyuka st. 19, 84100 Sloviansk, Ukraine
e-mail: nataliy_v@yahoo.com

V. A. Sadovnichiy and M. Z. Zgurovsky (eds.), *Contemporary Approaches and Methods in Fundamental Mathematics and Mechanics*, Understanding Complex Systems,
https://doi.org/10.1007/978-3-030-50302-4_9

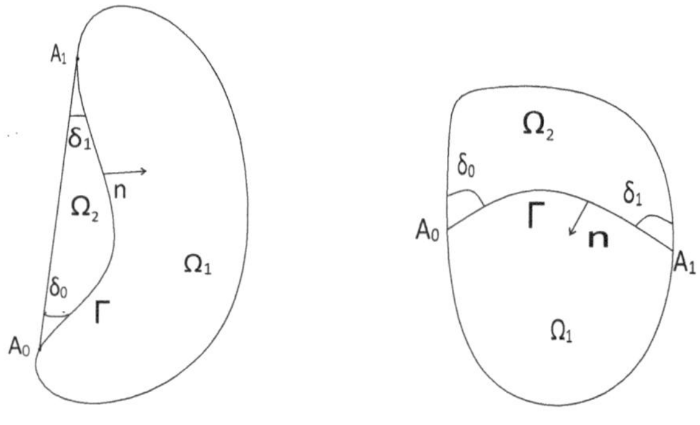

Fig. 9.1 Typical domain configuration with a smooth interface Γ

and moreover, Γ intersects with $\partial\Omega$ at two points A_0 and A_1, i.e. $\partial\Omega \cap \Gamma = \{A_0, A_1\}$ (see Fig. 9.1 for geometric setting). We assume that in vicinities of points A_0 and A_1 the curves Γ and $\partial\Omega$ form the angles δ_0 and δ_1, $\delta_i \in (0, \pi)$, $i = 0, 1$. Recall that the angle between curves is defined as the angle between their tangents at the point of intersection.

For a fixed positive k, $k \neq 1$, we analyze the Poisson equations in the unknown functions u_1 and u_2, $u_i = u_i(y_1, y_2) : \Omega_i \to \mathbb{R}$, $i = 1, 2$,

$$\Delta_y u_i = \mathscr{F}_{0i}(y), \tag{9.1}$$

subject the transmission conditions on Γ:

$$u_1 - u_2 = \mathscr{F}(y) \quad \text{and} \quad \frac{\partial u_1}{\partial n} - k\frac{\partial u_2}{\partial n} = \mathscr{F}_1(y), \tag{9.2}$$

and the Dirichlet boundary conditions on the rest part of $\partial\Omega_i$

$$u_1 = \mathscr{F}_2(y) \quad \text{on} \quad \partial\Omega_1 \backslash \Gamma, \quad u_2 = \mathscr{F}_3(y) \quad \text{on} \quad \partial\Omega_2 \backslash \Gamma, \tag{9.3}$$

where the functions $\mathscr{F}_{0i}$, $\mathscr{F}$, $\mathscr{F}_1$, $\mathscr{F}_2$, $\mathscr{F}_3$ are prescribed, n is the unit normal to Γ directed to Ω_1. In this work we examine two different cases of the interface Γ, the first one assumes Γ being from $C^{2+\beta}$ (as shown on Fig. 9.1), while the second option admits finite numbers of the corner points on Γ.

In our days, regularity results for elliptic transmission problems in domains with smooth boundaries are well known [12, 22, 27, 28, 30–33] (see also references therein). In particular, Schechter [31] and Sheftel' [33] investigated even-order elliptic equations with smooth coefficients in a bounded domain with a smooth boundary. Schechter proved existence of weak solutions and obtained the corresponding estimates. His strategy involved reduction of the interface problem to a mixed boundary value problem for a system of equations. Meanwhile, Sheftel' [32] provided the L_p estimates of a solution to the transmission problems. Exploiting approximate equations, Oleinik [27, 28] studied weak solutions of the elliptic equations of the second order with discontinuous coefficients. Then, in [22], these problems were treated in Sobolev and Hölder classes. In contrast to Schechter's technique, Ladyzhenskaya and Ural'tseva used cleverly chosen test functions to derive a priori estimates.

Starting in the 1970s a lot of mathematicians have studied linear transmission problems in non-smooth domains [8, 13, 14, 16, 18, 20, 25, 26, 29, 34] (see also references there). In particular, Kellogg [20], Ben M'Barek [8], Lemrabet [25] described the behavior of solutions of the transmission-Dirichlet and transmission-Neumann problems for the Laplace operator in a vicinity of a corner point in the case of two-dimensional domains. Dauge and Nicaise [13, 14, 26] extended some of Kondratiev-Maz'ya-Grisvard results, concerning the singular behavior of the weak solution to the transmission problems for the Laplace operator on a two-dimensional topological network. They obtained index formulae, calculated the dimension of the kernel and decomposed the weak solution into regular part (belonging to the appropriate space $W^{l,p}$) and singular part and then, based on the obtained results, established higher regularity results. Moreover, in [26], the solvability, regularity and asymptotic representation were stated in the weighted Sobolev spaces. Finaly, we quote the book [9], where the author described the behavior of weak solutions to the linear and quasilinear elliptic transmission problems in a neighborhood of boundary singularities: angular and conic points or edges. As for investigations of the transmission problems for elliptic equations with a singular interface in the smooth classes, we refer the paper [6], where in the case of a simple closed curve Γ (i.e. $\Gamma \cap \partial\Omega = \oslash$) having a corner point, authors proved the one-valued classical solvability of problem like (9.1)–(9.3) in the weighted Hölder classes $E_s^{2+\beta}$ with the weight s depending only on the angle.

The purpose of the present paper is to find the sufficiently conditions on the parameters of the model under which mixed Dirichlet-transmission problem (9.1)–(9.3) is solvable in the weighted Hölder classes. It should be noted that a main direction in investigation of boundary value problems with nonsmooth boundaries is related to asymptotic behavior of a solution in the neighborhood of the boundary singularities. The weighted Hölder classes used in the paper allow us to obtain the decay degree of the solution to (9.1)–(9.3) near corner points. On the other hand, the motivation in study of problem (9.1)–(9.3) in the classes $E_s^{2+\beta}$ arises from investigations of the two-phase free boundary problem for elliptic equations (contact two-phase Hele-Shaw problem or contact Muskat problem) in the case of corner points on the moving boundary. In fact, in this case, the solution (u_1, u_2) of problem (9.1)–(9.3) describes the initial pressure of the fluids in the contact Muskat problem. We recall that the Muskat problem describes the evolution of an interface between two immiscible incompressible fluids; see [7] for the derivation details. It is worth noting that the existence of waiting time phenomena is a main direction in the study of free boundary problems with a singular moving boundary. As follows from works [5, 7, 35, 36], solvability of the free boundary problems in the weighted Hölder spaces allows one to obtain the sufficiently conditions under which the waiting time exists.

In order to prove the one-valued classical solvability of problem (9.1)–(9.3) in the corresponding weighted Hölder classes, we, first, prove solvability of (9.1)–(9.3) in the weighted Sobolev spaces $W_\gamma^{2,p}$ with the appropriately chosen weight γ. On this route, we essential use the results from [26]. Then, taking into account these results, we estimate of the maximum modules of the solution u_i. After that, we increase the regularity of the solution (u_1, u_2) and show that u_i belongs to $E_s^{2+\beta}$, where the weight s, unlike [6], depends not only on the size of corners but also on the parameter k in transmission condition (9.2). The main ingredient in this study is the Schauder approach based on investigation of corresponding transmission problems in the plain corners. Namely, the main analytical difficulties are related to study of these problems.

Outline of the paper. In the next Sect. 9.2 we introduce the functional spaces and describe their useful properties. The main Theorems 9.1 and 9.2 of the paper, along with the general hypothesis, are stated in Sect. 9.3. Section 9.4 is devoted to the proof of Theorems 9.1 and 9.2 in the special case of Ω_i, namely, Ω_i, $i = 1, 2$, are plane corners. The principal results of Sect. 9.4 are stated in Theorems 9.3–9.8. It is worth mentioning that the main approach in this section consist of two steps. The first one is related with the obtaining integral representation of the solution. Then one has to produce a coercive estimates for this solution. In the final Sect. 9.5, based on Theorems 9.3–9.5 and results from [26], we prove Theorems 9.1 and 9.2 in the general case and discuss the conditions under which there are smoother solutions than indicated in the main theorems. The last result is formulated in Theorem 9.6. Appendix contains the proofs of some auxiliary assertions playing a central role in the course of the investigations in Sect. 9.4.

9.2 Functional Setting and Notations

Throughout this work, the symbol C will denote a generic positive constant, depending only on the structural quantities of the problem. We will carry out our analysis in the framework of the weighted Hölder spaces.

Let D be a given domain in $\mathbb{R}^2$ and ∂D have finite numbers of the corner points $A_0, A_1, \dots A_q$. Till the end of the paper, let

$$\beta \in (0, 1), \quad s \in \mathbb{R}$$

be arbitrarily fixed and for each $y, \bar{y} \in \bar{D}$, we put

$$r_i(y) = |A_i - y|, \quad r_i(\bar{y}) = |A_i - \bar{y}|, \quad i = 0, 1, \dots, q,$$

$$r(y) = \min\{r_0(y), r_1(y), \dots, r_q(y)\}, \quad r(\bar{y}) = \min\{r_0(\bar{y}), r_1(\bar{y}), \dots, r_q(\bar{y})\},$$

$$r(y, \bar{y}) = \min\{r(y), r(\bar{y})\}.$$

For any nonnegative integer l we introduce the class $E_s^{l+\beta}(\bar{D})$. Denoting

$$\langle v\rangle^{(\beta)}_{y,s,D} = \sup\left\{r^{-s+\beta}(y,\bar{y})\frac{|v(y)-v(\bar{y})|}{|y-\bar{y}|^{\beta}}:\quad y,\bar{y}\in\bar{D},\quad 0<|y-\bar{y}|<\frac{r(y,\bar{y})}{2}\right\}$$

we give the following definition.

Definition 9.1 A function $v = v(y)$ belongs to the class $E_s^{l+\beta}(\bar{D})$, if the norm here below is finite

$$\|v\|_{E_s^{l+\beta}(\bar{D})} = \sum_{|m|=0}^{l} \sup_{\bar{D}} r^{-s+|m|}(y)|D_y^m v(y)| + \sum_{|m|=l} \langle D_y^m v\rangle^{(\beta)}_{y,s-l-D}$$

where $|m| = m_1 + m_2$.

In a similar way, we define the space $E_s^{l+\beta}(\partial D)$.

If the domain D does not contain any corner points, the space $E_s^{l+\beta}(\bar{D})$ boils down to the usual Hölder space $\mathbb{C}^{l+\beta}(\bar{D})$.

The straightforward calculations arrive at the following properties of these spaces which will be used in Sect. 9.5.

Corollary 9.1 *Let D be a bounded domain, $0 < s_1 < s_2$, and let $u \in E_{s_2}^{l+\beta}(\bar{D})$. Then $u \in E_{s_1}^{l+\beta}(\bar{D})$ and the estimate holds*

$$\|u\|_{E_{s_1}^{l+\beta}(\bar{D})} \leq C\|u\|_{E_{s_2}^{l+\beta}(\bar{D})}$$

with the constant depending only on $|D|^{s_2-s_1}$, where $|D|$ is the Lebesgue measure of D.

If, for some fixed $\delta \in [0, \pi/2)$, D consists of two plain corners: $D = G_1^+ \cup G_2^+$ or $D = G_1^- \cup G_2^-$,

$$\begin{aligned} G_1^{\pm} &= \{(y_1, y_2):\quad y_1 > 0,\quad y_2 < \pm y_1 \tan\delta\},\\ G_2^{\pm} &= \{(y_1, y_2):\quad y_1 > 0,\quad y_2 > \pm y_1 \tan\delta\}, \end{aligned} \tag{9.4}$$

we will use the equivalent definition of $E_s^{l+\beta}(\bar{D})$. To this end, we define new independent variables

$$x_1 = \ln\sqrt{y_1^2 + y_2^2},\quad x_2 = \arctan(y_2/y_1). \tag{9.5}$$

It is apparent that, this mapping transforms the plane corners $G_1^{\pm}$ and $G_2^{\pm}$ to the finite strips $B_1^{\pm}$ and $B_2^{\pm}$:

$$\begin{aligned} B_1^{\pm} &= \{(x_1, x_2):\quad x_1 \in \mathbb{R},\quad -\pi/2 < x_2 < \pm\delta\},\\ B_2^{\pm} &= \{(x_1, x_2):\quad x_1 \in \mathbb{R},\quad \pm\delta < x_2 < \pi/2\}. \end{aligned}$$

Then direct calculations lead to the assertion.

Proposition 9.1 *For* $i = 1, 2$, *there is the equivalence*

$$v_i(y) \in E_s^{l+\beta}(\bar{G}_i^{\pm}) \ \ \textit{iff} \ \ v_i y(x)e^{-sx_1} \in \mathbb{C}^{l+\beta}(\bar{B}_i^{\pm}).$$

The similar result holds for the functions $v_i(y)$ defined on $\partial G_i^{\pm}$, $i = 1, 2$.

Along the paper, we will also encounter the usual Sobolev space $W^{l,p}(D)$ with the norm

$$\|v\|_{l,p;D} = \Big(\sum_{|m|\leq l} \int_D |D_y^m v(y)|^p dy\Big)^{1/p}$$

for the integer l, and the weighted Sobolev space $W_s^{l,p}(D)$:

$$\|v\|_{l,p;s,D} = \Big(\sum_{|m|\leq l} \int_D r^{p(s+|m|-l)}(y)|D_y^m v(y)|^p dy\Big)^{1/p}.$$

The definition of these spaces and their properties in the case of noninteger l can be found, for instance, in [1, 26].

It is apparent that $W^{0,p}(D) = L^p(D)$ and $W_s^{0,p}(D) = L_s^p(D)$.

9.3 The Main Results

First, we state our general hypotheses on the structural terms in the model.

h1 (Condition on the weight) We assume that s, k, δ_0 and δ_1 satisfy one of four conditions:

(i) $s \in (0, 1)\backslash\Big\{\frac{\pi}{2\pi-2\delta_0}, \frac{\pi}{2\pi-2\delta_1}\Big\}$, $k > 1$ and $\delta_0, \delta_1 \in (0, \pi/2]$;

(ii) $s \in \Big(1, \min\Big\{2, \frac{\pi}{\pi-2\delta_0}, \frac{\pi}{\pi-2\delta_1}\Big\}\Big)\backslash\Big\{\frac{\pi}{\pi-\delta_0}, \frac{\pi}{\pi-\delta_1}, \frac{\pi}{2\delta_0}, \frac{\pi}{2\delta_1}\Big\}$, $k < 1$ and $\delta_0, \delta_1 \in (0, \pi/2]$;

(iii) $s \in (0, 1)\backslash\Big\{\frac{\pi}{2\delta_0}, \frac{\pi}{2\delta_1}\Big\}$, $k < 1$ and $\delta_0, \delta_1 \in (\pi/2, \pi)$;

(iv) $s \in \Big(1, \min\Big\{2, \frac{\pi}{|\pi-2\delta_0|}, \frac{\pi}{|\pi-2\delta_1|}\Big\}\Big)\backslash\Big\{\frac{\pi}{\delta_0}, \frac{\pi}{\delta_1}, \frac{\pi}{2\pi-2\delta_0}, \frac{\pi}{2\pi-2\delta_1}\Big\}$, $k > 1$ and $\delta_0, \delta_1 \in (\pi/2, \pi)$.

h2 (Condition on the given functions) The following inclusions hold

$$\mathscr{F}_{0i} \in E_{s-2}^{\beta}(\bar{\Omega}_i), \quad i = 1, 2,$$
$$\mathscr{F} \in E_s^{2+\beta}(\Gamma), \quad \mathscr{F}_1 \in E_{s-1}^{1+\beta}(\Gamma),$$
$$\mathscr{F}_2 \in E_s^{2+\beta}(\partial\Omega_1\backslash\Gamma), \quad \mathscr{F}_3 \in E_s^{2+\beta}(\partial\Omega_2\backslash\Gamma).$$

We are now in the position to state our main result.

Theorem 9.1 *Let $\Gamma \in \mathbb{C}^{2+\beta}$, assumptions* ***h1*** *and* ***h2*** *hold. Then transmission problem* (9.1)–(9.3) *admits a unique classical solution* $(u_1(y_1, y_2), u_2(y_1, y_2))$, *satisfying the regularity*

$$u_1 \in E_s^{2+\beta}(\bar{\Omega}_1) \quad \textit{and} \quad u_2 \in E_s^{2+\beta}(\bar{\Omega}_2).$$

In addition, the following estimate holds

$$\|u_1\|_{E_s^{2+\beta}(\bar{\Omega}_1)} + \|u_2\|_{E_s^{2+\beta}(\bar{\Omega}_2)} \leq C\Big[\|\mathscr{F}_{01}\|_{E_{s-2}^{\beta}(\bar{\Omega}_1)} + \|\mathscr{F}_{02}\|_{E_{s-2}^{\beta}(\bar{\Omega}_2)} + \|\mathscr{F}\|_{E_s^{2+\beta}(\Gamma)}$$
$$+ \|\mathscr{F}_1\|_{E_{s-1}^{1+\beta}(\Gamma)} + \|\mathscr{F}_2\|_{E_s^{2+\beta}(\partial\Omega_1\backslash\Gamma)} + \|\mathscr{F}_3\|_{E_s^{\beta+2}(\partial\Omega_2\backslash\Gamma)}\Big]$$

for some C independent of the right-hand sides in (9.1)–(9.3) *and* u_i, $i = 1, 2$.

As follows from Theorem 9.1 and definition of the space $E_s^{2+\beta}$, the solution of problem (9.1)–(9.3) near corner points A_0 and A_1 behaves like $|A_0 - y|^s$ and $|A_1 - y|^s$, correspondingly. In fact, it can be shown that for sufficiently small δ_0 and δ_1 the solution near corner points vanishes faster than this theorem provides. Thus, it is possible to obtain more smooth solution than it is stated above (see forthcoming Theorem 9.6 in Sect. 9.5.2).

Coming to the case of a nonsmooth interface Γ, we first make additional assumptions:

h3 (Condition on the curve Γ) We assume that Γ has rth corner points: $A_2, A_3, \ldots, A_{r+1}$, and in the vicinity of each A_j the curve Γ consists of two intersecting segments with an angular opening $\alpha_j \in (0, \pi)$, $j = 2, \ldots, r+1$, and $\Gamma \in \mathbb{C}^{2+\beta}$ outside of these neighborhoods.

h4 (Condition on the weight) Denoting $\alpha = \min\{\alpha_2, \alpha_3, \ldots, \alpha_{r+1}\}$, we require that $\alpha_j \in (0, \pi)$, and parameters: s, k, δ_0, δ_1 and α_j, satisfy one of two conditions

(i) $s \in \left(\frac{1}{2}, \frac{\pi}{2\pi-\alpha}\right)\backslash\left\{\frac{\pi}{2\pi-2\delta_0}, \frac{\pi}{2\pi-2\delta_1}\right\}$, $k > 1$ and $\delta_0, \delta_1 \in (0, \pi/2]$;

(ii) $s \in \left(\frac{1}{2}, \frac{\pi}{2\pi-\alpha}\right)\backslash\left\{\frac{\pi}{2\delta_0}, \frac{\pi}{2\delta_1}\right\}$, $k < 1$ and $\delta_0, \delta_1 \in (\pi/2, \pi)$.

Theorem 9.2 *Under assumptions* ***h2–h4*** *transmission problem* (9.1)–(9.3) *admits a unique classical solution* $(u_1(y_1, y_2), u_2(y_1, y_2))$, *satisfying the regularity*

$$u_1 \in E_s^{2+\beta}(\bar{\Omega}_1) \quad \textit{and} \quad u_2 \in E_s^{2+\beta}(\bar{\Omega}_2).$$

Besides, the following estimate holds

$$\|u_1\|_{E_s^{2+\beta}(\bar{\Omega}_1)} + \|u_2\|_{E_s^{2+\beta}(\bar{\Omega}_2)} \leq C\Big[\|\mathscr{F}_{01}\|_{E_{s-2}^{\beta}(\bar{\Omega}_1)} + \|\mathscr{F}_{02}\|_{E_{s-2}^{\beta}(\bar{\Omega}_2)} + \|\mathscr{F}\|_{E_s^{2+\beta}(\Gamma)}$$
$$+ \|\mathscr{F}_1\|_{E_{s-1}^{1+\beta}(\Gamma)} + \|\mathscr{F}_2\|_{E_s^{2+\beta}(\partial\Omega_1\backslash\Gamma)} + \|\mathscr{F}_3\|_{E_s^{\beta+2}(\partial\Omega_2\backslash\Gamma)}\Big]$$

for some C independent of the right-hand sides in (9.1)–(9.3) *and u_i, $i = 1, 2$.*

Here we prove in detail Theorems 9.1 and 9.2 if vicinities of corner points A_j, $j = 0, 1, \ldots, r+1$, consist of segments. The general case is obtained with the same arguments and the transformation (A.13) in [35] .

9.4 Solvability of Transmission Problem (9.1)–(9.3) in $G_1^{\pm} \cup G_2^{\pm}$

We begin our analysis with the proof of Theorem 9.1 if Ω_1 and Ω_2 are plane corners defined with (9.4), i.e. $\Omega_1 = G_1^+$ and $\Omega_2 = G_2^+$. The consideration in the case of $\Omega_1 = G_1^-$ and $\Omega_2 = G_2^-$ is almost identical and is left to the interested reader.

Additionally to assumptions **h1** and **h2**, we require that the right-hand sides of (9.1)–(9.3) are finite functions.

Introducing

$$g = \{(y_1, y_2) : \quad y_1 \geq 0, \quad y_2 = y_1 \tan\delta\},$$

we first dwell on the special case in (9.1)–(9.3) when

$$\mathscr{F} = \mathscr{F}_2 = \mathscr{F}_3 \equiv 0. \tag{9.6}$$

Namely, (9.1)–(9.3) is replaced by the simpler conditions

$$\begin{cases} \Delta_y u_i = \mathscr{F}_{0i} \quad \text{in} \quad G_i^+, \; i = 1, 2, \\ u_1 - u_2 = 0 \quad \text{and} \quad \frac{\partial u_1}{\partial n} - k\frac{\partial u_2}{\partial n} = \mathscr{F}_1 \quad \text{on} \quad g, \\ u_1 = 0 \quad \text{on} \quad \partial G_1^+\backslash g, \quad u_2 = 0 \quad \text{on} \quad \partial G_2^+\backslash g. \end{cases} \tag{9.7}$$

In a further step, we will show how to reduce the general case to this special one.

Introducing new variables (9.5) and new functions

$$v_i(x) = e^{-sx_1}u_i(x),$$
$$f_{0i}(x) = e^{(2-s)x_1}\mathscr{F}_{0i}(x),$$
$$f_1(x) = -e^{(1-s)x_1}\mathscr{F}_1(x),$$

where the value s will be chosen later, we rewrite problem (9.7) in new form

$$\begin{cases} \Delta_x v_i + 2s\frac{\partial v_i}{\partial x_1} + s^2 v_i = f_{0i}(x) \quad \text{in} \quad B_i^+,\ i = 1, 2; \\ v_1 - v_2 = 0 \quad \text{and} \quad \frac{\partial v_1}{\partial n} - k\frac{\partial v_2}{\partial n} = f_1 \quad \text{on} \quad b; \\ v_1 = 0 \quad \text{on} \quad \partial B_1^+ \backslash b, \quad \text{and} \quad v_2 = 0 \quad \text{on} \quad \partial B_2^+ \backslash b. \end{cases} \tag{9.8}$$

Here b denotes the image of the boundary g after transformation (9.5), i.e.

$$b = \{(x_1, x_2) : \quad x_1 \in \mathbb{R}, \quad x_2 = \delta\}.$$

We will look for a solution $(v_1(x_1, x_2), v_2(x_1, x_2))$ of problem (9.8) in the Hölder classes $\mathbb{C}^{2+\beta}(\bar{B}_i^+)$ if $f_{0i} \in \mathbb{C}^{\beta}(\bar{B}_i^+)$ and $f_1 \in \mathbb{C}^{1+\beta}(b)$. To this end, we represent solution (9.8) as

$$v_i = v_i^0(x) + v_i^1, \quad i = 1, 2, \tag{9.9}$$

where (v_1^1, v_2^1) solves (9.8) with the homogenous equations (i.e. $f_{0i} \equiv 0$), while (v_1^0, v_2^0) is a solution of (9.8) with the homogenous boundary conditions (i.e. $f_1 \equiv 0$). On the next step we construct integral representations of v_i^0 and v_i^1, $i = 1, 2$, and then obtain the coercive estimates of the constructed solutions.

9.4.1 *Integral Representation of* (v_1^0, v_2^0)

We begin to analyze (9.8) in the case of homogenous boundary conditions (i.e. $f_1 = 0$). Let $\hat{w}(\lambda, x_2)$ be the Fourier transformation of the function $w(x_1, x_2)$ with respect to x_1, i.e.

$$\hat{w}(\lambda, x_2) = \int_{-\infty}^{+\infty} w(x_1, x_2) e^{-i\lambda x_1} dx_1.$$

Applying the Fourier transformation to (9.8) (where $f_1 \equiv 0$) derives to the problem

$$\begin{cases} \frac{d^2\hat{v}_1^0}{dx_2^2} + [2s\lambda i + s^2 - \lambda^2]\hat{v}_1^0 = \hat{f}_{01}(\lambda, x_2), \quad x_2 \in (-\pi/2, \delta), \\ \frac{d^2\hat{v}_2^0}{dx_2^2} + [2s\lambda i + s^2 - \lambda^2]\hat{v}_2^0 = \hat{f}_{02}(\lambda, x_2), \quad x_2 \in (\delta, \pi/2), \\ \hat{v}_1^0(\lambda, \delta) = \hat{v}_2^0(\lambda, \delta) \quad \text{and} \quad \frac{d\hat{v}_1^0}{dx_2} - k\frac{d\hat{v}_2^0}{dx_2}\Big|_{x_2=\delta} = 0, \\ \hat{v}_1^0(\lambda, -\pi/2) = 0, \quad \hat{v}_2^0(\lambda, \pi/2) = 0. \end{cases} \tag{9.10}$$

Next we denote by $\rho = i\lambda + s$. Then straightforward calculations lead to the integral representation of the solution

$$\hat{v}_1^0 = c_1^1 \sin \rho x_2 + c_2^1 \cos \rho x_2 + \int\limits_{-\pi/2}^{x_2} \frac{\hat{f}_{01}(z,\lambda)}{\rho} \sin \rho(x_2 - z)dz, \tag{9.11}$$

$$\hat{v}_2^0 = c_1^2 \sin \rho x_2 + c_2^2 \cos \rho x_2 + \int\limits_{\delta}^{x_2} \frac{\hat{f}_{02}(z,\lambda)}{\rho} \sin \rho(x_2 - z)dz,$$

where c_1^i and c_2^i, $i = 1, 2$, are arbitrary constants.

In order to find the unknown coefficients c_i^j, $i, j = 1, 2$, it is necessary to substitute (9.11) to the boundary conditions in (9.10) and then to solve the linear system of the algebraic equations. In summary, after tedious calculations, we obtain the determinant of this system:

$$\mathscr{D}(\lambda) = \frac{1}{2}[(k+1)\sin \rho\pi + (k-1)\sin 2\delta\rho]. \tag{9.12}$$

In next assertion, we give sufficiently conditions on the parameters: s, k and δ, when $\mathscr{D}(\lambda)$ does not vanish for any $\lambda \in \mathbb{R}$. It is worth mentioning that these conditions ensure the well-posedness of problem (9.10).

Proposition 9.2 *Let $\delta \in [0, \pi/2)$, m and n be nonnegative integer numbers. Then*

$$\mathscr{D}(\lambda) \neq 0 \quad \textit{for each} \quad \lambda \in \mathbb{R},$$

if the parameters: s, k and δ, satisfy one of the following conditions:

(i)

$$s \neq \pm n, \quad n = 0, 1, 2, \ldots, \quad \delta = 0, \quad \textit{and} \quad k > 0,\ k \neq 1;$$

(ii)

$$0 < s < 1, \quad \delta \in [0, \pi/2), \quad \textit{and} \quad k > 1;$$

(iii)

$$\max\left\{2n; \frac{\pi m}{\delta}\right\} < s < \min\left\{1 + 2n; \frac{\pi(1+2m)}{2\delta}\right\}, \quad k > 1,$$

$$\textit{and} \quad \delta \in \left(\frac{\pi m}{1+2n}; \min\left\{\frac{\pi}{2}; \frac{\pi(1+2m)}{4n}\right\}\right), \quad m < n + 1/2;$$

(iv)

$$\max\left\{1 + 2n; \frac{\pi(1+2m)}{2\delta}\right\} < s < \min\left\{2 + 2n; \frac{\pi(1+m)}{\delta}\right\}, \quad k > 1,$$

$$\textit{and} \quad \delta \in \left(\frac{\pi(1+2m)}{4(1+n)}; \min\left\{\frac{\pi}{2}; \frac{\pi(1+m)}{1+2n}\right\}\right), \quad m < n + 1/2;$$

(v)

$$1 < s < \min\left\{2; \frac{\pi}{2\delta}\right\}, \quad \delta \in [0, \pi/2), \quad \textit{and} \quad 0 < k < 1;$$

(vi)

$$\max\left\{1 + 2n; \frac{\pi m}{\delta}\right\} < s < \min\left\{2 + 2n; \frac{\pi(1 + 2m)}{2\delta}\right\}, \quad 0 < k < 1$$

$$\textit{and} \quad \delta \in \left(\frac{\pi m}{2(1 + n)}; \min\left\{\frac{\pi}{2}; \frac{\pi(1 + 2m)}{2(1 + 2n)}\right\}\right), \quad m < n + 1;$$

(vii)

$$\max\left\{2n; \frac{\pi(1 + 2m)}{2\delta}\right\} < s < \min\left\{1 + 2n; \frac{\pi(1 + m)}{\delta}\right\}, \quad 0 < k < 1,$$

$$\textit{and} \quad \delta \in \left(\frac{\pi(1 + 2m)}{2(1 + 2n)}; \min\left\{\frac{\pi}{2}; \frac{\pi(1 + m)}{2n}\right\}\right), \quad m < n.$$

Proof We will carry out the detailed proof of Proposition 9.2 in the case $k > 1$. The arguments in the case of $k < 1$ is almost identical, and left to the interested reader. It is apparent that the function $\mathscr{D}(\lambda)$ vanishes iff

$$\begin{cases} Im\, \mathscr{D}(\lambda) = 0, \\ Re\, \mathscr{D}(\lambda) = 0. \end{cases}$$

This system is equivalent to relations

$$\begin{cases} \sinh \pi\lambda \cos \pi s + q \sinh 2\delta\lambda \cos 2\delta s = 0, \\ \cosh \pi\lambda \sin \pi s + q \cosh 2\delta\lambda \sin 2\delta s = 0, \end{cases} \tag{9.13}$$

where

$$q = \frac{k - 1}{k + 1}.$$

After that, we find enough conditions on s when system (9.13) is unsolvable for each $\lambda \in \mathbb{R}$. It is apparent that if $\delta = 0$, system (9.13) can be rewritten in the compact form

$$\begin{cases} \sinh \pi\lambda \cos \pi s = 0, \\ \cosh \pi\lambda \sin \pi s = 0. \end{cases}$$

It is easy to verify that sufficient conditions of the insolubility to this system is the inequality

$$\sin \pi s \neq 0.$$

This means that

$$s \neq \pm n, \quad n = 0, 1, 2, \ldots,$$

which proves statement (i).

Next, we assume that $\delta \neq 0$, $\sin \pi s \neq 0$, and rewrite the second equation in (9.13) as

$$\frac{\cosh \pi \lambda}{\cosh 2\delta\lambda} = -q \frac{\sin 2\delta s}{\sin \pi s}. \tag{9.14}$$

In virtue of the positivity of both the left-hand side in (9.14) and q, we can conclude that Eq. (9.14) does not have any solutions if the inequality holds

$$\frac{\sin 2\delta s}{\sin \pi s} \geq 0.$$

Finally, straightforward calculations lead to the conditions (ii)–(iv) on the weight s, the angle δ and k under which the last inequality holds, other words these assumptions provide nonexistence of solutions in (9.13). That finishes the proof of Proposition 9.2. □

9.4.2 Coercive Estimates of (v_1^0, v_2^0)

At this point, we begin to describe the behavior of the solution (v_1^0, v_2^0) on the boundary b. To this end, we introduce a new unknown function $M(x_1)$:

$$M(x_1) = v_1^0(x_1, \delta) = v_2^0(x_1, \delta).$$

The Fourier transformation in this relation provides

$$\hat{M}(\lambda) = \hat{v}_1^0(\lambda, \delta) = \hat{v}_2^0(\lambda, \delta).$$

Then, keeping in mind boundary conditions in (9.10), we obtain representation

$$\hat{M}(\lambda) = \hat{M}_1(\lambda) + \hat{M}_2(\lambda), \tag{9.15}$$

where we put

$$\hat{M}_1(\lambda) = -\frac{1}{\rho\mathscr{D}_1(\lambda)} \int\limits_{-\pi/2}^{\delta} \hat{f}_{01}(\lambda, z)\frac{\sin\rho(z+\pi/2)}{\sin\rho(\delta+\pi/2)}dz,$$

$$\hat{M}_2(\lambda) = -\frac{1}{\rho\mathscr{D}_1(\lambda)} \int\limits_{\delta}^{\pi/2} \hat{f}_{02}(\lambda, z)\frac{\sin\rho(\pi/2-z)}{\sin\rho(\pi/2-\delta)}dz,$$

$$\mathscr{D}_1(\lambda) = \cot\rho(\delta+\pi/2) + k\cot\rho(\pi/2-\delta)$$
$$= \frac{\mathscr{D}(\lambda)}{\sin\rho(\delta+\pi/2)\sin\rho(\pi/2-\delta)}.$$

Then, straightforward calculations with aid Proposition 9.2 lead to the statement.

Remark 9.1 Let $\delta \in [0, \pi/2)$. Under conditions of Proposition 9.2, the functions $\mathscr{D}_1(\lambda)\sin\rho(\frac{\pi}{2}\pm\delta)$ and $\mathscr{D}_1(\lambda)$ do not vanish for each $\lambda \in \mathbb{R}$. Besides, there is the asymptotic representation of $\mathscr{D}_1(\lambda)$:

$$\mathscr{D}_1(\lambda) \approx \begin{cases} \mp i(k+1), & \lambda \to \pm\infty, \\ \cot s(\delta+\pi/2) + k\cot(\pi/2-\delta)s, & |\lambda| < 1. \end{cases}$$

Coming to the function $\hat{M}(\lambda)$, we apply the inverse Fourier transformation in (9.15) and then obtain

$$M(x_1) = M_1(x_1) + M_2(x_1),$$

$$M_1(x_1) = \int\limits_{-\pi/2}^{\delta} dz \int\limits_{-\infty}^{+\infty} f_{01}(x_1-\xi, z)\mathscr{G}_1(\xi, z)d\xi,$$

$$M_2(x_1) = \int\limits_{\delta}^{\pi/2} dz \int\limits_{-\infty}^{+\infty} f_{02}(x_1-\xi, z)\mathscr{G}_2(\xi, z)d\xi,$$

where

$$\mathscr{G}_1(\xi, z) = -\frac{1}{2\pi}\int\limits_{-\infty}^{+\infty} \frac{e^{i\lambda\xi}}{\rho\mathscr{D}_1(\lambda)}\frac{\sin\rho(z+\pi/2)}{\sin\rho(\delta+\pi/2)}d\lambda,$$

$$\mathscr{G}_2(\xi, z) = -\frac{1}{2\pi}\int\limits_{-\infty}^{+\infty} \frac{e^{i\lambda\xi}}{\rho\mathscr{D}_1(\lambda)}\frac{\sin\rho(\pi/2-z)}{\sin\rho(\pi/2-\delta)}d\lambda.$$

Note that the functions $M_1(x_1)$ and $M_2(x_1)$ have the similar properties. Thus, we will carry out here the detailed estimating of $M_1(x_1)$.

Introducing the function

$$J(z) = \int_0^{+\infty} \frac{\mu_1 e^{-\mu_1(\delta - z)}}{\mu_1^2 + (s-\mu_2)^2} d\mu_1,$$

where the positive quantity μ_2, $0 < \mu_2 < s$, is chosen such that the value $s - \mu_2$ has the properties of the weight s from Proposition 9.2, we describe below the properties of the kernel $\mathcal{G}_1(\xi, z)$.

Lemma 9.1 *Let $\delta \in [0, \pi/2)$, $\xi \in \mathbb{R}$, $z \in [-\pi/2, \delta]$ and ε be some positive numbers, conditions of Proposition 9.2 hold. Then the functions $\mathcal{G}_1(\xi, z)$ and $J(z)$ satisfy inequalities:*

(i)

$$|\mathcal{G}_1(\xi, z)| \leq Ce^{-\mu_2|\xi|}(1 + J(z)),$$

(ii)

$$\int_{-\pi/2}^{\delta} J(z)dz \leq C,$$

(iii)

$$\left|\frac{\partial \mathcal{G}_1}{\partial \xi}(\xi, z)\right| \leq Ce^{-\mu_2|\xi|}\left(1 + J(z) + \frac{\delta - z + |\xi|}{\xi^2 + (\delta - z)^2}\right),$$

(iv)

$$\left|\int_{-\pi/2}^{\delta} \frac{\partial \mathcal{G}_1}{\partial \xi}(\pm\varepsilon, z)dz\right| \leq C,$$

(v)

$$\left|\frac{\partial^l \mathcal{G}_1}{\partial \xi^l}(\xi, z)\right| \leq Ce^{-\mu_2|\xi|}\left(1 + J(z) + \frac{1 + \delta - z + |\xi|}{\xi^2 + (\delta - z)^2}\right), \quad l = 2, 3,$$

where the positive constant C depends on the value of δ but is bounded for each fixed $\delta \in [0, \pi/2)$.

Proof First, we prove statement (i). For the sake of definiteness we assume $\xi > 0$. In order to evaluate the function $\mathcal{G}_1(\xi, z)$, we calculate the integral along the shifted contour: $\lambda = \mu_1 + i\mu_2$. Thus, we have

$$\mathcal{G}_1(\xi, z) = -\frac{1}{2\pi} \int_{-\infty}^{+\infty} \frac{e^{i\mu_1\xi - \mu_2\xi}}{(s - \mu_2 + i\mu_1)\mathcal{D}_1(\mu_1 + i\mu_2)} \frac{\sin(i\mu_1 + s - \mu_2)(\pi/2 + z)}{\sin(i\mu_1 + s - \mu_2)(\pi/2 + \delta)} d\mu_1,$$

where the denominator of the integrand is nonzero due to s and $s - \mu_2$ satisfy assumptions of Proposition 9.2. The direct calculations together with Remark 9.1 provide the following asymptotic representation

$$\frac{\sin(i\mu_1 + s - \mu_2)(\frac{\pi}{2} + z)}{(s - \mu_2 + i\mu_1)\mathscr{D}_1(\mu_1 + i\mu_2)\sin(i\mu_1 + s - \mu_2)(\frac{\pi}{2} + \delta)} \tag{9.16}$$
$$\approx \begin{cases} \pm\frac{ie^{(-|\mu_1|\pm i(s-\mu_2))(\delta - z)}}{(k+1)(i\mu_1+s-\mu_2)}, & \mu_1 \to \pm\infty, \\ \frac{2\sin(s-\mu_2)(\frac{\pi}{2}+z)\sin(s-\mu_2)(\frac{\pi}{2}-\delta)}{(s-\mu_2)[(k+1)\sin(s-\mu_2)\pi+(k-1)\sin 2\delta(s-\mu_2)]}, & |\mu_1| < 1. \end{cases}$$

After that, introducing

$$\mathscr{I}(\xi, z, \mu_1, \mu_2) = \frac{e^{i\mu_1\xi}\sin(i\mu_1 + s - \mu_2)(\pi/2 + z)}{(s - \mu_2 + i\mu_1)\mathscr{D}_1(\mu_1 + i\mu_2)\sin(i\mu_1 + s - \mu_2)(\pi/2 + \delta)},$$

we rewrite the function $\mathscr{G}_1(\xi, z)$ in more comfortable form

$$\mathscr{G}_1(\xi, z) = \sum_{j=1}^{3} \mathscr{G}_{1j}(\xi, z)e^{-\mu_2 z}.$$

Here, for some positive fixed N, we put

$$\mathscr{G}_{11}(\xi, z) = -\frac{1}{2\pi}\int_{-N}^{N} \mathscr{I}(\xi, z, \mu_1, \mu_2)d\mu_1, \quad \mathscr{G}_{12}(\xi, z) = -\frac{1}{2\pi}\int_{N}^{+\infty} \mathscr{I}(\xi, z, \mu_1, \mu_2)d\mu_1,$$
$$\mathscr{G}_{13}(\xi, z) = -\frac{1}{2\pi}\int_{-\infty}^{+\infty} \mathscr{I}(\xi, z, \mu_1, \mu_2)d\mu_1.$$

Then, we estimate each terms $\mathscr{G}_{1j}$ separately.
• By asymptotic (9.16), we have

$$|\mathscr{G}_{11}(\xi, z)| \le C,$$

where the positive quantitative C depends only on the structural parameters of the model.
• Since the function $\mathscr{I}(\xi, z, \mu_1, \mu_2)$ defined in $\mu_1 \in (-\infty, -N)$ and $\mathscr{I}(\xi, z, \mu_1, \mu_2)$ defined in $\mu_1 \in (N, \infty)$ share the same properties, the estimates of $\mathscr{G}_{12}$ and $\mathscr{G}_{13}$ are similar. Thus, we will confine ourselves to the consideration of the case $\mathscr{G}_{12}$.

By virtue of asymptotic (9.16), the estimating of $\mathscr{G}_{12}$ is equivalent to evaluating the integral

$$\bar{\mathscr{G}}_{12} = -\frac{i}{2\pi(k+1)}\int_{N}^{+\infty} \frac{e^{\mu_1[i\xi - \delta + z]}e^{i(s-\mu_2)(\delta - z)}}{i\mu_1 + s - \mu_2}d\mu_1.$$

After straightforward calculations, we have

$$|\bar{\mathscr{G}}_{12}| \leq \left| \int_0^N \frac{e^{\mu_1[i\xi-\delta+z]}e^{i(s-\mu_2)(\delta-z)}}{i\mu_1+s-\mu_2} d\mu_1 \right| + \left| \int_0^{+\infty} \frac{(s-\mu_2)e^{\mu_1[i\xi-\delta+z]}e^{i(s-\mu_2)(\delta-z)}}{\mu_1^2+(s-\mu_2)^2} d\mu_1 \right|$$

$$+ \left| \int_0^{+\infty} \frac{i\mu_1 e^{\mu_1[i\xi-\delta+z]}e^{i(s-\mu_2)(\delta-z)}}{\mu_1^2+(s-\mu_2)^2} d\mu_1 \right| \leq C + J(z).$$

Collecting the last inequality with the estimate of $\mathscr{G}_{11}(\xi, z)$, we derive to statement (i) of this lemma. It is worth noting that the statement (ii) follows immediately from the easily verified equality

$$\int_{-\pi/2}^{\delta} e^{-\mu_1(\delta-z)} dz = \frac{1-e^{-\mu_1(\delta+\pi/2)}}{\mu_1}.$$

Indeed, based on this relation, we conclude

$$\int_{-\pi/2}^{\delta} J(z)dz = \int_0^{+\infty} \frac{1-e^{-\mu_1(\delta+\pi/2)}}{\mu_1^2+(s-\mu_2)^2} d\mu_1 \leq C.$$

Coming to inequality (iii), we first represent $\frac{\partial \mathscr{G}_1}{\partial \xi}(\xi, z)$ as

$$\frac{\partial \mathscr{G}_1}{\partial \xi}(\xi, z) = s\mathscr{G}_1(\xi, z) - \frac{1}{2\pi} \int_{-\infty}^{+\infty} \frac{e^{i\mu_1\xi-\mu_2\xi}}{\mathscr{D}_1(\mu_1+i\mu_2)} \frac{\sin(i\mu_1+s-\mu_2)(z+\pi/2)}{\sin(i\mu_1+s-\mu_2)(\delta+\pi/2)} d\mu_1.$$

Then, using asymptotic (9.16), simple albeit tedious calculations entail

$$\left| \int_{-\infty}^{+\infty} \frac{e^{i\mu_1\xi-\mu_2\xi}}{\mathscr{D}_1(\mu_1+i\mu_2)} \frac{\sin(i\mu_1+s-\mu_2)(z+\pi/2)}{\sin(i\mu_1+s-\mu_2)(\delta+\pi/2)} d\mu_1 \right|$$

$$\leq Ce^{-\mu_2\xi}\left[1 + \left| \int_{-\infty}^{0} e^{i\mu_1\xi} e^{(\delta-z)[\mu_1-i(s-\mu_2)]} d\mu_1 - \int_0^{+\infty} e^{i\mu_1\xi} e^{(\delta-z)[-\mu_1+i(s-\mu_2)]} d\mu_1 \right|\right]$$

$$= Ce^{-\mu_2\xi}\left[1 + \left| \frac{e^{-i(\delta-z)(s-\mu_2)}}{i\xi+\delta-z} + \frac{e^{i(\delta-z)(s-\mu_2)}}{i\xi-\delta+z} \right|\right]$$

$$\leq Ce^{-\mu_2\xi}\left[1 + \frac{2|(\delta-z)\sin(s-\mu_2)(\delta-z)+\xi\cos(s-\mu_2)(\delta-z)|}{\xi^2+(\delta-z)^2}\right].$$

Finally, keeping in mind the representation of $\frac{\partial \mathscr{G}_1}{\partial \xi}$, we collect the last inequality with the estimate of $\mathscr{G}_1$ and deduce the statement (iii) of Lemma 9.1.

Note that, the last inequality of Lemma 9.1 is the easy consequence of statements (i)–(iii). Thus we are left here to prove the statement (iv). To this end, taking into account the representation of $\frac{\partial \mathscr{G}_1}{\partial \xi}$ and statements (i)-(iii), we should estimate the function

$$\mathscr{L} = e^{-\mu_2 \varepsilon} \int_{-\pi/2}^{\delta} \Big[\int_{-\infty}^{0} e^{i\mu_1\varepsilon} e^{(\delta - z)[\mu_1 - i(s-\mu_2)]} d\mu_1 - \int_{0}^{\infty} e^{i\mu_1\varepsilon} e^{(\delta - z)[-\mu_1 + i(s-\mu_2)]} d\mu_1 \Big] dz$$

$$= e^{-\mu_2 \varepsilon} \Big[\int_{-\infty}^{0} \frac{e^{(\delta+\pi/2)(\mu_1 - i(s-\mu_2))} - 1}{\mu_1 - i(s-\mu_2)} e^{i\mu_1\varepsilon} d\mu_1 - \int_{0}^{\infty} \frac{e^{(\delta+\pi/2)(-\mu_1 + i(s-\mu_2))} - 1}{-\mu_1 + i(s-\mu_2)} e^{i\mu_1\varepsilon} d\mu_1 \Big]$$

for some positive number ε.

Next we rewrite the function $\mathscr{L}$ in the more convenient form:

$$\mathscr{L} \equiv \sum_{j=1}^{3} \mathscr{L}_j e^{-\mu_2 \varepsilon},$$

where we put

$$\mathscr{L}_1 = \int_{-\infty}^{0} \frac{e^{-(\delta+\pi/2)(s-\mu_2)i} e^{i\mu_1\varepsilon} e^{(\pi/2+\delta)\mu_1}}{\mu_1 - i(s-\mu_2)} d\mu_1,$$

$$\mathscr{L}_2 = -\int_{0}^{\infty} \frac{e^{(\delta+\pi/2)(s-\mu_2)i} e^{i\mu_1\varepsilon} e^{-(\pi/2+\delta)\mu_1}}{i(s-\mu_2) - \mu_1} d\mu_1,$$

$$\mathscr{L}_3 = \int_{0}^{\infty} \frac{e^{i\mu_1\varepsilon}}{i(s-\mu_2) - \mu_1} d\mu_1 + \int_{0}^{\infty} \frac{e^{-i\mu_1\varepsilon}}{i(s-\mu_2) + \mu_1} d\mu_1.$$

Due to $s - \mu_2 > 0$, the direct calculations allow us to obtain the following inequalities

$$|\mathscr{L}_1| + |\mathscr{L}_2| \le C,$$

$$|\mathscr{L}_3| = 2\Big| Im \int_{0}^{\infty} \frac{e^{-i\varepsilon\mu_1}(\mu_1 - i(s-\mu_2))}{\mu_1^2 + (s-\mu_2)^2} d\mu_1 \Big|$$

$$\le 2(s-\mu_2)\Big| \int_{0}^{+\infty} \frac{\cos(\mu_1\varepsilon)}{\mu_1^2 + (s-\mu_2)^2} d\mu_1 \Big| + \Big| \int_{0}^{+\infty} \frac{\mu_1 \sin(\mu_1\varepsilon)}{\mu_1^2 + (s-\mu_2)^2} d\mu_1 \Big|$$

$$\leq C\Big[1+\Big|\int\limits_{\pi/2}^{+\infty}\frac{\mu_1\sin\mu_1 d\mu_1}{\mu_1^2+\varepsilon^2(s-\mu_2)^2}\Big|+\Big|\int\limits_{0}^{\pi/2}\frac{\mu_1\sin\mu_1 d\mu_1}{\mu_1^2+\varepsilon^2(s-\mu_2)^2}\Big|\Big]$$

$$\leq C\Big[1+\Big|\int\limits_{\pi/2}^{+\infty}\frac{\mu_1\sin\mu_1 d\mu_1}{\mu_1^2+\varepsilon^2(s-\mu_2)^2}\Big|+\Big|\int\limits_{0}^{\pi/2}\frac{\sin\mu_1 d\mu_1}{\mu_1}\Big|\Big]$$

$$\leq C\Big[1+\Big|\int\limits_{\pi/2}^{+\infty}\frac{\mu_1\sin\mu_1 d\mu_1}{\mu_1^2+\varepsilon^2(s-\mu_2)^2}\Big|\Big].$$

Finally, integrating by parts in the second term provides the estimate

$$|\mathscr{L}_3|\leq C.$$

Collecting estimates of $\mathscr{L}_i$, $i=1,2,3$, with the representation of $\mathscr{L}$ arrives at

$$|\mathscr{L}|\leq C.$$

This finishes the proof of Lemma 9.1. □

Now we are ready to estimate the functions M_1.

Proposition 9.3 *Let k, δ and s meet requirements of Proposition 9.2, and let the finite function $f_{01}\in\mathbb{C}^{\beta}(\bar{B}_1^+)$. Then $M_1\in\mathbb{C}^{2+\beta}(\mathbb{R})$ and the estimate holds*

$$\|M_1\|_{\mathbb{C}^{2+\beta}(\mathbb{R})}\leq C\|f_{01}\|_{\mathbb{C}^{\beta}(\bar{B}_1^+)},$$

where the positive constant C depends only of the structural parameters of the model.

Proof It is worth mentioning that the estimate of $\sup\limits_{\mathbb{R}}|M_1|$ is the simple consequence of statements (i) and (ii) of Lemma 9.1. Then, in order to prove Proposition 9.3, it should be obtained the estimate

$$\|M_1''\|_{\mathbb{C}^{\beta}(\mathbb{R})}\leq C\|f_{01}\|_{\mathbb{C}^{\beta}(\bar{B}_1^+)}. \tag{9.17}$$

Namely, keeping in mind this inequality with the corresponding estimate of $\sup\limits_{\mathbb{R}}|M_1|$ and applying the interpolation inequality, we conclude that

$$\|M_1'\|_{\mathbb{C}^{\beta}(\mathbb{R})}\leq C\|f_{01}\|_{\mathbb{C}^{\beta}(\bar{B}_1^+)}.$$

Coming to inequality (9.17), we first evaluate $\sup\limits_{\mathbb{R}}|M_1''|$. To this end, we represent M_1'' in more convenient form

$$
\begin{aligned}
M_1'' &= \mathscr{M}_{11} + \mathscr{M}_{12} + \mathscr{M}_{13},\\
\mathscr{M}_{11} &= \int_{-\pi/2}^{\delta} dz \int_{|x_1-\xi|\geq\varepsilon} f_{01}(\xi, z)\frac{\partial^2 \mathscr{G}_1}{\partial(x_1-\xi)^2} d\xi,\\
\mathscr{M}_{12} &= \int_{-\pi/2}^{\delta} dz \int_{|x_1-\xi|\leq\varepsilon} [f_{01}(\xi, z) - f_{01}(x_1, \delta)]\frac{\partial^2 \mathscr{G}_1}{\partial(x_1-\xi)^2} d\xi,\\
\mathscr{M}_{13} &= -f_{01}(x_1,\delta)\int_{-\pi/2}^{\delta} dz \frac{\partial \mathscr{G}_1}{\partial \xi}\Big|_{|\xi|=\varepsilon}
\end{aligned}
$$

with some fixed number $\varepsilon > 0$.

At this point, we estimate each $\mathscr{M}_{1i}$ separately.

- By statement (iv) in Lemma 9.1,

$$
|\mathscr{M}_{13}| \leq C \underset{\bar{B}_1^+}{|} f_{01}|.
$$

- Coming the term $\mathscr{M}_{11}$, we apply Lemma 9.1 and obtain

$$
\begin{aligned}
|\mathscr{M}_{11}| &\leq C\sup_{\bar{B}_1^+}|f_{01}| \int_{-\pi/2}^{\delta} dz \int_{\varepsilon}^{+\infty} e^{-\mu_2\xi}\Big[1 + J(z) + \frac{1+\delta-z+\xi}{\xi^2+(\delta-z)^2}\Big] d\xi\\
&\leq C\sup_{\bar{B}_1^+}|f_{01}|\Big[1 + \int_{-\pi/2}^{\delta} dz \int_{\varepsilon}^{+\infty} e^{-\mu_2\xi}(1+\delta-z+\xi)\varepsilon^{-2} d\xi\Big].
\end{aligned}
$$

Then, the direct calculations entail

$$
|\mathscr{M}_{11}| \leq C\sup_{\bar{B}_1^+}|f_{01}|.
$$

- As for $\mathscr{M}_{12}$, we rewrite it in more comfortable form

$$
\begin{aligned}
\mathscr{M}_{12} &= \int_{-\pi/2}^{\delta} dz \int_{-\varepsilon}^{\varepsilon} [f_{01}(x_1-\xi, z) - f_{01}(x_1, z)]\frac{\partial^2 \mathscr{G}_1}{\partial\xi^2} d\xi\\
&+ \int_{-\pi/2}^{\delta} dz \int_{-\varepsilon}^{\varepsilon} [f_{01}(x_1, z) - f_{01}(x_1, \delta)]\frac{\partial^2 \mathscr{G}_1}{\partial\xi^2} d\xi.
\end{aligned}
$$

Then, keeping in mind the smoothness of f_{01} and statements (ii), (iv) of Lemma 9.1, we arrive at

$$
\begin{aligned}
|\mathscr{M}_{12}| &\leq C\Big[\langle f_{01}\rangle^{(\beta)}_{x_1,B_1^+} + \langle f_{01}\rangle^{(\beta)}_{x_2,B_1^+}\Big]\Big[1 + \int_{-\pi/2}^{\delta} dz \int_0^{\varepsilon} \xi^\beta \frac{1+\delta-z+\xi}{\xi^2+(\delta-z)^2} d\xi\\
&+ \int_{-\pi/2}^{\delta} dz \int_0^{\varepsilon} (z-\delta)^\beta \frac{1+\delta-z+\xi}{\xi^2+(\delta-z)^2} d\xi\Big].
\end{aligned}
$$

Next, we take advantage of the straightforward inequalities:

$$\int_{-\pi/2}^{\delta} dz \int_0^{\varepsilon} \frac{\xi^\beta(\delta - z)}{\xi^2 + (\delta - z)^2} d\xi \le C \int_{-\pi/2}^{\delta} dz \int_0^{\varepsilon} \xi^{\beta-1} d\xi,$$
$$\int_{-\pi/2}^{\delta} dz \int_0^{\varepsilon} \frac{\xi(\delta - z)^\beta}{\xi^2 + (\delta - z)^2} d\xi \le C \int_{-\pi/2}^{\delta} dz \int_0^{\varepsilon} (\delta - z)^{\beta-1} d\xi,$$
$$\int_{-\pi/2}^{\delta} dz \int_0^{\varepsilon} \frac{\xi^\beta}{\xi^2 + (\delta - z)^2} d\xi = \int_0^{\varepsilon} \xi^{\beta-1} \arctan \frac{\delta - z}{\xi}\Big|_{z=-\pi/2}^{z=\delta} \le C,$$
$$\int_{-\pi/2}^{\delta} dz \int_0^{\varepsilon} \frac{(z - \delta)^\beta}{\xi^2 + (\delta - z)^2} d\xi \le C \int_{-\pi/2}^{\delta} (z - \delta)^\beta dz \le C.$$

Therefore, we end up with

$$|\mathscr{M}_{12}| \le C\Big[\langle f_{01}\rangle_{x_1, B_1^+}^{(\beta)} + \langle f_{01}\rangle_{x_2, B_1^+}^{(\beta)}\Big].$$

Finally, collecting the estimates of $\mathscr{M}_{1j}$ provides

$$\sup_{\mathbb{R}} |M_1''| \le C \| f_{01} \|_{\mathcal{C}^\beta(\bar{B}_1^+)}. \tag{9.18}$$

At last, to complete the proof of this proposition, we are left to evaluate $\langle M_1''\rangle_{x,\mathbb{R}}^{(\beta)}$. To this end, we study the difference for $x_1, \bar{x}_1 \in \mathbb{R}$

$$M_1''(x_1) - M_1''(\bar{x}_1) = \sum_{j=1}^{4} S_j,$$

where

$$S_1 = [f_{01}(x_1, \delta) - f_{01}(\bar{x}_1, \delta)] \int_{-\pi/2}^{\delta} \frac{\partial \mathscr{G}_1}{\partial x_1}(\bar{x}_1 - \xi, z)\Big|_{|x_1-\xi|=2|x_1-\bar{x}_1|},$$
$$S_2 = \int_{-\pi/2}^{\delta} dz \int_{|x_1-\xi|\le 2|x_1-\bar{x}_1|} [f_{01}(\xi, z) - f_{01}(x_1, \delta)] \frac{\partial^2 \mathscr{G}_1}{\partial x_1^2}(x_1 - \xi, z) d\xi,$$
$$S_3 = \int_{-\pi/2}^{\delta} dz \int_{|x_1-\xi|\le 2|x_1-\bar{x}_1|} [f_{01}(\xi, z) - f_{01}(\bar{x}_1, \delta)] \frac{\partial^2 \mathscr{G}_1}{\partial \bar{x}_1^2}(x_1 - \xi, z) d\xi,$$
$$S_4 = \int_{-\pi/2}^{\delta} dz \int_{|x_1-\xi|\ge 2|x_1-\bar{x}_1|} [f_{01}(\xi, z) - f_{01}(x_1, \delta)] \Big[\frac{\partial^2 \mathscr{G}_1}{\partial x_1^2}(x_1 - \xi, z) - \frac{\partial^2 \mathscr{G}_1}{\partial x_1^2}(\bar{x}_1 - \xi, z)\Big] d\xi.$$

It is apparent that, the smoothness of f_{01} and estimate (iv) in Lemma 9.1 provide

$$|S_1| \le C\langle f_{01}\rangle_{x_1, B_1^+}^{(\beta)} |x_1 - \bar{x}_1|^\beta.$$

We preliminarily observe that the estimate for S_3 is the same as the one for S_2. Reason why, we evaluate here only S_2. To this end we rewrite S_2 as

$$S_2 = \int_{-\pi/2}^{\delta} dz \int_{|x_1-\xi|\leq 2|x_1-\bar{x}_1|} [f_{01}(\xi, z) - f_{01}(x_1, z)] \frac{\partial^2 \mathscr{G}_1}{\partial x_1^2} d\xi$$
$$+ \int_{-\pi/2}^{\delta} dz \int_{|x_1-\xi|\leq 2|x_1-\bar{x}_1|} [f_{01}(x_1, z) - f_{01}(x_1, \delta)] \frac{\partial^2 \mathscr{G}_1}{\partial x_1^2} d\xi.$$

After that, recasting the arguments leading to estimate (9.18), we end up with

$$|S_2| \leq C[\langle f_{01}\rangle^{(\beta)}_{x_2, B_1^+} + \langle f_{01}\rangle^{(\beta)}_{x_1, B_1^+}]|x_1 - \bar{x}_1|^\beta.$$

Finally, we are left to tackle the term S_4. Applying, first, the mean value theorem to the difference $\left[\frac{\partial^2 \mathscr{G}_1}{\partial x_1^2}(x_1 - \xi, z) - \frac{\partial^2 \mathscr{G}_1}{\partial x_1^2}(\bar{x}_1 - \xi, z)\right]$, and then statement (v) of Lemma 9.1 give

$$|S_4| \leq C\|f_{01}\|_{\mathbb{C}^\beta(\bar{B}_1^+)} \left[1 + \int_{-\pi/2}^{\delta} dz \int_{2|\bar{x}_1 - x_1|}^{+\infty} e^{-\mu_2 \xi} \frac{\xi^\beta + (\delta - z)^\beta}{\xi^2 + (\delta - z)^2} d\xi \right]$$
$$\leq C\|f_{01}\|_{\mathbb{C}^\beta(\bar{B}_1^+)} \left[1 + \int_{2|\bar{x}_1 - x_1|}^{+\infty} d\xi e^{-\mu_2 \xi} \xi^{\beta-1} \int_{-\pi/2}^{\delta} \frac{dz}{\xi[1 + (\delta - z)^2 \xi^{-2}]}\right.$$
$$\left. + \int_{-\pi/2}^{\delta} dz (\delta - z)^{\beta-1} \int_{2|\bar{x}_1 - x_1|}^{+\infty} \frac{e^{-\mu_2 \xi}}{(\xi^2 (\delta - z)^{-2} + 1)} \frac{d\xi}{(z - \delta)}\right]$$
$$\leq C|x_1 - \bar{x}_1|^\beta \|f_{01}\|_{\mathbb{C}^\beta(\bar{B}_1^+)}.$$

Collecting estimates of S_j, $j = \overline{1, 4}$, we reach the desired inequality

$$\langle M_1''\rangle^{(\beta)}_{x_1, \mathbb{R}} \leq C\|f_{01}\|_{\mathbb{C}^\beta(\bar{B}_1^+)},$$

which together with (9.18) provides estimate (9.17).

Thus, this finishes the proof of Proposition 9.3. □

It is apparent that the same result holds to the function M_2. Thus, taking into account representation (9.15), Propositions 9.2 and 9.3, we deduce that

$$\|M\|_{\mathbb{C}^{2+\beta}(\mathbb{R})} \leq \sum_{j=1}^{2} \|M_j\|_{\mathbb{C}^{2+\beta}(\mathbb{R})} \tag{9.19}$$
$$\leq C[\|f_{01}\|_{\mathbb{C}^\beta(\bar{B}_1^+)} + \|f_{02}\|_{\mathbb{C}^\beta(\bar{B}_2^+)}].$$

Using the definition of the function M and relations (9.10), we conclude that v_i^0, $i = 1, 2$, solves the Dirichlet problem

$$\begin{cases} \Delta_x v_i^0 + 2s\frac{\partial v_i^0}{\partial x_1} + s^2 v_i^0 = f_{0i}(x) \text{ in } B_i^+, \ i = 1, 2, \\ v_i^0 = M \quad \text{if } x_2 = \delta, \quad x_1 \in \mathbb{R}, \\ v_1^0(x_1, -\pi/2) = 0 \quad \text{and } v_2^0(x_1, \pi/2) = 0, \quad x_1 \in \mathbb{R}. \end{cases}$$

After that, recasting, with minor changes, the arguments of Chap. 3 [22] and using estimate (9.19) entail the assertion.

Lemma 9.2 *Let the finite functions* $f_{0i} \in \mathbb{C}^{\beta}(\bar{B}_i^+)$, $i = 1, 2$, $\delta \in [0, \pi/2)$. *Under assumptions of Proposition 9.2, there exists a unique classical solution* (v_1^0, v_2^0) *of problem* (9.8) *with* $f_1 \equiv 0$, *satisfying the regularity*

$$v_1^0 \in \mathbb{C}^{2+\beta}(\bar{B}_1^+) \quad \textit{and} \quad v_2^0 \in \mathbb{C}^{2+\beta}(\bar{B}_2^+).$$

Besides, the solution (v_1^0, v_2^0) *fulfills the estimate*

$$\|v_1^0\|_{\mathbb{C}^{2+\beta}(\bar{B}_1^+)} + \|v_2^0\|_{\mathbb{C}^{2+\beta}(\bar{B}_2^+)} \leq C(\|f_{01}\|_{\mathbb{C}^{\beta}(\bar{B}_1^+)} + \|f_{02}\|_{\mathbb{C}^{\beta}(\bar{B}_2^+)}),$$

where the quantity C *is independent of the right-hand sides in* (9.8).

9.4.3 Solvability of Problem (9.8) with $f_{0i} \equiv 0$, $i = 1, 2$, and Coercive Estimates

Analysis of problem (9.8) in the case of homogenous equations and the finite function $f_1 \in \mathbb{C}^{1+\beta}(b)$ repeats the steps of Sect. 9.4.2. Therefore, here we present only the key points.

It is easy to verify that the solution of this problem is represented after the Fourier transformation as

$$\hat{v}_1^1(\lambda, x_2) = \hat{N}(\lambda)\frac{\sin\rho(x_2 + \pi/2)}{\sin\rho(\delta + \pi/2)}, \quad \hat{v}_2^1(\lambda, x_2) = \hat{N}(\lambda)\frac{\sin\rho(\pi/2 - x_2)}{\sin\rho(\pi/2 - \delta)},$$

where the new unknown function $\hat{N}(\lambda)$ meets the requirements

$$\hat{v}_1^1(\lambda, \delta) = \hat{N}(\lambda) = \hat{v}_2^1(\lambda, \delta).$$

Next, the direct calculations provide the equality

$$\hat{N}(\lambda) = \hat{K}(\lambda)\hat{f}_1(\lambda) \quad \text{with} \quad \hat{K}(\lambda) = \frac{1}{\rho\mathscr{D}_1(\lambda)}.$$

Then, applying the inverse Fourier transformation in these relations, we obtain

$$K(\xi) = \frac{1}{2\pi} \int_{-\infty}^{+\infty} \frac{e^{i\lambda\xi}}{\rho\mathscr{D}_1(\lambda)} d\lambda \quad \text{and} \quad N(x_1) = \int_{-\infty}^{+\infty} f_1(x_1 - \xi) K(\xi) d\xi. \tag{9.20}$$

The main properties of the kernel $K(\xi)$ are described with the following statement.

Proposition 9.4 *Let δ, s and k satisfy conditions of Proposition 9.2. We assume that the positive value μ_2, $\mu_2 < s$, is chosen such that the quantity $(s - \mu_2)$ has the properties of the weight s from Proposition 9.2. Then the following estimates hold.*

(i)

$$|K(\xi)| + \left| \frac{\partial K(\xi)}{\partial \xi} \right| \leq C e^{-\mu_2 |\xi|} (1 + |\xi|^{-1}),$$

(ii)

$$\left| \frac{\partial^2 K(\xi)}{\partial \xi^2} \right| \leq C e^{-\mu_2 |\xi|} (1 + |\xi|^{-1} + \xi^{-2}),$$

(iii)

$$\left| \int_{-\varepsilon}^{\varepsilon} \frac{\partial K(\xi)}{\partial \xi} d\xi \right| \leq C,$$

for some positive quantity ε, and the positive constant C depends only on the structure parameters of the model.

The proof of this result is technically tedious, so that we represent it in the Appendix (see Sect. 9.6).

After that, recasting the arguments from Sect. 9.4.2 and using Proposition 9.4 entail.

Proposition 9.5 *Let the finite function $f_1 \in \mathbb{C}^{1+\beta}(\mathbb{R})$, δ, k and s satisfy conditions of Proposition 9.2. Then there is the estimate*

$$\|N\|_{\mathbb{C}^{2+\beta}(\mathbb{R})} \leq C \|f_1\|_{\mathbb{C}^{1+\beta}(\mathbb{R})}.$$

Then, repeating the arguments of Lemma 9.2 and taking advantage of Proposition 9.5, we assert.

Lemma 9.3 *Let $f_{0i} \equiv 0$, the finite function $f_1 \in \mathbb{C}^{1+\beta}(\mathbb{R})$ and s, δ and k satisfy conditions of Proposition 9.2. Then there exists a unique solution (v_2^1, v_2^2) of problem (9.8) satisfying regularity*

$$v_1^1 \in \mathbb{C}^{2+\beta}(\bar{B}_1^+) \quad \textit{and} \quad v_2^1 \in \mathbb{C}^{2+\beta}(\bar{B}_2^+),$$

Besides, the solution (v_2^1, v_2^2) fulfills the estimate

$$\|v_1^1\|_{\mathbb{C}^{2+\beta}(\bar{B}_1^+)} + \|v_2^1\|_{\mathbb{C}^{2+\beta}(\bar{B}_2^+)} \leq C \|f_1\|_{\mathbb{C}^{1+\beta}(\mathbb{R})}.$$

9.4.4 Solvability of Problem (9.7)

It is worth mentioning that Lemmas 9.2 and 9.3 together with representation (9.9) guarantee the one-to-one solvability of transmission problem (9.8) and the coercive estimate

$$\|v_1\|_{\mathbb{C}^{2+\beta}(\bar{B}_1^+)} + \|v_2\|_{\mathbb{C}^{2+\beta}(\bar{B}_2^+)} \leq C[\|f_1\|_{\mathbb{C}^{1+\beta}(\mathbb{R})} + \sum_{j=1}^{2} \|f_{0i}\|_{\mathbb{C}^{\beta}(\bar{B}_j^+)}].$$

Then, based on Proposition 9.1 and mapping (9.5), we come from the functions v_1 and v_2 defined in B_1^+ and B_2^+ to the functions u_1 and u_2 defined in G_1^+ and G_2^+, correspondingly. In summary, we obtain the following result.

Theorem 9.3 *Let $\delta \in [0, \pi/2)$ and δ, k, s meet the requirements of Proposition 9.2. We assume that $\mathscr{F}_{0i}$ and $\mathscr{F}_1$ are finite functions satisfying regularity $\mathscr{F}_{0i} \in E_{s-2}^{\beta}(\bar{G}_i^+)$, $i = 1, 2$, $\mathscr{F}_1 \in E_{s-1}^{1+\beta}(\bar{g})$. Under restriction* (9.6), *problem* (9.7) *admits a unique classical solution (u_1, u_2) satisfying regularity:*

$$u_1 \in E_s^{2+\beta}(\bar{G}_1^+) \quad \textit{and} \quad u_2 \in E_s^{2+\beta}(\bar{G}_2^+).$$

Besides, the solution fulfills the estimate

$$\sum_{i=1}^{2} \|u_i\|_{E_s^{2+\beta}(\bar{G}_i^+)} \leq C\Big[\sum_{i=1}^{2} \|\mathscr{F}_{0i}\|_{E_{s-2}^{\beta}(\bar{G}_i^+)} + \|\mathscr{F}_1\|_{E_{s-1}^{1+\beta}(\bar{g})}\Big].$$

To remove restriction (9.6), we look for solution of (9.1)–(9.3) in the form

$$u_i = U_i + \bar{u}_i,$$

where the unknown functions U_1 and U_2 solve the Dirichlet problems

$$\begin{cases} \Delta U_i = 0 \quad \text{in} \quad G_i^+,\ i = 1, 2, \\ U_1|_{\partial G_1^+ \backslash g} = \mathscr{F}_2, \quad U_1|_g = 0, \\ U_2|_{\partial G_2^+ \backslash g} = \mathscr{F}_3, \quad U_2|_g = -\mathscr{F}, \end{cases} \tag{9.21}$$

and the functions $\bar{u}_1$ and $\bar{u}_2$ satisfy conditions (9.7) with the new function $\mathscr{F}_1 := \mathscr{F}_1 - \frac{\partial U_1}{\partial n} + k\frac{\partial U_2}{\partial n}$. It is easy to see that Theorem 9.3 hold in this case.

Then Lemma 5.1 in [36] ensures the one-valued solvability of (9.21) with regularity

$$U_1 \in E_{\bar{s}}^{2+\beta}(\bar{G}_1^+), \quad U_2 \in E_{\bar{s}}^{2+\beta}(\bar{G}_2^+),$$

if

$$\mathscr{F}_2 \in E_{\bar{s}}^{2+\beta}(\partial G_1^+ \backslash g), \quad \mathscr{F}_3 \in E_{\bar{s}}^{2+\beta}(\partial G_2^+ \backslash g), \quad \mathscr{F} \in E_{\bar{s}}^{2+\beta}(g),$$

where

$$\bar{s} \neq \frac{2\pi n}{\pi \pm 2\delta}, \quad \frac{\pi(2n+1)}{\pi \pm 2\delta}, \quad \frac{\pi(2n-1)}{\pi \pm 2\delta}, \quad n = 1, 2, ... \tag{9.22}$$

Finally, the obtained results for the functions U_i and Theorem 9.3 allow us to remove restriction (9.6).

Theorem 9.4 *Let assumptions of Theorem 9.3 and condition* (9.22) *with* $\bar{s} = s$ *hold. We assume that the right-hand sides in* (9.1)–(9.3) *are finite functions. Then problem* (9.1)–(9.3) *in* $G_1^+ \cup G_2^+$ *admits a unique classical solution*

$$u_1 \in E_s^{2+\beta}(\bar{G}_1^+) \quad \text{and} \quad u_2 \in E_s^{2+\beta}(\bar{G}_2^+).$$

Besides, the estimate holds

$$\|u_1\|_{E_s^{2+\beta}(\bar{G}_1^+)} + \|u_2\|_{E_s^{2+\beta}(\bar{G}_2^+)} \leq C\big[\|\mathscr{F}_{01}\|_{E_{s-2}^{\beta}(\bar{G}_1^+)} + \|\mathscr{F}_{02}\|_{E_{s-2}^{\beta}(\bar{G}_2^+)} + \|\mathscr{F}_1\|_{E_{s-1}^{1+\beta}(\bar{g})}$$
$$+ \|\mathscr{F}\|_{E_s^{2+\beta}(\bar{g})} + \|\mathscr{F}_2\|_{E_s^{2+\beta}(\partial G_1^+\backslash g)} + \|\mathscr{F}_3\|_{E_s^{2+\beta}(\partial G_2^+\backslash g)}\big].$$

The results of Theorem 9.4 can be extended on the case $\Omega_1 = G_1^-$ and $\Omega_2 = G_2^-$. To this end, we recast step-by-step all the arguments of Sects. 9.4.1–9.4.4 and state.

Theorem 9.5 *Let* $\delta \in (0, \pi/2]$, *the right-hand sides in* (9.1)–(9.3) *be finite functions, and conditions* (9.22) *hold with* $\bar{s} = s$. *We assume that parameters:* s, k, δ, *satisfy one of two conditions*

(i)

$$0 < s < 1, \quad \delta \in (0, \pi/2] \quad \text{and} \quad k < 1;$$

(ii)

$$1 < s < \min\left\{2; \frac{\pi}{2\delta}\right\}, \quad \delta \in (0, \pi/2] \quad \text{and} \quad 1 < k.$$

Then, under assumption **h2**, *transmission problem* (9.1)–(9.3) *in* $G_1^- \cup G_2^-$ *admits a unique classical solution* $(u_1, u_2) : u_i \in E_s^{2+\beta}(G_i^-)$, $i = 1, 2$, *which satisfies the estimate*

$$\|u_1\|_{E_s^{2+\beta}(\bar{G}_1^-)} + \|u_2\|_{E_s^{2+\beta}(\bar{G}_2^-)} \leq C\big[\|\mathscr{F}_{01}\|_{E_{s-2}^{\beta}(\bar{G}_1^-)} + \|\mathscr{F}_{02}\|_{E_{s-2}^{\beta}(\bar{G}_2^-)} + \|\mathscr{F}_1\|_{E_{s-1}^{1+\beta}(\bar{g})}$$
$$+ \|\mathscr{F}\|_{E_s^{2+\beta}(\bar{g})} + \|\mathscr{F}_2\|_{E_s^{2+\beta}(\partial G_1^-\backslash g)} + \|\mathscr{F}_3\|_{E_s^{2+\beta}(\partial G_2^-\backslash g)}\big].$$

9.5 The Proof of the Main Results

9.5.1 *Proof of Theorems 9.1 and 9.2*

First of all, we prove Theorem 9.1. Our strategy of the proof consists in two main steps. The first is related to a solvability of (9.1)–(9.3) in the weighted Sobolev spaces $W_{\gamma}^{2,p}$ in the case of $\mathscr{F} \equiv 0$. On this step, we will obtain the weighted estimate of the maximum module of u_i, $i = 1, 2$. The second step deals with increasing of the smoothness of the functions u_i, $i = 1, 2$. To reach this, we use Schauder approach related with partition of unity together with local estimates in the corresponding model problems.

• Coming to solvability (9.1)–(9.3), we assume that $\mathscr{F} \equiv 0$. Denoting by η_q^1 and η_q^0, $q = 1, 2, \dots$, eigenvalues of the following spectral problems in the unknown functions Z_i^j, $i = 1, 2$, and $j = 0, 1$,

$$\begin{cases} \frac{d^2 Z_1^j}{d\varphi^2} + \eta^2 Z_1^j = 0, \quad \varphi \in (\delta_j, \pi), \\ \frac{d^2 Z_2^j}{d\varphi^2} + \eta^2 Z_2^j = 0, \quad \varphi \in (0, \delta_j), \\ Z_1^j(\delta_j) - Z_2^j(\delta_j) = 0, \quad \frac{dZ_1^j(\delta_j)}{d\varphi} - k\frac{dZ_2^j(\delta_j)}{d\varphi} = 0, \\ Z_1^j(\pi) = 0, \quad Z_2^j(0) = 0. \end{cases}$$

These problems arise if one looks for the nontrivial solutions to homogenous problem like (9.7) after change of variables and putting there $\delta = \pi/2 - \delta_j$, $j = 0, 1$. Lemma 1.58 [26] provides existence of the countable increasing set of η_q^j, $q = 1, 2, \dots$,

$$0 < \eta_1^j \le \eta_2^j \le \dots \le \eta_q^j < \dots, \quad 0 < \eta_1^j < 1, \quad \eta_q^j > C^\star q - 1$$

with positive constant $C^\star$.

Denoting $\eta_1 = \min\{\eta_1^0, \eta_1^1\}$, we assert the following.

Lemma 9.4 *Let* $\mathscr{F} \equiv 0$, $\gamma \in (\max\{2 - 2/p - \eta_1, 2 - s - 1/p\}, \min\{2 - 2/p, 2 - s\})$, $\delta_0, \delta_1 \in (0, \pi)$, *and* s, p *satisfy one of the following conditions:*
(i) $s \in (0, \eta_1]$, $p > 1/s$;
(ii) $s \in (\eta_1, 1)$, $p \in (1/s, 2/(s - \eta_1))$;
(iii) $s \in \left(1, \min\left\{2; \frac{\pi}{|\pi - 2\delta_0|}, \frac{\pi}{|\pi - 2\delta_1|}\right\}\right)$, $p \in (1, 2/(2 - \eta_1))$.
Then, under condition **h2**, *problem* (9.1)–(9.3) *admits a unique weak solution* (u_1, u_2), $u_i \in W^{1,2}(\Omega_i)$. *Besides, the estimates hold*

$$\|u_1\|_{2,p;\gamma,\Omega_1} + \|u_2\|_{2,p;\gamma,\Omega_2} \le C\Big[\|\mathscr{F}_{01}\|_{E_{s-2}^{\beta}(\bar{\Omega}_1)} + \|\mathscr{F}_{02}\|_{E_{s-2}^{\beta}(\bar{\Omega}_2)} + \|\mathscr{F}_1\|_{E_{s-1}^{1+\beta}(\Gamma)} \tag{9.23}$$

$$+ \|\mathscr{F}_2\|_{E_s^{2+\beta}(\partial\Omega_1\setminus\Gamma)} + \|\mathscr{F}_3\|_{E_s^{\beta+2}(\partial\Omega_2\setminus\Gamma)}\Big]$$
$$\equiv C\Xi_s(\mathscr{F}_{01}, \mathscr{F}_{02}, \mathscr{F}_1, \mathscr{F}_2, \mathscr{F}_3),$$

$$\max_{\bar{\Omega}_1} r^{2-s}(y)|u_1| + \max_{\bar{\Omega}_2} r^{2-s}(y)|u_2| \le C\Xi_s(\mathscr{F}_{01}, \mathscr{F}_{02}, \mathscr{F}_1, \mathscr{F}_2, \mathscr{F}_3). \tag{9.24}$$

Proof If

$$\eta_q^j \ne 2-\gamma-2/p, \quad q = 1, 2, \ldots, j = 0, 1, \quad 1 \ne 2-\gamma-2/p > 0, \quad p > 1,$$

and $\mathscr{F}_{0i} \in L_\gamma^p(\Omega_i)$, $\mathscr{F}_1 \in W_\gamma^{1-1/p,p}(\Gamma)$, $\mathscr{F}_2 \in W_\gamma^{2-1/p,p}(\partial\Omega_1\setminus\Gamma)$, $\mathscr{F}_3 \in W_\gamma^{2-1/p,p}(\partial\Omega_2\setminus\Gamma)$, then Theorem 3.12 [26] guarantees existence of a unique weak solution (u_1, u_2) to (9.1)–(9.3) which is presented as

$$u_i(y) = u_{0i}(y) + \sum_{\eta_q^j \in (0, 2-\gamma-2/p)} \mathscr{T}_q(\mathscr{F}_{01}, \mathscr{F}_{02}, \mathscr{F}_1, \mathscr{F}_2, \mathscr{F}_3) S_q(y), \quad i = 1, 2,$$

where $u_{0i} \in W_\gamma^{2,p}(\Omega_i)$ and $\mathscr{T}_q$ is the bounded functional on the corresponding space, $S_q(y)$ is some given functions which contains the terms $|y_1^2 + y_2^2|^{\frac{\eta_q^j}{2}}$ (see their definitions (2.50), (3.76) in [26]). Moreover, there is the estimate

$$\sum_{i=1}^{2} \|u_{0i}\|_{2,p;\gamma,\Omega_i} + \max_{\bar{\Omega}_i} \sum_{\eta_q^j \in (0, 2-\gamma-2/p)} |\mathscr{T}_q S_q| \le C\mathscr{R}_{p,\gamma}(\mathscr{F}_{01}, \mathscr{F}_{02}, \mathscr{F}_1, \mathscr{F}_2, \mathscr{F}_3), \quad i = 1, 2, \tag{9.25}$$

where we set

$$\mathscr{R}_{p,\gamma}(\mathscr{F}_{01}, \mathscr{F}_{02}, \mathscr{F}_1, \mathscr{F}_2, \mathscr{F}_3) = \sum_{i=1}^{2} \|\mathscr{F}_{0i}\|_{0,p;\gamma,\Omega_i} + \|\mathscr{F}_1\|_{1-1/p,p;\gamma,\Gamma}$$
$$+ \|\mathscr{F}_2\|_{2-1/p,p;\gamma,\partial\Omega_1\setminus\Gamma} + \|\mathscr{F}_3\|_{2-1/p,p;\gamma,\partial\Omega_2\setminus\Gamma}.$$

After that, choosing s, p and γ subject the conditions of Lemma 9.4, we conclude that there is not any η_q^j belongs to $(0, 2-\gamma-2/p)$, and that is why

$$u_i(y) = u_{0i}(y) \in W_\gamma^{2,p}(\Omega_i).$$

Finally, we remark that

$$\mathscr{R}_{p,\gamma}(\mathscr{F}_{01}, \mathscr{F}_{02}, \mathscr{F}_1, \mathscr{F}_2, \mathscr{F}_3) \le C\Xi_s(\mathscr{F}_{01}, \mathscr{F}_{02}, \mathscr{F}_1, \mathscr{F}_2, \mathscr{F}_3)$$

for $\gamma > 2-s-1/p$. The last inequality together with (9.25) complete the proof of inequality (9.23).

As for inequality (9.24), the direct calculations with aid the embedding theorem and the definition of the space $W^{2,p}_{\gamma}$ arrive at estimates

$$\max_{\bar{\Omega}_i} r^{2-s}(y)|u_i| \leq C\|r^{2-s}u_i\|_{2,p;\Omega_i} \leq C\|u_i\|_{2,p;\gamma,\Omega_i}, \quad i=1,2.$$

After that, using (9.23), we end up with

$$\max_{\bar{\Omega}_i} r^{2-s}(y)|u_i| \leq C\Xi_s(\mathscr{F}_{01}, \mathscr{F}_{02}, \mathscr{F}_1, \mathscr{F}_2, \mathscr{F}_3).$$

This completes the proof of Lemma 9.4. □

At this point we show that the weak solution (u_1, u_2) constructed in Lemma 9.4 belongs to the class $E_s^{2+\beta}$. To this end, we apply the partition of the unity together with estimate (9.24) and the local estimates from Theorems 9.3–9.5 with $\delta = \pi/2 - \delta_0$ and $\delta = \pi/2 - \delta_1$, correspondingly. Note that the smoothness of the functions $u_i(y)$, $i = 1, 2$, outside of the neighborhoods of the corner points A_0, A_1 is provided with the general theory for the elliptic boundary value problems [22, 30, 33]. Thus, the proof of Theorem 9.1 is completed in the case of $\mathscr{F} \equiv 0$.

To remove this restriction, we look for solution (9.1)–(9.3) in the form

$$u_1 = \mathscr{U}_1 \quad \text{and} \quad u_2 = \mathscr{U}_2 + \mathscr{V}_2$$

where the unknown functions $\mathscr{U}_i$, $i = 1, 2$, solve (9.1)–(9.3) with $\mathscr{F} \equiv 0$, and $\mathscr{V}_2$ is a solution of the Dirichlet problem

$$\begin{cases} \Delta\mathscr{V}_2 = 0 & \text{in} \quad \Omega_2, \\ \mathscr{V}_2|_{\partial\Omega_2\setminus\Gamma} = 0, & \mathscr{V}_2|_{\Gamma} = \mathscr{F}. \end{cases}$$

After that, Lemma 5.1 in [36] and Sect. 6 in [19] guarantee the existence of a unique solution $\mathscr{V}_2 \in E_s^{2+\beta}(\bar{\Omega}_2)$ which satisfies the bound

$$\|\mathscr{V}_2\|_{E_s^{2+\beta}(\bar{\Omega}_2)} \leq C\|\mathscr{F}\|_{E_s^{2+\beta}(\Gamma)}.$$

This finishes the proof of Theorem 9.1. □

Coming to the proof of Theorem 9.2, we repeat step-by-step the arguments above and use, additionally Lemmas 2.1 and 2.2 with Remark 3.1 in [6]. This completes the proof of the main results. □

9.5.2 More Regular Solutions of (9.1)–(9.3)

It is worth mentioning that the similar statements to Theorems 9.1 and 9.2 take place in the case of the greater value s.

Introducing

$$\delta = \max\{\delta_0, \delta_1\}$$

we assert the following

Theorem 9.6 *Let $\Gamma \in C^{2+\beta}$ and parameters: s, k, δ_0 and δ_1, satisfy one of two conditions for $n = 1, 2, ...$:*

i

$$k > 1, \quad \delta_0, \delta_1 \in \Big(0, \frac{\pi}{4n+2}\Big), \quad s \in \Big(\frac{2\pi n}{\pi - 2\delta}, 1 + 2n\Big);$$

ii

$$k < 1, \quad \delta_0, \delta_1 \in \Big(0, \frac{\pi}{2n+2}\Big), \quad s \in \Big(\max\Big\{1 + 2n, \frac{2\pi n}{\pi - 2\delta}\Big\}, 2 + 2n\Big)\backslash\Big\{\frac{\pi}{2\delta_1}, \frac{\pi}{2\delta_0}\Big\}.$$

Then, under assumption **h2**, *transmission problem* (9.1)–(9.3) *admits a unique classical solution* (u_1, u_2) *satisfying the estimate*

$$\sum_{i=1}^{2} \|u_i\|_{E_s^{2+\beta}(\bar{\Omega}_i)} \leq C[\sum_{i=1}^{2} \|\mathscr{F}_{0i}\|_{E_{s-2}^{\beta}(\bar{\Omega}_i)} + \|\mathscr{F}\|_{E_s^{2+\beta}(\Gamma)} + \|\mathscr{F}_1\|_{E_{s-1}^{1+\beta}(\Gamma)}$$
$$+ \|\mathscr{F}_2\|_{E_s^{2+\beta}(\partial\Omega_1\backslash\Gamma)} + \|\mathscr{F}_3\|_{E_s^{2+\beta}(\partial\Omega_2\backslash\Gamma)}]. \tag{9.26}$$

If Γ has rth corner points, and parameters: s, k, δ_0 and δ_1, α_i, $i = 1, 2, \ldots, r$, satisfy one of the two conditions:

iii

$$k > 1, \quad \delta_0, \delta_1 \in \Big(0, \frac{\pi}{4n+2}\Big), \quad \alpha_i \in \Big(0, \frac{\pi}{4n+2}\Big), \quad \delta < \frac{\pi + 2n\alpha}{8n+2},$$
$$\max\Big\{2n + 1/2, \frac{2\pi n}{\pi - 2\delta}\Big\} < s < \frac{\pi(4n+1)}{2\pi - \alpha};$$

iv

$$k < 1, \quad \delta_0, \delta_1 \in \Big(0, \frac{\pi}{2n+2}\Big), \quad \alpha_i \in \Big(0, \frac{\pi}{4n+3}\Big), \quad \delta < \frac{3\pi + 2n\alpha}{8n+6},$$
$$\max\Big\{2n + 3/2, \frac{2\pi n}{\pi - 2\delta}\Big\} < s < \frac{\pi(4n+3)}{2\pi - \alpha},$$

and assumptions **h2** *and* **h3** *hold, then there exists a unique classical solution of* (9.1)–(9.3), *and inequality* (9.26) *is performed.*

Proof In virtue of the proof of Theorem 9.1, it is enough to validity the statements of Theorem 9.6 only in the case of $\mathscr{F} \equiv 0$. The general case is reduced to this special one with the arguments similar to the end of the proof of Theorem 9.1.

Let us analyze the case of a smooth curve Γ, and prove the theorem for $k > 1$ (i.e. when assumption **(i)** holds). The arguments in the case of $k < 1$ is almost identical and left to the interested reader.

At the beginning, we consider the case of condition **(i)** with $n = 1$, which means the parameters: s, δ_0, δ_1 and k, meet the following requirements:

$$k > 1, \quad s = s_1 \in \Big(\frac{2\pi}{\pi - 2\delta}, 3\Big), \quad \delta_0, \delta_1 \in (0, \pi/6).$$

To obtain the one-valued solvability of problem (9.1)–(9.3) in this case, we set $s_0 = s_1 - 2$ and then, taking into account the inclusion: $(0, \pi/6) \subset (0, \pi/2)$, and Corollary 9.1, we apply Theorem 9.1 with $s = s_0$ and $\delta_0, \delta_1 \in (0, \pi/6)$, $k > 1$, to problem (9.1)–(9.3). Thus, we obtain the unique solution (u_1, u_2) satisfying the estimate

$$\|u_1\|_{E^{2+\beta}_{s_0}(\bar{\Omega}_1)} + \|u_2\|_{E^{2+\beta}_{s_0}(\bar{\Omega}_2)} \leq C\Xi_{s_0}(\mathscr{F}_{01}, \mathscr{F}_{02}, \mathscr{F}_1, \mathscr{F}_2, \mathscr{F}_3) \leq C\Xi_{s_1}(\mathscr{F}_{01}, \mathscr{F}_{02}, \mathscr{F}_1, \mathscr{F}_2, \mathscr{F}_3).$$

After that the straightforward calculations and estimates above arrive at the inequalities

$$\begin{aligned}\max_{\bar{\Omega}_i} r^{2-s_1}(y)\,|u_i| &= \max_{\bar{\Omega}_i} r^{2-s_1+s_0}(y)\,|u_i| r^{-s_0}(y)\\ &= \max_{\bar{\Omega}_i} r^{-s_0}(y)|u_i|\\ &\leq C\Xi_{s_1}(\mathscr{F}_{01}, \mathscr{F}_{02}, \mathscr{F}_1, \mathscr{F}_2, \mathscr{F}_3), \quad i = 1, 2. \end{aligned} \tag{9.27}$$

Finally, applying the partition of unity together with estimate (9.27), Theorem 9.4 and the general theory for the elliptic boundary value problems [30, 31], we deduce that the constructed solution on the previous step belongs to the class $E^{2+\beta}_{s_1}$ and u_1, u_2 satisfy estimate (9.26) with $s = s_1$. Thus we complete the proof of Theorem 9.6 in the case of **(i)** with $n = 1$.

To extend the obtained results to the case arbitrary n, $n > 1$, we apply mathematical induction method. Accordingly, for $j = 1, 2, \ldots, n$, we consider the following monotonically increasing sequence s_j, $s_j \in \Big(\frac{2\pi j}{\pi - 2\delta}, 1 + 2j\Big)$, and the family of nested intervals $\Big(0, \frac{\pi}{4j+2}\Big)$. Indeed, it is easy to verify that in the considered case, there are

$$2j < \frac{2\pi j}{\pi - 2\delta} < s_j < 2j+1, \quad \frac{2\pi(j-1)}{\pi - 2\delta} + \frac{4\delta}{\pi - 2\delta} < s_j - 2 < 2j - 1,$$
$$2 < \frac{2\pi}{\pi - 2\delta} < s_1 < s_2 < \cdots < s_j < s_{j+1} < \cdots < s_n = s,$$
$$\Big(0, \frac{\pi}{2}\Big) \supset \Big(0, \frac{\pi}{6}\Big) \supset \cdots \supset \Big(0, \frac{\pi}{4j+2}\Big) \supset \Big(0, \frac{\pi}{4j+6}\Big) \supset \cdots \supset \Big(0, \frac{\pi}{4n+2}\Big). \tag{9.28}$$

It is apparent that the validity of the theorem for $j = 1$ is examined with the arguments written above for the case $n = 1$. Next, for $j = n - 1$, we assume that problem (9.1)–(9.3) with $\mathscr{F} \equiv 0$ is solvable and its unique solution (u_1, u_2) satisfies estimate (9.26) with $s = s_{n-1}$ and $\delta_0, \delta_1 \in \Big(0, \frac{\pi}{4n-2}\Big)$. After that we have to prove Theorem 9.6 in the case of $j = n$, other words if the parameters s, k, δ_0, δ_1 satisfy the assumption

$$k > 1, \quad s = s_n \in \Big(\frac{2\pi n}{\pi - 2\delta}, 1 + 2n\Big), \quad \delta_0, \delta_1 \in \Big(0, \frac{\pi}{4n+2}\Big).$$

To this end, choosing $s_{n-1} = s_n - 2$ and keeping in mind properties (9.28) and Corollary 9.1, we recast the arguments of the step $j = 1$ for the solution $u_i \in E^{2+\beta}_{s_{n-1}}(\bar{\Omega}_i)$, $i = 1, 2$. Thus, we come to the estimate

$$\sum_{i=1}^{2} \max_{\bar{\Omega}_i} r^{2-s_n}(y)\, |u_i| = \sum_{i=1}^{2} \max_{\bar{\Omega}_i} r^{-s_{n-1}}(y) |u_i|$$
$$\leq C \Xi_{s_n}(\mathscr{F}_{01}, \mathscr{F}_{02}, \mathscr{F}_1, \mathscr{F}_2, \mathscr{F}_3).$$

As a result, Theorem 9.4 and a partition of unity, we arrive at

$$\sum_{i=1}^{2} \|u_i\|_{E^{2+\beta}_{s_n}(\bar{\Omega}_i)} \leq C\Big[\sum_{i=1}^{2} \|\mathscr{F}_{0i}\|_{E^{\beta}_{s_n-2}(\bar{\Omega}_i)} + \|\mathscr{F}_1\|_{E^{1+\beta}_{s_n-1}(\Gamma)}$$
$$+ \|\mathscr{F}_2\|_{E^{2+\beta}_{s_n}(\partial\Omega_1 \setminus \Gamma)} + \|\mathscr{F}_3\|_{E^{2+\beta}_{s_n}(\partial\Omega_2 \setminus \Gamma)}\Big].$$

That completes the proof of Theorem 9.6 in the case of the smooth interface Γ.

To prove Theorem 9.6 in the case of a singular interface under conditions **(iii)** and **(iv)**, we repeat here the arguments written above with applying Theorem 9.2 instead of Theorem 9.1 and using Remark 3.1 [6]. Thus, the proof of Theorem 9.6 is now finished. □

9.6 Appendix: Proof of Proposition 9.4

Asymptotic $\mathscr{D}(\lambda)$ obtained in Remark 9.1 provides the following behavior of $\hat{K}(\lambda)$ for small and large $|\lambda|$:

$$\hat{K}(\lambda) \approx \begin{cases} \pm\frac{i(1+k)^{-1}}{i\lambda+s}, & \lambda \to \pm\infty, \\ ([\cot s(\delta+\pi/2)+k\cot s(\pi/2-\delta)]s)^{-1}, & |\lambda|<1. \end{cases}$$

Keeping in mind the asymptotic of $\hat{K}(\lambda)$ for large value of $|\lambda|$, and the absence of poles in $\hat{K}(\lambda)$ for s satisfying assumptions in Proposition 9.2, we change the variable $\lambda = \mu_1 + i\mu_2$ to calculate the function $K(\xi)$.

For the sake of clarity, we consider here the case $\xi \geq 0$, then

$$K(\xi) = \frac{e^{-\mu_2\xi}}{2\pi} \int_{-\infty}^{+\infty} \frac{e^{i\mu_1\xi}d\mu_1}{(i\mu_1+s-\mu_2)\mathscr{D}_1(\mu_1+i\mu_2)}$$

with $\mu_2 > 0$ and $s-\mu_2 > 0$, where μ_2 is chosen such that the quantity $s-\mu_2$ has the same properties as s.

Based on this relation and the asymptotic representation of $\hat{K}(\lambda)$, we can conclude that the proof of Proposition 9.4 is equivalent to the verification of statements (i)–(iii) to the function

$$K_1(\xi) = \frac{e^{-\mu_2\xi}}{2\pi}\sum_{j=1}^{5} K_{1j},$$

where we put

$$K_{11} = \int_{-N}^{N} \frac{e^{i\mu_1\xi}[\cot(s-\mu_2)(\delta+\pi/2)+k\cot(s-\mu_2)(\pi/2-\delta)]^{-1}d\mu_1}{(s-\mu_2)(i\mu_1+s-\mu_2)},$$

$$K_{12} = -\int_{0}^{N} \frac{i(1+k)^{-1}e^{i\mu_1\xi}d\mu_1}{i\mu_1+s-\mu_2}, \quad K_{13} = \int_{-N}^{0} \frac{i(1+k)^{-1}e^{i\mu_1\xi}d\mu_1}{i\mu_1+s-\mu_2},$$

$$K_{14} = \int_{0}^{+\infty} \frac{i(1+k)^{-1}e^{i\mu_1\xi}d\mu_1}{i\mu_1+s-\mu_2}, \quad K_{15} = -\int_{-\infty}^{0} \frac{i(1+k)^{-1}e^{i\mu_1\xi}d\mu_1}{i\mu_1+s-\mu_2}.$$

After that we estimate each term K_{1j}.

- Keeping in mind that s and $s-\mu_2$ meet requirements of Proposition 9.2, we deduce that

$$\sum_{j=1}^{3}\left|\frac{\partial^l K_{1j}}{\partial\xi^l}\right| \leq C \quad \text{for} \quad l = 0, 1, 2. \tag{9.29}$$

- Coming to K_{14} and K_{15}, we have

$$
\begin{aligned}
|K_{14}+K_{15}| &= C\Big|Im\int_0^{+\infty}\frac{e^{i\xi\mu_1}d\mu_1}{i\mu_1+s-\mu_2}\Big| \\
&= C\Big|\int_0^{+\infty}\frac{(s-\mu_2)\sin\mu_1\xi}{\mu_1^2+(s-\mu_2)^2}d\mu_1-\int_0^{+\infty}\frac{\mu_1\cos\mu_1\xi}{\mu_1^2+(s-\mu_2)^2}d\mu_1\Big| \\
&\leq C\Big[1+\Big|\int_0^{+\infty}\frac{\mu_1\cos\mu_1\xi}{\mu_1^2+(s-\mu_2)^2}d\mu_1\Big|\Big].
\end{aligned}
$$

Then, integrating by parts in the second term arrives at the inequality

$$
|K_{14}+K_{15}|\leq C(1+\xi^{-1}). \tag{9.30}
$$

Summing up the estimates to K_{1j}, $j=\overline{1,5}$, we deduce

$$
|K_1|\leq Ce^{-\mu_2\xi}(1+\xi^{-1}). \tag{9.31}
$$

Thus, in order to prove statement (i), we are left to evaluate $\frac{\partial K_1}{\partial\xi}$. The direct calculations lead to the representation

$$
\frac{\partial K_1}{\partial\xi}=-\mu_2K_1(\xi)+\frac{e^{-\mu_2\xi}}{2\pi}\sum_{j=1}^{5}\frac{\partial K_{1j}}{\partial\xi}.
$$

Inequalities (9.29) and (9.31) ensure the corresponding estimate of the first fourth terms in the equality above. Let us consider the sum of the last two terms

$$
\begin{aligned}
\frac{\partial K_{14}}{\partial\xi}+\frac{\partial K_{15}}{\partial\xi} &= -C\,Im\Big[\int_0^{+\infty}e^{i\xi\mu_1}d\mu_1+\int_0^{+\infty}\frac{s-\mu_2}{i\mu_1+s-\mu_2}e^{i\xi\mu_1}d\mu_1\Big] \\
&= -C\,Im\,[\pi\delta(-\xi)+i\xi^{-1}]+C(s-\mu_2)[K_{14}+K_{15}],
\end{aligned}
$$

where $\delta(\xi)$ is the delta-function. Note that, in order to obtain the last equality, we use Lemma 5.3 in [4]. Next, this representation with aid inequality (9.30) derives

$$
\Big|\frac{\partial K_{14}}{\partial\xi}+\frac{\partial K_{15}}{\partial\xi}\Big|\leq C(1+\xi^{-1}).
$$

Collecting this inequality with (9.31), we deduce statement (i).

It is apparent that, the same approach allows us to obtain estimate (ii). Finally, to complete the proof of Proposition 9.4, we integrate the function $\left(\frac{\partial K_{14}}{\partial\xi}+\frac{\partial K_{15}}{\partial\xi}\right)$ and conclude

$$\int_{-\varepsilon}^{\varepsilon} \frac{\partial(K_{14}+K_{15})}{\partial\xi} d\xi = C\, Im \int_{0}^{+\infty} \frac{d\mu_1}{i\mu_1 + s - \mu_2} \int_{-\varepsilon}^{+\varepsilon} i\mu_1 e^{i\xi\mu_1} d\xi$$

$$= C \int_{0}^{+\infty} \frac{\sin(\mu_1\varepsilon) d\mu_1}{\mu_1^2 + (s-\mu_2)^2} \le C. \qquad \square$$

References

1. Adams, R.A.: Sobolev Spaces. Academic, New York (1975)
2. Aharoni, A.: Introduction to the Theory of Ferromagnetism. Oxford University Press, Oxford (1996)
3. Arjman, D., Engblom, S., Kreiss, G.: Temporal upscaling in micromagnetism via heterogeneous multiscale methods. J. Comput. Appl. Math. **345**, 99–113 (2019)
4. Bazaliy, B.V., Friedman, A.: The Hele-Shaw problem with surface tension in a half-plane: a model problem. J. Differ. Equ. **216**, 387–435 (2005)
5. Bazaliy, B.V., Friedman, A.: The Hele-Shaw problem with surface tension in a half-plane. J. Differ. Equ. **216**, 439–469 (2005)
6. Bazaliy, B.V., Vasylyeva, N.: The transmission problem in domain with a corner point for the Laplace operator in weighted Hölder spaces. J. Differ. Equ. **249**, 2476–2499 (2010)
7. Bazaliy, B.V., Vasylyeva, N.: The two-phase Hele-Shaw problem with a nonregular initial interface and without surface tension. J. Math. Phys. Anal. Geom. **10**(1), 3–43 (2014)
8. Ben M'Barek, A., Mérigot, M.: Régularité de la solution d'un probléme de transmission. C.R. Acad. Sci. Paris série A **280**, 1591–1593 (1975)
9. M. Borsuk, Transmission problems for elliptic second-order equations in non-smooth domains, Spriner Basel AG, (2010)
10. Ciarlet, P.G.: Plates and Junctions in Elastic Multi-structures, RMA 14. Masson, Paris (1990)
11. Cimrák, I.: A survey on the numeric and computations for the Landau-Lifshitz equation of micromagnetism. Arch. Comput. Meth. Eng. **15**(3), 277–309 (2008)
12. Costabel, M., Dauge, M., Nicaise, S.: Corner singularities and analytic regularity for linear elliptic systems. Part I: Smooth domains, https://hal.archives-ouvertes.fr/hal-00453934/en/ online version of Chapters 1 to 5 (2010)
13. Costabel, M., Dauge, M., Nicaise, S.: Analytic regularity for linear elliptic systems in polygons and polyhedra. Math. Models Methods Appl. Sci. **22**(8), 1250015 (2012)
14. Dauge, M., Nicaise, S.: Oblique derivative and interface problems on polygonal domains and networks. Commun. Partial Differ. Equ. **14**, 1147–1192 (1989)
15. Destuynder, P.: Une théorie asymptotique des plaques minces en élasticité linéaire, RMA 2. Masson, Paris (1986)
16. Druet, P-E.: Regularity of second derivatives in elliptic transmission problems near an interior regular multiple line of contact. Math. Meth. Appl. Sci. **41**(16), 6457–6479 (2018)
17. Escher, J., Matioc, A.-V.: Analysis of a two-phase model describing the growth of solid tumors. Eur. J. Appl. Math. **24**, 25–48 (2013)
18. Eskin, G.: Transmission problems for equations of principal type with two variables. Trans. Moscow Math. Soc. **21**, 245–292 (1970)
19. Grisvard, P.: Elliptic problems in nonsmooth domains, Monographs and Studies in Mathematics 21. Pitman, Boston (1985)
20. Kellogg, R.B.: Higher order singularities for interface problems, The Math. foundations of finite el. method with appl. to P.D.E., Ed.Aziz, Acad. Press, (1972)
21. Knees, D., Sändig, A.-M.: Regularity of elastic fields in composites. Multifield Problems in Solid and Fluid Mechanics. Lecture Notes in Applied and Computational Mechanics, vol. 28, pp. 331–360. Springer, Berlin (2006)

22. O.A. Ladyzhenskaia, N.N. Ural'tseva, *Linear and quasilinear elliptic equations*, Nauka, Moscow, 1964. (English translation: O.A. Ladyzhenskaia, N.N. Ural'tseva, *Linear and quasilinear elliptic equations,* Academies Press, New York, 1968.)
23. Le Dret, H.: Folded plates revisited. Comput. Mech. **5**, 345–365 (1989)
24. Le Dret, H.: Problémes variationnels dans les multi-domaines. RMA, vol. 19. Springer, Masson (1999)
25. Lemrabet, K.: Régularité de la solution d'un probléme de transmission. J. Math. Pures et Appl. **56**, 1–38 (1977)
26. Nicaise, S.: Polygonal interface problems. Peter D. Lang Verlag, Frankfurt a. M. (1993)
27. Oleinik, O.A.: Solution of fundamental boundary value problems for second order equations with discontinuous coefficients. Akad. Nauk SSSR **124**(6), 1219–1222 (1959)
28. Oleinik, O.A.: Boundary-value problems for linear equations of elliptic parabolic type with discontinuous coefficients. Izv. Akad. Nauk SSSR Ser. Math. **25**(1), 3–20 (1961)
29. Perfekt, H-M.: The transmission problem on a three-dimensional wedge. Arc. Ration. Mech. Anal. **231**(3), 1745–1780 (2019)
30. Roitberg, J.A., Sheftel', Z.G.: General boundary value problems for elliptic equations with discontinuous coefficients. Soviet. Math. Dokl. **4**, 231–234 (1963). (in Russian)
31. Schechter, M.A.: A generalization of the problem of transmission. Ann. Sc. Norm. Super. Pisa Cl. Sci. (3) **14**, 207–236 (1960)
32. Sheftel', Z.G.: Estimates in L_p of solutions of elliptic equations with discontinuous coefficients and satisfying general boundary conditions and conjugacy conditions. Soviet Math. Dokl. **4**, 321–324 (1963)
33. Sheftel', Z.G.: Energy inequalities and general boundary problems for elliptic equations with discontinuous coefficients. Sibirsk. Math. Zh. **6**, 636–668 (1965). (Russian)
34. Vasilyev, V.B.: On some transmission problems in a lane corner. Tatra Mt. Math. Publ. **63**, 291–301 (2015)
35. Vasylyeva, N.: On the solvability of the Hele-Shaw problem in the case of nonsmooth initial data in weighted Hölder classes. Ukr. Math. Bull. **2**(3), 323–349 (2005)
36. Vasylyeva, N.: The Mullins-Sekerka problem with surface tension and a nonregular initial interface. Nonlinear Bound. Value Prob. **21**, 165–204 (2012)

Part II
Determinism and Stochasticity in Modeling in Real Phenomena

Chapter 10
Convergence Rate of Random Attractors for 2D Navier–Stokes Equation Towards the Deterministic Singleton Attractor

Hongyong Cui and Peter E. Kloeden

Abstract In this paper we study the long-time behavior of a 2D Navier–Stokes equation. It is shown that under small forcing intensity the global attractor of the equation is a singleton. When endowed with additive or multiplicative white noise no sufficient evidence was found that the random attractor keeps the singleton structure, but the estimate of the convergence rate of the random attractor towards the deterministic singleton attractor as stochastic perturbation vanishes is obtained.

10.1 Introduction

Attractors, which are usually expected to be compact sets that are invariant and attract some kinds of sets (e.g., bounded subsets of the state space), are important objects to study the long-time behavior of dynamical systems generated by dissipative evolution equations [9, 11, 20]. When an evolution equation is perturbed by stochastic noises, it is useful to study the random attractor of the system [3, 13, 17]. A random attractor is defined to be pathwise pullback attracting, using the past information of the system, but is also forward attracting in probability [15, 26], and so has also the desired ability to depict the future of the system.

The effects of noises on attractors are often of good interest [4]. One question is the stability of the attractor under stochastic perturbations, i.e., the convergence of the random attractors $\mathscr{A}_\varepsilon = \{A_\varepsilon(\omega)\}_{\omega\in\Omega}$ towards the deterministic attractor $\mathscr{A}$ as the perturbation intensity ε vanishes. Since each attractor is generally a set rather than a single point, two kinds of convergences differ: one is the so-called upper semicontinuity

H. Cui (✉)
School of Mathematics and Statistics, Huazhong University of Science & Technology, Wuhan 430074, China
e-mail: h.cui@outlook.com

P. E. Kloeden
Mathematisches Institut, Universität Tübingen, 72076 Tübingen, Germany
e-mail: kloeden@na-uni.tuebingen.de

V. A. Sadovnichiy and M. Z. Zgurovsky (eds.), *Contemporary Approaches and Methods in Fundamental Mathematics and Mechanics*, Understanding Complex Systems,
https://doi.org/10.1007/978-3-030-50302-4_10

$$\lim_{\varepsilon \to 0^+} \operatorname{dist}\big(\mathscr{A}_\varepsilon(\omega), \mathscr{A}\big) = 0,$$

and, the second, the lower semi-continuity

$$\lim_{\varepsilon \to 0^+} \operatorname{dist}\big(\mathscr{A}, \mathscr{A}_\varepsilon(\omega)\big) = 0,$$

where dist denotes the Hausdorff semi-distance between non-empty sets $\operatorname{dist}(A, B) = \sup_{a \in A} \inf_{b \in B} d(a, b)$. The upper semicontinuity is relatively easy to realize in applications, requiring only general study on the dynamical systems and the attractors, so not too restrictive conditions could be sufficient [5, 6, 25, 28, 30]. Nevertheless, the lower semi-continuity is more complicated and needs more detailed study either on the structure of the unperturbed attractor [7, 8, 10, 14, 19, 21] or on the equi-attraction of the family of the perturbed random attractors [2, 14, 22–24], etc. Both approaches to the lower semi-continuity are technical in general applications.

In addition to the qualitative study mentioned above, some quantitive results in terms of the convergence rate of the convergences are also available in the literature, see for instance [2, 8] for deterministic perturbations of attractors where the convergence rate was obtained under the condition that the perturbed attractors attract exponentially. However, in general applications the attraction rate, which usually relates to the structure of the attractor, is another technical subject.

In this paper we study the convergence rate of the random attractors under additive or multiplicative white noise towards the deterministic (unperturbed) attractor, for a particular case that the deterministic attractor is a singleton. We consider the 2D Navier–Stokes equation

$$\frac{\mathrm{d}u}{\mathrm{d}t} - \nu \Delta u + (u \cdot \nabla)u + \nabla p = f(x), \quad \nabla \cdot u = 0,$$

where $\nu > 0$ and $f \in H$. It is shown that when the forcing intensity is sufficiently small ($\frac{|f|}{\nu^2 \lambda} < \frac{1}{c_1}$, see Sect. 10.2.2), the deterministic global attractor $\mathscr{A}$ is a singleton. In addition, under additive or multiplicative white noise, the random attractors $\mathscr{A}_\varepsilon = \{A_\varepsilon(\omega)\}_{\omega \in \Omega}$ converge both upper and lower semi-continuously towards the global attractor $\mathscr{A}$ at the rate of the same order as the perturbation intensity ε vanishes, i.e., more precisely, we have

$$\operatorname{dist}_H\big(A_\varepsilon(\omega), \mathscr{A}\big) \leqslant \rho_\varepsilon(\omega)$$

where $\operatorname{dist}_H(A, B) = \max\{\operatorname{dist}(A, B), \operatorname{dist}(B, A)\}$, and each ρ_ε is a random variable such that $\rho_\varepsilon(\omega) \sim \varepsilon$ as $\varepsilon \to 0^+$.

The qualitative study of the upper semicontinuity for the equation was given by Caraballo, Langa & Robinson [6], but since we are here interested in the convergence rate, more careful analysis is needed. Thanks to the small forcing intensity condition, it is proved that the distance between the solutions of the perturbed equation and the solution of the original system is eventually bounded by the random variable $\rho_\varepsilon(\cdot)$

mentioned above as the initial data pulled back to $-\infty$ (see Lemmas 10.9 and 10.11), and then the convergence rate of the attractors follows.

Note finally that, though the deterministic attractor under small forcing intensity is a singleton, the random attractors are generally not. Essentially, this is because that the random perturbation as an additional random forcing makes the "effective" forcing intensity no longer controllable. In this sense, the additive and multiplicative noises destroy the singleton structure of the undisturbed attractor.

10.2 Singleton Attractor of 2D Navier–Stokes Equation with Small Forcing Intensity

10.2.1 Preliminaries

We consider the following viscous incompressible 2D Navier–Stokes (NS) equation

$$\begin{aligned} \frac{\mathrm{d}u}{\mathrm{d}t} - \nu\Delta u + (u\cdot\nabla)u + \nabla p = f(x), \\ \nabla\cdot u = 0, \end{aligned} \tag{10.1}$$

defined on a bounded smooth domain $\mathscr{O} := [0, L]^2 \subset \mathbb{R}^2$ and endowed with initial value and periodic boundary conditions

$$u(t,x)|_{t=0} = u_0(x), \quad u(t, x + Le_j) = u(t,x), \tag{10.2}$$

where $\nu > 0$ represents the kinematic viscosity, f the volume forces that are applied to the fluid and e_j, $j = 1, 2$, are unit vectors parallel to the coordinate axes.

Keeping in mind that the state function u of Eq. (10.1) is vector-valued, we define in a standard way the space of smooth function that are periodic and divergence-free by

$$\dot{\mathbb{C}}_p^\infty = \left\{u \in \left[\dot{C}_p^\infty(\mathscr{O})\right]^2 : \nabla\cdot u = 0\right\},$$

that is, each member is 2-component divergence-free and each component is in $\dot{C}_p^\infty(\mathscr{O})$, where the subscript "$_p$" denotes the periodicity. In the same say we use the notations

$$\mathbb{L}^2(\mathscr{O}) = [L^2(\mathscr{O})]^2, \quad \mathbb{H}_p^k = [H_p^k(\mathscr{O})]^2,$$

where the norm is given, e.g., by

$$\|u\|_{\mathbb{L}^2(\mathscr{O})} = \sum_{j=1}^{2} |u_j|^2_{L^2(\mathscr{O})}.$$

The periodic boundary condition facilitates the analysis. For instance, we can expand u by Fourier expansion

$$u = \sum_{k \in \mathbb{Z}^2} e^{2\pi i k \cdot x/L} u_k$$

with $u_k \in \mathbb{R}^2$, which gives

$$-\Delta u = \frac{4\pi^2}{L^2} \sum_{k \in \mathbb{Z}^2} e^{2\pi i k \cdot x/L} |k|^2 u_k.$$

Denote by $Au := -\Delta u$ for $u \in D(A) = \{u \in \dot{\mathbb{H}}_p^2(\mathscr{O}) : \nabla \cdot u = 0\}$. Then in this case $(A, D(A))$ coincides with the Stokes operator [27, p237], and it is easy to observe that for any $s \geqslant 1$ we can define analogously the $s/2$-power of A with $D(A^{s/2}) = \{\dot{\mathbb{H}}_p^s(\mathscr{O}) : \nabla \cdot u = 0\}$ and that $|A^{s/2}u| = C\|u\|_{\dot{\mathbb{H}}_p^s(\mathscr{O})}$. In addition, since the inverse A^{-1} is a compact self-adjoint operator on $H := \{u \in \dot{\mathbb{L}}^2(\mathscr{O}) : \nabla \cdot u = 0\}$ we have a complete family of orthonormal eigenfunctions $\{w_j\}_{j \in \mathbb{N}}$ of A to expand H and such w_j can be chosen from $\dot{C}_p^\infty(\mathscr{O})$ [27, 239]. It is clear then that the Poincaré inequality holds for some $\lambda > 0$

$$\lambda |u|^2 \leqslant |A^{\frac{1}{2}} u|^2. \tag{10.3}$$

With the above preparations, Eq. (10.1) can be rewritten in the abstract form

$$\frac{\mathrm{d}u}{\mathrm{d}t} + \nu A u + B(u, u) + \nabla p = f, \tag{10.4}$$

where $B(u, v) := (u \cdot \nabla)v$.

Define

$$V = \{u \in \dot{\mathbb{H}}_p^1(\mathscr{O}) : \nabla \cdot u = 0\}.$$

Lemma 10.1 ([27, Proposition 9.2]) *In the 2-dimensional case it holds the following for some $c_1, c_2 > 0$*

$$|b(u, v, w)| \leqslant \begin{cases} c_1 |u|^{\frac{1}{2}} |A^{\frac{1}{2}} u|^{\frac{1}{2}} |A^{\frac{1}{2}} v| |w|^{\frac{1}{2}} |A^{\frac{1}{2}} w|^{\frac{1}{2}}, & u, v, w \in V, \\ c_2 |u|^{\frac{1}{2}} |A^{\frac{1}{2}} u|^{\frac{1}{2}} |A^{\frac{1}{2}} v|^{\frac{1}{2}} |Av|^{\frac{1}{2}} |w|, & u \in V, v \in D(A), w \in H. \end{cases} \tag{10.5}$$

Remark 10.1 The constants c_1 and c_2 are absolute constants independent of the 2d NS equation. For instance, c_1, which is crucial for our latter analysis, can be taken as $c_1 = (2 + \sqrt{2}/\pi + 1/4\pi^2)^{1/2}$ in our space-periodic case, see [18, Appendix].

As a classical and important model, the 2D NS equation (10.1) has been much investigated. The following results are well-known, see, e.g., [27, 29].

Lemma 10.2 *Let $f \in V^*$, the dual of V. Then the 2D Navier–Stokes equation* (10.4) *has a unique weak solution*

$$u \in C^0([0, \infty); H) \cap L^2_{loc}(0, \infty; V)$$

which is continuous in initial data u_0 in H.

Lemma 10.3 *Let $f \in H$. Then the dynamical system generated by the 2D Navier–Stokes equation* (10.4) *has a bounded absorbing set in V and a compact global attractor $\mathscr{A}$ in H.*

10.2.2 Small Grashof Number and the Singleton Attractor

Consider the non-dimensional Grashof number G which measures the intensity of the force f against the kinematic viscosity ν

$$G = \frac{|f|}{\nu^2 \lambda}.$$

This number was introduced by Foias et al. [18] in a study of determining modes of 2D Navier–Stokes equations. It is closely related to the so-called Reynolds number, see Temam [29],

$$Re = \frac{|f|^{1/2}}{\nu \lambda^{1/2}}.$$

In this section, we consider small forcing intensity, setting

$$G = \frac{|f|}{\nu^2 \lambda} < \frac{1}{c_1}, \tag{10.6}$$

or, equivalently,

$$\varepsilon_* := \nu \lambda - \frac{c_1^2 |f|^2}{\nu^3 \lambda} > 0, \tag{10.7}$$

where c_1 is the constant involved in inequality (10.5). We will show that under the condition (10.6) the global attractor $\mathscr{A}$ of the NS equation (10.1) is a singleton. We begin with a useful lemma.

Lemma 10.4 *For any bounded set $B \subset V$ and $\varepsilon > 0$ there exists a $T_{B,\varepsilon} > 0$ such that any solution u with initial data in B satisfies*

$$|A^{\frac{1}{2}} u(t)|^2 \leqslant \frac{|f|^2}{\nu^2 \lambda} + \varepsilon, \quad \forall t \geqslant T_{B,\varepsilon}.$$

Proof Taking the inner product of (10.1) with Au we have

$$\frac{1}{2}\frac{\mathrm{d}}{\mathrm{d}t}|A^{\frac{1}{2}}u|^2 + \nu|Au|^2 = (f, Au) \leqslant \frac{1}{2\nu}|f|^2 + \frac{\nu}{2}|Au|^2,$$

where we have used the fact that $b(u, u, Au) = 0$ since we have 2D periodic boundary conditions (cf. [29, Lemma 3.1]). This along with Poincaré's inequality (10.3) gives

$$\frac{\mathrm{d}}{\mathrm{d}t}|A^{\frac{1}{2}}u|^2 + \nu\lambda|A^{\frac{1}{2}}u|^2 \leqslant \frac{\mathrm{d}}{\mathrm{d}t}|A^{\frac{1}{2}}u|^2 + \nu|Au|^2 \leqslant \frac{1}{\nu}|f|^2.$$

Then by Gronwall's lemma we have

$$|A^{\frac{1}{2}}u(t)|^2 \leqslant e^{-\nu\lambda t}|A^{\frac{1}{2}}u(0)|^2 + \int_0^t e^{\nu\lambda(\eta-t)}\frac{|f|^2}{\nu}\,\mathrm{d}\eta \leqslant e^{-\nu\lambda t}|A^{\frac{1}{2}}u(0)|^2 + \frac{|f|^2}{\nu^2\lambda},$$

from which the lemma follows. □

Denote the difference of two weak solutions by $w = u - v$. Then

$$\frac{\mathrm{d}w}{\mathrm{d}t} + \nu Aw = -B(w, u) - B(v, w). \tag{10.8}$$

Lemma 10.5 *Let* (10.6) *hold. For any bounded set* $B \subset V$ *there exists a* $T = T_B > 0$ *such that the difference* $w(t) = u(t) - v(t)$ *of any two solutions with initial data in* B *satisfies*

$$|w(t)|^2 \leqslant e^{-\frac{\varepsilon_*}{2}(t-T)}|w(T)|^2, \quad \forall t \geqslant T,$$

where ε_* *is the constant given by* (10.7).

Proof Taking the inner product of (10.8) with w, by $b(v, w, w) = 0$ and (10.5) we obtain

$$\frac{1}{2}\frac{\mathrm{d}}{\mathrm{d}t}|w|^2 + \nu|A^{\frac{1}{2}}w|^2 = -b(w, u, w) \leqslant c_1|w||A^{\frac{1}{2}}w||A^{\frac{1}{2}}u| \leqslant \frac{\nu}{2}|A^{\frac{1}{2}}w|^2 + \frac{c_1^2}{2\nu}|w|^2|A^{\frac{1}{2}}u|^2,$$

which by Poincaré inequality is reformulated as

$$\frac{\mathrm{d}}{\mathrm{d}t}|w|^2 + \left(\nu\lambda - \frac{c_1^2}{\nu}|A^{\frac{1}{2}}u|^2\right)|w|^2 \leqslant \frac{\mathrm{d}}{\mathrm{d}t}|w|^2 + \nu|A^{\frac{1}{2}}w|^2 - \frac{c_1^2}{\nu}|A^{\frac{1}{2}}u|^2|w|^2 \leqslant 0.$$

Since, by (10.7), $\varepsilon_* = \nu\lambda - \frac{c_1^2|f|^2}{\nu^3\lambda} > 0$, from Lemma 10.4 there is a $T = T_{B,\frac{\nu\varepsilon_*}{c_1^2}} > 0$ such that

$$\left(\nu\lambda - \frac{c_1^2}{\nu}|A^{\frac{1}{2}}u(t)|^2\right) \geqslant \frac{\varepsilon_*}{2}, \quad \forall t \geqslant T. \tag{10.9}$$

Hence, we have

$$\frac{\mathrm{d}}{\mathrm{d}t}|w|^2 + \frac{\varepsilon_*}{2}|w|^2 \leqslant 0, \quad \forall t \geqslant T,$$

and the lemma follows from Gronwall's lemma. □

Theorem 10.1 *Let* (10.6) *hold. Then the global attractor* $\mathscr{A}$ *of 2D Navier–Stokes equation* (10.1) *is a singleton* $\mathscr{A} = \{a_*\}$ *with* $a_* \in D(A)$.

Proof (*Sketch of proof*) Take arbitrarily two points $x, y \in \mathscr{A}$. Since the attractor $\mathscr{A}$ is bounded in V and consists of bounded complete trajectories in V, there exist two complete trajectories ξ and η such that $x = \xi(0)$ and $y = \eta(0)$. Hence, by Lemma 10.5 we see that there is a $T = T_{\mathscr{A}} > 0$ such that

$$
\begin{aligned}
|x-y|^2 = |\xi(0)-\eta(0)|^2 &= |u(t,\xi(-t)) - u(t,\eta(-t))|^2 \\
&\leqslant e^{-\frac{\varepsilon_*}{2}(t-T)}|u(T,\xi(-t)) - u(T,\eta(-t))|^2 \\
&\leqslant e^{-\frac{\varepsilon_*}{2}(t-T)}[\text{Diameter }(\mathscr{A})]^2 \to 0, \quad \text{as } T < t \to \infty.
\end{aligned}
$$

□

Corollary 10.1 *The global attractor* $\mathscr{A}$ *of 2D Navier–Stokes equation* (10.1) *with* $f = 0$ *is a singleton* $\mathscr{A} = \{0\}$.

10.3 Additive Noise Case

In the following two sections we shall study the 2D NS equation endowed with additive and multiplicative white noises, respectively. The existence of the random attractors for these situations have been established by quite many researchers, see, e.g., [13, 17], and the upper semi-continuity for theory and applications see, e.g., [6, 28, 30]. Now we are interested in the convergence rate of the corresponding random attractors towards the deterministic singleton attractor as the noise intensity approaches zero.

10.3.1 Preliminaries

Let us endow the 2D Navier–Stokes equation (10.1) with additive white noise, considering

$$
\begin{aligned}
\frac{du_\varepsilon}{dt} - \nu\Delta u_\varepsilon + (u_\varepsilon \cdot \nabla)u_\varepsilon + \nabla p = f(x) + \varepsilon h(x)\frac{dW}{dt}, \\
\nabla \cdot u = 0.
\end{aligned}
\tag{10.10}
$$

As the deterministic Eq. (10.1), initial and periodic boundary condition (10.2) is considered. The term $\varepsilon h(x)$ denotes the perturbation intensity with $\varepsilon \in (0, 1]$,

$h(x) \in D(A)$. $W(t, \omega)$ is the standard scalar Wienner process on probability space $(\Omega, \mathfrak{F}, \mathscr{P})$, where

$$\Omega = \{\omega \in C(\mathbb{R}; \mathbb{R}) : \omega(0) = 0\},$$

endowed with the compact-open topology given by the complete metric

$$d_\Omega(\omega, \omega') := \sum_{n=1}^{\infty} \frac{1}{2^n} \frac{\|\omega - \omega'\|_n}{1 + \|\omega - \omega'\|_n}, \quad \|\omega - \omega'\|_n := \sup_{-n \leqslant t \leqslant n} |\omega(t) - \omega'(t)|,$$

and $\mathfrak{F}$ the Borel sigma-algebra induced by the compact-open topology of Ω, $\mathscr{P}$ the two-sided Wiener measure on $(\Omega, \mathfrak{F})$. It is widely used the translation-operator group $\{\theta_t\}_{t\in\mathbb{R}}$ on Ω defined by

$$\theta_t \omega = \omega(\cdot + t) - \omega(t), \quad \forall t \in \mathbb{R}, \omega \in \Omega,$$

under which the measure $\mathscr{P}$ is ergodic and invariant [17]. This operator is usually known as Wiener shift operator.

Consider for some $\alpha > 0$ (which will be specified latter such that (10.28) holds)

$$z(\omega) = -\int_{-\infty}^{0} e^{\alpha\tau} \mathrm{d}W(\tau), \quad \omega \in \Omega, \tag{10.11}$$

which is a stationary solution of the one-dimensional Ornstein-Uhlenbeck equation

$$\mathrm{d}z(\theta_t\omega) + \alpha z(\theta_t\omega)\mathrm{d}t = \mathrm{d}W(t). \tag{10.12}$$

It is known from [1, 12, 16] that there exists a θ-invariant subset $\tilde{\Omega} \subset \Omega$ of full measure such that $z(\theta_t\omega)$ is continuous in t for every $\omega \in \tilde{\Omega}$, and

$$\mathbb{E}\left(e^{\delta \int_s^{s+t} |z(\theta_r\omega)|\, \mathrm{d}r}\right) \leqslant e^{\frac{\delta t}{\sqrt{\alpha}}}, \quad \forall s \in \mathbb{R},\ \alpha^3 \geqslant \delta^2 \geqslant 0,\ t \geqslant 0, \tag{10.13}$$

$$\mathbb{E}\left(|z(\theta_s\omega)|^r\right) = \frac{\Gamma\left(\frac{1+r}{2}\right)}{\sqrt{\pi\alpha^r}}, \quad \forall r > 0,\ s \in \mathbb{R}, \tag{10.14}$$

$$\lim_{t\to\pm\infty} \frac{|z(\theta_t\omega)|}{|t|} = 0, \quad \lim_{t\to\pm\infty} \frac{1}{t}\int_0^t z(\theta_t\omega)\, \mathrm{d}s = 0, \tag{10.15}$$

$$\lim_{t\to\infty} e^{-\delta t}|z(\theta_{-t}\omega)| = 0, \quad \forall \delta > 0, \tag{10.16}$$

where Γ is the Gamma function. Hereafter, we will not distinguish between $\tilde{\Omega}$ and Ω.

Property (10.16) is called the tempered property of the random variable $z(\omega)$ in the literature. Analogously, for a random set $D(\omega)$ in Banach space X, i.e., a nonempty set-valued mapping from Ω to $2^X \setminus \emptyset$, $\omega \mapsto D(\omega)$, which is measurable in the sense that for any $x \in X$ the distance $\mathrm{dist}(x, D(\omega))$ is $(\mathfrak{F}, \mathscr{B}(\mathbb{R}))$-measurable in ω, is called tempered if for any $\delta > 0$

$$\lim_{t\to\infty} e^{-\delta t}\|D(\theta_{-t}\omega)\|_X = \lim_{t\to\infty} e^{-\delta t}\Big(\sup_{x\in D(\theta_{-t}\omega)} \|x\|_X\Big) = 0.$$

To study the random attractor of the NS equation we define $\mathscr{D}$ as the class of tempered random sets in H, i.e.,

$$\mathscr{D} = \big\{D:\ D \text{ is a tempered random set in } H\big\}.$$

The following lemma will be useful.

Lemma 10.6 *For any $T, k > 0$,*

$$\lim_{t\to\infty} \frac{\int_{-t}^{T-t} |z(\theta_s\omega)|^k\, \mathrm{d}s}{t-T} = 0, \quad \forall \omega \in \Omega.$$

Proof By ergodic theory we have

$$\lim_{t\to\infty} \frac{1}{t}\int_{-t}^{0} |z(\theta_s\omega)|^k\, \mathrm{d}s = \mathbb{E}(|z|^k).$$

Hence,

$$\lim_{t\to\infty} \frac{\int_{-t}^{T-t} |z(\theta_s\omega)|^k\, \mathrm{d}s}{t-T} = \lim_{t\to\infty}\left(\frac{\int_{-t}^{0} |z(\theta_s\omega)|^k\, \mathrm{d}s}{t}\cdot\frac{t}{t-T} - \frac{\int_{T-t}^{0} |z(\theta_s\omega)|^k\, \mathrm{d}s}{t-T}\right) = 0.$$

10.3.2 Uniform Estimates of Solutions

Define $v_\varepsilon(t,\omega,v_0) = u_\varepsilon(t,\omega,u_0) - \varepsilon z(\theta_t\omega)h(x)$ with $v_{\varepsilon,0} = u_{\varepsilon,0} - \varepsilon z(\omega)h(x)$ where u_ε is the solution of (10.10). Then v_ε satisfies

$$\begin{cases} \dfrac{\mathrm{d}v_\varepsilon}{\mathrm{d}t} + \nu A v_\varepsilon + B(v_\varepsilon + \varepsilon z(\theta_t\omega)h, v_\varepsilon + \varepsilon z(\theta_t\omega)h) + \nabla p \\ \qquad = f(x) + \varepsilon\alpha z(\theta_t\omega)h - \varepsilon\nu z(\theta_t\omega)Ah, \\ \nabla\cdot v_\varepsilon = 0. \end{cases} \tag{10.17}$$

Now we obtain some careful estimates of the solutions.

Lemma 10.7 *Let $f, v_\varepsilon(0) \in H$. Then for any $t \geqslant T > 0$, the solution v_ε of* (10.17) *satisfies*

$$
\begin{aligned}
&|v_\varepsilon(T, \theta_{-t}\omega, v_\varepsilon(0))|^2 + \int_0^T e^{\int_T^\eta (\frac{\varepsilon_*}{8} - \varepsilon^2 c_3 |z(\theta_{s-t}\omega)|^2) \mathrm{d}s} |A^{\frac{1}{2}} v_\varepsilon(\eta, \theta_{-t}\omega, v_\varepsilon(0))|^2 \, \mathrm{d}\eta \\
&\quad \leqslant c e^{-\frac{\varepsilon_*}{8} T + \varepsilon^2 c_3 \int_{-t}^{T-t} |z(\theta_s\omega)|^2 \mathrm{d}s} |v_\varepsilon(0)|^2 \\
&\qquad + e^{\frac{\varepsilon_*}{8}(t-T)} \int_{-t}^{T-t} c e^{\frac{\varepsilon_*}{8}\eta + \varepsilon^2 c_3 \int_\eta^{T-t} |z(\theta_s\omega)|^2 \mathrm{d}s} \left(\varepsilon^2 |z(\theta_\eta\omega)|^4 + 1\right) \mathrm{d}\eta, \qquad (10.18)
\end{aligned}
$$

where $c_3 := \max\left\{16\lambda c_1^2 |A^{\frac{1}{2}} h|^2/\varepsilon_*, 8c_2^2 |Ah|^2/\nu\right\}$, *and* $c > 0$ *is an absolute constant.*

Proof Taking the inner product of (10.17) with v_ε we have

$$
\begin{aligned}
&\frac{1}{2}\frac{\mathrm{d}}{\mathrm{d}t}|v_\varepsilon|^2 + \nu |A^{\frac{1}{2}} v_\varepsilon|^2 + b(v_\varepsilon + \varepsilon z(\theta_t\omega)h, v_\varepsilon + \varepsilon z(\theta_t\omega)h, v_\varepsilon) \\
&\quad = (f, v_\varepsilon) + (\varepsilon\alpha z(\theta_t\omega)h, v_\varepsilon) - (\varepsilon\nu z(\theta_t\omega)Ah, v_\varepsilon) \qquad (10.19) \\
&\quad \leqslant c|f|^2 + \varepsilon^2 c |z(\theta_t\omega)|^2 + \frac{\varepsilon_*}{64}|v_\varepsilon|^2.
\end{aligned}
$$

Note that

$$
\begin{aligned}
&|b(v_\varepsilon + \varepsilon z(\theta_t\omega)h, v_\varepsilon + \varepsilon z(\theta_t\omega)h, v_\varepsilon)| = |b(v_\varepsilon + \varepsilon z(\theta_t\omega)h, \varepsilon z(\theta_t\omega)h, v_\varepsilon)| \\
&\quad \leqslant |b(v_\varepsilon, \varepsilon z(\theta_t\omega)h, v_\varepsilon)| + \varepsilon^2 c |z(\theta_t\omega)|^4 + \frac{\varepsilon_*}{64}|v_\varepsilon|^2 \\
&\quad \leqslant \varepsilon \frac{c_1}{\sqrt{\lambda}} |A^{\frac{1}{2}} h| |z(\theta_t\omega)| |A^{\frac{1}{2}} v_\varepsilon|^2 + \varepsilon^2 c |z(\theta_t\omega)|^4 + \frac{\varepsilon_*}{64}|v_\varepsilon|^2 \\
&\quad \leqslant \left(\frac{\varepsilon^2 16\lambda c_1^2 |A^{\frac{1}{2}} h|^2 |z(\theta_t\omega)|^2}{2\lambda\varepsilon_*} + \frac{\varepsilon_*}{32\lambda}\right) |A^{\frac{1}{2}} v_\varepsilon|^2 + \varepsilon^2 c |z(\theta_t\omega)|^4 + \frac{\varepsilon_*}{64}|v_\varepsilon|^2 \\
&\quad \leqslant \left(\frac{\varepsilon^2 c_3 |z(\theta_t\omega)|^2}{2\lambda} + \frac{\varepsilon_*}{32\lambda}\right) |A^{\frac{1}{2}} v_\varepsilon|^2 + \varepsilon^2 c |z(\theta_t\omega)|^4 + \frac{\varepsilon_*}{64}|v_\varepsilon|^2,
\end{aligned}
$$

where, for latter purpose, c_3 is a constant defined by

$$
c_3 := \max\left\{\frac{16\lambda c_1^2 |A^{\frac{1}{2}} h|^2}{\varepsilon_*}, \frac{8c_2^2 |Ah|^2}{\nu}\right\}. \qquad (10.20)
$$

Hence, we have

$$
\begin{aligned}
&\frac{1}{2}\frac{\mathrm{d}}{\mathrm{d}t}|v_\varepsilon|^2 + \left(\nu - \frac{\varepsilon_*}{8\lambda} + \frac{\varepsilon_*}{8\lambda}\right)|A^{\frac{1}{2}} v_\varepsilon|^2 - \left(\frac{\varepsilon^2 c_3 |z(\theta_t\omega)|^2}{2\lambda} + \frac{\varepsilon_*}{32\lambda}\right) |A^{\frac{1}{2}} v_\varepsilon|^2 - \frac{\varepsilon_*}{32}|v_\varepsilon|^2 \\
&\quad \leqslant c|f|^2 + \varepsilon^2 c |z(\theta_t\omega)|^2 + \varepsilon^2 c |z(\theta_t\omega)|^4,
\end{aligned}
$$

and then, by Poincaré's inequality,

$$\begin{aligned}&\frac{\mathrm{d}}{\mathrm{d}t}|v_\varepsilon|^2+\Big(2\nu-\frac{\varepsilon_*}{4\lambda}\Big)|A^{\frac12}v_\varepsilon|^2+\Big(\frac{\varepsilon_*}{8}-\varepsilon^2c_3|z(\theta_t\omega)|^2\Big)|v_\varepsilon|^2\\&\leqslant c|f|^2+\varepsilon^2c\left(|z(\theta_t\omega)|^4+1\right).\end{aligned}\tag{10.21}$$

Applying Gronwall's lemma to (10.21) we obtain

$$\begin{aligned}&|v_\varepsilon(t)|^2+\Big(2\nu-\frac{\varepsilon_*}{4\lambda}\Big)\int_0^t e^{\int_t^\eta(\frac{\varepsilon_*}{8}-\varepsilon^2c_3|z(\theta_s\omega)|^2)\mathrm{d}s}|A^{\frac12}v_\varepsilon(\eta)|^2\,\mathrm{d}\eta\\&\leqslant e^{-\int_0^t(\frac{\varepsilon_*}{8}-\varepsilon^2c_3|z(\theta_s\omega)|^2)\mathrm{d}s}|v_\varepsilon(0)|^2+\int_0^t ce^{\int_t^\eta(\frac{\varepsilon_*}{8}-\varepsilon^2c_3|z(\theta_s\omega)|^2)\mathrm{d}s}\left(\varepsilon^2|z(\theta_\eta\omega)|^4+1\right)\mathrm{d}\eta.\end{aligned}$$

Therefore, for any $t\geqslant T>0$,

$$\begin{aligned}&|v_\varepsilon(T,\theta_{-t}\omega,v_\varepsilon(0))|^2\\&\quad+\Big(2\nu-\frac{\varepsilon_*}{4\lambda}\Big)\int_0^T e^{\int_T^\eta(\frac{\varepsilon_*}{8}-\varepsilon^2c_3|z(\theta_{s-t}\omega)|^2)\mathrm{d}s}|A^{\frac12}v_\varepsilon(\eta,\theta_{-t}\omega,v_\varepsilon(0))|^2\,\mathrm{d}\eta\\&\leqslant e^{-\int_0^T(\frac{\varepsilon_*}{8}-\varepsilon^2c_3|z(\theta_{s-t}\omega)|^2)\mathrm{d}s}|v_\varepsilon(0)|^2\\&\quad+\int_0^T ce^{\int_T^\eta(\frac{\varepsilon_*}{8}-\varepsilon^2c_3|z(\theta_{s-t}\omega)|^2)\mathrm{d}s}\left(\varepsilon^2|z(\theta_{\eta-t}\omega)|^4+1\right)\mathrm{d}\eta\\&=e^{-\frac{\varepsilon_*}{8}T+\varepsilon^2c_3\int_{-t}^{T-t}|z(\theta_s\omega)|^2\mathrm{d}s}|v_\varepsilon(0)|^2\\&\quad+e^{\frac{\varepsilon_*}{8}(t-T)}\int_{-t}^{T-t}ce^{\frac{\varepsilon_*}{8}\eta+\varepsilon^2c_3\int_\eta^{T-t}|z(\theta_s\omega)|^2\mathrm{d}s}\left(\varepsilon^2|z(\theta_\eta\omega)|^4+1\right)\mathrm{d}\eta.\end{aligned}$$

Since $2\nu-\frac{\varepsilon_*}{4\lambda}>0$, the lemma follows. □

Lemma 10.8 *Let $f,v_\varepsilon(0)\in H$. Then for any $t\geqslant\eta>0$, the solution v_ε of* (10.17) *satisfies*

$$\begin{aligned}&|A^{\frac12}v_\varepsilon(\eta,\theta_{-t}\omega,v_\varepsilon(0))|^2\leqslant ce^{-\frac{\varepsilon_*}{8}\eta+\varepsilon^2c_3\int_{-t}^{\eta-t}|z(\theta_s\omega)|^2\mathrm{d}s}|v_\varepsilon(0)|^2\\&\quad+e^{\frac{\varepsilon_*}{8}(t-\eta)}\int_{-t}^{\eta-t}ce^{\frac{\varepsilon_*}{8}\eta+\varepsilon^2c_3\int_\xi^{\eta-t}|z(\theta_s\omega)|^2\mathrm{d}s}\Big(\varepsilon^4|z(\theta_\xi\omega)|^4+1\Big)\mathrm{d}\xi,\end{aligned}\tag{10.22}$$

where c_3 is the constant given by (10.20)*, and c is an absolute constant.*

Proof Take the inner product of (10.17) with Av_ε to obtain

$$\begin{aligned}&\frac12\frac{\mathrm{d}}{\mathrm{d}t}|A^{\frac12}v_\varepsilon|^2+\nu|Av_\varepsilon|^2+b(v_\varepsilon+\varepsilon z(\theta_t\omega)h,v_\varepsilon+\varepsilon z(\theta_t\omega)h,Av_\varepsilon)\\&\quad=(f,Av_\varepsilon)+(\varepsilon\alpha z(\theta_t\omega)h,Av_\varepsilon)-(\varepsilon\nu z(\theta_t\omega)Ah,Av_\varepsilon)\\&\quad\leqslant\frac{\nu}{8}|Av_\varepsilon|^2+c\left(1+\varepsilon^2|z(\theta_t\omega)|^2\right).\end{aligned}$$

Since $b(u_\varepsilon,u_\varepsilon,Au_\varepsilon)=0$, by (10.5) we have

$$
\begin{aligned}
&|b(v_\varepsilon + \varepsilon z(\theta_t\omega)h, v_\varepsilon + \varepsilon z(\theta_t\omega)h, Av_\varepsilon)| \\
&\quad = |b(v_\varepsilon + \varepsilon z(\theta_t\omega)h, v_\varepsilon + \varepsilon z(\theta_t\omega)h, \varepsilon z(\theta_t\omega)Ah)| \\
&\quad \leqslant |b(v_\varepsilon, v_\varepsilon, \varepsilon z(\theta_t\omega)Ah)| + |b(\varepsilon z(\theta_t\omega)h, v_\varepsilon, \varepsilon z(\theta_t\omega)Ah)| \\
&\qquad + |b(v_\varepsilon, \varepsilon z(\theta_t\omega)h, \varepsilon z(\theta_t\omega)Ah)| + \varepsilon^3 c|z(\theta_t\omega)|^3 \\
&\quad \leqslant c_2|A^{\frac{1}{2}}v_\varepsilon||Av_\varepsilon||\varepsilon z(\theta_t\omega)Ah| + \varepsilon^2 c|z(\theta_t\omega)|^2|Av_\varepsilon| \\
&\qquad + \varepsilon^2 c_0|z(\theta_t\omega)|^2|A^{\frac{1}{2}}v_\varepsilon| + \varepsilon^3 c|z(\theta_t\omega)|^3 \\
&\quad \leqslant \frac{\nu}{8}|Av_\varepsilon|^2 + \frac{\varepsilon^2 8c_2^2|Ah|^2}{2\nu}|z(\theta_t\omega)|^2|A^{\frac{1}{2}}v_\varepsilon|^2 + \frac{c_4}{2}|A^{\frac{1}{2}}v_\varepsilon|^2 + \varepsilon^4 c|z(\theta_t\omega)|^4 + c \\
&\quad \leqslant \frac{\nu}{8}|Av_\varepsilon|^2 + \frac{\varepsilon^2 c_3}{2}|z(\theta_t\omega)|^2|A^{\frac{1}{2}}v_\varepsilon|^2 + \frac{c_4}{2}|A^{\frac{1}{2}}v_\varepsilon|^2 + \varepsilon^4 c|z(\theta_t\omega)|^4 + c \ \text{(by (10.20))},
\end{aligned}
$$

where c_3 is given in (10.20) and for latter purpose c_4 is taken as

$$
c_4 := \nu\lambda - \frac{\varepsilon_*}{8} \left(= \frac{7\nu\lambda}{8} + \frac{c_1^2|f|^2}{8\nu^3\lambda} \right).
$$

Hence, as $\frac{\nu\lambda}{2} - \frac{c_4}{2} = \frac{\varepsilon_*}{16}$,

$$
\begin{aligned}
&\frac{1}{2}\frac{\mathrm{d}}{\mathrm{d}t}|A^{\frac{1}{2}}v_\varepsilon|^2 + \frac{\nu}{4}|Av_\varepsilon|^2 + \left(\frac{\varepsilon_*}{16} - \frac{\varepsilon^2 c_3}{2}|z(\theta_t\omega)|^2\right)|A^{\frac{1}{2}}v_\varepsilon|^2 \\
&\quad = \frac{1}{2}\frac{\mathrm{d}}{\mathrm{d}t}|A^{\frac{1}{2}}v_\varepsilon|^2 + \frac{\nu}{4}|Av_\varepsilon|^2 + \frac{\nu\lambda}{2}|A^{\frac{1}{2}}v_\varepsilon|^2 - \left(\frac{c_4}{2} + \frac{\varepsilon^2 c_3}{2}|z(\theta_t\omega)|^2\right)|A^{\frac{1}{2}}v_\varepsilon|^2 \\
&\quad \leqslant \varepsilon^4 c|z(\theta_t\omega)|^4 + c.
\end{aligned}
$$

Applying Gronwall's lemma, for any $t \geqslant \rho \geqslant 0$ we have

$$
\begin{aligned}
&|A^{\frac{1}{2}}v_\varepsilon(t)|^2 + \frac{\nu}{2}\int_\rho^t e^{\int_t^\eta (\frac{\varepsilon_*}{8} - \varepsilon^2 c_3|z(\theta_s\omega)|^2)\mathrm{d}s}|Av_\varepsilon(\eta)|^2\,\mathrm{d}\eta \\
&\quad \leqslant e^{\int_t^\rho (\frac{\varepsilon_*}{8} - \varepsilon^2 c_3|z(\theta_s\omega)|^2)\mathrm{d}s}|A^{\frac{1}{2}}v_\varepsilon(\rho)|^2 \\
&\qquad + \int_0^t c e^{\int_t^\eta (\frac{\varepsilon_*}{8} - \varepsilon^2 c_3|z(\theta_s\omega)|^2)\mathrm{d}s}\left(\varepsilon^4|z(\theta_\eta\omega)|^4 + 1\right)\mathrm{d}\eta.
\end{aligned}
$$

Hence, for any $t \geqslant \eta > \rho > 0$,

$$
\begin{aligned}
|A^{\frac{1}{2}}v_\varepsilon(\eta, \theta_{-t}\omega, v_\varepsilon(0))|^2 &\leqslant e^{\int_\eta^\rho (\frac{\varepsilon_*}{8} - \varepsilon^2 c_3|z(\theta_{s-t}\omega)|^2)\mathrm{d}s}|A^{\frac{1}{2}}v_\varepsilon(\rho, \theta_{-t}\omega, v_\varepsilon(0))|^2 \\
&\quad + \int_0^\eta c e^{\int_\eta^\xi (\frac{\varepsilon_*}{8} - \varepsilon^2 c_3|z(\theta_{s-t}\omega)|^2)\mathrm{d}s}\left(\varepsilon^4|z(\theta_{\xi-t}\omega)|^4 + 1\right)\mathrm{d}\xi.
\end{aligned}
$$

For $t \geqslant \eta > \rho > \eta - 1 \geqslant 0$, integrate the above inequality w.r.t. ρ over $(\eta - 1, \eta)$ to obtain

$$
\begin{aligned}
|A^{\frac{1}{2}} v_\varepsilon(\eta, \theta_{-t}\omega, v_\varepsilon(0))|^2 &\leqslant \int_{\eta-1}^{\eta} e^{\int_\eta^\rho (\frac{\varepsilon_*}{8} - \varepsilon^2 c_3 |z(\theta_{s-t}\omega)|^2) \mathrm{d}s} |A^{\frac{1}{2}} v_\varepsilon(\rho)|^2 \,\mathrm{d}\rho \\
&+ \int_0^\eta c e^{\int_\eta^\xi (\frac{\varepsilon_*}{8} - \varepsilon^2 c_3 |z(\theta_{s-t}\omega)|^2) \mathrm{d}s} \Big(\varepsilon^4 |z(\theta_{\xi-t}\omega)|^4 + 1 \Big) \mathrm{d}\xi \\
&\leqslant c e^{-\frac{\varepsilon_*}{8}\eta + \varepsilon^2 c_3 \int_{-t}^{\eta-t} |z(\theta_s\omega)|^2 \mathrm{d}s} |v_\varepsilon(0)|^2 \\
&+ e^{\frac{\varepsilon_*}{8}(t-\eta)} \int_{-t}^{\eta-t} c e^{\frac{\varepsilon_*}{8}\eta + \varepsilon^2 c_3 \int_\xi^{\eta-t} |z(\theta_s\omega)|^2 \mathrm{d}s} \Big(\varepsilon^4 |z(\theta_\xi\omega)|^4 + 1 \Big) \mathrm{d}\xi,
\end{aligned}
$$

where the last inequality is by (10.18). Hence, the lemma follows. □

10.3.3 Perturbation Radius of the Singleton Attractor Under Additive Noise

Consider the difference $w_\varepsilon = v_\varepsilon - u$ of the unique solutions v, u of (10.17) and (10.1), respectively. Then w_ε satisfies the following equation in the weak solution sense

$$
\begin{aligned}
\frac{\mathrm{d}w_\varepsilon}{\mathrm{d}t} + \nu A w_\varepsilon + B(u_\varepsilon, u_\varepsilon) - B(u, u) &= \varepsilon \alpha z(\theta_t\omega) h - \varepsilon \nu z(\theta_t\omega) A h, \\
\nabla \cdot w_\varepsilon &= 0.
\end{aligned}
\tag{10.23}
$$

Lemma 10.9 *Let $f \in H$. Suppose that v_ε and u are solutions of* (10.17) *and* (10.1) *corresponding to initial data $v_\varepsilon(0) \in D \in \mathscr{D}$ and u_0 with $|A^{\frac{1}{2}} u_0|^2 \leqslant R$, respectively. Then there exists a random variable $\gamma(\omega)$ depending only on D and R such that the difference w_ε of the two solutions satisfies*

$$
\limsup_{t\to\infty} |w_\varepsilon(t, \theta_{-t}\omega, w_\varepsilon(0))|^2 \leqslant \varepsilon^2 \gamma(\omega), \quad \forall \varepsilon \in (0, 1].
$$

Proof Take the inner product of (10.23) with w_ε in H to obtain

$$
\begin{aligned}
&\frac{1}{2}\frac{\mathrm{d}}{\mathrm{d}t} |w_\varepsilon|^2 + \nu |A^{\frac{1}{2}} w_\varepsilon|^2 \\
&\quad \leqslant |b(u_\varepsilon, u_\varepsilon, w_\varepsilon) - b(u, u, w_\varepsilon)| + \big(\varepsilon \alpha z(\theta_t\omega) h - \varepsilon \nu z(\theta_t\omega) A h, w_\varepsilon \big) \\
&\quad \leqslant |b(u_\varepsilon, u_\varepsilon, w_\varepsilon) - b(u, u, w_\varepsilon)| + \varepsilon^2 c |z(\theta_t\omega)|^2 + \frac{\varepsilon_*}{24\lambda} |A^{\frac{1}{2}} w_\varepsilon|^2.
\end{aligned}
$$

By (10.5) we have

$$
\begin{aligned}
&|b(u_\varepsilon, u_\varepsilon, w_\varepsilon) - b(u, u, w_\varepsilon)| \\
&\quad = |b(u_\varepsilon, \varepsilon z(\theta_t\omega) h, w_\varepsilon) + b(w_\varepsilon, u, w_\varepsilon) + b(\varepsilon z(\theta_t\omega) h, u, w_\varepsilon)|
\end{aligned}
$$

$$\leqslant c_1|u_\varepsilon|^{\frac{1}{2}}|A^{\frac{1}{2}}u_\varepsilon|^{\frac{1}{2}}|A^{\frac{1}{2}}h\varepsilon z(\theta_t\omega)||w_\varepsilon|^{\frac{1}{2}}|A^{\frac{1}{2}}w_\varepsilon|^{\frac{1}{2}}+c_1|w_\varepsilon||A^{\frac{1}{2}}w_\varepsilon||A^{\frac{1}{2}}u|$$
$$+c_1|\varepsilon z(\theta_t\omega)h|^{\frac{1}{2}}|A^{\frac{1}{2}}h\varepsilon z(\theta_t\omega)|^{\frac{1}{2}}|A^{\frac{1}{2}}u||w_\varepsilon|^{\frac{1}{2}}|A^{\frac{1}{2}}w_\varepsilon|^{\frac{1}{2}}.$$

Carefully estimating the three terms in the right-hand side using Young's inequality, we obtain

$$\begin{aligned}
&\left|b(u_\varepsilon,u_\varepsilon,w_\varepsilon)-b(u,u,w_\varepsilon)\right|\\
&\leqslant\left(\varepsilon^2c|z(\theta_t\omega)|^2|A^{\frac{1}{2}}u_\varepsilon|^2+\frac{\varepsilon_*}{24\lambda}|A^{\frac{1}{2}}w_\varepsilon|^2\right)+\left(\frac{c_1^2}{2\nu}|w_\varepsilon|^2|A^{\frac{1}{2}}u|^2+\frac{\nu}{2}|A^{\frac{1}{2}}w_\varepsilon|^2\right)\\
&\quad+\left(\varepsilon^2c|z(\theta_t\omega)|^2|A^{\frac{1}{2}}u|^2+\frac{\varepsilon_*}{24\lambda}|A^{\frac{1}{2}}w_\varepsilon|^2\right)\\
&=\varepsilon^2c|z(\theta_t\omega)|^2\left(|A^{\frac{1}{2}}u_\varepsilon|^2+|A^{\frac{1}{2}}u|^2\right)+\frac{c_1^2}{2\nu}|w_\varepsilon|^2|A^{\frac{1}{2}}u|^2+\left(\frac{\nu}{2}+\frac{\varepsilon_*}{12\lambda}\right)|A^{\frac{1}{2}}w_\varepsilon|^2.
\end{aligned}$$

Hence,

$$\begin{aligned}
&\frac{1}{2}\frac{\mathrm{d}}{\mathrm{d}t}|w_\varepsilon|^2+\left(\frac{\nu}{2}-\frac{\varepsilon_*}{8\lambda}\right)|A^{\frac{1}{2}}w_\varepsilon|^2\\
&\quad\leqslant\varepsilon^2c|z(\theta_t\omega)|^2\left(|A^{\frac{1}{2}}u_\varepsilon|^2+|A^{\frac{1}{2}}u|^2+1\right)+\frac{c_1^2}{2\nu}|w_\varepsilon|^2|A^{\frac{1}{2}}u|^2,
\end{aligned}$$

i.e.,

$$\begin{aligned}
&\frac{\mathrm{d}}{\mathrm{d}t}|w_\varepsilon|^2+\left(\nu-\frac{\varepsilon_*}{4\lambda}\right)|A^{\frac{1}{2}}w_\varepsilon|^2-\frac{c_1^2}{\nu}|w_\varepsilon|^2|A^{\frac{1}{2}}u|^2\\
&\quad\leqslant\varepsilon^2c|z(\theta_t\omega)|^2\left(|A^{\frac{1}{2}}u_\varepsilon(t)|^2+|A^{\frac{1}{2}}u(t)|^2+1\right),
\end{aligned}$$

and then, by Poincaré's inequality,

$$\begin{aligned}
&\frac{\mathrm{d}}{\mathrm{d}t}|w_\varepsilon|^2+\left(\nu\lambda-\frac{\varepsilon_*}{4}-\frac{c_1^2}{\nu}|A^{\frac{1}{2}}u|^2\right)|w_\varepsilon|^2\\
&\quad\leqslant\varepsilon^2c|z(\theta_t\omega)|^2\left(|A^{\frac{1}{2}}u_\varepsilon(t)|^2+|A^{\frac{1}{2}}u(t)|^2+1\right).
\end{aligned}\tag{10.24}$$

Now, consider big lapsing time $T=T_{B,\frac{\nu\varepsilon_*}{c_1^2}}>0$ such that (10.9) holds, i.e.,

$$\left(\nu\lambda-\frac{c_1^2}{\nu}|A^{\frac{1}{2}}u(t)|^2\right)\geqslant\frac{\varepsilon_*}{2},\quad\forall t\geqslant T.$$

Then from (10.24) it follows

$$\frac{\mathrm{d}}{\mathrm{d}t}|w_\varepsilon|^2+\frac{\varepsilon_*}{4}|w_\varepsilon|^2\leqslant\varepsilon^2c|z(\theta_t\omega)|^2\left(|A^{\frac{1}{2}}u_\varepsilon(t)|^2+1\right),\quad t\geqslant T.\tag{10.25}$$

Apply Gronwall's lemma to (10.25) over (T,t) to obtain

$$|w_\varepsilon(t,\omega,w_\varepsilon(0))|^2 \leqslant e^{-\frac{\varepsilon_*}{4}(t-T)}|w_\varepsilon(T,\omega,w_\varepsilon(0))|^2 + \varepsilon^2 c\int_T^t e^{\frac{\varepsilon_*}{4}(\eta-t)}|z(\theta_\eta\omega)|^2\Big(|A^{\frac12}u_\varepsilon(\eta)|^2+1\Big)\mathrm{d}\eta,\quad t>T,$$

so, replacing ω with $\theta_{-t}\omega$ we have

$$\begin{aligned}|w_\varepsilon(t,\theta_{-t}\omega,w_\varepsilon(0))|^2 &\leqslant e^{-\frac{\varepsilon_*}{4}(t-T)}|w_\varepsilon(T,\theta_{-t}\omega,w_\varepsilon(0))|^2\\ &\quad+\varepsilon^2 c\int_T^t e^{\frac{\varepsilon_*}{4}(\eta-t)}|z(\theta_{\eta-t}\omega)|^2|A^{\frac12}u_\varepsilon(\eta,\theta_{-t}\omega,u_\varepsilon(0))|^2\,\mathrm{d}\eta\\ &\quad+\varepsilon^2 c\int_{T-t}^0 e^{\frac{\varepsilon_*}{4}\eta}|z(\theta_\eta\omega)|^2\,\mathrm{d}\eta,\quad t>T.\end{aligned}\tag{10.26}$$

For the first term on the right-hand side of (10.26), by (10.18) we have

$$\begin{aligned}&e^{-\frac{\varepsilon_*}{4}(t-T)}|w_\varepsilon(T,\theta_{-t}\omega,w_\varepsilon(0))|^2 \leqslant e^{-\frac{\varepsilon_*}{4}(t-T)}\Big(|v_\varepsilon(T,\theta_{-t}\omega,v_\varepsilon(0))|^2+|u(T,u_0)|^2\Big)\\ &\leqslant e^{-\frac{\varepsilon_*}{4}(t-T)+\varepsilon c\int_{-t}^{T-t}|z(\theta_s\omega)|\mathrm{d}s}|v_\varepsilon(0)|^2\\ &\quad+e^{-\frac{\varepsilon_*}{8}(t-T)+\varepsilon c\int_{-t}^{T-t}|z(\theta_s\omega)|\mathrm{d}s}\int_{-\infty}^0 ce^{\frac{\varepsilon_*}{8}\eta}\left(\varepsilon^2|z(\theta_\eta\omega)|^4+1\right)\mathrm{d}\eta\\ &\quad+e^{-\frac{\varepsilon_*}{4}(t-T)}|u(T,u_0)|^2\to 0,\quad \text{as } t\to\infty,\end{aligned}\tag{10.27}$$

where Lemma 10.6 and tempered properties were used in taking the limit.

Now we consider the second term on the right-hand side of (10.26). By (10.22) we have

$$\begin{aligned}&\int_T^t e^{\frac{\varepsilon_*}{4}(\eta-t)}|z(\theta_{\eta-t}\omega)|^2|A^{\frac12}v_\varepsilon(\eta,\theta_{-t}\omega,v_\varepsilon(0))|^2\,\mathrm{d}\eta\\ &\leqslant\int_T^t ce^{\frac{\varepsilon_*}{4}(\eta-t)}|z(\theta_{\eta-t}\omega)|^2\Bigg[e^{-\frac{\varepsilon_*}{8}\eta+\varepsilon^2c_3\int_{-t}^{\eta-t}|z(\theta_s\omega)|^2\mathrm{d}s}|v_\varepsilon(0)|^2\\ &\quad+e^{-\frac{\varepsilon_*}{8}(\eta-t)}\int_{-t}^{\eta-t}ce^{\frac{\varepsilon_*}{8}\xi+\varepsilon^2c_3\int_\xi^{\eta-t}|z(\theta_s\omega)|^2\mathrm{d}s}\Big(\varepsilon^4|z(\theta_\xi\omega)|^4+1\Big)\mathrm{d}\xi\Bigg]\mathrm{d}\eta\quad\text{(by (10.22))}\\ &=e^{-\frac{\varepsilon_*}{8}t}\int_{T-t}^0 e^{\frac{\varepsilon_*}{8}\eta+\varepsilon^2c_3\int_{-t}^{\eta}|z(\theta_s\omega)|^2\mathrm{d}s}|z(\theta_\eta\omega)|^2|v_\varepsilon(0)|^2\,\mathrm{d}\eta\\ &\quad+\int_{T-t}^0 e^{\frac{\varepsilon_*}{8}\eta}|z(\theta_\eta\omega)|^2\int_{-t}^{\eta}ce^{\frac{\varepsilon_*}{8}\xi+\varepsilon^2c_3\int_\xi^{\eta}|z(\theta_s\omega)|^2\mathrm{d}s}\Big(\varepsilon^4|z(\theta_\xi\omega)|^4+1\Big)\mathrm{d}\xi\mathrm{d}\eta\\ &\leqslant e^{-\frac{\varepsilon_*}{8}t+c_3\int_{-t}^0|z(\theta_s\omega)|^2\mathrm{d}s}\int_{T-t}^0 e^{\frac{\varepsilon_*}{8}\eta}|z(\theta_\eta\omega)|^2|v_\varepsilon(0)|^2\,\mathrm{d}\eta\\ &\quad+\int_{T-t}^0 e^{\frac{\varepsilon_*}{8}\eta}|z(\theta_\eta\omega)|^2\mathrm{d}\eta\int_{-t}^0 ce^{\frac{\varepsilon_*}{8}\xi+c_3\int_\xi^0|z(\theta_s\omega)|^2\mathrm{d}s}\Big(|z(\theta_\xi\omega)|^4+1\Big)\mathrm{d}\xi.\end{aligned}$$

By (10.14) we take α large enough such that

$$c_3\mathbb{E}(|z|^2) < \frac{\varepsilon_*}{8} \quad \left(\text{i.e., } \alpha > \frac{8c_3\Gamma(\frac{3}{2})}{\varepsilon_*\sqrt{\pi}}\right). \tag{10.28}$$

Then by ergodic theory and the temperedness of $s \mapsto |z(\theta_s\omega)|^k$ as $s \to -\infty$, $k \in \mathbb{N}$, for the second term we have

$$\limsup_{t\to\infty} \int_T^t e^{\frac{\varepsilon_*}{4}(\eta-t)}|z(\theta_{\eta-t}\omega)|^2|A^{\frac{1}{2}}v_\varepsilon(\eta,\theta_{-t}\omega,v_\varepsilon(0))|^2\,\mathrm{d}\eta \leqslant r(\omega), \tag{10.29}$$

where $r(\omega)$ is the random variable (independent of ε) given by

$$r(\omega) := c\int_{-\infty}^0 e^{\frac{\varepsilon_*}{8}\eta}|z(\theta_\eta\omega)|^2\mathrm{d}\eta \int_{-\infty}^0 e^{\frac{\varepsilon_*}{8}\xi + c_3\int_\xi^0 |z(\theta_s\omega)|^2\mathrm{d}s}\Big(|z(\theta_\xi\omega)|^4+1\Big)\mathrm{d}\xi.$$

Hence, by (10.26), (10.27) and (10.29) we conclude the lemma. □

Theorem 10.2 *Let $f \in H$. There exists a random variable $R(\omega)$ (independent of ε) such that for every $\varepsilon \in (0, 1]$ the additive noise random attractor $\mathscr{A}_\varepsilon$ and the deterministic attractor $\mathscr{A}$ satisfy*

$$\mathrm{dist}_H(\mathscr{A}_\varepsilon(\omega), \mathscr{A}) \leqslant \varepsilon R(\omega), \quad \forall \omega \in \Omega,$$

where $\mathrm{dist}_H(A, B) = \max\{\mathrm{dist}(A, B), \mathrm{dist}(B, A)\}$.

Proof Take arbitrarily $a(\omega) \in \mathscr{A}_\varepsilon(\omega)$. Then by the invariance of the random attractor $a(\omega) = v(t, \theta_{-t}\omega, a(\theta_{-t}\omega))$ for some $a(\theta_{-t}\omega) \in \mathscr{A}(\theta_{-t}\omega)$, $t \geqslant 0$. Hence, since the deterministic attractor $\mathscr{A} = \{a_*\}$ is a singleton, we have

$$\|a(\omega) - a_*\| = \|v(t, \theta_{-t}\omega, a(\theta_{-t}\omega)) - u(t, a_*)\|, \quad \forall t \geqslant 0.$$

Hence, by Lemma 10.9, the theorem follows by taking the limit as $t \to \infty$. □

10.4 Multiplicative Noise Case

10.4.1 Preliminaries

Consider the stochastic 2D Navier–Stokes equation with multiplicative white noise

$$\begin{aligned} \frac{\mathrm{d}u_\varepsilon}{\mathrm{d}t} - \nu\Delta u_\varepsilon + (u_\varepsilon\cdot\nabla)u_\varepsilon + \nabla p = f(x) + \varepsilon u_\varepsilon \circ \frac{\mathrm{d}W}{\mathrm{d}t}&, \\ \nabla\cdot u_\varepsilon = 0&, \end{aligned} \tag{10.30}$$

endowed also with the periodic boundary condition (10.2), where $f \in H$ and $\circ$ represents the Stratonovich integral. We also make use of the random variable z

defined by (10.11) with $\alpha > 0$, satisfying (10.12), i.e.,

$$\mathrm{d}z(\theta_t\omega) + \alpha z(\theta_t\omega)\mathrm{d}t = \mathrm{d}W(t).$$

Let $v_\varepsilon(t) = e^{-\varepsilon z(\theta_t\omega)}u_\varepsilon(t)$. Then

$$\begin{aligned}\frac{\mathrm{d}v_\varepsilon}{\mathrm{d}t} - \nu\Delta v_\varepsilon + e^{\varepsilon z(\theta_t\omega)}(v_\varepsilon\cdot\nabla)v_\varepsilon + \nabla p e^{-\varepsilon z(\theta_t\omega)} = f e^{-\varepsilon z(\theta_t\omega)} + \varepsilon\alpha z(\theta_t\omega)v_\varepsilon,\\ \nabla\cdot v_\varepsilon = 0.\end{aligned} \tag{10.31}$$

10.4.2 Uniform Estimates of Solutions

We first derive some estimates which will be useful later.

Lemma 10.10 *Let $f \in H$. Then for any $\varepsilon \in (0, 1]$ the solution v_ε of (10.31) with initial data $v_\varepsilon(0) \in H$ satisfies*

$$\begin{aligned}|A^{\frac{1}{2}}v_\varepsilon(s, \theta_{-t}\omega, v_\varepsilon(0))|^2 &\leqslant ce^{-\frac{\varepsilon_*}{8}s+\int_{-t}^{s-t}2\varepsilon\alpha z(\theta_\eta\omega)\mathrm{d}\eta}|v_\varepsilon(0)|^2\\ &+ ce^{\frac{\varepsilon_*}{8}(t-s)}\int_{-t}^{s-t} e^{\frac{\varepsilon_*}{8}\eta-\int_{s-t}^{\eta}2\varepsilon\alpha z(\theta_\xi\omega)\mathrm{d}\xi-2\varepsilon z(\theta_\eta\omega)}\mathrm{d}\eta, \quad t \geqslant s > 1.\end{aligned} \tag{10.32}$$

Proof Taking the inner product of (10.31) with v_ε we have

$$\begin{aligned}\frac{1}{2}\frac{\mathrm{d}}{\mathrm{d}t}|v_\varepsilon|^2 + \left(\frac{\nu}{4}+\frac{3\nu}{4}\right)|A^{\frac{1}{2}}v_\varepsilon|^2 &= e^{-\varepsilon z(\theta_t\omega)}(f, v_\varepsilon) + \varepsilon\alpha z(\theta_t\omega)|v_\varepsilon|^2\\ &\leqslant e^{-2\varepsilon z(\theta_t\omega)}\frac{|f|^2}{\nu\lambda} + \frac{\nu\lambda}{4}|v_\varepsilon|^2 + \varepsilon\alpha z(\theta_t\omega)|v_\varepsilon|^2.\end{aligned}$$

Hence, by Poincaré's inequality,

$$\frac{\mathrm{d}}{\mathrm{d}t}|v_\varepsilon|^2 + \frac{\nu}{2}|A^{\frac{1}{2}}v_\varepsilon|^2 + \big(\nu\lambda - 2\varepsilon\alpha z(\theta_t\omega)\big)|v_\varepsilon|^2 \leqslant e^{-2\varepsilon z(\theta_t\omega)}\frac{2|f|^2}{\nu\lambda},$$

and thereby, since $\frac{\varepsilon_*}{8} < \nu\lambda$ by (10.7),

$$\frac{\mathrm{d}}{\mathrm{d}t}|v_\varepsilon|^2 + \frac{\nu}{2}|A^{\frac{1}{2}}v_\varepsilon|^2 + \left(\frac{\varepsilon_*}{8} - 2\varepsilon\alpha z(\theta_t\omega)\right)|v_\varepsilon|^2 \leqslant e^{-2\varepsilon z(\theta_t\omega)}\frac{2|f|^2}{\nu\lambda}. \tag{10.33}$$

By Gronwall's lemma we have

$$|v_\varepsilon(s,\omega,v_\varepsilon(0))|^2+\frac{\nu}{2}\int_0^s e^{\int_s^\eta(\frac{\varepsilon_*}{8}-2\varepsilon\alpha z(\theta_\xi\omega))\mathrm{d}\xi}|A^{\frac{1}{2}}v_\varepsilon(\eta,\omega,v_\varepsilon(0))|^2\,\mathrm{d}\eta$$
$$\leqslant e^{-\int_0^s(\frac{\varepsilon_*}{8}-2\varepsilon\alpha z(\theta_\eta\omega))\mathrm{d}\eta}|v_\varepsilon(0)|^2+\frac{2|f|^2}{\nu\lambda}\int_0^s e^{\int_s^\eta(\frac{\varepsilon_*}{8}-2\varepsilon\alpha z(\theta_\xi\omega))\mathrm{d}\xi-2\varepsilon z(\theta_\eta\omega)}\mathrm{d}\eta,$$

for all $s\geqslant 0$. Hence, for any $t\geqslant s\geqslant 0$ we have

$$|v_\varepsilon(s,\theta_{-t}\omega,v_\varepsilon(0))|^2+\frac{\nu}{2}\int_0^s e^{\int_s^\eta(\frac{\varepsilon_*}{8}-2\varepsilon\alpha z(\theta_{\xi-t}\omega))\mathrm{d}\xi}|A^{\frac{1}{2}}v_\varepsilon(\eta,\theta_{-t}\omega,v_\varepsilon(0))|^2\,\mathrm{d}\eta$$
$$\leqslant e^{-\int_0^s(\frac{\varepsilon_*}{8}-2\varepsilon\alpha z(\theta_{\eta-t}\omega))\mathrm{d}\eta}|v_\varepsilon(0)|^2+c\int_0^s e^{\int_s^\eta(\frac{\varepsilon_*}{8}-2\varepsilon\alpha z(\theta_{\xi-t}\omega))\mathrm{d}\xi-2\varepsilon z(\theta_{\eta-t}\omega)}\mathrm{d}\eta \tag{10.34}$$
$$=e^{-\frac{\varepsilon_*}{8}s+\int_{-t}^{s-t}2\varepsilon\alpha z(\theta_\eta\omega))\mathrm{d}\eta}|v_\varepsilon(0)|^2$$
$$+ce^{\frac{\varepsilon_*}{8}(t-s)}\int_{-t}^{s-t}e^{\frac{\varepsilon_*}{8}\eta-\int_{s-t}^\eta 2\varepsilon\alpha z(\theta_\xi\omega)\mathrm{d}\xi-2\varepsilon z(\theta_\eta\omega)}\mathrm{d}\eta. \tag{10.35}$$

Taking the inner product of (10.31) with Av_ε we have (recall that $b(u,u,Au)=0$ since we are in the 2D periodic case)

$$\frac{1}{2}\frac{\mathrm{d}}{\mathrm{d}t}|A^{\frac{1}{2}}v_\varepsilon|^2+\nu|Av_\varepsilon|^2=e^{-\varepsilon z(\theta_t\omega)}(f,Av_\varepsilon)+\varepsilon\alpha z(\theta_t\omega)|A^{\frac{1}{2}}v_\varepsilon|^2$$
$$\leqslant e^{-2\varepsilon z(\theta_t\omega)}\frac{|f|^2}{2\nu}+\frac{\nu}{2}|Av_\varepsilon|^2+\varepsilon\alpha z(\theta_t\omega)|A^{\frac{1}{2}}v_\varepsilon|^2.$$

Hence,

$$\frac{\mathrm{d}}{\mathrm{d}t}|A^{\frac{1}{2}}v_\varepsilon|^2+\big(\nu\lambda-2\varepsilon\alpha z(\theta_t\omega)\big)|A^{\frac{1}{2}}v_\varepsilon|^2\leqslant e^{-2\varepsilon z(\theta_t\omega)}\frac{|f|^2}{\nu},$$

and then

$$\frac{\mathrm{d}}{\mathrm{d}t}|A^{\frac{1}{2}}v_\varepsilon|^2+\Big(\frac{\varepsilon_*}{8}-2\varepsilon\alpha z(\theta_t\omega)\Big)|A^{\frac{1}{2}}v_\varepsilon|^2\leqslant e^{-2\varepsilon z(\theta_t\omega)}\frac{|f|^2}{\nu},\quad\forall t>0.$$

Following the same argument as dealing with (10.33) to obtain (10.34), for $t\geqslant s\geqslant\rho>0$ we have

$$|A^{\frac{1}{2}}v_\varepsilon(s,\theta_{-t}\omega,v_\varepsilon(0))|^2\leqslant e^{-\int_\rho^s(\frac{\varepsilon_*}{8}-2\varepsilon\alpha z(\theta_{\eta-t}\omega))\mathrm{d}\eta}|A^{\frac{1}{2}}v_\varepsilon(\rho,\theta_{-t}\omega,v_\varepsilon(0))|^2$$
$$+\int_\rho^s ce^{\int_s^\eta(\frac{\varepsilon_*}{8}-2\varepsilon\alpha z(\theta_{\xi-t}\omega))\mathrm{d}\xi-2\varepsilon z(\theta_{\eta-t}\omega)}\mathrm{d}\eta$$
$$\leqslant e^{-\int_\rho^s(\frac{\varepsilon_*}{8}-2\varepsilon\alpha z(\theta_{\eta-t}\omega))\mathrm{d}\eta}|A^{\frac{1}{2}}v_\varepsilon(\rho,\theta_{-t}\omega,v_\varepsilon(0))|^2$$
$$+\int_0^s ce^{\int_s^\eta(\frac{\varepsilon_*}{8}-2\varepsilon\alpha z(\theta_{\xi-t}\omega))\mathrm{d}\xi-2\varepsilon z(\theta_{\eta-t}\omega)}\mathrm{d}\eta.$$

Integrate the above inequality w.r.t. ρ over $(s-1,s)$ for $s>1$ to obtain

$$
\begin{aligned}
|A^{\frac{1}{2}}v_\varepsilon(s,\theta_{-t}\omega,v_\varepsilon(0))|^2 \leqslant & \int_{s-1}^{s} e^{-\int_\rho^s(\frac{\varepsilon_*}{8}-2\varepsilon\alpha z(\theta_{\eta-t}\omega))\mathrm{d}\eta}|A^{\frac{1}{2}}v_\varepsilon(\rho,\theta_{-t}\omega,v_\varepsilon(0))|^2\,\mathrm{d}\rho \\
& + \int_0^s ce^{\int_s^\eta(\frac{\varepsilon_*}{8}-2\varepsilon\alpha z(\theta_{\xi-t}\omega)\mathrm{d}\xi-2\varepsilon z(\theta_{\eta-t}\omega)}\mathrm{d}\eta \\
\leqslant & \; ce^{-\int_0^s(\frac{\varepsilon_*}{8}-2\varepsilon\alpha z(\theta_{\eta-t}\omega))\mathrm{d}\eta}|v_\varepsilon(0)|^2 \\
& + \int_0^s ce^{\int_s^\eta(\frac{\varepsilon_*}{8}-2\varepsilon\alpha z(\theta_{\xi-t}\omega))\mathrm{d}\xi-2\varepsilon z(\theta_{\eta-t}\omega)}\mathrm{d}\eta \quad \text{(by (10.34))} \\
= & \; ce^{-\frac{\varepsilon_*}{8}s+\int_{-t}^{s-t}2\varepsilon\alpha z(\theta_\eta\omega)\mathrm{d}\eta}|v_\varepsilon(0)|^2 \\
& + ce^{\frac{\varepsilon_*}{8}(t-s)}\int_{-t}^{s-t} e^{\frac{\varepsilon_*}{8}\eta-\int_{s-t}^{\eta}2\varepsilon\alpha z(\theta_\xi\omega)\mathrm{d}\xi-2\varepsilon z(\theta_\eta\omega)}\mathrm{d}\eta,
\end{aligned}
$$

for $t \geqslant s > 1$. □

10.4.3 *Perturbation Radius of the Singleton Attractor Under Multiplicative Noise*

Now we estimate the Hausdorff distance between the random attractor and the deterministic attractor in terms of the perturbation intensity ε. The difference $w_\varepsilon := v_\varepsilon - u$ of any two weak solutions satisfies

$$
\begin{aligned}
\frac{\mathrm{d}w_\varepsilon}{\mathrm{d}t} - \nu\Delta w_\varepsilon + e^{\varepsilon z(\theta_t\omega)}(v_\varepsilon\cdot\nabla)v_\varepsilon - (u\cdot\nabla)u = fe^{-\varepsilon z(\theta_t\omega)} - f + \varepsilon\alpha z(\theta_t\omega)v_\varepsilon, \\
\nabla\cdot w_\varepsilon = 0.
\end{aligned}
\tag{10.36}
$$

Lemma 10.11 *Let $f \in H$. Then for any $\varepsilon \in (0,1]$ there exists a random variable $\gamma_\varepsilon(\omega)$ such that*

$$
\limsup_{t\to\infty}|w_\varepsilon(t,\theta_{-t}\omega,w_\varepsilon(0))|^2 \leqslant \gamma_\varepsilon(\omega), \quad \forall\varepsilon\in(0,1],\ \omega\in\Omega.
$$

Moreover, as a mapping of ε, $\gamma_\varepsilon(\omega) \sim \varepsilon^2$ as $\varepsilon \to 0$ for every $\omega \in \Omega$.

Proof Taking the inner product of (10.36) with w_ε to obtain

$$
\begin{aligned}
&\frac{1}{2}\frac{\mathrm{d}}{\mathrm{d}t}|w_\varepsilon|^2 + \nu|A^{\frac{1}{2}}w_\varepsilon|^2 + e^{\varepsilon z(\theta_t\omega)}b(v_\varepsilon,v_\varepsilon,w_\varepsilon) - b(u,u,w_\varepsilon) \\
&= (e^{-\varepsilon z(\theta_t\omega)}-1)(f,w_\varepsilon) + (\varepsilon\alpha z(\theta_t\omega)v_\varepsilon,w_\varepsilon) \\
&= (e^{-\varepsilon z(\theta_t\omega)}-1)(f,w_\varepsilon) + \varepsilon\alpha z(\theta_t\omega)|w_\varepsilon|^2 + (\varepsilon\alpha z(\theta_t\omega)u,w_\varepsilon) \\
&\leqslant c|e^{-\varepsilon z(\theta_t\omega)}-1|^2 + \left(\frac{\varepsilon_*}{16}+\varepsilon\alpha z(\theta_t\omega)\right)|w_\varepsilon|^2 + \varepsilon^2c|z(\theta_t\omega)|^2|u|^2.
\end{aligned}
\tag{10.37}
$$

Since

$$\begin{aligned}&\left|e^{\varepsilon z(\theta_t\omega)}b(v_\varepsilon,v_\varepsilon,w_\varepsilon)-b(u,u,w_\varepsilon)\right|=\left|(e^{-\varepsilon z(\theta_t\omega)}-1)b(v_\varepsilon,v_\varepsilon,w_\varepsilon)+b(w_\varepsilon,v_\varepsilon,w_\varepsilon)\right|\\&\quad=\left|(e^{-\varepsilon z(\theta_t\omega)}-1)b(v_\varepsilon,u,w_\varepsilon)+b(w_\varepsilon,u,w_\varepsilon)\right|\\&\quad\leqslant c|e^{-\varepsilon z(\theta_t\omega)}-1||A^{\frac12}v_\varepsilon||A^{\frac12}u||A^{\frac12}w_\varepsilon|+c_1|A^{\frac12}u||w_\varepsilon||A^{\frac12}w_\varepsilon|\\&\quad\leqslant c|e^{-\varepsilon z(\theta_t\omega)}-1|^2|A^{\frac12}v_\varepsilon|^2|A^{\frac12}u|^2+\frac{\varepsilon_*}{16\lambda}|A^{\frac12}w_\varepsilon|^2\\&\quad+\left(\frac{c_1^2}{2\nu}|A^{\frac12}u|^2|w_\varepsilon|^2+\frac{\nu}{2}|A^{\frac12}w_\varepsilon|^2\right),\end{aligned}$$

we have

$$\begin{aligned}&\frac12\frac{\mathrm d}{\mathrm dt}|w_\varepsilon|^2+\left(\frac{\nu}{2}-\frac{\varepsilon_*}{16\lambda}\right)|A^{\frac12}w_\varepsilon|^2\leqslant\left(\varepsilon\alpha z(\theta_t\omega)+\frac{c_1^2}{2\nu}|A^{\frac12}u|^2+\frac{\varepsilon_*}{16}\right)|w_\varepsilon|^2\\&\quad+\varepsilon^2c|z(\theta_t\omega)|^2|u|^2+c\left|e^{-\varepsilon z(\theta_t\omega)}-1\right|^2\left(|A^{\frac12}v_\varepsilon|^2|A^{\frac12}u|^2+1\right).\end{aligned}$$

Hence, by Poincaré's inequality,

$$\begin{aligned}&\frac{\mathrm d}{\mathrm dt}|w_\varepsilon|^2+\left(\nu\lambda-\frac{c_1^2}{\nu}|A^{\frac12}u|^2-\frac{\varepsilon_*}{4}-2\varepsilon\alpha z(\theta_t\omega)\right)|w_\varepsilon|^2\\&\quad\leqslant\varepsilon^2c|z(\theta_t\omega)|^2|u|^2+c\left|e^{-\varepsilon z(\theta_t\omega)}-1\right|^2\left(|A^{\frac12}v_\varepsilon|^2|A^{\frac12}u|^2+1\right),\quad t\geqslant0.\end{aligned}$$

Now by (10.9) again we have a $T=T_{B,\frac{\nu\varepsilon_*}{c_1^2}}>0$ such that

$$\begin{aligned}&\frac{\mathrm d}{\mathrm dt}|w_\varepsilon|^2+\left(\frac{\varepsilon_*}{4}-2\varepsilon\alpha z(\theta_t\omega)\right)|w_\varepsilon|^2\\&\quad\leqslant\varepsilon^2c|z(\theta_t\omega)|^2+c\left|e^{-\varepsilon z(\theta_t\omega)}-1\right|^2\left(|A^{\frac12}v_\varepsilon|^2+1\right),\quad t\geqslant T.\end{aligned}$$

By Gronwall's lemma, we have

$$\begin{aligned}&|w_\varepsilon(t,\omega,w_\varepsilon(0))|^2\leqslant e^{\int_s^t-(\frac{\varepsilon_*}{4}-2\varepsilon\alpha z(\theta_\eta\omega))\mathrm d\eta}|w_\varepsilon(s,\omega,w_\varepsilon(0))|^2\\&+\int_s^t ce^{\int_\xi^t-(\frac{\varepsilon_*}{4}-2\varepsilon\alpha z(\theta_\eta\omega))\mathrm d\eta}\left(\varepsilon^2|z(\theta_\xi\omega)|^2+\left|e^{-\varepsilon z(\theta_\xi\omega)}-1\right|^2\left(|A^{\frac12}v_\varepsilon(\xi)|^2+1\right)\right)\mathrm d\xi,\end{aligned}$$

for $t\geqslant s\geqslant T$. Replacing ω with $\theta_{-t}\omega$, for $t\geqslant s\geqslant T$ we have

$$\begin{aligned}&|w_\varepsilon(t,\theta_{-t}\omega,w_\varepsilon(0))|^2\leqslant e^{-\frac{\varepsilon_*}{4}(t-s)+\int_{s-t}^0 2\varepsilon\alpha z(\theta_\eta\omega)\mathrm d\eta}|w_\varepsilon(s,\theta_{-t}\omega,w_\varepsilon(0))|^2\\&+\varepsilon^2c\int_{s-t}^0 e^{\frac{\varepsilon_*}{4}\xi+\int_\xi^0 2\varepsilon\alpha z(\theta_\eta\omega)\mathrm d\eta}|z(\theta_\xi\omega)|^2\,\mathrm d\xi\end{aligned}$$

$$+\int_s^t ce^{-\frac{\varepsilon_*}{4}(t-\xi)+\int_{\xi-t}^0 2\varepsilon\alpha z(\theta_\eta\omega)\mathrm{d}\eta}\left|e^{-\varepsilon z(\theta_{\xi-t}\omega)}-1\right|^2\left(|A^{\frac{1}{2}}v_\varepsilon(\xi,\theta_{-t}\omega,v_\varepsilon(0))|^2+1\right)\mathrm{d}\xi$$
$$:= I_1(\varepsilon,t)+I_2(\varepsilon,t)+I_3(\varepsilon,t). \tag{10.38}$$

For the first term of the right-hand side of (10.38), by (10.35) we have

$$\begin{aligned}
I_1(\varepsilon,t) &\leqslant e^{-\frac{\varepsilon_*}{4}(t-s)+\int_{s-t}^0 2\varepsilon\alpha z(\theta_\eta\omega)\mathrm{d}\eta}\left(|v_\varepsilon(s,\theta_{-t}\omega,v_\varepsilon(0))|^2+|u(s,u_0)|^2\right)\\
&\leqslant ce^{-\frac{\varepsilon_*}{4}(t-s)+\int_{s-t}^0 2\varepsilon\alpha z(\theta_\eta\omega)\mathrm{d}\eta}\Bigg(e^{-\frac{\varepsilon_*}{8}s+\int_{-t}^{s-t}2\varepsilon\alpha z(\theta_\eta\omega))\mathrm{d}\eta}|v_\varepsilon(0)|^2\\
&\quad+e^{\frac{\varepsilon_*}{8}(t-s)}\int_{-t}^{s-t}e^{\frac{\varepsilon_*}{8}\eta-\int_{s-t}^{\eta}2\varepsilon\alpha z(\theta_\xi\omega))\mathrm{d}\xi-2\varepsilon z(\theta_\eta\omega)}\mathrm{d}\eta+c_R\Bigg)\\
&\leqslant ce^{-\frac{\varepsilon_*}{8}(t-s)-\frac{\varepsilon_*}{8}t+\int_{-t}^0 2\varepsilon\alpha z(\theta_\eta\omega)\mathrm{d}\eta}|v_\varepsilon(0)|^2\\
&\quad+ce^{-\frac{\varepsilon_*}{8}(t-s)}\int_{-t}^{s-t}e^{\frac{\varepsilon_*}{8}\eta+\int_\eta^0 2\varepsilon\alpha z(\theta_\xi\omega))\mathrm{d}\xi-2\varepsilon z(\theta_\eta\omega)}\mathrm{d}\eta\\
&\quad+c_Re^{-\frac{\varepsilon_*}{4}(t-s)+\int_{s-t}^0 2\varepsilon\alpha z(\theta_\eta\omega)\mathrm{d}\eta}\xrightarrow{t\to\infty}0,
\end{aligned}$$

where c_R is taken as the uniform H-bound of solutions corresponding to initial data from the R-ball of H. For the second term, it is trivial to obtain

$$\limsup_{t\to\infty} I_2(\varepsilon,t)\leqslant \varepsilon^2c\int_{-\infty}^0 e^{\frac{\varepsilon_*}{4}\xi+\int_\xi^0 2\varepsilon\alpha z(\theta_\eta\omega)\mathrm{d}\eta}|z(\theta_\xi\omega)|^2\mathrm{d}\xi.$$

For the third term, we have

$$\begin{aligned}
&I_3(\varepsilon,t)\\
&=\int_s^t ce^{-\frac{\varepsilon_*}{4}(t-\xi)+\int_{\xi-t}^0 2\varepsilon\alpha z(\theta_\eta\omega)\mathrm{d}\eta}\left|e^{-\varepsilon z(\theta_{\xi-t}\omega)}-1\right|^2\left(|A^{\frac{1}{2}}v_\varepsilon(\xi,\theta_{-t}\omega,v_\varepsilon(0))|^2+1\right)\mathrm{d}\xi\\
&\leqslant\int_s^t ce^{-\frac{\varepsilon_*}{4}(t-\xi)+\int_{\xi-t}^0 2\varepsilon\alpha z(\theta_\eta\omega)\mathrm{d}\eta}\left|e^{-\varepsilon z(\theta_{\xi-t}\omega)}-1\right|^2\Bigg(e^{-\frac{\varepsilon_*}{8}\xi+\int_{-t}^{\xi-t}2\varepsilon\alpha z(\theta_\eta\omega)\mathrm{d}\eta}|v_\varepsilon(0)|^2\\
&\quad+ce^{\frac{\varepsilon_*}{8}(t-\xi)}\int_{-t}^{\xi-t}e^{\frac{\varepsilon_*}{8}\eta-\int_{\xi-t}^{\eta}2\varepsilon\alpha z(\theta_\rho\omega)\mathrm{d}\rho-2\varepsilon z(\theta_\eta\omega)}\mathrm{d}\eta+1\Bigg)\mathrm{d}\xi\quad(\text{by }(10.32))\\
&=\int_s^t ce^{-\frac{\varepsilon_*}{8}(t-\xi)-\frac{\varepsilon_*}{8}t+\int_{-t}^0 2\varepsilon\alpha z(\theta_\eta\omega)\mathrm{d}\eta}\left|e^{-\varepsilon z(\theta_{\xi-t}\omega)}-1\right|^2|v_\varepsilon(0)|^2\,\mathrm{d}\xi\\
&\quad+\int_s^t ce^{-\frac{\varepsilon_*}{8}(t-\xi)}\left|e^{-\varepsilon z(\theta_{\xi-t}\omega)}-1\right|^2\int_{-t}^{\xi-t}e^{\frac{\varepsilon_*}{8}\eta+\int_\eta^0 2\varepsilon\alpha z(\theta_\rho\omega)\mathrm{d}\rho-2\varepsilon z(\theta_\eta\omega)}\mathrm{d}\eta\mathrm{d}\xi\\
&\quad+\int_s^t ce^{-\frac{\varepsilon_*}{4}(t-\xi)+\int_{\xi-t}^0 2\varepsilon\alpha z(\theta_\eta\omega)\mathrm{d}\eta}\left|e^{-\varepsilon z(\theta_{\xi-t}\omega)}-1\right|^2\mathrm{d}\xi\\
&\leqslant ce^{-\frac{\varepsilon_*}{8}t+\int_{-t}^0 2\varepsilon\alpha z(\theta_\eta\omega)\mathrm{d}\eta}\int_{s-t}^0 e^{\frac{\varepsilon_*}{8}\xi}\left|e^{-\varepsilon z(\theta_\xi\omega)}-1\right|^2|v_\varepsilon(0)|^2\,\mathrm{d}\xi
\end{aligned}$$

$$+c\int_{s-t}^{0} e^{\frac{\varepsilon_*}{8}\xi}\left|e^{-\varepsilon z(\theta_\xi\omega)}-1\right|^2 \mathrm{d}\xi \int_{-t}^{0} e^{\frac{\varepsilon_*}{8}\eta+\int_\eta^0 2\varepsilon\alpha z(\theta_\rho\omega)\mathrm{d}\rho-2\varepsilon z(\theta_\eta\omega)}\,\mathrm{d}\eta$$
$$+\int_{s-t}^{0} ce^{\frac{\varepsilon_*}{4}\xi+\int_\xi^0 2\varepsilon\alpha z(\theta_\eta\omega)\mathrm{d}\eta}\left|e^{-\varepsilon z(\theta_\xi\omega)}-1\right|^2 \mathrm{d}\xi.$$

Let us take $\alpha > 0$ large enough such that, by (10.14),

$$\mathbb{E}(2|z|) < \frac{\varepsilon_*}{8}. \tag{10.39}$$

Then

$$\limsup_{t\to\infty} I_3(\varepsilon,t) \leqslant \int_{-\infty}^{0} ce^{\frac{\varepsilon_*}{8}\xi}\left|e^{-\varepsilon z(\theta_\xi\omega)}-1\right|^2 \mathrm{d}\xi \int_{-\infty}^{0} e^{\frac{\varepsilon_*}{8}\eta+\int_\eta^0 2\varepsilon\alpha z(\theta_\rho\omega)\mathrm{d}\rho-2\varepsilon z(\theta_\eta\omega)}\mathrm{d}\eta$$
$$+\int_{-\infty}^{0} ce^{\frac{\varepsilon_*}{4}\xi+\int_\xi^0 2\varepsilon\alpha z(\theta_\eta\omega)\mathrm{d}\eta}\left|e^{-\varepsilon z(\theta_\xi\omega)}-1\right|^2 \mathrm{d}\xi =: \gamma_\varepsilon^{(1)}(\omega).$$

Since it is elementary to see that $\gamma_\varepsilon^{(1)}(\omega) \sim \varepsilon^2$ as $\varepsilon \to 0$, the proof is complete. □

Analogously to Theorem 10.2 for the additive noise case, by Lemma 10.11 we have

Theorem 10.3 *Let $f \in H$. Then for any $\varepsilon \in (0, 1]$ there exists a random variable $\gamma_\varepsilon(\omega)$ such that*

$$\mathrm{dist}_H(\mathscr{A}_\varepsilon(\omega), \mathscr{A}) \leqslant \sqrt{\gamma_\varepsilon(\omega)},$$

where $\mathrm{dist}_H(A, B) = \max\{\mathrm{dist}(A, B), \mathrm{dist}(B, A)\}$. *Moreover, as a mapping of ε, $\sqrt{\gamma_\varepsilon(\omega)} \sim \varepsilon$ as $\varepsilon \to 0$ for every $\omega \in \Omega$.*

Acknowledgements H. Cui was partially funded by NSFC Grant 11801195 and the Fundamental Research Funds for the Central Universities 5003011026.

References

1. Arnold, L.: Random Dynamical Systems. Springer, Berlin (1998)
2. Babin, A.V., Vishik, M.I.: Attractors of Evolution Equations. North-Holland (1992)
3. Bates, P.W., Lu, K., Wang, B.: Random attractors for stochastic reaction-diffusion equations on unbounded domains. J. Differ. Equ. **246**, 845–869 (2009)
4. Caraballo, T., Crauel, H., Langa, J.A., Robinson, J.C.: The effect of noise on the "chafee-infante" equation: a nonlinear case study. Proc. Am. Math. Soc. **135**, 373–382 (2007)
5. Caraballo, T., Langa, J.A.: On the upper semicontinuity of cocycle attractors for non-autonomous and random dynamical systems. Dyn. Contin. Discret. Impuls. Syst. Ser. A **10**, 491–514 (2003)
6. Caraballo, T., Langa, J.A., Robinson, J.C.: Upper semicontinuity of attractors for small random perturbations of dynamical systems. Commun. Partial Differ. Equ. **23**, 1557–1581 (1998)

7. Carvalho, A.N., Langa, J.A.: Non-autonomous perturbation of autonomous semilinear differential equations: Continuity of local stable and unstable manifolds. J. Differ. Equ. **233**, 622–653 (2007)
8. Carvalho, A.N., Langa, J.A., Robinson, J.C.: On the continuity of pullback attractors for evolution processes. Nonlinear Anal.: Theory Methods Appl. **71**, 1812–1824 (2009)
9. Carvalho, A.N., Langa, J.A., Robinson, J.C.: Attractors for Infinite-Dimensional Non-autonomous Dynamical Systems, vol. 182. Springer, Berlin (2013)
10. Carvalho, A.N., Langa, J.A., Robinson, J.C., Suárez, A.: Characterization of non-autonomous attractors of a perturbed infinite-dimensional gradient system. J. Differ. Equ. **236**, 570–603 (2007)
11. Chepyzhov, V.V., Vishik, M.I.: Attractors for Equations of Mathematical Physics, vol. 49. American Mathematical Society, Providence (2002)
12. Chueshov, I.: Monotone Random Systems Theory and Applications, vol. 1779, Springer Science & Business Media, Berlin (2002)
13. Crauel, H., Flandoli, F.: Attractors for random dynamical systems. Probab. Theory Relat. Fields **100**, 365–393 (1994)
14. Cui, H., Freitas, M.M., Langa, J.A.: On random cocycle attractors with autonomous attraction universes. Discret. Contin. Dyn. Syst. - Ser. B **22**, 3379–3407 (2017)
15. Cui, H., Langa, J.A.: Uniform attractors for non-autonomous random dynamical systems. J. Differ. Equ. **263**, 1225–1268 (2017)
16. Fan, X.: Attractors for a damped stochastic wave equation of the sine-Gordon type with sublinear multiplicative noise. Stoch. Anal. Appl. **24**, 767–793 (2006)
17. Flandoli, F., Schmalfuss, B.: Random attractors for the 3D stochastic Navier-Stokes equation with multiplicative white noise. Stoch. Stoch. Rep. **59**, 21–45 (1996)
18. Foias, C., Manley, O., Temam, R., Treve, Y.: Asymptotic analysis of the Navier-Stokes equations. Phys. D: Nonlinear Phenom. **9**, 157–188 (1983)
19. Hale, J.K., Raugel, G.: Lower semicontinuity of attractors of gradient systems and applications. Ann. Mat. Pura ed Appl. **154**, 281–326 (1989)
20. Kloeden, P.E., Rasmussen, M.: Nonautonomous Dynamical Systems, vol. 176. American Mathematical Society, Providence (2011)
21. Langa, J.A., Robinson, J.C., Suárez, A., Vidal-López, A.: The stability of attractors for non-autonomous perturbations of gradient-like systems. J. Differ. Equ. **234**, 607–625 (2007)
22. Li, D., Kloeden, P.: Equi-attraction and the continuous dependence of attractors on parameters. Glasg. Math. J. **46**, 131–141 (2004)
23. Li, D., Kloeden, P.: Equi-attraction and the continuous dependence of pullback attractors on parameters. Stoch. Dyn. **4**, 373–384 (2004)
24. Li, D., Kloeden, P.: Equi-attraction and continuous dependence of strong attractors of set-valued dynamical systems on parameters. Set-Valued Anal. **13**, 405–416 (2005)
25. Li, Y., Gu, A., Li, J.: Existence and continuity of bi-spatial random attractors and application to stochastic semilinear Laplacian equations. J. Differ. Equ. **258**, 504–534 (2015)
26. Ochs, G.: Weak Random Attractors, Citeseer (1999)
27. Robinson, J.C.: Infinite-Dimensional Dynamical Systems: An Introduction to Dissipative Parabolic PDEs and the Theory of Global Attractors, vol. 28. Cambridge University Press, Cambridge (2001)
28. Robinson, J.C.: Stability of random attractors under perturbation and approximation. J. Differ. Equ. **186**, 652–669 (2002)
29. Temam, R.: Infinite Dimensional Dynamical Systems in Mechanics and Physics, 2nd edn. Springer, New York (1997)
30. Wang, B.: Upper semicontinuity of random attractors for non-compact random dynamical systems. Electron. J. Differ. Equ. **2009**, 1–18 (2009)

Chapter 11
The Dynamics of Periodic Switching Systems

Jose S. Cánovas

Abstract Switching systems have been recently used to model phenomena from Biology, Economy, Physics, etc. They consist in the iteration of a finite number of maps which can be viewed as the dynamics of a class of triangular maps. We make a special emphasis to periodic switching by showing that the dynamics of these systems can be analyzed from associated dynamical systems. Hence, we introduce the basic background of regular enough piecewise monotone maps and then, as an application, we study the dynamics of two periodic models coming from Biology.

11.1 Introduction

Let X be a compact metric space with metric d and let $f_i : X \to X$ be continuous maps for $i = 1, ..., k$. A switching system can be seen as the triangular or skew product map $F : \mathscr{S} \times X \to \mathscr{S} \times X$, where $\mathscr{S}$ is a shift invariant subset of $\{1, ..., k\}^{\mathbb{N}}$. Namely, if σ denotes the shift map $\sigma : \{1, ..., k\}^{\mathbb{N}} \to \{1, ..., k\}^{\mathbb{N}}$, given by $\sigma(s_n) = (s_{n+1})$, then $\sigma(\mathscr{S}) = \mathscr{S}$. Given $x \in X$ and $(s_n) \in \mathscr{S}$, its orbit is given by the sequence $Orb(x, (s_n), F) = (\pi(F^n(s_n, x)))$, where $\pi : \mathscr{S} \times X \to X$ denotes the natural projection.

We are interested in periodic switching, that is, when the set $\mathscr{S}$ is finite and contains exactly a periodic orbit. Hence, the map F can be viewed as a periodic sequence $(f_1, ..., f_k, f_1, ..., f_k, ...)$, which we will denote by $[f_1, ..., f_k]$.

For $x \in X$, its orbit, denoted by $Orb(x, [f_1, ..., f_k])$, fulfills that

$$Orb(x, [f_1, ..., f_k]) = Orb(x, f_k \circ \cdots \circ f_1) \cup \cdots \cup Orb((f_{k-1} \circ \cdots \circ f_1)(x), f_{k-1} \circ \cdots \circ f_1 \circ f_k).$$

Therefore, its ω–limit set, that is, the set of limit points of the orbits, satisfies the relationship

J. S. Cánovas (✉)
Departamento de Matemática Aplicada y Estadística, Universidad Politécnica de Cartagena, C/Dr. Fleming sn, 30202 Cartagena, Spain
e-mail: jose.canovas@upct.es

V. A. Sadovnichiy and M. Z. Zgurovsky (eds.), *Contemporary Approaches and Methods in Fundamental Mathematics and Mechanics*, Understanding Complex Systems,
https://doi.org/10.1007/978-3-030-50302-4_11

$$\omega(x, [f_1, ..., f_k]) = \omega(x, f_k \circ \cdots \circ f_1) \cup ... \cup \omega((f_{k-1} \circ \cdots \circ f_1)(x), f_{k-1} \circ \cdots \circ f_1 \circ f_k).$$

Hence, the dynamics of $[f_1, ..., f_k]$ can be studied from the dynamics of the maps $f_k \circ \cdots \circ f_1, ..., f_{k-1} \circ \cdots \circ f_1 \circ f_k$, that is, it can be analyzed as the dynamics of k associated dynamical systems. It is also easy to see that

$$(f_{i-1} \circ \cdots \circ f_1)(Orb(x, f_k \circ \cdots \circ f_1)) = Orb((f_{i-1} \circ \cdots \circ f_1)(x), f_{i-1} \circ \cdots \circ f_k \circ f_1 \circ \cdots \circ f_i)$$

for $i = 1, ..., k-1$. So, it is reasonable that the dynamics of $f_k \circ \cdots \circ f_1$ will be similar to that of $f_{i-1} \circ \cdots \circ f_k \circ f_1 \circ \cdots \circ f_i$, $i = 1, ..., k-1$. In general, the last assertion is not true; for instance, the topological sequence entropy of two non-commuting maps can be different [3], but for many dynamical notions, say e.g. periodicity, topological entropy, attractors, this is the case.

Then, in many cases the dynamics of periodic switching systems can be analyzed from that of the map $f_k \circ \cdots \circ f_1$. However, in practice, we will show that some technical difficulties appear when the number of maps increases.

To fix ideas, let us assume that $X = I = [a, b]$, $a, b \in \mathbb{R}$, $a < b$, and the maps f_i are unimodal, that is, continuous interval maps for which there is $c_i \in I$ such that f_i is strictly increasing in $[a, c_i]$ and strictly decreasing in $[c_i, b]$. We call c_i the turning point of f_i. Then, the composition $f_k \circ \cdots \circ f_1$ can have at most $2^k - 1$ turning points. Since turning points are the key for analyzing the dynamics of multimodal maps, more than three might represent a technical problem for analyzing the model.

Below, as a motivation for studying the periodic systems, we introduce some periodically switched models coming from Economics, Physics and Biology.

1.1 Economic models. Periodically switched models appear in several economic models. The so-called duopolies, in which two firms compete in a closed market, are the first example. Sometimes these models can be stated as a discrete dynamical system given by the antitriangular map

$$F(x, y) = (f(y), g(x)), \quad x, y \geq 0.$$

For instance, we consider an isoelastic inverse demand function

$$p = \frac{1}{x + y},$$

where x and y are the firms' production and p is the price, and a linear cost function for each firm

$$C_1(x) = c_1 x, \quad C_2(y) = c_2 y, \quad c_1, c_2 > 0.$$

Then, under naive expectations for future productions the map

$$F(x, y) = (f(y), g(x)) = \left(\max\left\{0, \sqrt{y/c_1} - y\right\}, \max\left\{0, \sqrt{x/c_2} - x\right\}\right)$$

defines the Puu duopoly (see [29]). It is easy to realize that the dynamics of F is strongly related to that of the periodic sequence $(f, g, f, g, ...)$. Models defined by the same kind of functions can be found in [21, 22, 30].

The periodic iteration of maps can be also found in oligopoly models where firms have two different production strategies depending on whether the firms do invest or not in that period. Then, for each firm we have a reaction function in the investment period (long run) and another one when it does not invest (short run). These models can be found in [10].

The switching process can be also found in models where firms change their production strategies. For instance, from naive expectations to adaptive expectations or Cournot (see [7, 12, 14, 15]).

Finally, in [11, 20], we find economic models with seasonality assumptions, which are constructed by the periodic switching of two maps.

1.2 Physical models. Switching models can be found when modeling electrical circuits. For instance, in [40], it can be seen that the model can be described by the iteration of the periodic sequence $(f_a, f_b, f_b, f_b, f_a, ...)$ of period 4 where $f_a(x) = ax(1-x)$ is the well-known logistic model.

This kind of models can be also found in the so-called deformation of nonlinear maps (see [16, 27, 33]). These models can be seen as the iteration of a map, for instance the logistic map f_a, and an homeomorphism φ. Then, we have the two periodic sequence $(\varphi, f_a, \varphi, f_a, ...)$, and the associated maps $f_a \circ \varphi$ and $\varphi \circ f_a$.

1.3 Biological models. Finally, population growth models also offer several examples of switching. We will consider a biological model consisting of one species whose population x_n follows the difference equation

$$x_{n+1} = x_n g(x_n),$$

where g is known as production function per capita. If $g(x) = a(1-x)$, we have the classical logistic model introduced above, but there are other examples in the literature. For instance, when

$$g(x) = \frac{a}{1+bx}$$

with $a, b > 0$, we obtain the Beverton–Holt compensatory model (see [6]). If

$$g(x) = e^{r-x}$$

with $r > 0$, we have the over-compensatory Ricker model (see [31]). Finally, if

$$g(x) = \frac{\beta x}{1+\beta x} e^{r(1-x)},$$

with $\beta, r > 0$, we obtain the Ricker-Schreiber depensatory model, which presents Allee effect (see [32]), that is, extinction when the population is not big enough.

Switching models can appear when seasonality is added (see e.g. [24, 28] or [34]). Then, individual growth models are combined with different parameters according to seasonal characteristics. A simple example is obtained when two different parameters r_1 and r_2 are considered for the Ricker model and construct the sequence $(f_1, f_2, f_1, f_2, ...)$, where $f_i(x) = xe^{r_i(1-x)}$, $i = 1, 2$.

This type of models also appears when harvesting or hunting phenomena play some role (see [17, 18]). Assuming them proportional to the current population say $\phi(x) = (1-\gamma)x$, $\gamma \in (0, 1)$, then, the sequence $(f, \phi, f, \phi, ...)$ gives us models where growth and hunting are combined. One could expect that hunting reduces the model complexity. The dynamics is given by the maps $\phi \circ f$ or $f \circ \phi$, which are conjugated. In these models, it is interesting to investigate whether the census time have some relevance in order to plan the hunting or harvesting time.

The aim of this paper is to survey some techniques and ideas that are useful to analyze systems defined by periodic sequences of good enough interval maps. They will consist of results of multimodal maps which allow us to characterize metric attractors and chaotic dynamics. Next, we show that these results are effective to analyze the dynamics of two Biological models, which depend on several parameters.

The rest of the paper is organized as follows. Next, we introduce the basic mathematical background which we will use in the last section, where the two Biological models are studied.

11.2 One-Dimensional Dynamics Background

The aim of this section is to give some guidelines for the practical analysis of periodically switched sequences of interval maps. Let $I = [a, b] \subset \mathbb{R}$ and consider $f : I \to I$, a continuous interval map. We will assume that it is piecewise monotone, which means that there are $a = a_1 < a_2 < ... < a_k = b$ such that $f|_{[a_i, a_{i+1}]}$ is strictly monotone for $i = 1, 2, ..., k-1$. The $a_i' s$ are called turning points of f. We will always assume that the map f is smooth enough, in our case C^3, and non flat on the turning points a_i, that is, for x close to a_i, $i = 1, 2, ..., k$,

$$f(x) = \pm|\phi_i(x)|^{\beta_i} + f(a_i),$$

where ϕ_i is C^3, $\phi_i(a_i) = 0$ and $\beta_i > 0$.

Following [25], a metric attractor is a subset $A \subset [0, 1]$ such that $f(A) \subseteq A$, $O(A) = \{x : \omega(x, f) \subset A\}$ has positive Lebesgue measure, and there is no proper subset $A' \subsetneq A$ with the same properties. The set $O(A)$ is called the *basin* of the attractor.

By [39], the regularity properties of f imply that there are three possibilities for its metric attractors:

(A1) A periodic orbit.

(A2) A minimal Cantor set containing at least one recurrent turning point. Recall that a set is minimal if it is the ω–limit set of all the orbits, and a point is recurrent if belongs to its ω–limit set. This class contains the solenoidal attractors, which are basically Cantor sets where the dynamic is quasi periodic.

(A3) A union of periodic intervals $J_1, ..., J_k$, with $f^k(J_i) = J_i$ and $f^{j-i}(J_i) = J_j$, $1 \leq i < j \leq k$, and such that f^k is topologically mixing. Topologically mixing property implies the existence of dense orbits on each periodic interval (under the iteration of f^k).

Moreover, if f has an attractor of type (A2) or (A3), then they must contain the orbit of a turning point. Therefore, the number of non-periodic attractors is bounded by the cardinality of the set of turning points of f.

Recall that the Schwarzian derivative is given by

$$S(f)(x) = \frac{f'''(x)}{f'(x)} - \frac{3}{2}\left(\frac{f''(x)}{f'(x)}\right)^2.$$

If the map f holds that $S(f)(x) < 0$, then the number of attractors is bounded by the number of turning points since each attractor must attract their orbits[1]. Notice that, if a finite orbit $P = \{x_0, ..., x_{n-1}\}$ is an attractor, then $|(f^n)'(x_0)| \leq 1$. Since the Schwarzian derivative is negative, this finite attractor must attract the orbit of the turning point. Hence, the Lyapunov exponent [26] at the turning point x_M, given by

$$\text{lyex}(x_M) = \lim_{n\to\infty} \frac{1}{n} \log |(f^n)'(f(x_M))| = \lim_{n\to\infty} \frac{1}{n} \sum_{j=1}^{n} \log |f'(f^j(x_M))|,$$

must be lower than or equal to zero.

Now, we discuss on topological entropy, which we use to define topological chaos. It was introduced in the setting of continuous maps on compact topological spaces by Adler, Konheim and McAndrew [1] and for metric spaces by Bowen [8].[2] For continuous interval maps the definition read as follows. Given $\varepsilon > 0$, we say that a set $E \subset I$ is *(n, ε, f)–separated* if for any $x, y \in E$, $x \neq y$, there exists $k \in \{0, 1, ..., n-1\}$ such that $|f^k(x) - f^k(y)| > \varepsilon$. Denote by $s(n, \varepsilon, f)$ the biggest cardinality of any maximal (n, ε, f)–separated set in I. Then, the topological entropy of f is

$$h(f) = \lim_{\varepsilon\to 0} \limsup_{n\to\infty} \frac{1}{n} \log s(n, \varepsilon, f).$$

There is an equivalent definition using spanning sets as follows. We say that a set $F \subset I$ *(n, ε, f)–spans* I if for any $x \in I$, there exists $y \in F$ such that $|f^i(x) - f^i(y)| < \varepsilon$ for any $i \in \{0, 1, ..., n-1\}$. Denote by $r(n, \varepsilon, f)$ the smallest cardinality of any

[1] With the exception of a fixed point or a two periodic orbit at the interval endpoints.

[2] Dinaburg [13] gave simultaneously a Bowen like definition for continuous maps on a compact metric space.

minimal (t, ε, f)–spanning set in X. Then, topological entropy can be computed as

$$h(f) = \lim_{\varepsilon \to 0} \limsup_{n \to \infty} \frac{1}{n} \log r(n, \varepsilon, f).$$

On the other hand, for one dimensional dynamics the topological entropy is a useful tool to check the dynamical complexity of a map because it is strongly connected with the notion of horseshoe (see [2, page 205]). We say that the map $f : I \to I$ has a k-horseshoe, $k \in \mathbb{N}$, $k \geq 2$, if there are k disjoint subintervals J_i, $i = 1, ..., k$, such that $J_1 \cup ... \cup J_k \subseteq f(J_i)$, $i = 1, ..., k$[3]. It was proved that if $h(f) > 0$, then there is an integer l such that f^l has a horseshoe. This result links positive topological entropy with topological chaos for continuous interval maps. The problem is that Bowen's definitions of topological entropy are not suitable for working with families of interval maps depending on parameters. Then, in practice, some numerical algorithms are needed to make the computations.

For unimodal maps (see [4]), it is possible to make the computations by using the turning point. Let f be an unimodal map with maximum (turning point) at c. Let $k(f) = (k_1, k_2, k_3, ...)$ be its kneading sequence given by the rule

$$k_i = \begin{cases} R & \text{if } f^i(c) > c, \\ C & \text{if } f^i(c) = c, \\ L & \text{if } f^i(c) < c. \end{cases}$$

We fix that $L < C < R$. For two different unimodal maps f_1 and f_2, we fix their kneading sequences $k(f_1) = (k_n^1)$ and $k(f_2) = (k_n^2)$. We say that $k(f_1) \leq k(f_2)$ provided there is $m \in \mathbb{N}$ such that $k_i^1 = k_i^2$ for $i < m$ and either an even number of k_i^1's are equal to R and $k_m^1 < k_m^2$ or an odd number of k_i^1's are equal to R and $k_m^2 < k_m^1$. In [4], it is proved that, if $k(f_1) \leq k(f_2)$, then $h(f_1) \leq h(f_2)$. In addition, if $k_m(f)$ denotes the first m symbols of $k(f)$, then $h(f_1) \leq h(f_2)$, if $k_m(f_1) < k_m(f_2)$.

The algorithm for computing the topological entropy is based on the fact that the tent family

$$g_k(x) = \begin{cases} kx & if\ x \in [0, 1/2], \\ -kx + k & if\ x \in [1/2, 1], \end{cases}$$

with $k \in [1, 2]$, holds that $h(g_k) = \log k$. The idea of the algorithm is to bound the topological entropy of a unimodal maps between the topological entropies of two tent maps. The algorithm is divided into four steps:

Step 1. Fix $\varepsilon > 0$ (fixed accuracy) and an integer n such that $\delta = 1/n < \varepsilon$.
Step 2. Find the least positive integer m such that $k_m(g_{1+i\delta})$, $0 \leq i \leq n$, are distinct kneading sequences.
Step 3. Compute $k_m(f)$ for a fixed unimodal map f.

[3] Since Smale's work (see [36]), horseshoes have been in the core of chaotic dynamics, describing what we could call random deterministic systems.

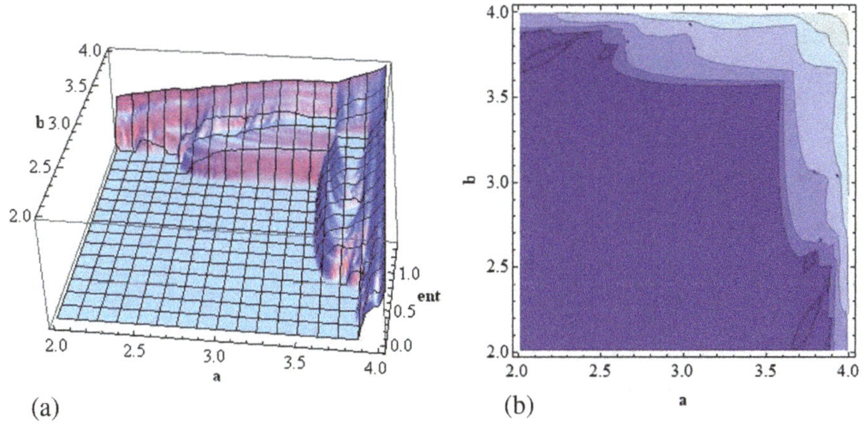

Fig. 11.1 **a** Topological entropy of $f_a \circ f_b$ when the parameter values a and b ranges from 2 to 4, with accuracy 10^{-6}. **b** Level curves of the topological entropy of $f_a \circ f_b$ for the same parameters range

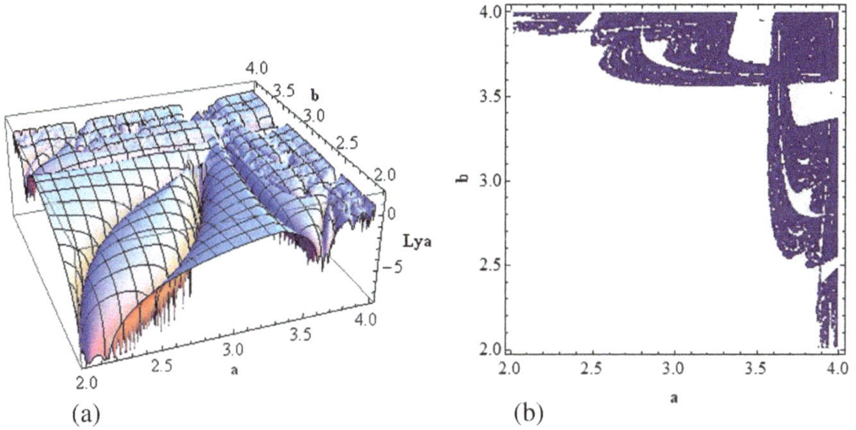

Fig. 11.2 **a** Estimations maximal Lyapunov exponent at the turning points of $f_a \circ f_b$ when the parameter values a and b ranges from 2 to 4. **b** In dark, the region where the maximal Lyapunov exponent at the turning points of $f_a \circ f_b$ is positive for the same parameters range

Step 4. Find r the largest integer such that $k_m(g_{1+r\delta}) < k_m(f)$. Hence $\log(1 + r\delta) \leq h(f) \leq \log(1 + (r + 2)\delta)$.

The algorithm is easily programmed. We usually use Mathematica, which has the advantage of computing the kneading invariants of tent maps without round off errors, improving in practice the accuracy of the method. It is remarkable that we can have positive topological entropy, and hence topological chaos, while the attractor is a periodic orbit. Then, topological chaos is not physically observable since it is hidden in a set of zero Lebesgue measure.

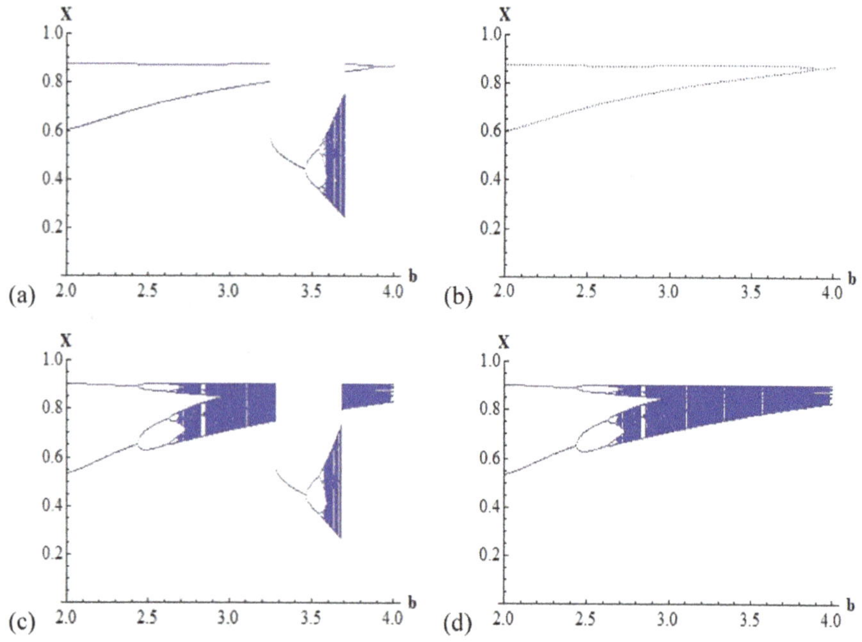

Fig. 11.3 **a** Bifurcation diagram of $f_a \circ f_b$ when $a = 3.5$ and b ranges from 2 to 4. The initial condition for each iteration is the minimum $1/2$. **b** Bifurcation diagram of $f_a \circ f_b$ when $a = 3.5$ and b ranges from 2 to 4. The initial condition for each iteration is the maximum $\frac{b-\sqrt{b^2-2b}}{2b}$. **c** Bifurcation diagram of $f_a \circ f_b$ when $a = 3.6$ and b ranges from 2 to 4. The initial condition for each iteration is the minimum $1/2$. **d** Bifurcation diagram of $f_a \circ f_b$ when $a = 3.6$ and b ranges from 2 to 4. The initial condition for each iteration is the maximum $\frac{b-\sqrt{b^2-2b}}{2b}$. We check that two possible diagrams may appear. Note also the existence of period doubling bifurcations (see [38] and references therein)

11.3 Applications

11.3.1 The Periodic Logistic Map

The well-known logistic map $f_a(x) = ax(1-x)$ (see [23]) was proposed as a model of population growth. It has been extensively studied from the mathematical point of view (see e.g. [37]).

In [34], this model was considered by adding some seasonality, in such a way that the logistic map is periodically switched. First, a two periodic switched was considered, and then, a twelve periodic one. In the two periodic case, we can compute the topological entropy of the map $f_a \circ f_b, a, b \in [0, 4]$. This map is unimodal when $b \leq 2$. Otherwise, it is trimodal, that is, with one minimum at $\frac{1}{2}$ and two maxima, $\frac{b-\sqrt{b^2-2b}}{2b}$ and $\frac{b+\sqrt{b^2-2b}}{2b}$, with the same image.

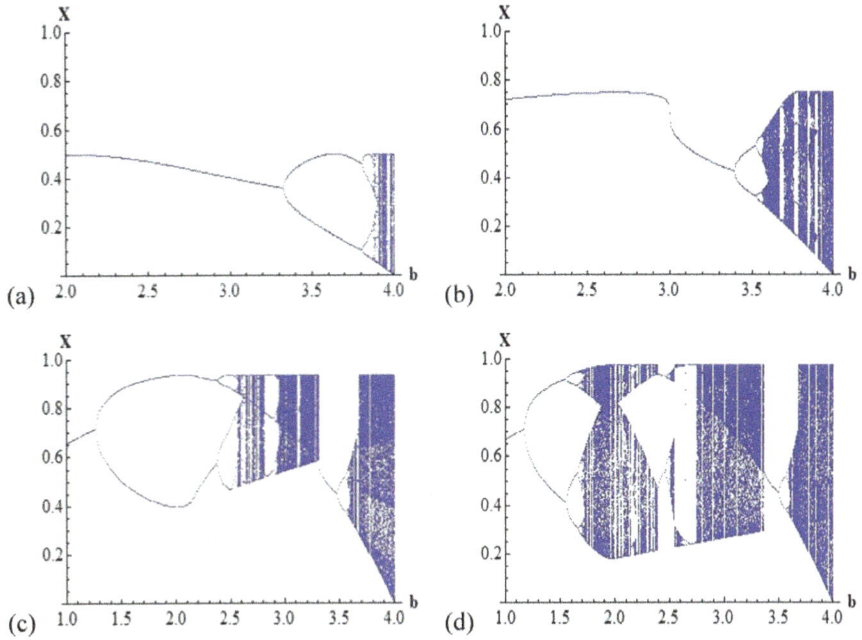

Fig. 11.4 **a** Bifurcation diagram of $f_a \circ f_b$ when $a = 2$ and b ranges from 2 to 4. The initial condition for each iteration is the minimum 1/2. **b** Bifurcation diagram of $f_a \circ f_b$ when $a = 3$ and b ranges from 2 to 4. The initial condition for each iteration is the minimum 1/2. **c** Bifurcation diagram of $f_a \circ f_b$ when $a = 3.75$ and b ranges from 2 to 4. The initial condition for each iteration is the minimum 1/2. **d** Bifurcation diagram of $f_a \circ f_b$ when $a = 3.9$ and b ranges from 2 to 4. The initial condition for each iteration is the minimum 1/2 but the same diagrams are obtained if we change the initial conditions. We note that the greater a, the greater the complexity of the diagram. Note also the existence of period doubling bifurcations (see [38] and references therein)

When it is unimodal we can use the above described algorithm, but when it is trimodal we take a variation of the algorithm from [5] for bimodal maps (with two extrema) which can be found in [9]. This algorithm is more technical but it is based on the same ideas of the unimodal case. Here, we consider both a and b greater than 2 and therefore, the unimodal algorithm cannot be applied. In Figs. 11.1 and 11.2, we show the computation of topological entropy and the estimation of Lyapunov exponents.

Below, we show some one dimensional bifurcation diagrams in which we fix the parameter a while b ranges the interval [2, 4]. It is easy to see that the map $f_a \circ f_b$ has negative Schwarzian derivative because $S(f_a \circ f_b) = (S(f_a) \circ f_b) \cdot (f_b')^2) + S(f_b)$. Since it has either one maximum or one minimum and two maxima with the same image, and taking into account that the end points of the interval are not attractors, we have that at most two attractors can be found. Then, we obtain two types of bifurcation diagrams due to the existence of a minimum and two maxima in $f_a \circ f_b$

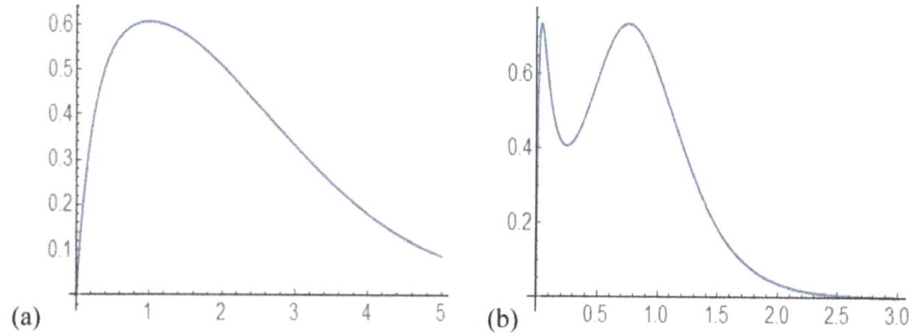

Fig. 11.5 **a** Graph of $f \circ m$ when the parameter values are $a = 0.5$ and $r = 1$. **b** Graph of $f \circ m$ when the parameter values are $a = 0.5$ and $r = 4$

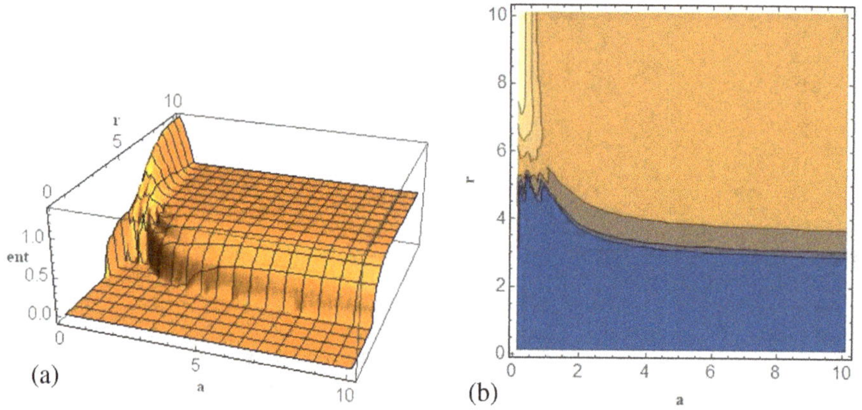

Fig. 11.6 **a** Topological entropy of $f \circ m$ when the parameter values a and r ranges from 0 to 10, with accuracy 10^{-6}. **b** Level curves of the topological entropy of $f \circ m$ when the parameter values a and r ranges from 0 to 10

(see Fig. 11.3). We also show that only one diagram is obtained, that is, there is only one attractor. Figure 11.4 shows the bifurcation diagrams in this case.

11.3.2 The Jonzén-Lundberg Model

We consider the Jonzén-Lundberg model (see [17, 18]). This model takes a year divided into a reproductive season (summer), a non-reproductive one (winter) and a hunting period. The reproductive period is given by the Ricker model stated with the map $f(x) = xe^{r(1-x)}$, the hunting one by the map $c(x) = (1-\gamma)x$, and finally, the mortality map is $m(x) = xe^{-ax}$. Then, we can consider two essentially different

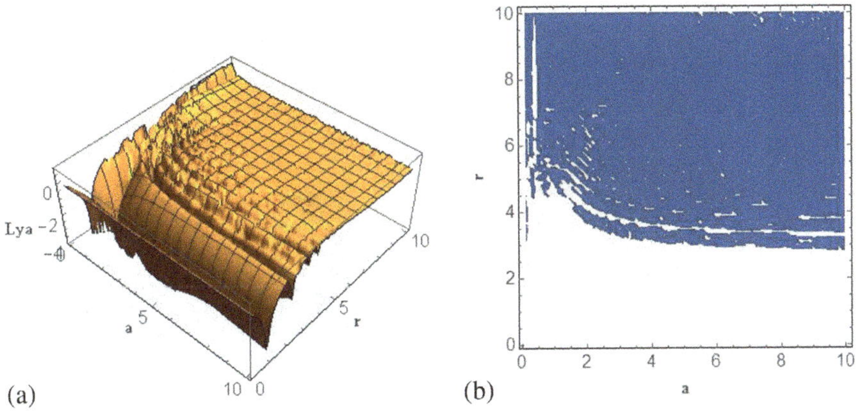

(a) (b)

Fig. 11.7 **a** Estimations of Lyapunov exponents at the turning points of $f \circ m$ when the parameter values a and r ranges from 0 to 10. of the map $f \circ m$. **b** In dark the region where the maximal Lyapunov exponent at the turning points of $f \circ m$ is positive when the parameter values a and r ranges from 0 to 10

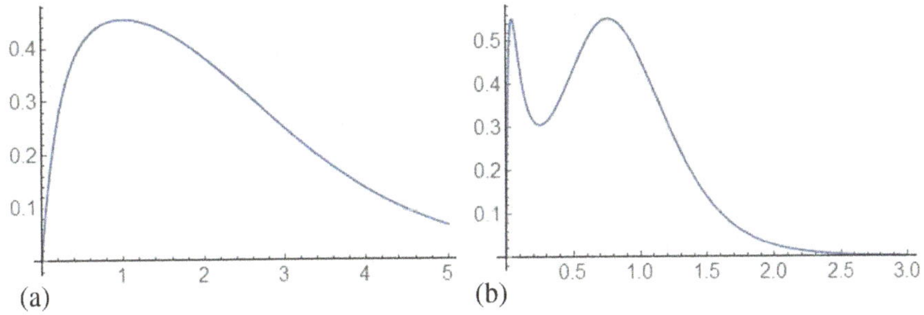

(a) (b)

Fig. 11.8 **a** Graph of $c \circ f \circ m$ when the parameter values are $\gamma = 0.25$, $a = 0.5$ and $r = 1$. **b** Graph of $c \circ f \circ m$ when the parameter values are $\gamma = 0.25$, $a = 0.5$ and $r = 4$

cases: the sequence $[m, c, f]$ means that the hunting period is spring, while for the sequence $[c, m, f]$ the hunting period is autumn.

Notice that we have three different parameters, namely, γ, a and r. This is a problem for presenting the results. So, we have chosen several values of γ, that is, we fix the rate of hunting and check the differences between the two different cases $[m, c, f]$ and $[m, f, c]$. In other words, we are going to analyze the dynamics of $c \circ f \circ m$ and $c \circ m \circ f$.

First, we study the case $\gamma = 0$. In this case $c(x) = x$. We have to analyze the associated maps $f \circ m$ and $m \circ f$. We compute the topological entropy of both maps. Note that, by the formula $h(f \circ m) = h(m \circ f)$ (see [19]), we can simplify the computations. When the map is unimodal we use the algorithm explained in Sect. 11.2, while if it is trimodal, we use the algorithm from [9], which is a variation of the algorithm from [5] for bimodal maps. We choose the map $m \circ f$ and note that

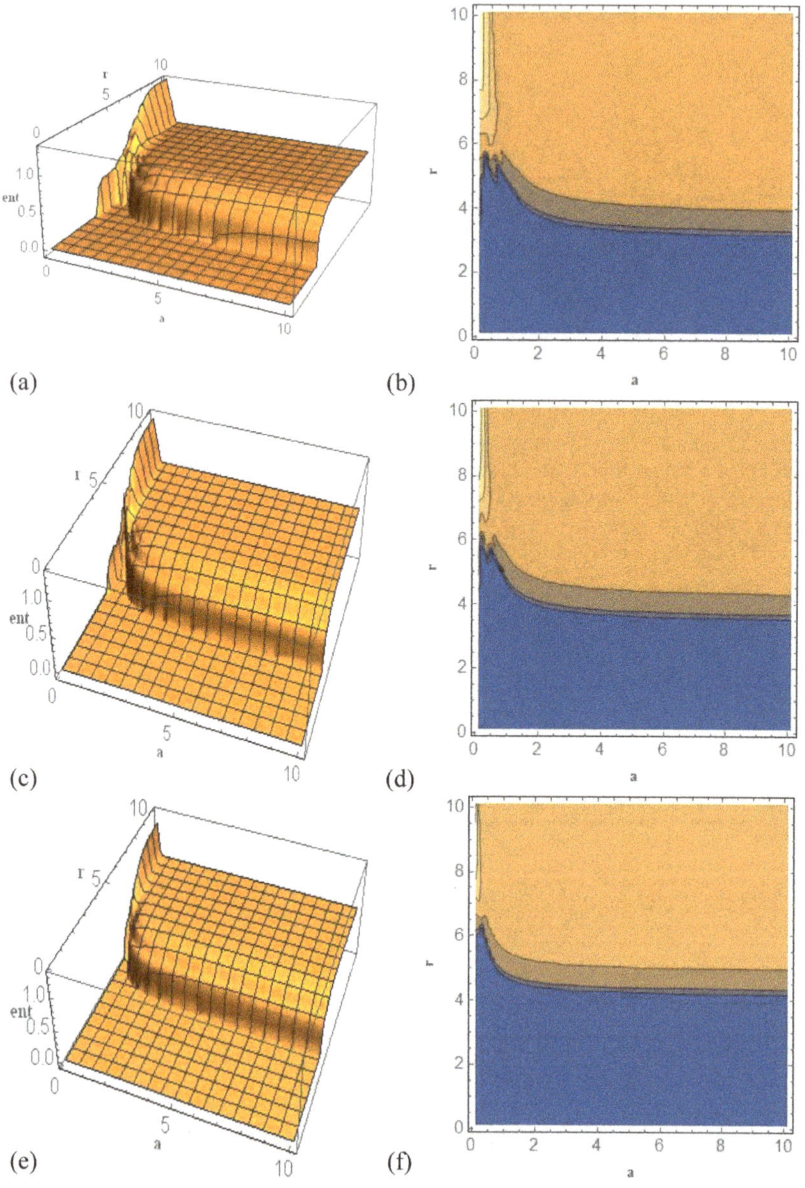

Fig. 11.9 **a** Topological entropy of $c \circ f \circ m$ when $\gamma = 0.25$ and the parameter values a and r ranges from 0 to 10. **b** Level curves of the topological entropy of $c \circ f \circ m$ when $\gamma = 0.25$ and the parameter values a and r ranges from 0 to 10. **c** Topological entropy of $c \circ f \circ m$ when $\gamma = 0.5$ and the parameter values a and r ranges from 0 to 10. **d** Level curves of the topological entropy of $c \circ f \circ m$ when $\gamma = 0.5$ and the parameter values a and r ranges from 0 to 10. **e** Topological entropy of $c \circ f \circ m$ when $\gamma = 0.75$ and the parameter values a and r ranges from 0 to 10. **f** Level curves of the topological entropy of $c \circ f \circ m$ when $\gamma = 0.75$ and the parameter values a and r ranges from 0 to 10

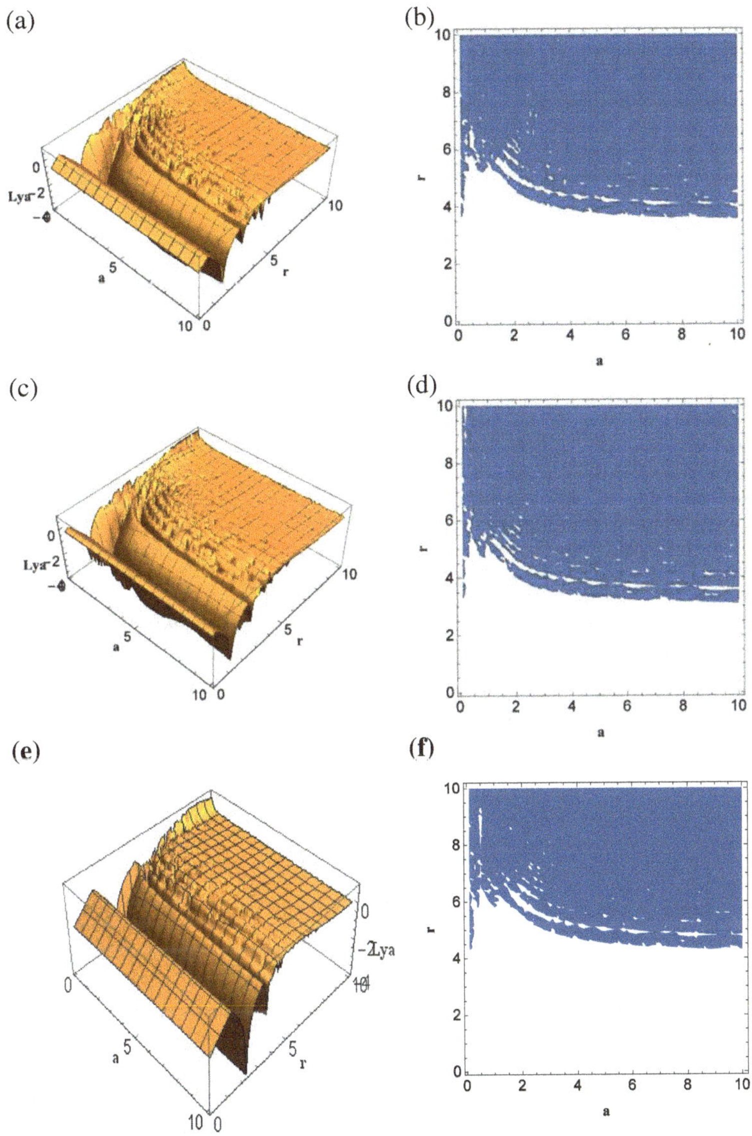

Fig. 11.10 **a** Maximal Lyapunov exponent at the turning points of $c \circ f \circ m$ when $\gamma = 0.25$ and the parameter values a and r ranges from 0 to 10. **b** In dark the region where the maximal Lyapunov exponent at the turning points of $c \circ f \circ m$ when $\gamma = 0.25$ and the parameter values a and r ranges from 0 to 10. **c** Maximal Lyapunov exponent at the turning points of $c \circ f \circ m$ when $\gamma = 0.5$ and the parameter values a and r ranges from 0 to 10. **d** In dark the region where the maximal Lyapunov exponent at the turning points of $c \circ f \circ m$ when $\gamma = 0.5$ and the parameter values a and r ranges from 0 to 10. **e** Maximal Lyapunov exponent at the turning points of $c \circ f \circ m$ when $\gamma = 0.75$ and the parameter values a and r ranges from 0 to 10. **f** In dark the region where the maximal Lyapunov exponent at the turning points of $c \circ f \circ m$ when $\gamma = 0.75$ and the parameter values a and r ranges from 0 to 10

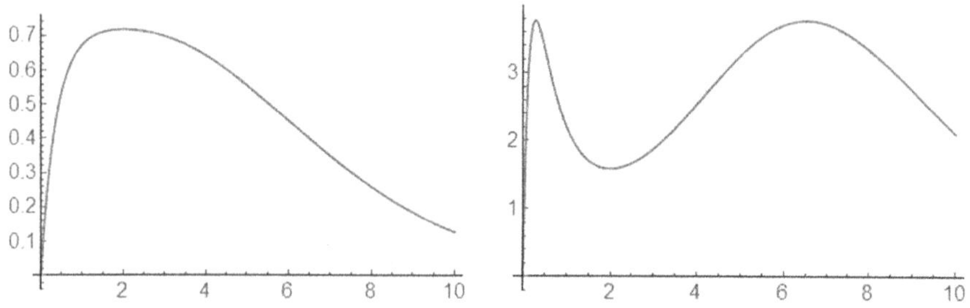

(a) (b)

Fig. 11.11 **a** Graph of $c \circ m \circ f$ when the parameter values are $\gamma = 0.25$, $a = 0.5$ and $r = 1$. **b** Graph of $c \circ m \circ f$ when the parameter values are $\gamma = 0.25$, $a = 0.5$ and $r = 4$

from $f'(x) = 0$ we get that $x = \frac{1}{r}$ is a turning point. The map can have two additional turning points provided $m'(f(x)) = 0$, which is equivalent to $xe^{r(1-x)} = \frac{1}{a}$. This equations must be solved numerically, having solution whenever $e^{r-1} \geq \frac{r}{a}$. So, the map $m \circ f$ is unimodal with a maximum turning point $1/r$ when $e^{r-1} < \frac{r}{a}$ and it is trimodal with two maximum turning points having the same trajectory and a minimum at $1/r$. Figure 11.5 shows different graphs of the map $f \circ m$.

Figures 11.6 and 11.7 show the topological entropy and the maximal Lyapunov exponent of the map $f \circ m$, respectively.

Now, let us analyze the case $\gamma \neq 0$ to check its influence in the dynamics complexity. For the first map $c \circ f \circ m$, we see from the chain rule that

$$(c \circ f \circ m)'(x) = c'((f \circ m)(x))f'(m(x))m'(x) = 0$$

From $m'(x) = 0$ we get that $x = \frac{1}{a}$ is a turning point. The map can have two additional turning points provided $f'(m(x)) = 0$, which is equivalent to $xe^{-ax} = \frac{1}{r}$. This equation must be solved numerically and has solution whenever $\frac{r}{a} \geq e$. No other extrema are possible because the map $c(x)$ is strictly increasing. So, the map $c \circ f \circ m$ is unimodal with a maximum turning point $1/a$ when $r/a < e$. Otherwise, it is trimodal with two maximum turning points having the same trajectory and a minimum at $1/a$. Figure 11.8 shows different graphics of this map.

Next, we take some values of γ, namely 0.25, 0.5 and 0.75 and check how topological entropy and Lyapunov exponents change. The result can be seen in Figs. 11.9 and 11.10. From them, we can see that the region for which the dynamic is complex decreases when the parameter γ increases, as one might expect.

For the map $c \circ m \circ f$ we find a similar scenario. The chain rule is now

$$(c \circ m \circ f)'(x) = c'((m \circ f)(x))m'(f(x))f'(x) = 0.$$

From $f'(x) = 0$ we get that $x = \frac{1}{r}$ is a turning point. The map can have two additional turning points provided $m'(f(x)) = 0$, which is equivalent to $xe^{r(1-x)} = \frac{1}{a}$. Again,

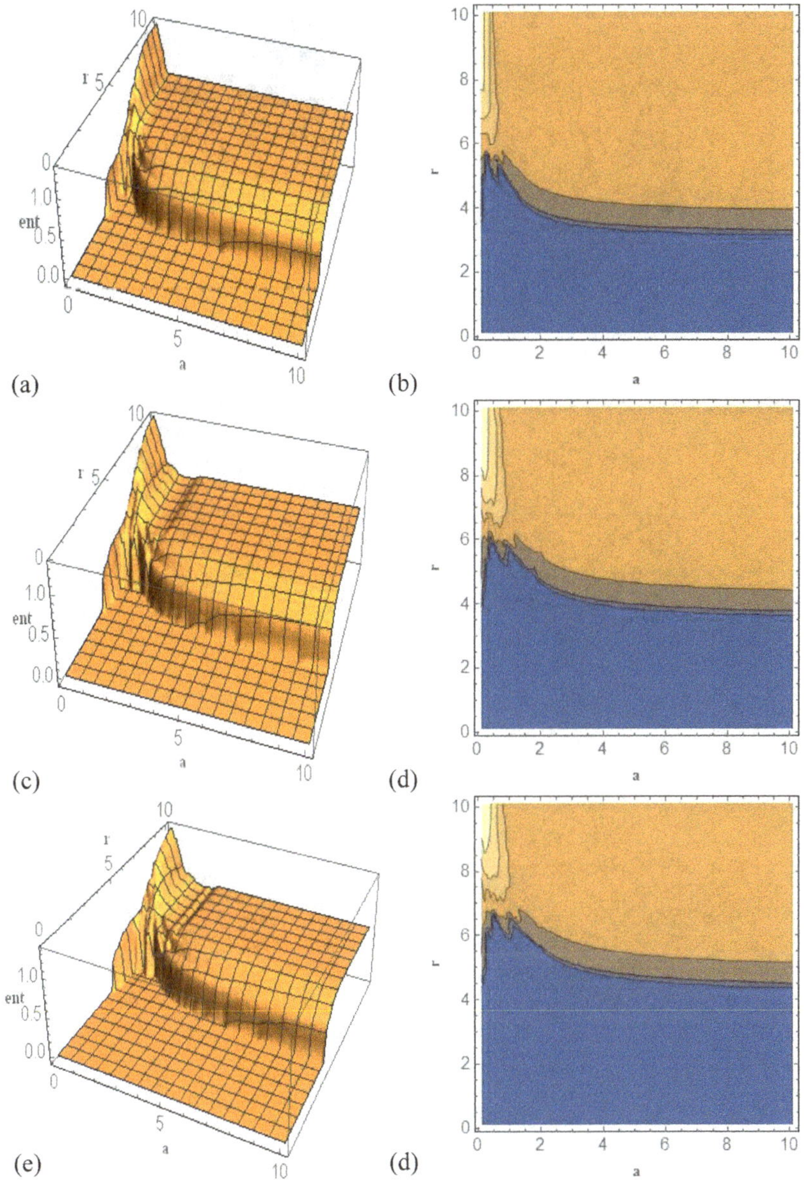

Fig. 11.12 **a** Topological entropy of $c \circ m \circ f$ when $\gamma = 0.25$ and the parameter values a and r ranges from 0 to 10. **b** Level curves of the topological entropy of $c \circ m \circ f$ when $\gamma = 0.25$ and the parameter values a and r ranges from 0 to 10. **c** Topological entropy of $c \circ m \circ f$ when $\gamma = 0.5$ and the parameter values a and r ranges from 0 to 10. **d** Level curves of the topological entropy of $c \circ m \circ f$ when $\gamma = 0.5$ and the parameter values a and r ranges from 0 to 10. **e** Topological entropy of $c \circ m \circ f$ when $\gamma = 0.75$ and the parameter values a and r ranges from 0 to 10. **f** Level curves of the topological entropy of $c \circ m \circ f$ when $\gamma = 0.75$ and the parameter values a and r ranges from 0 to 10

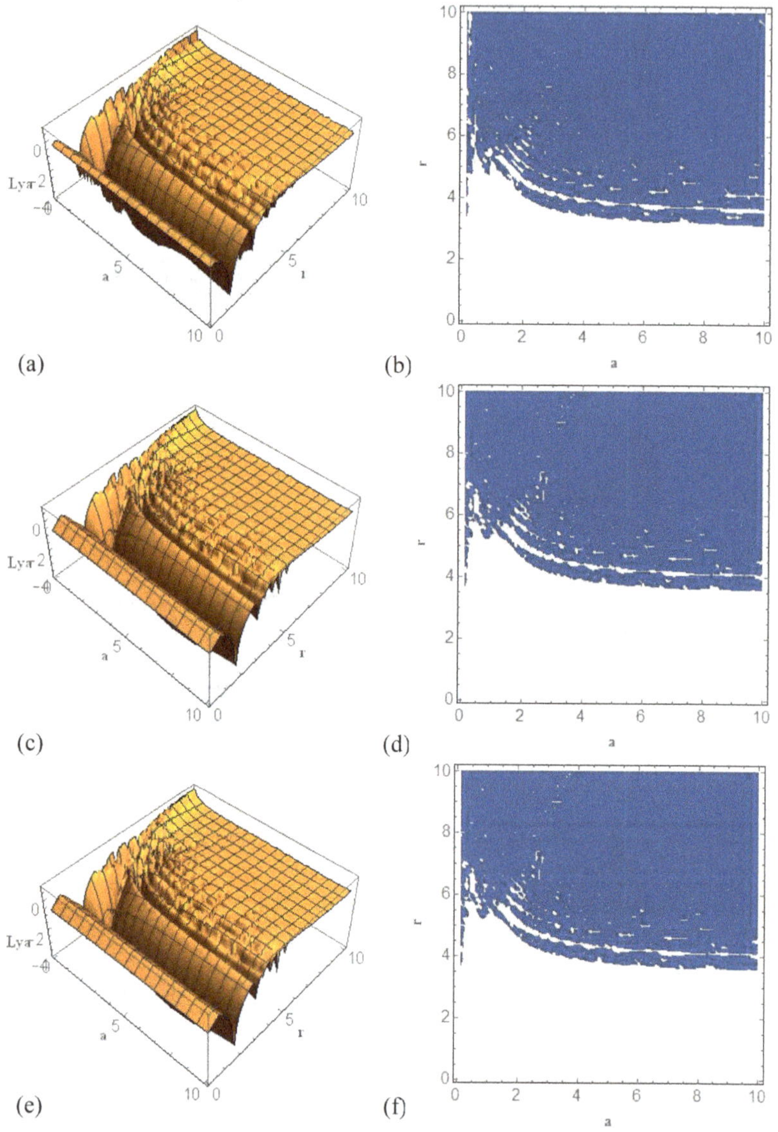

Fig. 11.13 **a** Maximal Lyapunov exponent at the turning points of $c \circ m \circ f$ when $\gamma = 0.25$ and the parameter values a and r ranges from 0 to 10. **b** In dark the region where the maximal Lyapunov exponent at the turning points of $c \circ m \circ f$ when $\gamma = 0.25$ and the parameter values a and r ranges from 0 to 10. **c** Maximal Lyapunov exponent at the turning points of $c \circ m \circ f$ when $\gamma = 0.5$ and the parameter values a and r ranges from 0 to 10. **d** In dark the region where the maximal Lyapunov exponent at the turning points of $c \circ m \circ f$ when $\gamma = 0.5$ and the parameter values a and r ranges from 0 to 10. **e** Maximal Lyapunov exponent at the turning points of $c \circ m \circ f$ when $\gamma = 0.75$ and the parameter values a and r ranges from 0 to 10. **f** In dark the region where the maximal Lyapunov exponent at the turning points of $c \circ m \circ f$ when $\gamma = 0.75$ and the parameter values a and r ranges from 0 to 10

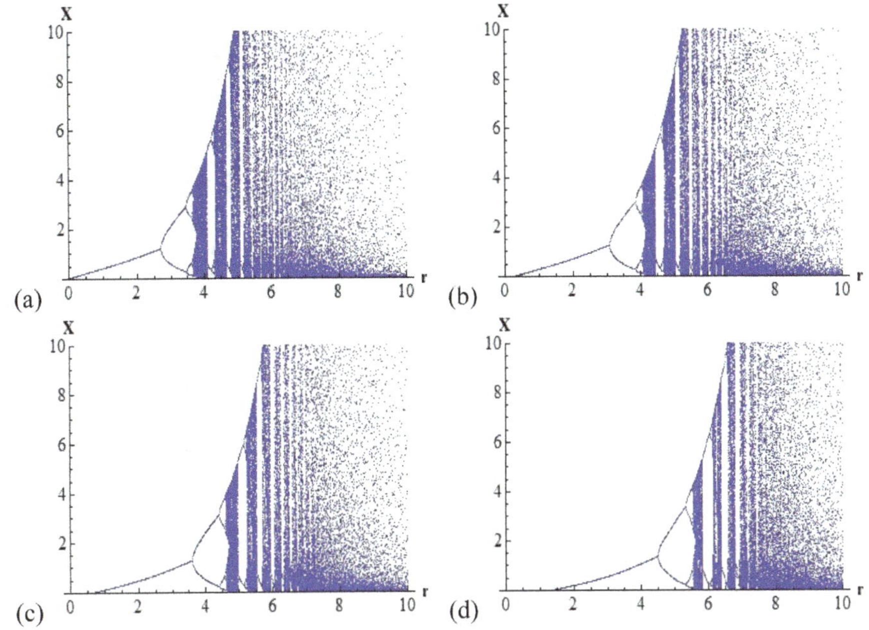

Fig. 11.14 **a** Bifurcation diagram of $c \circ f \circ m$ when $\gamma = 0$, $a = 2$ and r ranges from 0 to 10. **b** Bifurcation diagram of $c \circ f \circ m$ when $\gamma = 0.25$, $a = 2$ and r ranges from 0 to 10. **c** Bifurcation diagram of $c \circ f \circ m$ when $\gamma = 0.5$, $a = 2$ and r ranges from 0 to 10. **d** Bifurcation diagram of $c \circ f \circ m$ when $\gamma = 0.75$, $a = 2$ and r ranges from 0 to 10. Note the existence of period doubling bifurcations (see [38] and references therein)

we must solve numerically this equation having solution whenever $e^{r-1} \geq \frac{r}{a}$. So, we have that $c \circ m \circ f$ is unimodal with a maximum turning point $1/r$ when $e^{r-1} < \frac{r}{a}$. Otherwise, it is trimodal with two maximum turning points having the same trajectory and a minimum at $1/r$. Figure 11.11 shows different graphs of this map.

As in the previous case, we take values of γ equal to 0.25, 0.5 and 0.75, respectively. We check how topological entropy and Lyapunov exponents change. The result can be seen in Figs. 11.12 and 11.13. From them, we can check that the region for which the dynamic is complex decreases when the parameter γ increases, as one might expect.

It is easy to see that all the maps have negative Schwarzian derivative, and therefore at most two attractors appear. We show that for $a = 0$ fixed, according to our simulations, only one attractor appears. We also see that when γ increases, the Allee

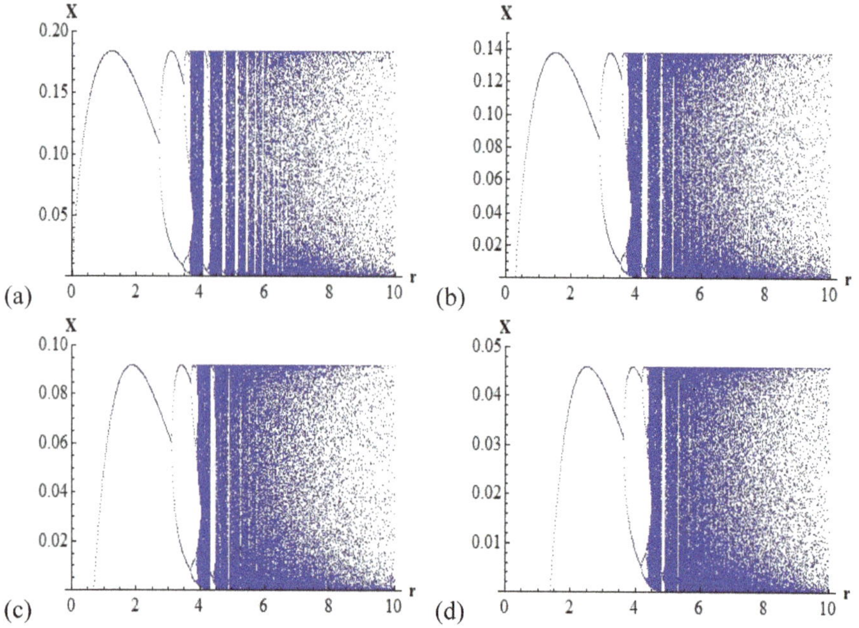

Fig. 11.15 (a) Bifurcation diagram of $c \circ m \circ f$ when $\gamma = 0$, $a = 2$ and r ranges from 0 to 10. (b) Bifurcation diagram of $c \circ m \circ f$ when $\gamma = 0.25$, $a = 2$ and r ranges from 0 to 10. (c) Bifurcation diagram of $c \circ m \circ f$ when $\gamma = 0.5$, $a = 2$ and r ranges from 0 to 10. (d) Bifurcation diagram of $c \circ m \circ f$ when $\gamma = 0.75$, $a = 2$ and r ranges from 0 to 10. Note the existence of period doubling bifurcations (see [38] and references therein)

effect appears (see e.g. [32]), that is, the species goes to extinction when the population is not big enough. We remark the difference of populations in each case shown in Figs. 11.14 and 11.15.

11.4 Conclusions

We have shown that periodically switching systems can be used to model some phenomena of Biology, Economy and Physics. We can study many dynamical properties of these periodic systems by using associated discrete dynamical systems. For one dimensional systems, we have introduced some techniques that are useful to analyze the dynamics. These techniques are valid for regular enough piecewise monotone maps. However, in practice, the analysis of some models has some technical difficulties, as for example the number of parameters, or the number of extrema of the piecewise monotone maps associated to the periodic system.

As an application, we study the dynamics of two biological models. We compute their topological entropies, estimate their Lyapunov exponents, and draw their bifurcation diagrams. Then, we find the parameter values for which the systems are topologically chaotic, when this topological chaos is physically observable, and how is the transition from simple dynamics to chaotic one. The existence of multiple attractors is also shown.

Acknowledgements I wish to thank to the anonymous referees for their advices to improve the paper. This work has been supported by the grant MTM 2017-84079-P Agencia Estatal de Investigación (AEI) y Fondo Europeo de Desarrollo Regional (FEDER).

References

1. Adler, R.L., Konheim, A.G., McAndrew, M.H.: Topological entropy. Trans. Am. Math. Soc. **114**, 309–319 (1965)
2. Alsedà, Ll., Llibre, J., Misiurewicz, M.: Combinatorial Dynamics and Entropy in Dimension One. Advances Series in Nonlinear Dynamics, vol. 5. World Scientific Publishing Co. Inc., River Edge (1993)
3. Balibrea, F., Cánovas, J.S., Jiménez López, V.: Commutativity and non-commutativity of topological sequence entropy. Ann. de l'Institut Fourier **49**, 1693–1709 (1999)
4. Block, L., Keesling, J., Li, S.H., Peterson, K.: An improved algorithm for computing topological entropy. J. Stat. Phys. **55**, 929–939 (1989)
5. Block, L., Keesling, J.: Computing the topological entropy of maps of the interval with three monotone pieces. J. Stat. Phys. **66**, 755–774 (1992)
6. Beverton, R.J.H., Holt, S.J.: On the Dynamics of Exploited Fish Populations. Fisheries investment Series 2, 19. Her Majesty's Stationary Office, London (1957)
7. Bischi, G.I., Lamantia, F., Radi, D.: An evolutionary Cournot model with limited market knowledge. J. Econ. Behav. Organ. **116**, 219–238 (2015)
8. Bowen, R.: Entropy for group endomorphism and homogeneous spaces. Trans. Am. Math. Soc. **153**, 401–414 (1971)
9. Cánovas, J.S., Muñoz-Guillermo, M.: Computing topological entropy for periodic sequences of unimodal maps. Commun. Nonlinear Sci. Numer. Simul. **19**, 3119–3127 (2014)
10. Cánovas, J.S., Panchuk, A., Puu, T.: Asymptotic dynamics of a piecewise smooth map modelling a competitive market. Math. Comput. Simul. **117**, 20–38 (2015)
11. Cavalli, F., Naimzada, A.: Monopoly models with time-varying demand function. Commun. Nonlinear Sci. Numer. Simul. **58**, 15–35 (2018)
12. Cerboni Baiardi, L., Lamantia, F., Radi, D.: Evolutionary competition between boundedly behavioral rules in oligopoly games. Chaos Solitons Fractals **79**, 204–225 (2015)
13. Dinaburg, E.I.: The relation between topological entropy and metric entropy. Sov. Math. **11**, 13–16 (1970)
14. Droste, E., Hommes, C.H., Tuinstra, J.: Endogeneous fluctuations under evolutionary pressure in Cournot competition. Games Econ. Behav. **40**, 232–269 (2002)
15. Hommes, C.H., Ochea, M.I., Tuinstra J.: On the stability of the Cournot equilibrium: An evolutionary approach. (Preprints, CeNDEF Working Paper, no 11-10) Universiteit van Amsterdam, Amsterdam (2011)
16. Jaganathan, R., Sinha, S.: A q-deformed nonlinear map. Phys. Lett. A **338**, 277–287 (2005)
17. Jonzén, N., Lundberg, P.: Temporally structured density dependence and population management. Ann. Zool. Fennici. **36**, 39–44 (1999)
18. Liz, E.: Effects of strength and timing of harvest on seasonal population models: stability switches and catastrophic shifts. Theor. Ecol. **10**, 235–244 (2017)
19. Kolyada, S., Snoha, L.: Topological entropy of nonautononous dynamical systems. Random Comput. Dyn. **4**, 205–233 (1996)

20. Kollias, I., Camouzis, E., Leventides, J.: Global analysis of solutions on the Cournot-Theocharis duopoly with variable marginal costs. J. Dyn. Games **4**, 25–39 (2017)
21. Kopel, M.: Simple and complex adjustment dynamics in Cournot duopoly models. Chaos Solitons Fractals **7**, 2031–2048 (1996)
22. Matsumoto, A., Nonaka, N.: Statistical dynamics in a chaotic Cournot model with complementary goods. J. Econ. Behav. Organ. **61**, 769–783 (2006)
23. May, R.M.: Simple mathematical models with very complicated dynamics. Nature **261**, 459–467 (1976)
24. Mendoza, S.A., Peacock-López, E.: Switching induced oscillations in discrete one-dimensional systems. Chaos Solitons Fractals **115**, 35–44 (2018)
25. Milnor, J.: On the concept of attractor. Commun. Math. Phys. **99**, 177–195 (1985)
26. Oseledets, V.I.: A multiplicative ergodic theorem. Lyapunov characteristic numbers for dynamical systems. Trans. Moscow Math. Soc. **19**, 197–231. Moscov. Mat. Obsch. **19**(1968), 179–210 (1968)
27. Patidar, V., Sud, K.K.: A comparative study on the co-existing attractors in the Gaussian map and its q-deformed version. Commun. Nonlinear Sci. Numer. Simul. **14**, 827–828 (2009)
28. Peacock-López, E.: Seasonality as a Parrondian game. Phys. Lett. A **375**, 3124–3129 (2011)
29. Puu, T.: Chaos in duopoly pricing. Chaos Solitons Fractals **1**, 573–581 (1991)
30. Puu, T., Norin, A.: Cournot duopoly when the competitors operate under capacity constraints. Chaos Solitons Fractals **18**, 577–592 (2003)
31. Ricker, W.E.: Stock and recruitment. J. Fisheries Res. Board Can. **11**, 559–623 (1954)
32. Schreiber, S.J.: Allee effects, extinctions, and chaotic transients in simple population models. Theor. Popul. Biol. **64**, 201–209 (2003)
33. Shrimali, M.D., Banerjee, S.: Delayeed q-deformed logisitc map. Commun. Nonlinear Sci. Numer. Simul. **18**, 3126–3133 (2013)
34. Silva, E., Peacock-Lopez, E.: Seasonality and the logisitic map. Chaos Solitons Fractals **95**, 152–156 (2017)
35. Singer, D.: Stable orbits and bifurcation of maps of the interval. SIAM J. App. Math. **35**, 260–267 (1978)
36. Smale, S.: Differentiable dynamical systems. Bull. Am. Math. Soc. **73**, 747–817 (1967)
37. Thunberg, H.: Periodicity versus chaos in one-dimensional dynamics. SIAM Rev. **43**, 3–30 (2001)
38. Tresser, C., Coullet, P., de Faria, E.: Period doubling. Scholarpedia **9**(6), 3958 (2014)
39. van Strien, S., Vargas, E.: Real bounds, ergodicity and negative Schwarzian for multimodal maps. J. Am. Math. Soc. **17**, 749–782 (2004)
40. Zhang, C., Shen, B., Yu, Y.: Induced complex behaviors via parameter-switching scheme in a standard logistic circuit, preprint (2019)

Chapter 12
Co-jumps and Markov Counting Systems in Random Environments

Carles Bretó

Abstract Motivated by the analysis of multi-strain infectious disease data, we provide closed-form transition rates for continuous-time Markov chains that arise from subjecting Markov counting systems to correlated environmental noises. Noise correlation induces co-jumps or counts that occur simultaneously in several counting processes. Such co-jumps are necessary and sufficient for infinitesimal correlation between counting processes of the system. We analyzed such infinitesimal correlation for a specific infectious disease model by randomizing time of Kolmogorov's Backward system of differential equations based on appropriate stochastic integrals.

12.1 Introduction

Continuous-time stochastic processes have proved to be useful for research in many areas of science. Such processes are often defined in terms of infinitesimal parameters, like infinitesimal transition rates in the case of continuous-time Markov chains. Assuming that such infinitesimal rates are subjected to external noises has proved particularly useful for data analysis and has been referred to as environmental stochasticity or, more generally, as random environments. Subjecting infinitesimal rates to such noises is known to result in new continuous-time Markov chains defined by new infinitesimal rates. While these new rates have been derived in closed form for the case of independent noises, facilitating the understanding of the properties of the resulting new continuous-time Markov chains, they have not been derived for the case of dependent noises. This lack of closed-form rates for the case of correlated noises creates uncertainty about the properties of the correlated system and can make applied researchers reluctant to take advantage of the modeling approaches of environmental stochasticity and random environments. To prevent such reticence, this paper considers introducing correlated external noises to the rates of Markov counting systems and provides closed-form expressions for the new rates, which capture

C. Bretó (✉)
Departament d'Anàlisi Econòmica, Universitat de València, Valencia, Spain
e-mail: carles.breto@uv.es

V. A. Sadovnichiy and M. Z. Zgurovsky (eds.), *Contemporary Approaches and Methods in Fundamental Mathematics and Mechanics*, Understanding Complex Systems,
https://doi.org/10.1007/978-3-030-50302-4_12

the effect of noise correlation. These new rates are the main contribution of the paper and are based on novel closed-form expressions for the infinitesimal covariances of Markov counting systems. Furthermore, this unusual focus of this paper on system covariances provides a characterization of correlated environmental noises in terms of simultaneous transitions in non-random environments.

The study of properties of counting processes has benefited among others the fields of epidemiology and ecology, which have relied on Markov processes both in deterministic and stochastic environments (the latter sometimes being favoured by empirical evidence). Research in these two disciplines has taken advantage of theoretical investigation of Markov processes both historically [5, 30] and more recently [18, 24]. Such Markov processes are often systems of interacting Markov counting processes, which are termed Markov counting systems by [16] and include networks of queues [10] and compartmental models [28, 37]. Markov counting systems are continuous-time Markov chains and can be defined by transition rates [10]. Noisy transition rates are often referred to as environmental stochasticity in epidemiology and ecology [14, 22]. The role of such stochasticity has been extensively studied, including in the context of deterministic ODE skeletons driven by diffusions [21, 26, 27, 31] and driven by Lévy processes [6, 33]. The role of stochastic environments has also been studied in the context of Markov counting systems, both paying attention to the system probabilistic properties e.g., [9, 15, 36, 47] and focusing on the biological implications from empirical studies that favour random environments over non-random ones e.g., [35, 40, 44, 46].

Epidemiological applications have come to consider multiple interacting pathogens and to study them based on counting systems subjected to correlated environmental noise, fitting in the framework provided by Bretó et al. [15] and Bretó and Ionides [16] who formalize the transition rates of the system subjected to noise. Pathogen interaction has received attention for some time now [23, 29], both without considering the role of external noises [1, 17, 41] and considering it [15, 44]. In particular, Shrestha et al. [44] consider a Markov counting system corresponding to a compartmental model of the susceptible-infectious-recovered type (a simpler version of which is considered by [15] and reproduced in Fig. 12.1). In [44], two pathogens co-exist but there are more than two possible different types of infection (depending on the history of past infections of individuals). The rate at which these different types of infection occur are assumed to be subjected to a single (common to all infection types) external white noise, making all infection rates correlated. Such rate randomization has been formalized by Bretó and Ionides [16], showing that the system subjected to noises is a new Markov counting system not only when the noises are independent (in which case Bretó et al. [15] provide closed-form rates) but also when the noises are correlated, on which this paper focuses.

The problem we take up in this paper is providing closed-form transition rates that define Markov counting systems accounting for correlated noises and is made difficult by the lack of closed-form transition probabilities of general systems. Closed-form transition probabilities are readily available for basic systems, e.g., those of a Poisson process correspond to a Poisson distribution and those of a linear pure death process to a binomial distribution [7]. However, they are not available for general compartmental models, including the system of interacting birth-death processes considered by [44] or that represented in Fig. 12.1. If such closed-form transition probabilites for general systems without noises were available, they could be used to obtain the desired closed-form transition rates with noises by integrating out the noise and taking limits.

A key downside of lacking closed-form rates is that promising scientific results and genuine biological rationale behind correlated noises may be outweighted by uncertainty about the properties of the model subjected to correlated noises and about the interpretation of empirical results. For example, [44] show that it is feasible to arrive at correct and precise biological conclusions regarding pathogen interaction based on their Markov counting systems with correlated noises. In addition, a heuristic biological justification for correlated noises could be as follows: while localized environmental variations need not affect all types of infection, changes at a larger scale in the environment should be expected to, like heat or cold waves. However, unless the properties of the model after subjecting it to correlated noise are clear and appealing, there is the risk that these models are avoided in actual applications.

The main contribution of this paper is to provide closed-form transition rates for Markov counting systems subjected to correlated noises based on the system infinitesimal covariances and to provide an illustration in the context of biological analysis of multi-strain pathogen dynamics. The provided closed-form expressions apply to a broad range of cases considered in the applied literature. They reduce the uncertainty about the properties of the model with correlated noises by giving a precise definition of the system as formalized in Sects. 12.2 and 12.3. Another motivation for pursuing the closed-form transition rates we provide is that they can be used to obtain closed-form infinitesimal covariances using our results from Sect. 12.4.

These covariances allow circumventing the problem of unavailable transition probabilities from which to directly integrate out the noise. Although our focus on infinitesimal covariances is unusual in the context of Markov counting systems, it is as pertinent as in the context of multivariate diffusions and leads to the novel closed-form expressions provided in Theorem 12.1. These expressions show that correlated noises induce simultaneous counts and that these in turn induce stronger correlations within the system. Hence, if additional correlation is demanded by data, it could be modelled with random environments. In this case, these environments could be interpreted as devices that generate the needed correlation in a non-random environment, instead of as actual random changes in parameters, analogously to interpreting parameter randomization as a device to generate overdispersion [38]. This is illustrated in Sect. 12.5, where the rates and interpretation of the role of correlated noises for Fig. 12.1 are given.

12.2 Markov Counting Systems Without External Noise

Markov counting systems are defined as continuous-time Markov chains driven by a collection of interacting counting processes that fully characterizes the transition rates of the system [15] and such definition can often be formalized in a diagram (similar to that in Fig. 12.1). Before formally defining Markov counting systems, we introduce their key aspects. First, consider a population whose members are at any point in time in one (and only one) of C possible stages (or compartments) of their lives, with stages belonging to finite collection $\mathscr{C}$. Next, let the number of population members that are at stage c at time t define integer-valued random variables $X_c(t)$, which make up the system $\{\boldsymbol{X}(t)\} \equiv \{X_c(t) : c \in \mathscr{C}\}$. Then, let the number of population members that have transitioned from stage i to stage j by time t define non-decreasing, integer-valued random variables $N_{ij}(t)$, which in turn, for all pairs (i, j) belonging to a collection of allowed transitions $\mathscr{T}$, define the collection of counting processes $\{\boldsymbol{N}(t)\} \equiv \{N_{ij}(t) : (i, j) \in \mathscr{T}\}$. Next, let the collection $\{\boldsymbol{N}(t)\}$ drive the dynamics of the system $\{\boldsymbol{X}(t)\}$ via the "conservation of mass" identity

$$X_c(t) = X_c(0) + \sum_{i:(i,c)\in\mathscr{T}} N_{ic}(t) - \sum_{j:(c,j)\in\mathscr{T}} N_{cj}(t), \tag{12.1}$$

with counts initialized at zero, i.e., $\boldsymbol{N}(0) = \boldsymbol{0}$, so that changes in $\{\boldsymbol{X}(t)\}$ are the result of changes in $\{\boldsymbol{N}(t)\}$. Mass conservation identity (12.1) restricts the transitions that can occur in $\{\boldsymbol{X}(t)\}$ as follows. Let $\mathbb{N}_0$ $(\mathbb{N})$ be the natural numbers including (excluding) zero. For any given increments of the collection of counts $\boldsymbol{\ell} \equiv \{\ell_{ij} : (i, j) \in \mathscr{T}\} \in \left\{\mathbb{N}_0^{\mathscr{T}} - (0, \ldots, 0)\right\}$, the system $\{\boldsymbol{X}(t)\}$ must make transitions $\boldsymbol{u} \equiv \{u_c : c \in \mathscr{C}\} \in \mathbb{Z}^{\mathscr{C}}$ with $u_c = \sum_{(i,c)\in\mathscr{T}} \ell_{ic} - \sum_{(c,j)\in\mathscr{T}} \ell_{cj}$. Finally, let the following transition rates define the continuous-time Markov chain $\{\mathbf{X}(t), \mathbf{N}(t)\}$

$$q(\boldsymbol{x}, \boldsymbol{\ell}) \equiv \lim_{h\downarrow 0} \frac{P\Big(\boldsymbol{N}(t+h)=\boldsymbol{n}+\boldsymbol{\ell},\ \boldsymbol{X}(t+h)=\boldsymbol{x}+\boldsymbol{u}\ |\boldsymbol{N}(t)=\boldsymbol{n}, \boldsymbol{X}(t)=\boldsymbol{x}\Big)}{h}. \tag{12.2}$$

Since the left hand side of (12.2) is assumed to only depend on $\boldsymbol{x}$ (and not $\boldsymbol{n}$) and on $\boldsymbol{\ell}$, $\{\boldsymbol{X}(t)\}$ is itself a continuous-time Markov chain and we call it a Markov counting system,[1] which we illustrate with the following example.

Figure 12.1 defines a Markov counting system by relying on the concepts of marginal transition rates and of pairwise transition rates, which are necessary for its interpretation and which are also key to studying the effect of correlated external noise. Consider the rate at which k population members simultaneously undergo a transition of the ij-type (regardless of whether other members undergo other transitions), which can be defined as $q_{ij}(\boldsymbol{x}, k) \equiv \sum_{\boldsymbol{\ell}:\ell_{ij}=k} q(\boldsymbol{x}, \boldsymbol{\ell})$ for $k \in \mathbb{N}$ and which we call the (i, j) marginal transition rate. Marginal transition rates of size one $q_{ij}(\boldsymbol{x}, 1)$ are the labels on the arrows in Fig. 12.1. Marginal rates of sizes greater than one do not appear in Fig. 12.1 because they are assumed to be zero. Another assumption needed to interpret Fig. 12.1 is that there are no co-jumps of different types. To formalize this second assumption, consider the rate at which $\boldsymbol{k} = (k_{ij}, k_{i'j'})$ population members simultaneously undergo transitions of the (i, j) and (i', j') types (regard-

[1] The transition rates in (12.2) are time homogeneous, since its left hand side does not depend on t. This homogeneity adds clarity to the concepts, results and proofs but can readily be relaxed.

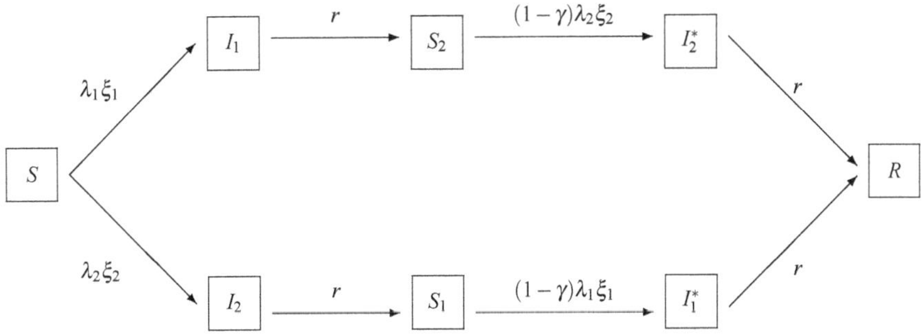

Fig. 12.1 Multi-strain SIR-type compartmental model of [15]. This model will be used in Sect. 12.5 to illustrate our results. Each individual falls in one compartment: S, susceptible to both strains ; I_1, infected with strain 1; I_2, infected with strain 2; S_1, susceptible to strain 1 (but immune to strain 2); S_2, susceptible to strain 2 (but immune to strain 1); I_1^*, infected with strain 1 (but immune to strain 2); I_2^*, infected with strain 2 (but immune to strain 1); and R, immune to both strains. Regarding demography, births enter S from compartment B (not plotted), at rate $b(t)$ driven by birth data (which is treated as a covariate), and all individuals have a common mortality rate m at which they leave each compartment in the diagram into D (not plotted). Regarding disease dynamics, r is the recovery rate from infection; γ measures the strength of cross-immunity between strains; and λ_i is the per-capita infection rate of strain i with ξ_i being the stochastic noise on this rate. Moreover, $\lambda_i = \beta(t)(I_i(t) + I_i^*(t))^\alpha / P(t) + \omega$, where $0 \leq \beta(t)$ is parameterized with a trend and a smooth seasonal component, $0 \leq \omega$ models infections from an environmental reservoir and $0 \leq \alpha \leq 1$ captures inhomogeneous mixing of the population

less of whether other members undergo other transitions), which can be defined as $q_{ij,i'j'}(\boldsymbol{x}, \boldsymbol{k}) \equiv \sum_{\boldsymbol{\ell}:\ell_{ij}=k_{ij},\ell_{i'j'}=k_{i'j'}} q(\boldsymbol{x}, \boldsymbol{\ell})$ for $\boldsymbol{k} \in \left\{\mathbb{N}_0^2 - (0,0)\right\}$ and which we call the $(i, j) - (i', j')$ pairwise transition rate. Requiring all pairwise transition rates to satisfy $q_{ij,i'j'}(\boldsymbol{x}, (1, 0)) = q_{ij}(\boldsymbol{x}, 1)$ and $q_{ij,i'j'}(\boldsymbol{x}, (0, 1)) = q_{i'j'}(\boldsymbol{x}, 1)$ guarantees no co-jumps and allows interpreting figures such as Fig. 12.1 (letting ξ_i be constants) as continuous-time Markov chains see [2, 10, 28]. Such interpretation also assumes that all non-zero rates $q_{ij}(\boldsymbol{x}, 1)$ are deterministic functions of the chain state $\boldsymbol{x}$ and not subjected to external noise.

12.3 Markov Counting Systems with External Noise

The addition of independent external noises to the Markov counting systems of Sect. 12.2 has been extensively studied in the literature, which we briefly review here from a theoretical perspective that focuses on infinitesimal overdispersion, compound processes and time randomization.

Infinitesimal equidispersion of Markov counting systems is the property that the infinitesimal mean and variance of increments of each counting process of the system are the same [11, 16]. Infinitesimal equidispersion implies that the counting processes of the system must be simple (i.e., increase by no more than one count at a time; [20]). Such dispersion constraints of the counting processes involved, $\{\boldsymbol{N}(t)\}$, which have equal infinitesimal means and variances, are inherited by the Markov counting systems of Sect. 12.2, $\{\boldsymbol{X}(t)\}$. These dispersion constraints are the result of combining simple processes and exponential inter-event times. Both simpleness and exponential times hold for mixed processes. In mixed processes, parameters (e.g., the rate in a Poisson process) are drawn at random from the mixing distribution. Hence, parameters remain constant over time but can take different values for different process realizations; for a formal definition, see, e.g., [20, 45]. This randomness in the parameter value generates additional variability between the trajectories of multiple realizations of the mixed process. However, for any given process realization, the randomness in the paramenter value fails to generate additional variability within the trajectory of that given realization as argued in more detail in, e.g., [25]. As a consequence, mixed processes remain infinitesimally equidispersed [16]. To obtain infinitesimal overdispersion, it is necessary to allow for the possibility of multiple transitions occurring at the same time (something that might approximate clustered transitions relative to exponential inter-event times).

Simultaneous transitions have been studied see, e.g., [12, 32, 34] in the context of a change of time by a stochastic process or clock see, e.g., [4, 8], e.g., by a gamma or Poisson clock [13]. Such time-changed Markov counting processes can be characterized as Markov counting processes in random environments.

Consider randomizing the rate of the counting process corresponding to a simple linear death process, $\{N(t)\}$, with death rate $\lambda(n) = (x - n)\rho$. This process has increments that follow a binomial distribution [42] and satisfies Kolmogorov's Backward Differential System (KBDS), as formally defined in, e.g., [10]. KBDS plays a key role in our approach to randomizing rates of counting processes. For completeness, we summarize such approach here and point the interested reader to Appendix B of the original paper [16]. Define a new process $\{N^\star(t)\}$ for which the same KBDS holds but with $\lambda(n)$ replaced by a new, noisy rate $\lambda(n)\xi(t)$, i.e., adding multiplicative gamma continuous-time white noise $\xi(t) = d\Gamma(t)/dt$ to the original rate. This noise addition defines a stochastic version of the original KBDS. Appropriately defining stochastic integration[2] against gamma processes, the stochastic version of the KBDS is satisfied by time-changed process $\big\{N\big(\Gamma(t)\big)\big\}$. A Taylor series expansion of the mean and variance of the binomial increments confirms that a gamma time-change produces infinitesimal overdispersion, something that need not occur in general e.g., changing time by stationary Ornstein-Uhlenbeck processes, as in [36, 39, 47]. These ideas regarding time change and simultaneous transitions extend to Markov counting systems used in applications.This extension can be made more straightforward by directly defining systems of interacting univariate time-changed counting processes as in [11] instead of randomizing time of continuous-time Markov chains more general than the basic death binomial process.

---

[2]For stochastic integration, [16] use the Marcus canonical stochastic integral with Marcus map $\Phi(u, x, y) = \pi^{M(\lambda)}_{m,m+k}(x + uy)$, developed in the context of Lévy calculus [3]. Unlike the Itô integral, it satisfies a chain rule of the Newton-Leibniz type, becoming the Stratonovich integral in the case of continuous Lévy processes.

12.4 Markov Counting Systems with Correlated External Noise

12.4.1 Correlated External Noise in Bivariate Death Markov Counting Systems

Introducing correlated noise to the rates is easier if one considers two independent death processes, in which case such correlated noise results in co-jumps and infinitesimal covariance as stated in Proposition 12.1 below, which considers a common multiplicative gamma external noise, a common death rate, and which will later be useful when considering general Markov counting systems. Proposition 12.1 was proved in [16] and we state it here to make the paper self-contained (since it is helpful in Sect. 12.5). We state it in terms of our notation for Markov counting systems, after introducing our notation for the noise. The noise affecting the individual death rates is assumed to be continuous-time white noise obtained from a gamma process, which is also the choice of noise in [15, 44]. Gamma white noise is defined as $\{\xi(t)\} \equiv \{d\Gamma(t)/dt\}$ with $\Gamma(t) \sim \text{Gamma}\,(t/\tau, \tau)$, $E[\Gamma(t)] = t$, and $V[\Gamma(t)] = \tau t$, so that τ parameterizes the magnitude of the noise.

Proposition 12.1 (Proposition 7 of [16] [3]) *Consider the bivariate Markov counting system* $\{\boldsymbol{Y}(t)\} \equiv \left\{\left(Y_1(t), Y_2(t)\right)\right\}$ *defined by counting processes* $\left\{\left(N_{Y_1D}(t), N_{Y_2D}(t)\right)\right\}$ *through mass conservation equations*

$$Y_i(t) = Y_i(0) - N_{Y_iD}(t),$$

and by transition rates $q_{Y_iD}(y_i, 1) = \delta y_i\, \mathbb{I}\{0 < y_i\}$ *i.e., two independent linear death processes having equal individual death rate* $\delta \in R^+$ *and initial population sizes* $Y_i(0)$. *Consider subjecting both* $q_{Y_iD}(y_i, 1)$ *to a common gamma white noise, which defines counting processes* $\{N_{\tilde{Y}_i\tilde{D}}(t)\} = \{N_{Y_iD}\left(\Gamma(t)\right)\}$ *and the corresponding Markov counting system* $\{\tilde{\boldsymbol{Y}}(t)\}$, *i.e., two death processes each having stochastic rate* $\delta\xi(t)$. *Then, the transition rates of* $\{\tilde{\boldsymbol{Y}}(t)\}$ *correspond to pairwise transition rates, letting* $k_i = k_{\tilde{Y}_i\tilde{D}} \in \left\{\mathbb{N}_0^2 - (0,0) : k_i \leq \tilde{y}_i\right\}$,

$$q_{\tilde{Y}_1\tilde{D},\tilde{Y}_2\tilde{D}}\left((\tilde{y}_1, \tilde{y}_2), (k_1, k_2)\right) = \binom{\tilde{y}_1}{k_1}\binom{\tilde{y}_2}{k_2}\sum_{j=0}^{k_1+k_2}\binom{k_1+k_2}{j}(-1)^{k_1+k_2-j+1}\tau^{-1}\ln\left(1 + \delta\tau(\tilde{y}_1 + \tilde{y}_2 - j)\right).$$

Furthermore, the infinitesimal covariance between $\{N_{\tilde{Y}_1\tilde{D}}(t)\}$ *and* $\{N_{\tilde{Y}_2\tilde{D}}(t)\}$ *is*

---

[3] In Proposition 12.1, $\mathscr{C} = \{Y_1, Y_2, D\}$ so that, following our notation from (12.1), $\{\boldsymbol{X}(t)\} = \{(X_{Y_1}(t), X_{Y_2}(t), X_D(t))\}$, which we write as $\{(Y_1(t), Y_2(t))\}$ for simplicity. Furthermore, transition rates $q_{\tilde{Y}_1\tilde{D},\tilde{Y}_2\tilde{D}}$ are defined in (12.2) and the paragraph that follows it.

$$\lim_{h\downarrow 0} h^{-1} Cov\left[N_{\tilde{Y}_1\tilde{D}}(t+h) - N_{\tilde{Y}_1\tilde{D}}(t),\ N_{\tilde{Y}_2\tilde{D}}(t+h) - N_{\tilde{Y}_2\tilde{D}}(t)\ \middle|\ \tilde{\boldsymbol{Y}}(t) = \tilde{\boldsymbol{y}}\right]$$

$$= \tilde{y}_1\tilde{y}_2\tau^{-1}\ln\left(\frac{(1+\delta\tau)^2}{1+2\delta\tau}\right) > 0.$$

The common rate assumption of Proposition 12.1 can be relaxed at the cost of more complex closed-form expressions for the covariance and for the pairwise rates. Although different death rates are assumed in the interacting death processes of Fig. 12.1, they will be assumed to be equal for the sake of simplicity when we illustrate in Sect. 12.5 our results for general Markov counting systems (which hold regardless of whether individual death rates are equal).

12.4.2 Correlated External Noise in General Markov Counting Systems

Consider generalizing Proposition 12.1 to general Markov counting systems, which will lead us to defining infinitesimal covariances of such general systems. Consider a general Markov counting system as defined by transition rates (12.2) that satisfies the standard assumptions to interpret Fig. 12.1 of neither multiple jumps nor co-jumps and denote such system by $\{\boldsymbol{W}(t)\}$. Consider now subjecting some (or all) transition rates of $\{\boldsymbol{W}(t)\}$ to a collection of (possibly correlated, not necessarily gamma) white noises derived from $\{\boldsymbol{\Gamma}(t)\}$ (analogously to Sect. 12.4.1) and call the resulting process $\{\tilde{\boldsymbol{W}}(t)\}$. The transitions rates of $\{\tilde{\boldsymbol{W}}(t)\}$ could be obtained by integrating out the noises $\{\boldsymbol{\Gamma}(t)\}$ from the (now randomized) transition probabilities appearing inside the limit in (12.2) as follows. Consider the collection of randomized time increments $\boldsymbol{H} \equiv \{H_{ij} : (i,j) \in \mathscr{T}\}$, necessarily with $E[H_{ij}] = h$. The nature of H_{ij} depends on whether $q_{ij}(\boldsymbol{w}, 1)$ is subjected to noise: if yes, then H_{ij} is the corresponding noise random variable with density $f_{H_{ij}}$; if not, then it is the degenerate random variable $H_{ij} = h$. Then, provided they exist, the transition rates of $\{\tilde{\boldsymbol{W}}(t)\}$ are

$$q(\tilde{\boldsymbol{w}}, \boldsymbol{\ell}) = \lim_{h\downarrow 0} \int \frac{P\Big(\boldsymbol{N}(t+s) = \boldsymbol{n}+\boldsymbol{\ell},\ \boldsymbol{W}(t+s) = \tilde{\boldsymbol{w}}+\boldsymbol{u}\ |\boldsymbol{N}(t) = \boldsymbol{n},\ \boldsymbol{W}(t) = \tilde{\boldsymbol{w}}\Big)}{s} f_{\boldsymbol{H}}(s)ds \tag{12.3}$$

with $\boldsymbol{u}$ and $\boldsymbol{l}$ defined as in Sect. 12.2. While such direct integration of the noise in equation (12.3) was straightforward for the bivariate death process of Proposition 12.1, it is not so straightforward for more sophisticated models, like the one in Fig. 12.1 [15] or similar models [44]. Hence, instead of obtaining the new transition rates by direct integration, we propose constructing such new rates by directly specifying transition rates that produce the same infinitesimal covariance as that produced by introducing noise to appropriate simpler systems, assuming such simpler systems exist. In the case of Fig. 12.1, noises $\xi_i(t)$ affect the rate of death processes $\{N_{SI_i}\}$ and $\{N_{S_i I_i^*}\}$, for which appropriate simpler processes do exist, e.g., the bivariate death process of Proposition 12.1. Note that covariances corresponding to bivariate processes other than the gamma-binomial of Proposition 1 could be used, e.g., that of a Poisson-binomial process, e.g., [12], or beta-binomial [16]. For Fig. 12.1, our proposal amounts to specifying a new set of transition rates for the system represented by Fig. 12.1 such that the infinitesimal covariance between $\{N_{SI_i}\}$ and $\{N_{S_i I_i^*}\}$ matches the covariance given by Proposition 12.1 between $\{N_{Y_1 D}\}$ and $\{N_{Y_2 D}\}$. To do this, we first derive closed-form expressions for the infinitesimal covariances between two counting processes involved in a Markov counting system.

12.4.3 Infinitesimal Covariance of Markov Counting Systems

Define the infinitesimal covariances of a general Markov counting system $\{\boldsymbol{X}(t)\}$ as defined in Sect. 12.2 as the collection

$$\{\boldsymbol{\sigma}_{dX}(\boldsymbol{x})\} \equiv \left\{\sigma_{dX}^{ij,i'j'}(\boldsymbol{x}) : (i,j) \neq (i',j') \in \mathscr{T}\right\}$$

of infinitesimal covariances between counting processes $\{N_{ij}(t)\}$ and $\{N_{i'j'}(t)\}$:

$$\sigma_{dX}^{ij,i'j'}(\boldsymbol{x}) \\ \equiv \lim_{h\downarrow 0} h^{-1} Cov\Big[N_{ij}(t+h) - N_{ij}(t),\ N_{i'j'}(t+h) - N_{i'j'}(t) \ \Big|\ \boldsymbol{X}(t) = \boldsymbol{x}\Big]. \tag{12.4}$$

Our closed-form expression (12.8) below requires two moment existence conditions: $(P3^{\star})$ and $(P4^{\star})$. Similar conditions were required by Theorem 1 of [16] to provide closed-form expressions for the infinitesimal mean and variance of Markov counting processes. Since [16] call their condition for the mean $(P1^{\star})$ and that for the variance $(P2^{\star})$, we shall call our conditions for covariances $(P3^{\star})$ and $(P4^{\star})$. $(P3^{\star})$ and $(P4^{\star})$ are related to the number of transitions occurring in the Markov counting system over a time interval. This number of transitions is related not only to the sizes of the increments of each counting process $\{N_{ij}(t)\}$ but also to the overall rate at which these increments occur, which we call the rate function of the Markov counting system and define as

$$\lambda(\boldsymbol{x}) \equiv \lim_{h\downarrow 0} \frac{1 - P\left(N_{ij}(t+h) - N_{ij}(t) = 0 \ \text{ for all } (i,j) \in \mathscr{T} \ \middle|\ \boldsymbol{X}(t) = \boldsymbol{x}\right)}{h}. \tag{12.5}$$

This quantity is also know as the intensity of the process in the point process literature [20]. If the rate function satisfies that $\lambda(\boldsymbol{x}) = \sum_{\boldsymbol{\ell}} q(\boldsymbol{x}, \boldsymbol{\ell}) < \infty$ for all $\boldsymbol{x}$, then the process is said to be stable and conservative.

Consider stochastically bounding the rate function and the increment of each pair of counting processes over $[t, t+\bar{h}]$ by:

$$\bar{\Lambda}(t) \equiv \sup_{t\le s\le t+\bar{h}} \lambda\Big(\boldsymbol{X}(s)\Big) \tag{12.6}$$

$$\bar{Z}_{ij,i'j'}(t) \equiv \max\Big\{ \sup_{t\le s\le t+\bar{h}} dN_{ij}(s),\ \sup_{t\le s\le t+\bar{h}} dN_{i'j'}(s)\Big\}. \tag{12.7}$$

A combination of these two bounds gives the following property:

P3*. For each t, $\boldsymbol{x}$ and $(i, j) \neq (i', j')$ there is some $\bar{h} > 0$ such that

$$E\Big[\bar{Z}^2_{ij,i'j'}(t)\bar{\Lambda}(t)|\boldsymbol{X}(t) = \boldsymbol{x}\Big] < \infty,$$

P4*. For each t, $\boldsymbol{x}$ and $(i, j) \neq (i', j')$ there is some $\bar{h} > 0$ such that

$$E\Big[\bar{Z}^2_{ij,i'j'}(t)\bar{\Lambda}(t)^2|\boldsymbol{X}(t) = \boldsymbol{x}\Big] < \infty.$$

Properties $(P3^{\star})$ and $(P4^{\star})$ require that the Markov counting system does not have an explosive behaviour and holds, for example, for SIR-type compartmental models like that of Fig. 12.1, as shown in Sect. 12.5. $(P3^{\star})$ and $(P4^{\star})$ suffice to guarantee that infinitesimal covariances exist and that are given by the expression in Theorem 12.1 below.

Theorem 12.1 (Infinitesimal covariances of a Markov counting system). *Let $\{\boldsymbol{X}(t)\}$ be a time homogeneous Markov counting system defined by counting processes $\{\boldsymbol{N}(t)\}$ and by transition rates $q(\boldsymbol{x}, \boldsymbol{\ell})$ as in* (12.2) *that is stable and conservative. Supposing (P3*) and (P4*), the infinitesimal covariance between $\{N_{ij}(t)\}$ and $\{N_{i'j'}(t)\}$ is*

$$\sigma_{dX}^{ij,i'j'}(\boldsymbol{x}) = \sum_{\boldsymbol{k}} k_{ij}k_{i'j'}q_{ij,i'j'}(\boldsymbol{x}, \boldsymbol{k}). \tag{12.8}$$

Theorem 12.1 generalizes Theorem 1 of [16] to covariances and is proved in Appendix A.

12.5 Transition Rates of SIR-type Models Subjected to External Correlated Noises

Theorem 12.1 can be used to show that, after minimal simplifications to add clarity to our contribution, the system represented in Fig. 12.1 can be defined by transition rates that reproduce the effects (identified in Proposition 12.1) of correlated noises. First, consider the continuous-time Markov chain $\{\boldsymbol{Z}(t)\} \equiv \Big\{\big(S(t), I_1(t), I_2(t), S_1(t),$

$S_2(t), I_1^*(t), I_2^*(t), R(t))\}$ specified by Fig. 12.1. The standard interpretation of Fig. 12.1 gives the transition rates for $\{\boldsymbol{Z}(t)\}$ in Table 12.1 (e.g., letting the ξ_i be deterministic constants). Next, before considering adding noise to $\{\boldsymbol{Z}(t)\}$, we make the following simplifications to $\{\boldsymbol{Z}(t)\}$ so that the contributions in this paper can be presented more clearly (our approach works without these simplifications but results are not as straightforward): (i) instead of a time-inhomogeneous birth rate $b(t)$, births compensate deaths so that the total population size remains constant and is equal to $P < \infty$, as in [44]; (ii) instead of a time-inhomogeneous infection rate $\beta(t)$ within λ_i, this rate is constant and equal to β; and (iii) instead of different transition rates from S to I_i and from S_i to I_i^*, these rates are both $\lambda_i\xi_i(t)$, i.e., $\gamma = 0$. Simplifications (i) and (ii) impose time-homogeneity, which allows for a simpler notation in the rest of the paper. Simplification (iii) allows for simpler closed-form expressions for both transition rates and covariances. Now, let stable, conservative continuous-time Markov chain $\{\tilde{\boldsymbol{Z}}(t)\}$ be defined by the rates of Table 12.1 modified according to (i)–(iii) and by non-zero pairwise rates of transitions involving ξ_i equal to those in Proposition 12.1 as follows:

$$q_{\tilde{S}\tilde{I}_i,\tilde{S}_i\tilde{I}_i^*}(\tilde{z},\boldsymbol{k}) = \binom{\tilde{s}}{k_1}\binom{\tilde{s}_i}{k_2}\sum_{j=0}^{k_1+k_2}\binom{k_1+k_2}{j}(-1)^{k_1+k_2-j+1}\tau^{-1}\ln\left(1+\lambda_i\tau(\tilde{s}+\tilde{s}_i-j)\right).$$

System $\{\tilde{\boldsymbol{Z}}(t)\}$ defined by these rates satisfies $(P3^\star)$ and $(P4^\star)$, since letting the fixed population size be $\tilde{P}$,

$$\begin{aligned}\lambda\left(\tilde{\boldsymbol{Z}}(t)\right) &= m(\tilde{P}-\tilde{S}(t)) + r\left(\tilde{I}_1(t)+\tilde{I}_2(t)+\tilde{I}_1^*(t)+\tilde{I}_2^*(t)\right) + \sum_{\boldsymbol{k}} q_{\tilde{S}\tilde{I}_1,\tilde{S}_1\tilde{I}_1^*}\left(\tilde{\boldsymbol{Z}}(t),\boldsymbol{k}\right) + \\ &\quad + \sum_{\boldsymbol{k}} q_{\tilde{S}\tilde{I}_2,\tilde{S}_2\tilde{I}_2^*}\left(\tilde{\boldsymbol{Z}}(t),\boldsymbol{k}\right) \\ &\le \left(m + r + \lambda_1 + \lambda_2\right)\tilde{P} \end{aligned} \tag{12.9}$$

where the inequality follows by bounding all compartments by $\tilde{P}$ and because

$$\sum_{\boldsymbol{k}} q_{\tilde{S}\tilde{I}_i,\tilde{S}_i\tilde{I}_i^*}\left(\tilde{\boldsymbol{Z}}(t),\boldsymbol{k}\right) = \tau^{-1}\ln\left(1+\tau\lambda_i\left(\tilde{S}(t)+\tilde{S}_i\right)\right) \le \lambda_i\tilde{P}$$

(as follows from the properties of the binomial gamma process of [16]). Since the right hand side of (12.9) is not time-varying, it also bounds $\bar{\Lambda}(t)$ in $(P3^\star)$ and $(P4^\star)$. Similarly, $\tilde{P}$ is an upper bound for the increments: $\bar{Z}_{ij,i'j'}(t) \le \tilde{P}$, so that

$$\begin{aligned} E\left[\bar{Z}_{ij,i'j'}^2(t)\bar{\Lambda}(t) \mid \tilde{\boldsymbol{Z}}(t)=\tilde{z}\right] &\le \left(m+r+\lambda_1+\lambda_2\right)\tilde{P}^3, \\ E\left[\bar{Z}_{ij,i'j'}^2(t)\bar{\Lambda}^2(t) \mid \tilde{\boldsymbol{Z}}(t)=\tilde{z}\right] &\le \left(m+r+\lambda_1+\lambda_2\right)^2\tilde{P}^4 \end{aligned}$$

Table 12.1 Transition rates according the standard interpretation of Fig. 12.1 as a continuous-time Markov chain with rate function $\lambda_Z(z) \equiv \sum_{(i,j)\in\mathcal{T}} q_{ij}(z, 1)$ and with all marginal rates $q_{ij}(z, k)$ for $k > 1$ and all pairwise transition rates $q_{ij,i'j'}(z, \boldsymbol{k})$ assumed to be zero

N_{ij}	N_{SI_i}	$N_{I_iS_i}$	$N_{S_iI_i^*}$	$N_{I_i^*R}$	N_{BS}	N_{SD}	N_{I_iD}	N_{S_iD}	$N_{I_i^*D}$	N_{RD}
$q_{ij}(z, 1)$	$\lambda_i\xi_i$	r	$(1-\gamma)\lambda_i\xi_i$	r	$b(t)$	m	m	m	m	m

$(P3^\star)$ and $(P4^\star)$ can be analogously verified for the bivariate system $\{\tilde{\boldsymbol{Y}}(t)\}$ of Proposition 12.1 by assuming, for example, that the initial population sizes are deterministic, i.e., for fixed $y_i(0) = \tilde{y}_i(0)$

$$\lambda\left(\tilde{Y}(t)\right) = \sum_{\boldsymbol{k}} q_{\tilde{Y}_1\tilde{D},\tilde{Y}_2\tilde{D}}\left(\tilde{Y}(t), \boldsymbol{k}\right) = \tau^{-1}\ln\left(1 + \tau\delta\left(\tilde{Y}_1(t) + \tilde{Y}_2(t)\right)\right) \leq \delta(\tilde{y}_1(0) + \tilde{y}_2(0)).$$

Hence, since all conditions stated in Theorem 2 hold for $\{\tilde{\boldsymbol{Z}}(t)\}$ and for $\{\tilde{\boldsymbol{Y}}(t)\}$ from Proposition 12.1, it follows that

$$\sigma_{d\tilde{Z}}^{\tilde{S}\tilde{I}_i,\tilde{S}_i\tilde{I}_i^*}(\tilde{z}) = \tilde{s}\tilde{s}_i\tau^{-1}\ln\left(\frac{(1+\lambda_i\tau)^2}{1+2\lambda_i\tau}\right) > 0, \tag{12.10}$$

since rates $q_{\tilde{S}\tilde{I}_i,\tilde{S}_i\tilde{I}_i^*}(\tilde{z}, \boldsymbol{k})$ involved in the infinite sum of the right hand side of (12.10) have been defined to match rates $q_{\tilde{Y}_1\tilde{D},\tilde{Y}_2\tilde{D}}(\tilde{\boldsymbol{y}}, \boldsymbol{k})$ in Proposition 12.1. Equation (12.10) can be interpreted as follows. First, the effects of introducing correlated noises in the bivariate system of Proposition 12.1 can be reproduced in more general systems, as long as simpler (e.g., bivariate) systems appropriate for these more general systems exist and are analytically tractable. This is the case for most SIR-type models, which usually involve a combination of birth and death processes (for which bivariate Poisson, binomial or negative binomial processes are appropriate and mixtures with gamma processes or Poisson processes are often analytically tractable). Second, it permits an alternative interpretation of correlated noises $\xi_i(t)$ in a non-random environment context. These noises have effectively been integrated out in $\{\tilde{\boldsymbol{Z}}(t)\}$, which can be seen as a regular continuous-time Markov chain in a non-random environment, with the caveat that it now allows for simultaneous co-transitions that drive the new infinitesimal correlations.

Acknowledgements This work was supported by the Spanish Ministry of Science, Innovation and Universities and the Spanish Agency of Research, co-funded with FEDER funds, project ECO2017-87245-R. I thank an anonymous referee for suggestions that led to substantial improvements in the paper.

Appendix 1: Proof of Theorem 12.1

Proof First, we prove that the infinitesimal covariance equals the infinitesimal cross-product under conditions $(P3^\star)$ and $(P4^\star)$. Let $\Delta N_{ij}(t) \equiv N_{ij}(t+h) - N_{ij}(t)$ and analogously for all other subindices and counting processes in this proof. Since $(P3^\star)$ implies $(P1^\star)$ in Theorem 1 of [11], it follows that

$$E\Big[\Delta N_{ij}(t) \mid \boldsymbol{X}(t) = \boldsymbol{x}\Big] = h \sum_{k} k q_{ij}(\boldsymbol{x}, k) + o(h).$$

Then

$$\begin{aligned}
&\lim_{h \downarrow 0} h^{-1} Cov\Big[\Delta N_{ij}(t),\ \Delta N_{i'j'}(t) \mid \boldsymbol{X}(t) = \boldsymbol{x}\Big] \\
&= \lim_{h \downarrow 0} h^{-1} \Big\{ E\Big[\Delta N_{ij}(t) \Delta N_{i'j'}(t) \mid \boldsymbol{X}(t) = \boldsymbol{x}\Big] - o(h) \Big\} \\
&= \lim_{h \downarrow 0} h^{-1} E\Big[\Delta N_{ij}(t) \Delta N_{i'j'}(t) \mid \boldsymbol{X}(t) = \boldsymbol{x}\Big].
\end{aligned}$$

The rest of this proof follows closely the proof of Theorem 1 in [11]. While that theorem provided expressions for the infinitesimal mean and variance of $\{\boldsymbol{X}(t)\}$, this one provides them for its covariances.

All probabilities and expectations in this proof are conditional on $\boldsymbol{X}(t) = \boldsymbol{x}$ (in addition to other conditioning, where appropriate). Define the following: (i) let $\{\bar{N}(t)\}$ be a process such that, conditional on $\bar{\Lambda}(t)$ and $\bar{Z}(t) \equiv \bar{Z}_{ij,i'j'}(t)$, realizations of $\{\bar{N}(t)\}$ are those of a compound Poisson process [19] with Poisson event rate $\bar{\Lambda}(t)$ and degenerate jump or batch size distribution [20] with mass one at $\bar{Z}(t)$, i.e., a process with jumps arriving according to the Poisson process and for which the size of the jumps is $\bar{Z}(t)$; and (ii) let S be the event that there is exactly one transition time occurring in the interval $[t, t+h]$ in the Markov counting system $\{\boldsymbol{X}(t)\}$. Then,

$$\begin{aligned}
E\left[\Delta N_{ij}(t) \Delta N_{i'j'}(t)\right] = {} & E\left[\Delta N_{ij}(t) \Delta N_{i'j'}(t)\, \mathbb{I}\{S\}\right] \\
& + E\left[\Delta N_{ij}(t) \Delta N_{i'j'}(t)\, \mathbb{I}\{S^c\}\right].
\end{aligned} \tag{12.11}$$

Consider the first term on the right hand side of (12.11). Let $S_{ij,i'j'} \subset S$ be the event that there is exactly one transition time occurring in the interval $[t, t+h]$ in the Markov counting system $\{\boldsymbol{X}(t)\}$ and that this transition increases both the $\{N_{ij}(t)\}$ and the $\{N_{i'j'}(t)\}$ processes (and possibly other processes). Then, in (12.11), letting $\boldsymbol{k} \equiv (k_{ij}, k_{i'j'})$

$$
\begin{aligned}
E[\Delta N_{ij}(t)\Delta N_{i'j'}(t)\,\mathbb{I}\{S\}\,] &= E[\Delta N_{ij}(t)\Delta N_{i'j'}(t)|S_{ij,i'j'}] \;\times\; P(S_{ij,i'j'}|S) \;\times\; P(S) \\
&= \sum_{\boldsymbol{k}} k_{ij}k_{i'j'}\frac{q_{ij,i'j'}(\boldsymbol{x},\boldsymbol{k})}{\sum_{\boldsymbol{k}} q_{ij,i'j'}(\boldsymbol{x},\boldsymbol{k})} \times \frac{\sum_{\boldsymbol{k}} q_{ij,i'j'}(\boldsymbol{x},\boldsymbol{k})}{\lambda(\boldsymbol{x})} \times \\
&\quad \times \Big[h\lambda(\boldsymbol{x}) + o(h)\Big] \qquad (12.12)\\
&= h\sum_{\boldsymbol{k}} k_{ij}k_{i'j'}q_{ij,i'j'}(\boldsymbol{x},\boldsymbol{k}) + o(h) \qquad (12.13)
\end{aligned}
$$

where $P(S)$ in (12.12) follows by a standard result on continuous-time Markov chains; see for example [43], p. 492.

To finish the proof, we show that the second term on the right hand side of (12.11) disappears infinitesimally. Let $\bar{S}$ be the event that there is exactly one transition time occurring in the interval $[t, t+h]$ in the compound Poisson process $\{\bar{N}_{ij}(t)\}$. Since the random variable $\Delta N_{ij}(t)\Delta N_{i'j'}(t)$ is stochastically smaller than $\left(\Delta\bar{N}(t)\right)^2$,

$$
\begin{aligned}
E\left[\Delta N_{ij}(t)\Delta N_{i'j'}(t)\,\mathbb{I}\{S^c\}\right] &\le E\left[\left(\Delta\bar{N}(t)\right)^2\mathbb{I}\{\bar{S}^c\}\right] \\
&= E\left[E\left[\left(\Delta\bar{N}(t)\right)^2 \mid \bar{\Lambda}(t), \bar{Z}(t)\right]\right] - \\
&\quad - E\left[E\left[\left(\Delta\bar{N}(t)\right)^2\mathbb{I}\{\bar{S}\} \mid \bar{\Lambda}(t), \bar{Z}(t)\right]\right] \qquad (12.14)\\
&= \left(E\left[\bar{Z}^2(t)\bar{\Lambda}(t)h\right] + \underbrace{E\left[\bar{Z}^2(t)\bar{\Lambda}^2(t)\right]h^2}_{=o(h)}\right) - \\
&\quad - E\left[\bar{Z}^2(t)\bar{\Lambda}(t)h\exp\{-h\bar{\Lambda}(t)\}\right] \qquad (12.15)\\
&= E\left[\bar{Z}^2(t)\bar{\Lambda}(t)h\Big(1-\exp\big\{-h\bar{\Lambda}(t)\big\}\Big)\right] + o(h)
\end{aligned}
$$

where (12.14) follows as in (12.11), and (12.15) follows by the properties of the compound Poisson distribution. The second term of (12.15) is $o(h)$ by $(P4^{\star})$. Since $\bar{z}^2\bar{\lambda}\Big(1-\exp\big\{-h\bar{\lambda}\big\}\Big) \le \bar{z}^2\bar{\lambda}$ and, by $(P3^{\star})$, $E\left[\bar{Z}^2(t)\bar{\Lambda}(t)\right]$ is assumed finite (note that the distribution of $\bar{Z}^2(t)\bar{\Lambda}(t)$ depends on $\bar{h}$ and not h), it follows by dominated convergence that

$$
\begin{aligned}
&\lim_{h\downarrow 0}\frac{E\left[\bar{Z}^2(t)\bar{\Lambda}(t)h\Big(1-\exp\big\{-h\bar{\Lambda}(t)\big\}\Big)\right]}{h} \\
&= E\left[\lim_{h\downarrow 0}\bar{Z}^2(t)\bar{\Lambda}(t)\Big(1-\exp\big\{-h\bar{\Lambda}(t)\big\}\Big)\right] = 0.
\end{aligned}
$$

References

1. Aguiar, M., Stollenwerk, N., Kooi, B.W.: The Stochastic Multi-strain Dengue Model: Analysis of the Dynamics. In: Simos, T. E., Psihoyios, G., Tsitouras, C., Anastassi, Z. (Eds.), American Institute of Physics Conference Series. Volume 1389 of American Institute of Physics Conference Series, pp. 1224–1227 (2011)
2. Anderson, R.M., May, R.M.: Infectious Diseases of Humans. Oxford University Press, Oxford (1991)
3. Applebaum, D.: Lévy Processes and Stochastic Calculus. Cambridge University Press, Cambridge (2004)
4. Barndorff-Nielsen, O.E., Shiryaev, A.: Change of time and change of measure. World Scientific Publishing, Singapore (2010)
5. Bartlett, M.S.: Deterministic and stochastic models for recurrent epidemics. In: Proceedings of the Third Berkeley Symposium on Mathematical Statistics and Probability, pp. 81–109. University California Press, Berkeley (1956)
6. Bhadra, A., Ionides, E.L., Laneri, K., Pascual, M., Bouma, M., Dhiman, R.C.: Malaria in Northwest India: Data analysis via partially observed stochastic differential equation models driven by Lévy noise. J. Amer. Stat. Ass. **106**, 440–451 (2011)
7. Bharucha-Reid, A.T.: Elements of the Theory of Markov Processes and their Applications. McGraw-Hill, New York (1960)
8. Bochner, S.: Diffusion equation and stochastic processes. Proc. Nat. Acad. Sci. U. S. A. **35**, 368–370 (1949)
9. Braumann, C.A.: Environmental Versus Demographic Stochasticity in Population growth, pp. 37–52. Springer, Berlin (2010)
10. Brémaud, P.: Markov Chains: Gibbs Fields, Monte Carlo Simulation, and Queues. Springer, New York (1999)
11. Bretó, C.: On infinitesimal dispersion of multivariate Markov counting systems. Stat. Probab. Lett. **82**, 720–725 (2012a)
12. Bretó, C.: Time changes that result in multiple points in continuous-time Markov counting processes. Stat. Probab. Lett. **82**, 2229–2234 (2012b)
13. Bretó, C.: Trajectory composition of poisson time changes and markov counting systems. Stat. Probab. Lett. **88**, 91–98 (2014)
14. Bretó, C.: Modeling and inference for infectious disease dynamics: a likelihood-based approach. Stat. Sci. **33**(1), 57–69 (2018)
15. Bretó, C., He, D., Ionides, E., King, A.: Time series analysis via mechanistic models. Ann. Appl. Stat. **3**, 319–348 (2009)
16. Bretó, C., Ionides, E.: Compound Markov counting processes and their applications to modeling infinitesimally over-dispersed systems. Stoch. Proc. Appl. **121**, 2571–2591 (2011)
17. Buckee, C.O., Recker, M., Watkins, E.R., Gupta, S.: Role of stochastic processes in maintaining discrete strain structure in antigenically diverse pathogen populations. Proc. Nat. Acad. Sci. **108**(37), 15504–15509 (2011)
18. Cauchemez, S., Ferguson, N.M.: Likelihood-based estimation of continuous-time epidemic models from time-series data: application to measles transmission in London. J. R. Soc. Interface **5**(25), 885–897 (2008)
19. Cox, D., Isham, V.: Point Processes. Chapman & Hall, New York (1980)
20. Daley, D., Vere-Jones, D.: An Introduction to the Theory of Point Processes. Volume I: Elementary Theory and Methods. Springer, Berlin (2003)
21. Dureau, J., Kalogeropoulos, K., Baguelin, M.: Capturing the time-varying drivers of an epidemic using stochastic dynamical systems. Biostatistics **14**(3), 541–555 (2013)
22. Engen, S., Bakke, O., Islam, A.: Demographic and environmental stochasticity: concepts and definitions. Biometrics **54**, 840–846 (1998)
23. Fenton, A., Pedersen, A.: Community epidemiology framework for classifying disease threats. Emer. Interface Dis. **11**, 1815–1821 (2005)

24. He, D., Ionides, E., King, A.: Plug-and-play inference for disease dynamics: measles in large and small populations as a case study. J. R. Soc. Interface **7**, 271–283 (2010)
25. Hougaard, P., Lee, M.-L.T., Whitmore, G.A.: Analysis of overdispersed count data by mixtures of poisson variables and poisson processes. Biometrics **53**(4), 1225–1238 (1997)
26. Hu, G., Wang, K.: Stability in distribution of competitive Lotka-Volterra system with Markovian switching. Appl. Math. Modell. **35**(7), 3189–3200 (2011)
27. Ionides, E.L., Bretó, C., King, A.A.: Inference for nonlinear dynamical systems. Proc. Nat. Acad. Sci. USA **103**, 18438–18443 (2006)
28. Jacquez, J.A.: Compartmental Analysis in Biology and Medicine, 3rd edn. BioMedware, Ann Arbor (1996)
29. Kamo, M., Sasaki, A.: The effect of cross-immunity and seasonal forcing in a multi-strain epidemic model. Phys. D **165**, 228–241 (2002)
30. Kermack, W.O., McKendrick, A.G: A contribution to the mathematical theory of epidemics. Proc. R. Soc. Lon., Ser. A **115**, 700–721 (1927)
31. King, A., Ionides, E., Pacual, M., Bouna, M.: Inapparent infections and cholera dynamics. Nature **454**, 877–880 (2008)
32. Kozubowski, T.J., Podgórski, K.: Distributional properties of the negative binomial Lévy process. Probab. Math. Stat. **29**, 43–71 (2009)
33. Laneri, K., Bhadra, A., Ionides, E., Bouma, M., Dhiman, R., Yadav, R.S., Pascual, M.: Forcing versus feedback: epidemic malaria and monsoon rains in Northwest India. PLoS Comput. Biol. 6 (2010)
34. Lee, M.-L.T., Whitmore, G.A.: Stochastic processes directed by randomized time. J. Appl. Probab. **30**(2), 302–314 (1993)
35. Magpantay, F.M.G., Domenech de Cellès, M., Rohani, P., King, A.A.: Pertussis immunity and epidemiology: mode and duration of vaccine-induced immunity. Parasitology **143**, 835–849 (2016)
36. Marion, G., Renshaw, E.: Stochastic modelling of environmental variation for biological populations. Theor. Popul. Biol. **57**, 197–217 (2000)
37. Matis, J.H., Kiffe, T.R.: Stochastic Population Models. A Compartmental Perpective. Springer, Berlin (2000)
38. McCullagh, P., Nelder, J.A.: Generalized Linear Models, 2nd edn. Chapman and Hall, London (1989)
39. Nisbet, R., Gurney, W.: Modelling Fluctuating Populations. Wiley Interscience publication, Wiley, New York (1982)
40. O'Dea Eamon, B., Park, A.W., Dark, J.: Estimating the distance to an epidemic threshold. J. R. Soc. Interface **15** (2018)
41. Reich, N.G., Shrestha, S., King, A.A., Rohani, P., Lessler, J., Kalayanarooj, S., Yoon, I.-K., Gibbons, R.V., Burke, D.S., Cummings, D.A.T.: Interactions between serotypes of dengue highlight epidemiological impact of cross-immunity. J. R. Soc. Interface **10**(86), (2013)
42. Renshaw, E.: Modelling Biological Populations in Space and Time. Cambridge University Press, Cambridge (1991)
43. Ross, S.: Stochastic Processes, 2nd edn. Wiley, New York (1996)
44. Shrestha, S., King, A., Rohani, P.: Statistical inference for multi-pathogen systems. PLoS Comput. Biol. **7**(8), e1002135 (2011)
45. Snyder, D., Miller, M.: Random Point Processes in Time and Space. Springer, Berlin (1991)
46. Stocks, T., Britton, T., Höhle, M.: Model selection and parameter estimation for dynamic epidemic models via iterated filtering: application to rotavirus in Germany. Biostatistics (2018)
47. Varughese, M., Fatti, L.: Incorporating environmental stochasticity within a biological population model. Theor. Popul. Biol. **74**, 115–129 (2008)

Chapter 13
On Fractal Dimension of Global and Exponential Attractors for Dissipative Higher Order Parabolic Problems in $\mathbb{R}^N$ with General Potential

Jan W. Cholewa and Radosław Czaja

Abstract We consider a Cauchy problem for a dissipative fourth order parabolic equation in $\mathbb{R}^N$ with a general potential. Using the method by Chueshov and Lasiecka we estimate from above fractal dimension of a global attractor. We also show that it is contained in a finite dimensional exponential attractor.

13.1 Introduction

We study a Cauchy problem

$$\begin{cases} u_t + \Delta^2 u = f(x,u), \quad t > 0,\ x \in \mathbb{R}^N, \\ u(0,x) = u_0(x),\ x \in \mathbb{R}^N, \end{cases} \tag{13.1}$$

for initial data $u_0 \in H^2(\mathbb{R}^N)$ with the right-hand side of the form

$$f(x,u) = g(x) + m(x)u + f_0(x,u),\ x \in \mathbb{R}^N,\ u \in \mathbb{R}. \tag{13.2}$$

The potential $m\colon \mathbb{R}^N \to \mathbb{R}$ satisfies integrability condition

$$\sup_{y \in \mathbb{R}^N} \|m\|_{L^r(B(y,1))} < \infty \ \text{ for some } \ \max\left\{\frac{N}{4}, 1\right\} < r \le \infty, \tag{13.3}$$

which is uniform for arbitrarily centered unit balls $B(y,1) \subset \mathbb{R}^N$, whereas

J. W. Cholewa · R. Czaja (✉)
Institute of Mathematics, University of Silesia in Katowice, Bankowa 14, 40-007 Katowice, Poland
e-mail: radoslaw.czaja@us.edu.pl

J. W. Cholewa
e-mail: jan.cholewa@us.edu.pl

V. A. Sadovnichiy and M. Z. Zgurovsky (eds.), *Contemporary Approaches and Methods in Fundamental Mathematics and Mechanics*, Understanding Complex Systems,
https://doi.org/10.1007/978-3-030-50302-4_13

$$g \in L^2(\mathbb{R}^N) \cap L^\infty(\mathbb{R}^N). \tag{13.4}$$

The map $f_0\colon \mathbb{R}^N \times \mathbb{R} \to \mathbb{R}$ is locally Lipschitz in $u \in \mathbb{R}$ uniformly for $x \in \mathbb{R}^N$ having derivative $\frac{\partial f_0}{\partial u}$ and satisfying

$$f_0(x,0) = 0, \quad \frac{\partial f_0}{\partial u}(x,0) = 0, \ x \in \mathbb{R}^N, \tag{13.5}$$

and a growth condition

$$|f_0(x,u) - f_0(x,v)| \le c|u-v|(1 + |u|^{\rho-1} + |v|^{\rho-1}), \ x \in \mathbb{R}^N, \ u, v \in \mathbb{R} \tag{13.6}$$

for some positive constant c with

$$\rho > 1 \text{ arbitrary if } N \le 4 \text{ or } 1 < \rho < 1 + \frac{8}{N-4} \text{ when } N \ge 5. \tag{13.7}$$

Concerning dissipativeness we assume a structure condition

$$uf(x,u) \le C(x)u^2 + D(x)|u|, \quad x \in \mathbb{R}^N, \ u \in \mathbb{R}, \tag{13.8}$$

with some $C, D\colon \mathbb{R}^N \to \mathbb{R}$ satisfying

$$\sup_{y \in \mathbb{R}^N} \|C\|_{L^r(B(y,1))} < \infty \quad \text{with r as in (13.3)}, \tag{13.9}$$

$$0 \le D \in L^s(\mathbb{R}^N) \ \text{ for some } \ \max\left\{\frac{2N}{N+4}, 1\right\} \le s \le 2 \quad (s > 1 \text{ if } N = 4), \tag{13.10}$$

and such that solutions of the linear problem

$$\begin{cases} v_t + \Delta^2 v = C(x)v, & t > 0, \ x \in \mathbb{R}^N, \\ v(0) = v_0 \in L^2(\mathbb{R}^N) \end{cases} \tag{13.11}$$

decay exponentially as $t \to \infty$ so that (13.11) enjoys uniform exponential stability property.

Higher order parabolic problems have been of wide interest in the past decades and various aspects of the theory, involving both bounded and unbounded domains as well as specific models, have been studied not only in a large number of papers (see e.g. [15, 16, 26, 27]), but also in many monographs (see e.g. [19, 23, 24, 29]). In this context we also mention the theory of higher order elliptic problems and analytic semigroups, involving fractional powers and interpolation-extrapolation phase spaces (see e.g. [1, 18, 20, 30]).

In this paper we deal with a semigroup of global solutions defined by (13.1) in an energy space. Our goal is to estimate from above fractal dimension of its global attractor and also show that it is contained in a finite dimensional exponential attractor. Although this is known to be a relevant information about a dynamical system (see e.g. the exposition of [28, Sects. 13.5 and 14.6] and [29, Chaps. V and VI]), to the best of our knowledge for higher order parabolic problems in large unbounded domains as in (13.1)–(13.2) such a result is not available as yet in the literature.

Concerning second order reaction-diffusion equations in $\mathbb{R}^N$, considered in the Lebesgue phase spaces of initial data for which existence of attractors has been extensively studied (see e.g. [3, 17, 32] and references therein), finite dimensionality of a global attractor under a general dissipativeness mechanism has been exhibited in [4] using a control of trace of the linearized operator from the equation, which remains in the vein of the approach in [29].

As for the dissipative higher order parabolic problems in bounded domains, fractal dimension of attractor was estimated in [7] following approach of [33], see also [25] and the remarks on the subject in [2].

Coming back to the problem (13.1) we recall from [10] that with the above mentioned assumptions (13.1) defines in $H^2(\mathbb{R}^N)$ a C^0 semigroup $\{S(t)\colon t \geq 0\}$ possessing a global attractor $\mathbf{A}$ in the sense of [21], that is, $\mathbf{A}$ compact in $H^2(\mathbb{R}^N)$, invariant under $\{S(t)\colon t \geq 0\}$ and attracting bounded sets in the sense that for each bounded subset B of $H^2(\mathbb{R}^N)$

$$d(S(t)B, \mathbf{A}) = \sup_{v \in B} \inf_{w \in \mathbf{A}} \|S(t)v - w\|_{H^2(\mathbb{R}^N)} \to 0 \text{ as } t \to \infty.$$

We remark that in [11], with a different approach than in [10], the existence of a global attractor was proved in a critical case $\rho = \frac{N+4}{N-4}$ or even in supercritical cases $\rho > \frac{N+4}{N-4}$ for $N \geq 5$. This required, however, a restriction on the range of integrability of the potential in (13.3) and some additional assumptions involving a structure condition on f (see [11, (1.12) and (1.14)]). This is a matter we will not pursue here as the critical and supercritical cases are rather subject of separate studies.

Below our concern is to estimate fractal dimension of the global attractor $\mathbf{A}$. To accomplish this task we rely on the method developed by Chueshov and Lasiecka (see [12, Theorem 3.1.21] and [13, Theorem 2.14]), which involves in particular suitable estimates for the semigroup on the attractor as in (13.15)–(13.16).

Theorem 13.1 *Let* (13.2)–(13.10) *hold and the solutions of the linear problem* (13.11) *decay exponentially as $t \to \infty$. Suppose also that*

$$\frac{\partial f_0}{\partial u}(x, u) \to 0 \text{ as } u \to 0 \text{ uniformly for } x \in \mathbb{R}^N, \tag{13.12}$$

and assume that

$$D \in L^p(\mathbb{R}^N) \text{ for some } p \geq \max\left\{\frac{N}{4}, 1\right\} \quad (p > 1 \text{ if } N = 4) \tag{13.13}$$

or

$$D(x) \to 0 \quad as \ |x| \to \infty. \tag{13.14}$$

Then for the semigroup $\{S(t) \colon t \geq 0\}$ *defined by* (13.1) *in* $H^2(\mathbb{R}^N)$ *and its global attractor* $\mathbf{A}$ *the following statements hold.*

(i) Given any $u_0, v_0 \in \mathbf{A}$, *the functions* $u = S(\cdot)u_0$ *and* $v = S(\cdot)v_0$ *for each* $T > 0$ *belong to the space* $Z := C^1([0,T]; L^2(\mathbb{R}^N)) \cap C([0,T]; H^2(\mathbb{R}^N))$ *equipped with the norm* $\|\cdot\|_Z = \|\cdot\|_{H^1(0,T;L^2(\mathbb{R}^N)) \cap L^2(0,T;H^2(\mathbb{R}^N))}$, *and satisfy for each* $T > 0$ *the estimates*

$$\|u - v\|_{H^1(0,T;L^2(\mathbb{R}^N)) \cap L^2(0,T;H^2(\mathbb{R}^N))} \leq \kappa_T \|u_0 - v_0\|_{H^2(\mathbb{R}^N)}, \tag{13.15}$$

$$\|u(T) - v(T)\|_{H^2(\mathbb{R}^N)} \leq \eta_T \|u_0 - v_0\|_{H^2(\mathbb{R}^N)} + \mu_T \|u - v\|_{L^2(0,T;L^{2r'}(B_{R^*}))}, \tag{13.16}$$

for some positive constants $\kappa_T, \eta_T, \mu_T, R^*$, *where* B_{R^*} *is a ball in* $\mathbb{R}^N$ *of radius* R^* *around zero and* r' *is Hölder's conjugate to* r *in* (13.3). *Furthermore, there exists* $\mu^* > 0$ *such that*

$$\mu_T = \mu^* \, for \ T \ large \ enough \ and \ \eta_T \to 0 \, as \ T \to \infty. \tag{13.17}$$

(ii) Given any T *large enough and any* $\delta \in (0, 1 - \eta_T)$, *fractal dimension of* $\mathbf{A}$ *in* $H^2(\mathbb{R}^N)$ *is estimated by*

$$dim_f(\mathbf{A}) \leq \frac{\ln \mathrm{m}_Z^{\mathscr{B}_\delta}}{\ln \frac{1}{\eta_T + \delta}}, \tag{13.18}$$

where $\mathrm{m}_Z^{\mathscr{B}_\delta}$ *denotes the maximal number of elements* z_j *in a closed ball* $\mathscr{B}_\delta$ *in* Z *of radius* $2\mu^* \kappa_T \delta^{-1}$ *around zero with the property that*

$$\|z_{j|_{B_{R^*}}} - z_{l|_{B_{R^*}}}\|_{L^2(0,T;L^{2r'}(B_{R^*}))} > 1 \ for \ j \neq l,$$

$\|\cdot\|_{L^2(0,T;L^{2r'}(B_{R^*}))}$ *being a compact seminorm in* Z *in the sense of Definition 13.1.*

We make the proof of Theorem 13.1 using the fact that D has at least one of the properties (13.13) or (13.14). Note however that (13.13) is automatically satisfied for $N \leq 4$ by taking $p = s$ with s from (13.10), whereas (13.14) is a consequence of (13.10) whenever D is uniformly continuous. Observe also that (13.13) is a counterpart of the assumption on D considered for a second order problem in [4, Theorem 4.4]. Moreover, in [4, (4.2)] it was required that $\left|\frac{\partial^2 f_0}{\partial u^2}(x,u)\right| \leq h(R)$ whenever $|u| \leq R$ and $x \in \mathbb{R}^N$, which together with the assumption $\frac{\partial f_0}{\partial u}(x,0) = 0$ in [4, (1.6)] implies (13.12).

Having the global attractor $\mathbf{A}$ we also know that it is contained in a closed and bounded in $H^2(\mathbb{R}^N)$ positively invariant set $\mathbf{B}$, which absorbs bounded sets of $H^2(\mathbb{R}^N)$. We remark that due to the smoothing action of the semigroup the set $\mathbf{B}$ can be chosen here bounded in the space of bounded continuous functions in $\mathbb{R}^N$

(see Proposition 13.2(iv)). We then obtain the quasi-stability of the semigroup on the absorbing set **B** in the sense of [12, Definition 3.4.1], and combining this with dissipativeness and Hölder continuity of the semigroup with respect to t we prove the existence of an exponential attractor **M** as stated in Theorem 13.2.

Theorem 13.2 *Under the assumptions of Theorem 13.1, the semigroup* $\{S(t): t \geq 0\}$ *in* $H^2(\mathbb{R}^N)$ *has an exponential attractor, that is there exists* $\mathbf{M} \subset \mathbf{B}$ *such that*

(i) $\mathbf{M}$ *is compact in* $H^2(\mathbb{R}^N)$,

(ii) $S(t)\mathbf{M} \subset \mathbf{M}$ *for every* $t \geq 0$,

(iii) $\mathbf{M}$ *exponentially attracts bounded subsets of* $H^2(\mathbb{R}^N)$ *in the sense that given any bounded set* $B \subset H^2(\mathbb{R}^N)$

$$d(S(t)B, \mathbf{M}) \leq C_B e^{-\gamma t}, \ t \geq t_B$$

for some positive constants C_B, γ, t_B *(*γ *being independent of* B*),*

(iv) $\mathbf{M}$ *has finite fractal dimension in* $H^2(\mathbb{R}^N)$.

A brief description of this work is as follows. In Sect. 13.2 we summarize some known results for (13.1) concerning dissipativeness mechanism and the existence of a global attractor. In Sect. 13.3 we derive the estimates for the semigroup on the attractor **A** and on the absorbing set **B** proving also compactness of the considered seminorm (see Lemma 13.8). Once this is known, the estimate of fractal dimension of **A** follows by Proposition 13.4 and hence we complete the proof of Theorem 13.1. The details concerning the proof of Theorem 13.2 are given in Sect. 13.4 based on dissipativeness and quasi-stability properties already mentioned above. In particular, we obtain an upper bound of fractal dimension of **M** as specified in (13.61).

Not being too exhaustive we remark that the approach can be applied also for second order reaction-diffusion equations in $\mathbb{R}^N$. It was actually used for discretized reaction-diffusion equations in [6], where fractal dimension of the global attractor for a lattice problem was estimated and the existence of an exponential attractor was proved in the presence of a general dissipativeness mechanism.

13.2 Some Known Results Concerning Global Attractor for (13.1)

We use Bessel spaces $H^s_p(\mathbb{R}^N)$ as in [30], which for $p = 2$ are denoted $H^s(\mathbb{R}^N)$. Since f in (13.1) involves a potential possessing merely some mild integrability properties (see (13.3)), we first recall from [9] the results concerning linear problem (13.11), that is,

$$\begin{cases} v_t + \Delta^2 v = C(x)v, & t > 0, \ x \in \mathbb{R}^N, \\ v(0) = v_0 \subset L^2(\mathbb{R}^N). \end{cases}$$

Henceforth, given a potential C as in (13.9), i.e.,

$$\sup_{y\in\mathbb{R}^N} \|C\|_{L^r(B(y,1))} < \infty \text{ for some } \max\left\{\frac{N}{4}, 1\right\} < r \leq \infty,$$

we denote

$$A_C = \Delta^2 - C(x)I. \tag{13.19}$$

Proposition 13.1 *If $C\colon \mathbb{R}^N \to \mathbb{R}$ satisfies* (13.9), *then*

(i) A_C in (13.19) *is a selfadjoint operator in $L^2(\mathbb{R}^N)$, its domain $D(A_C)$ is contained in $H^2(\mathbb{R}^N)$ and*

$$\int_{\mathbb{R}^N} A_C\phi\psi = \int_{\mathbb{R}^N} \Delta\phi\Delta\psi - \int_{\mathbb{R}^N} C(x)\phi\psi = \int_{\mathbb{R}^N} \phi A_C\psi \text{ for all } \phi, \psi \in D(A_C);$$

also $C\phi^2 \in L^1(\mathbb{R}^N)$ if $\phi \in H^2(\mathbb{R}^N)$,

(ii) A_C is a sectorial operator in $L^2(\mathbb{R}^N)$ and $-A_C$ generates a C^0 analytic semigroup $\{e^{-A_C t}: t \geq 0\}$ in $L^2(\mathbb{R}^N)$ which, given any $-\beta^* \leq \sigma \leq \xi \leq \beta^*$ and $T > 0$, satisfies

$$\|e^{-A_C t}\|_{\mathscr{L}(H^{4\sigma}(\mathbb{R}^N), H^{4\xi}(\mathbb{R}^N))} \leq M_T t^{\sigma-\xi}, \quad 0 < t \leq T,$$

for some constant $M_T \geq 1$, where

$$\beta^* = 1 + \left(\frac{N}{8} - \frac{N}{4r}\right)_- > \frac{1}{2};$$

in particular $e^{-A_C t}v_0$ is a solution of (13.11),

(iii) the solutions of (13.11) *decay exponentially as $t \to \infty$ so that* (13.11) *enjoys uniform exponential stability property if and only if for some $\omega_0 > 0$*

$$\int_{\mathbb{R}^N} (|\Delta\phi|^2 - C(x)\phi^2) \geq \omega_0 \|\phi\|^2_{L^2(\mathbb{R}^N)}, \quad \phi \in H^2(\mathbb{R}^N). \tag{13.20}$$

There also exists a continuous real valued function $\omega(\zeta)$ defined in a certain interval $[0, \zeta_0]$ and satisfying

$$\lim_{\zeta\to 0^+} \omega(\zeta) = \omega(0) = \omega_0,$$

such that

$$\int_{\mathbb{R}^N} ((1-\zeta)|\Delta\phi|^2 - C(x)\phi^2) \geq \omega(\zeta)\|\phi\|^2_{L^2(\mathbb{R}^N)} \text{ for } \phi \in H^2(\mathbb{R}^N),\ \zeta \in [0, \zeta_0]. \tag{13.21}$$

Proof The results are due to [9, Lemma 2.9 and Corollaries 2.5, 2.7, 2.10, 2.11]. □

Remark 13.1 Note that $\beta^* = 1$ if and only if $r \geq 2$ and if $r \geq 2$ then $D(A_C) = H^4(\mathbb{R}^N)$ (see [9, Corollary 2.6]).

We now give more detailed information concerning the solutions of the nonlinear problem (13.1) and the existence of a global attractor. We also prove that the absorbing set for the associated semigroup can be chosen bounded in the space of bounded continuous functions in $\mathbb{R}^N$.

Proposition 13.2 *If* (13.2)–(13.10) *hold then*
(i) given $u_0 \in H^2(\mathbb{R}^N)$ there exists a unique global solution u of (13.1) *satisfying*

$$u \in C([0,\infty); H^2(\mathbb{R}^N)) \cap C((0,\infty); H^{4\beta^*}(\mathbb{R}^N)) \cap C^1((0,\infty); L^2(\mathbb{R}^N)),$$
$$u \in C((0,\infty); D(A_m))$$

where $A_m = \Delta^2 - m(x)I$; in particular

$$S(t)u_0 = u(t; u_0),\ t \geq 0,$$

is a C^0 semigroup in $H^2(\mathbb{R}^N)$ satisfying Duhamel's formula

$$S(t)u_0 = e^{-A_m t}u_0 + \int_0^t e^{-A_m(t-s)}\big(f_0(\cdot, S(s)u_0) + g\big)ds,\ t \geq 0. \tag{13.22}$$

If, in addition, the solutions of (13.11) *decay exponentially as $t \to \infty$, then*
(ii) the orbits of bounded sets in $H^2(\mathbb{R}^N)$ for $\{S(t)\colon t \geq 0\}$ are almost immediately bounded in $H^\sigma_q(\mathbb{R}^N)$ for every $2 \leq q < \infty$, $\sigma \leq 4 + \left(\frac{N}{q} - \frac{N}{r}\right)_-$, i.e., for any B bounded in $H^2(\mathbb{R}^N)$ and any $\varepsilon > 0$ there exists $R_{B,\varepsilon} > 0$ such that

$$\|S(t)u_0\|_{H^\sigma_q(\mathbb{R}^N)} \leq R_{B,\varepsilon},\quad u_0 \in B,\ t \geq \varepsilon,$$

(iii) the semigroup $\{S(t)\colon t \geq 0\}$ has a global attractor $\mathbf{A}$, which is compact in $H^2(\mathbb{R}^N)$, invariant under $\{S(t)\colon t \geq 0\}$, bounded in $H^\sigma_q(\mathbb{R}^N)$, and attracting in $H^\sigma_q(\mathbb{R}^N)$ bounded sets of $H^2(\mathbb{R}^N)$ for every $2 \leq q < \infty$, $\sigma < 4 + \left(\frac{N}{q} - \frac{N}{r}\right)_-$,
(iv) the global attractor $\mathbf{A}$ is contained in a positively invariant closed and bounded absorbing set $\mathbf{B}$ in $H^2(\mathbb{R}^N)$, and both $\mathbf{A}$ and $\mathbf{B}$ are bounded in the space of bounded and continuous functions; hence all values of all elements of $\mathbf{B}$ (and $\mathbf{A}$) belong to a certain interval $I_{\mathbf{B}} = [-R_{\mathbf{B}}, R_{\mathbf{B}}]$.

Proof Part (i) follows from the theory of abstract parabolic problems (see [9, Sect. 3, $(3.29)_{|_{\gamma(\varepsilon)=\frac{1}{2}}}$]) and from [10, Theorem 1.2].

As for the statements in parts (ii) and (iii), they are consequences of [10, Theorems 3.9 and 4.4].

Concerning part (iv), in the role of $\mathbf{B}$ we take the closure in $H^2(\mathbb{R}^N)$ of the set $\{S(t)u_0\colon t \geq 1,\ u_0 \in \mathcal{O}_{\mathbf{A}}\}$, where $\mathcal{O}_{\mathbf{A}}$ is a neighborhood in $H^2(\mathbb{R}^N)$ of the attractor $\mathbf{A}$. Using (ii) with $q = \sigma = 2$ we observe that such set is bounded in $H^2(\mathbb{R}^N)$. Applying (ii) with $\sigma = 4 + \left(\frac{N}{q} - \frac{N}{r}\right)_-$ and $q > r$ where r is as in (13.3) we then observe via Lemma 13.10 that such set is also bounded in the space of bounded and continuous functions in $\mathbb{R}^N$. Here note that for the prescribed parameters σ and q we have $H^\sigma_q(\mathbb{R}^N) \hookrightarrow C^{\mu_0}(\mathbb{R}^N)$ for some positive μ_0 (see [30, Sect. 2.8.1]). □

Remark 13.2 (i) Recall from Proposition 13.2 that the values of elements of **B** belong to some interval $I_{\mathbf{B}} = [-R_{\mathbf{B}}, R_{\mathbf{B}}]$. Since $\mathbf{A} \subset \mathbf{B}$, it is also true for elements of **A**, although we could possibly find a shorter interval for elements belonging only to **A**.

(ii) Note that if we let $u, v \in I_{\mathbf{B}}$ and use (13.6) then we have

$$|f_0(x,u) - f_0(x,v)| \le L_{\mathbf{B}}|u - v| \text{ whenever } x \in \mathbb{R}^N,\ u, v \in I_{\mathbf{B}}, \tag{13.23}$$

where $L_{\mathbf{B}}$ is a positive constant depending only on $R_{\mathbf{B}}$ in (i) above and c, ρ in (13.6).

13.3 Estimate of Fractal Dimension of the Attractor

Our main concern here is to prove Theorem 13.1. This will be carried out in the following three subsections.

13.3.1 Estimates for the Semigroup on the Attractor and the Absorbing Set

In Lemmas 13.1–13.7 we use the assumptions of Theorem 13.1. Given $R > 0$ we also denote $B_R = \{x \in \mathbb{R}^N \colon |x| < R\}$ and $B_R^c = \mathbb{R}^N \setminus B_R$, whereas **B** is the absorbing set as in Proposition 13.2(iv).

Lemma 13.1 *For arbitrarily fixed $q \in [2, \infty)$, given any $\varepsilon > 0$ there exist certain $t_\varepsilon > 0$ and $R_\varepsilon > 0$ such that*

$$\|S(t)u_0\|_{L^q(B^c_{R_\varepsilon})} \le \varepsilon \text{ for any } t \ge t_\varepsilon,\ u_0 \in \mathbf{B}. \tag{13.24}$$

Moreover, we have

$$\|\phi\|_{L^q(B^c_{R_\varepsilon})} \le \varepsilon \text{ for each } \phi \in \mathbf{A}. \tag{13.25}$$

Proof The first part follows from [10, Lemma 4.1]. If $\phi \in \mathbf{A}$, by invariance of **A** we have $\phi = S(t_\varepsilon)u_0$ for some $u_0 \in \mathbf{A} \subset \mathbf{B}$ and hence we get (13.25). □

Lemma 13.2 *If $u_0, v_0 \in \mathbf{B}$ and $u = S(\cdot)u_0$, $v = S(\cdot)v_0$, then $U = u - v$ satisfies*

$$\sup_{t\in[0,T]} \|U(t)\|_{L^2(\mathbb{R}^N)} \le c_T\|U(0)\|_{L^2(\mathbb{R}^N)},\ T \ge 0, \tag{13.26}$$

$$\sup_{t\in[0,T]} t^{\frac{1}{2}}\|U(t)\|_{H^2(\mathbb{R}^N)} \leq c_T\|U(0)\|_{L^2(\mathbb{R}^N)},\ T\geq 0, \tag{13.27}$$

$$\sup_{t\in[0,T]} \|U(t)\|_{H^2(\mathbb{R}^N)} \leq c_T\|U(0)\|_{H^2(\mathbb{R}^N)},\ T\geq 0, \tag{13.28}$$

for some positive constant c_T.

In particular, we have

$$\begin{aligned}\|U\|_{L^2(0,T;L^2(\mathbb{R}^N))} &\leq c_T T^{\frac{1}{2}}\|U(0)\|_{L^2(\mathbb{R}^N)},\ T>0,\\ \|U\|_{L^2(0,T;H^2(\mathbb{R}^N))} &\leq c_T T^{\frac{1}{2}}\|U(0)\|_{H^2(\mathbb{R}^N)},\ T>0.\end{aligned} \tag{13.29}$$

Proof Given $u_0, v_0 \in \mathbf{B}$ and using Duhamel's formula (13.22) we have

$$U(t) = e^{-A_m t}U(0) + \int_0^t e^{-A_m(t-s)}\big(f_0(\cdot,u) - f_0(\cdot,v)\big)ds,\ t\geq 0, \tag{13.30}$$

where $A_m = \Delta^2 - m(x)I$. From (13.23) we infer that

$$\|f_0(\cdot,u) - f_0(\cdot,v)\|_{L^2(\mathbb{R}^N)} \leq L_{\mathbf{B}}\|U(t)\|_{L^2(\mathbb{R}^N)} \leq L_{\mathbf{B}}\|U(t)\|_{H^2(\mathbb{R}^N)}, \tag{13.31}$$

whereas due to part (ii) of Proposition 13.1 we have

$$\begin{aligned}\|e^{-A_m t}\|_{\mathscr{L}(H^2(\mathbb{R}^N))} &\leq M_T,\ t\in[0,T],\\ \|e^{-A_m t}\|_{\mathscr{L}(L^2(\mathbb{R}^N))} &\leq M_T,\ t\in[0,T],\\ \|e^{-A_m t}\|_{\mathscr{L}(L^2(\mathbb{R}^N),H^2(\mathbb{R}^N))} &\leq M_T t^{-\frac{1}{2}},\ t\in(0,T].\end{aligned} \tag{13.32}$$

Combining (13.30)–(13.32) we get

$$\|U(t)\|_{L^2(\mathbb{R}^N)} \leq M_T\|U(0)\|_{L^2(\mathbb{R}^N)} + M_T L_{\mathbf{B}}\int_0^t \|U(s)\|_{L^2(\mathbb{R}^N)}ds,\ t\in[0,T],$$

$$\|U(t)\|_{H^2(\mathbb{R}^N)} \leq M_T\|U(0)\|_{H^2(\mathbb{R}^N)} + M_T L_{\mathbf{B}}\int_0^t (t-s)^{-\frac{1}{2}}\|U(s)\|_{H^2(\mathbb{R}^N)}ds,\ t\in(0,T],$$

$$\begin{aligned}\|U(t)\|_{H^2(\mathbb{R}^N)} &\leq M_T t^{-\frac{1}{2}}\|U(0)\|_{L^2(\mathbb{R}^N)}\\ &\quad + M_T L_{\mathbf{B}}\int_0^t (t-s)^{-\frac{1}{2}}\|U(s)\|_{H^2(\mathbb{R}^N)}ds,\ t\in(0,T],\end{aligned}$$

and the results follow applying [8, Lemma 1.2.9]. □

Lemma 13.3 *Given any $\omega_0 > 0$ there exists $\alpha_0 > 0$ such that*

$$m(x) - C(x) \leq \frac{\omega_0}{4} + \alpha_0 D(x)\ \textit{for}\ x\in\mathbb{R}^N. \tag{13.33}$$

Proof From (13.8) we infer that $|g(x)| \leq D(x)$ in $\mathbb{R}^N$ and

$$m(x) - C(x) \leq -\frac{f_0(x,u)}{u} + \frac{2D(x)}{|u|}, \; x \in \mathbb{R}^N, \; u \neq 0. \tag{13.34}$$

Due to the mean value theorem and (13.5), (13.12) there is a certain $u_{\omega_0} > 0$ such that

$$\sup_{x\in\mathbb{R}^N} \left| \frac{f_0(x, u_{\omega_0})}{u_{\omega_0}} \right| = \sup_{x\in\mathbb{R}^N} \left| \frac{\partial f_0}{\partial u}(x, \theta_x u_{\omega_0}) \right| \leq \frac{\omega_0}{4},$$

where $\theta_x \in [0, 1]$. Substituting $u = u_{\omega_0}$ in (13.34) we get (13.33) with $\alpha_0 = \frac{2}{u_{\omega_0}}$. □

Lemma 13.4 *Given any $\omega_0 > 0$ there exists $c_* > 0$ such that*

$$|f_0(x,u) - f_0(x,v)| \leq \left(\frac{\omega_0}{8} + c_*(|u|^{\rho-1} + |v|^{\rho-1})\right) |u - v|, \; u, v \in \mathbb{R}, \; x \in \mathbb{R}^N. \tag{13.35}$$

Proof Since the mean value theorem yields

$$f_0(x,u) - f_0(x,v) = \frac{\partial f_0}{\partial u}\big(x, \theta_{x,u,v}u + (1 - \theta_{x,u,v})v\big)(u - v), \; x \in \mathbb{R}^N, \; u, v \in \mathbb{R},$$

where $\theta_{x,u,v} \in [0, 1]$, due to (13.12) one can choose a certain $\varepsilon_{\omega_0} > 0$ such that

$$\sup_{u,v\in[-\varepsilon_{\omega_0},\varepsilon_{\omega_0}]} \sup_{x\in\mathbb{R}^N} |f_0(x,u) - f_0(x,v)| \leq \frac{\omega_0}{8}|u - v|.$$

Having chosen such ε_{ω_0} we define

$$f_{01}(x,v) = \begin{cases} f_0(x,v), & x \in \mathbb{R}^N, \; |v| \leq \varepsilon_{\omega_0}, \\ f_0(x,\varepsilon_{\omega_0}), & x \in \mathbb{R}^N, \; v > \varepsilon_{\omega_0}, \\ f_0(x,-\varepsilon_{\omega_0}), & x \in \mathbb{R}^N, \; v < -\varepsilon_{\omega_0} \end{cases}$$

and

$$f_{02}(x,v) = f_0(x,v) - f_{01}(x,v), \; x \in \mathbb{R}^N, \; v \in \mathbb{R}.$$

With the above set-up we infer that

$$|f_{01}(x,u) - f_{01}(x,v)| \leq \frac{\omega_0}{8}|u - v|, \; x \in \mathbb{R}^N, \; u, v \in \mathbb{R},$$

and

$$|f_{02}(x,u) - f_{02}(x,v)| \leq c_*|u - v|(|u|^{\rho-1} + |v|^{\rho-1}), \; x \in \mathbb{R}^N, \; u, v \in \mathbb{R},$$

where constant $c_* > 0$ depends only on ε_{ω_0} and c, ρ from (13.6). Hence we get (13.35). □

Lemma 13.5 *There exist constants $a, b, R^*, T^* > 0$ such that for any $u_0, v_0 \in \mathbf{B}$ the function $U(t) = S(t)u_0 - S(t)v_0$, $t \geq 0$, satisfies the estimate*

$$\|U(t)\|_{L^2(\mathbb{R}^N)} \leq e^{-a(t-T^*)} \left\|U(T^*)\right\|_{L^2(\mathbb{R}^N)} + b\|U\|_{L^2(T^*,t;L^{2r'}(B_{R^*}))}, \quad t \geq T^*. \tag{13.36}$$

Moreover, if $u_0, v_0 \in \mathbf{A}$ then

$$\|U(t)\|_{L^2(\mathbb{R}^N)} \leq e^{-at} \|U(0)\|_{L^2(\mathbb{R}^N)} + b\|U\|_{L^2(0,t;L^{2r'}(B_{R^*}))}, \quad t \geq 0. \tag{13.37}$$

Proof Given $u_0, v_0 \in \mathbf{B}$ and $u = S(\cdot)u_0$, $v = S(\cdot)v_0$, we have from (13.1) and (13.2)

$$U_t + (\Delta^2 - m(x))U = f_0(x, u) - f_0(x, v)$$

and after multiplying by U in $L^2(\mathbb{R}^N)$

$$\frac{1}{2}\frac{d}{dt}\|U\|^2_{L^2(\mathbb{R}^N)} + \int_{\mathbb{R}^N} (|\Delta U|^2 - m(x)U^2) = \int_{\mathbb{R}^N} (f_0(x, u) - f_0(x, v))U. \tag{13.38}$$

We first consider assumption (13.13) and complete the proof in this case. Then we will complete the proof under assumption (13.14).

Proof under assumption (13.13). On the left-hand side of (13.38) we replace m by $C + m - C$. Then, given $\omega_0 > 0$ such that (13.20) holds, we use (13.21) with $\zeta > 0$ such that $\omega(\zeta) > \frac{3\omega_0}{4}$ and apply (13.33) to get for any $R > 0$

$$\begin{aligned}
\int_{\mathbb{R}^N} (|\Delta U|^2 - m(x)U^2) &= \int_{\mathbb{R}^N} (\zeta + 1 - \zeta)|\Delta U|^2 - \int_{\mathbb{R}^N} (C + m - C)(x)U^2 \\
&\geq \zeta\|\Delta U\|^2_{L^2(\mathbb{R}^N)} + \frac{3\omega_0}{4}\|U\|^2_{L^2(\mathbb{R}^N)} - \Big(\int_{B_R} + \int_{B_R^c}\Big)(m - C)(x)U^2 \\
&\geq \zeta\|\Delta U\|^2_{L^2(\mathbb{R}^N)} + \frac{\omega_0}{2}\|U\|^2_{L^2(\mathbb{R}^N)} - \int_{B_R} (m - C)(x)U^2 - \alpha_0 \int_{B_R^c} D(x)U^2.
\end{aligned} \tag{13.39}$$

Concerning the last term on the right-hand side above, observe that using embedding $H^2(\mathbb{R}^N) \hookrightarrow L^{2p'}(\mathbb{R}^N)$ with p as in (13.13) and using in $H^2(\mathbb{R}^N)$ equivalent norm[1] $\|(-\Delta + I)(\cdot)\|_{L^2(\mathbb{R}^N)}$ we obtain with some constant $b_p > 0$

$$\|U\|_{L^{2p'}(\mathbb{R}^N)} \leq b_p(\|U\|_{L^2(\mathbb{R}^N)} + \|\Delta U\|_{L^2(\mathbb{R}^N)}), \tag{13.40}$$

so that by the Hölder inequality

[1]The equivalence in $H^2(\mathbb{R}^N)$ of the standard $\|\cdot\|_{H^2(\mathbb{R}^N)}$ norm and the norm $\|(-\Delta + I)(\cdot)\|_{L^2(\mathbb{R}^N)}$ follows from [30, (2) and Step 3 in Sect. 2.5.3].

$$\begin{aligned}\int_{B_R^c} D(x)U^2 &\le \|D\|_{L^p(B_R^c)}\|U^2\|_{L^{p'}(B_R^c)} \le \|D\|_{L^p(B_R^c)}\|U\|^2_{L^{2p'}(\mathbb{R}^N)} \\ &\le 2b_p^2\|D\|_{L^p(B_R^c)}(\|U\|^2_{L^2(\mathbb{R}^N)} + \|\Delta U\|^2_{L^2(\mathbb{R}^N)}).\end{aligned} \tag{13.41}$$

Focusing below on $R = R_0$ chosen so large that

$$2\alpha_0 b_p^2\|D\|_{L^p(B_{R_0}^c)} \le \min\left\{\frac{\zeta}{2}, \frac{\omega_0}{8}\right\}, \tag{13.42}$$

which is possible due to the fact that $D \in L^p(\mathbb{R}^N)$, we thus have

$$\begin{aligned}\int_{\mathbb{R}^N}(|\Delta U|^2 - m(x)U^2) &\ge \zeta\|\Delta U\|^2_{L^2(\mathbb{R}^N)} + \frac{\omega_0}{2}\|U\|^2_{L^2(\mathbb{R}^N)} - \int_{B_{R_0}}(m-C)(x)U^2 \\ &\quad - \min\left\{\frac{\zeta}{2}, \frac{\omega_0}{8}\right\}(\|U\|^2_{L^2(\mathbb{R}^N)} + \|\Delta U\|^2_{L^2(\mathbb{R}^N)}) \\ &\ge \frac{\zeta}{2}\|\Delta U\|^2_{L^2(\mathbb{R}^N)} + \frac{3\omega_0}{8}\|U\|^2_{L^2(\mathbb{R}^N)} - \int_{B_{R_0}}(m-C)(x)U^2.\end{aligned}$$

Since due to (13.3), (13.9) there exists a constant $k_0 = k_0(R_0) > 0$ such that

$$\|m - C\|_{L^r(B_{R_0})} \le k_0,$$

by the Hölder inequality we get

$$\int_{B_{R_0}}(m(x) - C(x))U^2 \le k_0\|U\|^2_{L^{2r'}(B_{R_0})}, \tag{13.43}$$

where the exponent r' is Hölder's conjugate to r. Consequently, we have

$$\int_{\mathbb{R}^N}(|\Delta U|^2 - m(x)U^2) \ge \frac{\zeta}{2}\|\Delta U\|^2_{L^2(\mathbb{R}^N)} + \frac{3\omega_0}{8}\|U\|^2_{L^2(\mathbb{R}^N)} - k_0\|U\|^2_{L^{2r'}(B_{R_0})}. \tag{13.44}$$

To estimate the right-hand side of (13.38), we use (13.23) on B_R and (13.35) on B_R^c to get

$$\begin{aligned}\int_{\mathbb{R}^N}(f_0(x,u) - f_0(x,v))U &\le \Big(\int_{B_R} + \int_{B_R^c}\Big)|f_0(x,u) - f_0(x,v)||U| \\ &\le L_{\mathbf{B}}\int_{B_R} U^2 + \frac{\omega_0}{8}\int_{B_R^c} U^2 + c_*\int_{B_R^c}(|u|^{\rho-1} + |v|^{\rho-1})U^2.\end{aligned} \tag{13.45}$$

We fix arbitrarily $q \ge \max\{\frac{N}{4}, \frac{2}{\rho-1}, 1\}$ (also $q > 1$ if $N = 4$) and obtain (13.40) with q instead of p. For $\varepsilon > 0$ so small that

$$4b_q^2 c_* \varepsilon^{\rho-1} \leq \min\left\{\frac{\zeta}{2}, \frac{\omega_0}{8}\right\}, \tag{13.46}$$

we will apply (13.24) with exponent $q(\rho-1)$. For this exponent and above ε we choose $t_\varepsilon > 0$ and $R_\varepsilon > 0$ according to (13.24) and also such that

$$R_\varepsilon \geq R_0, \tag{13.47}$$

where R_0 was fixed before to satisfy (13.42). Proceeding in a similar manner as in (13.41) and using that $\|u\|_{L^{q(\rho-1)}(B^c_{R_\varepsilon})} \leq \varepsilon$ according to (13.24), we get

$$\begin{aligned}
\int_{B^c_{R_\varepsilon}} (|u|^{\rho-1} + |v|^{\rho-1})U^2 &\leq \||u|^{\rho-1} + |v|^{\rho-1}\|_{L^q(B^c_{R_\varepsilon})} \|U^2\|_{L^{q'}(B^c_{R_\varepsilon})} \\
&\leq (\|u\|^{\rho-1}_{L^{q(\rho-1)}(B^c_{R_\varepsilon})} + \|v\|^{\rho-1}_{L^{q(\rho-1)}(B^c_{R_\varepsilon})}) \|U\|^2_{L^{2q'}(\mathbb{R}^N)} \\
&\leq 4b_q^2 \varepsilon^{\rho-1} (\|U\|^2_{L^2(\mathbb{R}^N)} + \|\Delta U\|^2_{L^2(\mathbb{R}^N)}), \ t \geq t_\varepsilon.
\end{aligned} \tag{13.48}$$

Note that if $u_0, v_0 \in \mathbf{A}$ then, according to (13.25), we can take $t_\varepsilon = 0$.

From (13.38), (13.44), (13.45) and (13.48) we have

$$\begin{aligned}
\frac{1}{2}\frac{d}{dt}\|U\|^2_{L^2(\mathbb{R}^N)} &+ \left(\frac{\zeta}{2} - 4b_q^2 c_* \varepsilon^{\rho-1}\right) \|\Delta U\|^2_{L^2(\mathbb{R}^N)} + \left(\frac{\omega_0}{4} - 4b_q^2 c_* \varepsilon^{\rho-1}\right) \|U\|^2_{L^2(\mathbb{R}^N)} \\
&- k_0 \|U\|^2_{L^{2r'}(B_{R_0})} \leq L_{\mathbf{B}} \int_{B_{R_\varepsilon}} U^2, \ t \geq t_\varepsilon,
\end{aligned}$$

which for ε, R_ε as in (13.46), (13.47) gives

$$\frac{1}{2}\frac{d}{dt}\|U\|^2_{L^2(\mathbb{R}^N)} + \frac{\omega_0}{8}\|U\|^2_{L^2(\mathbb{R}^N)} - k_0\|U\|^2_{L^{2r'}(B_{R_\varepsilon})} \leq L_{\mathbf{B}} \int_{B_{R_\varepsilon}} U^2, \ t \geq t_\varepsilon.$$

Denoting now by α_ε the $\frac{1}{r}$-power of the volume of the ball in $\mathbb{R}^N$ of radius R_ε (respectively, letting $\alpha_\varepsilon = 1$ when $r = \infty$) we obtain by the Hölder inequality

$$\int_{B_{R_\varepsilon}} U^2 \leq \alpha_\varepsilon \|U\|^2_{L^{2r'}(B_{R_\varepsilon})}.$$

Hence we conclude that

$$\frac{1}{2}\frac{d}{dt}\|U\|^2_{L^2(\mathbb{R}^N)} + \frac{\omega_0}{8}\|U\|^2_{L^2(\mathbb{R}^N)} \leq (k_0 + L_{\mathbf{B}}\alpha_\varepsilon)\|U\|^2_{L^{2r'}(B_{R_\varepsilon})}, \ t \geq t_\varepsilon,$$

which yields

$$\|U(t)\|^2_{L^2(\mathbb{R}^N)} \le e^{-\frac{\omega_0}{4}(t-t_\varepsilon)} \|U(t_\varepsilon)\|^2_{L^2(\mathbb{R}^N)} + 2(k_0 + L_{\mathbf{B}}\alpha_\varepsilon) \int_{t_\varepsilon}^t \|U(s)\|^2_{L^{2r'}(B_{R_\varepsilon})} ds, \ t \ge t_\varepsilon,$$

which is a counterpart of (13.36) with $T^* = t_\varepsilon$ and $R^* = R_\varepsilon$, where ε is as in (13.46). Since $T^* = t_\varepsilon = 0$ when $u_0, v_0 \in \mathbf{A}$, we also get (13.37).

Proof under assumption (13.14). In this case using (13.33) there is $R_0 > 0$ such that

$$m(x) - C(x) \le \frac{3\omega_0}{8} \quad \text{for } |x| \ge R_0. \tag{13.49}$$

We proceed in a similar way as in the previous case, that is, we replace m by $C + m - C$ and use (13.21) with $\zeta > 0$ such that $\omega(\zeta) > \frac{3\omega_0}{4}$. However, instead of (13.33) we now use (13.49) and hence instead of (13.39) we get

$$\int_{\mathbb{R}^N} (|\Delta U|^2 - m(x)U^2) \ge \zeta \|\Delta U\|^2_{L^2(\mathbb{R}^N)} + \frac{3\omega_0}{8} \|U\|^2_{L^2(\mathbb{R}^N)} - \int_{B_{R_0}} (m - C)(x)U^2,$$

which together with (13.43) leads to (13.44). The rest is then the same as under the assumption (13.13) and we get (13.36)–(13.37) as before. □

Lemma 13.6 *If $u_0, v_0 \in \mathbf{B}$ then the function $U(t) = S(t)u_0 - S(t)v_0$, $t \ge 0$, satisfies for every $\tau > 0$ and any $t \ge T^*$ the estimate*

$$\|U(t+\tau)\|_{H^2(\mathbb{R}^N)} \le \frac{c_\tau c_{T^*}}{\tau^{\frac{1}{2}}} e^{-a(t-T^*)} \|U(0)\|_{H^2(\mathbb{R}^N)} + \frac{bc_\tau}{\tau^{\frac{1}{2}}} \|U\|_{L^2(T^*,t;L^{2r'}(B_{R^*}))}, \tag{13.50}$$

where c_τ, c_{T^*} come from Lemma 13.2 and $a, b, R^*, T^* > 0$ are as in Lemma 13.5.

Moreover, if $u_0, v_0 \in \mathbf{A}$ then

$$\|U(t+\tau)\|_{H^2(\mathbb{R}^N)} \le \frac{c_\tau}{\tau^{\frac{1}{2}}} e^{-at} \|U(0)\|_{H^2(\mathbb{R}^N)} + \frac{bc_\tau}{\tau^{\frac{1}{2}}} \|U\|_{L^2(0,t;L^{2r'}(B_{R^*}))}, \ \tau > 0, \ t \ge 0.$$

Hence, given $T > 0$ and letting $\tau = \min\{\frac{T}{2}, 1\}$ and $t = T - \tau$ we get for $u_0, v_0 \in \mathbf{A}$

$$\|U(T)\|_{H^2(\mathbb{R}^N)} \le \eta_T \|U(0)\|_{H^2(\mathbb{R}^N)} + \mu_T \|U\|_{L^2(0,T;L^{2r'}(B_{R^*}))}, \ T > 0,$$

where $\eta_T = c_1 e^{-a(T-1)}$ and $\mu_T = bc_1 = \mu^*$ for $T \ge 2$ so that (13.17) *holds.*

Proof Observe that, due to positive invariance of $\mathbf{B}$, $S(t)u_0, S(t)v_0 \in \mathbf{B}$ for every $t \ge 0$. Therefore using the semigroup property and (13.27) we get for $\tau > 0$

$$\begin{aligned} \|U(t+\tau)\|_{H^2(\mathbb{R}^N)} &= \|S(\tau)S(t)u_0 - S(\tau)S(t)v_0\|_{H^2(\mathbb{R}^N)} \\ &\le c_\tau \tau^{-\frac{1}{2}} \|S(t)u_0 - S(t)v_0\|_{L^2(\mathbb{R}^N)} = c_\tau \tau^{-\frac{1}{2}} \|U(t)\|_{L^2(\mathbb{R}^N)}, \ t \ge 0. \end{aligned} \tag{13.51}$$

Combining (13.51) with (13.26) and (13.36) or (13.37) respectively, and using the inequality $\|U(0)\|_{L^2(\mathbb{R}^N)} \leq \|U(0)\|_{H^2(\mathbb{R}^N)}$ we get the result. □

Remark 13.3 (i) Recall from Proposition 13.2 that

$$u, v \in C^1((0,\infty); L^2(\mathbb{R}^N)),\ A_m u, A_m v \in C((0,\infty); L^2(\mathbb{R}^N)) \text{ for } u_0, v_0 \in \mathbf{B},$$

where $A_m = \Delta^2 - m(x)I$.

(ii) Since the solutions enter $D(A_m)$, due to invariance of the attractor we have $\mathbf{A} \subset D(A_m)$ and therefore we conclude from [31, Theorem I.1] that

$$u, v \in C^1([0,\infty); L^2(\mathbb{R}^N)),\ A_m u, A_m v \in C([0,\infty); L^2(\mathbb{R}^N)) \text{ for } u_0, v_0 \in \mathbf{A}.$$

Lemma 13.7 *If $u_0, v_0 \in \mathbf{B}$ and $u = S(\cdot)u_0$, $v = S(\cdot)v_0$, then $U = u - v$ satisfies*

$$\|U\|_{H^1(0,T;L^2(\mathbb{R}^N))\cap L^2(0,T;H^2(\mathbb{R}^N))} \leq \kappa_T \|U(0)\|_{H^2(\mathbb{R}^N)},\ T > 0, \tag{13.52}$$

for some positive constant κ_T.

Proof Let λ_0 be so large that a sectorial selfadjoint operator $\mathscr{A} = \Delta^2 - (m(x) - \lambda_0)I$ in $L^2(\mathbb{R}^N)$ is positive. Given $u_0, v_0 \in \mathbf{B}$ observe from (13.1)–(13.2) that $U = u - v$ satisfies

$$U_t + \mathscr{A}U = f_0(\cdot, u) - f_0(\cdot, v) + \lambda_0 U =: \mathscr{F}(t), \tag{13.53}$$

where due to (13.26) and (13.31)

$$\sup_{t\in[0,T]} \|\mathscr{F}(t)\|_{L^2(\mathbb{R}^N)} \leq c_T(L_{\mathbf{B}} + \lambda_0)\|U(0)\|_{L^2(\mathbb{R}^N)}. \tag{13.54}$$

Since we have already proved (13.29), in order to obtain (13.52) it is enough to show that $\|U_t\|_{L^2(0,T;L^2(\mathbb{R}^N))}$ is bounded by a multiple of $\|U(0)\|_{H^2(\mathbb{R}^N)}$. This actually follows from maximal regularity result, see e.g. [5, Corollary 3]. However, for the completeness of argument we present a direct proof.

Multiplying (13.53) by $\mathscr{A}U$ in $L^2(\mathbb{R}^N)$ and using the Cauchy inequality we get

$$\frac{1}{2}\frac{d}{dt}\|\mathscr{A}^{\frac{1}{2}}U\|^2_{L^2(\mathbb{R}^N)} + \|\mathscr{A}U\|^2_{L^2(\mathbb{R}^N)} \leq \frac{1}{2}\|\mathscr{F}(t)\|^2_{L^2(\mathbb{R}^N)} + \frac{1}{2}\|\mathscr{A}U\|^2_{L^2(\mathbb{R}^N)}.$$

Using this and (13.54) we obtain

$$\frac{d}{dt}\|\mathscr{A}^{\frac{1}{2}}U\|^2_{L^2(\mathbb{R}^N)} + \|\mathscr{A}U\|^2_{L^2(\mathbb{R}^N)} \leq c_T^2(L_{\mathbf{B}} + \lambda_0)^2\|U(0)\|^2_{L^2(\mathbb{R}^N)},$$

which ensures that

$$\|\mathscr{A}U\|^2_{L^2(0,T;L^2(\mathbb{R}^N))} \leq Tc_T^2(L_{\mathbf{B}} + \lambda_0)^2\|U(0)\|^2_{L^2(\mathbb{R}^N)} + \|\mathscr{A}^{\frac{1}{2}}U(0)\|^2_{L^2(\mathbb{R}^N)}.$$

Since $D(\mathscr{A}^{\frac{1}{2}}) = H^2(\mathbb{R}^N)$ (see [9, Corollary 2.7]) and $\|U(0)\|_{L^2(\mathbb{R}^N)}$ is estimated by $\|U(0)\|_{H^2(\mathbb{R}^N)}$, we infer that for some $k_T > 0$

$$\|\mathscr{A}U\|_{L^2(0,T;L^2(\mathbb{R}^N))} \le k_T \|U(0)\|_{H^2(\mathbb{R}^N)}. \tag{13.55}$$

Now we connect (13.53), (13.54) and (13.55) to get

$$\begin{aligned}\|U_t\|_{L^2(0,T;L^2(\mathbb{R}^N))} &\le \|\mathscr{A}U\|_{L^2(0,T;L^2(\mathbb{R}^N))} + \|\mathscr{F}\|_{L^2(0,T;L^2(\mathbb{R}^N))} \\ &\le k_T\|U(0)\|_{H^2(\mathbb{R}^N)} + c_T T^{\frac{1}{2}}(L_{\mathbf{B}} + \lambda_0)\|U(0)\|_{L^2(\mathbb{R}^N)}.\end{aligned}$$

This and (13.29) give (13.52) with κ_T depending only on T, c_T, k_T, λ_0 and $L_{\mathbf{B}}$. □

13.3.2 Compactness of Associated Seminorm

The following lemma concerns compactness property of the seminorm which is used in the proof of finite dimensionality of attractor.

Lemma 13.8 *Given $T > T^* \ge 0$ let*

$$Z := C^1([T^*, T]; L^2(\mathbb{R}^N)) \cap C([T^*, T]; H^2(\mathbb{R}^N)) \tag{13.56}$$

be equipped with the norm

$$\|\cdot\|_Z = \|\cdot\|_{H^1(T^*,T;L^2(\mathbb{R}^N))} + \|\cdot\|_{L^2(T^*,T;H^2(\mathbb{R}^N))}, \tag{13.57}$$

and consider the map

$$\mathrm{n}_Z\colon Z \to [0,\infty), \qquad \mathrm{n}_Z(z) = \mu\left(\int_{T^*}^{T} \|(Pz)(s)\|^2_{L^{2r'}(B_{R^*})} ds\right)^{\frac{1}{2}}, \quad z \in Z,$$

where μ and R^* are positive constants, r' is Hölder's conjugate to r from (13.3)*, and the operator P is defined as follows:*

$$Z \ni z \xrightarrow{P} z_{|B_{R^*}} \in C^1([T^*, T]; L^2(B_{R^*})) \cap C([T^*, T]; H^2(B_{R^*})).$$

Then

$$P\,is\,compact\ from\ Z\,into\,L^2(T^*, T; L^{2r'}(B_{R^*})) =: Y,$$

that is, n_Z is a compact seminorm in Z in the sense of Definition 13.1.

Proof Given $z \in Z$ and writing at any fixed $t \in [T^*, T]$ the difference quotient for Pz in $\|\cdot\|_{L^2(B_{R^*})}$ norm observe that in $L^2(B_{R^*})$ there exists $\frac{d(Pz)}{dt} = P\frac{dz}{dt}$. Now, if

$\{z_n\} \subset Z$ is bounded in the norm given in (13.57) then, in particular, $\{Pz_n\}$ is bounded in the space

$$W = \left\{ w \in L^2\left(T^*, T; H^2(B_{R^*})\right), \frac{dw}{dt} \in L^2\left(T^*, T; L^2(B_{R^*})\right) \right\},$$

where

$$H^2(B_{R^*}) \overset{compact}{\hookrightarrow} L^{2r'}(B_{R^*}) \hookrightarrow L^2(B_{R^*}),$$

because $r > \max\{\frac{N}{4}, 1\}$ and $\frac{1}{r} + \frac{1}{r'} = 1$.

Due to [22, Theorem 5.1], there then exists a subsequence of $\{Pz_n\}$ convergent in Y, which proves compactness of P from Z into the space Y. □

13.3.3 Estimate from Above of Fractal Dimension of the Attractor

Having derived in Lemmas 13.6 and 13.7 the estimates (13.15)–(13.16) for the semigroup on the attractor **A** and having established compactness of the seminorm as in Lemma 13.8 we now estimate fractal dimension of **A**.

In particular, applying Proposition 13.4 we specify:

- map $V = S(T)$ with large enough T for which $\eta_T < 1$ and $\mu_T = \mu^*$,
- spaces $X = H^2(\mathbb{R}^N)$, $Y = L^2(0, T; L^{2r'}(B_{R^*}))$ with standard norms and space $Z = C^1([0, T]; L^2(\mathbb{R}^N)) \cap C([0, T]; H^2(\mathbb{R}^N))$ with the norm

$$\|\cdot\|_Z = \|\cdot\|_{H^1(0,T;L^2(\mathbb{R}^N))} + \|\cdot\|_{L^2(0,T;H^2(\mathbb{R}^N))},$$

- seminorm $\mathrm{n}_Z(z)$ of the form $\mu^* \|Pz\|_{L^2(0,T;L^{2r'}(B_{R^*}))}$ defined for $z \in Z$ with the aid of operator $Pz = z_{|B_{R^*}} \in C^1([0, T]; L^2(B_{R^*})) \cap C([0, T]; H^2(B_{R^*}))$,
- mapping K such that $\mathbf{A} \ni u_0 \overset{K}{\longrightarrow} S(\cdot)u_0 \in Z$.

With the above set-up we obtain from (13.62) the estimate of fractal dimension of **A** as in (13.18). The proof of Theorem 13.1 is thus complete.

13.4 Existence of Exponential Attractor

In this section we prove Theorem 13.2 beginning with the following auxiliary result.

Proposition 13.3 *Under the assumptions of Theorem 13.1, let* **B** *be the absorbing positively invariant set for the semigroup* $\{S(t): t \geq 0\}$ *in* $H^2(\mathbb{R}^N)$ *as in Proposition 13.2(iv). Then there are times* $T > T^* > 0$ *and* $\eta \in (0, 1)$ *such that for any* $\delta \in (0, 1 - \eta)$ *there exists a compact set* **E** $\subset$ **B** *in* $H^2(\mathbb{R}^N)$ *satisfying*

$$S(T)\mathbf{E} \subset \mathbf{E}, \qquad d(S(nT)\mathbf{B}, \mathbf{E}) \leq \Lambda\,(\eta+\delta)^n\,,\ n \in \mathbb{N},$$

for some $\Lambda > 0$*, and also such that fractal dimension of* $\mathbf{E}$ *in* $H^2(\mathbb{R}^N)$ *is finite and estimated by*

$$dim_f(\mathbf{E}) \leq \frac{\ln \mathrm{m}_Z^{\mathscr{B}_\delta}}{\ln \frac{1}{\eta+\delta}}, \tag{13.58}$$

where the space Z *is given in* (13.56)–(13.57) *and* $\mathrm{m}_Z^{\mathscr{B}_\delta}$ *denotes the maximal number of elements* z_j *in a closed ball* $\mathscr{B}_\delta$ *in* Z *of radius* $2\mu^*\kappa_T\delta^{-1}$ *around zero with the property that* $\|z_{j|_{B_{R^*}}} - z_{l|_{B_{R^*}}}\|_{L^2(T^*,T;L^{2r'}(B_{R^*}))} > 1$ *for* $j \neq l$*. Here constants* R^*, T^*, μ^* *and* κ_T *are as in Lemmas 13.6 and 13.7.*

Proof Based on (13.50) with $\tau = 1$ and (13.52) we directly apply [12, Theorem 3.2.1] specifying:

- map $V = S(T)$ with $T \geq T^* + 1$ such that $\eta := c_1 c_{T^*} e^{-a(T-T^*-1)} < 1$,
- spaces $X = H^2(\mathbb{R}^N)$, $Y = L^2(T^*, T; L^{2r'}(B_{R^*}))$ with standard norms and space Z given in (13.56)–(13.57),
- seminorm $\mathrm{n}_Z(z)$ of the form $\mu^*\|Pz\|_{L^2(T^*,T;L^{2r'}(B_{R^*}))}$ defined for $z \in Z$ with the aid of operator $Pz = z_{|_{B_{R^*}}} \in C^1([T^*, T]; L^2(B_{R^*})) \cap C([T^*, T]; H^2(B_{R^*}))$,
- mapping K such that

$$\mathbf{B} \ni u_0 \stackrel{K}{\longrightarrow} S(\cdot)u_0 \in Z. \tag{13.59}$$

Observe that (13.50) with $t = T - 1$, $\tau = 1$ is the counterpart of [12, (3.2.1)] on $\mathbf{B}$, the seminorm n_Z is compact in Z by Lemma 13.8, and K in (13.59) is Lipschitz continuous from $\mathbf{B}$ into Z by (13.52), as required in [12, Theorem 3.2.1]. □

Now we prove Hölder continuity of $t \to S(t)u_0$ for $t \in [T, 2T]$ uniformly for $u_0 \in \mathbf{B}$, which will be used below to estimate the fractal dimension of an exponential attractor in the proof of Theorem 13.2.

Lemma 13.9 *Under the assumptions of Theorem 13.1 for arbitrarily fixed* $\zeta \in (0, 1)$ *and any* $T > 0$ *we have*

$$\|S(t_1)u_0 - S(t_2)u_0\|_{H^2(\mathbb{R}^N)} \leq \chi_T\,|t_1 - t_2|^\zeta\,,\ t_1, t_2 \in [T, 2T],\ u_0 \in \mathbf{B} \tag{13.60}$$

for some positive constant χ_T*;* $\mathbf{B}$ *being the absorbing set as in Proposition 13.2(iv).*

Proof The proof follows the lines of the proof of [8, (2.2.3)] using Lipschitz properties of f_0 (see [9, (3.18) and Proposition 3.4]) and the characterization of the two-sided fractional power scale generated by A_m in $L^2(\mathbb{R}^N)$ (see [9, Corollary 2.7]). □

We finally conclude the existence of an exponential attractor completing the proof of Theorem 13.2.

Proof of Theorem 13.2. Let $T > T^*$ and $\eta \in (0, 1)$ be as in Proposition 13.3. For a given $\delta \in (0, 1-\eta)$ we use $\mathbf{E}$ from Proposition 13.3 to define $\mathbf{M} = \bigcup_{t\in[T,2T]} S(t)\mathbf{E}$, which is compact in $H^2(\mathbb{R}^N)$, because $S(t)u_0$ is continuous in $H^2(\mathbb{R}^N)$ with respect to a pair of arguments $(t, u_0) \in [0, \infty) \times H^2(\mathbb{R}^N)$. Moreover, $S(t)\mathbf{M} \subset \mathbf{M}$ for $t \geq 0$, since $S(T)\mathbf{E} \subset \mathbf{E}$.

Due to (13.28) and (13.60) fractal dimension of $\mathbf{M}$ in $H^2(\mathbb{R}^N)$ is estimated from above by $\frac{1}{\zeta}\left(1 + dim_f \mathbf{E}\right)$ (see [12, Proposition 3.1.13]). Since this holds for any $\zeta \in (0, 1)$, we get from (13.58)

$$dim_f(\mathbf{M}) \leq 1 + \frac{\ln \mathrm{m}_Z^{\mathscr{B}_\delta}}{\ln \frac{1}{\eta+\delta}}, \tag{13.61}$$

where $\mathrm{m}_Z^{\mathscr{B}_\delta}$ is as in Proposition 13.3.

Note that if B is bounded in $H^2(\mathbb{R}^N)$, then there exists $T_B \geq 0$ such that $S(T_B)B \subset \mathbf{B}$. For $t \geq T_B + 2T$, we have $t - T_B = n_t T + T + r_t$ with $r_t \in [0, T]$, $n_t \in \mathbb{N}$, and using (13.28), we obtain

$$\begin{aligned} d(S(t)B, \mathbf{M}) &= d(S(t - T_B)S(T_B)B, \mathbf{M}) \leq d(S(T + r_t)S(n_t T)\mathbf{B}, S(T + r_t)\mathbf{E}) \\ &\leq c_{2T} d(S(n_t T)\mathbf{B}, \mathbf{E}) \leq c_{2T} \Lambda e^{-\xi n_t} \leq C_B e^{-\frac{\xi}{T} t}, \end{aligned}$$

where $\xi = -\ln(\eta + \delta) > 0$ and $C_B = c_{2T} \Lambda e^{2\xi + \frac{\xi T_B}{T}} > 0$. Hence we get the result. □

Acknowledgements The authors are grateful to the anonymous referees for the positive comments and helpful suggestions which improved the final version of the paper. Jan W. Cholewa was partially supported by the grant MTM2016-75465 from MINECO, Spain.

13.5 Appendix: Abstract Estimate of Fractal Dimension

Recall that if $\mathbb{M}$ is a compact subset of a given metric space then its *fractal dimension* is

$$dim_f(\mathbb{M}) = \limsup_{\varepsilon \to 0} \log_{\frac{1}{\varepsilon}} n(\mathbb{M}, \varepsilon),$$

where $n(\mathbb{M}, \varepsilon)$ denotes the smallest number of ε–balls needed to cover $\mathbb{M}$ in the considered space. The relevance of this notion has been satisfactorily addressed in many references, see e.g. [28, p. 351], [4, p. 1083], [12, p. 97] and references therein.

In what follows we include the estimate of fractal dimension from the monograph by Chueshov [12, Theorem 3.1.21] (see also [13, Theorem 2.14]), which we use in the proof of Theorem 13.1. We remark that this involves consideration of a compact seminorm as in the definition below.

Definition 13.1 n_Z is a compact seminorm on a normed space Z if and only if n_Z satisfies

$$\mathrm{n}_Z(x+y) \leq \mathrm{n}_Z(x) + \mathrm{n}_Z(y), \quad \mathrm{n}_Z(\lambda x) = |\lambda|\, \mathrm{n}_Z(x), \ x, y \in Z, \ \lambda \in \mathbb{R},$$

and each sequence $\{z_n\}$ bounded in Z contains a subsequence $\{z_{n_j}\}$ such that

$$\mathrm{n}_Z(z_{n_j} - z_{n_l}) \to 0 \ \text{as} \ j, l \to \infty.$$

Proposition 13.4 *Assume that $\mathbb{M}$ is a bounded and closed subset of a Banach space X, which is negatively invariant under $V : \mathbb{M} \to X$, that is, $\mathbb{M} \subset V\mathbb{M}$. Assume also that there exists a normed space Z, a map $K : \mathbb{M} \to Z$ and a compact seminorm n_Z on Z such that for some constants $\eta \in [0, 1)$ and $\kappa > 0$ we have*

$$\|Kx - Ky\|_Z \leq \kappa \, \|x - y\|_X \,, \ x, y \in \mathbb{M},$$

$$\|Vx - Vy\|_X \leq \eta \, \|x - y\|_X + \mathrm{n}_Z(Kx - Ky), \ x, y \in \mathbb{M}.$$

Then $\mathbb{M}$ is compact in X, its fractal dimension $dim_f(\mathbb{M})$ in X is finite and, given any $\delta \in (0, 1 - \eta)$,

$$dim_f(\mathbb{M}) \leq \frac{\ln \mathrm{m}_Z^{\mathscr{B}_\delta}}{\ln \frac{1}{\eta+\delta}},$$

where $\mathrm{m}_Z^{\mathscr{B}_\delta}$ is the maximal number of points z_j in a closed ball $\mathscr{B}_\delta$ in Z of radius $2\kappa\delta^{-1}$ around zero possessing the property that $\mathrm{n}_Z(z_j - z_l) > 1$ whenever $j \neq l$.

Remark 13.4 (i) The seminorm in Proposition 13.4 can be defined in particular as

$$\mathrm{n}_Z(z) = \mu \, \|Pz\|_Y \,, \ z \in Z,$$

where μ is a positive constant and $P \colon Z \to Y$ is a linear compact operator with values in a normed space Y (cf. [14, Theorem 2.1]).

(ii) With the seminorm as in part (i) above the fractal dimension of $\mathbb{M}$ in Proposition 13.4 can be estimated for any $\delta \in (0, 1 - \eta)$ by

$$dim_f(\mathbb{M}) \leq \frac{\ln \mathrm{m}_Z^{\mathscr{B}_\delta}}{\ln \frac{1}{\eta+\delta}}, \tag{13.62}$$

$\mathrm{m}_Z^{\mathscr{B}_\delta}$ being now the maximal number of elements z_j in a closed ball $\mathscr{B}_\delta$ in Z of radius $2\mu\kappa\delta^{-1}$ around zero with the property that $\left\| P(z_j - z_l) \right\|_Y > 1$ for $j \neq l$.

13.6 Appendix: Boundedness of Absorbing Set in Supremum Norm

Lemma 13.10 *Assume that $\sigma > 0$ and $q > 1$ are such that $H^{\sigma}_{q}(\mathbb{R}^N)$ is embedded in the Hölder space $C^{\mu_0}(\mathbb{R}^N)$ for some $\mu_0 > 0$ and denote the constant from this embedding by $c_{\sigma,q}$.*

If a subset B of $H^2(\mathbb{R}^N)$ is contained in a ball in $H^{\sigma}_{q}(\mathbb{R}^N)$ of radius R centered at zero, then $cl_{H^2(\mathbb{R}^N)}B$ can be viewed as a subset of a closed ball of radius $c_{\sigma,q}R$ around zero in the space of bounded and continuous functions in $\mathbb{R}^N$.

Proof Consider $\phi \in cl_{H^2(\mathbb{R}^N)}B$ and a sequence $\{\phi_n\} \subset B$ convergent to ϕ in $H^2(\mathbb{R}^N)$. Denoting by $\overline{B}_k$ a closed ball in $\mathbb{R}^N$ centered at zero of radius $k \in \mathbb{N}$ we have

$$\|\phi_n\|_{C^{\mu_0}(\overline{B}_k)} \leq \|\phi_n\|_{C^{\mu_0}(\mathbb{R}^N)} \leq c_{\sigma,q}R \text{ for all } k, n \in \mathbb{N}. \tag{13.63}$$

If $k \in \mathbb{N}$ is fixed then due to (13.63) and Arzela-Ascoli theorem a certain subsequence of $\{\phi_n\}$ converges to ϕ uniformly in $\overline{B}_k$. Therefore, a certain (diagonal type) subsequence of $\{\phi_n\}$ actually converges to ϕ uniformly in $\overline{B}_k$ for every $k \in \mathbb{N}$. If such a subsequence is chosen, then ϕ can be viewed as an almost uniform in $\mathbb{R}^N$ limit of this subsequence, which together with (13.63) implies that $|\phi(x)| \leq c_{\sigma,q}R$ for each $|x| \leq k$ and any $k \in \mathbb{N}$, which proves the claim. □

References

1. Amann, H.: Linear and Quasilinear Parabolic Problems. Birkhäuser, Basel (1995)
2. Anguiano, M., Haraux, A.: The ε-entropy of some infinite dimensional compact ellipsoids and fractal dimension of attractors. Evol. Equ. Control Theory **6**, 345–356 (2017)
3. Arrieta, J.M., Cholewa, J.W., Dlotko, T., Rodríguez-Bernal, A.: Asymptotic behavior and attractors for reaction diffusion equations in unbounded domains. Nonlinear Anal. **56**, 515–554 (2004)
4. Arrieta, J.M., Moya, N., Rodríguez-Bernal, A.: On the finite dimension of attractors of parabolic problems in $\mathbb{R}^N$ with general potentials. Nonlinear Anal. **68**, 1082–1099 (2008)
5. Haak, B.H., El Maati Ouhabaz: Maximal regularity for non-autonomous evolution equations. Math. Ann. **363**, 1117–1145 (2015)
6. Cholewa, J.W., Czaja, R.: Lattice dynamical systems: dissipative mechanism and fractal dimension of global and exponential attractors. J. Evol. Equ. **20**, 485–515 (2020)
7. Cholewa, J.W., Czaja, R., Mola, G.: Remarks on the fractal dimension of bi-space global and exponential attractors. Boll. Un. Math. Ital. (9) **1**, 121–145 (2008)
8. Cholewa, J.W., Dlotko, T.: Global Attractors in Abstract Parabolic Problems. Cambridge University Press, Cambridge (2000)
9. Cholewa, J.W., Rodríguez-Bernal, A.: Linear and semilinear higher order parabolic equations in $\mathbb{R}^N$. Nonlinear Anal. **75**, 194–210 (2012)
10. Cholewa, J.W., Rodríguez-Bernal, A.: Dissipative mechanism of a semilinear higher order parabolic equation in $\mathbb{R}^N$. Nonlinear Anal. **75**, 3510–3530 (2012)
11. Cholewa, J.W., Rodríguez-Bernal, A.: Critical and supercritical higher order parabolic problems in $\mathbb{R}^N$. Nonlinear Anal. **104**, 50–74 (2014)

12. Chueshov, I.: Dynamics of Quasi-stable Dissipative Systems. Springer, Heidelberg (2015)
13. Chueshov, I., Lasiecka, I.: Long-Time Behavior of Second Order Evolution Equations with Nonlinear Damping, Memoirs of the American Mathematical Society 195 (912). AMS, Providence, RI (2008)
14. Czaja, R., Efendiev, M.: A note on attractors with finite fractal dimension. Bull. Lond. Math. Soc. **40**, 651–658 (2008)
15. Efendiev, M.: Infinite-dimensional exponential attractors for fourth-order nonlinear parabolic equations in unbounded domains. Math. Methods Appl. Sci. **34**, 939–949 (2011)
16. Efendiev, M., Miranville, A., Zelik, S.: Exponential attractors for a singularly perturbed Cahn-Hilliard system. Math. Nachr. **272**, 11–31 (2004)
17. Efendiev, M., Zelik, S.: The attractor for a nonlinear reaction-diffusion system in an unbounded domain. Commun. Pure Appl. Math. **54**, 625–688 (2001)
18. Friedman, A.: Partial Differential Equations. Holt, Rinehart and Winston, New York (1969)
19. Galaktionov, V.A., Mitidieri, E.L., Pohozaev, S.I.: Blow-up for Higher-Order Parabolic, Hyperbolic, Dispersion and Schrödinger Equations. CRC Press, Boca Raton (2015)
20. Gazzola, F., Grunau, H.-C., Sweers, G.: Polyharmonic Boundary Value Problems. Positivity Preserving and Nonlinear Higher Order Elliptic Equations in Bounded Domains. Springer, Berlin (2010)
21. Hale, J.K.: Asymptotic Behavior of Dissipative Systems, Mathematical Surveys and Monographs 25. AMS, Providence (1988)
22. Lions, J.L.: Quelques Méthodes de Résolution des Problèmes aux Limites non Linéaires. Dunod, Paris (1969)
23. Lions, J.L., Magenes, E.: Problèmes aux Limites non Homogènes et Applications, vol. I. Dunod, Paris (1968)
24. Lunardi, A.: Analytic Semigroups and Optimal Regularity in Parabolic Problems. Birkhäuser, Basel (1995)
25. Málek, J., Pražák, D.: Large time behavior via the method of ℓ-trajectories. J. Differ. Equ. **181**, 243–279 (2002)
26. Miranville, A., Zelik, S.: Robust exponential attractors for Cahn-Hilliard type equations with singular potentials. Math. Methods Appl. Sci. **27**, 545–582 (2004)
27. Savostianov, A., Zelik, S.: Finite dimensionality of the attractor for the hyperbolic Cahn-Hilliard-Oono equation in $\mathbb{R}^3$. Math. Methods Appl. Sci. **39**, 1254–1267 (2016)
28. Robinson, J.C.: Infinite-Dimensional Dynamical Systems. An Introduction to Dissipative Parabolic PDEs and the Theory of Global Attractors. Cambridge University Press, Cambridge (2001)
29. Temam, R.: Infinite-Dimensional Dynamical Systems in Mechanics and Physics, 2nd edn. Springer, New York (1997)
30. Triebel, H.: Interpolation Theory. Function Spaces, Differential Operators. VEB Deutscher, Berlin (1978)
31. von Wahl, W.: Global solutions to evolution equations of parabolic type. In: Favini, A., Obrecht, E. (eds.) Differential Equations in Banach Spaces, pp. 254–266. Springer, Berlin (1986)
32. Wang, B.: Attractors for reaction-diffusion equations in unbounded domains. Phys. D **2199**, 1–12 (1999)
33. Zelik, S.: The attractor for a nonlinear reaction-diffusion system with a supercritical nonlinearity and its dimension. Rend. Accad. Naz. Sci. XL Mem. Mat. Appl. (5) **24**, 1–25 (2000)

Chapter 14
Ergodicity of Stochastic Hydrodynamical-Type Evolution Equations Driven by α-Stable Noise

Jianhua Huang, Tianlong Shen, and Yuhong Li

Abstract The present paper is devoted to the ergodicity of stochastic 2D hydrodynamical-type evolution equation driven by α-stable noise with $\alpha \in (\frac{3}{2}, 2)$, which covers stochastic Navier–Stokes equation, magneto-hydrodynamic equation, Boussinesq equation, magnetic Benard equation and so on. The existence and uniqueness of the invariant measure of this stochastic system are established by the strong Feller property and accessibility of the transition semigroup. The novel to overcome those difficulties caused by the trajectory discontinuity and lower regularity of the corresponding Ornstein–Uhlenbeck process for α-stable noise. As applications of the abstract result, the existence and uniqueness of the invariant measure for the stochastic Boussinesq equation and stochastic 2D Magneto-hydrodynamic equation are given.

14.1 Introduction

In recent years, stochastic partial differential equations driven by Lévy noise have attracted a lot of attention, see [2, 6, 10, 11, 16] and references therein. But most of them assumed that the Lévy noise are square integrable, which clearly rules out the interesting α-stable noise. As the α-stable noise do provide useful models for physics, queueing theory, mathematical finances and others, there have been many studies on the stochastic PDEs driven by α-stable noise (see [5, 11–13, 17–19]). Sun and Xie

---

J. Huang (✉)
College of Liberal Arts and Science, National University of Defense Technology, Changsha 410073, China
e-mail: jhhuang32@nudt.edu.cn

T. Shen
National Innovation Institute of Defense Technology, Academy of Military Sciences, Beijing 100071, China
e-mail: shentianlong303@163.com

Y. Li
School of Hydropower and Information Engineer, Huazhong University of Science and Technology, Wuhan 430074, China
e-mail: liyuhong@hust.edu.cn

V. A. Sadovnichiy and M. Z. Zgurovsky (eds.), *Contemporary Approaches and Methods in Fundamental Mathematics and Mechanics*, Understanding Complex Systems,
https://doi.org/10.1007/978-3-030-50302-4_14

in [15] studied the ergodicity of stochastic dissipative equation with m-dissipative function driven by α-stable process with $\alpha \in (1,2)$. Priola and Zabczyk investigated the structural properties of solutions to nonlinear stochastic equations driven by cylindrical stable processes with bounded and Lipschitz nonlinearities in [11]. While Xu in [18] studied the Ornstein–Uhlenbeck α-stable processes with $\alpha \in (1,2)$ and prove that the system is ergodic if the perturbation is small. The exponential mixing of SPDEs driven by α-stable noise has been established in [12, 13, 18]. Dong and Xie in [6] established the existence of the invariant measures for 2D stochastic Navier–Stokes equation forced by α-stable noise with $\alpha \in (1,2)$. Recently, Dong et al. in [5] proved the exponential ergodicity and strong Feller of stochastic Burgers equations driven by $\frac{\alpha}{2}$-subordinated cylindrical Brownian motions with $\alpha \in (1,2)$. Specially, Xu in [17] studied the ergodicity of stochastic real Ginzburg–Landau equation driven by α-stable noise with $\alpha \in (\frac{3}{2},2)$ and established a maximal inequality for stochastic α-stable convolution, which is useful for studying other SPDEs forced by α-stable noise.

The aim of this article is to investigate the ergodicity of the following abstract stochastic hydro-dynamical-type evolution equation on some Hilbert space H,

$$du(t) + [Au + B(u,u) + f(u)]dt = dL_t, \tag{14.1}$$

where L_t denotes a cylindrical α-stable noise with $\alpha \in (\frac{3}{2},2)$, f is a linear bounded and Lipschitz operator in H. We will specify some hypothesis on the linear operator A and the bilinear mapping $B(\cdot,\cdot)$ later. Stochastic 2D hydrodynamical-type equation (14.1) covers a lot of hydrodynamical partial differential equation such as stochastic Navier–Stokes equation, magneto-hydrodynamic equation, Boussinesq equation and magnetic Benard equation. It is worthy to mention that Chueshov and Millet in [3] proved the well-posedness property for the stochastic model (14.1) driven by multiplicative Gaussian noise, and presented the Wentzell–Freidlin type large deviation principle by a weak convergence method. The existence of martingale solutions for abstract stochastic hydrodynamical-type evolution equation driven by Lévy noise in certain Fréchet space is proved by Motyl in [9].

Motivated by [3, 9, 17], we consider the ergodicity of the abstract stochastic hydrodynamical-type evolution equation (14.1) with $\alpha \in (\frac{3}{2},2)$. The maximal inequality for the stochastic α-stable convolution developed by Xu in [17], vorticity transformation and Banach fixed point theorem are applied to show the existence and uniqueness of the global mild solution. By a priori estimates for the solution of the Galerkin approximation equation and the classical Bogoliubov–Krylov theorem, the existence of the invariant measures for the stochastic system is proved. Due to the trajectories discontinuity and gradient term in bilinear term $B(\cdot,\cdot)$, we truncate the bilinear term $B(\cdot,\cdot)$ to prove the strong Feller property by a gradient estimate for the associated transition semigroup developed by Priola and Zabczyk in [11], and prove the uniqueness of the invariant measure for abstract stochastic hydrodynamical-type evolution equation by the strong Feller property and the accessibility to zero instead of the normal irreducibility. Finally, we apply the results to stochastic Boussinesq equation and stochastic 2D Magneto-hydrodynamic equation to obtain the existence and uniqueness of the invariant measure.

The rest of the paper is organized as follows. Section 14.2 is devoted to introduce the functional setting and maximal inequality for stochastic α-stable convolution. In Sect. 14.3, we apply Banach fixed point theorem and vorticity transformation to show the well-posedness of the mild solution for stochastic abstract stochastic hydrodynamical-type evolution equations (14.1). In Sect. 14.4, some a priori estimates for the Galerkin approximation equations are presented and the existence of the invariant measures for stochastic systems (14.1) is obtained. Section 14.5 is devoted to the uniqueness of the invariant measure by the strong Feller property and the accessibility to zero. Finally, the abstract results are applied to stochastic 2D Boussinesq equation and stochastic 2D Magneto-hydrodynamic equation to obtain the corresponding uniqueness of the invariant measure respectively.

14.2 Preliminaries

In this section, we will introduce some notations and functional setting, and present the regularity for the Ornstein–Uhlenbeck process with α-stable noise.

For $D = [0, 1] \times [0, 1] \subset R^2$, the space H defined by

$$H = \{u \in (L^2(D, R))^2 : \nabla \cdot u = 0, \int_D u(s)ds = 0\},$$

which is a separable Hilbert space with the inner product

$$\langle u, v\rangle_H = \int_D u(x)v(x)\,dx, \qquad u, v \in H,$$

and a Banach space with the norm

$$\|u\|_H^2 = \langle u, u\rangle_H = \int_D |u|^2 dx.$$

Denote $Z_* := Z \setminus \{0\}$, for $m, n \in Z_*$, define

$$e_{m,n} = \sqrt{\frac{2}{\pi}} \sin(m\pi)\sin(n\pi), \qquad \lambda_{m,n} = \pi^2(m^2 + n^2).$$

It is well known that $\{e_{m,n} : m, n \in Z_*\}$ is an orthonormal basis of H, and for an unbounded self-adjoint linear operator A on H, $(e_{m,n}, \lambda_{m,n})$ are the eigenvectors and eigenvalues of A, i.e.

$$Ae_{m,n} = \lambda_{m,n}e_{m,n}, \qquad m, n \in Z_*.$$

Assume that A generate a compact analytic C_0-semigroup $(e^{-tA})_{t\geq 0}$ of negative type on H.

For $x = \sum_{m,n\in Z_*} x_{m,n}e_{m,n} \in H, x_{m,n} \in R, m, n \in Z_*$, define

$$A^{\frac{1}{2}}x := \sum_{m,n\in Z_*} \lambda_{m,n}^{\frac{1}{2}} x_{m,n}e_{m,n}.$$

Set

$$V := \{u : u = \sum_{m,n\in Z_*} u_{m,n}e_{m,n} \in H, m, n \in Z_*, \|u\|_V = \|A^{\frac{1}{2}}u\|_H := \left(\sum_{m,n} u_{m,n}^2 \lambda_{m,n}\right)^{\frac{1}{2}} < \infty\},$$

and let V' denote the dual of V with respect to the inner product $\langle \cdot, \cdot \rangle_H$ of H. Thus we have the Gelfand triple $V \subset H \subset V'$.

We assume that $B : V \times V \to V'$ is a continuous mapping satisfying the following antisymmetry and bound conditions:

(**C1**) $B : V \times V \to V'$ is a bilinear continuous mapping.
(**C2**) For $u_i \in V, i = 1, 2, 3$,

$$\langle B(u_1, u_2), u_3\rangle = -\langle B(u_1, u_3), u_2\rangle.$$

(**C3**) There exists a Banach interpolation space $\mathscr{H}$ satisfying the following conditions:

(C31) $V \subset \mathscr{H} \subset H$,
(C32) there exists a constant C such that for $u_i \in V, i = 1, 2, 3$,

$$|\langle B(u_1, u_2), u_3\rangle| \leq C\|u_1\|_H\|u_2\|_V\|u_3\|_{\mathscr{H}}.$$

Let $L_t = \Sigma_{m,n\in Z_*}\beta_{m,n}l_{m,n}(t)e_{m,n}$ be a cylindrical α-stable process on H with $\alpha \in (\frac{3}{2}, 2)$ and $\{l_{m,n}(t)\}_{m,n\in Z_*}$ are independent, real valued, normalized, symmetric α-stable processes defined on a fixed stochastic basis. Moreover, $(\beta_{m,n})$ is a sequence such that, for some $\frac{1}{2} + \frac{1}{2\alpha} < \beta < \frac{3}{2} - \frac{1}{\alpha}$,

there exist $C_1, C_2 > 0$ satisfying

$$C_1\lambda_{m,n}^{-\beta} \leq |\beta_{m,n}| \leq C_2\lambda_{m,n}^{-\beta}.$$

Remark 14.1 As $\alpha \in (\frac{3}{2}, 2)$ and $\frac{1}{2} + \frac{1}{2\alpha} < \beta < \frac{3}{2} - \frac{1}{\alpha}$, we have $\frac{3}{4} < \beta < 1$. This assumption guarantees that the corresponding Ornstein–Uhlenbeck process Z_t with drift $-A$ are in some spaces between $D(A^{\frac{3}{4}})$ and $D(A)$, which is just the so-called reproducing kernel Hilbert space.

Remark 14.2 (i) For stochastic 2-D Navier–Stokes equation driven by α-stable noise with Dirichlet boundary conditions:

$$\begin{cases} du(t) + [-\nu\Delta u + u \cdot \nabla u + \nabla p]dt = dL_t, \quad x \in D, \\ \nabla \cdot u = 0, \\ u = 0, \quad x \in \partial D, \end{cases} \tag{14.2}$$

where D is a bounded domain in R^2, $u = (u^1, u^2)$ is the velocity of a fluid, p is the pressure, ν is the kinematic viscosity.

Denote

$$H^1(D) = \{u \in L^2(D), D_i u \in L^2(D), i = 1, 2\},$$
$$H_0^1(D) = \text{the closure of } C_0^\infty(D) \text{ in } H^1(D),$$
$$H_1(D) = \{u \in [L^2(D)]^2 : \nabla \cdot u = 0, \quad u = 0 \quad on \quad \partial D\}.$$

Stochastic system (14.2) can be rewritten in the form (14.1) with $f = 0$ in the space $H = H_1$, where $A = -\nu\Delta$ is the Stokes operator and $V = V_1 = [H_0^1(D)]^2 \bigcap H_1$. The map $B = B_1 : V_1 \times V_1 \to V_1'$ is defined by

$$\langle B_1(u_1, u_2), u_3 \rangle = \int_D [u_1 \cdot \nabla u_2] u_3 dx = \sum_{i,j=1}^{2} \int_D u_1^j \partial_j u_2^i u_3^i dx, \quad u_i \in V_1. \tag{14.3}$$

We can choose the interpolation space $\mathscr{H} = [L^\infty(D)]^2 \bigcap H$ such that $V \subset \mathscr{H} \subset H$.

(ii) For stochastic 2D Boussinesq equation driven by α-stable noise in a 2D bounded domain $D \subset R^2$:

$$\begin{cases} du(t) + [u \cdot \nabla u - \Delta u + \nabla p - (0, \theta)dt = dL_{1t}, \\ d\theta(t) + [u \cdot \nabla\theta - \Delta\theta]dt = dL_{2t}, \\ \nabla \cdot u = 0, \\ u(0) = u_0, \quad \theta(0) = \theta_0, \\ u = 0, \quad \theta = 0, \quad x \in \partial D, \end{cases} \tag{14.4}$$

where $u(t, x) = (u_1, u_2) \in R^2$ represents the velocity vector field, $\theta(t, x) \in R$ represents the scalar temperature, $p = p(x, t) \in R$ is the scalar pressure, and $L_t = \begin{pmatrix} L_{1t} \\ L_{2t} \end{pmatrix}$ is cylindrical α-stable noise. Denote $H_2 = L^2(D)$ and $V_2 = H_0^1(D)$. Define the mapping B_2 by

$$\langle B_2(u_1, \theta_2), \theta_3 \rangle = \sum_{j=1}^{2} \int_D u_1^j \partial_j \theta_2 \theta_3 dx, \quad u_1 \in V_1, \theta_2, \theta_3 \in V_2. \tag{14.5}$$

Stochastic system (14.4) can be rewritten in the form (14.1) with $f(u, \theta) = ((0, \theta), 0)$ in the space $H = H_1 \times H_2$, $V = V_1 \times V_2$ and $A(u, \theta) = (\Delta u, \Delta\theta)$.

Define $B : V \times V \to V'$ by

$$\langle B(z_1, z_2), z_3\rangle = \langle B_1(u_1, u_2), u_3\rangle + \langle B_2(u_1, \theta_2), \theta_3\rangle, \quad z_i = (u_i, b_i) \in V. \tag{14.6}$$

We can choose the interpolation space $\mathscr{H}$ to be $\mathscr{H} = ([L^\infty(D)]^2 \times [L^\infty(D)]) \bigcap H$ such that $V \subset \mathscr{H} \subset H$.

(iii) For stochastic 2D Magneto-hydrodynamic equation driven by α-stable noise in a 2D bounded domain $D \subset R^2$:

$$\begin{cases} du + [\Delta u + b \cdot \nabla b - u \cdot \nabla u - \nabla p]dt = dL_{1t}, \\ db + [\Delta b + b \cdot \nabla u - u \cdot \nabla b]dt = dL_{2t}, \\ \nabla \cdot u = 0, \quad \nabla \cdot b = 0, \\ u(0, x) = u_0, \quad b(0, x) = b_0, \\ u = 0, \quad b = 0, \quad x \in \partial D, \end{cases} \tag{14.7}$$

where $u = (u_1, u_2) \in R^2$ and $b = (b_1, b_2) \in R^2$ denote the velocity field and magnetic field respectively, $p \in R$ is a scalar pressure and $L_t = \begin{pmatrix} L_{1t} \\ L_{2t} \end{pmatrix}$ is cylindrical α-stable noise. Stochastic system (14.7) can be rewritten in the form (14.1) with $f = 0$ in the space $H = H_1 \times H_1$, $V = V_1 \times V_1$ and $A(u, b) = (\Delta u, \Delta b)$.

Define $B : V \times V \to V'$ by

$$\langle B(z_1, z_2), z_3\rangle = \langle B_1(u_1, u_2), u_3\rangle - \langle B_1(b_1, b_2), u_3\rangle + \langle B_1(u_1, b_2), b_3\rangle - \langle B_1(b_1, u_2), b_3\rangle, \tag{14.8}$$

where $z_i = (u_i, b_i) \in V$. We can choose the interpolation space $\mathscr{H} = ([L^\infty(D)]^2 \times [L^\infty(D)]^2) \bigcap H$ such that $V \subset \mathscr{H} \subset H$.

(iv) For stochastic magnetic Benard equation driven by α-stable noise in a 2D bounded domain $D \subset R^2$:

$$\begin{cases} du + [\Delta u + b \cdot \nabla b - u \cdot \nabla u - \nabla p + (0, \theta)]dt = dL_{1t}, \\ d\theta + [\Delta\theta - u \cdot \nabla\theta]dt = dL_{2t}, \\ db + [\Delta b + b \cdot \nabla u - u \cdot \nabla b]dt = dL_{3t}, \\ \nabla \cdot u = 0, \quad \nabla \cdot b = 0, \\ u(0, x) = u_0, \quad \theta(0) = \theta_0, \quad b(0, x) = b_0, \\ u = 0, \quad \theta = 0, \quad b = 0, \quad x \in \partial D. \end{cases} \tag{14.9}$$

Stochastic system (14.9) can be rewritten in the form (14.1) with $f(u, \theta, b) = ((0, \theta), 0, 0)$ in the space $H = H_1 \times H_2 \times H_1$, $V = V_1 \times V_2 \times V_1$ and $A(u, \theta, b) = (\Delta u, \Delta\theta, \Delta b)$.

Define $B : V \times V \to V'$ by the relation

$$\langle B(z_1, z_2), z_3\rangle = \langle B_1(u_1, u_2), u_3\rangle - \langle B_1(b_1, b_2), u_3\rangle \\ + \langle B_1(u_1, b_2), b_3\rangle - \langle B_1(b_1, u_2), b_3\rangle + \langle B_2(u_1, \theta_2), \theta_3\rangle, \tag{14.10}$$

where $z_i = (u_i, \theta_i, b_i) \in V$. We can choose the interpolation space $\mathscr{H} = ([L^\infty(D)]^2 \times [L^\infty(D)] \times [L^\infty(D)]^2) \bigcap H$ such that $V \subset \mathscr{H} \subset H$.

14.3 Well-Posedness of the Mild Solution

In this section, we will apply Banach fixed point theorem to show the well-posedness of the mild solution for the stochastic hydrodynamical-type evolution equations (14.1) with $f = 0$.

Consider the following linear equation

$$\begin{cases} dZ_t + AX_t\,dt = dL_t, & t \geq 0, \\ Z_0 = 0. \end{cases} \tag{14.11}$$

Then under the above assumptions on A, and L_t, we have

$$Z_t = \int_0^t e^{-A(t-s)} dL_s = \sum_{m,n\geq 1} \Big(\int_0^t e^{-\lambda_{m,n}(t-s)} \beta_{m,n}\, dl_{m,n}(s) \Big) e_{m,n}. \tag{14.12}$$

The following lemmas play a crucial role in proving the well-posedness, strong Feller and accessibility for the solution of stochastic equation (14.1), which are adapted from [17].

Lemma 14.1 ([17]) *Let $T > 0$, then for all $0 \leq \kappa < \beta - \frac{1}{2\alpha}$ and all $0 < p < \alpha$, we have $Z_t = (Z_t)_{t\in[0,T]}$ is a random variable with values in $L^p(0, T; H)$ and*

$$\mathbf{E} \sup_{0\leq t\leq T} \|A^\kappa Z_t\|_H^p \leq C T^{\frac{p}{\alpha}},$$

where C depends on α, κ, β, p.

Lemma 14.2 ([17]) *Let $\kappa \in [0, \beta - \frac{1}{2\alpha})$, $T > 0$ and $\varepsilon > 0$, then we have*

$$\mathbf{P}(\sup_{0\leq t\leq T} \|A^\kappa Z_t\|_H \leq \varepsilon) > 0. \tag{14.13}$$

Definition 14.1 Let $I = [a, b] \subset R^+$, a H-valued stochastic process g is right continuous if

$$g(t+) := \lim_{s\downarrow t, s,t\in I} g(s) = g(t), \qquad t \in I.$$

g is called to have left limit if the limit $\lim_{s\in I, s\uparrow t} g(s) =: g(t-)$ exists.

We say that g is càdlàg if g is right continuous and has left limit. Let $\mathbf{D}([0,T],H)$ be the space of all *càdlàg* paths from $[0,T]$ into H.

For all $\omega \in \Omega$, let $v(t,\omega) = u(t,\omega) - Z(t,\omega)$, then from the equation (14.1), with $f = 0$, we have

$$\begin{cases} dv + [Av + B(v + Z_t, v + Z_t)]dt = 0, & t \geq 0, \\ v(0,x,\omega) = u_0. \end{cases} \tag{14.14}$$

For each $T > 0$, define

$$K_T(\omega) := \sup_{0 \leq t \leq T} \|Z_t(\omega)\|_V, \quad \omega \in \Omega. \tag{14.15}$$

Lemma 14.2 implies that for every $k \in \mathbf{N}$, there exists some set $N_k \subset \Omega$ such that $\mathbf{P}(N_k) = 0$ and

$$K_k(\omega) < \infty, \quad \omega \notin N_k. \tag{14.16}$$

Define $N = \bigcup_{k \geq 1} N_k$, it is easy to see $\mathbf{P}(N) = 0$ and that for all $T > 0$

$$K_T(\omega) < \infty, \quad \omega \notin N. \tag{14.17}$$

Lemma 14.3 *Assume that conditions (C1)–(C3), and the above assumptions on A, L_t hold. Then we have the following statements:*

(B1) *For $u_0 \in H$, and $\omega \notin N$, there exists a random variable $T(\omega)$ depending on $\|u_0\|_H$ and $K_1(\omega)$ such that $0 < T(\omega) \leq 1$, and Eq. (14.14) has a unique mild solution $v(\cdot,\omega) \in C([0,T]; H)$ satisfying*

$$\sup_{0 \leq t \leq T(\omega)} \|A^{\frac{1}{2}} v(t,\omega)\|_H \leq C(t^{-\frac{\theta}{2}} + 1), \quad t \in (0, T(\omega)], \quad \theta \in \left(1, \frac{3}{2}\right), \tag{14.18}$$

where C is a constant depending on $\|u_0\|_H$ and $K_1(\omega)$.

(B2) *For $u_0 \in V$, and $\omega \notin N$, there exists a random variable $\widehat{T}(\omega)$ depending on $\|u_0\|_H$ and $K_1(\omega)$ such that $0 < \widehat{T}(\omega) \leq 1$, and Eq. (14.14) has a unique mild solution $v(\cdot,\omega) \in C([0,\widehat{T}]; V)$ satisfying*

$$\sup_{0 \leq t \leq \widehat{T}(\omega)} \|A^{\frac{1}{2}} v(t,\omega)\|_H \leq 1 + \|A^{\frac{1}{2}} u_0\|_H. \tag{14.19}$$

Proof For the notational simplicity, we will omit the variable ω. Firstly, we are going to prove (B1). Let $0 < T \leq 1$ and $\tilde{B} > 0$ be some constants to be determined later, and let $\theta \in (1, \frac{3}{2})$,

Define

$$S = \{v \in C([0,T], H) : v_0 = u_0, v(t) \in V, t \in (0,T], \sup_{0 \leq t \leq T} t^{\frac{\theta}{2}} \|A^{\frac{1}{2}} v\|_H + \sup_{0 \leq t \leq T} \|v\|_H \leq \tilde{B}\}.$$

For any $v, v' \in S$, define

$$d(v, v') = \sup_{0\le t\le T} t^{\frac{\theta}{2}} \|A^{\frac{1}{2}}(v - v')\|_H + \sup_{0\le t\le T} \|(v - v')\|_H,$$

then (S, d) is a closed metric space.

Define a map $\mathscr{F} : S \to C([0, T]; H)$ as the following: for any $v \in S$,

$$\mathscr{F}v = e^{-At}u_0 + \int_0^t e^{-A(t-s)} B(v + Z_s, v + Z_s)ds. \tag{14.20}$$

We will prove that there exist $T_0 > 0$ and $B_0 > 0$ such that for $t \in (0, T_0]$ and $\tilde{B} \ge B_0$

(a) $\mathscr{F}v \in S$ for $v \in S$.
(b) $d(\mathscr{F}v, \mathscr{F}v') \le \frac{1}{2}d(v, v')$ for $v, v' \in S$.

We can derive that

$$\|A^{\frac{1}{2}}\mathscr{F}v\|_H \le \|A^{\frac{1}{2}}e^{-At}u_0\|_H + \int_0^t \|A^{\frac{1}{2}}e^{-A(t-s)}B(v + Z_s, v + Z_s)\|_H := I_1 + I_2. \tag{14.21}$$

Due to the fact $\|e_k\|_{\mathscr{H}} \le 1$, we have

$$\begin{aligned} I_1^2 &= \|A^{\frac{1}{2}}e^{-At}u_0\|_H^2 \\ &= \sum_{m,n=1}^{\infty} \langle A^{\frac{1}{2}}e^{-At}u_0, e_{m,n}\rangle^2 = \sum_{m,n=1}^{\infty} \langle u_0, \lambda_{m,n}^{\frac{1}{2}}e^{-\lambda_{m,n}t}e_{m,n}\rangle^2 \\ &\le C\sum_{m,n=1}^{\infty} (m+n)^2e^{-2\pi t(m+n)^2}\|u_0\|_H^2 \le C\sum_{m,n=1}^{\infty} \pi(m+n)^{2-2\theta}t^{-\theta}\|u_0\|_H^2. \end{aligned} \tag{14.22}$$

It can be seen $I_1 < \infty$ with $\theta > \frac{3}{2}$.

It follows from assumptions (C1)-(C3) that

$$\begin{aligned} \|A^{\frac{1}{2}}e^{-A(t-s)}B(v + Z_s, v + Z_s)\|_H^2 &= \sum_{m,n=1}^{\infty} \langle A^{\frac{1}{2}}e^{-A(t-s)}B(v + Z_s, v + Z_s), e_{m,n}\rangle^2 \\ &= \sum_{m,n=1}^{\infty} \langle B(v + Z_s, v + Z_s), \lambda_{m,n}^{\frac{1}{2}}e^{-\lambda_{m,n}(t-s)}e_{m,n}\rangle^2 \\ &\le C\sum_{m,n=1}^{\infty} (m+n)^2e^{-2(t-s)\pi(m+n)^2}\langle B(v + Z_s, v + Z_s), e_k\rangle^2 \\ &\le C\sum_{m,n=1}^{\infty} (m+n)^{2-2\theta}(t-s)^{-\theta}\|v + Z_s\|_H^2\|v + Z_s\|_V^2\|e_k\|_{\mathscr{H}}^2. \end{aligned} \tag{14.23}$$

It suffices to choose $\frac{3}{2} < \theta < 2$ such that $I_2 < \infty$. Thus, we have

$$
\begin{aligned}
I_2 &\le C\int_0^t (t-s)^{-\frac{\theta}{2}}\|v+Z_s\|_H\|v+Z_s\|_V ds \\
&\le C\int_0^t (t-s)^{-\frac{\theta}{2}}(\tilde{B}^2 s^{-\frac{\theta}{2}} + K_1\tilde{B} + K_1\tilde{B}s^{-\frac{\theta}{2}} + K_1^2)ds. \quad (14.24)
\end{aligned}
$$

Combining (14.22) and (14.24) gives

$$
\|A^{\frac{1}{2}}\mathscr{F}v\|_H \le Ct^{-\frac{\theta}{2}}\|u_0\|_H + C\int_0^t (t-s)^{-\frac{\theta}{2}}(\tilde{B}^2 s^{-\frac{\theta}{2}} + K_1\tilde{B} + K_1\tilde{B}s^{-\frac{\theta}{2}} + K_1^2)ds. \tag{14.25}
$$

Hence,

$$
t^{\frac{\theta}{2}}\|A^{\frac{1}{2}}\mathscr{F}v\|_H \le C\|u_0\|_H + Ct^{\frac{\theta}{2}}\int_0^t (t-s)^{-\frac{\theta}{2}}(\tilde{B}^2 s^{-\frac{\theta}{2}} + K_1\tilde{B} + K_1\tilde{B}s^{-\frac{\theta}{2}} + K_1^2)ds.
$$

Furthermore,

$$
\|\mathscr{F}v\|_H \le C\|u_0\|_H + C\int_0^t (\tilde{B}^2 s^{-\frac{\theta}{2}} + K_1\tilde{B} + K_1\tilde{B}s^{-\frac{\theta}{2}} + K_1^2)ds. \tag{14.26}
$$

By a similar calculation as the above prove, the continuity of $\mathscr{F}v$ can be verified. As $T > 0$ is sufficiently small and B is sufficiently large, it follows from (14.25) and (14.26) that (a) holds.

For any $v, v' \in S$, it follows from assumptions (C1)–(C3) and $\|e_k\|_{\mathscr{H}} \le 1$ that

$$
\begin{aligned}
&\|A^{\frac{1}{2}}\mathscr{F}v - A^{\frac{1}{2}}\mathscr{F}v'\|_H \\
\le &\int_0^t \|A^{\frac{1}{2}}e^{-A(t-s)}(B(v+Z_s, v+Z_s) - B(v'+Z_s, v'+Z_s))\|_H ds \\
\le &\int_0^t \|A^{\frac{1}{2}}e^{-A(t-s)}B(v+Z_s, v-v')\|_H + \|A^{\frac{1}{2}}e^{-A(t-s)}B(v-v', v+Z_s)\|_H ds \\
:= &\int_0^t (J_1 + J_2)ds.
\end{aligned}
$$

Similarly to (14.23),

$$
J_1^2 \le C\sum_{k=1}^{\infty}(\pi k)^{2-2\theta}(t-s)^{-\theta}\|v+Z_s\|_H^2\|v-v'\|_V^2, \tag{14.27}
$$

and

$$
J_2^2 \le C\sum_{k=1}^{\infty}(\pi k)^{2-2\theta}(t-s)^{-\theta}\|v-v'\|_H^2\|v'+Z_s\|_V^2. \tag{14.28}
$$

Then it follows from $\frac{3}{2} < \theta < 2$,

$$\begin{aligned}
&\sup_{0\le t\le T} t^{\frac{\theta}{2}}\|A^{\frac{1}{2}}\mathscr{F}v - A^{\frac{1}{2}}\mathscr{F}v'\|_H \\
&\le CT^{\frac{\theta}{2}}(\tilde{B}+K_1)\int_0^t (t-s)^{-\frac{\theta}{2}}s^{-\frac{\theta}{2}}[s^{\frac{\theta}{2}}\|v-v'\|_V^2]ds \\
&+ CT^{\frac{\theta}{2}}\int_0^t (t-s)^{-\frac{\theta}{2}}(s^{-\frac{\theta}{2}}\tilde{B}+K_1)\|v-v'\|_H ds \\
&\le CT^{1-\frac{\theta}{2}}(\tilde{B}+K_1)\sup_{0\le t\le T}(t^{\frac{\theta}{2}}\|v-v'\|_V) \\
&+ C(\tilde{B}T^{1-\frac{\theta}{2}}+K_1T)\|v-v'\|_H.
\end{aligned} \tag{14.29}$$

Similarly,

$$\begin{aligned}
\sup_{0\le t\le T}\|\mathscr{F}v-\mathscr{F}v'\|_H \le CT^{1-\frac{\theta}{2}}(\tilde{B}+K_1)&\sup_{0\le t\le T}(t^{\frac{\theta}{2}}\|v-v'\|_V) \\
+ C(\tilde{B}T^{1-\frac{\theta}{2}}+K_1T)&\sup_{0\le t\le T}\|v-v'\|_H.
\end{aligned} \tag{14.30}$$

Combining (14.30) and (14.30) gives

$$d(\mathscr{F}v,\mathscr{F}v') \le C(\tilde{B}T^{1-\frac{\theta}{2}}+K_1T^{1-\frac{\theta}{2}}+K_1T)d(v,v'). \tag{14.31}$$

Choosing T small enough, we can get (b).

It follows from (a) and (b) that there exists a unique solution in S for Eq. (14.14) by the Banach fixed point theorem. Let $v \in S$ be the unique solution in S for Eq. (14.14), then

$$\|A^{\frac{1}{2}}v\|_H \le Ct^{-\frac{\theta}{2}}\|u_0\|_H + C\int_0^t (t-s)^{-\frac{\theta}{2}}(\tilde{B}^2s^{-\frac{\theta}{2}}+K_1\tilde{B}+K_1\tilde{B}s^{-\frac{\theta}{2}}+K_1^2)ds,$$

which means the desired inequality can be easily verified.

Let us move to the prove the uniqueness. Let $v, v' \in C([0,T];H)$ be two solutions satisfying the specified inequality, then

$$\begin{aligned}
&\sup_{0\le t\le T}\|v(t)-v'(t)\|_H \\
&= \sup_{0\le t\le T}\|\int_0^t e^{-A(t-s)}(B(v+Z_s,v+Z_s)-B(v'+Z_s,v'+Z_s))\,ds\| \\
&\le CT^{1-\frac{\theta}{2}}(\tilde{B}+K_1)\sup_{0\le t\le T}(t^{\frac{\theta}{2}}\|v-v'\|_V)+C(\tilde{B}T^{1-\frac{\theta}{2}}+K_1T)\sup_{0\le t\le T}\|v-v'\|_H,
\end{aligned} \tag{14.32}$$

which implies that $\|v-v'\|_H = 0$ for all $t \in [0,T]$.

Next, we move to prove (B2). Let $0 < \widehat{T} \leq 1$ which will be specified later. Define

$$S' = \{ v \in C([0,T],V) : v_0 = u_0,\ \sup_{0\leq t\leq T} \|A^{\frac{1}{2}}v\|_H \leq 1 + \|A^{\frac{1}{2}}u_0\|_H\}. \tag{14.33}$$

For any $v, v' \in S'$, define

$$d(v,v') = \sup_{0\leq t\leq T} \|A^{\frac{1}{2}}(v-v')\|_H, \tag{14.34}$$

then (S', d) is a closed metric space.

Define the map $\mathscr{F} : S' \to C([0,\widehat{T}];V)$ as the following: for any $v \in S'$.

$$\mathscr{F}v = e^{-At}u_0 + \int_0^t e^{-A(t-s)} B(v(s)+z_s, v(s)+z_s)\,ds.$$

Similar to the proof of (i), we can verify that for any $v \in S'$, $\mathscr{F}v \in S'$ and

$$d(\mathscr{F}v, \mathscr{F}v') \leq \frac{1}{2} d(v,v').$$

Then the Banach fixed point theorem is applied to get the unique solution, which finish the proof of (B2).

Lemma 14.4 *Assume that conditions (C1)–(C3) hold, and the above assumptions on A, L_t hold. Then*

(E1) *For $u_0 \in H$, and $\omega \notin N$, Eq. (14.14) has a unique global solution $v(\cdot,\omega) \in C([0,\infty);H) \cap C((0,\infty);V)$.*

(E2) *For $u_0 \in V$, and $\omega \notin N$, $v(\cdot,\omega) \in C([0,\infty);V)$.*

Proof **(E1).** Multiplying Eq. (14.14) with $v \in V$, then integrating over D leads to

$$\frac{1}{2}\frac{d}{dt}\|v\|_H^2 + \langle Av, v\rangle + \langle B(v+Z_t, v+Z_t), v\rangle_H = 0. \tag{14.35}$$

It follows from assumptions (C1)–(C3) that

$$\begin{aligned}\langle B(v+Z_t, v+Z_t), v\rangle_H &\leq \|v+Z_t\|_H \|v+Z_t\|_V \|v\|_{\mathscr{H}} \\ &\leq \frac{1}{2}\|v+Z_t\|_H^2 \|v+Z_t\|_V^2 + \frac{1}{2}\|v\|_{\mathscr{H}}^2.\end{aligned} \tag{14.36}$$

However, as $V \subset \mathscr{H} \subset H$, by the Sobolev embedding theorem, $\|v\|_{\mathscr{H}}^2 \leq C\|v\|_V^2$. Then we have

$$\frac{1}{2}\frac{d}{dt}\|v\|_H^2 + \frac{1}{2}\|v\|_V^2 \leq \frac{1}{2}\|Z_t\|_V^2\|v\|_H^2 + C\|Z_t\|_V^4. \tag{14.37}$$

It follows from *Poincaré* inequality $2\pi\|x\|_H^2 \le \|x\|_V^2$ that

$$\frac{d}{dt}\|v\|_H^2 \le (\varepsilon^2 - 2\pi)\|v\|_H^2 + C\varepsilon^4. \tag{14.38}$$

Gronwall inequality implies that

$$\|v\|_H^2 \le e^{-(2\pi-\varepsilon^2)t}\|u_0\|_H^2 + c\varepsilon^4. \tag{14.39}$$

It follows from $\varepsilon \le \sqrt{2\pi}$ that $\|v\|_H^2 < \infty$ when $t \to \infty$.

It follows from (B1) of Lemma 14.3 that Eq. (14.14) has a unique local solution v such that for some $T > 0$, $v(\cdot, \omega) \in C([0, T]; H) \bigcap C((0, T]; V)$. By (14.39), we can extend this solution $v(\cdot, \omega) \in C([0, \infty); H) \bigcap C((0, \infty); V)$.

Next, we need to prove the uniqueness of $v(\cdot, \omega) \in C([0, \infty); H) \bigcap C((0, \infty); V)$. Suppose there exist two solutions $v(\cdot, \omega), \widehat{v}(\cdot, \omega) \in C([0, \infty); H) \bigcap C((0, \infty); V)$, by the local uniqueness on $[0, T]$, $v(t, \omega) = \widehat{v}(t, \omega)$ almost surely for all $t \in [0, T]$. For any $T_0 \ge T$ and $t \in [T, T_0]$, we have

$$\begin{aligned} &\|v(t) - \widehat{v}(t)\|_V \\ &= \|\int_T^t e^{-A(t-s)}(B(v + Z_s, v + Z_s) - B(\widehat{v} + Z_s, \widehat{v} + Z_s))\,ds\| \\ &\le CT_0^{1-\frac{\theta}{2}}(B + K_1)\sup_{T\le t\le T_0}(t^{\frac{\theta}{2}}\|v - \widehat{v}\|_V) + C(BT_0^{1-\frac{\theta}{2}} + K_1 T_0)\sup_{T\le t\le T_0}\|v - \widehat{v}\|_H, \end{aligned} \tag{14.40}$$

By the continuity of $v, \widehat{v}$, it can be verified that $v(t) = \widehat{v}(t)$ for all $t \in [T, T_0]$. Since T_0 is arbitrary, we proved the uniqueness of the solution $v(\cdot, \omega) \in C([0, \infty); H) \cap C((0, \infty); V)$.

(E2). If $u_0 \in V$, it follows from (E1) that Eq. (14.14) has a unique solution $v(\cdot, \omega) \in C([0, \infty); H) \cap C((0, \infty); V)$. By (B2) of Lemma 14.3 that Eq. (14.14) has a unique local solution $\widehat{v}(\cdot, \omega) \in C([0, \hat{T}]; V)$ for some $\hat{T} > 0$. Since $C([0, \hat{T}]; V)$ is a subset of $C([0, \hat{T}]; H) \bigcap C((0, \hat{T}]; V)$, $\widehat{v} = v$ for all $t \in [0, \hat{T}]$. Hence, $v(\cdot, \omega) \in C([0, \infty); V)$. Thus, the proof of Lemma 14.4 is complete.

Theorem 14.1 *Assume that conditions (C1)–(C3) hold, and the above assumptions on A, L_t hold. Then the following statements hold.*

(D1) For $u_0 \in H$, and $\omega \in \Omega$ a.s., Eq. (14.1) possesses a unique mild solution $u(\cdot, \omega) \in \mathbf{D}([0, \infty); H) \cap \mathbf{D}((0, \infty); V)$. Moreover, $u(\cdot, \omega)$ has the following form:

$$u = e^{-At}u_0 + \int_0^t e^{-A(t-s)}B(u, u)ds + \int_0^t e^{-A(t-s)}dL_s(\omega). \tag{14.41}$$

(D2) u is a Markov process.

(D3) For every $u_0 \in V$ and $\omega \in \Omega$ a.s., we have $u(\cdot, \omega) \in D([0, \infty); V)$. For every $T > 0$,

$$\sup_{0\le t\le T} \|u(t,\omega)\|_V \le C,$$

where C is some constant depending on T, α, β and ω.

Proof It follows from Lemma 3.3 of [17] that $Z(t) \in \mathbf{D}([0,\infty); V)$. By Lemma 14.4, $v + Z(\omega)$ is the unique solution to Eq. (14.1) in $\mathbf{D}([0,\infty); H) \cap \mathbf{D}((0,\infty); V)$. The Markov property follows from the uniqueness. (D3) follows from (E2) of Lemma 14.4.

14.4 The Existence of the Invariant Measure

We follow the method in [4] to prove the existence of invariant measures. First, we consider the Galerkin approximation of Equation (14.1).

Let $\{e'_k\}_{k\in Z}$ be an orthonormal basis of H and define

$$H_m := span\{e'_k;\ |k| \le m\}.$$

It is known that H_m is a finite dimensional Hilbert space equipped with the norm adopted from H. For any $m > 0$, let $\pi_m : H \to H_m$ be the projection from H to H_m.

The Galerkin approximation of (14.1) has the following form

$$\begin{cases} du^m = [Au^m + B^m(u^m, u^m)]dt + dL_t^m, & t \ge 0, \\ u^m(0,x) = u_0^m, \end{cases} \tag{14.42}$$

where $u^m = \pi_m u$, $B^m(u,u) = \pi_m B(u^m, u^m)$, $L_t^m = \sum_{|k|\le m} \beta_k l_k(t) e'_k$.

Lemma 14.5 *Assume that conditions (C1)–(C3) hold, and the above assumptions on A, L_t hold. Then the following statement hold.*

(G1) For $u_0 \in W$ with $W = H, V$ and $\omega \in \Omega$ a.s., there exists a unique mild solution $v^m(\cdot,\omega) \in \mathbf{D}([0,\infty); W_m)$ satisfying

$$\sup_{0\le t\le T} \|v^m(t,\omega)\|_W \le C, \quad T > 0, \tag{14.43}$$

where C is some constant depending on $\|u_0\|_W$, T and $K_T(\omega)$.

(G2) For $u_0 \in W$ with $W = H, V$ and $\omega \in \Omega$ a.s., we have

$$\lim_{m\to\infty} \|v^m(t,\omega) - v(t,\omega)\|_W = 0, \quad t \ge 0. \tag{14.44}$$

Proof (G1) can be immediately shown by the same method in Theorem 14.1. Let's show (G2). We only show the case $W = V$ since the case $W = H$ can be proved by the same method. For $t > 0$, it follows from Theorem 14.1 that

$$\sup_{0\le s\le t}\|v(s)\|_V\le \hat{C},\quad \sup_{0\le s\le t}\|v^m(s)\|_V\le \hat{C},$$

where $\widehat{C}>0$ depends on $\|u_0\|_V$, t and K_t.

It is clear that

$$\begin{aligned}u^m-u&=e^{-At}(u_0^m-u_0)+(Z_t-Z_t^m)+\int_0^t e^{-A(t-s)}(I-\pi_m)B(u,u)\\&+\int_0^t e^{-A(t-s)}[B^m(u_m,u_m)-B^m(u,u)]=:I_1+I_2+I_3+I_4.\end{aligned}$$

Let $m\to\infty$, then $\|I_1\|_V\to 0,\quad \|I_2\|_V\to 0$. Similarly to (14.23),

$$\|I_3\|_V\le C\int_0^t \|I-\pi_m\|(t-s)^{-\frac{\theta}{2}}\|u(s)\|_H^2\|u(s)\|_V^2ds\to 0,\quad m\to\infty,\tag{14.45}$$

and

$$\begin{aligned}\|I_4\|_V&\le C\widehat{K}\int_0^t(t-s)^{-\frac{\theta}{2}}\|u^m(s,\omega)-u(s,\omega)\|_Vds\\&\le C\widehat{K}\int_0^t((t-s)^{-\frac{p\theta}{2}}ds)^{\frac{1}{p}}(\int_0^t\|u^m(s,\omega)-u(s,\omega)\|_V^qds)^{\frac{1}{q}},\end{aligned}\tag{14.46}$$

where $\widehat{K}=\sup_{0\le s\le t,m}(\|u(s)\|_V^2+\|u^m(s)\|_V^2)\le \hat{C}$, $\frac{3}{2}<\theta<2$, $\frac{1}{p}+\frac{1}{q}=1$ and $1<p<\frac{2}{\theta}$.

Recalling that $\sup_{0\le s\le t}\|u(s)-u^m(s)\|_V\le 2\widehat{C}$, it derives from Fatou's theorem that

$$\begin{aligned}\lim_{m\to\infty}\sup_m\|u-u^m\|_V&\le C\widetilde{K}t^{\frac{1}{p}-\frac{\theta}{2}}\lim_{m\to\infty}\sup_m[(\int_0^t\|u^m-u\|_V^qds)^{\frac{1}{q}}]\\&\le C\widetilde{K}t^{\frac{1}{p}-\frac{\theta}{2}}[(\int_0^t\lim_{m\to\infty}\sup_m\|u^m-u\|_V^qds)^{\frac{1}{q}}],\end{aligned}$$

which implies

$$\lim_{m\to\infty}\sup_m\|u-u^m\|_V\to 0.\tag{14.47}$$

Thus, the proof of Lemma 14.5 is complete.

Before proving Theorem 14.2, let's recall the definition on the pure jump Lévy processes. Let $\{(l_j(t))_{t\ge 0}, j\in Z_*\}$ be a sequence of independent one dimensional purely jump Lévy processes with the same characteristic function, i.e.,

$$\mathbf{E}e^{i\varepsilon l_j(t)}=e^{-t\Psi(\varepsilon)},\quad \forall t\ge 0, j\in Z_*,\tag{14.48}$$

where $\Psi(\varepsilon)$ is a complex valued function called the Lévy symbol given by

$$\Psi(\varepsilon) = \int_{R\backslash 0} (e^{i\varepsilon y} - 1 - i\varepsilon y 1_{|y|\leq 1})\nu(dy), \tag{14.49}$$

where ν is the Lévy measure and satisfies that

$$\int_{R\backslash 0} 1 \wedge |y|^2 \nu(dy) < +\infty. \tag{14.50}$$

For $t > 0$ and $\Gamma \in \mathscr{B}(R \setminus 0)$, the Poisson random measure associated with $l_j(t)$ is defined by

$$N^{(j)}(t, \Gamma) := \Sigma_{s\in(0,t]} 1_\Gamma(l_j(s) - l_j(s-)). \tag{14.51}$$

The compensated Poisson random measure is given by

$$N^{(j)}(t, \Gamma) := N^{(j)}(t, \Gamma) - t\nu(\Gamma). \tag{14.52}$$

It follows from Lévy-It$\widehat{o}$'s decomposition [1] that

$$l_j(t) = \int_{|x|\leq 1} x\widetilde{N}^{(j)}(t, dx) + \int_{|x|>1} xN^{(j)}(t, dx). \tag{14.53}$$

Theorem 14.2 *Assume that conditions (C1)–(C3) hold, and the above assumptions on A, L_t hold. Then the solution u of stochastic system (14.1) admits at least one invariant measure. The invariant measures are supported on V.*

Proof We follow the method in [4]. Define

$$f(u) := (\|u\|_H^2 + 1)^{\frac{1}{2}}, \quad u \in H_m. \tag{14.54}$$

Applying Itô formula [1] gives

$$f(u^m) =: f(u_0^m) - G_1^m(t) + G_2^m(t) + G_3^m(t) + G_4^m(t), \tag{14.55}$$

where

$$G_1^m(t) := \int_0^t \frac{\|u^m\|_V^2}{(1 + \|u^m\|_H^2)^{\frac{1}{2}}} ds + \int_0^t \frac{\langle B(u_m, u_m), u_m\rangle}{(1 + \|u^m\|_H^2)^{\frac{1}{2}}} ds,$$

$$G_2^m(t) := \sum_{|j|<m} \int_0^t \int_{|y|\leq 1} [f(u^m + \gamma\beta_j e_j) - f(u^m)]\widetilde{N}^{(j)}(ds, dy),$$

$$G_3^m(t) := \sum_{|j|<m} \int_0^t \int_{|y|\le 1} [f(u^m + \gamma\beta_j e_j) - f(u^m)] - \frac{< u^m, y\beta_j e_j >}{\|u^m\|_H^2 + 1} \nu(dy)ds,$$

$$G_4^m(t) := \sum_{|j|<m} \int_0^t \int_{|y|>1} [f(u^m + \gamma\beta_j e_j) - f(u^m)] N^{(j)}(ds, dy).$$

Obviously,

$$\langle B(u_m, u_m), u_m \rangle = 0.$$

We can show the following estimates

$$\mathbf{E}[\sup_{0<t\le T} |I_2^m(t)|] \le CT^{\frac{1}{2}}, \quad \mathbf{E}[\sup_{0<t\le T} |I_3^m(t)|] \le CT, \quad \mathbf{E}[\sup_{0<t\le T} |I_4^m(t)|] \le CT.$$

Thus, we obtain

$$\mathbf{E}[\sup_{0<t\le T} (\|u_m\|_H^2 + 1)^{\frac{1}{2}}] + \mathbf{E}\int_0^t \frac{\|u^m\|_V^2}{(1 + \|u^m\|_H^2)^{\frac{1}{2}}} ds \le (\|u_0\|_H^2 + 1)^{\frac{1}{2}} + CT + CT^{\frac{1}{2}}.$$

Lemma 14.5 leads to

$$\lim_{m\to\infty} \|u^m\|_H = \|u\|_H, \quad \lim_{m\to\infty} \|u^m\|_V = \|u\|_V.$$

Fatou's Lemma gives

$$\mathbf{E}[\sup_{0<t\le T} (\|u\|_H^2 + 1)^{\frac{1}{2}}] + \mathbf{E}\int_0^t \frac{\|u\|_V^2}{(1 + \|u\|_H^2)^{\frac{1}{2}}} ds \le (\|u_0\|_H^2 + 1)^{\frac{1}{2}} + CT + CT^{\frac{1}{2}}.$$

By Young's inequality, we have

$$\begin{aligned}
\mathbf{E}\left(\int_0^T \|u\|_V ds\right) &\le \mathbf{E}\left(\int_0^T \frac{\|u\|_V(\|u\|_H + 1)}{(\|u\|_H^2 + 1)^{\frac{1}{2}}} ds\right) \\
&\le C\mathbf{E}\left(\int_0^T \frac{\|u\|_V^2 + 1}{(\|u\|_H^2 + 1)^{\frac{1}{2}}} ds\right) \\
&\le C(1 + \|u_0\|_H + T),
\end{aligned}$$

which with the classical Bogoliubov–Krylov's argument implies the existence of the invariant measures and that the support of the invariant measure is V. The proof of Theorem 14.2 is complete.

14.5 The Uniqueness of the Invariant Measure

In this section, we will follow the idea of [17] to show the uniqueness of the invariant measure. Due to the discontinuity of trajectories, the uniqueness of the invariant measure is obtained by the strong Feller property and the accessibility to zero instead of the irreducibility.

14.5.1 *Strong Feller Property*

Denote by $B_b(E)$ the space of bounded measurable function $f : E \to R$. For all $f \in B_b(H)$, define

$$P_t f(u_0) = \mathbf{E}[f(u)],$$

where $t \geq 0$ and $u_0 \in H$. It follows from Theorem 14.1 that $(P_t)_{t\geq 0}$ is a Markov semigroup on $B_b(H)$.

The noise $(L_t)_{t\geq 0}$ under the norm $\|\cdot\|_V$ needs to get a gradient estimate for the O-U process corresponding to $(Z_t)_{t\geq 0}$ to show the strong Feller property of the semigroup $(P_t)_{t\geq 0}$ on $B_b(V)$.

As the bilinear operator B is not bounded, we consider the equation with truncated bilinearity as follows:

$$du^\rho = [Au^\rho + B^\rho(u^\rho, u^\rho)]dt + dL_t, \quad t \geq 0, \tag{14.56}$$

where $\rho > 0$, $B^\rho(x, x) = B(x, x)\chi(\frac{\|x\|_V}{\rho})$ for all $x \in V$ and $\chi : R \to [0, 1]$ is a smooth function such that

$$\chi(z) = 1, \quad |z| < 1, \quad \chi(z) = 0, \quad |z| \geq 2.$$

For all $x \in V$, it follows from conditions C that

$$\|B^\rho(x, x)\|_{V'} \leq C\|x\|_V^2 \chi\left(\frac{\|x\|_V}{\rho}\right) \leq C\rho^2,$$

and

$$\|B^\rho(x, x) - B^\rho(y, y)\|_{V'} \leq C(1 + \rho^2)\|x - y\|_V,$$

which implies that B^ρ is a Lipschitz function from V to V'.

So there exists a unique solution $u^\rho \in \mathbf{D}([0,\infty); V)$ for Eq. (14.56). For every $f \in B_b(V)$, define

$$P_t^\rho f(u_0) = \mathbf{E}[f(u^\rho)], \quad t \geq 0, \quad u_0 \in V,$$

$(P_t^\rho)_{t\geq 0}$ is a Markov semigroup.

Define the derivative of $f \in C_b^1(V)$:

$$D_h f(x) :- \lim_{\varepsilon\to 0} \frac{f(x+\varepsilon h) - f(x)}{\varepsilon}.$$

It follows from Riesz representation theorem that there exists some $Df(x) \in V$ such that

$$D_h f(x) = \langle Df(x), h\rangle_V, \quad h \in V.$$

Define

$$\|D_h f(x)\|_\infty = \sup_{x\in V} \|Df(x)\|_V.$$

It follows from B^ρ is a bounded Lipschitz function, Lemma 5.9 of [11] and $\beta < \frac{3}{2} - \frac{1}{\alpha}$ that we have the following proposition.

Proposition 14.1 *For $\alpha \in (\frac{3}{2}, 2)$, there exists some $\gamma \in [\frac{1}{\alpha}, 1)$ such that*

$$\|DP_t^\rho f\|_\infty \leq Ct^{-\gamma}\|f\|_\infty, \quad t > 0,$$

where $f \in B_b(V)$ and $C > 0$ depends on ρ, α and γ.

Theorem 14.3 *The semigroup $(P_t)_{t\geq 0}$ is strong Feller on $B_b(V)$.*

Proof Assume $\|f\|_\infty = 1$. For $T_0 > 0$, it suffices to show that for all $t \in (0, T_0]$

$$\lim_{\|u_0'-u_0\|_V=0} P_t f(u_0') = P_t f(u_0).$$

Recall

$$K_{T_0}(\omega) := \sup_{0\leq t\leq T_0} \|Z_t(\omega)\|_V, \quad \omega \in \Omega,$$

it follows from Lemma 14.2 and the Markov inequality that

$$\mathbf{P}\left[K_{T_0}(\omega) > \frac{\rho}{2}\right] \leq \frac{C}{\rho}.$$

Let ρ be large enough such that $\|u_0\|_V < \sqrt{\rho}$ and $A := \{K_{T_0} \leq \frac{\rho}{2}\}$.

By Lemma 14.3, there exists some $0 < t_0 \leq T_0$ depending on ρ such that for all $\omega \in A$,

$$\sup_{0 \leq t \leq t_0} \|u(\omega)\|_V \leq 1 + \|u_0\|_V. \tag{14.57}$$

We have

$$\begin{aligned}\mathbf{P}(\sup_{0\leq t\leq t_0} \|u\|_V \geq \rho) &\leq \mathbf{P}(\sup_{0\leq t\leq t_0} \|v\|_V + \sup_{0\leq t\leq t_0} \|Z_t\|_V \geq \rho) \\ &\leq \mathbf{P}(K_{T_0} > \frac{\rho}{2}) + \mathbf{P}(\sup_{0\leq t\leq t_0} \|v\|_V > \frac{\rho}{2}, A).\end{aligned}$$

It follows from $\omega \in A$ and (14.57) that

$$\sup_{0\leq t\leq t_0} \|v(\omega)\|_V \leq 1 + \|u_0\|_V \leq 1 + \sqrt{\rho} \leq \frac{\rho}{2}.$$

Thus, we derive that

$$\mathbf{P}\left(\sup_{0\leq t\leq t_0} \|v\|_V \geq \frac{\rho}{2}, A\right) = 0,$$

which leads to

$$\mathbf{P}(\sup_{0\leq t\leq t_0} \|u\|_V \geq \rho) \leq \mathbf{P}\left(K_{T_0} > \frac{\rho}{2}\right) \leq \frac{C}{\rho}.$$

Define

$$\tau = \inf\{t > 0, \|u\|_V \geq \rho\}.$$

For any $t \in [0, t_0]$, we obtain

$$\mathbf{P}(\tau \leq t) = \mathbf{P}(\sup_{0\leq t\leq t_0} \|u\|_V \geq \rho) \leq \frac{C}{\rho}. \tag{14.58}$$

Hence, for $t \in [0, \tau)$, stochastic systems (14.1) and stochastic systems (14.56) both have a unique mild solution, $u^\rho = u$.

Let $u' \in V$ such that $\|u - u'\|_V \leq 1$ and ρ be so large that $\|u\|_V, \|u'\|_V \leq \sqrt{\rho}$. For any $t \in (0, t_0]$,

$$|P_t f(u_0) - P_t f(u_0')| = |\mathbf{E}[f(u)] - \mathbf{E}[f(u')]| = I_1 + I_2 + I_3,$$

where

$$I_1 := |\mathbf{E}[f(u)1_{\tau_1>t}] - \mathbf{E}[(f(u')1_{\tau_1>t}]|, \quad I_2 := |\mathbf{E}[f(u)1_{\tau_1\le t}]|, \quad I_3 := |\mathbf{E}[f(u')1_{\tau_1\le t}]|.$$

It turns out from (14.58) that

$$I_2 \le \frac{C}{\rho}, \quad I_3 \le \frac{C}{\rho}.$$

We can derive from Proposition 14.1 and (14.58) that

$$\begin{aligned} I_1 &= |\mathbf{E}[f(u)1_{\tau_1>t}] - \mathbf{E}[f(u')1_{\tau_1>t}]| \\ &\le |\mathbf{E}[f(u)] - \mathbf{E}[f(u')]| + |\mathbf{E}[f(u)1_{\tau_1\le t}]| + |\mathbf{E}[f(u')1_{\tau_1>t}]| \\ &\le Ct^{-\gamma}\|u_0 - u_0'\|_V + \frac{2C}{\rho}. \end{aligned}$$

Therefore, we obtain

$$|P_t f(u_0) - P_t f(u_0')| \le Ct^{-\gamma}\|u_0 - u_0'\|_V + \frac{4C}{\rho}.$$

We can choose suitable ρ and δ such that for all $\varepsilon > 0$,

$$\rho \ge \max\left\{\frac{12C}{\varepsilon}, \quad 2\|u_0\|_V^2 + 2\right\}, \quad \delta \le \frac{\varepsilon t^\gamma}{2C}.$$

Then, it follows that for $\|u_0 - u_0'\|_V < \delta$,

$$|P_t f(u_0) - P_t f(u_0')| \le \varepsilon, \quad t \in (0, t_0].$$

Thus, we deduce that for $t_0 < t \le T_0$,

$$P_t f(u_0) - P_t f(u_0') = P_{t_0}[P_{t-t_0} f]u_0 - P_{t_0}[P_{t-t_0} f]u_0' \to 0,$$

as $\|u_0 - u_0'\|_V \to 0$, which finish the proof of Theorem 14.3.

Theorem 14.4 *The semigroup $(P_t)_{t\ge 0}$ is strong Feller on $B_b(H)$.*

Proof For any $T_0 > 0$, it suffices to show that for all $t \in (0, T_0]$ and $u_0 \in V$

$$\lim_{\|u_0' - u_0\|_{H=0}} P_t f(u_0') = P_t f(u_0).$$

Let $\Omega_N := \{\omega \in \Omega : \sup_{0\le t_0\le T_0} \|Z(t)\|_V \le N\}$ and it follows from Lemma 14.2 and Chebyshev's inequality that

$$P(\Omega_N^c) \le \frac{c}{N}. \tag{14.59}$$

Observe that

$$u - u' = I_1 + I_2, \tag{14.60}$$

where

$$I_1 = e^{-At}u_0 - e^{-At}u_0', \quad I_2 = \int_0^t e^{-A(t-s)}[B(u,u) - B(u',u')]ds.$$

Similarly to (14.22) and (14.23),

$$\|I_1\|_V \le Ct^{-\frac{\theta}{2}}\|u_0 - u_0'\|_H,$$
$$\|I_2\|_V \le C\int_0^t (t-s)^{-\frac{\theta}{2}}(\|u\|_V + \|u'\|_V)\|u - u'\|_V ds.$$

Theorem 14.1 and $u_0 \in V$ gives

$$\|u\|_V \le \|v\|_V + \|Z\|_V \le C.$$

Hence,

$$\|I_2(t)\|_V \le C\int_0^t (t-s)^{-\frac{\theta}{2}}\|u - u'\|_V ds.$$

For $r \in (0, t_0]$, define

$$\Phi_r = \sup_{0\le t\le r} t^{\frac{\theta}{2}}\|u - u'\|_V.$$

Theorem 14.1 yields that

$$\Phi_r \le \sup_{0\le t\le r} t^{\frac{\theta}{2}}(\|v\|_V + \|v'\|_V) + 2r^{\frac{1}{2}}N < \infty.$$

We deduce from (14.60) that

$$\Phi_r \le C\|u_0 - u_0'\|_H + C\sup_{0\le t\le r}\left[t^{\frac{\theta}{2}}\int_0^t (t-s)^{-\frac{\theta}{2}}s^{-\frac{\theta}{2}}ds\right]\Phi_r \le C\|u_0 - u_0'\|_H + Cr^{1-\frac{\theta}{2}}\Phi_r.$$

Choosing r small enough such that $Cr^{1-\frac{\theta}{2}} \leq \frac{1}{2}$, we get

$$\Phi_r \leq C\|u_0 - u_0'\|_H,$$

which implies that

$$\|u - u'\|_V \leq Ct^{-\frac{\theta}{2}}\|u_0 - u_0'\|_H. \tag{14.61}$$

For all $0 < t \leq T_0$, by the Markov property, we obtain

$$|P_t f(u_0) - P_t f(u_0')| \leq |\mathbf{E}[P_{t-s} f(u) - P_{t-s} f(u')]| \leq H_1 + H_2,$$

where $s = \frac{t}{2} \wedge r$ and

$$H_1 = |\mathbf{E}[P_{t-s} f(u) - P_{t-s} f(u')]1_{\Omega_N^c}|, \quad H_2 = |\mathbf{E}[P_{t-s} f(u) - P_{t-s} f(u')]1_{\Omega_N}|.$$

Equation (14.59) implies that

$$H_1 \leq 2c\frac{\|f\|_\infty}{N}.$$

Applying Theorem 14.3, dominated convergence theorem and inequality (14.61), we deduce that

$$H_2 \to 0, \quad \|u_0 - u_0'\|_H \to 0.$$

14.5.2 The Accessibility

As we can't get the irreducibility, we fail to get the ergodicity through the classical Doob's Theorem. Alternatively, a criterion in [8] can be applied to prove the corresponding ergodicity.

Definition 14.2 (*Accessibility*) Let $(X_t)_{t\geq 0}$ be a metric space E-valued stochastic process and let $(P_t(x,.))_{x\in E}$ be the transition probability family. $(X_t)_{t\geq 0}$ is said to be accessible to $x_0 \in E$ if the resolvent $\mathscr{R}_\lambda$ satisfies

$$\mathscr{R}_\lambda(x, \mathscr{U}) := \int_0^\infty e^{-\lambda t} P_t(x, \mathscr{U})dt > 0,$$

for all $x \in E$ and all neighborhoods $\mathscr{U}$ of x_0, where $\lambda > 0$ is arbitrary.

Theorem 14.5 ([8]) *If $(X_t)_{t\geq 0}$ is strong Feller at an accessible point $x \in E$, then it can have at most one invariant measure.*

Theorem 14.6 *The solution u of stochastic hydrodynamical-type evolution equation (14.1) driven by an* α*-stable noise* L_t *with* $\alpha \in (\frac{3}{2}, 2)$ *satisfying the above mentioned conditions has a unique invariant measure.*

Proof For $t > 0$ and $\varepsilon > 0$, define

$$\Omega_{\varepsilon,t} := \{\omega \in \Omega : \sup_{1 \le s \le t} \|Z_s\| \le \varepsilon\},$$

it follows from Lemma 14.2 that

$$\mathbf{P}(\Omega_{\varepsilon,t}) > 0. \tag{14.62}$$

It follows from (14.39) that for $\omega \in \Omega_{\varepsilon,t}$

$$\|v\|_H^2 \le e^{-(2\pi-\varepsilon^2)t}\|u_0\|_H^2 + c\varepsilon^4. \tag{14.63}$$

For all $r > 0$, define

$$B_H(r) := \{u_0 \in H; \|u_0\|_H \le r\}. \tag{14.64}$$

For all $R > 0$, let $T := T_{R,\delta}$ be sufficiently large and $\varepsilon := \varepsilon_{R,\delta}$ be sufficiently small. It follows from (14.63) that for all $\delta > 0$,

$$\|u\|_H \le \|v\|_H + \|Z_t\|_H \le e^{-(\pi-\frac{\varepsilon^2}{2})t}\|u_0\|_H + c(\varepsilon^2+\varepsilon) < \delta, \quad t \ge T, \tag{14.65}$$

for $x \in B_H(r)$ and $\omega \in \Omega_{\varepsilon,t}$.

By (14.62), we have for all $u_0 \in B_H(r)$,

$$P(t, x, B_H(\delta)) > 0, \quad t \ge T, \tag{14.66}$$

which implies that

$$\mathscr{R}_\lambda(x, B_H(\delta)) > 0. \tag{14.67}$$

Since $R > 0$ is arbitrary, inequality (14.67) is true for all $u_0 \in H$. Thus, $\{u\}_{t\ge 0}$ is accessible to 0.

14.6 Applications

In this section, we apply the former abstract results to stochastic 2D Boussinesq equation and stochastic 2D Magneto-hydrodynamic equation to obtain the uniqueness of the invariant measure.

14.6.1 Stochastic 2D Boussinesq Equation

Consider stochastic Boussinesq equation (14.68) driven by α-stable noise in a two dimensional bounded domain $D \subset R^2$:

$$\begin{cases} du(t) + [u \cdot \nabla u - \Delta u + \nabla p - (0, \theta)]dt = \dot{L}_{1t}, \\ d\theta(t) + [u \cdot \nabla \theta - \Delta \theta]dt = \dot{L}_{2t}, \\ \nabla \cdot u = 0, \\ u(0) = u_0, \quad \theta(0) = \theta_0, \\ u = 0, \quad \theta = 0, \quad x \in \partial D. \end{cases} \tag{14.68}$$

Denote the function space $H = H_1 \times H_2$, $V = V_1 \times V_2$, where

$$H_1 = \{u \in [L^2(D)]^2 : \nabla \cdot u = 0, \quad u = 0 \quad on \quad \partial D\}, \quad H_2 = L^2(D),$$

$$V_1 = [H_0^1(D)]^2 \bigcap H_1, \quad V_2 = H_0^1.$$

Define the operators $A(u, \theta)$ and $B : V \times V \to V'$ by

$$A(u, \theta) = (\Delta u, \Delta \theta)$$
$$\langle B(z_1, z_2), z_3 \rangle = \langle B_1(u_1, u_2), u_3 \rangle + \langle B_2(u_1, \theta_2), \theta_3 \rangle, \quad z_i = (u_i, b_i) \in V, i = 1, 2,$$

where

$$\langle B_1(u_1, u_2), u_3 \rangle = \int_D [u_1 \cdot \nabla u_2] u_3 dx = \sum_{i,j=1}^{2} \int_D u_1^j \partial_j u_2^i u_3^i dx, \quad u_i \in V_1, i = 1, 2,$$

$$\langle B_2(u_1, \theta_2), \theta_3 \rangle = \sum_{j=1}^{2} \int_D u_1^j \partial_j \theta_2 \theta_3 dx, \quad u_1 \in V_1, \quad \theta_2, \theta_3 \in V_2.$$

Then stochastic system (14.68) can be rewritten in the form (1) with $f(u, \theta) = ((0, \theta), 0)$.

Lemma 14.6 *The assumptions (C1), (C2) and (C3) hold for stochastic Boussinesq equations (14.68).*

Proof It suffices to check the assumption (C3). In fact, denote the interpolation space $\mathscr{H}$ by

$$\mathscr{H} = ([L^\infty(D)]^2 \times [L^\infty(D)]) \bigcap H.$$

By using Hölder inequality, we can obtain

$$\begin{aligned} |\langle B_1(u_1, u_2), u_3 \rangle| &\le \|u_1\|_{H_1} \|u_2\|_{V_1} \|u_3\|_{[L^\infty(D)]^2}, \\ |\langle B_2(u_1, \theta_2), \theta_3 \rangle| &\le \|u_1\|_{H_1} \|\theta_2\|_{V_2} \|\theta_3\|_{[L^\infty(D)]}. \end{aligned} \tag{14.69}$$

Then

$$\begin{aligned}\langle B(z_1, z_2), z_3\rangle &\le \|u_1\|_{H_1}(\|u_2\|_{V_1}\|u_3\|_{[L^\infty(D)]^2} + \|\theta_2\|_{V_2}\|\theta_3\|_{[L^\infty(D)]})\\ &\le \|z_1\|_H\|z_2\|_V\|z_3\|_{\mathscr{H}},\end{aligned} \tag{14.70}$$

which implies assumption (C3) holds. Thus, the proof of Lemma 14.6 is completed.

For all $\omega \in \Omega$, denote

$$v_t(\omega) := u_t(\omega) - Z_{1t}(\omega), \qquad \eta_t(\omega) := \theta_t(\omega) - Z_{2t}(\omega),$$

then

$$\begin{cases} dv + [Av + B_1(v + Z_{1t}, v + Z_{1t}) - (0, \eta + Z_{2t})]dt = 0, \quad t \ge 0,\\ d\eta + [A\eta + B_2(v + Z_{1t}, \theta + Z_{2t})]dt = 0, \quad t \ge 0,\\ v(0, x, \omega) = u_0, \quad \theta(0, x, \omega) = \theta_0. \end{cases} \tag{14.71}$$

Let $e_{k\{k \in Z_*\}}$ be an orthonormal basis of H and define

$$H_m := span\{e'_k;\ |k| \le m\}.$$

It is known that H_m is a finite dimensional Hilbert space equipped with the norm adopted from H. For any $m > 0$, let $\pi_m : H \to H_m$ be the projection from H to H_m. Then we have the following Galerkin approximation equations

$$\begin{cases} du^m + [Au^m + B_1^m(u^m, u^m) - (0, \theta^m)]dt = dL_t^m, \quad t \ge 0,\\ d\theta^m + [A\theta^m + B_2^m(u^m, \theta^m)]dt = dL_t^m, \quad t \ge 0,\\ \nabla \cdot u^m = 0, \quad \nabla \cdot b^m = 0,\\ u^m(0, x) = u_0, \quad \theta^m(0, x) = \theta_0, \end{cases} \tag{14.72}$$

where $u^m = \pi_m u$, $B_1^m(u^m, u^m) = \pi_m B(u^m, u^m)$, $L_t^m = \sum_{|k| \le m} \beta_k l_k(t) e'_k$.

For each $T > 0$, define

$$K_T(\omega) := \max\{\sup_{0 \le t \le T} \|Z_{1t}(\omega)\|, \sup_{0 \le t \le T} \|Z_{2t}(\omega)\|\}, \quad \omega \in \Omega. \tag{14.73}$$

Lemma 14.2 yields that for every $k \in \mathbf{N}$, there exists some set $N_k \in \Omega$ such that $\mathbf{P}(N_k) = 0$ and

$$K_k(\omega) < \infty, \quad \omega \notin N_k. \tag{14.74}$$

Let $N = \bigcup_{k \ge 1} N_k$, it is easy to see $\mathbf{P}(N) = 0$ and that for all $T > 0$

$$K_T(\omega) < \infty, \quad \omega \notin N. \tag{14.75}$$

Let $0 < T \le 1$ and $B > 0$ be some constants to be determined later. Define

$$S = \{\ \psi = (v, \eta) \in C([0, T], H) : \psi_0 = (u_0, \theta_0), \psi(t) \in V, t \in (0, T], \tag{14.76}$$

$$\sup_{0 \le t \le T} t^{\frac{\delta}{2}} \|A^{\frac{1}{2}} v\|_{H_1} + \sup_{0 \le t \le T} \|v\|_{H_1} + \sup_{0 \le t \le T} t^{\frac{\delta}{2}} \|A^{\frac{1}{2}} \eta\|_{H_2} + \sup_{0 \le t \le T} \|\eta\|_{H_2} \le B\}. \tag{14.77}$$

For any $\psi, \psi' \in S$, define

$$\begin{aligned} d(\psi, \psi') = & \sup_{0 \le t \le T} t^{\frac{\delta}{2}} \|A^{\frac{1}{2}} (v - v')\|_{H_1} + \sup_{0 \le t \le T} \|(v - v')\|_{H_1} \\ & + \sup_{0 \le t \le T} t^{\frac{\delta}{2}} \|A^{\frac{1}{2}} (\eta - \eta')\|_{H_2} + \sup_{0 \le t \le T} \|(\eta - \eta')\|_{H_2}, \end{aligned}$$

then (S, d) is a closed metric space.

Define a map $\mathscr{F} : S \to C([0, T]; H)$ as the following: for any $\psi \in S$,

$$\begin{aligned} \mathscr{F}\psi = (\ & e^{-At} u_0 + \textstyle\int_0^t e^{-A(t-s)} B_1(v + Z_{1s}, v + Z_{1s}) ds - \int_0^t e^{-A(t-s)} (0, \eta + Z_{2s}) ds, \\ & e^{-At} \theta_0 + \textstyle\int_0^t e^{-A(t-s)} B_2(v + Z_{1s}, \theta + Z_{2s}) ds). \end{aligned}$$

By similar arguments in Sect. 14.4, we can prove the following Theorem 14.7.

Theorem 14.7 *Assume that the assumptions on the noise process L_t hold, then the following statements hold.*

(1) For $(u_0, \theta_0) \in H$, and $\omega \in \Omega$, stochastic system (14.68) possesses a unique mild solution $\phi(\omega) = (u(\omega), \theta(\omega)) \in \mathbf{D}([0, \infty); H) \cap \mathbf{D}([0, \infty); V)$. Moreover, $\phi(\omega)$ has the following form:

$$\begin{aligned} \phi(\omega) = \Big(& e^{-At} u_0 + \int_0^t e^{-A(t-s)} B_1(u, u) ds - \int_0^t e^{-A(t-s)} (0, \theta) ds + \int_0^t e^{-A(t-s)} dL_{1s}(\omega), \\ & e^{-At} \theta_0 + \int_0^t e^{-A(t-s)} B_2(u, \theta) ds + \int_0^t e^{-A(t-s)} dL_{2s}(\omega) \Big). \end{aligned} \tag{14.78}$$

(2) ϕ is a Markov process.

(3) For every $(u_0, \theta_0) \in V$ and $\omega \in \Omega$ a.s., we have $\phi(\omega) \in D([0, \infty); V)$. For every $T > 0$,

$$\sup_{0 \le t \le T} \|\phi(\omega)\|_V \le C,$$

where C is some constant depending on T, α, β and ω.

Applying Theorem 14.6 to stochastic 2D Boussinesq equation (14.68), we have

Theorem 14.8 *Assume that the assumptions on the noise process L_t hold, then the solution u of stochastic 2D Boussinesq equation (14.68) driven by α-stable noise with $\alpha \in (\frac{3}{2}, 2)$ has a unique invariant measure.*

14.6.2 Stochastic 2D Magneto-Hydrodynamic Equation

Consider the stochastic 2D Magneto-hydrodynamic equation driven by α-stable noise in a 2D bounded domain D:

$$\begin{cases} du = [\Delta u + b \cdot \nabla b - u \cdot \nabla u - \nabla p]dt + dL_t, \\ db = [\Delta b + b \cdot \nabla u - u \cdot \nabla b]dt + dL_t, \\ \nabla \cdot u = 0, \quad \nabla \cdot b = 0, \\ u(0, x) = u_0, \quad b(0, x) = b_0, \\ u = 0, \quad b = 0, \quad x \in \partial D, \end{cases} \tag{14.79}$$

where $u = (u_1, u_2)$ and $b = (b_1, b_2)$ denote the velocity field and magnetic field respectively, p is a scalar pressure and L_t is cylindrical α-stable noise.

Denote

$$H = H_1 \times H_1, \quad V = V_1 \times V_1, \quad A(u, b) = (\Delta u, \Delta b),$$

and define $B : V \times V \to V'$ by

$$\langle B(z_1, z_2), z_3 \rangle = \langle B_1(u_1, u_2), u_3 \rangle - \langle B_1(b_1, b_2), u_3 \rangle + \langle B_1(u_1, b_2), b_3 \rangle - \langle B_1(b_1, u_2), b_3 \rangle, \tag{14.80}$$

where $z_i = (u_i, b_i) \in V$. Then stochastic system (14.79) can be rewritten in the form (14.1) with $f = 0$.

Lemma 14.7 *The assumptions (C1), (C2) and (C3) hold for stochastic Magneto-hydrodynamic equation (14.79).*

Proof It suffices to check the assumption (C3). In fact, denote the interpolation space $\mathscr{H}$ by

$$\mathscr{H} = ([L^\infty D)]^2 \times [L^\infty(D)]^2) \bigcap H.$$

We can derive from Hölder inequality that

$$\begin{aligned} |\langle B_1(u_1, u_2), u_3 \rangle| &\le \|u_1\|_{H_1} \|u_2\|_{V_1} \|u_3\|_{[L^\infty(D)]^2}, \\ |\langle B_1(b_1, b_2), u_3 \rangle| &\le \|b_1\|_{H_1} \|b_2\|_{V_1} \|u_3\|_{[L^\infty(D)]^2}, \\ |\langle B_1(u_1, b_2), b_3 \rangle| &\le \|u_1\|_{H_1} \|b_2\|_{V_1} \|b_3\|_{[L^\infty(D)]^2}, \\ |\langle B_1(b_1, u_2), b_3 \rangle| &\le \|b_1\|_{H_1} \|u_2\|_{V_1} \|b_3\|_{[L^\infty(D)]^2}. \end{aligned} \tag{14.81}$$

Then

$$\begin{aligned}\langle B(z_1, z_2), z_3\rangle &\leq \|u_1\|_{H_1}(\|u_2\|_{V_1}\|u_3\|_{[L^\infty(D)]^2} + \|b_2\|_{V_1}\|b_3\|_{[L^\infty(D)]^2}) \\ &\quad + \|b_1\|_{H_1}(\|b_2\|_{V_1}\|u_3\|_{[L^\infty(D)]^2} + \|u_2\|_{V_1}\|b_3\|_{[L^\infty(D)]^2}) \\ &\leq \|z_1\|_H\|z_2\|_V\|z_3\|_{\mathscr{H}},\end{aligned}$$

which implies assumption (C3) holds. Thus, the proof of Lemma 14.7 is complete.

Applying Theorem 14.6 to stochastic 2D Magneto-hydrodynamic equation (14.79), we have

Theorem 14.9 *Assume that the assumptions on the noise process L_t hold, then the solution u of stochastic 2D Magneto-hydrodynamic equation (14.79) driven by α-stable noise with $\alpha \in (\frac{3}{2}, 2)$ has a unique invariant measure.*

Acknowledgements The authors appreciate the referee's valuable comments and suggestions, which are very important to improve the quality of the manuscript. The first author was supported by the NSF of China (No.11771449), The second author was supported by the NSF of China (No.11801563), and the third author was supported by the Fundamental Research Funds for the Central Universities (HUST No.2016YXMS226).

References

1. Applebaum, D.: Lévy Processes and Stochastic Calculus. Cambridge University Press, Cambridge (2009)
2. Albeverio, S., Wu, J., Zhang, T.: Parabolic spdes driven by Poisson white noise. Stoch. Process. Their Appl. **74**, 21–36 (1998)
3. Chueshov, I., Millet, A.: Stochastic 2D hydrodynamical type systems: well posedness and large deviations. Appl. Math. Optim. **61**, 379–420 (2010)
4. Dong, Z., Xu, L., Zhang, X.: Invariance measures of stochastic 2D Navier-Stokes equations driven by a-stable processes. Electron. Commun. Probab. **16**, 678–688 (2011)
5. Dong, Z., Xu, L., Zhang, X.: Exponential ergodicity of stochastic Burgers equations. J. Stat. Phys. **154**, 929–949 (2014)
6. Dong, Z., Xie, Y.: Ergodicity of stochastic 2D Navier-Stokes equations with Levy noise. J. Differ. Equ. **251**, 196–222 (2011)
7. Feireisl, E., Bohdan, M., Antoin, N.: Compressible fluid flows driven by stochastic forcing. J. Differ. Equ. **254**, 1342–1358 (2013)
8. Hairer, M.: Ergodicity for stochastic PDEs (2008). http://www.haier.org/notes/Imperial.pdf
9. Motyl, E.: Stochastic hydrodynamic-type evolution equations driven by Lévy noise in 3D unbounded domains-abstract framework and applications. Stoch. Process. Their Appl. **124**, 2052–2097 (2014)
10. Peszat, S., Zabczyk, J.: Stochastic Partial Differential Equations with Lévy Noise: An Evolution Equation Approach, vol. 113. Cambridge University Press, Cambridge (2007)
11. Priola, E., Zabczyk, J.: Structrual properties of semilinear SPDEs driven by cylindrical stable processes. Probab. Theory Relat. Fields **149**, 97–137 (2011)
12. Priola, E., Xu, L., Zabczyk, J.: Exponential mixing for some SPDEs with Lévy noise. Stoch. Dyn. **11**, 521–534 (2011)

13. Priola, E., Shirkyan, A., Xu, L., Zabczyk, J.: Exponential ergodicity and regularity for equations with Lévy noise. Stoch. Process. Their Appl. **122**, 106–133 (2012)
14. Sermange, M., Temam, R.: Some mathematical questions related to the MHD equations. Commun. Pure Appl. Math. **36**, 635–664 (1983)
15. Sun, X., Xie, Y.: Ergodicity of stochastic dissipative equations driven by a-stable process. Stoch. Anal. Appl. **32**, 61–76 (2014)
16. Wang, F.: Gradient estimate for Ornstein-Uhlenbeck jump processes. Stoch. Process. Their Appl. **121**, 466–478 (2011)
17. Xu, L.: Ergodicity of the stochastic real Ginzburg-Landau equation driven by a-stable noise. Stoch. Process. Their Appl. **123**, 3710–3736 (2013)
18. Xu, L., Zegarlinski, B.: Ergodicity of the finite and infinite dimensional a-stable systems. Stoch. Anal. Appl. **27**, 797–824 (2009)
19. Xu, L.: Exponential mixing of 2D SDEs forced by degenerate Lévy noise. J. Evol. Equ. **14**, 249–272 (2014)

Part III
Advances in Control and Optimization

Chapter 15
Uniform Global Attractor for a Class of Nonautonomous Evolution Hemivariational Inequalities with Multidimensional "Reaction-Velocity" Law

Michael Z. Zgurovsky, Ciro D'Apice, Umberto De Maio, Nataliia V. Gorban, Pavlo O. Kasyanov, Oleksiy V. Kapustyan, Olha V. Khomenko, and José Valero

Abstract We consider non-autonomous evolution inclusions and hemivariation inequalities with possibly non-monotone multidimensional "reaction-velocity" law. The dynamics of all weak solutions defined on the positive semi-axis of time is investigated. We prove the existence global attractor. New properties of complete trajectories are justified. The pointwise behavior of such problem solutions on attractor is described in the autonomous case.

M. Z. Zgurovsky
National Technical University of Ukraine "Igor Sikorsky Kyiv Politechnic Institute", Peremogy ave. 37, build. 1, Kyiv 03056, Ukraine
e-mail: zgurovsm@hotmail.com

C. D'Apice
Dipartimento di Science Aziendali-Management e Innovation Systems, University of Salerno, Via Giovanni Paolo II, 132, Fisciano, SA, Italy
e-mail: cdapice@unisa.it

U. De Maio
Dipartimento di Matematica e Applicazioni "R. Caccioppoli", Università degli Studi di Napoli "Federico II", Complesso Monte S. Angelo, via Cintia, 80126 Napoli, Italy
e-mail: udemaio@unina.it

N. V. Gorban · P. O. Kasyanov (✉) · O. V. Khomenko
Institute for Applied System Analysis, National Technical University of Ukraine "Igor Sikorsky Kyiv Polytechnic Institute", Peremogy ave. 37, build 35, Kyiv 03056, Ukraine
e-mail: kasyanov@i.ua

N. V. Gorban
e-mail: nata_gorban@i.ua

O. V. Khomenko
e-mail: olgkhomenko@ukr.net

V. A. Sadovnichiy and M. Z. Zgurovsky (eds.), *Contemporary Approaches and Methods in Fundamental Mathematics and Mechanics*, Understanding Complex Systems,
https://doi.org/10.1007/978-3-030-50302-4_15

15.1 Introduction

Let $N = 2, 3$ and Ω be a bounded domain in $\mathbb{R}^N$. Assume that the boundary $\partial\Omega$ of the set Ω is regular and it is divided into three parts: Γ_D, Γ_N and Γ_C. Let Γ_D, Γ_N and Γ_C be disjoint sets and $\partial\Omega = \overline{\Gamma_D} \cup \overline{\Gamma_N} \cup \overline{\Gamma_C}$. Assume that $\Gamma_C \subset \partial\Omega$ is an open subset with positive surface measure (see, for example, [36, p. 196]). In this paper we establish the qualitative asymptotic behaviour (as time $t \to \infty$) of all weak solutions of nonautonomous evolution hemivariational inequalities with pointwise pseudomonotone multivalued mappings. As a motivational example we consider the following hemivariational inequality of hyperbolic type with the multidimensional "reaction-velocity" law [9, 32, 35, 37]:

$$\sigma_{ij}(u) = \sum_{k,h=1}^{N} b_{ijhk}\varepsilon_{kh}(u) + \sum_{k,h=1}^{N} a_{ijhk}\frac{\partial \varepsilon_{kh}(u)}{\partial t}, \quad i, j = 1, \ldots, N, \tag{15.1}$$

$$\sum_{j=1}^{N} \frac{\partial \sigma_{ij}(u)}{\partial x_j} + f_i = \frac{\partial^2 u_i}{\partial t^2} \quad \text{in} \quad \Omega \times (0, \infty), \quad i = 1, \ldots, N, \tag{15.2}$$

$$- f^2(x, t) \in \partial\psi(x, t, u_t(x, t)) \text{ in } \Omega \times (0, \infty), \tag{15.3}$$

$$u_i = 0 \text{ on } \Gamma_D \times (0, \infty), \quad i, j = 1, \ldots, N, \tag{15.4}$$

$$S_i = \sum_{j=1}^{N} \sigma_{ij} n_j = F_i(x, t) \text{ on } \Gamma_N \times (0, \infty), \quad i, j = 1, \ldots, N, \tag{15.5}$$

$$- S \in \partial j(x, t, u_t) \text{ on } \Gamma_C \times (0, \infty), \tag{15.6}$$

where $u = (u_1, u_2, \ldots, u_N)^T : \Omega \times (0, \infty) \mapsto \mathbb{R}^N$ is the displacement field, $\sigma = \{\sigma_{ij}(u)\}_{i,j=1}^N$ is the stress tensor, $\varepsilon = \{\varepsilon_{hk}(u)\}_{h,k=1}^N$ is the strain tensor, $\varepsilon_{hk}(u) = \frac{1}{2}(u_{k,h} + u_{h,k})$, $u_{k,h} := \frac{\partial u_k}{\partial x_h}$, $u_{h,k} := \frac{\partial u_h}{\partial x_k}$, $k, h = 1, 2, \ldots, N$, $\{a_{ijhk}\}_{ijhk}$ are the viscosity coefficients, $\{b_{ijhk}\}_{ijhk}$ are the elasticity coefficients, $S = \{S_i\}_{i=1}^N$ is the stress vector on Γ_N, and $n = \{n_i\}_{i=1}^N$ is the outward unit normal to $\partial\Omega$, $f = \{f_i\}_{i=1}^N = f^1 + f^2$ is the density of body force, $f^1 : \Omega \times (0, \infty) \mapsto \mathbb{R}^N$ is the exter-

O. V. Kapustyan
Taras Shevchenko National University of Kyiv, Institute for Applied System Analysis, National Technical University of Ukraine "Igor Sikorsky Kyiv Polytechnic Institute", Kyiv, Ukraine
e-mail: kapustyanav@gmail.com

J. Valero
Centro de Investigación Operativa, Universidad Miguel Hernandez de Elche, Avda. Universidad s/n, 03202 Elche, Spain
e-mail: jvalero@umh.es

nal loading, $f^2 : \Omega \times (0, \infty) \mapsto \mathbb{R}^N$ is the reaction of constrains introducing the skin effects (here f^2 is a possibly multivalued function of the velocity $u_t := \frac{\partial u}{\partial t}$), $F = \{F_i\}_{i=1}^N : \Gamma_N \times (0, \infty) \mapsto \mathbb{R}^N$ is the vector of surface traction, the functions $\psi : \Omega \times (0, \infty) \times \mathbb{R}^d \mapsto \mathbb{R}$ and $j : \Gamma_C \times (0, \infty) \times \mathbb{R}^N \mapsto \mathbb{R}$ are locally Lipschitz functions in u_t, and the multivalued mappings $\partial\psi$ and ∂j represent their Clarke's subdifferentials with respect to u_t.

To state the assumptions on parameters of Problem (15.1)–(15.6) we need to introduce the following constructions. Let $\gamma \geq 1$ and $\mathscr{E}$ be a real separable Banach space. As $L_\gamma^{\rm loc}(0, \infty; \mathscr{E})$ we consider the Fréchet space of all locally integrable functions with values in $\mathscr{E}$, that is, $\varphi \in L_\gamma^{\rm loc}(0, \infty; \mathscr{E})$ if and only if for any finite interval $[\tau, T] \subset (0, \infty)$ the restriction of φ on $[\tau, T]$ belongs to the space $L_\gamma(\tau, T; \mathscr{E})$. If $\mathscr{E} \subset L_1(D)$ for some bounded open set $D \subset \mathbb{R}^m$, $m = 1, 2, \ldots$, then any function φ from $L_\gamma^{\rm loc}(0, \infty; \mathscr{E})$ can be considered as a measurable mapping that acts from $\Omega \times (0, \infty)$ into $\mathbb{R}$. Further, we write $\varphi(x, t)$, when we consider this mapping as a function from $\Omega \times (0, \infty)$ into $\mathbb{R}$, and $\varphi(t)$, if this mapping is considered as an element from $L_\gamma^{\rm loc}(0, \infty; \mathscr{E})$; [2, 8, 12, 40, 46, 47] and references therein. Further we denote that

$$T(h)y(\cdot) = \Pi_{0,\infty}\, y(\cdot + h), \quad y \in L_\gamma^{\rm loc}(0, \infty; \mathscr{E}),\ h \geq 0,$$

where $\Pi_{0,\infty}$ is the restriction operator to the time interval $[0, \infty)$. A function $\varphi \in L_\gamma^{\rm loc}(0, \infty; \mathscr{E})$ is called *translation compact* in $L_\gamma^{\rm loc}(0, \infty; \mathscr{E})$, if the set $\{T(h)\varphi : h \geq 0\}$ is precompact in $L_\gamma^{\rm loc}(0, \infty; \mathscr{E})$; [2]. A translation compact in $L_\gamma^{\rm loc}(0, \infty; \mathscr{E})$ function $\varphi \in L_\gamma^{\rm loc}(0, \infty; \mathscr{E})$ is called *univocal*, if a net $\{T(h)\varphi\}_{h\geq 0}$ converges in $L_\gamma^{\rm loc}(0, \infty; \mathscr{E})$ as $h \to \infty$. A function $\varphi \in L_\gamma^{\rm loc}(0, \infty; \mathscr{E})$ is called *translation bounded* in $L_\gamma^{\rm loc}(0, \infty; \mathscr{E})$, if

$$\sup_{t\geq 0} \int_t^{t+1} \|\varphi(s)\|_{\mathscr{E}}^\gamma ds < \infty. \tag{15.7}$$

Reference [40, p. 105]. A function $\varphi \in L_1^{\rm loc}(0, \infty; \mathscr{E})$ is called *translation uniform integrable (t.u.i.)* in $L_1^{\rm loc}(0, \infty; \mathscr{E})$, if

$$\lim_{K\to\infty} \sup_{t\geq 0} \int_t^{t+1} \|\varphi(s)\|_{\mathscr{E}}\, \chi_{\{\|\varphi(s)\|\geq K\}} ds = 0. \tag{15.8}$$

In particular, a function $\varphi \in L_1^{\rm loc}(D \times (0, \infty))$ is called *t.u.i.* in $L_1^{\rm loc}(D \times (0, \infty))$, if

$$\lim_{K\to\infty} \sup_{t\geq 0} \int_t^{t+1} \int_D |\varphi(x, s)|\, \chi_{\{|\varphi(x,s)|\geq K\}} dx ds = 0. \tag{15.9}$$

Dunford–Pettis compactness criterion provides that a function $\varphi \in L_1^{\rm loc}(0, \infty; \mathscr{E})$ is t.u.i. in $L_1^{\rm loc}(0, \infty; \mathscr{E})$ if and only if for every sequence of elements $\{\tau_n\}_{n\geq 1} \subset (0, \infty)$ the sequence $\{T(\tau_n)\varphi(\cdot)\}_{n\geq 1}$ contains a subsequence which converges weakly in

$L_1^{\text{loc}}(0,\infty;\mathscr{E})$. According to [12], we note that for any $\gamma > 1$ Hölder's and Chebyshev's inequalities imply that every translation bounded in $L_\gamma^{\text{loc}}(0,\infty;\mathscr{E})$ function is t.u.i. in $L_1^{\text{loc}}(0,\infty;\mathscr{E})$. Moreover, every translation compact in $L_\gamma^{\text{loc}}(0,\infty;\mathscr{E})$ function is translation bounded in $L_\gamma^{\text{loc}}(0,\infty;\mathscr{E})$.

Let the following three assumptions hold:

Assumption 15.1 Let $\psi : \Omega \times (0,\infty) \times \mathbb{R}^N \mapsto \mathbb{R}$ satisfy the following properties:

(i) $\psi(\cdot,\xi)$ is measurable for each $\xi \in \mathbb{R}^N$ and $\psi(\cdot,\bar{0}) \in L_1(\Omega \times (\tau,T))$ for each finite time interval $(\tau,T) \subset (0,\infty)$;

(ii) $\psi(x,t,\cdot)$ is locally Lipschitz for a.e. $(x,t) \in \Omega \times (0,\infty)$;

(iii) there exist a constant $\gamma_1 > 0$ and a nonnegative t.u.i. in $L_1^{\text{loc}}((0,\infty) \times \Omega)$ function $C_1 : \Omega \times (0,\infty) \to \mathbb{R}$ such that $\|\eta\|^2 \le C_1(x,t) + \gamma_1\|\xi\|^2$ for a.e. $(x,t) \in \Omega \times (0,\infty)$ and for each $\eta \in \partial\psi(x,t,\xi)$;

(iv) $\psi^0(x,t,\xi;-\xi) \le 0$ for each $\xi \in \mathbb{R}^N$, $x \in \Omega$, and $t > 0$, where $\psi^0(x,t,\xi;\eta)$ is the directional derivative of $\psi(x,t,\cdot)$ at the point $\xi \in \mathbb{R}^N$ in the direction $\eta \in \mathbb{R}^N$.

Assumption 15.2 Let $j : \Gamma_C \times (0,\infty) \times \mathbb{R}^N \mapsto \mathbb{R}$ satisfy the following conditions:

(i) $j(\cdot,\xi)$ is a measurable function for each $\xi \in \mathbb{R}^N$ and $j(\cdot,\bar{0}) \in L_1(\Gamma_C \times (\tau,T))$ for each finite time interval $(\tau,T) \subset (0,\infty)$;

(ii) $j(x,t,\cdot)$ is a locally Lipschitz function for each $x \in \Gamma_C$ and $t > 0$;

(iii) there exist a constant $\gamma_2 > 0$ and a nonnegative t.u.i. in $L_1^{\text{loc}}((0,\infty) \times \Gamma_C)$ function $\bar{c} : \Gamma_C \times (0,\infty) \mapsto \mathbb{R}$ such that $\|\eta\|_{\mathbb{R}^N}^2 \le \bar{c}(x,t) + \gamma_2\|\xi\|_{\mathbb{R}^N}^2$ for each $x \in \Gamma_C$, $t > 0$, $\xi \in \mathbb{R}^N$, and $\eta \in \partial j(x,t,\xi)$;

(iv) $j^0(x,t,\xi;-\xi) \le 0$ for each $\xi \in \mathbb{R}^N$, $x \in \Gamma_N$, and $t > 0$, where $j^0(x,t,\xi;\eta)$ is the directional derivative of $j(x,t,\cdot)$ at the point $\xi \in \mathbb{R}^N$ in the direction $\eta \in \mathbb{R}^N$.

Assumption 15.3 Assume that $f^1 \in L_2^{\text{loc}}(0,\infty;L_2(\Omega;\mathbb{R}^N))$ and $F \in L_2^{\text{loc}}(0,\infty;L_2(\Gamma_N;\mathbb{R}^N))$ are univocally translation compact in $L_2^{\text{loc}}(0,\infty;L_2(\Omega;\mathbb{R}^N))$ and $L_2^{\text{loc}}(0,\infty;L_2(\Gamma_N;\mathbb{R}^N))$ respectively.

Note that all nonconvex superpotential graphs from [35, Chap. 4.6], in particular, the functions j, defined as a minimum and as a maximum of quadratic convex functions, satisfy Assumption 15.2.

For the variational formulation of Problem (15.1)–(15.6) we set: $H := L_2(\Omega;\mathbb{R}^N)$, $Z = H^\delta(\Omega;\mathbb{R}^N)$, $V := \{v \in H^1(\Omega;\mathbb{R}^N) : v_i = 0 \text{ on } \Gamma_D\}$, where $\delta \in \left(\frac{1}{2};1\right)$; see [32] for details. Let $\langle\cdot,\cdot\rangle_V : V^* \times V \mapsto \mathbb{R}$ be the pairing in $V^* \times V$ that coincides on $H \times V$ with the inner product $(\cdot,\cdot)$ in Hilbert space H. For each $u, v \in V$ we set:

$$\langle f_0, v\rangle_V := \sum_{i=1}^N \int_\Omega f_i v_i dx + \sum_{i=1}^N \int_{\Gamma_N} F_i v_i d\sigma(x),$$

$$a(u,v) := \sum_{i,j,k,h=1}^{N} \int_{\Omega} a_{ijhk}\varepsilon_{ij}(u)\varepsilon_{kh}(v)dx, \quad b(u,v) := \sum_{i,j,k,h=1}^{N} \int_{\Omega} b_{ijhk}\varepsilon_{ij}(u)\varepsilon_{kh}(v)dx,$$

$\bar{\gamma} : Z \mapsto L_2(\partial\Omega; \mathbb{R}^N)$ be the *trace operator*, $\bar{\gamma}^* : L_2(\partial\Omega; \mathbb{R}^N) \mapsto Z^*$ be its *dual*, that is,

$$\bar{\gamma}^* u(z) = \int_{\partial\Omega} u(x)\bar{\gamma}z(x)d\sigma(x), \quad z \in Z, \; u \in L_2(\partial\Omega; \mathbb{R}^N).$$

For each $t > 0$ let us consider a locally Lipschitz functional $J[t] : L_2(\Gamma_C; \mathbb{R}^N) \mapsto \mathbb{R}$,

$$J[t](z) = \int_{\Gamma_C} j(x,t,z(x))d\sigma(x), \quad z \in L_2(\Gamma_C; \mathbb{R}^N).$$

Then the operators A_1, A_2, and B_0 are defined as follows:

$$\langle A_1 u, v\rangle_V = a(u,v), \quad \langle B_0 u, v\rangle_V = b(u,v), \; A_0(t,u) = A_1 u + A_2(t,u), \quad u, v \in V, \quad t > 0;$$
$$A_2(t,z) = \bar{\gamma}^* \left(\partial J[t](\bar{\gamma}z)\right) + \partial\eta[t](z), \quad z \in Z,$$

where for each $T > 0$ the function $\eta[t] : L_2(\Omega; \mathbb{R}^N) \mapsto \mathbb{R}$ is defined as follows:

$$\eta[t](v) = \int_{\Omega} \psi(x,t,v(x))dx, \quad v \in L_2(\Omega; \mathbb{R}^N). \tag{15.10}$$

We note that

$$\partial(J \circ \bar{\gamma})[t](z) \subset \bar{\gamma}^* \left(\partial J[t](\bar{\gamma}z)\right)$$

in the general case. Moreover, the equality holds if J is regular [46, 47]. According to [32, 46, 47], the following properties hold:

(H_1) V, Z, H are Hilbert spaces; $H^* \equiv H$, and the following embeddings are dense and compact:

$$V \subset Z \subset H \equiv H^* \subset Z^* \subset V^*;$$

(H_2) $f_0 \in L_2^{\mathrm{loc}}(0,\infty; V^*)$ is univocally translation compact in $L_2^{\mathrm{loc}}(0,\infty; V^*)$;

(A_1) there exist a positive constant c and a nonnegative t.u.i. in $L_1^{\mathrm{loc}}(0,\infty)$ function $c : (0,\infty) \mapsto (0,\infty)$ such that

$$\|d\|_{V^*}^2 \le c(t) + c\|u\|_V^2,$$

for each $t > 0$, $u \in V$, and $d \in A_0(t,u)$;

(A_2) there exist a positive constant α such that

$$\alpha\|u\|_V^2 \le \langle d, u\rangle_V,$$

for each $t > 0$, $u \in V$, and $d \in A_0(t, u)$;

(A_3) $A_0 = A_1 + A_2$, where $A_1 : V \mapsto V^*$ is a linear self-adjoint (symmetric) continuous operator and $A_2 : (0, \infty) \times V \rightrightarrows V^*$ satisfies the following conditions:

(a) there exists a Hilbert space Z such that the embedding $V \subset Z$ is dense and compact and the embedding $Z \subset H$ is dense and continuous;
(b) for each $t > 0$ and $u \in Z$ the set $A_2(t, u)$ is nonempty, convex, and weakly compact in Z^*;
(c) for each $t > 0$ the mapping $A_2(t, \cdot) : Z \rightrightarrows Z^*$ is bounded, that is, A_2 transforms bounded sets from Z into bounded sets in the space Z^*;
(d) for each $t > 0$ the mapping $A_2(t) : Z \rightrightarrows Z^*$ is a demiclosed map, that is, if $u_n \to u$ in Z and $d_n \to d$ weakly in Z^* as $n \to \infty$, and $d_n \in A_2(t, u_n)$ for each $n = 1, 2, \dots,$ then $d \in A_2(t, u)$;
(e) the multivalued mapping $t \rightrightarrows A_2(t, \cdot)$ is measurable (see [32]);

(B_1) $B_0 : V \mapsto V^*$ is is a linear self-adjoint (symmetric) continuous operator;
(B_2) there exists a positive constant γ such that

$$\gamma \|u\|_V^2 \le \langle B_0 u, u\rangle_V \quad \text{for each } u \in V.$$

The variational setting of Problem (15.1)–(15.6) has the following formulation:

$$y''(t) + A_0(t, y'(t)) + B_0(y(t)) \ni f_0(t), \tag{15.11}$$

where parameters of this problem satisfy conditions (H_1), (H_2), (A_1)–(A_3) and (B_1)–(B_2). As a *weak solution* of Problem (15.11) on the finite time interval $[\tau, T]$ we consider the pair of elements $(y, y')^T \in L_2(\tau, T; V \times V)$ such that for some $d \in L_2(\tau, T; V^*)$

$$\begin{aligned} &d(t) \in A_0(t, y'(t)) \quad \text{for a.e. } t \in (\tau, T), \\ &-\int_\tau^T (\zeta'(t), y'(t))dt + \int_\tau^T \langle d(t), \zeta(t)\rangle_V dt \\ &+\int_\tau^T \langle B_0 y(t), \zeta(t)\rangle_V dt = \int_\tau^T \langle f_0(t), \zeta(t)\rangle_V \quad \forall \zeta \in C_0^\infty([\tau, T]; V), \end{aligned} \tag{15.12}$$

where y' is the derivative of y in the sense of $\mathscr{D}^*([\tau, T]; V^*)$. As a *generalized solution* of Problem (15.1)–(15.6) on the finite time interval $[\tau, T] \subset (0, \infty)$ we consider the weak solution of the respective Problem (15.11) on $[\tau, T]$. This definition corresponds to Definition 3 from [32]. Note that several optimization and control problems meet this formulation [3–7, 13, 16–23, 26–30, 33, 34, 38].

The main purpose of this paper is to establish the long-time dynamics (as time $t \to \infty$) of all globally defined on $(0, \infty)$ weak solutions of nonautonomous Problem (15.11), where parameters of this problem satisfy conditions (H_1), (H_2), (A_1)–(A_3), (B_1), and (B_2).

We remark that the abstract existence results for Problem (15.11) under the similar conditions on their parameters were considered in [11, 14, 32, 41, 42, 44, 45]. In this paper we establish the long-time behaviour of all weak solutions as time $t \to \infty$ in the natural phase space $V \times H$. We remark also that a such problem in the autonomous case (under the same conditions) was established in [43, 46].

15.2 Main Result

Let us consider Problem (15.11) under Assumptions (H_1), (H_2), (A_1)–(A_3), (B_1), (B_2). Further, to simplify the conclusions, we consider the equivalent norm $\|u\|_V = \sqrt{\langle B_0 u, u\rangle_V}$, $u \in V$, on the space V. This norm is generated by the following inner product $(u, v)_V = \langle B_0 u, v\rangle_V$, $u, v \in V$. For fixed $\tau < T$ let us set:

$$\begin{aligned}
&X_{\tau,T} = L_2(\tau, T; V), \quad X^*_{\tau,T} = L_2(\tau, T; V^*), \quad W_{\tau,T} = \{u \in X_{\tau,T} \ : \ u' \in X^*_{\tau,T}\},\\
&\mathscr{A}_{\tau,T}(y) = \{d \in X^*_{\tau,T} \ : \ d(t) \in A_0(t, y(t)) \text{ for a.e. } t \in (\tau, T)\}, \quad y \in X_{\tau,T},\\
&B_{\tau,T}(y)(t) = B_0(y(t)) \text{ for a.e. } t \in (\tau, T), \ \ y \in X_{\tau,T},\\
&f_{\tau,T}(t) = f_0(t) \text{ for a.e. } t \in (\tau, T).
\end{aligned}$$

Note, that $\mathscr{A}_{\tau,T} : X_{\tau,T} \rightrightarrows X^*_{\tau,T}$, $B_{\tau,T} : X_{\tau,T} \mapsto X^*_{\tau,T}$, and $f_{\tau,T} \in X^*_{\tau,T}$. Moreover, the space $W_{\tau,T}$ is the Hilbert space with the graph norm of the derivative (see [44, 45]):

$$\|u\|^2_{W_{\tau,T}} = \|u\|^2_{X_{\tau,T}} + \|u'\|^2_{X^*_{\tau,T}}, \quad u \in W_{\tau,T}. \tag{15.13}$$

Let $\langle \cdot, \cdot\rangle_{X_{\tau,T}} : X^*_{\tau,T} \times X_{\tau,T} \to \mathbb{R}$ be the pairing on $X^*_{\tau,T} \times X_{\tau,T}$ which restriction to $L_2(\tau, T; H) \times X_{\tau,T}$ coincides with the inner product on $L_2(\tau, T; H)$. According to [8, Theorem IV.1.17, P. 177]), the embedding $W_{\tau,T} \subset C([\tau, T]; H)$ is continuous and dense. Moreover,

$$(u(T), v(T)) - (u(\tau), v(\tau)) = \int_\tau^T \Big[\langle u'(t), v(t)\rangle_V + \langle v'(t), u(t)\rangle_V\Big]dt. \tag{15.14}$$

for each $u, v \in W_{\tau,T}$.

The definition of the derivative in the sense of $\mathscr{D}^*([\tau, T]; V^*)$ and equality (15.12) implies that each weak solution $(y, y')^T$ of Problem (15.11) on $[\tau, T]$ belongs to $C([\tau, T]; V) \times W_{\tau,T}$. Moreover

$$y'' + \mathscr{A}_{\tau,T}(y') + B_{\tau,T}(y) \ni f_{\tau,T}. \tag{15.15}$$

Vice versa, if $y \in C([\tau, T]; V)$, $y' \in W_{\tau,T}$, and y satisfies (15.15), then $(y, y')^T$ is a weak solution of (15.11) on $[\tau, T]$.

For each $a \in V$ and $b \in H$ let us consider the initial data:

$$y(\tau) = a, \quad y'(\tau) = b. \tag{15.16}$$

According to [32, Theorem 11, p. 523], for each $a \in V$ and $b \in H$ there exists a weak solution of Problem (15.11) on $[\tau, T]$ satisfying initial conditions (15.16). Moreover, each weak solution $(y, y')^T$ of Cauchy problem (15.11), (15.16) on $[\tau, T]$ belongs to $C([\tau, T]; V) \times W_{\tau,T}$ and y satisfies (15.15). Therefore, since $W_{\tau,T} \subset C([\tau, T]; H)$, initial data (15.16) have sense.

For each $\tau \geq 0$ and a separable real Hilbert space E we consider the Fréchet spaces

$$C^{\mathrm{loc}}([\tau, \infty); E) := \{y : [\tau, \infty) \to E \;:\; \Pi_{t_1,t_2} y \in C([t_1, t_2]; E) \text{ for each } [t_1, t_2] \subset [\tau, \infty)\},$$
$$W^{\mathrm{loc}}([\tau, \infty)) := \{y : [\tau, \infty) \to V \;:\; \Pi_{t_1,t_2} y \in W_{t_1,t_2} \text{ for each } [t_1, t_2] \subset [\tau, \infty)\},$$

where Π_{t_1,t_2} is the restriction operator to the finite time interval $[t_1, t_2]$. We recall that the sequence $\{f_n\}_{n\geq 1}$ converges in $W^{\mathrm{loc}}([\tau, \infty))$ (in $C^{\mathrm{loc}}([\tau, \infty); E)$ respectively) to $f \in W^{\mathrm{loc}}([\tau, \infty))$ (to $f \in C^{\mathrm{loc}}([\tau, \infty); E)$ respectively) as $n \to \infty$ if and only if the sequence $\{\Pi_{t_1,t_2} f_n\}_{n\geq 1}$ converges in W_{t_1,t_2} (in $C([t_1, t_2]; E)$ respectively) to $\Pi_{t_1,t_2} f$ as $n \to \infty$ for each finite time interval $[t_1, t_2] \subset [\tau, \infty)$.

According to Assumption (A_2) and Schwarz's inequality, each weak solution $(y, y')^T$ of Problem (15.11) on $[\tau, T]$ satisfies the following a priory estimate:

$$\begin{aligned} \|y'(t)\|_H^2 + \|y(t)\|_V^2 + \alpha \int_s^t \|y'(\xi)\|_V^2 d\xi \\ \leq \|y'(s)\|_H^2 + \|y(s)\|_V^2 + \frac{1}{\alpha} \int_s^t \|f(\xi)\|_{V^*}^2 d\xi, \end{aligned} \tag{15.17}$$

for each t and s satisfying $\tau \leq s \leq t \leq T$. Therefore, Assumption (H_2) implies that each weak solution of Problem (15.11) can be extended to a *global* one, defined on $[0, \infty)$.

Let $E := V \times H$ be the *phase space* and $C^{\mathrm{loc}}([\tau, \infty); E)$ be the *extended phase space* for Problem (15.11). Let $\mathscr{K}_\tau^+$ be the family of globally defined on $[\tau, \infty)$ weak solutions of Problem (15.11). According to [47], to characterize the uniform long-time behavior of all weak solutions of Problem (15.11), we consider the *united trajectory space* $\mathscr{K}_\cup^+$ for the family of solutions $\{\mathscr{K}_\tau^+\}_{\tau\geq 0}$ shifted to zero:

$$\mathscr{K}_\cup^+ := \bigcup_{\tau\geq 0} \left\{ T(h)y(\,\cdot + \tau\,) \;:\; y(\cdot) \in \mathscr{K}_\tau^+,\; h \geq 0 \right\}, \tag{15.18}$$

and the *extended united trajectory space* for the family $\{\mathscr{K}_\tau^+\}_{\tau\geq 0}$:

$$\mathscr{K}^+ := \mathrm{cl}_{C^{\mathrm{loc}}([0,\infty);E)} \left[\mathscr{K}_\cup^+ \right], \tag{15.19}$$

where $\mathrm{cl}_{C^{\mathrm{loc}}([0,\infty);E)}[\,\cdot\,]$ is the closure in $C^{\mathrm{loc}}([0, \infty); E)$. According to [47],

$$T(h)\mathscr{K}^+ \subset \mathscr{K}^+ \text{ for each } h \geq 0, \tag{15.20}$$

because $T(h)\mathscr{K}_{\mathrm{U}}^+ \subset \mathscr{K}_{\mathrm{U}}^+$ for each $h \geq 0$, and

$$\rho_{C^{\mathrm{loc}}([0,\infty);E)}(T(h)u, T(h)v) \leq \rho_{C^{\mathrm{loc}}([0,\infty);E)}(u, v) \text{ for each } u, v \in C^{\mathrm{loc}}([0,\infty); E),$$

where $\rho_{C^{\mathrm{loc}}([0,\infty);E)}$ is the standard metric on Fréchet space $C^{\mathrm{loc}}([0,\infty); E)$. Let us define the multi-valued semiflow (*m-semiflow*) $G : [0,\infty) \times E \to 2^E$:

$$G(t, y_0) := \{y(t) : y(\cdot) \in \mathscr{K}^+ \text{ and } y(0) = y_0\}, \quad t \geq 0, \ y_0 \in E. \tag{15.21}$$

Note that the set $G(t, y_0)$ is nonempty for each $t \geq 0$ and $y_0 \in E$. Moreover, the following two conditions hold:

(i) $G(0, \cdot) = I$ is the identity map;
(ii) $G(t_1 + t_2, y_0) \subset G(t_1, G(t_2, y_0))$, $\forall t_1, t_2 \in [0,\infty)$, $\forall y_0 \in E$,

where $G(t, D) = \bigcup_{y \in D} G(t, y)$, $D \subset E$. We denote by $\mathrm{dist}_E(C, D) = \sup_{c \in C} \inf_{d \in D} \rho(c, d)$ the *Hausdorff semidistance* between nonempty subsets C and D of the Hilbert space E. Recall that the set $\mathfrak{R} \subset E$ is a *global attractor* of the m-semiflow G if it satisfies the following conditions:

(i) $\mathfrak{R}$ attracts each bounded subset $B \subset E$, that is,

$$\mathrm{dist}_E(G(t, B), \mathfrak{R}) \to 0, \quad t \to \infty; \tag{15.22}$$

(ii) $\mathfrak{R}$ is negatively semi-invariant set, that is, $\mathfrak{R} \subset G(t, \mathfrak{R})$ for each $t \geq 0$;
(iii) $\mathfrak{R}$ is the minimal set among all nonempty closed subsets $C \subset E$ that satisfy (15.22).

We examine the uniform long-time behavior of solutions for Problem (15.11) in the strong topology of the natural phase space E (as time $t \to \infty$) in the sense of the existence of a compact global attractor for m-semiflow G generated by the family of solution sets $\{\mathscr{K}_\tau^+\}_{\tau \geq 0}$ and their shifts. The following theorem is the main result of the paper.

Theorem 15.1 *Let assumptions (H_1), (H_2), (A_1)–(A_3), (B_1), (B_2) hold. Then the m-semiflow G defined in (15.21) has a compact global attractor $\mathfrak{R}$ in the phase space E.*

15.3 Proof of Theorem 15.1

Before the proof of Theorem 15.1 we establish some auxiliary statements.

Lemma 15.4 *Let assumptions of Theorem 15.1 hold. Then,*

$$\|\varphi(t)\|_E^2 \le c_1 \|\varphi(s)\|_E^2 e^{-c_2(t-s)} + \sup_{h \ge 0} \int_s^t \psi(\xi + h) e^{c_2(\xi - t)} d\xi, \tag{15.23}$$

for each $\varphi \in \mathcal{K}^+$, *and* $t \ge s \ge 0$, *where*

$$c_1 := \frac{\sqrt{\lambda_1} + \varepsilon}{\sqrt{\lambda_1} - \varepsilon},\ c_2 := \varepsilon\rho,\ \varepsilon := \frac{\alpha\lambda_1}{3 + \lambda_1 c} < \min\left\{\frac{\alpha}{c}, \sqrt{\lambda_1}\right\},\ \rho := \frac{2\sqrt{\lambda_1}}{\sqrt{\lambda_1} - \varepsilon},$$
$$\psi(t) := \left(\frac{\rho}{2\alpha} + \frac{\varepsilon\rho}{2}\right) \|f_0(t)\|_{V^*}^2 + \frac{\varepsilon\rho}{2} c(t), \quad t > 0. \tag{15.24}$$

Proof According to (15.19), if inequality (15.23) holds for each $\varphi \in \mathcal{K}_{\cup}^+$, then it holds for each $\varphi \in \mathcal{K}^+$. The rest of the proof establishes inequality (15.19) for each $\varphi \in \mathcal{K}_{\cup}^+$.

For any $\varphi \in \mathcal{K}_{\cup}^+$, there exist $\tau, h \ge 0$ and $(y, y')^T \in \mathcal{K}_\tau^+$: $\varphi(\cdot) = (T(h)y(\cdot + \tau), T(h)y'(\cdot + \tau))^T$. For an arbitrary fixed $\varepsilon > 0$ we set

$$Y_\varepsilon(t) := \frac{1}{2}\|y'(t)\|_H^2 + \frac{1}{2}\|y(t)\|_V^2 + \varepsilon(y'(t), y(t)), \quad t \ge 0.$$

Note that

$$\frac{dY_\varepsilon(t)}{dt} = (y''(t), y'(t)) + \langle B_0 y(t), y'(t)\rangle_V + \varepsilon\|y'(t)\|_H^2 + \varepsilon(y''(t), y(t)),$$

for a.e. $t > 0$. Therefore, according to (15.12), and Assumptions (A_1), (A_2),

$$\frac{dY_\varepsilon(t)}{dt} \le \langle f_0(t + \tau + h), y'(t)\rangle_V - \alpha\|y'(t)\|_V^2 + \varepsilon\|y'(t)\|_H^2$$
$$+ \varepsilon\langle f_0(t + \tau + h), y(t)\rangle_V + \varepsilon\|y(t)\|_V \sqrt{c(t + \tau + h) + c\|y'(t)\|_V^2} - \varepsilon\|y(t)\|_V^2, \tag{15.25}$$

for a.e. $t > 0$. Let $\lambda_1 > 0$ be the first eigenvalue of A_1. Since

$$\lambda_1\|u\|_H^2 \le \|u\|_V^2 \quad \forall u \in V, \tag{15.26}$$

then

$$|(y'(t), y(t))| \le \|y'(t)\|_H \|y(t)\|_H$$
$$\le \frac{1}{\sqrt{\lambda_1}}\|y'(t)\|_H \|y(t)\|_V \le \frac{1}{2\sqrt{\lambda_1}}\left(\|y'(t)\|_H^2 + \|y(t)\|_V^2\right), \tag{15.27}$$

for a.e. $t > 0$, where the last inequality follows form Schwarz's inequality. Moreover, Schwarz's inequality imply:

$$\langle f_0(t+\tau+h), y'(t)\rangle_V \le \|f_0(t+\tau+h)\|_{V^*}\|y'(t)\|_V \le \frac{\alpha\|y'(t)\|_V^2}{2} + \frac{\|f_0(t+\tau+h)\|_{V^*}^2}{2\alpha},$$

$$\|y(t)\|_V\left(\|f_0(t+\tau+h)\|_{V^*} + \sqrt{c(t+\tau+h)+c\|y'(t)\|_V^2}\right)$$
$$\le \frac{1}{2}(\|y(t)\|_V^2 + \|f_0(t+\tau+h)\|_{V^*}^2 + c(t+\tau+h) + c\|y'(t)\|_V^2), \tag{15.28}$$

for a.e. $t > 0$. Let us apply the last two inequalities to the right-hand side of (15.25):

$$\begin{aligned}\frac{dY_\varepsilon(t)}{dt} &\le \left(-\frac{\alpha}{2}+\frac{\varepsilon c}{2}\right)\|y'(t)\|_V^2 + \varepsilon\|y'(t)\|_H^2 \\ &- \frac{\varepsilon}{2}\|y(t)\|_V^2 + \|f_0(t+\tau+h)\|_{V^*}^2\left(\frac{1}{2\alpha}+\frac{\varepsilon}{2}\right) + \frac{\varepsilon}{2}c(t+\tau+h),\end{aligned} \tag{15.29}$$

for a.e. $t > 0$. If $\varepsilon < \frac{\alpha}{c}$, then, according to (15.26), inequality (15.29) implies

$$\begin{aligned}\frac{dY_\varepsilon(t)}{dt} &\le \left(-\frac{\alpha\lambda_1}{2}+\frac{\varepsilon c\lambda_1}{2}+\varepsilon\right)\|y'(t)\|_H^2 \\ &- \frac{\varepsilon}{2}\|y(t)\|_V^2 + \left(\frac{1}{2\alpha}+\frac{\varepsilon}{2}\right)\|f_0(t+\tau+h)\|_{V^*}^2 + \frac{\varepsilon}{2}c(t+\tau+h),\end{aligned} \tag{15.30}$$

for a.e. $t > 0$. Let us choose $\varepsilon > 0$ such that

$$-\frac{\alpha\lambda_1}{2}+\frac{\varepsilon c\lambda_1}{2}+\varepsilon = -\frac{\varepsilon}{2},$$

that is, $\varepsilon = \dfrac{\alpha\lambda_1}{3+\lambda_1 c} < \dfrac{\alpha}{c}$. If we set

$$c_2 := \frac{\alpha\lambda_1^{\frac{3}{2}}}{(1+2\sqrt{\lambda_1})(3+\lambda_1 c)}, \quad \eta(t) := \left(\frac{1}{2\alpha}+\frac{\varepsilon}{2}\right)\|f_0(t+\tau+h)\|_{V^*}^2 + \frac{\varepsilon}{2}c(t+\tau+h),\ t>0,$$

then, according to (15.27), we obtain that

$$\frac{dY_\varepsilon(t)}{dt} + c_2 Y_\varepsilon(t) \le \eta(t),$$

for a.e. $t > 0$, because $c_2\left(1+\dfrac{1}{2\sqrt{\lambda_1}}\right) = \dfrac{\varepsilon}{2}$. Therefore,

$$Y_\varepsilon(t) \le Y_\varepsilon(s)e^{-c_2(t-s)} + \int_s^t \eta(\xi)e^{c_2(\xi-t)}d\xi, \tag{15.31}$$

for each $t \ge s \ge 0$. Since $\lambda_1 c - \sqrt{\lambda_1 c} + 3 > 0$, then

$$\sqrt{\lambda_1} > \frac{\sqrt{c}\lambda_1}{3+\lambda_1 c} \geq \frac{\alpha\lambda_1}{3+\lambda_1 c} = \varepsilon,$$

where the last inequality holds because, according to Assumptions (A_1) and (A_2), $\alpha^2 \leq c$. Thus, inequalities (15.27) and (15.31) imply

$$\|y'(t)\|_H^2 + \|y(t)\|_V^2 \leq c_1(\|y'(s)\|_H^2 + \|y(s)\|_V^2)e^{-c_2(t-s)} + \rho\int_s^t \eta(\xi)e^{c_2(\xi-t)}d\xi,$$

for each $t \geq s \geq 0$, that is, inequality (15.23) holds.

Corollary 15.1 *Let assumptions of Theorem 15.1 hold. Then, there exists $R > 0$ such that*

$$\|\varphi(t)\|_E^2 \leq c_1\|\varphi(s)\|_E^2 e^{-c_2(t-s)} + R, \tag{15.32}$$

for each $\varphi \in \mathscr{K}^+$, and $t \geq s \geq 0$, where constants $c_1 > 1$ and $c_2 > 0$ are defined in (15.24).

Proof Assumptions (H_2) and (A_1) imply that the function ψ defined in (15.24) is t.u.i. because the family of t.u.i. nonnegative real functions form a convex cone in the space $L_1^{\mathrm{loc}}(0,\infty)$. Therefore, according to (15.8), there exists $K > 0$ such that

$$\sup_{t\geq 0}\int_t^{t+1} \psi(s)\chi_{\{\psi(s)\geq K\}}ds \leq 1. \tag{15.33}$$

Therefore, inequality (15.23) implies that

$$\begin{aligned}
\|\varphi(t)\|_E^2 \leq & c_1\|\varphi(s)\|_E^2 e^{-c_2(t-s)} + e^{c_2(1-t)}\sup_{h\geq 0}\int_0^1 \psi(\xi+h)d\xi \\
& + e^{c_2(2-t)}\sup_{h\geq 0}\int_1^2 \psi(\xi+h)d\xi + \cdots + e^{c_2([t]-t)}\sup_{h\geq 0}\int_{[t]-1}^{[t]} \psi(\xi+h)d\xi \\
& + e^{c_2([t]+1-t)}\sup_{h\geq 0}\int_{[t]}^t \psi(\xi+h)d\xi \\
\leq & c_1\|\varphi(s)\|_E^2 e^{-c_2(t-s)} + \frac{e^{c_2}}{e^{c_2}-1}\sup_{\zeta\geq 0}\int_\zeta^{\zeta+1} \psi(\xi)d\xi \\
\leq & c_1\|\varphi(s)\|_E^2 e^{-c_2(t-s)} + \frac{e^{c_2}}{e^{c_2}-1}(K+1),
\end{aligned}$$

for each $\varphi \in \mathscr{K}^+$, and $t \geq s \geq 0$, that is, inequality (15.32) holds with $R := \dfrac{e^{c_2}}{e^{c_2}-1}(K+1)$.

Lemma 15.5 *Let $\{(y_n, y_n')^T\}_{n\geq 1} \subset \mathscr{K}^+$ be a sequence such that $(y_n(0), y_n'(0)) \to (y_0, y_0')$ weakly in $V \times H$ for some $(y_0, y_0') \in V \times H$. Then there exist a subsequence*

$\{y_{n_k}\}_{k\geq 1} \subset \{y_n\}_{n\geq 1}$ *and an element* $(y, y')^T \in \mathscr{K}^+$ *such that* $(y(0), y'(0))^T = (y_0, y_0')^T$, $y_n' \to y'$ *in* $C^{\mathrm{loc}}([0, \infty); H)$ *and for each finite time interval* $[\tau, T] \subset [0, \infty)$*:*

$$y_{n_k}(t) \to y(t) \text{ weakly in } V, \text{ uniformly on } [\tau, T], \tag{15.34}$$

as $k \to \infty$*. If, additionally,* $(y_{n_k}(0), y_{n_k}'(0))^T \to (y_0, y_0')^T$ *strongly in E as* $k \to \infty$*, then* $(y_{n_k}, y_{n_k}')^T \to (y, y')^T$ *strongly in* $C^{\mathrm{loc}}([0, \infty); E)$ *as* $k \to \infty$.

Proof Let $\{(y_n, y_n')^T\}_{n\geq 1} \subset \mathscr{K}^+$ be an arbitrary sequence such that $(y_n(0), y_n'(0)) \to (y_0, y_0')$ weakly in $V \times H$ for some $(y_0, y_0') \in V \times H$. Corollary 15.1 implies the boundedness of the sequence $\{(y_n, y_n')^T\}_{n\geq 1}$ in $L_\infty([0, \infty); E)$. Since $\mathscr{K}_\cup^+$ is a dense set in a Polish space $\mathscr{K}^+$ endowed with the topology induced from $C^{\mathrm{loc}}([0, \infty); E)$, then for each $n \geq 1$ there exists $(u_n, u_n')^T \in \mathscr{K}_\cup^+$ such that

$$\rho_{C^{\mathrm{loc}}([0,\infty); E)}((y_n, y_n')^T, (u_n, u_n')^T) \leq \frac{1}{n}, \text{ for each } n \geq 1. \tag{15.35}$$

Therefore, the rest of the proof establishes the statements of the lemma for the sequence $\{(u_n, u_n')^T\}_{n\geq 1} \subset \mathscr{K}_\cup^+$. For each $n \geq 1$ formula (15.18) provides the existence of $\tau_n, h_n \geq 0$ and $(z_n, z_n')^T \in \mathscr{K}_{\tau_n}^+$ such that $u_n(\cdot) = T(h_n)z_n(\cdot + \tau_n)$. Therefore, according to the definition of the global weak solution of Problem (15.11), for each $n = 1, 2, \ldots$ there exists $d_n \in L_2^{\mathrm{loc}}(0, \infty; V^*)$ such that

$$d_n(t) = g_n(t) - u_n''(t) - B_0 u_n(t) \in A_0(t + \tau_n + h_n, u_n'(t)) \text{ for a.e. } t > 0. \tag{15.36}$$

where $g_n(\cdot) := T(h_n) f_0(\cdot + \tau_n)$. Note that $(u_n, u_n')^T \in C([0, \infty); E)$, for each $n = 1, 2, \ldots$.

According to Corollary 15.1 and Assumption (A_1), for each $T > 0$ there exists $C_T > 0$ such that

$$\|u_n'\|_{X_{0,T}} + \|u_n''\|_{X_{0,T}^*} + \|u_n'\|_{C([0,T]; H)} + \|d_n\|_{X_{0,T}^*} + \|u_n\|_{C([0,T]; V)} \leq C_T, \tag{15.37}$$

for each $n = 1, 2, \ldots$. Setting

$$v_n(t) = \int_0^t u_n'(s) ds,$$

for each $t \geq 0$ and $n = 1, 2, \ldots$, we note that

$$u_n(t) = v_n(t) + u_n(0), \tag{15.38}$$

and

$$\|v_n(t) - v_n(s)\|_V < C|t - s|^{\frac{1}{2}}, \; v_n(0) = \bar{0}, \tag{15.39}$$

for each $n = 1, 2, \ldots,$ and $t, s \geq 0$. Therefore, according to (15.36)–(15.39), Assumptions (A_2), (H_2), the continuity of the embedding $W^{\mathrm{loc}}([0, \infty)) \subset C^{\mathrm{loc}}([0, \infty); H)$, the compactness of the embedding $W^{\mathrm{loc}}([0, \infty)) \subset L_2^{\mathrm{loc}}(0, \infty; H)$, the reflexivity of spaces $W_{0,T}$, $X_{0,T}$, $X_{0,T}^*$ for each $T > 0$, there exist a subsequence $\{u_{n_k}, d_{n_k}\}_{k=1,2,\ldots} \subset \{u_n, d_n\}_{n=1,2,\ldots}$ and elements $u \in C^{\mathrm{loc}}([0, \infty); V)$, $u' \in W^{\mathrm{loc}}([0, \infty))$, $d, g \in L_2^{\mathrm{loc}}(0, \infty; V^*)$ such that the following convergences hold:

$$
\begin{array}{ll}
v_{n_k} \to v \text{ in } C^{\mathrm{loc}}([0, \infty); V), & u_{n_k}(t) \to u(t) \text{ weakly in } V, \ \forall t \geq 0, \\
u'_{n_k} \to u' \text{ weakly in } L_2^{\mathrm{loc}}(0, \infty; V), & u''_{n_k} \to u'' \text{ weakly in } L_2^{\mathrm{loc}}(0, \infty; V^*), \\
d_{n_k} \to d \text{ weakly in } L_2^{\mathrm{loc}}(0, \infty; V^*), & u'_{n_k} \to u' \text{ weakly in } C^{\mathrm{loc}}([0, \infty); H), \\
u'_{n_k} \to u' \text{ in } L_2^{\mathrm{loc}}(0, \infty; H), & u'_{n_k}(t) \to u'(t) \text{ in } H \text{ for a.e. } t > 0, \\
g_{n_k} \to g \text{ in } L_2^{\mathrm{loc}}(0, \infty; V^*), &
\end{array}
\tag{15.40}
$$

as $k \to \infty$, where

$$
v(\cdot) = u(\cdot) - u(0), \quad u(0) = y_0, \ u'(0) = y'_0. \tag{15.41}
$$

Therefore,

$$
(u, u')^T \in \mathscr{K}^+, \tag{15.42}
$$

Moreover, (15.34) holds for each finite time interval $[\tau, T] \subset [0, \infty)$ (see [46, 47]).

To complete the proof we additionally assume that $(u_{n_k}(0), u'_{n_k}(0))^T \to (y_0, y'_0)^T$ strongly in E as $k \to \infty$, and prove that $(u_{n_k}, u'_{n_k})^T \to (u, u')^T$ strongly in $C^{\mathrm{loc}}([0, \infty); E)$ as $k \to \infty$. For this purpose it is sufficient to verify that

$$
u'_{n_k} \to u' \text{ strongly in } C^{\mathrm{loc}}([0, \infty); H), \tag{15.43}
$$

as $k \to \infty$ because $v_{n_k} \to v$ strongly in $C^{\mathrm{loc}}([0, \infty); V)$, $u_{n_k}(0) \to y_0$ strongly in V as $k \to \infty$, and (15.38), (15.41) hold.

On the contrary, if (15.43) does not hold, there exist constants $T, L, \varepsilon > 0$ and a subsequence $\{u_{k_j}\}_{j=1,2,\ldots} \subset \{u_{n_k}\}_{k=1,2,\ldots}$ such that

$$
\max_{[0,T]} \|u'_{k_j}(t) - u'(t)\|_H = \|u'_{k_j}(t_j) - u'(t_j)\|_H \geq L,
$$

for each $j = 1, 2, \ldots$ and some $\{t_j\}_{j \geq 1} \subset [0, T]$. Without loss of the generality we assume that $t_j \to t_0$ as $j \to \infty$ for some $t_0 \in [0, T]$, and, therefore,

$$
\varliminf_{j \to \infty} \|u'_{k_j}(t_j) - u'(t_0)\|_H \geq L, \tag{15.44}
$$

because $u' \in C^{\mathrm{loc}}([0, \infty); H)$.

On the other hand, since $u'_{n_k} \to u'$ weakly in $C^{\mathrm{loc}}([0, \infty); H)$ as $k \to \infty$, then

$$u'_{k_j}(t_j) \to u'(t_0) \text{ weakly in } H,\ \ j \to \infty. \tag{15.45}$$

If, additionally,

$$\overline{\lim_{j\to\infty}}\ \|u'_{k_j}(t_j)\|_H \le \|u'(t_0)\|_H, \tag{15.46}$$

then $u'_{k_j}(t_j) \to u'(t_0)$ strongly in H as $j \to \infty$ which contradicts to (15.44). Therefore, to finish the proof of (15.43) it is sufficient to establish inequality (15.46). Indeed, according to (15.17) and (15.36), the real functions $\{J_j\}_{j\ge1}$,

$$J_j(t) = \|u'_{k_j}(t)\|_H^2 + \|u_{k_j}(t)\|_V^2 - \frac{1}{\alpha}\int_0^t \|g_{k_j}(\xi)\|_{V^*}^2 d\xi \tag{15.47}$$

$t \in [0, T]$, are nonincreasing and continuous. Moreover, the real function J,

$$J(t) = \|u'(t)\|_H^2 + \|u(t)\|_V^2 - \frac{1}{\alpha}\int_0^t \|g(\xi)\|_{V^*}^2 d\xi, \tag{15.48}$$

$t \in [0, T]$ is nonincreasing and continuous. Indeed, (15.19) and (15.42) implies the existence of a sequence $\{(w_n, w'_n)^T\}_{n\ge1} \subset \mathcal{K}_U^+$ that converges to $(u, u')^T$ strongly in $C^{\text{loc}}([0,\infty); E)$. According to (15.18), for each $n \ge 1$ there exist $\tilde{\tau}_n, \tilde{h}_n \ge 0$ and a global weak solution $(\tilde{w}, \tilde{w}')^T$ of Problem (15.11) on $[\tilde{\tau}_n, \infty)$ such that $w_n(\cdot) = T(\tilde{h}_n)\tilde{w}_n(\cdot + \tilde{\tau}_n)$. Therefore, (15.17) implies that

$$\|w'_n(t)\|_H^2 + \|w_n(t)\|_V^2 - \frac{1}{\alpha}\int_0^t \|T(\tilde{h}_n)f(\xi + \tilde{\tau}_n)\|_{V^*}^2 d\xi$$
$$\le \|w'_n(s)\|_H^2 + \|w_n(s)\|_V^2 - \frac{1}{\alpha}\int_0^s \|T(\tilde{h}_n)f_0(\xi + \tilde{\tau}_n)\|_{V^*}^2 d\xi,$$

for each $t \ge s \ge 0$. Therefore, the strong convergence of $\{(w_n, w'_n)^T\}_{n\ge1}$ to $(u, u')^T$ in $C^{\text{loc}}([0,\infty); E)$ and Assumption (H_2) imply that $J(t) \le J(s)$ for each $t \ge s \ge 0$. The continuity of J follows from (15.42). Moreover, (15.40) and Assumption (H_2) imply that

$$J_j(t) \to J_0(t), \text{ as } j \to \infty, \tag{15.49}$$

for a.e. $t \in (0, T)$. Therefore, according to [24],

$$\overline{\lim_{j\to\infty}}\ J_j(t_j) \le J(t_0).$$

Therefore, according to (15.40), we obtain (15.46), that is, (15.43) holds.

Let $\beta(E)$ denotes the family of all nonempty bounded subsets of E.

Proof (*Proof of Theorem* 15.1) According to [31, Theorem 3], the m-semiflow G defined in (15.21) has a compact global attractor $\mathfrak{R}$ in the phase space E, if the following properties hold:

(i) the ball $B_0 = \{u \in E : \|u\|_E \leq \sqrt{R} + 1\}$ is an absorbing set, that is, for each $B \in \beta(E)$ there exists $T(B) > 0$ such that $G(t, B) \subset B_0$ for each $t \geq T(B)$, where $R > 0$ is a positive constant from Corollary 15.1;

(ii) for each $t \geq 0$ the mapping $G(t, \cdot) : E \to \beta(E)$ is upper semi-continuous and takes nonempty compact values;

(iii) G is asymptotically upper semicompact m-semiflow, that is, if $\forall B \in \beta(E)$ such that for some $T(B) \in \mathbb{R}_+$, $\gamma^+_{T(B)}(B) = \bigcup_{\tau \geq T(B)} G(\tau, B) \in \beta(E)$, any sequence $\xi_n \in G(t_n, B)$, $t_n \to \infty$, is precompact in E.

Note that property (i) implies that m-semiflow G is pointwise dissipative, that is, the bounded absorbing set B_0 attracts each point $\varphi_0 \in E$; see [31] for details. Moreover, the set $\cup_{t \geq 0} G(t, B)$ is bounded in E for each $B \in \beta(E)$. We remark also, a strict compact-valued mapping acting in real separable Hilbert space is upper semi-continuous if and only if its graphic is a closed set.

Let us establish properties (i)–(iii). Indeed, (i) and (ii) directly follow from Corollary 15.1 and Lemma 15.5 respectively. Therefore, to finish the proof it is sufficient to verify the asymptotic upper semi-compactness of the m-semiflow G.

Let $0 < t_1 < t_2 < \cdots < t_n < t_{n+1} \to \infty$ as $n \to \infty$, $B \in \beta(E)$, $\{v_n\}_{n \geq 1} \subset B$, and $\xi_n \in G(t_n, v_n)$, $n = 1, 2, \ldots$, be arbitrary fixed. To prove that the sequence $\{\xi_n\}_{n=1,2,\ldots}$ has a convergent subsequence in E we note that Corollary 15.1 and Lemma 15.5 imply the existence of a subsequence $\{\xi_{n_k}\}_{k \geq 1} \subset \{\xi_n\}_{n \geq 1}$ and an element $\xi \in E$ such that

$$\xi_{n_k} \to \xi \text{ weakly in } E, \text{ and } \|\xi_{n_k}\|_E \to a \geq \|\xi\|_E, \text{ as } k \to \infty. \tag{15.50}$$

If, additionally,

$$a \leq \|\xi\|_E, \tag{15.51}$$

then (15.50)–(15.51) imply that $\xi_{n_k} \to \xi$ strongly in E as $k \to \infty$.

Let us prove that (15.51) holds. Indeed, let us fix an arbitrary $T_0 > \sqrt{\lambda_1}$, where $\lambda_1 > 0$ is the constant from (15.26). Then for sufficiently large $k = 1, 2, \ldots$

$$G(t_{n_k}, v_{n_k}) \subset G(T_0, G(t_{n_k} - T_0, v_{n_k})) \subset G(T_0, B_0).$$

Therefore, $\xi_{n_k} \in G(T_0, \beta_{n_k})$, where $\beta_{n_k} \in G(t_{n_k} - T_0, v_{n_k})$ and $\|\beta_{k_j}\|_E^2 \leq R + 1$, $j = 1, 2, \ldots$; see (15.32) for details. According to Corollary 15.1, Lemma 15.5, and (15.50), there exist a subsequence $\{\xi_{k_j}, \beta_{k_j}\}_{j=1,2,\ldots} \subset \{\xi_{n_k}, \beta_{n_k}\}_{k=1,2,\ldots}$ and an element $\beta_{T_0} \in E$ such that

$$\xi \in G(T_0, \beta_{T_0}), \quad \beta_{k_j} \to \beta_{T_0} \text{ weakly in } E, \quad j \to \infty. \tag{15.52}$$

The definition of G implies the existence of a sequence $\{y_j\}_{j \geq 0} \subset \mathscr{K}_+$ such that

$$\xi_{k_j} = \begin{pmatrix} y_j(T_0) \\ y_j'(T_0) \end{pmatrix}, \ \beta_{k_j} = \begin{pmatrix} y_j(0) \\ y_j'(0) \end{pmatrix}, \ \xi = \begin{pmatrix} y_0(T_0) \\ y_0'(T_0) \end{pmatrix}, \ \beta_{T_0} = \begin{pmatrix} y_0(0) \\ y_0'(0) \end{pmatrix},$$

for each $j = 1, 2, \ldots$. Since $\mathcal{K}_{\cup}^{+}$ is the dense set in a Polish space $\mathcal{K}^{+}$ endowed with the topology induced from $C^{\text{loc}}([0,\infty); E)$, then for each $j \geq 1$ there exists $(u_j, u_j')^T \in \mathcal{K}_{\cup}^{+}$ such that

$$\rho_{C^{\text{loc}}([0,\infty);E)}((y_j, y_j')^T, (u_j, u_j')^T) \leq \frac{1}{j}, \text{ for each } j \geq 1. \tag{15.53}$$

Therefore, the rest of the proof establishes the statements of the lemma for the sequence $\{(u_j, u_j')^T\}_{j\geq 1} \subset \mathcal{K}_{\cup}^{+}$. For each $j \geq 1$ formula (15.18) provides the existence of $\tau_j, h_j \geq 0$ and $(z_j, z_j')^T \in \mathcal{K}_{\tau_j}^{+}$ such that $u_j(\cdot) = T(h_j)z_j(\cdot + \tau_j)$. Therefore, according to the definition of the global weak solution of Problem (15.11), for each $j = 1, 2, \ldots$ there exists $d_j \in L_2^{\text{loc}}(0,\infty; V^*)$:

$$d_j(t) = g_j(t) - u_j''(t) - B_0 u_j(t) \in A_0(t + \tau_j + h_j, u_j'(t)) \text{ for a.e. } t > 0. \tag{15.54}$$

where $g_j(\cdot) := T(h_j) f_0(\cdot + \tau_j)$. Note that $(u_j, u_j')^T \in C([0,\infty); E)$, for each $j = 0, 1, \ldots$. Similarly to the proof of Lemma 15.5, without loss of generality we claim that

$$\begin{array}{ll}
v_j \to v_0 \text{ in } C^{\text{loc}}([0,\infty); V), & u_j(t) \to y_0(t) \text{ weakly in } V,\ \forall t \geq 0, \\
u_j' \to y_0' \text{ weakly in } L_2^{\text{loc}}(0,\infty; V), & u_j'' \to y_0'' \text{ weakly in } L_2^{\text{loc}}(0,\infty; V^*), \\
d_j \to d_0 \text{ weakly in } L_2^{\text{loc}}(0,\infty; V^*), & u_j' \to y_0' \text{ weakly in } C^{\text{loc}}([0,\infty); H), \\
u_j' \to y_0' \text{ in } L_2^{\text{loc}}(0,\infty; H), & u_j'(t) \to y_0'(t) \text{ in } H \text{ for a.e. } t > 0, \\
g_j \to g_0 \text{ in } L_2^{\text{loc}}(0,\infty; V^*), & u_j(t) \to y_0(t) \text{ weakly in } V, \text{ uniformly on } [0, T_0], \\
T(h_j)c_j(\cdot + \tau_j) \to \bar{c}(\cdot), &
\end{array} \tag{15.55}$$

as $k \to \infty$, where $v_0(\cdot) = y_0(\cdot) - y_0(0)$, $\bar{c}(\cdot)$ is t.u.i. in $L_1^{\text{loc}}(0,\infty)$, $g_0 \in L_2^{\text{loc}}(0,\infty; V^*)$ is univocally translation compact in $L_2^{\text{loc}}(0,\infty; V^*)$, and d_0 satisfies inequalities from Assumptions (A_1) and (A_2) with $y(\cdot) = y_0(\cdot)$ and $c(\cdot) = \hat{c}(\cdot)$. Let us fix an arbitrary $\varepsilon \in (0, \sqrt{\lambda_1})$ and set:

$$Y_j(t) = \frac{1}{2}\left[\|y_j(t)\|_V^2 + \|y_j'(t)\|_H^2\right] + \varepsilon(y_j'(t), y_j(t)), \quad t \in [0, T_0],\ j \geq 0.$$

Repeating several lines of the proof of Theorem 2 from [43] (see also [46, pp. 89–92]), we obtain that for each $j \geq 0$ and for a.e. $t \in (0, T_0)$

$$\begin{aligned}
Y_j(T_0) &= Y_j(0)e^{-2\varepsilon T_0} + 2\varepsilon \int_0^{T_0} \|y_j'(t)\|_H^2 e^{-2\varepsilon(T_0-t)} dt \\
&\quad - \int_0^{T_0} \langle d_j(t), y_j'(t)\rangle_V e^{-2\varepsilon(T_0-t)} dt - \varepsilon \int_0^{T_0} \langle d_j(t), y_j(t)\rangle_V e^{-2\varepsilon(T_0-t)} dt \\
&\quad + 2\varepsilon^2 \int_0^{T_0} (y_j'(t), y_j(t)) e^{-2\varepsilon(T_0-t)} dt.
\end{aligned} \tag{15.56}$$

Moreover,

$$2\varepsilon\int_0^{T_0}\|y_j'(t)\|_H^2 e^{-2\varepsilon(T_0-t)}dt \to 2\varepsilon\int_0^{T_0}\|y_0'(t)\|_H^2 e^{-2\varepsilon(T_0-t)}dt, \quad j\to\infty; \tag{15.57}$$

$$\int_\varepsilon^{T_0}\langle d_j(t), y_j'(t)\rangle_V e^{-2\varepsilon(T_0-t)}dt \to \int_\varepsilon^{T_0}\langle d_0(t), y_0'(t)\rangle_V e^{-2\varepsilon(T_0-t)}dt, \quad j\to\infty; \tag{15.58}$$

$$\forall j\geq 0 \quad \left|\int_0^{\varepsilon}\langle d_j(t), y_j'(t)\rangle_V e^{-2\varepsilon(T_0-t)}dt\right| \leq c(1+\bar{R})\bar{R}e^{-2\varepsilon(T_0-\varepsilon)}\varepsilon. \tag{15.59}$$

In particular, Assumption (A_3) implies that for each $j\geq 1$ exists $z_j\in L_2(0,T_0;Z^*)$ such that $d_j(\cdot)=A_1y_j'(\cdot)+z_j(\cdot)$. According to Assumption (A_3), and (15.57), we obtain that $y_j\to y_0$ in $L_2(0,T_0;Z)$, $z_j\to z_0$ weakly in $L_2(0,T_0;Z^*)$, $j\to\infty$. Therefore, for some $\hat{R},\tilde{R}>0$

$$\begin{aligned}\overline{\lim_{j\to\infty}}&\left(-2\varepsilon\int_0^{T_0}\langle d_j(t), y_j(t)\rangle_V e^{-2\varepsilon(T_0-t)}dt\right)\\ &\leq -\varepsilon\int_0^{T_0}\langle d_0(t), y_0(t)\rangle_V e^{-2\varepsilon(T_0-t)}dt + \|A_1\|_{L(V;V^*)}(\tilde{R}^2e^{-2\varepsilon T_0}+2\bar{R}^2)\varepsilon;\end{aligned} \tag{15.60}$$

$$2\varepsilon^2\left|\int_0^{T_0}(y_j'(t), y_j(t))e^{-2\varepsilon(T_0-t)}dt\right| \leq \frac{\varepsilon}{2\sqrt{\lambda_1}}\bar{R}^2. \tag{15.61}$$

Finally, from (15.27), (15.56)–(15.61), according to [46, p. 92], if we pass $T_0\to\infty$ and, then, $\varepsilon\to 0+$, we obtain (15.51).

15.4 Auxiliary Properties of the Global Attractor in the Autonomous Case

Let A_0 do not depend on t. Assume that $j\equiv 0$, $Z=H$, and $f_0\in H$. Then each weak global solution $(u,u')^T$ became regular, that is, $u'(\cdot)\in C([\tau+\varepsilon,T];V)\cap L^2(\tau+\varepsilon,T;D\cap V)$ and $u''(\cdot)\in L^2(\tau+\varepsilon,T;H)$, for each $0\leq\tau<T<\infty$ and $\varepsilon\in(0,T-\tau)$, where

$$D:=\{w\in V\ :\ B_0w\in H\}$$

is endowed with the natural respective norm and inner product; cf. [10, 15, 25, 39, 47]. Under such assumptions the global attractor from Theorem 15.1 satisfy respective regularity (more smooth) conditions.

We say that the complete trajectory $\varphi\in\mathscr{K}$ is stationary if $\varphi(t)=z$ for all $t\in\mathbb{R}$ for some $z\in E$. We note that the m-semiflow G coincides with the usual multivalued semigroup generated by all weak solutions in the autonomous case. Following [1, p.486] we denote the set of rest points of G by $Z(G)$. Note that

$$Z(G) = \{(z, \bar{0}) |\; z \in B_0^{-1}(f_0 - A_0(\bar{0})\}.$$

Thus, $Z(G)$ is a convex, nonempty, weakly compact in $V \times V$ set.

For investigating the trajectory behaviour of solutions on the attractor $\Re$ it is necessary to consider similar definitions to [1, p. 486]:

Definition 15.1 A functional $\mathscr{V} : \Re \to \mathbb{R}$ is a Lyapunov type function for G on $\Re$ provided
(i) $\mathscr{V}$ is continuous;
(ii) $\mathscr{V}(\varphi(t)) \le \mathscr{V}(\varphi(s))$ whenever $\varphi \in \mathscr{K}$ and $t \ge s \ge 0$;
(iii) if $\mathscr{V}(\psi(t)) = constant$ for some $\psi \in \mathscr{K}$ and all $t \in \mathbb{R}$, then ψ is stationary.

As a consequence of the following theorem in the presence of a Lyapunov function the behavior of such complete orbits can be characterized.

Theorem 15.2 *Suppose that there exists a Lyapunov type function $\mathscr{V}$ for G on $\Re$. Then for each $\psi \in \mathscr{K}$ the limit sets*

$$\alpha(\psi) = \{z \in E |\; \psi(t_j) \to z \textit{ for some sequence } t_j \to -\infty\},$$

$$\omega(\psi) = \{z \in E |\; \psi(t_j) \to z \textit{ for some sequence } t_j \to \infty\}$$

are connected subsets of $Z(G)$ on which $\mathscr{V}$ is constant.

Remark 15.1 We note that the Lyapunov type function exists if $j \equiv 0$ and damping is sufficiently large.

Proof (*Proof of Theorem* 15.2) The proof follows from the proof of [1, Theorem 5.1], the asymptotic compactness of G and from the properties of solutions of problem (15.11).

15.5 Applications

As an application we consider the hemivariational inequality of hyperbolic type with multidimensional "reaction-velocity" law [32] (see Problem (15.1)–(15.6)). From the results of previous sections it follows the next result.

Corollary 15.2 *The m-semiflow G constructed on all generalized solutions of (15.1)–(15.6) has the compact global attractor $\Re$.*

If, for example, $j \equiv 0$ and the damping is sufficiently large, then there exists a Lyapunov type function $\mathscr{V} : \Re \to \mathbb{R}$ for G on $\Re$. Therefore, from Theorem 15.2 we obtain the following result.

Corollary 15.3 *For each $\psi \in \mathcal{K}$ the limit sets*

$$\alpha(\psi) = \{z \in E \mid \psi(t_j) \to z \text{ for some sequence } t_j \to -\infty\},$$

$$\omega(\psi) = \{z \in E \mid \psi(t_j) \to z \text{ for some sequence } t_j \to \infty\}$$

are connected subsets of $Z(G)$ on which $\mathcal{V}$ is constant.

Acknowledgements This research was partially supported by the National Academy of Sciences of Ukraine, Grant No. 10-02/05-2020.

References

1. Ball, J.M.: Continuity properties and global attractors of generalized semiflows and the Navier-Stokes equations. J. Nonlinear Sci. (1997). https://doi.org/10.1007/s003329900037
2. Chepyzhov, V.V., Vishik, M.I.: Evolution equations and their trajectory attractors. J. Math. Pures Appl. (1997). https://doi.org/10.1016/S0021-7824(97)89978-3
3. D'Apice, C., Kogut, P.I., Manzo, R.: On relaxation of state constrained optimal control problem for a PDE-ODE model of supply chains. Netw. Heterog. Media (NHM) **9**(3), 501–518 (2014). https://doi.org/10.3934/nhm.2014.9.501
4. D'Apice, C., D'Arienzo, M.P., Kogut, P.I., Manzo, R.: On boundary optimal control problem for an arterial system: existence of feasible solutions. J. Evol. Equ. **18**(4), 1745–1786 (2018). https://doi.org/10.1007/s00028-018-0460-4
5. D'Apice, C., Manzo, R., Piccoli, B.: Optimal input flows for a PDE-ODE model of supply chains. Commun. Math. Sci. **10**(36), 1226–1240 (2012). https://doi.org/10.4310/CMS.2012.v10.n4.a10
6. D'Apice, C., Manzo, R., Piccoli, B.: Numerical schemes for the optimal input flow of a supply-chain. SIAM J. Numer. Anal. (SINUM) **51**(5), 2634–2650 (2013). https://doi.org/10.1137/120889721
7. D'Apice, C., Manzo, R., Piccoli, B.: Existence of solutions to Cauchy problems for a mixed continuum-discrete model for supply chains and networks. J. Math. Anal. Appl. **362**(2), 374–386 (2010). https://doi.org/10.1016/j.jmaa.2009.07.058
8. Gajewski, H., Gröer, K., Zacharias, K.: Nichtlineare operatorgleichungen und operatordifferentialgleichungen. Akademie, Berlin (1974)
9. Goeleven, D., Miettinen, M., Panagiotopoulos, P.D.: Dynamic hemivariational inequalities and their applications. J. Optim. Theory Appl. (1999). https://doi.org/10.1023/A:1021783924105
10. Gluzman, M.O., Gorban, N.V., Kasyanov, P.O.: Lyapunov type functions for classes of autonomous parabolic feedback control problems and applications. Appl. Math. Lett. **39**, 19–21 (2015). https://doi.org/10.1016/j.aml.2014.08.006
11. Gorban, N.V., Kapustyan, O.V., Kasyanov, P.O., Paliichuk, L.S.: On global attractors for autonomous damped wave equation with discontinuous nonlinearity. Solid Mechanics and Its Applications, vol. 211, pp. 221–237 (2014)
12. Gorban, N.V., Kapustyan, O.V., Kasyanov, P.O.: Uniform trajectory attractor for non-autonomous reaction-diffusion equations with Caratheodory's nonlinearity. Nonlinear Anal. **98**, 13–26 (2014). https://doi.org/10.1016/j.na.2013.12.004
13. Gorban, N.V., Kapustyan, A.V., Kapustyan, E.A., Khomenko, O.V.: Strong global attractor for the three-dimensional Navier-Stokes system of equations in unbounded domain of channel type. J. Autom. Inf. Sci. **47**, 48–59 (2015)
14. Kapustyan, A.V., Pankov, A.V., Valero, J.: On global attractors of multivalued semiflows generated by the 3D Benard system. Set-Valued Var. Anal. **20**(3), 445–465 (2012)

15. Kapustyan, O.V., Kasyanov, P.O., Valero, J.: Structure and regularity of the global attractor of a reaction-diffusion equation with non-smooth nonlinear term. Discret. Contin. Dyn. Syst. **34**(10), 4155–4182 (2014). https://doi.org/10.3934/dcds.2014.34.4155
16. Kapustyan, E.A., Nakonechnyi, A.G.: Optimal bounded control synthesis for a parabolic boundary-value problem with fast oscillatory coefficients. J. Autom. Inf. Sci. **31**(12) (1999). https://doi.org/10.1615/JAutomatInfScien.v31.i12.50
17. Kapustyan, E.A., Nakonechnyi, A.G.: The minimax problems of pointwise observation for a parabolic boundary-value problem. J. Autom. Inf. Sci. **34**(5) (2002). https://doi.org/10.1615/JAutomatInfScien.v34.i5.60
18. Kapustian, O.A.: Approximate optimal regulator for distributed control problem with superposition functional and rapidly oscillating coefficients. Modern Mathematics and Mechanics, pp. 481–491 (2019)
19. Kapustian, O.A., Nakonechnyi, O.G., Podlipenko, Y.K.: Minimax estimation of solutions of the first order linear hyperbolic systems with uncertain data. Stat. Optim. Inf. Comput. **7**(4), 695–708 (2019)
20. Kapustian, O.A., Sobchuk, V.V.: Approximate homogenized synthesis for distributed optimal control problem with superposition type cost functional. Stat. Optim. Inf. Comput. **6**(2), 233–239 (2018)
21. Kapustyan, O.V., Kapustyan, O.A., Sukretna, A.V.: Approximate bounded synthesis for one weakly nonlinear boundary-value problem. Nonlinear Oscil. **3**(12), 297–304 (2009)
22. Kapustyan, O.V., Kapustian, O.A., Sukretna, A.V.: Approximate Bounded Synthesis for Distributed Systems. LAP LAMBERT Academic Publishing, Saarbrücken (2013), 236 p.
23. Kapustyan, V.O., Pyshnograiev, I.O., Kapustian, O.A.: Quasi-optimal control with a general quadratic criterion in a special norm for systems described by parabolic-hyperbolic equations with non-local boundary conditions. Discret. Contin. Dyn. Syst.-B, **24**(3) (2019). https://doi.org/10.3934/dcdsb.2019014
24. Kasyanov, P.O.: Multivalued dynamics of solutions of autonomous operator differential equations with pseudomonotone nonlinearity. Math. Notes **92**(1–2), 205–218 (2012)
25. Kasyanov, P.O., Toscano, L., Zadoianchuk, N.V.: Regularity of weak solutions and their attractors for a parabolic feedback control problem. Set-Valued Var. Anal. (2013). https://doi.org/10.1007/s11228-013-0233-8
26. Kogut, P.I., Manzo, R., Putchenko, A.O.: On approximate solutions to the Neumann elliptic boundary value problem with non linearity of exponential type. Bound. Value Probl. **2016**:208, 1–32 (2016). https://doi.org/10.1186/s13661-016-0717-1
27. Kupenko, O.P., Manzo, R.: Approximation of an optimal control problem in the coefficient for variational inequality with anisotropic p-Laplacian. Nonlinear Differ. Equ. Appl. NoDEA **23**(3), 18 (2016). Article 35. https://doi.org/10.1007/s00030-016-0387-9
28. Kupenko, O.P.: On an optimal control problem in coefficients for linear elliptic variational inequality. Abstr. Appl. Anal. **2013** (2013). Article ID 821964, 13 p. https://doi.org/10.1155/2013/821964
29. Kupenko, O.P., Manzo, R.: On optimal controls in coefficients for ill-posed non-linear elliptic Dirichlet bounday value problems. Discret. Contin. Dyn. Syst. Ser. B **23**(4), 1363–1393 (2018). https://doi.org/10.3934/dcdsb.2018155
30. Manzo, R.: On Nuemann boundary control problem for ill-posed strongly nonlinear elliptic equation with p-Laplace operator and type of nonlinearity. Ric. Mat. **68**(2), 769–802 (2019). https://doi.org/10.1007/s11587-019-00439-x
31. Melnik, V.S., Valero, J.: On attractors of multivalued semi-flows and differential inclusions. Set-Valued Anal. (1998). https://doi.org/10.1023/A:1008608431399
32. Migórski, S.: Boundary hemivariational inequalities of hyperbolic type and applications. J. Glob. Optim. (2005). https://doi.org/10.1007/s10898-004-7021-9
33. Nakonechnyi, A.G., Kapustyan, E.A., Chikriy, A.A.: Control of impulse systems in conflict situation. J. Autom. Inf. Sci. **51**(9) (2019). https://doi.org/10.1615/JAutomatInfScien.v51.i9.10

34. Nakonechnyi, O.G., Kapustia, O.A., Chikrii, A.O.: Approximate guaranteed mean square estimates of functionals on solutions of parabolic problems with fast oscillating coefficients under nonlinear observations. Cybern. Syst. Anal. **55**(5), 785–795 (2019)
35. Naniewicz, Z., Panagiotopoulos, P.D.: Mathematical Theory of Hemivariational Inequalities and Applications. Marcel Dekker Inc., New York (1995)
36. Panagiotopoulos, P.D.: Inequality Problems in Mechanics and Applications. Convex and Nonconvex Energy Functions. Birkhauser, Basel (1985)
37. Panagiotopoulos, P.D.: Hemivariational inequalities and Fan-variational inequalities. New applications and results. Atti Sem. Mat. Fis. Univ. Modena **XLIII**, 159–191 (1995)
38. Stanzhitskii, A., Dobrodzii, T.: Study of optimal control problems on the half-line by the averaging method. Differ. Equ. **47**(2), 264–277 (2011)
39. Valero, J., Gimenez, A., Kapustyan, O.V., Kasyanov, P.O., Amigo, J.M.: Convergence of equilibria for numerical approximations of a suspension model. Comput. Math. Appl. (2016). https://doi.org/10.1016/j.camwa.2016.05.034
40. Vishik, M., Chepyzhov, V.V.: Trajectory and global attractors of three-dimensional Navier-Stokes systems. Math. Notes (2002). https://doi.org/10.1023/A:1014190629738
41. Zadoyanchuk, N.V., Kas'yanov, P.O.: Faedo-Galerkin method for second-order evolution inclusions with W_λ-pseudomonotone mappings. Ukr. Math. J. (2009). https://doi.org/10.1007/s11253-009-0207-z
42. Zadoyanchuk, N.V., Kasyanov, P.O.: Analysis and control of second-order differential-operator inclusions with +-coercive damping. Cybern. Syst. Anal. (2010). https://doi.org/10.1007/s10559-010-9208-z
43. Zadoyanchuk, N.V., Kasyanov, P.O.: Dynamic of solutions of a class of the second-order autonomous evolution inclusions. Cybern. Syst. Anal. (2012). https://doi.org/10.1007/s10559-012-9421-z
44. Zgurovsky, M.Z., Mel'nik, V.S., Kasyanov, P.O.: Evolution Inclusions and Variation Inequalities for Earth Data Processing I. Springer, Heidelberg (2010). https://doi.org/10.1007/978-3-642-13837-9
45. Zgurovsky, M.Z., Mel'nik, V.S., Kasyanov, P.O.: Evolution Inclusions and Variation Inequalities for Earth Data Processing II. Springer, Heidelberg (2010). https://doi.org/10.1007/978-3-642-13878-2
46. Zgurovsky, M.Z., Kasyanov, P.O., Kapustyan, O.V., Valero, J., Zadoianchuk, N.V.: Evolution Inclusions and Variation Inequalities for Earth Data Processing III. Long-Time Behavior of Evolution Inclusions Solutions in Earth Data Analysis (English). Series: Advances in Mechanics and Mathematics, vol. 27. Springer, Berlin (2012), XLI, 330 p. ISBN 978-3-642-28511-0
47. Zgurovsky, M.Z., Kasyanov, P.O.: Qualitative and Quantitative Analysis of Nonlinear Systems: Theory and Applications. Series: Studies in Systems, Decision and Control, vol. 111. Springer International Publishing, Cham (2008), XXXIII+240 pp

Chapter 16
On a Lyapunov Characterization of Input-To-State Stability for Impulsive Systems with Unstable Continuous Dynamics

Petro Feketa, Alexander Schaum, and Thomas Meurer

Abstract This chapter studies the input-to-state stability (ISS) property for nonlinear control systems with impulsive jumps at fixed moments. Sufficient conditions for the ISS are formulated in terms of a candidate ISS-Lyapunov function equipped with nonlinear rate functions which characterize the evolution of this function along the discontinuous trajectories of the system. For the case of unstable continuous dynamics, we derive new sufficient conditions for ISS under average-type dwell-time that provide a lower bound for the frequency of stabilizing jumps sufficient for the ISS.

16.1 Introduction

Impulsive systems proved to be a convenient modeling framework for the mathematical description of processes that combine continuous and discontinuous behavior. This type of behavior is of great importance for many applications in logistics [34], robotics [31], population dynamics [26, 35], neuron models [12, 15], etc. The basis of mathematical theory of impulsive systems as well as fundamental results on the existence and stability of solutions were summarized in the late 1980s in [18, 27, 28].

Nearly at the same time the concept of input-to-state stability (ISS) for systems of ordinary differential equations with external inputs was introduced by Sontag [29]. ISS characterizes a behavior of solutions to control systems with respect to external

P. Feketa (✉) · A. Schaum · T. Meurer
Kiel University, Kaiserstraße 2, 24143 Kiel, Germany
e-mail: pf@tf.uni-kiel.de

A. Schaum
e-mail: alsc@tf.uni-kiel.de

T. Meurer
e-mail: tm@tf.uni-kiel.de

V. A. Sadovnichiy and M. Z. Zgurovsky (eds.), *Contemporary Approaches and Methods in Fundamental Mathematics and Mechanics*, Understanding Complex Systems,
https://doi.org/10.1007/978-3-030-50302-4_16

disturbances. The ISS property of impulsive systems and its Lyapunov characterization have been first proposed in [16, 17]. A candidate ISS-Lyapunov function proposed there, was parametrized by two real constants called rate coefficients. These constants were used to describe the evolution of the ISS-Lyapunov function along the trajectories of the impulsive system during flows (constant $c \in \mathbb{R}$) and impulsive jumps (constant $d \in \mathbb{R}$), see Remark 16.2. Positive values of the rate coefficients correspond to the case of a positive impact of flows/jumps on the ISS property and, vice versa, the negative sign of c (or d) means that the corresponding flows (or jumps) play against the stability. Additional relations called dwell-time conditions (DTC) which restrict the number/frequency of impulsive jumps in order to guarantee the ISS of impulsive systems have been introduced. These conditions balance continuous dynamics and discontinuous dynamics of the system to assure the desired stability property and they can be of two types: fixed or average dwell-time conditions. The fixed dwell-time conditions utilize the minimum/maximum distance between two consecutive jumps. The average ones provide a sufficient estimate of the average number of jumps in a unit time-interval in order to conclude ISS. Usually, fixed dwell-time conditions are more conservative compared to the average ones, however they are easier to be checked. Some generalizations of this approach involving exponential ISS-Lyapunov functions with multiple rate coefficients and the corresponding fixed and average dwell-time conditions have been proposed in [2, 3].

A more advanced technique to study the ISS of impulsive control systems has been utilized in [7–9, 21]. It relies on the concept of a candidate ISS-Lyapunov function with nonlinear rates $\varphi : [0, \infty) \to [0, \infty)$ and $\psi : [0, \infty) \to [0, \infty)$. The nonlinear rate functions φ and ψ can estimate the evolution of the ISS-Lyapunov function along nonlinear flows and jumps of the system in a much more precise manner than the linear ones with rate constants c and d. Hence, the resulting sufficient conditions are supposed to be less conservative.

While most of the recent works are primarily focused on extensions of pioneering results of [17] to other more complicated classes of impulsive systems (e.g. with delay [6, 19, 30, 33], in infinite-dimensional[7] and stochastic [25, 33, 36] settings, with switchings [19–21], with non-fixed and/or state-dependent moments of jumps [8], including hybrid dynamical systems [1, 22] to name a few), we consider a rather simple class of impulsive systems and aim to relax the sufficient conditions for the ISS. In this chapter, we extend recent results on the ISS [9] and global asymptotic stability [10] under average-type dwell-time to the systems with unstable continuous dynamics. Also, our results relax the sufficient conditions for ISS compared to the works [3, 7, 16, 17]. This is achieved by combining the advantages of nonlinear rate functions and average-type dwell-time conditions.

The rest of the chapter is organized as follows. Section 16.2 contains the problem formulation. In Sect. 16.3 we prove the main result of the paper which is a Lyapunov-based characterization of the input-to-state stability of impulsive systems with unstable continuous dynamics under average dwell-time. Finally, a short discussion on the derived results in Sect. 16.4 concludes the chapter.

16.2 Input-To-State Stability for Impulsive Control Systems

Let $\mathbb{T} = \{t_1, t_2, t_3, \ldots\}$ be a strictly increasing sequence of impulse times in (t_0, ∞) for some initial time $t_0 \in \mathbb{R}$ with no finite accumulation point, i.e., $\lim\limits_{i\to\infty} t_i = \infty$. The latter condition excludes Zeno behavior, which is characterized by infinitely many jumps over a finite interval of time and causes a bounded domain for solutions. Some recent approaches to prolong solutions beyond Zeno time are proposed in [4, 5, 13]. In this chapter, we consider Zeno-free impulsive system with external inputs

$$\begin{cases} \dot{x}(t) = f(x(t), u(t)), & t \notin \mathbb{T}, \\ x(t) = g(x^-(t), u^-(t)), & t \in \mathbb{T}, \end{cases} \tag{16.1}$$

where $x(t) \in \mathbb{R}^N$ is the state of system (16.1) at time $t \in \mathbb{R}$, $u(t) \in \mathbb{R}^M$ is a control input at time $t \in \mathbb{R}$, f and g are functions from $\mathbb{R}^N \times \mathbb{R}^M$ to $\mathbb{R}^N$, with f locally Lipschitz, $N, M \in \mathbb{N}$. The state x and the input u are assumed to be right-continuous, and to have left limits at all times. We denote by $(\cdot)^-$ the left-limit operator, i.e., $x^-(t) = \lim_{s\nearrow t} x(s)$. The space of admissible inputs is $U_c := PC([t_0, \infty), \mathbb{R}^M)$, i.e. the space of piecewise right-continuous functions from $[t_0, \infty)$ to $\mathbb{R}^M$ equipped with the norm $\|u\|_{U_c} := \sup_{t\geq t_0} |u(t)|$, where $|\cdot|$ stands for the Euclidean norm. For a given set $\mathbb{T}$, we denote by $x = \phi(t; t_0, x_0, u)$ a piecewise right-continuous and differentiable on (t_i, t_{i+1}), $i \in \mathbb{N} \cup \{0\}$ function from $[t_0, \infty)$ to $\mathbb{R}^N$ that is a solution to (16.1) with the initial data $(t_0, x_0) \in \mathbb{R} \times \mathbb{R}^N$ and input $u \in U_c$. The assumptions above guarantee the local existence and uniqueness of such solutions [28].

To introduce the notion of ISS, we recall the following standard definitions of comparison functions: a function $\alpha : [0, \infty) \to [0, \infty)$ is called positive definite, and we write $\alpha \in \mathscr{P}$, when α is continuous, $\alpha(0) = 0$, and $\alpha(r) > 0\ \forall r > 0$. A function $\alpha : [0, \infty) \to [0, \infty)$ is of class $\mathscr{K}$, and we write $\alpha \in \mathscr{K}$, when $\alpha \in \mathscr{P}$ and strictly increasing. If $\alpha \in \mathscr{K}$ is unbounded, then we say it is of class $\mathscr{K}_\infty$, and we write $\alpha \in \mathscr{K}_\infty$. A continuous function $\beta : [0, \infty) \times [0, \infty) \to [0, \infty)$ is of class $\mathscr{KL}$, and we write $\beta \in \mathscr{KL}$, when $\beta(\cdot, t)$ is of class $\mathscr{K}$ for each fixed $t \geq 0$, and for each fixed $r \geq 0$ function $\beta(r, \cdot)$ is strictly decreasing with $\lim\limits_{t\to\infty} \beta(r, t) = 0$.

Definition 16.1 For a given sequence $\mathbb{T}$ of impulse times we call system (16.1) input-to-state stable (ISS) if there exist functions $\beta \in \mathscr{KL}$ and $\gamma \in \mathscr{K}_\infty$, such that for every initial condition $(t_0, x_0) \in \mathbb{R} \times \mathbb{R}^N$ and every input $u \in U_c$, the corresponding solution to (16.1) exists globally and satisfies

$$|\phi(t; t_0, x_0, u)| \leq \beta(|x_0|, t - t_0) + \gamma(\|u\|_{U_c}) \quad \forall t \geq t_0. \tag{16.2}$$

In the sequel, we derive sufficient conditions for ISS of system (16.1) in terms of an auxiliary Lyapunov-like scalar function.

16.3 Lyapunov Characterization of the ISS Property

For the ISS analysis of impulsive systems we adapt the concept of a candidate ISS-Lyapunov function from [7], which is suitable to characterize the dynamics of impulsive systems with unstable continuous flow.

Definition 16.2 A Lipschitz continuous function $V : \mathbb{R}^N \to [0, \infty)$ is called a candidate ISS-Lyapunov function for system (16.1) if $\exists \alpha_1, \alpha_2 \in \mathscr{K}_\infty$ such that

$$\alpha_1(|x|) \leq V(x) \leq \alpha_2(|x|) \quad \forall x \in \mathbb{R}^N \tag{16.3}$$

holds and $\exists \chi \in \mathscr{K}_\infty$, $\psi \in \mathscr{P}$, and continuous function $\varphi : [0, \infty) \to \mathbb{R}$ with $\varphi(s) = 0 \Leftrightarrow s = 0$, such that $\forall x \in \mathbb{R}^N$ and $\forall \xi \in \mathbb{R}^M$ it holds that

$$\begin{aligned} V(x) \geq \chi(|\xi|) \quad \Rightarrow \quad \dot{V}_u(x) &\leq \varphi(V(x)), \\ V(g(x, \xi)) &\leq \psi(V(x)) \end{aligned} \tag{16.4}$$

$\forall u \in U_c$ with $u(t_0) = \xi$.

Remark 16.1 In Definition 16.2, for a given input $u \in U_c$ the Dini derivative $\dot{V}_u(x)$ is defined by

$$\dot{V}_u(x) = \lim_{t \to +0} \frac{V(\phi_c(t; 0, x, u)) - V(x)}{t},$$

where ϕ_c is a transition map that corresponds to the continuous part of system (16.1), i.e., $\phi_c(t; t_0, x, u)$ is the state of system (16.1) at time t if the state at time $t_0 := 0$ was x and no impulses occur. A locally Lipschitz continuous function can be recovered by the Riemann integral of its Dini derivative [14].

Remark 16.2 If in Definition 16.2 for some $c, d \in \mathbb{R}$

$$\varphi(s) = -cs \quad \text{and} \quad \psi(s) = e^{-d} s \quad \forall s \in [0, \infty),$$

then V is called a candidate exponential ISS-Lyapunov function with rate coefficients $c \in \mathbb{R}$ and $d \in \mathbb{R}$.

The existence of a candidate ISS-Lyapunov function V does not readily imply the ISS of system (16.1). Since the rate function φ may take values larger than zero, the continuous dynamics (flow) can play against the ISS of the system and destabilize it. However, the ISS still can be concluded if the discontinuous dynamics is stabilizing, i.e., $\psi(s) < s$ for all $s \in (0, \infty)$. Then, an additional condition that restricts the number/frequency of impulsive jumps is needed in order to conclude ISS.

Given a sequence $\mathbb{T}$ and a pair of times $s, t \in \mathbb{R}$ satisfying $t > s \geq t_0$, let $N(t, s)$ denote the number of impulsive times $t_i \in \mathbb{T}$ in the semi-open interval $(s, t]$. Now we are in position to state the main result of the chapter.

Theorem 16.1 (ISS under average dwell-time) *Let there exist a candidate ISS-Lyapunov function V for system* (16.1) *with rate functions* $\varphi, \psi \in \mathscr{P}$ *and a constant* $\rho > 0$ *such that the limit*

$$\lim_{t\to\infty} \frac{N(t,s)}{t-s} = \rho \quad \forall s \geq t_0 \tag{16.5}$$

exists uniformly with respect to $s \in [t_0, \infty)$ *and*

$$\int_1^\infty \frac{ds}{\varphi(s)} = \infty. \tag{16.6}$$

If for some $\delta > 0$ *and all* $a > 0$ *it holds that*

$$\int_{\psi(a)}^{a} \frac{ds}{\varphi(s)} \geq \frac{1}{\rho} + \delta, \tag{16.7}$$

then the impulsive system (16.1) *is ISS.*

Proof Fix arbitrary input $u \in U_c$ and initial data $(t_0, x_0) \in \mathbb{R} \times \mathbb{R}^N$. We shall prove ISS of system (16.1) by direct construction of the functions β and γ from Definition 16.1. For brevity we denote $x(\cdot) = \phi(\cdot; t_0, x_0, u)$ and $v(\cdot) := V(x(\cdot))$.

First, assume that $u \equiv 0$. We shall bound the solution to (16.1) from above by a function $\beta \in \mathscr{KL}$. Since, $u \equiv 0$, the following inequalities hold:

$$\begin{aligned} \dot{v}(t) &\leq \varphi(v(t)), \quad t \notin \mathbb{T}, \\ v(t) &\leq \psi(v^-(t)), \; t \in \mathbb{T}. \end{aligned} \tag{16.8}$$

From (16.8) it follows that

$$\frac{dv(t)}{\varphi(v(t))} \leq dt \tag{16.9}$$

for all times between impulses and any non-zero initial condition. Following the Wintner-Conti theorem for scalar differential equations [32], any solution to the Cauchy problem

$$\dot{s} = \varphi(s), \quad s(0) = s_0 \in [0, \infty)$$

is forward complete thanks to the condition (16.6). Then, integrating inequality (16.9) from t_0 to t_1, we obtain

$$\int_{v(t_0)}^{v^-(t_1)} \frac{ds}{\varphi(s)} \leq t_1 - t_0. \tag{16.10}$$

The proofs of the Lyapunov-based theorems for the ISS of impulsive systems which use nonlinear rates φ, ψ and fixed dwell-time aim to show that the sequence of values of the function v evaluated at the moments of jumps is strictly decreasing. However, this is not the case for our setting due to the average character of the dwell-time condition (16.7). Generally, the mentioned sequence is not decreasing under the conditions of Theorem 16.1. In order to overcome this difficulty we derive the estimates of v along the continuous parts of the trajectories until the time t_n for some $n \in \mathbb{N}$, taking into account the jumps of the value of function v:

$$\int_{\psi(v^-(t_1))}^{v^-(t_2)} \frac{ds}{\varphi(s)} \leq t_2 - t_1, \tag{16.11}$$

$$\vdots$$

$$\int_{\psi(v^-(t_{n-1}))}^{v^-(t_n)} \frac{ds}{\varphi(s)} \leq t_n - t_{n-1}. \tag{16.12}$$

By summation of the inequalities (16.10)–(16.12), we get

$$\int_{v(t_0)}^{v^-(t_1)} \frac{ds}{\varphi(s)} + \int_{\psi(v^-(t_1))}^{v^-(t_2)} \frac{ds}{\varphi(s)} + \cdots + \int_{\psi(v^-(t_{n-1}))}^{v^-(t_n)} \frac{ds}{\varphi(s)} \leq t_n - t_0.$$

The last inequality can be rewritten as

$$\int_{v(t_0)}^{v(t_n)} \frac{ds}{\varphi(s)} + \int_{\psi(v^-(t_1))}^{v^-(t_1)} \frac{ds}{\varphi(s)} + \cdots + \int_{\psi(v^-(t_{n-1}))}^{v^-(t_{n-1})} \frac{ds}{\varphi(s)} + \int_{\psi(v^-(t_n))}^{v^-(t_n)} \frac{ds}{\varphi(s)} \leq t_n - t_0.$$

Finally, utilizing the dwell-time condition (16.7), for any $n \in \mathbb{N}$ it holds that

$$\begin{aligned} \int_{v(t_0)}^{v(t_n)} \frac{ds}{\varphi(s)} &\leq - \left(\int_{\psi(v^-(t_1))}^{v^-(t_1)} \frac{ds}{\varphi(s)} + \cdots + \int_{\psi(v^-(t_n))}^{v^-(t_n)} \frac{ds}{\varphi(s)} \right) + t_n - t_0 \\ &\leq t_n - t_0 - n\left(\frac{1}{\rho} + \delta\right). \end{aligned} \tag{16.13}$$

Next, we make use of the existence of a uniform limit (16.5). This limit may be interpreted as the asymptotic frequency of impulsive jumps. The uniform convergence of $\frac{N(t,s)}{t-s}$ means that

$$\forall \varepsilon > 0 \quad \exists T = T(\varepsilon) > t_0 \quad \text{such that} \quad \forall t \geq T, \forall s \geq t_0$$
$$\left| \rho - \frac{N(t,s)}{t-s} \right| < \varepsilon.$$

Hence,
$$\frac{N(t,s)}{t-s} - \varepsilon < \rho < \frac{N(t,s)}{t-s} + \varepsilon \quad \forall t \geq T, \forall s \in [t_0, t).$$

This means, in particular, that for any $\varepsilon > 0$ there exists a sufficiently large number $n^* \in \mathbb{N}$ such that
$$\frac{N(t_{n^*}, t_0)}{t_{n^*} - t_0} - \varepsilon < \rho < \frac{N(t_{n^*}, t_0)}{t_{n^*} - t_0} + \varepsilon,$$

which yields
$$t_{n^*} - t_0 < \frac{n^*}{\rho - \varepsilon}.$$

Since the inequality (16.13) holds for any $n \in \mathbb{N}$, we get
$$\int\limits_{v(t_0)}^{v(t_{n^*})} \frac{ds}{\varphi(s)} < \frac{n^*}{\rho - \varepsilon} - n^* \left(\frac{1}{\rho} + \delta \right) = n^* \left(\frac{1}{\rho - \varepsilon} - \frac{1}{\rho} - \delta \right).$$

Now, taking $\varepsilon = \frac{\delta \rho^2}{2 + \delta \rho}$, we obtain
$$\int\limits_{v(t_0)}^{v(t_{n^*})} \frac{ds}{\varphi(s)} < -\frac{n^* \delta}{2} < 0. \tag{16.14}$$

The estimate (16.14) means that $v(t_{n^*}) < v(t_0)$ and there exists a finite decrease of the value of function v estimated in the points t_0 and t_{n^*} respectively. Since the limit (16.5) is uniform w.r.t the initial time, one may derive the estimates of the type (16.14) on every interval $(t_{n^*}^j, t_{n^*}^{j+1})$ for all $j \in \mathbb{N}$, where $t_{n^*}^1 := t_{n^*}$ and the length of every interval is bounded from above by $\frac{n^*}{\rho - \varepsilon}$. The existence of a function $\tilde{\beta} \in \mathscr{KL}$ that bounds the Lyapunov function evolution along the trajectories of (16.1), i.e.,
$$v(t) \leq \tilde{\beta}(v(t_0), t - t_0)$$

follows then using standard techniques (see, e.g., [7, 8, 10]).

Now, let $u \in U_c$ be an arbitrary admissible input. Define the set
$$\mathscr{A}_1 := \left\{ x \in \mathbb{R}^N : V(x) \leq \chi(\|u\|_{U_c}) \right\}.$$

For all $t : x(t) \notin \mathscr{A}_1$ the relations (16.4) hold and consequently

$$V(x(t)) \le \tilde{\beta}(V(x_0), t - t_0) \quad \forall t : x(t) \notin \mathcal{A}_1.$$

Let $t^* := \inf\{t : x(t) \in \mathcal{A}_1\}$. From the radial unboundedness of V we get

$$|\phi(t; t_0, x_0, u)| \le \beta(|x_0|, t - t_0) \quad \forall t \le t^*, \tag{16.15}$$

where $\beta(r, t) = \alpha_1^{-1}(\tilde{\beta}(\alpha_2(r), t))$. Functions $\alpha_1, \alpha_2 \in \mathcal{K}_\infty$ are taken from (16.3). The monotonicity and unboundedness of α_1, α_2 imply the existence of the inverse functions $\alpha_1^{-1}, \alpha_2^{-1} \in \mathcal{K}_\infty$. Next, we estimate $|\phi(t; t_0, x_0, u)|$ for $t > t^*$.

Condition (16.7) with $\varphi \in \mathcal{P}$ implies that all the jumps are stabilizing, i.e. $\psi(s) < s$ for all $s \in (0, \infty)$. Hence, the trajectory may leave the set $\mathcal{A}_1$ only along the flows. Denote by $y = \mu(t; r)$ a solution to the initial-value problem

$$\dot{y} = \varphi(y), \quad y(0) = r \ge 0, \quad y(t) \in [0, \infty).$$

For any $r \ge 0$ define a function $M : [0, \infty) \to [0, \infty)$ as follows: $M(r) = \mu\left(\frac{n^*}{\rho - \varepsilon}; r\right)$. Condition (16.6) implies the existence of $\mu(\bar{t}; r)$ for any initial condition $y(0) = r \ge 0$ and $M \in \mathcal{K}$. Let us define the function $\tilde{\psi} : [0, \infty) \to [0, \infty)$ by

$$\tilde{\psi}(r) := \max\{M(\chi(r)), \chi(r)\} \quad \forall r \in [0, \infty)$$

and introduce the set

$$\mathcal{A}_2 := \left\{x \in \mathbb{R}^N : V(x) \le \tilde{\psi}(\|u\|_{U_c})\right\} \supseteq \mathcal{A}_1.$$

The solution $x = \phi(t; t_0, x_0, u)$ remains inside the set $\mathcal{A}_2$ for $t \ge t^*$. Indeed, since $t_{n^*}^{j+1} - t_{n^*}^j \le \frac{n^*}{\rho - \varepsilon}$, there exists an upper bound

$$v(t) \le M(v(t_{n^*}^j)), \quad t \in [t_{n^*}^j, t_{n^*}^{j+1}] \quad \forall j \in \mathbb{N} \cup \{0\}, \tag{16.16}$$

where $t_{n^*}^j := t_0$. Since $v(t_{n^*}^j) < v(t_{n^*}^{j+1})$ for all $j \in \mathbb{N}$ as long as $x(t) \notin \mathcal{A}_1$, we have $v(t) \le \tilde{\psi}(\|u\|_{U_c})$ for all $t \ge t^*$, which implies

$$|\phi(t; t_0, x_0, u)| \le \alpha_1^{-1}(\tilde{\psi}(\|u\|_{U_c})) := \tilde{\gamma}(\|u\|_{U_c}) \quad \forall t \ge t^*. \tag{16.17}$$

The function $\tilde{\gamma}$ is positive definite and nondecreasing, hence it may be majorized by a $\mathcal{K}_\infty$ function γ. Combining (16.15) and (16.17), we obtain

$$|\phi(t; t_0, x_0, u)| \le \beta(|x_0|, t - t_0) + \gamma(\|u\|_{U_c}) \quad \forall t \ge t_0.$$

This completes the proof. □

Remark 16.3 Let function V from Theorem 16.1 be a candidate exponential ISS-Lyapunov function (see Remark 16.2) with rate coefficients $c < 0$ and $d > 0$. Then, condition (16.6) always holds true and condition (16.7) reads as $\frac{d}{|c|} \le \frac{1}{\rho} + \delta$.

Example 16.1 Let $\mathbb{T}$ be a sequence of impulsive jumps such that 3 impulses occur at three randomly picked moments within every interval $(2i, 2(i+1))$, $i \in \mathbb{N} \cup \{0\}$. Then, $\mathbb{T}$ satisfies condition (16.5) with $\rho = \frac{3}{2}$. Consider a system

$$\begin{cases} \dot{x} = \tanh(x) + u, & t \notin \mathbb{T}, \\ x = \frac{x^-}{2}, & t \in \mathbb{T}, \end{cases} \tag{16.18}$$

where $x(t), u(t) \in \mathbb{R}$. Taking radially unbounded function $V = |x|$ as a candidate ISS-Lyapunov function, we obtain

$$\begin{aligned} \dot{V} &= \mathtt{sign}(x)(\tanh(x) + u) \\ &\le \tanh(|x|) + |u| \\ &= (1+\varepsilon)\tanh V(x) - \varepsilon \tanh V(x) + |u| \\ &\le (1+\varepsilon)\tanh V(x) \end{aligned}$$

if $V(x) \ge \tanh^{-1} \frac{|u|}{\varepsilon}$ for any $\varepsilon > 0$. The evolution of the ISS-Lyapunov function along the jumps can be estimated as $V(\frac{x}{2}) = \frac{V(x)}{2}$. Summarizing, the gain functions from (16.4) are

$$\varphi(s) = (1+\varepsilon)\tanh(s), \quad \psi(s) = \frac{s}{2}, \quad \text{and} \quad \chi(s) = \tanh^{-1}\frac{s}{\varepsilon}.$$

Let us check conditions (16.6) and (16.7) of Theorem 16.1:

$$\frac{1}{1+\varepsilon}\int_1^{\infty} \frac{ds}{\tanh(s)} = \frac{1}{1+\varepsilon} \ln \sinh s|_1^{\infty} = \infty.$$

$$\frac{1}{1+\varepsilon}\int_{\frac{a}{2}}^{a} \frac{ds}{\tanh(s)} = \frac{1}{1+\varepsilon}\ln\left(\frac{\sinh a}{\sinh\frac{a}{2}}\right) = \frac{1}{1+\varepsilon}\ln\sqrt{2(\cosh a + 1)} \ge \frac{\ln 2}{1+\varepsilon}.$$

Then, from the dwell-time condition (16.7), we obtain

$$\frac{\ln 2}{1+\varepsilon} \ge \frac{1}{\rho} + \delta = \frac{2}{3} + \delta. \tag{16.19}$$

There always exist (possibly small) positive constants $\varepsilon, \delta > 0$ satisfying (16.19). Hence, system (16.18) is ISS with external ISS-Lyaponov gain $\chi(s) = \tanh^{-1}\frac{s}{\varepsilon}$. The constant $\varepsilon > 0$ characterizes a trade-off between the ISS-Lyapunov gain χ and dwell-time condition (16.19). Also, since the distance between two consecutive jumps in $\mathbb{T}$ may be up to $4 \gg \frac{1}{\rho} = \frac{2}{3}$, it is not possible to conclude the ISS of (16.18) based on the sufficient conditions with fixed dwell-time (see, e.g., [7, 8]). □

The ISS is a generalization of the global asymptotic stability (GAS) property of ODEs to the case of systems with input. By letting $u \equiv 0$, it can be seen directly from Definition 16.1 that ISS of (16.1) implies the GAS property of a so-called zero-system

$$\begin{cases} \dot{x}(t) = \tilde{f}(x(t)), & t \notin \mathbb{T}, \\ x(t) = \tilde{g}(x^-(t)), & t \in \mathbb{T}, \end{cases} \tag{16.20}$$

where $\tilde{f}(x) = f(x, 0), \tilde{g}(x) = g(x, 0)$ for all $x \in \mathbb{R}^N$. From Theorem 16.1 we obtain

Corollary 16.1 *Let there exist a continuously differentiable function* $V : \mathbb{R}^N \to [0, \infty)$ *such that for some* $\alpha_1, \alpha_2 \in \mathcal{K}_\infty$ *and* $\varphi, \psi \in \mathcal{P}$ *the following inequalities*

$$\alpha_1(|x|) \le V(x) \le \alpha_2(|x|),$$
$$\nabla V(x) \cdot \tilde{f}(x) \le -\varphi(V(x)), \quad V(\tilde{g}(x)) \le \psi(V(x))$$

hold for all $x \in \mathbb{R}^N$ *and*

$$\int_1^\infty \frac{ds}{\varphi(s)} = \infty.$$

If for some $\rho > 0$ *the limit*

$$\lim_{t\to\infty} \frac{N(t, s)}{t - s} = \rho \quad \forall s \ge t_0$$

exists uniformly w.r.t. $s \in [t_0, \infty)$ *and for some* $\delta > 0$ *and all* $a > 0$ *it holds that*

$$\int_{\psi(a)}^{a} \frac{ds}{\varphi(s)} \ge \frac{1}{\rho} + \delta,$$

then the impulsive system (16.20) *is GAS.*

16.4 Concluding Remarks

The main contribution of the chapter is a new Lyapunov-based theorem equipped with an average-type dwell-time condition that balances stable and unstable dynamics of an impulsive system to guarantee the input-to-state stability property. This theorem relaxes the previously known sufficient conditions for ISS by combining the advantages of nonlinear rates of the corresponding ISS-Lyapunov function and the average-type dwell-time condition.

Finally, we discuss the question of the existence of the limit (16.5):

$$\lim_{t\to\infty} \frac{N(t,s)}{t-s} = \rho \quad \forall s \geq t_0$$

uniformly with respect to s, which is used in Theorem 16.1. Impulsive systems with this property are quite natural. One of the simplest examples when limit (16.5) exists is the case of periodic impulsive perturbations. Then $\rho = \frac{1}{\theta}$, where θ is the distance between two consecutive jumps. Irregular impulses occurring within periodic time-windows [10] also fall into the class of (16.5). Another example that satisfies (16.5) comes from the mathematical theory of discontinuous multi-frequency oscillations, which can be modeled by differential equations that undergo impulsive jumps at the moments of intersection of a quasi-periodic trajectory on a surface of a torus with some predefined submanifold of the torus [11, 23, 24].

References

1. Cai, C., Teel, A.R.: Results on input-to-state stability for hybrid systems. In: Proceedings of the 44th IEEE Conference on Decision and Control and 2005 European Control Conference, pp. 5403–5408 (2005)
2. Dashkovskiy, S., Feketa, P.: Input-to-state stability of impulsive systems with different jump maps. IFAC-PapersOnLine **49**(18), 1073 – 1078 (2016); In: 10th IFAC Symposium on Nonlinear Control Systems NOLCOS 2016
3. Dashkovskiy, S., Feketa, P.: Input-to-state stability of impulsive systems and their networks. Nonlinear Anal.: Hybrid Syst. **26**, 190–200 (2017)
4. Dashkovskiy, S., Feketa, P.: Prolongation and stability of Zeno solutions to hybrid dynamical systems. IFAC-PapersOnLine 50(1), 3429–3434 (2017); In: 20th IFAC World Congress
5. Dashkovskiy, S., Feketa, P.: Asymptotic properties of Zeno solutions. Nonlinear Anal.: Hybrid Syst. **30**, 256–265 (2018)
6. Dashkovskiy, S., Kosmykov, M., Mironchenko, A., Naujok, L.: Stability of interconnected impulsive systems with and without time delays, using Lyapunov methods. Nonlinear Anal.: Hybrid Syst. **6**(3), 899–915 (2012)
7. Dashkovskiy, S., Mironchenko, A.: Input-to-state stability of nonlinear impulsive systems. SIAM J. Control Optim. **51**(3), 1962–1987 (2013)
8. Feketa, P., Bajcinca, N.: Stability of nonlinear impulsive differential equations with non-fixed moments of jumps. In: Proceedings of the 2018 European Control Conference (ECC), Limassol, Cyprus, pp. 900–905 (2018)
9. Feketa, P., Bajcinca, N.: Average dwell-time for impulsive control systems possessing ISS-Lyapunov function with nonlinear rates. In: Proceedings of the 2019 European Control Conference (ECC), Naples, Italy, pp. 3686–3691 (2019)
10. Feketa, P., Bajcinca, N.: On robustness of impulsive stabilization. Automatica **104**, 48–56 (2019)
11. Feketa, P., Perestyuk, Y.: Perturbation theorems for a multifrequency system with pulses. J. Math. Sci. **217**(4), 515–524 (2016)

12. Gluzman, M.O., Gorban, N.V., Kasyanov, P.O.: Lyapunov type functions for classes of autonomous parabolic feedback control problems and applications. Appl. Math. Lett. **39**, 19–21 (2015)
13. Goebel, R., Sanfelice, R.G.: Pointwise asymptotic stability in a hybrid system and well-posed behavior beyond Zeno. SIAM J. Control Optim. **56**(2), 1358–1385 (2018)
14. Hagood, J.W., Thomson, B.S.: Recovering a function from a Dini derivative. Am. Math. Mon. **113**(1), 34–46 (2006)
15. He, Z., Li, C., Chen, L., Cao, Z.: Dynamic behaviors of the FitzHugh-Nagumo neuron model with state-dependent impulsive effects. Neural Netw. **121**, 497–511 (2020)
16. Hespanha, J.P., Liberzon, D., Teel, A.R.: On input-to-state stability of impulsive systems. In: Proceedings of the 44th IEEE Conference on Decision and Control and 2005 European Control Conference, pp. 3992–3997 (2005)
17. Hespanha, J.P., Liberzon, D., Teel, A.R.: Lyapunov conditions for input-to-state stability of impulsive systems. Automatica **44**(11), 2735–2744 (2008)
18. Lakshmikantham, V., Bainov, D., Simeonov, P.S.: Theory of Impulsive Differential Equations. World Scientific, Singapore (1989)
19. Liu, J., Liu, X., Xie, W.-C.: Input-to-state stability of impulsive and switching hybrid systems with time-delay. Automatica **47**(5), 899–908 (2011)
20. Liu, J., Liu, X., Xie, W.-C.: Class-$\mathcal{KL}$ estimates and input-to-state stability analysis of impulsive switched systems. Syst. Control Lett. **61**(6), 738–746 (2012)
21. Mancilla-Aguilar, J., Haimovich, H.: Uniform input-to-state stability for switched and time-varying impulsive systems (2019). arXiv:1904.03440
22. Mironchenko, A., Yang, G., Liberzon, D.: Lyapunov small-gain theorems for networks of not necessarily ISS hybrid systems. Automatica **88**, 10–20 (2018)
23. Perestyuk, M., Feketa, P.: Invariant sets of impulsive differential equations with particularities in ω-limit set. Abstr. Appl. Anal. **2011**, Article ID 970469, 1–14 (2011)
24. Perestyuk, M.O., Feketa, P.V.: Invariant manifolds of one class of systems of impulsive differential equations. Nonlinear Oscil. **13**(2), 260–273 (2010)
25. Ren, W., Xiong, J.: Stability analysis of impulsive stochastic nonlinear systems. IEEE Trans. Autom. Control **62**(9), 4791–4797 (2017)
26. Rogovchenko, Y.V.: Nonlinear impulse evolution systems and applications to population models. J. Math. Anal. Appl. **207**(2), 300–315 (1997)
27. Samoilenko, A., Perestyuk, N.: Differential Equations with Impulse Effect. Visca Skola, Kiev (1987)
28. Samoilenko, A.M., Perestyuk, N.A.: Impulsive Differential Equations. World Scientific Series on Nonlinear Science, vol. 14. World Scientific Publishing Co., Inc, River Edge (1995)
29. Sontag, E.D.: Smooth stabilization implies coprime factorization. IEEE Trans. Autom. Control **34**(4), 435–443 (1989)
30. Sun, X.-M., Wang, W.: Integral input-to-state stability for hybrid delayed systems with unstable continuous dynamics. Automatica **48**(9), 2359–2364 (2012)
31. Tang, Y., Xing, X., Karimi, H.R., Kocarev, L., Kurths, J.: Tracking control of networked multi-agent systems under new characterizations of impulses and its applications in robotic systems. IEEE Trans. Ind. Electron. **63**(2), 1299–1307 (2015)
32. Wintner, A.: The non-local existence problem of ordinary differential equations. Am. J. Math. **67**(2), 277–284 (1945)
33. Wu, X., Tang, Y., Zhang, W.: Input-to-state stability of impulsive stochastic delayed systems under linear assumptions. Automatica **66**, 195–204 (2016)
34. Xu, W., Cui, X.: System optimization on distribution center of retail. Int. J. Nonlinear Sci. **6**(1), 79–85 (2008)
35. Yang, X., Peng, D., Lv, X., Li, X.: Recent progress in impulsive control systems. Math. Comput. Simul. **155**, 244–268 (2019)
36. Yao, F., Qiu, L., Shen, H.: On input-to-state stability of impulsive stochastic systems. J. Frankl. Inst. **351**(9), 4636–4651 (2014)

Chapter 17
Practical Stability of Discrete Systems: Maximum Sets of Initial Conditions Concept

V. V. Pichkur and Ya. M. Linder

Abstract In this chapter we consider practical stability of discrete systems on the basis of maximum sets of initial conditions concept. We propose results concerning nonlinear discrete systems including topological properties of the maximum sets of initial conditions for internal practical stability, the necessary and sufficient conditions of internal practical stability using the optimal Lyapunov function. Further we offer the analytical forms (such as the Minkowski function, the inverse Minkowski function, and the support function) of the maximum sets of initial conditions representation in linear case, optimal ellipsoidal estimations of practical stability domains, and optimal estimations of phase constraints. In the last section we consider the problem of external practical stability of discrete systems.

17.1 Introduction

In recent years, theory of practical stability of motion has been developed very intensively. For the first time, the concept of practical stability was proposed in [1, 2]. Since then the direct Lyapunov's method has been considered as the main research technique in this theory. Due to works [3–6] the second Lyapunov method has been generalized to this type of stability. Further necessary and sufficient conditions of practical stability in terms of the Lyapunov functions have been obtained for various classes of systems including the discrete ones [4, 7–17]. The technique based on properties of the maximum sets of initial conditions has been introduced in [9, 10, 18–20]. The topological properties of the optimal sets of initial conditions such as

V. V. Pichkur (✉) · Ya. M. Linder
Taras Shevchenko National University of Kyiv, Volodymyrska st., 64, Kyiv, Ukraine
e-mail: vpichkur@gmail.com

Ya. M. Linder
e-mail: yaroslav.linder@gmail.com

V. A. Sadovnichiy and M. Z. Zgurovsky (eds.), *Contemporary Approaches and Methods in Fundamental Mathematics and Mechanics*, Understanding Complex Systems,
https://doi.org/10.1007/978-3-030-50302-4_17

compactness, boundary and interior properties have been studied in [9, 10, 21–26]. Such sets have been effectively described in case of linear systems with convex phase constraints using convex analysis technique, in particular, using the support functions and the Minkowski functions properties.

In this paper we consider the problems of practical stability of discrete systems solutions. Nowadays, discrete systems are one of the most applicable tools in simulation and modelling. They are often used in combination with continuous systems (hybrid systems, systems with impulse impact). Different problems concerning qualitative analysis, stability theory and their applications in modelling have been presented in [27–31].

We introduce the following notations: $\mathbb{R}^n$ is an n-dimensional Euclidean space; $\langle x, y\rangle = x^*y$ is the usual inner product of $x, y \in \mathbb{R}^n$, where $*$ denotes the transpose sign; $\|x\| = \sqrt{\langle x, x\rangle}$; $K_r(a) = a + rK_1(0)$ denotes the closed ball with radius $r \geq 0$ and center at $a \in \mathbb{R}^n$; $S = \{x \in \mathbb{R}^n : \|x\| = 1\}$ stands for the unit sphere with center at 0; $int A$ is the set of inner points while ∂A denotes the boundary of a set $A \subset \mathbb{R}^n$; $comp(\mathbb{R}^n)$ ($conv(\mathbb{R}^n)$) is the set of nonempty (convex) compact subsets of $\mathbb{R}^n$; $c(A, \psi) = \sup_{x\in A} \langle x, \psi\rangle$ stands for the support function of $A \subset \mathbb{R}^n$, $\psi \in \mathbb{R}^n$. We use the symbol I to denote the identity matrix.

Let us consider a discrete system

$$x(k+1) = f_k(x(k)),\ k = 0, 1, \ldots, N-1, \tag{17.1}$$

where $x \in \mathbb{R}^n$ is a state vector, $f_k : D \to \mathbb{R}^n$, $D \subset \mathbb{R}^n$ is a bounded set.

Suppose that $f_k(0) = 0, k = 0, 1, \ldots, N-1$. This condition is equal to the statement that 0 is an equilibrium point of system (17.1). Denote by $x(k, x_0, s)$ the solution of system (17.1) corresponding to the Cauchy condition $x(s) = x_0$, $k \geq s$, $x(k, x_0) = x(k, x_0, 0)$, $k = 0, 1, \ldots, N$. Assume that the function f_k satisfies the local Lipschitz condition with respect to x, $k = 0, 1, \ldots, N-1$. It means that for every point $x \in D$ there exists a neighborhood $U_r(x)$ and a constant $C_k > 0$ such that for each $y \in U_r(x)$

$$\|f_k(x) - f_k(y)\| \leq C_k \|x - y\|, k = 0, 1, \ldots, N-1 \tag{17.2}$$

holds. In addition the inverse condition takes place i.e for every point $x \in D$ there exists a neighborhood $U_r(x)$ of this point and a constant $L_k > 0$ such that for each $y \in U(x)$

$$\|f_k(x) - f_k(y)\| \geq L_k \|x - y\| \tag{17.3}$$

takes place. Conditions (17.2), (17.3) can be applied to any compact subset of D. Inequality (17.3) makes the mapping $x(k, x_0)$ injective with respect to x_0 and the functions f_k open on the corresponding set. It means that on any compact set D there exists a function $\varphi(k, x)$ such that $x_0 = \varphi(k, x)$ if and only if $x = x(k, x_0)$. Moreover from (17.3) it follows that $\varphi(k, x)$ satisfies the Lipschitz condition with respect to x.

Let $M \subset D$ be a compact set. Then

$$x(k, M) = \cup_{x_0 \in M} x(k, x_0), \ \varphi(k, M) = \cup_{x \in M} \varphi(k, x)$$

are compact sets in $\mathbb{R}^n$, $k = 0, 1, \ldots, N$. Moreover the following statements take place:

1. $x(k, \partial M) = \partial x(k, M), x(k, int M) = int x(k, M), k = 0, 1, \ldots, N;$
2. $\varphi(k, \partial M) = \partial \varphi(k, M), \varphi(k, int M) = int \varphi(k, M), k = 0, 1, \ldots, N.$

Suppose that $G_0 \subset \mathbb{R}^n$, $\Phi(k) \subset D$ are compact sets such that $0 \in int\ G_0$, $0 \in int\ \Phi(k)$ for all $k = 0, 1, \ldots, N$.

Definition 17.1 The zero solution of system (17.1) is called $\{G_0, \Phi(k), 0, N\}$-stable if for every $x_0 \in G_0$ the solution $x(k, x_0)$ of system (17.1) belongs to the sets $\Phi(k)$ for all $k = 0, 1, \ldots, N$.

Note that $\{G_0, \Phi(k), 0, N\}$-stability of the zero solution of system (17.1) is called internal. Besides internal stability there exists external practical stability of the zero solution of system (17.1). Let $D_0 \subset \mathbb{R}^n$, $0 \in int D_0$ and the set D_0 contains points lying outside $\Phi(0)$.

Definition 17.2 The zero solution of system (17.1) is called external $\{D_0, \Phi(k), 0, N\}$-stable, if for any $x_0 \in D_0$ there exists $k \in \{0, 1, \ldots, N\}$ such that $x(k, x_0) \in \Phi(k)$.

On the basis of definition 17.1 we state the following problems:

1. given G_0, $\Phi(k)$, and $k = 0, 1, \ldots, N$. Verify whether the zero solution of system (17.1) is $\{G_0, \Phi(k), 0, N\}$-stable;
2. given $\Phi(k)$, $k = 0, 1, \ldots, N$. Find the maximum set $G_* \subseteq \Phi(0)$ such that the equilibrium position $x(k) = 0$ of discrete system (17.1) is $\{G_*, \Phi(k), 0, N\}$ - stable;
3. given $\Phi(k)$, $k = 0, 1, \ldots, N$, and the set of initial conditions $G_0(a) \subseteq \Phi(0)$ depends on some parameter $a \geq 0$. Find all parameter a values so that the solution $x(k) = 0$ of discrete system (17.1) is $\{G_0(a), \Phi(k), 0, N\}$ - stable. Suppose that the zero solution of discrete system (17.1) is $\{G_0(a), \Phi(k), 0, N\}$ - stable iff $a \in [0, a_*]$. In this case a_* is called the maximum estimation of practical stability;
4. given $G_0, k = 0, 1, \ldots, N$, the states constraints $\Phi(k, b)$ depend on some parameter $b \geq 0$, $k = 0, 1, \ldots, N$. Find all parameter b values so that the equilibrium position $x(k) = 0$ of discrete system (17.1) is $\{G_0, \Phi(k, b), 0, N\}$ - stable;
5. given G_0, $\Phi(k)$, $k = 0, 1, \ldots$. Find the maximum N such that the equilibrium position $x(k) = 0$ of discrete system (17.1) is $\{G_0, \Phi(k), 0, N\}$ - stable.

One can state the same problems for external modes of practical stability by similar way. It turns out that the problem of finding the maximum set is central in the following sense: the solution of other problems is based on the properties of the maximum set.

Bellow we discuss topological properties of the maximum sets of initial conditions for internal practical stability and the necessary and sufficient conditions of internal practical stability using the optimal Lyapunov function. Further we offer the analytical forms of the maximum sets of initial conditions representation in the linear discrete system case. In the last section we consider the similar problems of external practical stability.

17.2 Internal Practical Stability

In this section we describe properties of a set $G_* \subseteq \Phi(0)$. The set G_* consists of all points $x_0 \subseteq \Phi(0)$ such that any solution $x(\cdot, x_0)$ of discrete system (17.1) belongs to $\Phi(k)$ for all $k = 0, 1, \ldots, N$. We call such a set the maximum set of internal practical stability [7, 18]. One can observe that definition 17.1 takes place if and only if $G_0 \subseteq G_*$. Consider the following assertions [21].

Theorem 17.1 *The maximum set of practical stability of system* (17.1) *with phase constraints* $\Phi(k)$, $k = 0, 1, \ldots, N$ *is a compact set in* $\mathbb{R}^n$. *In addition*

$$G_* = \bigcap_{k=0}^{N} \varphi(k, \Phi(k)),$$

where $\varphi(k, \Phi(k)) = \cup_{x \in \Phi(k)} \varphi(k, x)$, $k = 0, 1, \ldots, N$.

Theorem 17.2 *The point* $x_0 \in \partial G_*$ *if and only if* $x(k, x_0) \in \Phi(k)$, $k = 0, 1, \ldots, N$ *and there exists* t *from the set* $\{0, 1, \ldots, N\}$ *such that* $x(t, x_0) \in \partial\Phi(t)$.

Corollary 17.1 *The point* $x_0 \in int G_*$ *if and only if* $x(k, x_0) \in int\Phi(k)$ *for all* $k = 0, 1, \ldots, N$.

Consider a linear discrete system

$$x(k + 1) = A_k x(k) + b_k, \tag{17.4}$$

where A_k is a nonsingular matrix of dimension $n \times n$, b_k is a vector of dimension n, $k = 0, 1, \ldots, N - 1$. The general solution of system (17.4) can be written down in the form

$$x(k) = \Theta(k)x_0 + r(k), k = 0, 1, \ldots, N,$$

where $\Theta(k, s) = A_{k-1}A_{k-2}...A_s$ is a nonsingular matrix, $\Theta(s, s) = I$, $\Theta(k) = \Theta(k, 0), r(k) = \sum_{i=0}^{k-1} \Theta(k, i)b_i, k, s = 0, 1, \ldots, N$. In particular, $\Theta(0) = I, r(0) = 0$. Assume that $r(k) \in int\Phi(k)$, $k = 0, 1, \ldots, N$. The following theorems are true.

Theorem 17.3 *If* $\Phi(k) \in conv(\mathbb{R}^n)$, $k = 0, 1, \ldots, N$, *then* $0 \in intG_*$ *and the maximum set of practical stability* $G_* \in conv(\mathbb{R}^n)$. *Then its Minkowski function can be written as*

$$m_*(x_0) = \inf\left\{\lambda > 0 : \frac{x_0}{\lambda} \in G_*\right\} = \max_{k=0,1,\ldots,N, \psi \in S} \frac{\langle \Theta(k)x_0, \psi \rangle}{c(\Phi(k), \psi) - \langle r(k), \psi \rangle},$$

where $x_0 \in \mathbb{R}^n$ *and*

$$G_* = \{x_0 \in \mathbb{R}^n : m_*(x_0) \leq 1\}.$$

Theorem 17.4 *The inverse Minkowski function of* G_* *can be expressed as*

$$d_*(x_0) = \min_{k=0,1,\ldots,N,\psi\in P(k)} \frac{c(\Phi(k),\psi) - \langle r(k),\psi\rangle}{\langle \Theta(k)x_0,\psi\rangle},$$

where $P(k) = \{\psi \in S : \langle \Theta(k)x_0,\psi\rangle > 0\}$, $x_0 \in \mathbb{R}^n$. *In this case*

$$G_* = \cup_{e\in S}\, [0, d_*(e)]\, e.$$

Theorem 17.5 *The support function of the set* G_* *is equal to*

$$c(G_*,\xi) = \overline{co}g(\xi),$$

where

$$g(\xi) = \min_{k=0,1,\ldots,N} \left(c(\Phi(k),(\Theta^*(k))^{-1}\xi) - \left\langle r(k),(\Theta^*(k))^{-1}\xi\right\rangle\right), \xi \in \mathbb{R}^n$$

and $\overline{co}g(\xi)$ *stands for convex closure of the function* $g(\xi)$. *In this case*

$$G_* = \cap_{\xi\in S} H(\xi),$$

where $H(\xi) = \{x \in \mathbb{R}^n : \langle x,\xi\rangle \leq g(\xi)\}$, $\xi \in \mathbb{R}^n$.

Theorem 17.6 *Any point* $x_0 \in \partial G_*$ *if and only if*

$$\max_{k=0,1,\ldots,N,\psi\in S} \frac{\langle \Theta(k)x_0,\psi\rangle}{c(\Phi(k),\psi) - \langle r(k),\psi\rangle} = 1.$$

Corollary 17.2 *If system* (17.4) *is homogeneous, i.e* $b_k = 0$, $k = 0, 1, \ldots, N-1$, *then the Minkowski function of the maximum set of practical stability is equal to*

$$m_*(x_0) = \max_{k=0,1,\ldots,N,\psi\in S} \frac{\langle \Theta(k)x_0,\psi\rangle}{c(\Phi(k),\psi)}, x_0 \in \mathbb{R}^n,$$

the inverse Minkowski function is written as

$$d_*(x_0) = \min_{k=0,1,\ldots,N,\psi\in P(k)} \frac{c(\Phi(k),\psi)}{\langle \Theta(k)x_0,\psi\rangle},$$

where $P(k) = \{\psi \in S : \langle \Theta(k)x_0,\psi\rangle > 0\}$, $x_0 \in \mathbb{R}^n$, *and the support function is expressed in the form*

$$c(G_*,\xi) = \overline{co}g(\xi),$$

where $g(\xi) = \min_{k=0,1,\ldots,N} c(\Phi(k), (\Theta^*(k))^{-1}\xi)$, $\xi \in \mathbb{R}^n$.

Example 17.1 Find the Minkowski function, the inverse Minkowski function, and the support function if system (17.4) is homogeneous, i. e $b_k = 0, k = 0, 1, \ldots, N - 1$. Let $\Phi(k) = p_k K_1(0)$, $p_k > 0$, $k = 0, 1, \ldots, N$. In such a case

$$m_*(x_0) = \max_{k=0,1,\ldots,N} \frac{\|\Theta(k)x_0\|}{p_k}, \; d_*(x_0) = \min_{k=0,1,\ldots,N} \frac{p_k}{\|\Theta(k)x_0\|},$$

$$c(G_*, \xi) = \overline{co} \min_{k=0,1,\ldots,N} p_k \left\|(\Theta^*(k))^{-1}\xi\right\|, \xi \in \mathbb{R}^n.$$

17.3 Practical Stability and Lyapunov Functions

In this section we use the Lyapunov functions technique in order to describe necessary and sufficient conditions of internal practical stability [8].

Theorem 17.7 *System* (17.1) *is* $\{G_0, \Phi(k), 0, N\}$*-stable if and only if there exist the Lyapunov functions* $V_k(x)$, $k = 0, 1, \ldots, N$ *such that the following statements are fulfilled:*

$$G_0 \subseteq \left\{x \in \mathbb{R}^n : V_0(x) \le 1\right\},$$

$$\left\{x \in \mathbb{R}^n : V_k(x) \le 1\right\} \subseteq \Phi(k), \; k = 0, 1, \ldots, N,$$

$$V_{k+1}(f_k(x)) \le V_k(x), \; k = 0, 1, \ldots, N - 1.$$

Proof of Theorem 17.7 is constructive. We choose a continuous function $\alpha_* : \mathbb{R}^n \to \mathbb{R}^1$ such that

$$\alpha_*(x) < 1, x \in int\, G_*; \alpha_*(x) = 1, x \in \partial G_*; \alpha_*(x) > 1, x \notin G_*.$$

The function α_* exists. For example

$$\alpha_*(x) = \begin{cases} 1 - \rho(x, \partial G_*), & x \in G_*, \\ 1 + \rho(x, \partial G_*), & x \notin G_*, \end{cases}$$

where $\rho(x, \partial G_*) = \min_{y \in \partial G_*} \|x - y\|$ is the distance from a point $x \in \mathbb{R}^n$ to the boundary of the set G_*. Hence

$$V_k(x) = \alpha_*(\varphi(k, x)), \; k = 0, 1, \ldots, N,$$

where $\varphi(k, y) = x(0, y, k)$ so that $y_0 = \varphi(k, y)$ if and only if $y = x(k, y_0)$. The Lyapunov function $V_k(x) = \alpha_*(\varphi(k, x))$ is called optimal since it is constructed using the maximum set of initial conditions.

The Lyapunov functions can be constructed taking into account the set $G_0 \subset \Phi(0)$. To do this, we choose a continuous function $\alpha_0 : \mathbb{R}^n \to \mathbb{R}^1$ such that

$$\alpha_0(x) < 1, x \in int\, G_0; \alpha_0(x) = 1, x \in \partial G_0; \alpha_0(x) > 1, x \notin G_0.$$

For example,

$$\alpha_0(x) = \begin{cases} 1 - \rho(x, \partial G_0), & x \in G_0, \\ 1 + \rho(x, \partial G_0), & x \notin G_0, \end{cases}$$

where $\rho(x, \partial G_0) = \min_{y \in \partial G_0} \|x - y\|$.

If the set $G_0 \subset \Phi(0)$ is convex, then the function α_0 can be defined using the support function of the set G_0

$$\alpha(x) = \max_{\psi \in S} \left(\langle x, \psi \rangle - c(G_0, \psi) + 1\right),\ k = 0, 1, \ldots, N.$$

Taking into account that $0 \in int G_0$, the function α_0 can be chosen as the Minkowski function of the set G_0 i.e

$$\alpha_0(x) = \inf \{\lambda > 0 : x \in \lambda G_0\} = \max_{\psi \in S} \frac{\langle x, \psi \rangle}{c(G_0, \psi)}.$$

In that case the Lyapunov function $V_k(x) = \alpha_0(\varphi(k, x)), k = 0, 1, \ldots, N$.

Theorem 17.7 is used to find the optimal estimations of the set of initial conditions and the sets of phase constraints. Consider system (17.4) assuming that it is homogeneous and hence can be written as

$$x(k+1) = A_k x(k),\ k = 0, 1, \ldots, N-1. \tag{17.5}$$

Here $b_k = 0,\ k = 0, 1, \ldots, N-1$. Let the set of initial conditions $G_0 = K_r(0)$, $r \geq 0$ and the sets of phase constraints $\Phi(k)$ be convex and compact, $0 \in int\Phi(k)$, $k = 0, 1, \ldots, N$. In this case the Lyapunov function

$$V_k(x) = \max_{\psi \in S} \left\langle \Theta^{-1}(k)x, \psi \right\rangle - r + 1 = \left\| \Theta^{-1}(k) \cdot x \right\| - r + 1,\ k = 0, 1, \ldots N.$$

Definition 17.3 The zero solution of system (17.5) is called $\{r, \Phi(k), 0, N\}$ - stable if for each $x_0 \in K_r(0)$ the corresponding solution $x(k, x_0)$ of system (17.5) belongs to $\Phi(k)$, $k = 0, 1, \ldots N$.

Definition 17.4 The maximal value $r > 0$ for which $\{r, \Phi(k), 0, N\}$ - stability of the zero solution of system (17.5) takes place is called the optimal estimation of the practical stability of system (17.5) in the class $\Omega = \{K_r(0) : r \geq 0\}$ with phase constraints $\Phi(k)$, $k = 0, 1, \ldots, N$.

Theorem 17.7 yields the following statement.

Theorem 17.8 *The optimal estimation of practical stability of system* (17.5) *in the class* Ω *with the phase constraints* $\Phi(k)$, $k = 0, 1, \ldots, N$ *is equal to*

$$r_* = \min_{k=0,1,\ldots,N} \min_{\psi \in S} \frac{c(\Phi(k), \psi)}{\sqrt{\langle B(k)\psi, \psi\rangle}},$$

where $B(k)$ *satisfies the matrix discrete equation*

$$B(k+1) = A_k B(k) A_k^T, \; k = 0, 1, \ldots, N-1 \tag{17.6}$$

with the initial condition $B(0) = I$.

Example 17.2 Find the optimal estimation of practical stability of system (17.5) in the class Ω if sets of phase constraints are balls $\Phi(k) = p_k K_1(0)$, $p_k > 0$, $k = 0, 1, \ldots, N$. Then

$$r_* = \min_{k=0,1,\ldots,N} \min_{\psi \in S} \frac{p_k}{\sqrt{\langle B(k)\psi, \psi\rangle}} = \min_{k=0,1,\ldots,N} \frac{p_k}{\sqrt{\lambda_*(B(k))}},$$

where $\lambda_*(B(k))$ is the maximal eigenvalue of the matrix $B(k)$.

For nonhomogeneous system (17.4) the Lyapunov function can be written as follows

$$V_k(x) = \left\| \Theta^{-1}(k) \cdot (x - r(k)) \right\| - r + 1, \; k = 0, 1, \ldots N.$$

The following theorem takes place.

Theorem 17.9 *The optimal estimation of practical stability of system* (17.4) *in the class* Ω *with phase constraints* $\Phi(k)$, $k = 0, 1, \ldots, N$ *is equal to*

$$r_* = \min_{k=0,1,\ldots,N} \min_{\psi \in S} \frac{c(\Phi(k), \psi) - \langle r(k), \psi\rangle}{\sqrt{\langle B(k)\psi, \psi\rangle}},$$

where $B(k)$ *satisfies* (17.6) *with the initial condition* $B(0) = I$.

Let the phase constraints are considered as compact polytopes so that

$$\Phi(k) = \left\{ x \in \mathbb{R}^n : \left\langle x, l_s^k \right\rangle \le 1, \; s = 1, 2, \ldots, M \right\}. \tag{17.7}$$

Here $l_s^k \in \mathbb{R}^n \setminus \{0\}$, $s = 1, 2, \ldots, M$, $k = 0, 1, \ldots, N$. The following theorems are true.

Theorem 17.10 *The optimal estimation of practical stability of system* (17.5) *in the class* Ω *with phase constraints* (17.7) *is equal to*

$$r_* = \min_{k=0,1,\ldots,N} \min_{s=1,2,\ldots,M} \frac{1}{\sqrt{\left\langle B(k) l_s^k, l_s^k \right\rangle}},$$

where $B(k)$ is a solution of matrix equation (17.6), $B(0) = I$.

Theorem 17.11 *The optimal estimation of the practical stability of system* (17.4) *in the class Ω with phase constraints* (17.7) *is equal to*

$$r_* = \min_{k=0,1,\ldots,N} \min_{s=1,2,\ldots,M} \frac{1 - \langle l_s^k, r(k)\rangle}{\sqrt{\langle B(k) l_s^k, l_s^k\rangle}},$$

where $B(k)$ satisfies matrix equation (17.6), $B(0) = I$.

Denote $E_r(0, Q) = \{x \in \mathbb{R}^n : \langle Qx, x\rangle \le r^2\}$, where Q is a positive definite symmetric matrix of dimension $n \times n$.

Definition 17.5 The zero solution of system (17.5) is called $\{r, Q, \Phi(k), 0, N\}$ - stable, if for each $x_0 \in E_r(0, Q)$ the corresponding solution $x(k, x_0)$ of system (17.5) belongs to $\Phi(k)$, $k = 0, 1, \ldots N$. In addition the maximal value $r > 0$ for which $\{r, Q, \Phi(k), 0, N\}$ - stability of the zero solution of system (17.5) takes place is called the optimal estimation of practical stability of system (17.5) in the class

$$\Sigma = \{E_r(0, Q) : r \ge 0\}$$

with the phase constraints $\Phi(k)$, $k = 0, 1, \ldots, N$.

Consider the optimal estimation of the initial conditions in $\Sigma = \{E_r(0, Q) : r \ge 0\}$ for system (17.5) solutions. In this case the Lyapunov functions are equal to

$$V_k(x) = \left\| Q^{\frac{1}{2}} \Theta^{-1}(k) \cdot x \right\| - r + 1, \; k = 0, 1, \ldots N.$$

Theorem 17.12 *The optimal estimation of practical stability of system* (17.5) *in the class Σ with phase constraints $\Phi(k)$, $k = 0, 1, \ldots, N$ satisfies*

$$r_* = \min_{k=0,1,\ldots,N} \min_{\psi \in S} \frac{c(\Phi(k), \psi)}{\sqrt{\langle B(k)\psi, \psi\rangle}},$$

where $B(k)$ is a solution of matrix equation (17.6) *with initial condition* $B(0) = Q^{-1}$.

Theorem 17.13 *The optimal estimation of practical stability of system* (17.5) *in the class Σ with the phase constraints* (17.7) *is equal to*

$$r_* = \min_{k=0,1,\ldots,N} \min_{s=1,2,\ldots,M} \frac{1}{\sqrt{\langle B(k) l_s^k, l_s^k\rangle}}.$$

Here $B(k)$ is a solution of matrix equation (17.6), $B(0) = Q^{-1}$.

If system (17.4) is not homogeneous, then the Lyapunov functions can be written in the form

$$V_k(x) = \left\| Q^{\frac{1}{2}} \Theta^{-1}(k) \cdot (x - r(k)) \right\| - r + 1,\ k = 0, 1, \ldots N.$$

In this case the following theorems are fulfilled.

Theorem 17.14 *The optimal estimation of practical stability in Σ with the phase constraints $\Phi(k)$, $k = 0, 1, \ldots, N$ is equal to*

$$r_* = \min_{k=0,1,\ldots,N} \min_{\psi \in S} \frac{c(\Phi(k), \psi) - \langle r(k), \psi \rangle}{\sqrt{\langle B(k)\psi, \psi \rangle}},$$

where $B(k)$ is a solution of matrix equation (17.6), $B(0) = Q^{-1}$.

Theorem 17.15 *The optimal estimation of practical in the class Σ with the phase constraints* (17.7) *satisfies*

$$r_* = \min_{k=0,1,\ldots,N} \min_{s=1,2,\ldots,M} \frac{1 - \left\langle l_s^k, r(k) \right\rangle}{\sqrt{\left\langle B(k) l_s^k, l_s^k \right\rangle}},$$

where $B(k)$ is a solution of matrix equation (17.6), $B(0) = Q^{-1}$.

Consider linear system (17.5). Given the set of initial conditions $G_0 = K_r(0)$ and the phase constraints belong to the class

$$W = \left\{ \Phi(k, p) \subset \mathbb{R}^n : p \geq 0,\ k = 0, 1, \ldots, N \right\}. \tag{17.8}$$

Here $\Phi(k, p) = p\Phi(k)$, $\Phi(k) \subset \mathbb{R}^n$ are convex compact sets, $0 \in int\Phi(k)$, $k = 0, 1, \ldots, N$. The problem is to obtain the minimal value of the parameter p such that all the solutions of system (17.5) with initial conditions in G_0 belongs to the phase constraints $\Phi(k, p)$, $k = 0, 1, \ldots, N$. The following theorem holds.

Theorem 17.16 *Let $G_0 = K_r(0)$. Then the optimal estimation of phase constraints of system* (17.5) *in class* (17.8) *is equal to*

$$p_* = r \max_{k=0,1,\ldots,N} \max_{\psi \in S} \frac{\sqrt{\langle B(k)\psi, \psi \rangle}}{c(\Phi(k), \psi)}. \tag{17.9}$$

where $B(k)$ is a solution of matrix equation (17.6), $B(0) = I$. *If $G_0 = E_r(0, Q)$, then the optimal estimation can be defined from* (17.9), *where* $B(0) = Q^{-1}$.

Assume that

$$\Phi(k, p) = \bigcap_{s=1}^{M} \left\{ x \in \mathbb{R}^n : \left\langle l_s^k, x \right\rangle \leq p \right\}, \tag{17.10}$$

where $l_s^k \in \mathbb{R}^n \backslash \{0\}$, $s = 1, 2, \ldots, M$, $k = 0, 1, \ldots, N$, $p \geq 0$. The following theorem takes place.

Theorem 17.17 *Let $G_0 = K_r(0)$. In this case the optimal estimation of phase constraints in class* (17.10) *is equal to*

$$p_* = r \max_{k=0,1,\ldots,N} \max_{s=1,2,\ldots,M} \sqrt{\langle B(k)l_s^k, l_s^k \rangle}. \tag{17.11}$$

Here $B(k)$ is a solution of matrix equation (17.6), *$B(0) = I$. If $G_0 = E_r(0, Q)$, then the optimal estimation can be obtained from* (17.11), *where $B(0) = Q^{-1}$.*

17.4 Maximum Sets of External Practical Stability

Consider discrete system (17.1), $\Phi(k) \subset D$ are compact sets, $0 \in int\Phi(k)$, $k = 0, 1, \ldots, N$ [22].

Definition 17.6 We say that E_* is the maximum set of external practical stability with respect to the zero solution of system (17.1) if it consists of all $x_0 \in \mathbb{R}^n$ such that there exists $k \in \{0, 1, \ldots, N\}$ so that $x(k, x_0) \in \Phi(k)$

Consider some properties of the set E_*.

Theorem 17.18 *The maximum set of external practical stability E_* is compact in $\mathbb{R}^n$ and*

$$E_* = \cup_{k=0}^{N} \varphi(k, \Phi(k)).$$

Here $\varphi(k, \Phi(k)) = \cup_{x \in \Phi(k)} \varphi(k, x)$, $k = 0, 1, \ldots, N$.

Theorem 17.19 *Assume that $x_0 \in \partial E_*$. Then there exists $t \in \{0, 1, \ldots, N\}$ such that $x(t, x_0) \in \partial\Phi(t)$, and $x(k, x_0) \notin int\Phi(k)$ for all $k = 0, 1, \ldots, N$.*

Theorem 17.20 *Suppose that $\Phi(k)$ is closure of some open simply connected bounded set, $k = 0, 1, \ldots, N$. We assume that for $x_0 \in D$ there exists $t \in \{0, 1, \ldots, N\}$ such that $x(t, x_0) \in \partial\Phi(t)$, and $x(k, x_0) \notin int\Phi(k)$ for all $k = 0, 1, \ldots, N$. Then $x_0 \in \partial E_*$.*

One can observe that under theorem 17.20 conditions any point $x_0 \in intE_*$ if and only if there exists $k \in \{0, 1, \ldots, N\}$ such that $x(k, x_0) \in int\Phi(k)$. Consider linear system (17.4) and assume that $r(k) \in int\Phi(k)$, $k = 0, 1, \ldots, N$.

Theorem 17.21 *If $\Phi(k) \in conv(\mathbb{R}^n)$, $k = 0, 1, \ldots, N$ and $0 \in intE_*$, then the maximum set of external practical stability E_* is starshaped. The Minkowski function of the set E_* is equal to*

$$m_*(x_0) = \inf\left\{\lambda > 0 : \frac{x_0}{\lambda} \in E_*\right\} = \min_{k=0,1,\ldots,N} \max_{\psi \in S} \frac{\langle \Theta(k)x_0, \psi \rangle}{c(\Phi(k), \psi) - \langle r(k), \psi \rangle},$$

$x_0 \in \mathbb{R}^n$, and $E_* = \{x_0 \in \mathbb{R}^n : m_*(x_0) \leq 1\}$.

Theorem 17.22 *The inverse Minkowski function of the set* E_* *satisfies*

$$d_*(x_0) = \max_{k=0,1,\dots,N} \min_{\psi \in P(k)} \frac{c(\Phi(k), \psi) - \langle r(k), \psi \rangle}{\langle \Theta(k)x_0, \psi \rangle},$$

where $P(k) = \{\psi \in S : \langle \Theta(k)x_0, \psi \rangle > 0\}$, $x_0 \in \mathbb{R}^n$. *In this case* $E_* = \cup_{e \in S} [0, d_*(e)]\, e$.

Theorem 17.23 *The support function of the set* E_* *can be written as follows*

$$c(E_*, \xi) = \max_{k=0,1,\dots,N} \left(c(\Phi(k), (\Theta^*(k))^{-1}\xi) - \left\langle r(k), (\Theta^*(k))^{-1}\xi \right\rangle \right), \ \xi \in \mathbb{R}^n.$$

In this case

$$coE_* = \cap_{\xi \in S} H(\xi),$$

where $H(\xi) = \{x \in \mathbb{R}^n : \langle x, \xi \rangle \le g(\xi)\}$, $\xi \in \mathbb{R}^n$.

Theorem 17.24 *A point* x_0 *belongs to* ∂E_* *if and only if*

$$\min_{k=0,1,\dots,N} \max_{\psi \in S} \frac{\langle \Theta(k)x_0, \psi \rangle}{c(\Phi(k), \psi) - \langle r(k), \psi \rangle} = 1.$$

Consider homogeneous system (17.5). In this case $b_k = 0$, $k = 0, 1, \dots, N - 1$.

Corollary 17.3 *The Minkowski function of the set* E_* *is equal to*

$$m_*(x_0) = \min_{k=0,1,\dots,N} \max_{\psi \in S} \frac{\langle \Theta(k)x_0, \psi \rangle}{c(\Phi(k), \psi)}, \ x_0 \in \mathbb{R}^n,$$

the inverse Minkowski function can be defined as follows

$$d_*(x_0) = \max_{k=0,1,\dots,N} \min_{\psi \in P(k)} \frac{c(\Phi(k), \psi)}{\langle \Theta(k)x_0, \psi \rangle},$$

where $P(k) = \{\psi \in S : \langle \Theta(k)x_0, \psi \rangle > 0\}$, $x_0 \in \mathbb{R}^n$, *and the support function of the set* E_* *is equal to*

$$c(E_*, \xi) = \max_{k=0,1,\dots,N} c(\Phi(k), (\Theta^*(k))^{-1}\xi), \ \xi \in \mathbb{R}^n.$$

Example 17.3 Consider the particular case $\Phi(k) = p_k K_1(0)$, $p_k > 0$, $k = 0, 1, \dots, N$. According to Corollary 17.3 we obtain

$$m_*(x_0) = \min_{k=0,1,\dots,N} \frac{\|\Theta(k)x_0\|}{p_k}, \ d_*(x_0) = \max_{k=0,1,\dots,N} \frac{p_k}{\|\Theta(k)x_0\|}, \ x_0 \in \mathbb{R}^n,$$

$$c(E_*, \xi) = \max_{k=0,1,\dots,N} p_k \left\| (\Theta^*(k))^{-1}\xi \right\|, \ \xi \in \mathbb{R}^n.$$

References

1. Chetaev, N.G.: On certain questions related to the problem of the stability of unsteady motion. J. Appl. Math. Mech. **24**, 6–19 (1960)
2. Lasalle, J., Lefshetz, S.: Stability by Lyapunov Direct Method and Application. Academic Press, Boston (1961)
3. Kirichenko, N.F.: Introduction to Stability Theory. Vyshcha Shkola, Kyiv (1978)
4. Bublik, B.N., Garashchenko, F.G., Kirichenko, N.F.: Structural - Parametric Optimization and Stability of Bunch Dynamics. Naukova dumka, Kyiv (1985)
5. Grippo, L., Lampariello, F.: Practical stability of discrete-time systems. J. Franklin Inst. **302**(3), 213–224 (1976)
6. Lakshmikantham, V., Leela, S., Martynyuk, A.A.: Practical Stability of Nonlinear Systems. World Scientific, Singapore (1990)
7. Bashnyakov, O.M., Garashchenko, F.G., Pichkur, V.V.: Practical Stability. Estimations and Optimization. Taras Shevchenko National University of Kyiv, Kyiv (2008)
8. Bashnyakov, A.N., Pichkur, V.V., Hitko, I.V.: Conditions of practical stability of discrete systems and Lyapunov functions. Syst. Res. Inf. Technol. **3**, 125–133 (2012)
9. Pichkur, V.: On practical stability of differential inclusions using Lyapunov functions. Discret. Contin. Dyn. Syst. Ser. B **22**, 1977–1986 (2017)
10. Pichkur, V.: Maximum sets of initial conditions in practical stability and stabilization of differential inclusions. In: Sadovnichiy, V.A., Zgurovsky, M. (eds.) Modern Mathematics and Mechanics. Fundamentals, Problems and Challenges, pp. 397–410. Springer, Berlin (2019)
11. Benabdallah, A., Ellouze, I., Hammami, M.A.: Practical stability of nonlinear time-varying cascade systems. J. Dyn. Control Syst. **15**(1), 45–62 (2015)
12. Ghanmi, B.: On the Practical h-stability of nonlinear systems of differential equations. J. Dyn. Control Syst. **25**(4), 691–713 (2019)
13. Cao, Y., Sun, J.: Practical stability of nonlinear measure differential equations. Nonlinear Anal.: Hybrid Syst. **30**, 163–170 (2018)
14. Wang, P., Wu, M., Wu, Y.: Practical stability in terms of two measures for discrete hybrid systems. Nonlinear Anal.: Hybrid Syst. **2**(1), 58–64 (2008)
15. Borysenko, S.D., Toscano, S.: Impulsive differential systems: the problem of stability and practical stability. Nonlinear Anal.: Theory, Methods Appl. **71**(12), e1843–e1849 (2009)
16. Yakar, C., Cicek, M., Gücen, M.: B: Practical stability, boundedness criteria and Lagrange stability of fuzzy differential systems. Comput. Math. Appl. **64**(6), 2118–2127 (2012)
17. Zhang, Y., Sun, J.: Practical stability of impulsive functional differential equations in terms of two measurements. Comput. Math. Appl. **48**(10–11), 1549–1556 (2004)
18. Garashchenko, F.G., Pichkur, V.V.: On optimal sets of practical stability. J. Autom. Inf. Sci. **31**(5), 1–12 (1999)
19. Garashchenko, F.G., Pichkur, V.V.: Analysis of optimal properties of practical stability of dynamic systems. Cybern. Syst. Anal. **38**(5), 703–715 (2002)
20. Garashchenko, F.G., Pichkur, V.V.: Properties of optimal sets of practical stability of differential inclusions. Part I. Part II. J. Autom. Inf. Sci. **38** (3), 1–19 (2006)
21. Bashnyakov, A.N., Pichkur, V.V., Hitko, I.V.: On maximal initial data set in problems of practical stability of discrete system. J. Autom. Inf. Sci. **43**(3), 1–8 (2011)
22. Garashchenko, F.G., Pichkur, V.V.: On properties of maximal set of external practical stability of discrete systems. J. Autom. Inf. Sci. **48**(3), 46–53 (2016)
23. Pichkur, V.V., Sasonkina, M.S.: Maximum set of initial conditions for the problem of weak practical stability of a discrete inclusion. J. Math. Sci. **194**, 414–425 (2013)
24. Pichkur, V.V., Sasonkina, M.S.: Practical stabilization of discrete control systems. Int. J. Pure Appl. Math. **81**(6), 877–884 (2012)
25. Linder, Ya.N., Pichkur, V.V.: Conditions of weak practical stability of differential inclusions with impulse impact. J. Autom. Inf. Sci. **43**(11), 57-65 (2011)
26. Tairova, M.S., Strakhov, Y.M.: Reverse algorithm for practical stabilization of discrete inclusions. Int. J. Pure Appl. Math. **116**(2), 89–499 (2017)

27. Kapustyan, A.V., Pankov, A.V., Valero, J.: On global attractors of multivalued semiflows generated by the 3D Bénard system. Set-Valued Var. Anal. **20**(3), 445–465 (2012)
28. Gorban, N.V., Kapustyan, O.V., Kasyanov, P.O., Paliichuk, L.S.: On global attractors for autonomous damped wave equation with discontinuous nonlinearity. Solid Mech. Its Appl. **211**, 221–237 (2014)
29. Perestyuk, N.A., Plotnikov, V.A., Samoilenko, A.M., Skripnik, N.V.: Differential Equations with Impulse Effects: Multivalued Right-Hand Sides with Discontinuities. Walter De Gruyter, Berlin (2011)
30. Gavrilyuk, I., Makarov, V., Vasylyk, V.: Exponentially Convergent Algorithms for Abstract Differential Equations. Frontiers in Mathematics. Birkhäuser, Springer, Basel (2011)
31. Dovgiy, S.O., Lyashko, S.I., Cherniy, D.I.: Algorithms of the discrete singularity method for computing technologies. Cybern. Syst. Anal. **53**(6), 950–962 (2017)

Chapter 18
Optimal Control for Systems of Differential Equations on the Infinite Interval of Time Scale

O. Stanzhytskyi, V. Mogylova, and O. Lavrova

Abstract In the given paper we are dealing with an optimal control problem on the semi-axis. We have stated the connection between solutions to optimal control problems on time scales and to the corresponding problem on real semi-axis. A new method is proposed for constructing a minimizing sequence for a problem on the semi-axis.

18.1 Introduction

The theory of differential equations on time scales has first arised in 1988, when Stefan Hilger has introduced the concept of Δ-derivative on any nonempty closed subset $\mathbb{T} \subset \mathbb{R}^1$ of the real numbers, named a time scale, in [1]. Since in the case of continuous time Δ-derivative coincides with the usual derivative, and in the case of discrete time—with the difference ratio, the theory of time scales allows us to unify continuous and discrete analysis. The basics of the theory of equations on time scales are given in [2].

The optimal control theory on time scales has developed intensively. For instance, several efforts have been made for spreading the maximum principle on time scales. Weak version of the maximum principle has been got in [3], strong version has been

O. Stanzhytskyi (✉) · O. Lavrova
Taras Shevchenko National University of Kyiv, Volodymyrska Street, 60, Kyiv 01601, Ukraine
e-mail: ostanzh@gmail.com

O. Lavrova
e-mail: lavrova_olia@ukr.net

V. Mogylova
National Technical University of Ukraine "Igor Sikorsky Kyiv Polytechnic Institute", Peremogy ave. 37, build 1, Kyiv 03056, Ukraine
e-mail: mogilova@gmail.com

V. A. Sadovnichiy and M. Z. Zgurovsky (eds.), *Contemporary Approaches and Methods in Fundamental Mathematics and Mechanics*, Understanding Complex Systems,
https://doi.org/10.1007/978-3-030-50302-4_18

obtained in [4] via using Ekeland's topological principle. In [5, 6] strong version of the maximum principle has been received in the form, convenient for using. The paper [7] is devoted to applying the dynamic programming method and viscosity solution for equations on time scales.

All the works, mentioned above, give, in the main, necessary conditions of optimality. Herewith it is important to have sufficient conditions for existence of optimal controls. Such conditions for some classes of equations have been obtained in terms of right parts of the system under investigation and the quality criterion in [8]. Let us note that similar conditions for systems of ordinary differential equations have been obtained in [9, 10], for functionally-differential equations—in [11], and for stochastic systems—in [12]. Qualitative behavior of control problems for infinite-dimensional systems was considered in [13, 14].

From the point of view of applications, it is important to approximate the optimal control problem with continuous time by the family of problems on time scales in the case, when the graininess function is tending to zero. This approach makes it possible to construct a minimizing sequence for optimal control problems for ordinary differential equations. This issue has been considered in [15] for a case of finite interval of time.

The aim of the given paper is to obtain similar results on semi-axis.

Let us note that the process of obtaining results on infinite intervals similar to results on finite intervals is a very non-trivial problem, which is confirmed, for example, from [16].

The given paper consists of introduction and two sections, and is organised as follows. In the Sect. 18.2 we introduce a few necessary results on the theory of time scales, needed in what follows, statement of the problem, and formulate our main results. Section 18.3 is devoted to proof of the theorems.

18.2 Preliminaries, Statement of the Problem and Main Results

Let $\mathbb{T}$ be a time scale, that is, any closed nonempty subset of $\mathbb{R}^1$. We assume that $\text{card}(\mathbb{T}) \geq 2$, $0 \in \mathbb{T}$, and $\sup \mathbb{T} = +\infty$. For every subset $A \subset \mathbb{R}^1$ we denote $A_{\mathbb{T}} = A \cap \mathbb{T}$.

The backward and forward jump operators $\rho, \sigma : \mathbb{T} \to \mathbb{T}$ are defined, respectively, by $\rho(t) = \sup\{s \in \mathbb{T} : s < t\}$ and $\sigma(t) = \inf\{s \in \mathbb{T} : s > t\}$ for every $t \in \mathbb{T}$.

A point $t \in \mathbb{T}$ is called *left-scattered* (respectively, *right-scattered*), if $\rho(t) < t$ ($\sigma(t) > t$). A point $t \in \mathbb{T}$ is called *left-dense*(respectively, *right-dense*), if $\rho(t) = t$ and $t > \inf \mathbb{T}$ ($\sigma(t) = t$).

The graininess function $\mu : \mathbb{T} \to \mathbb{R}^+$ is defined by $\mu(t) = \sigma(t) - t$. We denote by RS the set of all right-scattered points of $\mathbb{T}$, and by RD—the set of all right-dense points.

The function $f : \mathbb{T} \to \mathbb{R}^d$ is called Δ-differentiable at $t \in \mathbb{T}$, if the limit

$$f^{\Delta}(t) = \lim_{\substack{s\to t\\ s\in\mathbb{T}}} \frac{f(\sigma(t)) - f(s)}{\sigma(t) - s}$$

exists in $\mathbb{R}^d$.

Let μ_Δ be the Lebesgue Δ-measure on $\mathbb{T}$, defined in terms of Caratheodory's extension in [2, Ch. 5]. For all $c, d \in \mathbb{T}$ such that $c \le d$ one has $\mu_\Delta([c, d)_\mathbb{T}) = d - c$. Recall that $A \subset \mathbb{T}$ is μ_Δ-measurable set of $\mathbb{T}$ if and only if A is usual μ_L-measurable set of $\mathbb{R}^1$, where μ_L denotes the standard Lebesgue measure. Moreover, if $A \subset \mathbb{T}$, then

$$\mu_\Delta(A) = \mu_L(A) + \sum_{r\in A\cap \mathrm{RS}} \mu(r).$$

Let $A \subset \mathbb{T}$. Some property is said to hold Δ-almost everywhere (Δ-a. e.) on A, if it holds for every $t \in A\backslash A'$, where $A' \subset A$ is some μ_Δ-measurable subset $\mathbb{T}$ such that $\mu_\Delta(A') = 0$.

We consider a function f, defined on $A \subset \mathbb{T}$, with values in $\mathbb{R}^d$.

Let $\tilde{A} = A \cup (r, \sigma(r)), r \in A \cap \mathrm{RS}$, and let $\tilde{f}$ be the extension of f, defined μ_L-a. e. on $\tilde{A}$, defined as $\tilde{f}(t) = f(t)$ for $t \in A$ and as $\tilde{f}(t) = f(t)$, where $t \in (r, \sigma(r))$ for every $r \in A \cap \mathrm{RS}$. Recall that f is μ_Δ-measurable on A if and only if $\tilde{f}$ is μ_L-measurable on $\tilde{A}$. The functional space $L^p_\mathbb{T}(A, \mathbb{R}^d)$ is the set of all functions f, defined Δ-a. e. on A, with values in $\mathbb{R}^d$, that are μ_Δ-measurable on A, and such that $\int_A |f(t)|^p \,\Delta t < \infty$, $p \ge 1$. Here $|\cdot|$ is the standard Euclidean norm of $\mathbb{R}^d$, and $\|\cdot\|$—the standard matrix norm. This space, endowed with the norm $\|f\|_{L^p_\mathbb{T}} = \int_A |f(t)|^p \,\Delta t$, is a Banach space [17]. We recall here that if $f \in L^1_\mathbb{T}(A, \mathbb{R}^d)$, then

$$\int_A f(t)\Delta t = \int_A f(t)dt + \sum_{r\in A\cap \mathrm{RS}} \mu(r) f(r). \tag{18.1}$$

Let $\mathbb{T}_\lambda$ be the set of time scales, $\lambda \in \Lambda \subset \mathbb{R}^1$, 0—a limit point of Λ, $0 \in \mathbb{T}_\lambda$ for all $\mathbb{T}_\lambda$, and $\sup \mathbb{T}_\lambda = +\infty$ for all $\lambda \in \Lambda$. Denote $\mu_\lambda = \sup_{t\in\mathbb{T}_\lambda} \mu_\lambda(t)$. If $\mu_\lambda \to 0$, $\lambda \to 0$, then $\mathbb{T}_\lambda$ is tending to a continuous time scale $\mathbb{T}_0 = [0, \infty)$ (for example, in Hausdorff metric), that is, $\mathbb{T}_\lambda$ goes into semi-axis $t \ge 0$. Therefore the question about connection between optimal control problems on $\mathbb{T}_\lambda$ and such problems on $\mathbb{R}^+$ arises.

Let consider the following optimal control problems on each of time scales $\mathbb{T}_\lambda$

$$\begin{cases} x^\Delta = f_1(t, x) + f_2(t, x)u(t) \\ x(0) = x_0 \end{cases} \tag{18.2}$$

with quality criterion of the form

$$J_\lambda(u) = \int_{[0,\sigma(\tau))_{\mathbb{T}_\lambda}} [g(t)A(t, x(t)) + B(t, u(t))]\Delta t \to \inf \tag{18.3}$$

or

$$J_\lambda(u) = \int_{[0,\sigma(\tau))_{\mathbb{T}_\lambda}} [g(t)A(t,x(t)) + |u(t)|^2]\Delta t \to \inf . \tag{18.4}$$

Here $x \in D$ is some phase vector, $x_0 \in D$—some fixed vector, x^Δ—delta-derivative on time scale, D—some bounded domain in $\mathbb{R}^d$, ∂D—its boundary, $\overline{D} = D \cup \partial D$, τ—the moment of the first exit of $x(t)$ from D, that is, $\tau = \sup_{t\in\mathbb{T}_\lambda} \{\forall s \in [0,t]_{\mathbb{T}_\lambda} : x(s) \in D\}$.

The control parameter $u(t)$, $t \geq 0$, is m-measured, measurable vector-function such that for μ_L-a. e. $t \geq 0$ $u(t) \in V$ is convex closed set in $\mathbb{R}^m$, $0 \in V$.

For every such control a continuous (continuous from the right for $t \in \mathrm{RS}$) function $x(t)$ is called the corresponding solution (trajectory) to the system (18.2) on $\mathbb{T}_\lambda$ such that the next conditions are valid:

(1) $x(0) = x_0$;
(2) $(t, x(t)) \in D$ for $t \in \mathbb{T}_\lambda$ before the moment of its exit from D;
(3) $x(t)$ satisfies the integral equation of the form

$$x(t) = x_0 + \int_{[0,t)_{\mathbb{T}_\lambda}} [f_1(s,x(s)) + f_2(s,x(s))u(s)]\Delta s \tag{18.5}$$

for $t \in \mathbb{T}_\lambda$ before the moment of its exit from D.

The control $u(t)$ is regarded to be an admissible for (18.2)–(18.3) and (18.2), (18.4), if the following conditions are fulfilled:

(1) $u(t) \in L^p_{\mathbb{T}_\lambda}$;
(2) $u(t) \in V$ for μ_Δ-a. e. $t \in \mathbb{T}_\lambda$;
(3) the solution $x(t)$, corresponding to $u(t)$, exists on $[0;\sigma(\tau))_{\mathbb{T}_\lambda}$, $\sigma(\tau) > 0$;
(4) $J_\lambda(u) < \infty$.

We will denote the set of all admissible controls as U.

Concerning the system (18.2), let the following conditions be true:

(a1) the mappings $f_1(t,x) : [0,+\infty)_{\mathbb{T}} \times D \to \mathbb{R}^d$ and $f_2(t,x) : [0,+\infty)_{\mathbb{T}} \times D \to \mathbb{R}^d \times \mathbb{R}^m$ are defined and measurable with respect to all their arguments, and satisfy the linear growth and Lipschitz conditions with respect to x, that is, there exists some $K > 0$ such that for any $t \in [0,+\infty)_{\mathbb{T}}$ and $x \in D$ we have

$$|f_1(t,x)| + \|f_2(t,x)\| \leq K(1+|x|), \tag{18.6}$$

and for all $t \in [0,+\infty)_{\mathbb{T}}$, $x_1, x_2 \in D$, we have

$$|f_1(t,x_1) - f_1(t,x_2)| + \|f_2(t,x_1) - f_2(t,x_2)\| \leq K\,|x_1 - x_2|; \tag{18.7}$$

(a2) $g(t) \in L^p[0,+\infty)_{\mathbb{T}}$, $g(t) \in [0,1]$ for all $t \geq 0$;

(a3) the mapping $A : [0, +\infty)_{\mathbb{T}} \times D \to \mathbb{R}^1$, $A(t, x) \geq 0$ for $t \geq 0$, $x \in D$, is continuous with respect to all its arguments, and there exists some constant $K_A > 0$ such that

$$A(t, x) \leq K_A(1 + |x|); \tag{18.8}$$

(a4) the mapping $B : [0, +\infty)_{\mathbb{T}} \times V \to \mathbb{R}^1$ is measurable with respect to all its arguments, and there exist some constants $a > 0$, $a_1 > 0$, and $p > 1$ such that

$$a_1|z|^p \leq B(t, z) \leq a|z|^p \tag{18.9}$$

for all $t \geq 0$, $z \in V$;

(a5) for any $t \geq 0$ and $z \in V$ $B(t, z)$ has a derivative $\frac{\partial B}{\partial z}$, satisfying the estimate

$$\left\| \frac{\partial B}{\partial z} \right\| \leq a_2 \|z\|^{p-1} \tag{18.10}$$

for some $a_2 > 0$, and besides a_2 does not depend on t, z;

(a6) there exists a function $\varphi(T) \to 0$, $T \to \infty$, such that for all scales $\mathbb{T}_\lambda$ the following condition is true

$$\int_{[T,\infty)_{\mathbb{T}_\lambda}} |u(t)|^p \Delta t \leq \varphi(T) \tag{18.11}$$

for all admissible controls and

$$\sum_{r \in (T,\infty)_{\mathbb{T}_\lambda} \cap \mathrm{RS}} \mu_\lambda(r) \leq \varphi(T). \tag{18.12}$$

Let (18.2)–(18.3) and (18.2), (18.4) have solutions for all $\mathbb{T}_\lambda$, $\lambda \in \Lambda$, and $\lambda = 0$.

Let us denote by $V_\lambda(x_0)$ the Bellman function of the problems (18.2)–(18.3) and (18.2), (18.4). Wherein $V_0(x_0) = V(x_0)$ is the Bellman function for the problems on the real semi-axis, that is, for the following problems

$$\begin{cases} \frac{dx}{dt} = f_1(t, x) + f_2(t, x)u(t) \\ x(0) = x_0, \end{cases} \tag{18.13}$$

$$J_0(u) = J(u) = \int_0^\tau [g(t)A(t, x(t)) + B(t, u(t))]dt \to \inf, \tag{18.14}$$

$$J_0(u) = J(u) = \int_0^\tau [g(t)A(t, x(t)) + |u(t)|^2]dt \to \inf. \tag{18.15}$$

The main results of our paper are the next theorems.

Theorem 18.1 *If (a1)—(a6) are true, then we have*

$$V_\lambda(x_0) \to V(x_0),\ \lambda \to 0, \tag{18.16}$$

uniformly in $x_0 \in D$.

Let the solutions to (18.2)–(18.3), (18.2), (18.4), and (18.13)–(18.14), (18.13), (18.15) are noted as $(x_\lambda^*, u_\lambda^*)$ and (x^*, u^*), respectively.

Let $T > 0, T \in \mathbb{T}_\lambda$, is fixed. Let $(x_{\lambda,T}^*, u_{\lambda,T}^*)$ note solution to (18.2)–(18.3), (18.2), (18.4) on the segment $[0, T]_{\mathbb{T}_\lambda}$. For the problem on $\mathbb{T}_\lambda$ we put

$$u_{\lambda,T,\infty}^*(t) = \begin{cases} u_{\lambda,T}^*(t),\ t \in [0, T]_{\mathbb{T}_\lambda}, \\ 0, \quad t > T,\ t \in \mathbb{T}_\lambda, \end{cases} \tag{18.17}$$

and $x_{\lambda,T,\infty}^*(t)$ is the trajectory, corresponding to it. This control is obviously an admissible one for (18.2)–(18.3) and (18.2), (18.4). For continuous problems (18.13)–(18.14) and (18.13), (18.15) we construct control $\tilde{u}_{\lambda,T,\infty}^*(t)$ with the help of control $u_{\lambda,T,\infty}^*(t)$ and the formula

$$\tilde{u}_{\lambda,T,\infty}^*(t) = \begin{cases} u_{\lambda,T,\infty}^*(t),\ t \in \mathbb{T}_\lambda, \\ u_{\lambda,T,\infty}^*(r),\ t \in (r, \sigma(r)). \end{cases} \tag{18.18}$$

This control is, clearly, also an admissible control for these problems, and $\tilde{x}_{\lambda,T,\infty}^*(t)$ is the corresponding trajectory.

Theorem 18.2 *Let conditions (a1)—(a6) be fulfilled. Then there exist sequences* $\{\lambda_n\} \in \Lambda$, $\lambda_n \to 0$, $n \to \infty$, *and* $\{T_k\} \in \mathbb{T}_{\lambda_n}$, $T_k \to \infty$, $k \to \infty$, *such that the following conditions are true:*

(1) the sequence $\{\tilde{u}_{\lambda_n,T_k,\infty}^*\}$ *is minimising sequence for (18.13)–(18.14), that is,*

$$J(\tilde{u}_{\lambda_n,T_k,\infty}^*) \to V,\ n \to \infty,\ k \to \infty; \tag{18.19}$$

(2) for this sequence we have

$$\tilde{u}_{\lambda_n,T_k,\infty}^* \to u^*,\ n \to \infty,\ k \to \infty, \tag{18.20}$$

weakly in $L^p(0, \infty)$;

(3) for this sequence we have

$$\tilde{x}_{\lambda_n,T_k,\infty}^*(t) \to x^*(t),\ n \to \infty,\ k \to \infty, \tag{18.21}$$

pointwise, uniformly on every finite interval.

If (18.13)–(18.14) has a unique solution, then convergence in (18.19)–(18.21) is valid for any sequences $\{\lambda_n\} \in \Lambda$, $\lambda_n \to 0$, *and* $\{T_k\} \in \mathbb{T}_{\lambda_n}$, $T_k \to \infty$, $n \to \infty$, $k \to \infty$.

For the functional (18.15) all these statements are valid with the change of weak convergence in (18.20) to strong convergence in $L^2(0, +\infty)$.

18.3 Proof of the Theorems

18.3.1 Proof of Theorem 18.1

We will realise the proof only for (18.2)–18.3) and (18.13)–(18.14), because for (18.2), (18.4) and (18.13), (18.15) it is similar.

Let V_λ^T note the Bellman function of the problem (18.2)–(18.3) on $[0, T]_{\mathbb{T}_\lambda}$, and V^T — the Bellman function of the problem (18.13)–(18.14) on $[0, T]$.

It follows from [18] that

$$V^T(x_0) \to V(x_0), \quad T \to \infty. \tag{18.22}$$

From the mentioned above work we have

$$V_\lambda^T(x_0) \to V_\lambda(x_0), \quad T \to \infty. \tag{18.23}$$

Now we are going to demonstrate that convergence in (18.23) is uniform in λ. Indeed, from the definition of the Bellman function we get

$$\begin{aligned} V_\lambda \le J(u^*_{\lambda,T,\infty}) &= \int_{[0,\sigma(\tau_T))_{\mathbb{T}_\lambda}} [g(t)A(t, x^*_{\lambda,T,\infty}(t)) + B(t, u^*_{\lambda,T,\infty}(t))]\Delta t + \\ &+ \int_{[T,\sigma(\tau_T))_{\mathbb{T}_\lambda}} g(t)A(t, x^*_{\lambda,T,\infty}(t))\Delta t = \\ &= V_\lambda^T + \int_{[T,\sigma(\tau_T))_{\mathbb{T}_\lambda}} g(t)A(t, x^*_{\lambda,T,\infty}(t))\Delta t, \end{aligned} \tag{18.24}$$

where τ_T is the moment of exit of $x^*_{\lambda,T,\infty}$ from D.

Let us note that, since $u^*_{\lambda,T,\infty}(t) = u^*_{\lambda,T}(t)$ on $[0, T]_{\mathbb{T}_\lambda}$, $x^*_{\lambda,T,\infty}(t) = x^*_{\lambda,T}(t)$ on $[0, T]_{\mathbb{T}_\lambda}$ in virtue of uniqueness of solution to the Cauchy problem for (18.2), wherefrom we get (18.24). But, due to (18.7) and boundedness of D, we obtain

$$\int_{[T,\sigma(\tau_T))_{\mathbb{T}_\lambda}} g(t)A(t, x^*_{\lambda,T,\infty}(t))\Delta t \le C \int_{[T,\infty)_{\mathbb{T}_\lambda}} g(t)\Delta t \tag{18.25}$$

for some constant C, that does not depend on T and λ. Thus, in virtue of (18.1) and (a6), we have

$$\int_{[T,\infty)_{\mathbb{T}_\lambda}} g(t)\Delta t = \int_{[T,\infty)_{\mathbb{T}_\lambda}} g(t)dt + \sum_{r\in[T,\infty)_{\mathbb{T}_\lambda}\cap \mathrm{RS}} \mu(r)g(r) \leq$$
$$\leq \int_{[T,\infty)} g(t)dt + \varphi(T) \to 0, \quad T \to \infty. \tag{18.26}$$

Hence, the second term in (18.24) is tending to zero while $T \to \infty$ uniformly in λ. Let us note that $(x^*_{\lambda,T}, u^*_{\lambda,T})$ is the optimal couple also for the problem on $[0,\infty)_{\mathbb{T}_\lambda}$, if $\sigma(\tau_T) \leq T$. In this case $V^T_\lambda = V_\lambda$.

Let $\sigma(\tau_T) > T$ for some λ. Then

$$V_\lambda = J(u^*_\lambda) = \int_{[0,T)_{\mathbb{T}_\lambda}} [g(t)A(t,x^*_\lambda(t)) + B(t,u^*_\lambda(t))]\Delta t +$$
$$+ \int_{[T,\sigma(\tau_T))_{\mathbb{T}_\lambda}} [g(t)A(t,x^*_\lambda(t)) + B(t,u^*_\lambda(t))]\Delta t \geq$$
$$\geq V^T_\lambda + \int_{[T,\sigma(\tau_T))_{\mathbb{T}_\lambda}} [g(t)A(t,x^*_\lambda(t)) + B(t,u^*_\lambda(t))]\Delta t. \tag{18.27}$$

But

$$\int_{[T,\sigma(\tau_T))_{\mathbb{T}_\lambda}} [g(t)A(t,x^*_\lambda(t)) + B(t,u^*_\lambda(t))]\Delta t \leq C\int_{[T,\infty)_{\mathbb{T}_\lambda}} g(t)\Delta t +$$
$$+ a\int_{[T,\infty)_{\mathbb{T}_\lambda}} \left|u^*_\lambda(t)\right|^p \Delta t \leq C\int_{[T,\infty)} g(t)dt + a\int_{[T,\infty)_{\mathbb{T}_\lambda}} \left|u^*_\lambda(t)\right|^p \Delta t +$$
$$+ \sum_{r\in[T,\infty)_{\mathbb{T}_\lambda}\cap \mathrm{RS}} \mu(r)g(r) \leq C\int_{[T,\infty)} g(t)dt + 2\varphi(T) \to 0$$

while $T \to \infty$. Thus, on the one hand, we conclude

$$V_\lambda - V^T_\lambda \leq \int_{[T,\infty)} g(t)dt + \varphi(T),$$

and, from the other hand, we have (18.27).

Thereby there exists some function $\psi(T) \to 0$, $T \to \infty$, such that the following estimate is valid for all λ

$$\left|V_\lambda - V^T_\lambda\right| \leq \psi(T) \to 0, \ T \to \infty, \tag{18.28}$$

which means uniform convergence in (18.23).

Let us fix an arbitrary $\varepsilon > 0$ and choose rather big $T > 0$ such that

$$\left|V^T(x_0) - V(x_0)\right| < \frac{\varepsilon}{3} \tag{18.29}$$

and

$$\left|V_\lambda^T(x_0) - V_\lambda(x_0)\right| < \frac{\varepsilon}{3} \tag{18.30}$$

uniformly in all $\lambda \in \Lambda$. Such a choice is possible due to (18.28). For such a fixed T the problem of optimal control has been considered on the finite interval $[0, T]$. Thus, we can choose $\lambda_0 \in \Lambda$ from [15] such that

$$\left|V_\lambda^T - V^T\right| < \frac{\varepsilon}{3}, \tag{18.31}$$

if $\lambda < \lambda_0$. Then we have (18.16) from (18.29)–(18.31), thereby completing the proof.

18.3.2 Proof of Theorem 18.2

It follows from Theorem 18.1 that

$$V_\lambda^T(x) \to V(x_0), \quad T \to \infty, \quad \lambda \to 0.$$

It means that there exists some $L > 0$ such that

$$L \geq V_\lambda^{T_n} = J_\lambda(u^*_{\lambda_n,T_k}) \geq a_1 \int_{[0,T_k)_{\mathbb{T}_\lambda}} |u^*_{\lambda_n,T_k}|^p \Delta t = a_1 \int_0^\infty |\tilde{u}^*_{\lambda_n,T_k,\infty}(t)|^p dt.$$

Thus, the sequence of admissible controls is weakly compact in $L^p(0, \infty)$. Hereupon there exists some weakly convergent subsequence $\tilde{u}^*_{\lambda_n,T_n,\infty}$, which we will note as $\tilde{u}^*_{\lambda_n,T_n,\infty}$. Then

$$\tilde{u}^*_{\lambda_n,T_k,\infty} \to u^* \text{ in } L^p(0, \infty), \quad n, k \to \infty.$$

Let $x^*_{T_n}$ note the solution to (18.13), corresponding to the control $\tilde{u}^*_{\lambda_n,T_k,\infty}$, τ_{T_k}—the moment of the first exit of the solution $x^*_{T_n}$ to the boundary of D. Then

$$\begin{aligned} J(\tilde{u}^*_{\lambda_n,T_k,\infty}) = &\int_0^{T_n} [g(t)A(t, x^*_{T_k}(t)) + B(t, \tilde{u}^*_{\lambda_n,T_k,\infty}(t))]dt + \\ &+ \int_{T_n}^{\tau_{T_n}} [g(t)A(t, x^*_{T_k}(t)) + B(t, \tilde{u}^*_{T_k,\infty}(t))]dt. \end{aligned} \tag{18.32}$$

But $B(t, \tilde{u}^*_{T_k,\infty}(t)) = 0$ for $t \geq T_k$ due to construction of $\tilde{u}^*_{\lambda_n,T_n,\infty}$ and a4). Herewith from a3) we have

$$\int_{T_n}^{\tau_n} g(t)A(t, x^*_{T_k}(t))dt \leq \int_{T_k}^{\tau_n} g(t)K_A\left(1 + |x^*_{T_k}(t)|\right) dt < C \int_{T_k}^{\infty} g(t)dt \to 0 \tag{18.33}$$

for some positive constant C.

But we get from [15] that the first term in (18.32) is tending to V_{T_k} while $\lambda_n \to 0$. Due to (18.33), convergence of the second term in (18.32) to zero is uniform in λ_n. The first statement of theorem 18.2 follows from these conclusions.

Let $x^*(t)$ note the solution to (18.13), corresponding to the control $u^*(t)$. Such a solution exists and is unique, clearly, in virtue of the conditions on the coefficients of the system. Proof of the fact that $(x^*(t), u^*(t))$ is the optimal couple for (18.13)–(18.14) is realised in standard way with the help of passage to the limit in (18.5) and (18.14). Thereby we get statements (2) and (3).

If (18.13)–(18.14) has a unique solution, then it follows from the earlier conclusions that it is possible to choose a weakly convergent to the optimal control subsequence from any sequence $\{\tilde{u}_{\lambda_n,T_k,\infty}\}$. But such a control is unique. Thus, these convergences in (18.19)–(18.21) take place for any $T_n \to \infty$ and $\lambda_k \to 0$. In order to prove the last statement, let us note that it is not difficult to show existence of the limit

$$\lim_{\substack{n\to\infty \\ k\to\infty}} \int_0^\infty \left|\tilde{u}^*_{\lambda_n,T_k,\infty}(t)\right|^2 dt = \int_0^\infty \left|u^*\right|^2 dt.$$

It left to apply the Schur's lemma, thereby completing our proof.

References

1. Hilger, S.: Ein Maßkettenkalkül mit Anwendungen auf Zentrumsmannigfaltigkeiten. Ph.D thesis, Universität Würzburg (1988)
2. Bohner, M., Peterson, A.: Advances in Dynamic Equations on Time Scales. Birkhäuser Boston Inc., Boston (2003)
3. Hilscher, R., Zeidan, V.: Weak maximum principle and accessory problem for control problems on time scales. Nonlinear Anal. **70**(9), 3209–3226 (2009)
4. Bourdin, L., Trelat, E.: Pontryagin maximum principle for finite dimensional nonlinear optimal control problems on time scales. SIAM J. Control Optim. **51**(5), 3781–3813 (2013)
5. Bohner, M., Kenzhebaev, K., Stanzhytskyi, O., Lavrova, O.: Pontryagins maximum principle for dynamic systems on time scales. Differ. Equ. Appl. **23**(7), 1161–1189 (2017)
6. Bourdin, L., Stanzhytskyi, O., Trelat, E.: Addendum to Pontryagin's maximum principle for dynamic systems on time scales. Differ. Equ. Appl. **23**(10), 1760–1763 (2017)
7. Danilov, V.Y., Lavrova, O.E., Stanzhytskyi, O.M.: Viscous solutions of the Hamilton-Jacobi-Bellman equation on time scales. Ukr. Math. J. **69(7)**, 1085–1106 (2017)
8. Lavrova, O.E.: Conditions for the existence of optimal control for some classes of differential equations on time scales. J. Math. Sci. **222**(3), 276–295 (2017)
9. Stanzhytskyi, A.N., Samoilenko, E.A.: Coefficient conditions for existence of an optimal control for systems of differential equations. Sib. Math. J. **55**(1), 156–170 (2014)
10. Stanzhitskii, A.N., Dobrodzii, T.V.: Study of optimal control problems on the half-line by the averaging method. Differ. Equ. **47**(2), 264–277 (2011)
11. Kichmarenko, O., Stanzhytskyi, O.: Sufficient conditions for the existence of optimal controls for some classes of functional-differential equations. Nonlinear Dyn. Syst. Theory **18**(2), 196–211 (2018)
12. Stanzhytskyi, A.N., Samoilenko, E.A., Mogileva, V.V.: On the existence of an optimal feedback control for stochastic systems. Differ. Equ. **49**(11), 1456–1464 (2013)
13. Gorban, N.V., Kasyanov, P.O.: On regularity of all weak solutions and their attractors for reaction-diffusion inclusion in unbounded domain. Solid Mech. Its Appl. **211**, 205–220 (2014)

14. Gorban, N.V., Gluzman, M.O., Kasyanov, P.O., Tkachuk, A.M.: Long-time behavior of state functions for badyko models. Stud. Syst., Decis. Control **69**, 351–359 (2016)
15. Stanzhytskyi, O., Lavrova, O., Mogylova, V., Misiats, O.: Approximation of the optimal control problem on an interval with a family of optimization problems on time scales. Nonlinear Dyn. Syst. Theory **17**(3), 303–314 (2017)
16. Dzhumabaev, D.S., Uteshova, R.E.: A limit with weight solution in the singular point of a nonlinear ordinary differential equation and its property. Ukr. Math. J. **69**(12), 1717–1722 (2018)
17. Agarwal, R.P., Otero-Espinar, V., Perera, K., Vivero, D.R.: Basic properties of Sobolev's spaces on time scales. Adv. Differ. Equ. article 38121 (2006)
18. Kovalchuk, T.V., Lavrova, O.E., Mogylova, V.V.: Optymalne keruvannya deyakymy klasamy dynamichnyx rivnyan na neskinchennomu intervali chasovoi shkaly. Nelinijni kolyvannya., **2(3)**, 58–72 (2019) (in Ukrainian)

Chapter 19
Approximate Feedback Control for Hyperbolic Boundary-Value Problem with Rapidly Oscillating Coefficients in the Case of Non-convex Objective Functional

Olena Kapustian

Abstract The article deals with the optimal control problem consisting of the hyperbolic boundary-value problem with rapidly oscillating coefficients and non-convex objective functional. In the general case, finding an exact formula for optimal control in the feedback form for such class of distributed processes in micro-inhomogeneous media does not seem possible. We assume that the corresponding problem with homogenized parameters allows finding optimal control in the feedback form. The main result of the work is to prove that obtained regulator realizes an approximate optimal control for the original problem.

19.1 Introduction

In this work, we focus on the finding effective methods of control for complicated infinite-dimensional systems, initiated in the works [4, 5, 14]. Finding control in the feedback form plays important role here. Interesting application of feedback control theory and its connection with evolution inclusions and stability of invariant sets can be found in [9, 10, 15, 16, 19–22]. In [7, 8, 11–13] it was proposed and substantiated a procedure for constructing approximate optimal feedback control for a wide class of distributed processes in micro-inhomogeneous medium. We use some known facts on G-convergence theory from [1–6]. In this paper, we consider the optimal control problem in the feedback form for a wave equation with rapidly oscillating coefficients and non-convex objective functional with superposition type nonlinearity. For parabolic equation such problem was considered in [7]. In general, to find the exact formula of optimal feedback control is not possible for such a problem because we can not directly apply the Fourier method. But the transition to the homogenized

O. Kapustian (✉)
Taras Shevchenko National University of Kyiv, Volodymyrska Street, 60, Kyiv 01601, Ukraine
e-mail: olena.kap@gmail.com

V. A. Sadovnichiy and M. Z. Zgurovsky (eds.), *Contemporary Approaches and Methods in Fundamental Mathematics and Mechanics*, Understanding Complex Systems,
https://doi.org/10.1007/978-3-030-50302-4_19

parameters may greatly simplifies the structure of the problem. Assuming that the problem with the homogenized coefficients already admits optimal feedback control (see [17]), we ground approximate optimal control in the feedback form for the original problem. We give an example of linear-quadratic control problem for specific conditions in this paper.

19.2 Setting of the Problem

Let $\Omega \subset R^n$ be a bounded domain and let $\varepsilon \in (0, 1)$ be a small parameter. In the cylinder $Q = (0, T) \times \Omega$ controlled process $\{y, u\}$ is described by the problem

$$\begin{cases} y_{tt}(t, x) = A^\varepsilon y(t, x) + u(t, x), \quad (t, x) \in Q, \\ y|_{\partial\Omega} = 0, \\ y|_{t=0} = y_0^\varepsilon, \; y_t|_{t=0} = y_1^\varepsilon, \end{cases} \tag{19.1}$$

with an objective functional

$$J_\varepsilon(u) = \int_\Omega q_\varepsilon(x, y(T, x))y(T, x)dx + \int_Q u^2(t, x)dtdx \to \inf, \tag{19.2}$$

where

$$A^\varepsilon = div(a^\varepsilon \nabla), \quad a^\varepsilon(x) = a\left(\frac{x}{\varepsilon}\right),$$

a is measurable, symmetric, periodic matrix, satisfying conditions of uniform ellipticity and boundness: $\exists \nu_1 > 0, \nu_2 > 0 \; \forall \eta \in R^n \; \forall x \in R^n$

$$\nu_1 \sum_{i=1}^n \eta_i^2 \le \sum_{i,j=1}^n a_{i,j}(x)\eta_i\eta_j \le \nu_2 \sum_{i=1}^n \eta_i^2, \tag{19.3}$$

$q_\varepsilon : \Omega \times R \mapsto R$ is a Caratheodory function, and there exist functions $C_1 \in L^2(\Omega)$, $C_2 \in L^1(\Omega)$, and constant $C > 0$, independent of $\varepsilon \in (0, 1)$, such that for all $\xi \in R$ and almost all $x \in \Omega$ the following inequalities hold

$$|q_\varepsilon(x, \xi)| \le C|\xi| + C_1(x), q_\varepsilon(x, \xi)\xi \ge -C_2(x). \tag{19.4}$$

Under these conditions the operator $q_\varepsilon(x, \cdot) : L^2(\Omega) \mapsto L^2(\Omega)$ is continuous. Hence, by conditions (19.3), (19.4) and properties of solutions of the problem (19.1) we obtain that functional $J_\varepsilon : L^2(Q) \to R$ is weakly lower semicontinuous and, therefore [14], the problem (19.1), (19.2) has a unique solution $\{\bar{y}^\varepsilon, \bar{u}^\varepsilon\}$ (optimal process) in class $W(0, T) \times L^2(Q)$, where $W(0, T)$ is a class of functions $y \in L^2(0, T; H_0^1(\Omega))$, which have distributional derivatives with respect to t from class $L^2(0, T; L^2(\Omega))$ [14]. In general case, we are not able to find the exact optimal feedback law for the problem (19.1), (19.2). However, in many cases [6] a transition to homogenized parameters simplifies the structure of the problem. We will assume that the problem with homogenized coefficients already admits optimal feedback control of the form $u[t, x, y, y_t]$.

The main goal of this paper is to prove the fact that the form $u[t, x, y, y_t]$ realizes an approximate feedback control in initial problem (19.1), (19.2), i.e. for any $\eta > 0$

$$|J_\varepsilon(\bar{u}^\varepsilon) - J_\varepsilon(u[t, x, y_\varepsilon, y_{\varepsilon t}])| < \eta \tag{19.5}$$

for $\varepsilon > 0$ small enough, where y_ε is a solution of the problem (19.1), (19.2) with control $u[t, x, y_\varepsilon, y_{\varepsilon t}]$.

19.3 Main Results

We shall use $\|\cdot\|$ to denote the norm and $(\cdot, \cdot)$ to denote the inner product in $L^2(\Omega)$. Let us assume that there exists a Caratheodory function $q : \Omega \times R \mapsto R$, such that

$$\begin{array}{c} \forall r > 0 \quad q_\varepsilon(x, \xi) \to q(x, \xi) \text{ weakly in } L_2(\Omega) \\ \text{uniformly for } |\xi| \le r. \end{array} \tag{19.6}$$

We refer to the following problem

$$\begin{cases} y_{tt}(t, x) = A^0 y(t, x) + u(t, x), \quad (t, x) \in Q, \\ y|_{\partial\Omega} = 0, \\ y|_{t=0} = y_0, \ y_t|_{t=0} = y_1, \end{cases} \tag{19.7}$$

$$J(u) = \int_\Omega q(x, y(T, x))y(T, x)dx + \int_Q u^2(t, x)dtdx \to \inf, \tag{19.8}$$

as an homogenized one for the problem (19.1), (19.2). Here a constant matrix a^0 is homogenized for a^ε [6], $A^0 = div(a^0\nabla)$, $y_0 \in H_0^1(\Omega)$, $y_1 \in L^2(\Omega)$, such that

$$\begin{array}{l} y_0^\varepsilon \to y_0 \text{ weakly in } H_0^1(\Omega), \ \varepsilon \to 0, \\ y_1^\varepsilon \to y_1 \text{ weakly in } L^2(\Omega), \ \varepsilon \to 0. \end{array} \tag{19.9}$$

Due to (19.6) function $q(x, \xi)$ satisfies conditions (19.4). So optimal control problem (19.7), (19.8) has a unique solution $\{\overline{y}, \overline{u}\}$.

Let us assume that the following conditions hold:

there exists a measurable map $u : [0, T] \times \Omega \times H_0^1(\Omega) \times L^2(\Omega) \mapsto L^2(\Omega)$ such that

$$\overline{u}(t, x) = u\,[t, x, \overline{y}(t, x), \overline{y}_t(t, x)] \text{ a.e. on } Q; \tag{19.10}$$

$$\exists\, D_1 > 0,\ D_2 > 0,\ \text{such that}\ \forall\, t \in [0,T],\ \forall\, y \in H_0^1(\Omega),\ \forall\, z \in L^2(\Omega)$$

$$\begin{aligned} \|u\,[t,x,y,z]\| &\le D_1\left(\|y\|_{H_0^1} + \|z\| + 1\right), \\ \|u\,[t,x,y,z] - u\,[t,x,y_1,z_1]\| &\le D_2\left(\|y - y_1\|_{H_0^1} + \|z - z_1\|\right). \end{aligned} \tag{19.11}$$

$$\begin{aligned} &y_n \to y \text{ weakly in } H_0^1(\Omega),\ z_n \to z \text{ weakly in } L^2(\Omega) \Rightarrow \\ &\Rightarrow \forall\, t \in [0,T]\ u\,[t,x,y_n,z_n] \to u\,[t,x,y,z] \text{ in } L^2(\Omega) \end{aligned} \tag{19.12}$$

Before we formulate the main result, we give a typical example of the function $q_\varepsilon : \Omega \times R \mapsto R$, for which the conditions (19.4), (19.6), (19.10), (19.11), (19.12) hold.

Example 19.1 Let $q_\varepsilon(x,\xi) = g\left(\frac{x}{\varepsilon}\right)\xi$, where g is measurable, bounded, positive, periodic function with mean value $\langle g\rangle$ [6]. Then conditions (19.4), (19.6) hold for $q(x,\xi) = \langle g\rangle\,\xi$.

Let $\{X_i\}$, $\{\lambda_i\}$ be solutions of the spectrum problem

$$\begin{cases} A^0 X_i = -\lambda_i X_i, \\ X_i\,|_{\partial\Omega} = 0, \end{cases}$$

$\{X_i\} \subset H_0^1\,(\Omega)$ is an orthonormal basis in $L^2\,(\Omega)$, $0 < \lambda_1 \le \lambda_2 \le \cdots$, $\lambda_i \to \infty$, $i \to \infty$.

We are looking for solution of (19.7), (19.8) in the form

$$y(t,x) = \sum_{i=1}^{\infty} y_i(t) X_i(x),\quad u(t,x) = \sum_{i=1}^{\infty} u_i(t) X_i(x).$$

Then we have countable family of one-dimensional optimal control problems:

$$\begin{cases} \ddot{y}_i(t) = -\lambda_i y_i(t) + u_i(t), \\ y_i(0) = (y_0, X_i),\ \dot{y}_i(0) = (y_1, X_i), \\ (y_i(T))^2 + \langle g\rangle^{-1} \int\limits_0^T (u_i(t))^2 dt \to \inf. \end{cases} \tag{19.13}$$

The optimal feedback control in each of these problems is defined by the formula [4]

$$u_i(t) = K_{1i}(t) y_i(t) + K_{2i}(t) \dot{y}_i(t), \tag{19.14}$$

where coefficients $K_{1i}(t)$, $K_{2i}(t)$ are defined through λ_i, $\langle g\rangle$ and they satisfy the estimates

$$\exists\, K > 0\ \forall\, t \in [0,T],\ \forall\, i \ge 1\ |K_{1i}(t)| \le \frac{K}{\sqrt{\lambda_i}},\quad |K_{2i}(t)| \le \frac{K}{\lambda_i}. \tag{19.15}$$

Passing to the original problem we get optimal feedback control

$$u[t,x,y,y_t] = \sum_{i=0}^{\infty}\Big(K_{1i}(t)(y,X_i) + K_{2i}(t)(y_t,X_i)\Big)X_i(x). \tag{19.16}$$

From Parseval's equality we have

$$\|u[t,x,\phi,\psi]\|^2 \le 2\frac{K^2}{\lambda_1^2}(\|\phi\|^2_{H_0^1} + \|\psi\|^2), \tag{19.17}$$

$$\|u[t,x,\phi_1,\psi_1] - u^0[t,x,\phi_2,\psi_2]\|^2 \le 2\frac{K^2}{\lambda_1^2}(\|\phi_1-\phi_2\|^2_{H_0^1} + \|\psi_1-\psi_2\|^2). \tag{19.18}$$

From this we immediately obtain (19.11). Further, for weakly convergent sequences $y_n \to y$ in $H_0^1(\Omega)$, $z_n \to z$ in $L^2(\Omega)$, and for arbitrary $i \ge 1$ we can write

$$(u[t,x,y_n,z_n] - u[t,x,y,z], X_i) =$$

$$= K_{1i}(t)(y_n - y, X_i) + K_{2i}(t)(z_n - z, X_i) \to 0,\ n \to \infty.$$

This equality and (19.17) mean that convergence (19.12) takes place in weak sense. Then from estimate

$$\sum_{i=N}^{\infty}(u[t,x,y_n,z_n],X_i)^2 \le \frac{K^2}{\lambda_N^2}(\|y_n\|^2_{H_0^1} + \|z_n\|^2).$$

we deduce strong convergence in $L^2(\Omega)$.

Now let us come back to our main task. Using feedback law (19.10), we consider the problem

$$\begin{cases} y_{tt} = A^{\varepsilon}y + u[t,x,y,y_t], \\ y|_{\partial\Omega} = 0, \\ y|_{t=0} = y_0^{\varepsilon},\ y_t|_{t=0} = y_1^{\varepsilon}. \end{cases} \tag{19.19}$$

The main result of this article is the following.

Theorem 19.1 *Let conditions (19.3), (19.4), (19.6), (19.10), (19.11), (19.12) hold and, moreover, there exists a positive function $l \in L^{\infty}(\Omega)$, such that for all $\varepsilon \in (0,1)$*

$$|q_{\varepsilon}(x,\xi_1) - q_{\varepsilon}(x,\xi_2)| \le l(x)|\xi_1 - \xi_2|. \tag{19.20}$$

Then for arbitrary $\eta > 0$ there exists $\bar{\varepsilon} \in (0,1)$, such that $\forall \varepsilon \in (0,\bar{\varepsilon})$

$$|J_{\varepsilon}(\bar{u}^{\varepsilon}) - J_{\varepsilon}(u[t,x,y_{\varepsilon},y_{\varepsilon t}])| < \eta \tag{19.21}$$

where $\{\bar{y}^\varepsilon, \bar{u}^\varepsilon\}$ *is an optimal process for the problem (19.1), (19.2),* y_ε *is a solution of the problem (19.19), the control* $u[t, x, y_\varepsilon, y_{\varepsilon t}]$ *is defined from (19.10).*

Proof Under conditions (19.11) the problem (19.19) has a unique solution y_ε in the class $W(0, T)$ [18]. Moreover, the following equality takes place

$$\frac{1}{2}\frac{d}{dt}\left(\|y_{\varepsilon t}\|^2 + (a^\varepsilon \nabla y_\varepsilon, \nabla y_\varepsilon)\right) = (u[t, x, y_\varepsilon, y_{\varepsilon t}], y_{\varepsilon t}). \tag{19.22}$$

Then from Gronwall Lemma we get

$$\sup_{t\in[0,T]}\left(\|y_{\varepsilon t}(t)\|^2 + \|y_\varepsilon(t)\|^2_{H_0^1(\Omega)}\right) \le C_1. \tag{19.23}$$

Here and after constants C_i depend only on parameters of the problem (19.1), (19.2) and do not depend on ε.

Due to inequalities (19.10) and (19.23) we deduce that

$$\begin{aligned} &\{y_\varepsilon\} \text{ is bounded in } L^\infty(0, T; H_0^1(\Omega));\\ &\{y_{\varepsilon t}\} \text{ is bounded in } L^\infty(0, T; L^2(\Omega));\\ &\{y_{\varepsilon tt}\} \text{ is bounded in } L^\infty(0, T; H^{-1}(\Omega)). \end{aligned} \tag{19.24}$$

Then from Compactness theorem [18] we conclude that there exists function $z(t, x) \in W(0, T)$ such that up to subsequence

$$\begin{aligned} &y_\varepsilon \to z \text{ in } C([0, T]; L^2(\Omega)) \cap C([0, T]; H^1_{0w}(\Omega)),\\ &y_{\varepsilon t} \to z_t \text{ in } C([0, T]; H^{-1}(\Omega)) \cap C([0, T]; L^2_w(\Omega)). \end{aligned} \tag{19.25}$$

From (19.25), (19.12) and Lebesgue's dominated convergence theorem we derive that

$$u_\varepsilon(t, x) := u[t, x, y_\varepsilon, y_{\varepsilon t}] \to u[t, x, z, z_t] \text{ in } L^2(Q). \tag{19.26}$$

Now we use the following result about convergence of hyperbolic operators which is the consequence of G-convergence of A^ε to A^0.

Lemma 19.1 ([1]) *Let* $y_0^\varepsilon \to y_0$ *weakly in* $H_0^1(\Omega)$, $y_1^\varepsilon \to y_1$ *weakly in* $L^2(\Omega)$, $u_\varepsilon \to u$ *weakly in* $L^2(Q)$. *Then* $y^\varepsilon \to y$ *in* $C([0, T]; L^2(\Omega))$ *and weakly in* $L^2(0, T; H_0^1(\Omega))$, *where* y^ε *is a solution of the problem (19.1) with control* u_ε, y *is a solution of the problem (19.7) with control function* u.

From this results we deduce that z is a solution of the problem (19.19) with operator A^0 and initial data y^0, y^1.

Since the optimal control problem (19.7), (19.8) has a unique solution $\{\bar{y}, \bar{u}\}$ and formula $\bar{u}(t, x) = u[t, x, \bar{y}(t, x), \bar{y}_t(t, x)]$ is valid for control $\bar{u}$, then $\bar{y}$ is a solution of the problem (19.19) with operator A^0 and initial data y^0, y^1. However, this problem also has a unique solution, so $\bar{y} \equiv z$, and moreover, the convergence (19.25) hold as $\varepsilon \to 0$ (not only along subsequence).

To finish the first part of the proof we need the following result

Lemma 19.2 ([7]) *Let functions* q_ε *satisfy conditions (19.4), (19.6), (19.20) for* $\varepsilon \to 0$, *and* $y_\varepsilon \to y$ *in* $L^2(\Omega)$. *Then*

$$q_\varepsilon(x, y_\varepsilon) \to q(x, y) \; weakly \; in \; L^2(\Omega).$$

From this lemma and convergence (19.25), (19.26) we derive

$$J_\varepsilon(u_\varepsilon) = \int\limits_\Omega q_\varepsilon(x, y_\varepsilon(T, x)) y_\varepsilon(T, x) dx + \int\limits_Q u_\varepsilon^2(t, x) dt dx \to J(\bar{u}). \tag{19.27}$$

Now we consider the optimal process $\{\bar{y}^\varepsilon, \; \bar{u}^\varepsilon\}$ of the problem (19.1), (19.2). We have the inequality

$$-\int_\Omega C_2(x) dx + \int_Q \left(\bar{u}^\varepsilon\right)^2 (t, x) dt dx \le$$

$$\le J_\varepsilon \left(\bar{u}^\varepsilon\right) \le C \int_\Omega z_\varepsilon^2(T, x) dx + \int_\Omega C_1(x) z_\varepsilon(T, x) dx,$$

where z_ε is a solution of the problem (19.1) with control $u \equiv 0$. For z_ε estimate (19.23) holds, so, the sequence $\{\bar{u}^\varepsilon\}$ is bounded in $L^2(Q)$. Then, there exists $v \in L^2(Q)$, such that up to subsequence

$$\bar{u}^\varepsilon \to \; v \text{ weakly in } L^2(Q), \;\; \varepsilon \to 0.$$

By the boundedness of $\{\bar{u}^\varepsilon\}$ in $L^2(Q)$ and equality (19.22) we deduce the boundedness of the sequence $\{\bar{y}^\varepsilon\}$ in the sense of (19.24). So up to subsequence it tends to some function $y \in W(0, T)$ as $\varepsilon \to 0$ within the meaning of (19.25). Using Lemma 19.1, we obtain that y is a solution of the problem (19.7) with control v.

Let us show that the process $\{y, \; v\}$ is optimal in the problem (19.7), (19.8). From the optimality of $\{\bar{y}^\varepsilon, \; \bar{u}^\varepsilon\}$ for arbitrary $u \in L^2(Q)$ the following inequality holds

$$J_\varepsilon \left(\bar{u}^\varepsilon\right) \le J_\varepsilon(u) = \int\limits_\Omega q_\varepsilon(x, p_\varepsilon(T, x)) p_\varepsilon(T, x) dx, \tag{19.28}$$

where p_ε is a solution of the problem (19.1) with control u. It is clear, that $\{p^\varepsilon\}$ is bounded in in the sense of (19.24) and up to subsequence it tends to some function $p \in W(0, T)$ as $\varepsilon \to 0$ within the meaning of (19.25). Moreover, p is a solution of the problem (19.7) with control u. Then Lemma 19.2 implies

$$
\begin{aligned}
J_\varepsilon(u) &\to J(u) = \int_\Omega q(x, p(T,x))p(T,x)dx, \\
\liminf_{\varepsilon\to 0} J_\varepsilon(\bar{u}^\varepsilon) &\geq J(v) = \int_\Omega q(x, y(T,x))y(T,x)dx.
\end{aligned}
\tag{19.29}
$$

From (19.28), (19.29) we ultimately derive inequality

$$
J(v) \leq J(u),
$$

which means that $\{y, v\}$ is an optimal process in the problem (19.7), (19.8). By the uniqueness it follows $\{y, v\} \equiv \{\overline{y}, \overline{u}\}$. After that, repeating arguments from [7] we can prove that

$$
J_\varepsilon(\bar{u}^\varepsilon) \to J(\overline{u}), \quad \varepsilon \to 0. \tag{19.30}
$$

Combining (19.27) and (19.30) we obtain the statements of the theorem.

References

1. Colombini, F., Spagnolo, S.: On the convergence of solutions of hyperbolic equations. Commun. Partial Differ. Equ. **3**(1), 77–103 (1978)
2. Denkowski, Z., Migorski, S.: Control problems for parabolic and hyperbolic equations via the theory of G convergence. Annali di Matematica Pura ed Applicata **149**, 22–39 (1987)
3. Denkowski, Z., Mortola, S.: Asymptotic behavior of optimal solutions to control problems for systems described by differential inclusions corresponding to partial differential equations. J. Optim. Theory Appl. **78**, 365–391 (1993)
4. Egorov, A.I.: Optimal Control by Heat and Diffusion Processes. Nauka, Moscow (1978)
5. Fursikov, A.V.: Optimal Control of Distributed Systems. Theory and Applications. AMS, Providence (1999)
6. Jikov, V.V., Kozlov, S.M., Oleinik, O.A.: Homogenization of Differential Operators and Integral Functionals. Springer, Berlin (1994)
7. Kapustian, O.A., Sobchuk, V.V.: Approximate homogenized synthesis for distributed optimal control problem with superposition type cost fucntional. Stat. Optim. Inf. Comput. **6**, 233–239 (2018)
8. Kapustian, O.A., Sukretna, A.V.: Approximate averaged synthesis of the problem of optimal control for a parabolic equation. Ukr. Math. J. **56**(10), 1653–1664 (2004)
9. Kapustyan, A.V., Pankov, A.V., Valero, J.: On global attractors of multivalued semiflows generated by the 3D Bénard system. Set-Valued Var. Anal. **20**(3), 445–465 (2012)
10. Kapustyan, O.V., Shkundin, D.V.: Global attractor of one nonlinear parabolic equation. Ukr. Math. J. **55**(4), 446–455 (2003)
11. Kapustyan, O.V., Kapustian, O.A., Sukretna, A.V.: Approximate bounded synthesis for one weakly nonlinear boundary-value problem. Nonlinear Oscil. **12**, 297–304 (2009)
12. Kapustyan, O.V., Kapustian, O.A., Sukretna, A.V.: Approximate stabilization for a nonlinear parabolic boundary-value problem. Ukr. Math. J. **63**, 759–767 (2011)
13. Kapustyan, O.V., Kapustian, O.A., Mazur, O.K.: Problem of optimal control for the Poisson equation with nonlocal boundary conditions. J. Math. Sci. **201**, 325–334 (2014)
14. Lions, J.-L.: Optimal Control of Systems, Governed by Partial Differential Equations. Springer, Berlin (1971)
15. Pichkur, V.: On practical stability of differential inclusions using Lyapunov functions. Discret. Contin. Dyn. Syst. Ser. B **22**, 1977–1986 (2017)

16. Pichkur, V.: Maximum sets of initial conditions in practical stability and stabilization of differential inclusions. In: Sadovnichiy, V.A., Zgurovsky, M. (eds.) Modern Mathematics and Mechanics. Fundamentals Problems and Challenges, pp. 397–410 (2019)
17. Rusina, A.V., Pereguda, O.V.: Approximate synthesis of distributed optimal control for hyperbolic equation with fast oscillating coefficients. J. Math. Sci. **228**, 306–313 (2018)
18. Sell, G.R., You, Y.: Dynamics of Evolutionary Equations. Springer, Berlin (2002)
19. Stanzhitskii, A., Dobrodzii, T.: Study of optimal control problems on the half-line by the averaging method. Differ. Equ. **47**(2), 264–277 (2011)
20. Stanzhyts'kyi, O.: Investigation of exponential dichotomy of it stochastic systems by using quadratic forms. Ukr. Math. J. **53**(11), 1882–1894 (2001)
21. Stanzhyts'kyi, O.: Investigation of invariant sets of it stochastic systems with the use of Lyapunov functions. Ukr. Math. J. **53**(2), 323–327 (2001)
22. Zgurovsky, M.Z., Kasyanov, P.O., Valero, J.: Noncoercive evolution inclusions for Sk type operators. Int. J. Bifurc. Chaos **20**(9), 2823–2834 (2010)

Chapter 20
Decomposition of Intersections with Fuzzy Sets of Operands

S. O. Mashchenko and D. O. Kapustian

Abstract We investigate the operation of intersection of fuzzy sets with a fuzzy set of operands. This is a natural generalization of the corresponding operation which involves a crisp set of operands. The decomposition approach was used to study the intersections of fuzzy sets with a fuzzy set of operands. The result of this operation is a type-2 fuzzy set (T2FS). We prove several results which enable us to simplify constructing the type-2 membership function. It is shown that the resulting T2FS can be decomposed according to secondary membership grades into a finite collection of type-1 fuzzy sets. Each of these sets is the intersection of the original sets with a crisp set of operands. This crisp set is the corresponding α-cut of the fuzzy set of operands. Illustrative examples are given.

20.1 Introduction

In [25], Zadeh proved the α-cut decomposition theorem. It allows one to decompose fuzzy sets into a collection of crisp sets, in accordance with membership degrees. Together with the extension principle [25, 27] this theorem has become a powerful tool which helped to construct the theory of fuzzy sets. Yager [24] generalized the concept of α-cuts and the extension principle to interval-valued fuzzy sets. Zeng and Li [28] considered various mathematical formulations of the extension principle for interval fuzzy sets and established a classification theorem for interval-valued fuzzy sets. Zeng et al. [29] gave an axiomatic definition of the generalized extension principle for a lattice-valued fuzzy set and developed the concept of α-cut for it. In [30], Zeng and Shi investigated properties of α-cuts for interval-valued fuzzy sets.

S. O. Mashchenko (✉) · D. O. Kapustian
Taras Shevchenko National University of Kyiv, Volodymyrska Street, 60, Kyiv 01601, Ukraine
e-mail: s.o.mashchenko@gmail.com

D. O. Kapustian
e-mail: kapustian.darina404@gmail.com

V. A. Sadovnichiy and M. Z. Zgurovsky (eds.), *Contemporary Approaches and Methods in Fundamental Mathematics and Mechanics*, Understanding Complex Systems, https://doi.org/10.1007/978-3-030-50302-4_20

Wu and Mendel [22, 23] used α-cuts to determine the linguistic weighted average for an interval-valued fuzzy set.

Fuzzy sets can be vague even though they are the main tool for formalizing uncertainty. In [17], Mendel and John pointed out that there are (at least) four sources of uncertainty in fuzzy sets. These are: the meanings of words that are used in the description of fuzzy sets (words mean different things to different people); the ambiguity of the opinions of various experts; measurement noises and data noises. All these factors lead to an uncertainty of the fuzzy set membership function. The main instrument for describing such an uncertainty is a type-2 fuzzy set (T2FS) theory, introduced by Zadeh in 1971 [26]. T2FSs attracted significant attention of researchers both in theory and in applications. On the other hand, T2FSs are difficult to understand and their use is computationally complicated.

The need for research and practical application of T2FSs led to the development of representation theorems. Four basic representation theorems are known for T2FSs. They are about vertical slice, wavy slice [17], α-plane (or z-slice) [11, 20] and representations utilizing computational geometry [3]. These representations are mainly used to reduce the computational complexity of operations on T2FSs. Chen and Kawase [2], Tahayori et al. [19] proved representation theorems for operations on T2FSs. Liu et al. [11, 18] used the idea of the α-plane representation for T2FSs to construct the centroid type-reduction strategy. Wagner and Hagras [20, 21] exploited the idea of decomposing T2FSs into a collection of interval-valued fuzzy sets with the help of the z-slices. Hamrawi and Coupland [4, 5] introduced arithmetic operations for type-2 fuzzy numbers. In [6–8] Hamrawi, Coupland and John explored the use of the α-cuts concept and the expansion principle for T2FSs. They showed that operations on general T2FSs may decomposed into a collection of type-2 interval operations or crisp interval operations. Summarizing, the decomposition approaches to T2FSs are a powerful tool for their investigation and practical applications.

In [14], the intersection and union operations on sets with a fuzzy set of operands were defined. The result is a complicated mathematical object like T2FS. In the case of crisp sets a nice feature of these operations is that the form of the type-2 membership function of the resulting T2FS is rather simple. This facilitates an application of this operation to solving various problems [12, 13, 15]. In the general case, the computation of the type-2 membership functions of T2FSs of the intersection and union of fuzzy sets with a fuzzy set of operands is much more difficult. This makes it almost impossible to use these operations to solve applied problems. Therefore, the development of simple methods for their implementation is important. In this article, we investigate the operation of intersection with a fuzzy set of operands and consider the decomposition of the resulting T2FS according to secondary membership grades into a finite collection of classical fuzzy sets. Each of these sets is the intersection of the original sets with a crisp set of operands. This crisp set is the corresponding α-cut of the fuzzy set of operands.

20.2 Preliminaries

There are two important preliminaries to this approach. These are the notion of the intersection with a fuzzy set of operands and the concept of T2FSs.

20.2.1 Intersection of Sets with a Fuzzy Set of Operands

Assume that fuzzy sets F_j with the membership functions $\varphi_j(x)$, $j \in N$ are defined on a crisp set X, where $N = \{1, 2, \dots, n\}$ is the set of their indices and n is the cardinality of N. To begin with, we consider the standard intersection $\bigcap_{j \in N} F_j$ of fuzzy sets F_j, $j \in N$. It is defined as the fuzzy set $F = \{(x, \phi(x)) : x \in X\}$ with the membership function $\phi(x) = \min_{j \in N} \varphi_j(x)$, $x \in X$. Let $\tilde{N}$ be a fuzzy set on N with an arbitrary membership function $\eta(j)$, $j \in N$. The following problem was addressed in [14]: 'What does we obtain if we intersect fuzzy sets F_j, $j \in N$ with the fuzzy set $\tilde{N}$ of operands?' The operation in question was denoted there by $\tilde{F} = \tilde{\bigcap}_{(j,\eta(j)) \in \tilde{N}} F_j$.

A natural generalization of the classical intersection operation among fuzzy sets leads to the situation in which one interprets the crisp set N as a fuzzy set $\tilde{N}$. In this way, the set $\tilde{F} = \tilde{\bigcap}_{(j,\eta(j)) \in \tilde{N}} F_j$ is defined by the membership function

$$y(x) = \min_{(j,\eta(j)) \in \tilde{N}} \varphi_j(x), \quad x \in X. \tag{20.1}$$

In this case, for each fixed $\hat{x} \in X$, the value of the membership function $y(\hat{x})$ is given by the value of the objective function of a discrete fuzzy mathematical problem with variables $j \in N$:

$$y(\hat{x}) = \min \varphi_j(\hat{x}), \quad s.t. \quad (j, \eta(j)) \in \tilde{N} \tag{20.2}$$

According to [14] a solution to (20.2) is a fuzzy set that we denote by $\tilde{N}(\hat{x}) = \{(j, \tilde{\eta}(\hat{x}, j)) : j \in N\}$. The support of $\tilde{N}(\hat{x})$ is the set (denoted by $N^{PO}(\hat{x}) \subseteq N$) of Pareto optimal solutions to a two-criteria discrete optimization problem:

$$\varphi_j(\hat{x}) \to \min, \ \eta(j) \to \max, \ s.t. \ j \in N. \tag{20.3}$$

Thus, the objective function φ_j of (20.2) is being minimized, whereas the membership degree $\eta(j)$ of the fuzzy set $\tilde{N}$ is being maximized. The restriction of the membership function $\eta(j)$, $j \in N$ to the set $N^{PO}(\hat{x}) \subseteq N$ is the membership function $\tilde{\eta}$ of the fuzzy set $\tilde{N}(\hat{x})$. In other words, this membership function takes the following form:

$$\tilde{\eta}(\hat{x}, j) = \begin{cases} \eta(j), & j \in N^{PO}(\hat{x}); \\ 0, & i \notin N^{PO}(\hat{x}). \end{cases} \tag{20.4}$$

To formalize the set of Pareto optimal solutions to two-criteria problem (20.3), a relation of domination on the set N of indices was introduced. This relation is generated by the objective functions of problem (20.3).

An index $j \in N$ dominates an index $i \in N$ in the descending order with respect to a function φ and in the ascending order with respect to a function η for $\hat{x} \in X$ (notation $jS_{\hat{x}}i$) if the inequalities $\varphi_j(\hat{x}) \leq \varphi_i(\hat{x})$ and $\eta(j) \geq \eta(i)$ hold, and $(\varphi_j(\hat{x}), \eta(j)) \neq (\varphi_i(\hat{x}), \eta(i))$. We rewrite these conditions in a more convenient form: either $\varphi_j(\hat{x}) < \varphi_i(\hat{x})$ and $\eta(j) \geq \eta(i)$, or $\varphi_j(\hat{x}) = \varphi_i(\hat{x})$ and $\eta(j) > \eta(i)$. This notion enables us to define the set of Pareto optimal solutions to two-criteria problem (20.3) in the form

$$N^{PO}(\hat{x}) = \{i \in N : j \neg S_{\hat{x}} i \quad \forall j \in N\} =$$
$$\{i \in N : \{j \in N | \; ((\varphi_j(\hat{x}) < \varphi_i(\hat{x})) \wedge (\eta(j) \geq \eta(i))) \vee$$
$$((\varphi_j(\hat{x}) = \varphi_i(\hat{x})) \wedge (\eta(j) > \eta(i)))\} = \varnothing\}.$$

To the set of solutions to problem (20.2) (which is the fuzzy set $\tilde{N}(\hat{x}) = \{(j, \tilde{\eta}(\hat{x}, j)) : j \in N\}$) there corresponds a fuzzy set $\tilde{\Phi}(\hat{x})$ on $[0, 1]$ of optimal values of the objective function of problem (20.2) with the membership function

$$\varphi(\hat{x}, y) = \max_{j \in N} \{\tilde{\eta}(\hat{x}, j) : \; \varphi_j(\hat{x}) = y\}, \;\; y \in J_{\hat{x}} \subseteq [0, 1],$$

where $J_{\hat{x}} = \{y \in [0, 1] : \; \varphi_j(\hat{x}) = y, \;\; j \in N\}$.

Thus, for each fixed $x \in X$, the values $y(x)$ of membership function (20.1) of the fuzzy set $\tilde{F} = \tilde{\bigcap}_{(j,\eta(j)) \in \tilde{N}} F_j$ also form a fuzzy set $\tilde{\Phi}(x)$ on $J_x \subseteq [0, 1]$. Therefore, the set $\tilde{F}$ is a T2FS.

Definition 20.1 We call the intersection $\tilde{\bigcap}_{(j,\eta(j)) \in \tilde{N}} F_j$ of fuzzy sets F_j, $j \in N$ with a fuzzy set $\tilde{N}$ of operands the T2FS $\tilde{F} = \{((x, y), \varphi(x, y)) : x \in X, \;\; y \in J_x \subseteq [0, 1]\}$ with the type-2 membership function

$$\varphi(x, y) = \begin{cases} \max_{j \in N} \{\tilde{\eta}(x, j) : \; \varphi_j(x) = y\}, & y \in J_x; \\ 0, & y \notin J_x; \end{cases} \tag{20.5}$$

$x \in X$, $y \in [0, 1]$ where

$$J_x = \{y \in [0, 1] : \varphi_j(x) = y, \;\; j \in N\} \tag{20.6}$$

is the set of primary membership degrees.

20.2.2 Type-2 Fuzzy Sets

The T2FSs were introduced by Zadeh in 1971 (see [26]) as an extension of the type-1 fuzzy sets (T1FSs). The degree of membership of elements in a T1FS is given by a number from the interval [0, 1], whereas the degree of membership of elements in a T2FS is a fuzzy set on [0, 1].

The following definition, based on the ideas of Karnik and Mendel [10], was given by Mendel and John [17]. A T2FS, denoted by $\tilde{A}$, on a crisp set X is characterized by the type-2 membership function (T2MF) $\mu_{\tilde{A}}(x, u)$, that is,

$$\tilde{A} = \{((x, u), \mu_{\tilde{A}}(x, u)) : \ x \in X, \ u \in J_x \subseteq [0, 1]\},$$

where J_x is the set of primary membership degrees $u \in [0, 1]$ of $x \in X$ to $\tilde{A}$; $\mu_{\tilde{A}}(x, u)$ is a crisp number from the interval [0, 1], known as a secondary grade of pair (x, u) to $\tilde{A}$. The primary membership degree u is usually understood as the degree of manifestation of some property (that defines the given fuzzy set) for $x \in X$. The secondary grade is usually [16] associated with the degree of truth of the corresponding primary degree u of this property for x.

Remark 20.1 According to amendments of Harding, Walker and Walker [9] and Aisbett, Rickard and Morgenthaler [1], the set of primary membership degrees $J_x \subseteq [0, 1]$ can be understood as the support of a T2MF for $x \in X$. Therefore, if we consider the T2MF $\mu_{\tilde{A}}(x, u)$ for all $u \in [0, 1]$, then we have to assume that $\mu_{\tilde{A}}(x, u) = 0$ for all $u \notin J_x$.

For each T2FS $\tilde{A}$ on X with the T2MF $\mu_{\tilde{A}}(x, u)$, $u \in J_x \subseteq [0, 1]$, $x \in X$ one can define embedded T2FSs and T1FSs [17]. Assume that, for any $x \in X$, the unique primary membership degree $u_x \in J_x$ is given. An embedded T2FS $\tilde{A}^e$ in $\tilde{A}$ is defined by $\tilde{A}^e = \{((x, u_x), \mu_{\tilde{A}}(x, u_x)) : \ x \in X\}$. A collection $A^e = \{(x, u_x) : \ x \in X\}$ is called an embedded T1FS in the T2FS $\tilde{A}$. Its membership function has the form $\mu_{A^e}(x) = u_x, x \in X$.

These notions are an important research tool for T2FSs. In [17], Mendel and John gave the Wavy-Slice Representation of T2FSs which states that each T2FS can be represented as a collection of all possible embedded T2FSs. These authors interpret the collection as a classical union of its elements in the sense of T1FS.

20.3 An Example of Intersection with a Fuzzy Set of Operands

Since Definition 20.1 given in Sect. 20.2.1 is rather complicated, we now give an example intended to facilitate its understanding.

Example 20.1 A diet is comprised of three parts: proteins (part 1), fats (part 2) and carbohydrates (part 3). We denote by $N = \{1, 2, 3\}$ the universal set of diet parts. For simplicity, we only consider four foods: bread (product x_1), milk (x_2), meat (x_3) and fish (x_4).

Let $X = \{x_1, x_2, x_3, x_4\}$ be the universal set of foods. We define three fuzzy sets on X:

- protein-rich foods $F_1 = \{(x_1; 0, 25), (x_2; 0, 1), (x_3; 1), (x_4; 0, 8)\}$;
- fat-rich foods $F_2 = \{(x_1; 0), (x_2; 0, 1), (x_3; 1), (x_4; 0, 25)\}$;
- carbohydrate-rich foods $F_3 = \{(x_1; 1), (x_2; 0, 25), (x_3; 0), (x_4; 0)\}$

with membership functions $\varphi_1(x)$, $\varphi_2(x)$ and $\varphi_3(x)$, respectively. Also, we know the fuzzy set $\tilde{N} = \{(1; 0, 8), (2; 0, 3), (3; 0, 5)\}$ (with a membership function $\eta(i)$, $i = 1, 2, 3$) of diet parts which are suitable for a healthy lifestyle. Our purpose is to determine the set of foods which are suitable for a healthy lifestyle within the whole diet. To accomplish, we have to find the intersection of three fuzzy sets F_1, F_2 and F_3 with the fuzzy set $\tilde{N}$ of operands. To this end, we shall use Definition 20.1.

Table 20.1 gives the values of membership functions $\varphi_1(x)$, $\varphi_2(x)$ and $\varphi_3(x)$; the sets $N^{PO}(x)$, $x \in X$ of Pareto-optimal solutions to two-criteria problem (20.3) and the values of the membership function $\tilde{\eta}(x, i)$ which are calculated with the help of formula (20.4). The values of the T2MF $\varphi(x, y)$ found by formula (20.5) are represented in Table 20.2.

Thus, the T2FS of foods which are suitable for a healthy lifestyle products has the form

$$\tilde{F} = \{((x_1; 0); 0, 3), ((x_1; 0, 25); 0, 8), ((x_2; 0, 1); 0, 8),$$
$$((x_3; 0); 0, 5), ((x_3; 1); 0, 8), ((x_4; 0); 0, 5), ((x_4; 0, 8); 0, 8).\} \tag{20.7}$$

Table 20.1 The calculated data for Example 20.1

Functions and sets	x_1	x_2	x_3	x_4
$\varphi_1(x)$	0,25	0,1	1	0,8
$\varphi_2(x)$	0	0,1	1	0,25
$\varphi_3(x)$	1	0,25	0	0
$N^{PO}(x)$	$\{1, 2\}$	$\{1\}$	$\{1, 3\}$	$\{1, 3\}$
$\tilde{\eta}(x, 1)$	0,8	0,8	0,8	0,8
$\tilde{\eta}(x, 2)$	0,3	0	0	0
$\tilde{\eta}(x, 3)$	0	0	0,5	0,5
$\phi_{0,3}(x)$	0	0,1	0	0
$\phi_{0,5}(x)$	0,25	0,1	0	0
$\phi_{0,8}(x)$	0,25	0,1	1	0,8

Table 20.2 T2MF $\tilde{\varphi}(x, y)$

y	x_1	x_2	x_3	x_4
0	0,3	0	0,5	0,5
0,1	0	0,8	0	0
0,25	0,8	0	0	0
0,8	0	0	0	0,8
1	0	0	0,8	0

Each triple $((x, y); \varphi(x, y))$ in $\tilde{F}$ may be interpreted as follows: the food x (bread, milk, meat) has the degree of membership y with the degree of truth of this statement being equal to $\varphi(x, y)$.

We conclude that even in such a simple setting determining the intersection of sets with a fuzzy set of operands is not immediate. The main complication stems from the necessity of finding the set of Pareto optimal solutions to two-criteria problem (20.3) for each fixed $x \in X$. This motivates the purpose of this article, which consists in simplifying the computational complexity of the intersection operation and presenting it in a form which is convenient for understanding and usage. As said in the Introduction, the decomposition approach is one of the main tools for investigating fuzzy sets. We are going to apply it to intersections with fuzzy sets of operands.

20.4 Decomposition of Intersections with Fuzzy Sets of Operands

Assume that fuzzy sets F_j with the membership functions $\varphi_j(x)$, $j \in N$ are defined on a crisp set X, where $N = \{1, 2, \ldots, n\}$ is the set of their indices and n is the cardinality of N. Let $\tilde{N}$ be a fuzzy set on N with an arbitrary membership function $\eta(j) \in (0, 1]$, $j \in N$. We want to represent the intersection T2FS $\tilde{F} = \tilde{\bigcap}_{(j,\eta(j))\in\tilde{N}} F_j$ of the fuzzy sets F_j, $j \in N$ with the fuzzy set $\tilde{N}$ of operands as a collection of simpler (in some sense) sets.

Denote by $N_\alpha = \{j \in N : \eta(j) \geq \alpha\}$, $\alpha \in (0, 1]$ the α-cut of the fuzzy set $\tilde{N} = \bigcup_{j\in N}(j, \eta(j))$ of operands indices.

Remark 20.2 Let $A = \{\eta(1), \eta(2), \ldots, \eta(n)\}$ be the set of membership degrees values of the fuzzy set $\tilde{N} = \bigcup_{j\in N}(j, \eta(j))$ of operands indices. It is clear that while obtaining the α-cuts $N_\alpha = \{j \in N : \eta(j) \geq \alpha\}$ of the T1FS $\tilde{N}$ we can assume that $\alpha \in A$ rather than $\alpha \in (0, 1]$. In addition, we can take $\alpha = \eta(i)$ for $i \in N$. We shall call $N_{\eta(i)}$ the $\eta(i)$-level set of operands indices.

For fixed $x \in X$ and any $i \in N$, consider the T1FS

$$\Phi_{\eta(i)} = \{(x, \phi_{\eta(i)}(x)) : x \in X\} \tag{20.8}$$

with the membership function

$$\phi_{\eta(i)}(x) = \min_{j \in N_{\eta(i)}} \varphi_j(x_j). \tag{20.9}$$

It is equal to the intersection $\bigcap_{j \in N_{\eta(i)}} F_j$ of fuzzy sets F_j with indices j from the $\eta(i)$-level set

$$N_{\eta(i)} = \{j \in N : \eta(j) \geq \eta(i)\}. \tag{20.10}$$

We shall call the T1FS $\Phi_{\eta(i)} = \bigcap_{j \in N_{\eta(i)}} F_j$ the intersection of the level $\eta(i)$.

The following theorem gives a representation of the T2FS $\tilde{F} = \tilde{\bigcap}_{(j,\eta(j)) \in \tilde{N}} F_j$ type-2 membership function which is more convenient for calculations.

Theorem 20.1 *The T2FS* $\tilde{F} = \tilde{\bigcap}_{(j,\eta(j)) \in \tilde{N}} F_j$ *has the T2MF*

$$\phi(x, y) = \begin{cases} \max_{i \in N}\{\eta(i) : y = \phi_{\eta(i)}(x)\}, & y \in Y_x; \\ 0, & \text{otherwise}; \end{cases} \tag{20.11}$$

$x \in X$, $y \in [0, 1]$, *where*

$$Y_x = \{y \in [0, 1] : y = \phi_{\eta(i)}(x), \ i \in N\}. \tag{20.12}$$

Proof We check that $\phi(x, y) \equiv \varphi(x, y)$. Suppose first that

$$\varphi(x, y) > 0. \tag{20.13}$$

We intend to prove that $\varphi(x, y) = \phi(x, y) > 0$. According to (20.5) and (20.6), inequality (20.13) ensures that there exists $i^* \in N$ such that

$$y = \varphi_{i^*}(x) \tag{20.14}$$

and $\tilde{\eta}(x, i^*) > 0$.Then formula (20.4) implies that

$$i^* \in N^{PO}(x) \tag{20.15}$$

and

$$\varphi(x, y) = \eta(i^*). \tag{20.16}$$

We are going to show that

$$\varphi_{i^*}(x) = \min\{\varphi_j(x) : \ \eta(j) \geq \eta(i^*)\} \tag{20.17}$$

and

$$\eta(i^*) = \max\{\eta(j) : \ \varphi_j(x) = \varphi_{i^*}(x)\}. \tag{20.18}$$

Assume on the contrary that $\varphi_{i^*}(x) > \min\{\varphi_j(x) : \ \eta(j) \geq \eta(i^*)\}$ or $\eta(i^*) < \max\{\eta(j) : \ \varphi_j(x) = \varphi_{i^*}(x)\}$. Then there exists $j \in N$ such that either $\varphi_{i^*}(x) > \varphi_j(x)$ and $\eta(j) \geq \eta(i^*)$ or $\varphi_{i^*}(x) = \varphi_j(x)$ and $\eta(j) > \eta(i^*)$. In both cases we infer $j S_x i^*$ and thereupon $i^* \notin N^{PO}(x)$, a contradiction to (20.15).

Formulae (20.14), (20.17), (20.9) and (20.10) imply that $y = \varphi_{i^*}(x) = \phi_{\eta(i^*)}(x)$, whence $y \in Y_x$ and $\phi(x, y) = \eta(i^*) > 0$ by (20.11), (20.12) and (20.18). According to (20.16), we conclude that $\varphi(x, y) = \phi(x, y) > 0$.

Conversely, let us check that $\phi(x, y) = \varphi(x, y) > 0$ provided that

$$\phi(x, y) > 0. \tag{20.19}$$

According to (20.11) and (20.12), formula (20.19) implies that there exists $i^* \in N$ for which

$$\eta(i^*) = \phi(x, y) = \max_{i \in N}\{\eta(i) : \ y = \phi_{\eta(i)}(x)\}. \tag{20.20}$$

Then, in view of (20.9) and (20.10), there exists $i' \in N$ such that

$$y = \varphi_{i'}(x) = \phi_{\eta(i^*)}(x) = \min_{i \in N}\{\varphi_i(x) : \ \eta(i) \geq \eta(i^*)\} \tag{20.21}$$

and

$$\eta(i^*) = \max_{i \in N}\{\eta(i) : \ \varphi_i(x) = \varphi_{i'}(x)\}. \tag{20.22}$$

Formula (20.22) guarantees $\varphi_{i'}(x) = \varphi_{i^*}(x)$. With this at hand we obtain

$$y = \varphi_{i^*}(x) = \phi_{\eta(i^*)}(x) = \min_{i \in N}\{\varphi_i(x) : \ \eta(i) \geq \eta(i^*)\} \tag{20.23}$$

having utilized (20.21). Then (20.6) ensures that

$$\varphi_{i^*}(x) = y \in J_x \tag{20.24}$$

and, with the help of (20.5), we infer $\varphi(x, y) = \tilde{\eta}(x, i^*)$. We consider two cases separately:
A) if $\tilde{\eta}(x, i^*) > 0$, then formulae (20.4) and (20.5) imply that

$$\varphi(x, y) = \eta(i^*) > 0; \tag{20.25}$$

B) if $\tilde{\eta}(x, i^*) = 0$, then formula (20.4) guarantees $i^* \notin N^{PO}(x)$. As a consequence, there exists $j \in N$, such that the relation $j S_x i^*$ holds. There are two subcases. In the former subcase, $y = \varphi_{i^*}(x) > \varphi_j(x)$ and $\eta(j) \geq \eta(i^*)$, so that

$$y = \varphi_{i^*}(x) > \varphi_j(x) \geq \min\{\varphi_i(x) : \eta(i) \geq \eta(i^*)\},$$

a contradiction to (20.23). In the latter subcase, $y = \varphi_{i^*}(x) = \varphi_j(x)$ and $\eta(j) > \eta(i^*)$, so that

$$\eta(i^*) < \eta(j) \geq \max\{\eta(i) : \varphi_i(x) = \varphi_{i^*}(x)\},$$

a contradiction to (20.22). Thus, formulae (20.24) and (20.25) hold true and thereupon $\phi(x, y) = \varphi(x, y) > 0$ by (20.20).

Summarizing, if $\varphi(x, y) > 0$, then $\varphi(x, y) = \phi(x, y) > 0$ and conversely, if $\phi(x, y) > 0$, then $\phi(x, y) = \varphi(x, y) > 0$. Thus, $\varphi(x, y) = 0$ if, and only if, $\phi(x, y) = 0$.

The proof of Theorem 1 is complete.

Let us illustrate Theorem 20.1 with the help of Example 20.1. For each $i = 1, 2, 3$, we use (20.9) to find the membership function value $\phi_i(x)$ of the intersection $\Phi_{\eta(i)} = \bigcap_{j \in N_{\eta(i)}} F_j$ of the level $\eta(i)$. This enables us to add values $\phi_{0,8}(x) = \varphi_1(x)$, $\phi_{0,5}(x) = \min\{\varphi_1(x), \varphi_3(x)\}$ and $\phi_{0,3}(x) = \min\{\varphi_1(x), \varphi_2(x), \varphi_3(x)\}$ to Table 20.1. Appealing to (20.11) we construct the intersection T2MF $\varphi(x, y)$. It is easy to verify that the values $\varphi(x, y)$, $y \in [0, 1]$, $x \in X$ match the corresponding values from Table 20.2. We note that this requires much less computation in comparison to our original argument.

Now we construct the decomposition of the intersection T2FS according to secondary membership grades into a collection of T1FSs. Assume that, for each $x \in X$, a unique $y_x \in J_x \subseteq [0, 1]$ is given. We say that an embedded T2FS $\tilde{F}^e = \{((x, y_x), \varphi(x, y_x)) : x \in X\}$ in the intersection T2FS $\tilde{F}$ has a constant secondary grade $\eta(i) \in (0, 1]$, $i \in N$, if $\varphi(x, y_x) \equiv \eta(i)$. Denote this embedded T2FS by $\tilde{F}^e_{\eta(i)} = \{((x, y_x), \eta(i)) : x \in X\}$.

Remark 20.3 Obviously, there is a unique embedded T1FS $F^e_{\eta(i)} = \{(x, y_x) : x \in X\}$ for the embedded T2FS $\tilde{F}^e_{\eta(i)}$. This gives a representation of the form $\tilde{F}^e_{\eta(i)} = \{(F^e_{\eta(i)}, \eta(i))\}$.

Remark 20.4 We intend to construct a representation of the intersection T2FS $\tilde{F}$ as a collection of embedded T2FSs $F^e_{\eta(i)} = \{(x, y_x) : x \in X\}$ with constant secondary grades $\eta(i) \in (0, 1]$, $i \in [0, 1]$. As in the theory of Wavy-Slice Representation of T2FSs [17], we consider each element of a T2FS as a subset. Therefore, we interpret the collection as a classical union of its elements in the sense of T1FS. Repeated elements are counted only once like in any union. For a pair (x, y_x) with different secondary grades $\eta(i)$ and $\eta(j)$ only one member of the collection is taken into account which has a maximum secondary grade $\max\{\eta(i), \eta(j)\}$.

Theorem 20.2 points out a useful property of the T2FS $\tilde{F}$.

Theorem 20.2 *For each $i \in N$, let a T1FS $F^e_{\eta(i)} = \{(x, y_x) : x \in X\}$ be the intersection of the level $\eta(i)$, that is, the fuzzy set $F^e_{\eta(i)} \equiv \Phi_{\eta(i)} = \{(x, \phi_{\eta(i)}(x)) : x \in X\}$ which is equal to the intersection $\bigcap_{j \in N_{\eta(i)}} F_j$ of fuzzy sets F_j with indices j from the $\eta(i)$-level set $N_{\eta(i)}$ of operands indices. Then the intersection T2FS $\tilde{F} = \tilde{\bigcap}_{(j,\eta(j)) \in \tilde{N}} F_j$ is given by a collection $\{\tilde{F}^e_{\eta(i)} : i \in N\}$ of the embedded T2FSs $\tilde{F}^e_{\eta(i)} = \{(F^e_{\eta(i)}, \eta(i))\}$ with a constant secondary grade $\eta(i) \in (0, 1]$, $i \in N$. In other words,*

$$\tilde{F} = \{(\Phi_{\eta(i)}, \eta(i)) : i \in N\}, \tag{20.26}$$

where

$$\Phi_{\eta(i)} = \bigcap_{j \in N_{\eta(i)}} F_j, \ i \in N. \tag{20.27}$$

Proof According to Theorem 20.1, a T2FS $\tilde{F}$ has the form

$$\tilde{F} = \{((x, y), \phi(x, y)) : x \in X, \ y \in Y_x \subseteq [0, 1]\},$$

where

$$Y_x = \{y \in [0, 1] : y = \phi_{\eta(i)}(x), \ i \in N\}.$$

We first prove the inclusion $\tilde{F} \subseteq \{(\Phi_{\eta(i)}, \eta(i)) : i \in N\}$, where, according to (20.8), $\Phi_{\eta(i)} = \{(x, \phi_{\eta(i)}(x)) : x \in X\}$. Assume that $((\hat{x}, \hat{y}), \phi(\hat{x}, \hat{y})) \in \tilde{F}$, $\hat{x} \in X$, $\hat{y} \in Y_x \subseteq [0, 1]$. In view of $\hat{y} \in Y_x \subseteq [0, 1]$ and formula (20.12) there exists $\hat{i} \in N$ such that $\hat{y} = \phi_{\eta(\hat{i})}(\hat{x})$. Also, according to (20.11), the equality $\phi(\hat{x}, \hat{y}) = \eta(\hat{i})$ holds. Thus, $((\hat{x}, \hat{y}), \phi(\hat{x}, \hat{y})) = ((\hat{x}, \phi_{\eta(\hat{i})}(\hat{x})), \eta(\hat{i}))$ and thereupon

$$((\hat{x}, \hat{y}), \phi(\hat{x}, \hat{y})) \in \{((x, \phi_{\eta(i)}(x)), \eta(i)) : x \in X, \ i \in N\} =$$

$$\{(\{(x, \phi_{\eta(i)}(x)) : x \in X\}, \eta(i)) : i \in N\} = \{(\Phi_{\eta(i)}, \eta(i)) : i \in N\}.$$

Next, we prove the inclusion $\tilde{F} \supseteq \{(\Phi_{\eta(i)}, \eta(i)) : i \in N\}$. Assume that

$$(\Phi_{\eta(\hat{i})}, \eta(\hat{i})) \in \{(\Phi_{\eta(i)}, \eta(i)) : i \in N\}.$$

By virtue of (20.8), this leads to $((\hat{x}, \phi_{\eta(\hat{i})}(\hat{x})), \eta(\hat{i})) \in \{(\Phi_{\eta(i)}, \eta(i)) : i \in N\}$. Assume on the contrary that

$$((\hat{x}, \phi_{\eta(\hat{i})}(\hat{x})), \eta(\hat{i})) \notin \tilde{F} = \{((x, y), \phi(x, y)) : x \in X, \ y \in Y_x \subseteq [0, 1]\}.$$

Then, for $x = \hat{x}$ and $y = \phi_{\eta(\hat{i})}(\hat{x})$, in view of (20.12), $\eta(\hat{i}) > \phi(\hat{x}, \phi_{\eta(\hat{i})}(\hat{x}))$. Therefore, formula (20.11) implies that

$$\eta(\hat{i}) > \phi(\hat{x}, \phi_{\eta(\hat{i})}(\hat{x})) = \max_{i \in N}\{\eta(i) : \phi_{\eta(\hat{i})}(\hat{x}) = \phi_{\eta(i)}(\hat{x})\} \geq \eta(\hat{i}),$$

a contradiction. Thus, the inclusion $\tilde{F} \supseteq \{(\Phi_{\eta(i)}, \eta(i)) : i \in N\}$ holds true, whence $\tilde{F} = \{(\Phi_{\eta(i)}, \eta(i)) : i \in N\}$.

The proof of Theorem 20.2 is complete.

Thus, the intersection T2FS $\tilde{F}$ can be decomposed, according to secondary grades, into the corresponding collection of T1FSs. Let us discuss this property in the setting of Example 20.1. To this end, we use (20.27) to construct the intersections $\Phi_{0,8} = F_1$, $\Phi_{0,5} = F_1 \bigcap F_3$ and $\Phi_{0,3} = F_1 \bigcap F_2 \bigcap F_3$ of the corresponding levels. Then, invoking (20.26) yields

$$\{(\Phi_{0,3}; 0,3), (\Phi_{0,5}; 0,5), (\Phi_{0,8}; 0,8)\} =$$

$$\{((x_1; 0); 0,3), ((x_2; 0,1); 0,3), ((x_3; 0); 0,3), ((x_4; 0); 0,3),$$

$$((x_1; 0,25); 0,5), ((x_2; 0,1); 0,5), ((x_3; 0); 0,5), ((x_4; 0); 0,5),$$

$$((x_1; 0,25); 0,8), ((x_2; 0,1); 0,8), ((x_3; 1); 0,8), ((x_4; 0,5); 0,8)\},$$

whence, by Remark 20.4,

$$\{(\Phi_{0,3}; 0,3), (\Phi_{0,5}; 0,5), (\Phi_{0,8}; 0,8)\} =$$

$$\{((x_1; 0); 0,3), ((x_1; 0,25); 0,8), ((x_2; 0,1); 0,8), ((x_3; 0); 0,5),$$

$$((x_3; 1); 0,8), ((x_4; 0); 0,5), ((x_4; 0,5); 0,8)\}.$$

Comparing these with (20.7) makes it clear that

$$\tilde{F} = \{(\Phi_{0,3}; 0,3), (\Phi_{0,5}; 0,5), (\Phi_{0,8}; 0,8)\}.$$

Note that the T1FSs $\Phi_{0,8} = F_1$, $\Phi_{0,5} = F_1 \bigcap F_3$ and $\Phi_{0,3} = F_1 \bigcap F_2 \bigcap F_3$ can be thought of as fuzzy sets of protein-rich foods, foods which are rich in proteins and carbohydrates and foods which are rich in proteins, fats and carbohydrates, respectively. The resulting T2FS can be interpreted as follows. The set of foods which are suitable for a healthy lifestyle within the whole diet consists of:

- protein-rich foods with the degree of truth of this statement being equal to 0,8;
- foods which are rich in proteins and carbohydrates with the degree of truth of this statement being equal to 0,5;

– foods which are rich in proteins, fats and carbohydrates with the degree of truth of this statement being equal to 0,3.

Among other things, Theorem 1 facilitates finding the intersection of fuzzy sets with a fuzzy set of operands indices not only for discrete universes of discourse. To illustrate this point we consider yet another example.

Example 20.2 We want to find the intersection of three fuzzy sets F_1, F_2 and F_3. The membership functions of these sets are

$$\varphi_1(x) = \begin{cases} x/4, & x \in [0,4]; \\ 2 - x/4, & x \in [4,8]; \\ 0, & x \in [8,10]; \end{cases} \quad \varphi_2(x) = \begin{cases} x/3, & x \in [0,3]; \\ 2 - x/3, & x \in [3,6]; \\ 0, & x \in [6,10]; \end{cases}$$

$$\varphi_3(x) = \begin{cases} x/4 - 1/2, & x \in [2,6]; \\ 5/2 - x/4, & x \in [6,10]; \\ 0, & x \in [0,2]. \end{cases}$$

These have domain $X = [0, 10]$. In Fig. 20.1 the graphs of their membership functions are represented by the three lines (the solid line corresponds to $\varphi_1(x)$, the dashed line corresponds to $\varphi_2(x)$, the dotted line corresponds to $\varphi_3(x)$).

Let $\tilde{N}$ be a fuzzy set on the set $N = \{1, 2, 3\}$ of operands indices with the membership function $\eta(i)$ given by $\eta(1) = 0,8$ and $\eta(2) = 0,6$ and $\eta(3) = 0,4$. By Theorem 20.2, the T2FS $\tilde{F} = \tilde{\bigcap}_{(j,\eta(j))\in\tilde{N}} F_j$ is given by

$$\tilde{F} = \{(\Phi_{0,8}; 0,8), (\Phi_{0,6}; 0,6), (\Phi_{0,4}; 0,4)\}.$$

Here, the fuzzy set $\Phi_{0,4}$ is the intersection of the level 0,4 which is equal to the intersection of fuzzy sets with indices from the set $N_{0,4} = \{1, 2, 3\}$ of operands indices. According to (20.9), the membership function of $\Phi_{0,4}$ is

$$\phi_{0,4}(x) = \begin{cases} x/4 - 1/2, & x \in [2, 30/7]; \\ 3/2 - x/4, & x \in [30/7, 6]; \\ 0, & x \notin [2,6]. \end{cases}$$

Similarly, by (20.9),

$$\phi_{0,6}(x) = \begin{cases} x/4, & x \in [0, 24/7]; \\ 2 - x/3, & x \in [24/7, 6]; \\ 0, & x \in [6,10]; \end{cases}$$

is the membership function of the fuzzy set $\Phi_{0,6}$. This fuzzy set is the intersection of the level 0,6 which is equal to the to the intersection of fuzzy sets with indices from the set $N_{0,6} = \{1, 2\}$ of operands indices. Finally, corresponding to the fuzzy set $\Phi_{0,6} = F_1$ is the set $N_{0,8} = \{1\}$ of operands indices. In view of (20.9), $\Phi_{0,6}$ has the membership function $\phi_{0,8}(x) \equiv \varphi_1(x)$.

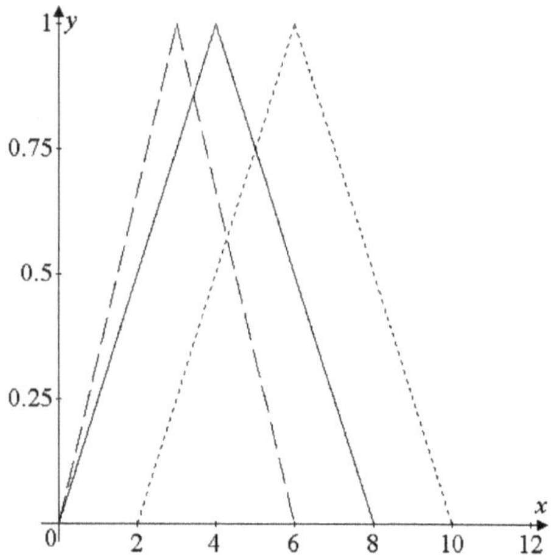

Fig. 20.1 The graphs of $\varphi_i(x), i = 1, 2, 3$

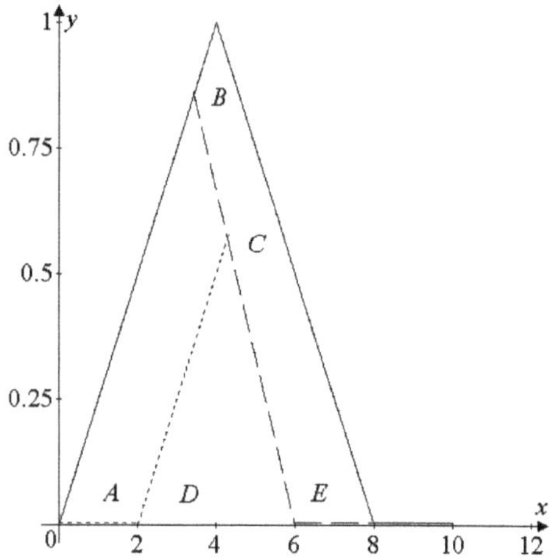

Fig. 20.2 The graph of the T2MF $\varphi(x, y)$

In Fig. 20.2, the graph of the T2MF $\varphi(x, y)$ of the T2FS $\tilde{F}$ is represented by the three lines (the solid line corresponds to $\varphi(x, y) = 0, 8$; the dashed line corresponds to $\varphi(x, y) = 0, 6$; the dotted line corresponds to $\varphi(x, y) = 0, 4$).

Observe that, according to Remark 20.4, the secondary membership grade is equal to 0,8 (maximum of 0.8 and 0.6) on the line segments AB and [8, 10] of Fig. 20.2. Similarly, it is equal to 0,6 (maximum of 0,6 and 0,4) on the line segments CE and [6,8].

20.5 Conclusion

The present article has confirmed that the decomposition approach is a powerful tool for studying fuzzy sets. The approach leads to a significant simplification of the description of fuzzy sets and operations on them. Further, it gives various representations of fuzzy sets which enable us to analyze them from different viewpoints and facilitate their understanding and interpretation. Here, the decomposition approach was applied to studying the operation of intersection of fuzzy sets with a fuzzy set of operands. The result of this operation is a T2FS, the mathematical object which is not easy to use and understand. Our findings offer a nice alternative: we decompose the resulting T2FS into a finite collection of T1FSs, thereby simplifying the construction of this set and giving a clear interpretation. We hope that wide possibilities are to be discovered for the use in the theory of fuzzy set of the intersection operation on fuzzy sets with a fuzzy set of operands. The idea of this work can be used to generalize other algebraic operations to the case of a fuzzy set of operands, most notably the union operation.

References

1. Aisbett, J., Rickard, J.T., Morgenthaler, D.G.: Type-2 fuzzy sets as functions on spaces. IEEE Trans. Fuzzy Syst. **18**(4), 841–844 (2010)
2. Chen, Q., Kawase, S.: On fuzzy-valued fuzzy reasoning. Fuzzy Sets Syst. **113**, 237–251 (2000)
3. Coupland, S., John, R.: Geometric type-1 and type-2 fuzzy logic systems. IEEE Trans. Fuzzy Syst. **15**(1), 3–15 (2007)
4. Hamrawi, H., Coupland, S.: Non-specificity measures for type-2 fuzzy sets. IEEE Trans. Fuzzy Syst. 732–737 (2009)
5. Hamrawi, H., Coupland, S.: Type-2 fuzzy arithmetic using alpha-planes. In: Proceedings of the IFSA-EUSFLAT, Portugal, pp. 606–611 (2009)
6. Hamrawi, H., Coupland, S., John, R.: Extending operations on type-2 fuzzy sets. In: Proceedings of the UKCI, Nottingham, UK (2009)
7. Hamrawi, H., Coupland, S., John, R.: A novel alpha-cut representation for type-2 fuzzy sets. IEEE Trans. Fuzzy Syst. 1–8 (2010)
8. Hamrawi, H., Coupland, S., John, R.: Type-2 fuzzy alpha-cuts. IEEE Trans. Fuzzy Syst. **25**(3), 682–692 (2017)
9. Harding, J., Walker, C., Walker, E.: The variety generated by the truth value algebra of T2FSs. Fuzzy Sets Syst. **161**, 735–749 (2010)
10. Karnik, N.N., Mendel, J.M.: Introduction to type-2 fuzzy logic systems. IEEE Trans. Fuzzy Syst. **2**, 915–920 (1998)
11. Liu, F.: An efficient centroid type-reduction strategy for general type-2 fuzzy logic system. Inf. Sci. **178**(9), 2224–2236 (2008)
12. Mashchenko, S.O.: Generalization of Germeyer's criterion in the problem of decision making under the uncertainty conditions with the fuzzy set of the states of nature. J. Autom. Inf. Sci. **44**(10), 26–34 (2012)
13. Mashchenko, S.O.: A mathematical programming problem with the fuzzy set of indices of constraints. Cybern. Syst. Anal. **49**(1), 62–70 (2013)
14. Mashchenko, S.: Intersections and unions of fuzzy sets of operands. Fuzzy Sets Syst. **352**, 12–25 (2018)

15. Mashchenko, S.O., Morenets, V.I.: Shapley value of a cooperative game with fuzzy set of feasible coalitions. Cybern. Syst. Anal. **53**(3), 432–440 (2017)
16. Mendel, J.M.: Type-2 fuzzy sets: some questions and answers. IEEE Connect. Newsl. IEEE Neural Netw. Soc. **1**, 10–13 (2003)
17. Mendel, J.M., John, R.: Type-2 fuzzy sets made simple. IEEE Trans. Fuzzy Syst. **10**(2), 117–127 (2002)
18. Mendel, J., Liu, F., Zhai, D.: α-plane representation for type-2 fuzzy sets: theory and applications. IEEE Trans. Fuzzy Syst. **17**(5), 1189–1207 (2009)
19. Tahayori, H., Tettamanzi, A., Antoni, G.: Approximated type-2 fuzzy set operations. In: Proceedings of IEEE World Congress on Computational Intelligence, Vancouver, Canada, pp. 1910–1917 (2006)
20. Wagner, C., Hagras, H.: zSlices towards bridging the gap between interval and general type-2 fuzzy Logic. IEEE Trans. Fuzzy Syst. 489–497 (2008)
21. Wagner, C., Hagras, H.: Toward general type-2 fuzzy logic systems based on z-slices. IEEE Trans. Fuzzy Syst. **18**(4), 637–660 (2010)
22. Wu, D., Mendel, J.M.: Aggregation using the linguistic weighted average and interval type-2 fuzzy sets. IEEE Trans. Fuzzy Syst. **15**(6), 1145 (2007)
23. Wu, D., Mendel, J.: Corrections to aggregation using the linguistic weighted average and interval type-2 fuzzy sets. IEEE Trans. Fuzzy Syst. **16**(6), 1664–1666 (2008)
24. Yager, R.R.: Level sets and the extension principle for interval valued fuzzy sets and its application to uncertain measures. Inf. Sci. **178**(18), 3565–3576 (2008)
25. Zadeh, L.A.: Fuzzy sets. Inf. Control **46**(3), 338–353 (1965)
26. Zadeh, L.A.: Quantitative fuzzy semantics. Inf. Sci. **3**, 159–176 (1971)
27. Zadeh, L.A.: The concept of a linguistic variable and its application to approximate reasoning. Inf. Sci. **8**(3), 199–249 (1975)
28. Zeng, W., Li, H.: Representation theorem of interval-valued fuzzy set. Int. J. Uncertain. Fuzziness Knowl.-Based Syst. **14**(3), 259–269 (2006)
29. Zeng, W., Li, H., Zhao, Y., Yu, X.: Extension principle of lattice-valued fuzzy set. In: Proceedings of the 4th International Conference on Fuzzy Systems and Knowledge Discovery, FSKD 2007, vol. 1, pp. 140–144 (2007)
30. Zeng, W., Shi, Y.: Note on interval-valued fuzzy set. Lect. Notes Artif. Intell. (Sub-Ser. Lect. Notes Comput. Sci.) **3613**, 20–25 (2005)

Chapter 21
Distribution of Values of Cantor Type Fractal Functions with Specified Restrictions

Oleg V. Barabash, Andrii P. Musienko, Valentyn V. Sobchuk, Nataliia V. Lukova-Chuiko, and Olga V. Svynchuk

Abstract We consider Q_s^*-representation of numbers $x \in [0, 1]$. It is an encoding of real numbers by means of the finite alphabet $A = \{0, 1, 2, \ldots, s-1\}$. The article is devoted to continuous non-monotonic singular functions of Cantor type defined in terms of a given Q_s^*-representation of numbers. We study their local and global properties: structural, variational, differential, integral, self-similar, and fractal. Level sets of functions as well as topological and metric properties of images of Cantor type sets are examined in detail. In this work we also study the distribution of random variable $Y = f(X)$, where f is a non-monotonic singular function of Cantor type and X is a random variable such that its distribution induced by distributions of digits of its Q_5^*-representation that are independent random variables.

21.1 Introduction

The space $C[0, 1]$ of continuous functions on the segment $[0, 1]$ with the uniform metric is rich in functions with locally complicated structure. These are nowhere monotonic functions, nowhere differentiable functions, singular functions (non-constant continuous functions whose derivatives are equal to zero almost everywhere in the sense of Lebesgue measure) and functions whose sets of constancy are of full Lebesgue measure.

O. V. Barabash (✉) · A. P. Musienko · V. V. Sobchuk · O. V. Svynchuk
State University of Telecommunications, Solomyanska Str., 7, Kyiv 03110, Ukraine
e-mail: bar64@ukr.net; 7011990@ukr.net

N. V. Lukova-Chuiko
Taras Shevchenko National University of Kyiv, Bogdan Gavrilishina Str., 24, Kyiv 04116, Ukraine

V. A. Sadovnichiy and M. Z. Zgurovsky (eds.), *Contemporary Approaches and Methods in Fundamental Mathematics and Mechanics*, Understanding Complex Systems,
https://doi.org/10.1007/978-3-030-50302-4_21

A set of non-constancy of function f(x) is the set S_f of all points x such that for any $\varepsilon > 0$ in ε-neighborhood $O_\varepsilon(x)$ of point x there exist points x_1 and x_2 such that $f(x_1) \neq f(x_2)$ [22]. If $f(x)$ is monotonic, then the set of non-constancy is a set of increasing points of the function.

A function of Cantor type is the function whose set of non-constancy is a nowhere dense set. If complement of a set of non-constancy is the set of full Lebesgue measure, then this function is called a *singular function of Cantor type*.

Singular functions of Cantor type forms a rich subclass in the class of singular probability distribution functions. They attract a special attention in stochastics. The non-monotonic functions of Cantor type have not been studied in detail so far. Nowhere non-monotonic functions as well as functions without monotonicity intervals except of constancy intervals did not attract attention even less. We study such objects in detail.

Singular functions are not only of theoretical interest [6, 14, 15, 19, 20, 26, 27, 29, 31] (they are related to complicated probabilistic problems [5]); they describe transient fractal radiation, appear in the problems of distributed systems management [1], analysis and design of antennas [7–10], in the field of information technology and telecommunications [2, 4, 5, 11, 16].

For constructing of a class of such functions, systems of encoding (representation) of real numbers with a finite alphabet can be used efficiently. We use Q_s^*-representation of the numbers [30], which is a self-similar generalization of the classic s-adic representation and Q_s-representation of real numbers [17].

Let $A_s = \{0, 1, \dots, s-1\}$ be an alphabet of s-adic numeral system and let $L = A_s \times A_s \times \cdots$ be a space of sequences of elements of the alphabet, and let $Q_s^* = \|q_{ij}\|$, $j \in N$, $i \in A_s$ be an infinite stochastic matrix with positive elements and the following properties:

1. $q_{0j} + q_{1j} + \cdots + q_{[s-1]j} = 1$;
2. $\prod\limits_{j=1}^{\infty} \max\{q_{0j}, q_{1j}, \dots, q_{[s-1]j}\} = 0$.

Theorem 21.1 ([17]) *For any $x \in [0, 1]$ there exists a sequence (α_k), $\alpha_k = \alpha_k(x) \in A_s$ such that*

$$x = \beta_{\alpha_1 1} + \sum_{k=2}^{\infty} \left[\beta_{\alpha_k k} \prod_{j=1}^{k-1} q_{\alpha_j j} \right] = \Delta^{Q_s^*}_{\alpha_1(x)\dots\alpha_k(x)\dots}, \tag{21.1}$$

where $\beta_{0j} = 0$ and

$$\beta_{ij} = \sum_{k=0}^{i-1} q_{kj}, i \in \overline{1, s-1}, j \in N.$$

The representation of the number x in the form of series (21.1) is called Q_s^*-*expansion* while the symbolic notation

$$x = \Delta^{Q_s^*}_{\alpha_1(x)\dots\alpha_k(x)\dots} \tag{21.2}$$

is called Q_s^**-representation*, and a number $\alpha_j(x)$ is called jth digit in the representation of x.

Each *irrational* number has a unique representation, but some rational numbers have two representations. These are numbers with the representations

$$\Delta^{Q_s^*}_{\alpha_1\ldots\alpha_{k-1}\alpha_k(0)} = \Delta^{Q_s^*}_{\alpha_1\ldots\alpha_{k-1}[\alpha_k-1](s-1)}.$$

They are called Q_s^**-rational*. By agreement we use only one of two representations of a rational number containing period (0). Then we have the uniqueness of the Q_s^*-representation of a number.

The notion of cylinder is important for the geometry of representations.

Definition 21.1 Let $(c_1, c_2, \ldots, c_m)$ be a fixed m-tuple of symbols $c_i \in A_s$. A set $\Delta^{Q_s^*}_{c_1c_2\ldots c_m}$ formed by all points $x \in [0, 1]$ with Q_s^*-representation such that $\alpha_j(x) = c_j$, $j = \overline{1, m}$ is called a cylinder of rank m with the base $c_1c_2\ldots c_m$.

Properties of Cylinders:

(1) a cylinder $\Delta^{Q_s^*}_{c_1c_2\ldots c_m}$ is a segment with the endpoints

$$a = \Delta^{Q_s^*}_{c_1\ldots c_m(0)} = \beta_{c_1 1} + \sum_{k=2}^{m}\left[\beta_{c_k k}\prod_{j=1}^{k-1} q_{c_j j}\right], \quad b = \Delta^{Q_s^*}_{c_1\ldots c_m(s-1)} = a + \prod_{i=1}^{m} q_{c_i i};$$

(2) $\Delta^{Q_s^*}_{c_1c_2\ldots c_m} = \bigcup_{i=0}^{s-1} \Delta^{Q_s^*}_{c_1c_2\ldots c_m i}$;

(3) $|\Delta^{Q_s^*}_{c_1c_2\ldots c_m}| = \prod_{i=1}^{m} q_{c_i i}$;

(4) $\max \Delta^{Q_s^*}_{c_1c_2\ldots c_m i} = \min \Delta^{Q_s^*}_{c_1c_2\ldots c_m [i+1]}$, $i = \overline{0, s-2}$;

(5) $\bigcap_{m=1}^{\infty} \Delta^{Q_s^*}_{c_1c_2\ldots c_m} = x = \Delta^{Q_s^*}_{c_1c_2\ldots c_m\ldots} \in [0, 1]$ for any sequence (c_i), $c_i \in A_S$.

If all columns of the matrix $\|q_{ij}\|$ are identical, then Q_s^*-representation is called Q_s*-representation*. In addition, if $q_i = \frac{1}{s}$, then the Q_s-representation is a classic *s-adic representation.*

21.2 Singular Functions as Solutions of Systems of Functional Equations

Let $\overline{q} = (q_0, q_1, \cdots, q_{s-1})$, $s > 1$ be a stochastic vector with non-negative coordinates ($q_i > 0, q_0 + q_1 + \cdots + q_{s-1} = 1$) and let $\overline{g} = (g_0, g_1, \ldots, g_{s-1})$ be a stochastic vector satisfying conditions: $|g_i| < 1$, $\delta_i = g_0 + g_1 + \cdots + g_{i-1} > 0$.

Theorem 21.2 *The system of s functional equations*

$$\begin{cases} f(q_0x) = g_0 f(x), \\ f(\beta_i + q_i x) = \delta_i + g_i f(x), \quad i = \overline{1, s-1}, \end{cases} \tag{21.3}$$

has a unique solution in a class of continuous functions on [0, 1]. *It is a continuous function*

$$f(x) = \delta_{\alpha_1(x)} + \sum_{k=2}^{\infty}\left(\delta_{\alpha_k(x)}\prod_{j=1}^{k-1} g_{\alpha_j(x)}\right). \tag{21.4}$$

Proof Let

$$\beta_i + q_i x = \beta_i + \beta_{\alpha_1} q_i + \beta_{\alpha_2} q_i q_{\alpha_1} + \cdots = \Delta^{Q_s}_{i\alpha_1(x)\alpha_2(x)\ldots\alpha_k(x)\ldots}, \quad i = \overline{0, s-1}.$$

Then

$$\begin{aligned} f(x) &= f\left(\Delta^{Q_s}_{\alpha_1\alpha_2\ldots\alpha_n\ldots}\right) \\ &= \delta_{\alpha_1} + g_{\alpha_1} f\left(\Delta^{Q_s}_{\alpha_2\alpha_3\ldots\alpha_n\ldots}\right) = \delta_{\alpha_1} + g_{\alpha_1}\left[\delta_{\alpha_2} + g_{\alpha_2} f\left(\Delta^{Q_s}_{\alpha_3\alpha_4\ldots\alpha_n\ldots}\right)\right] = \\ &= \delta_{\alpha_1} + \delta_{\alpha_2} g_{\alpha_1} + \cdots + \delta_{\alpha_m}\prod_{j=1}^{m-1} g_{\alpha_j} + \left(\prod_{j=1}^{m} g_{\alpha_j}\right) f\left(\Delta^{Q_s}_{\alpha_{m+1}\alpha_{m+2}\ldots\alpha_{m+k}\ldots}\right). \end{aligned}$$

Since $\left|\prod_{j=1}^{m} g_{\alpha_j}\right| < 1$ and $\left|\prod_{j=1}^{m} g_{\alpha_j}\right| \to 0$ as $m \to \infty$, $\left|f\left(\Delta^{Q_s}_{\alpha_{m+1}\alpha_{m+2}\ldots\alpha_{m+k}\ldots}\right)\right| \le C =$ $const$, we get $\left(\prod_{j=1}^{m} g_{\alpha_j}\right) f\left(\Delta^{Q_s}_{\alpha_{m+1}\alpha_{m+2}\ldots\alpha_{m+k}\ldots}\right) \to 0$ $(m \to \infty)$. Therefore the sequence $A_m = \delta_{\alpha_1} + \delta_{\alpha_2} g_{\alpha_1} + \cdots + \delta_{\alpha_m}\prod_{j=1}^{m-1} g_{\alpha_j}$ has a limit that is the value of f at x. So we get a unique expansion (21.4) for function f.

Now we prove the continuity of the function (21.4). Consider an arbitrary number x_0 of the segment [0, 1] and the difference

$$f(x) - f(x_0) = \left(\prod_{j=1}^{m-1} g_{\alpha_j}\right)\left(f\left(\Delta^{Q_s}_{\alpha_m(x)\alpha_{m+1}(x)\ldots\alpha_{m+k}(x)\ldots}\right) - f\left(\Delta^{Q_s}_{\alpha_m(x_0)\alpha_{m+1}(x_0)\ldots\alpha_{m+k}(x_0)\ldots}\right)\right),$$

where $\alpha_m(x) \neq \alpha_m(x_0)$, but $\alpha_i(x) = \alpha_i(x_0)$ for $i < m$.

If x_0 is the Q_s-irrational number, then the condition $x \to x_0$ is equivalent to $m \to \infty$ and

$$|f(x) - f(x_0)| \le C\prod_{j=1}^{m-1} g_{\alpha_j} \to 0, \quad m \to \infty.$$

That is, $\lim_{x \to x_0} f(x) = f(x_0)$, and the function $f(x)$ is continuous at x_0 by definition.

If x_0 is the Q_s-rational number, then we can use the above considerations, but, for the case $x \to x_0$ from the left, it is enough to use the second representation of a number with period $(s-1)$, and when x tends to x_0 from the right, we should use the first representation with period (0).

Corollary 21.1 *If* $q_0 = g_0 = \frac{1}{2}$, *then* $f(x) = x$.

We have an infinite family of continuous functions such that every function is defined by a finite set of parameters.

21.3 Constructive Generalization of a Class of Continuous Functions

Let $G_s^* = \|g_{ij}\|$, $i \in A_s$, $j \in N$ be an infinite stochastic matrix with the following properties:

1. $| g_{ij} |< 1, g_{0j} > 0, g_{(s-1)j} > 0$;
2. $g_{0j} + g_{1j} + \cdots + g_{[s-1]j} = 1$;
3. $\prod\limits_{j=1}^{\infty} \max\limits_{i}\{g_{ij}\} = 0$;
4. $\forall (i_j j), i_j \in A_s \sum\limits_{k=2}^{\infty} \prod\limits_{j=1}^{k-1} \{|g_{i_j j}|\} < \infty$.

Consider a function f given by the equality

$$f(x) = \delta_{\alpha_1(x)1} + \sum_{k=2}^{\infty} \left(\delta_{\alpha_k(x)k} \prod_{j=1}^{k-1} g_{\alpha_j(x)j} \right), \tag{21.5}$$

where $\delta_{0j} = 0, 0 < \delta_{ij} = \sum\limits_{k=0}^{i-1} g_{kj} < 1, i \in \overline{1, s-1}$.

If for all $i \in A_s$, $j \in N$ we have $g_{ij} = g_i$, that is, all columns of the matrix G_s^* are identical, then the G_s^*-representation is called the *G_s-representation*.

Let us prove that this function is *well defined*. Let us show that:

(1) the series (21.5) is convergent for any (α_k), $\alpha_k \in A_s$;

(2) $f(x)$ is defined at Q_s^*-rational points. That is, for different representations of the same Q_s^*-rational argument, values of the function $f(x)$ and $f(x')$ are the same.

1. We test if series (21.5) is absolutely convergent. Taking into account the initial conditions 3 and 4 we get

$$\sum_{k=2}^{\infty} \left| \delta_{\alpha_k k} \prod_{j=1}^{k-1} g_{\alpha_j j} \right| = \sum_{k=2}^{\infty} \left(\delta_{\alpha_k k} \prod_{j=1}^{k-1} |g_{\alpha_j j}| \right) < \sum_{k=2}^{\infty} \prod_{j=1}^{k-1} |g_{\alpha_j j}| < \infty,$$

whence we have that series (21.5) is absolutely convergent.

2. If $x \equiv \Delta^{Q_s^*}_{\alpha_1(x)\alpha_2(x)\ldots\alpha_k(x)(0)} = \Delta^{Q_s^*}_{\alpha_1(x)\alpha_2(x)\ldots[\alpha_k(x)-1](s-1)} \equiv x', \alpha_k(x) \neq 0$, then

$$f(x) - f(x') = \left(\prod_{j=1}^{k-1} g_{\alpha_j j}\right)\left(\delta_{\alpha_k k} - \delta_{[\alpha_k-1]k} - g_{[\alpha_k-1]k}\left(\delta_{[s-1][k+1]} + \delta_{[s-1][k+2]}g_{[s-1][k+1]} + \cdots\right)\right) =$$

$$= \left(\prod_{j=1}^{k-1} g_{\alpha_j j}\right)\left(\delta_{\alpha_k k} - \delta_{[\alpha_k-1]k} - g_{[\alpha_k-1]k}\right)$$

$$= \left(\prod_{j=1}^{k-1} g_{\alpha_j j}\right)\left(\delta_{\alpha_k k} + \left(\delta_{[\alpha_k-1]k} + g_{[\alpha_k-1]k}\right)\right) = 0.$$

The corresponding values $f(x)$ and $f(x')$ coincide for different representations of Q_s^*-rational point, hence the function f is well defined on the segment $[0; 1]$.

Theorem 21.3 *The function f is continuous at every point in the segment* $[0, 1]$. *It takes all values from that segment.*

Proof Let $x_0 \in [0, 1]$. By the definition of continuous function, we should prove that

$$\lim_{x \to x_0} |f(x) - f(x_0)| = 0.$$

1. Let x_0 be the Q_s^*-irrational point. For an arbitrary $x \in [0, 1]$, where $x \neq x_0$, there exists $m = m(x)$ such that

$$\begin{cases} \alpha_i(x) = \alpha_i(x_0), & i = \overline{1, m-1}, \\ \alpha_m(x) \neq \alpha_m(x_0). \end{cases}$$

Then

$$f(x) - f(x_0) = \left(\prod_{j=1}^{m-1} g_{\alpha_j(x_0)j}\right)\left(\delta_{\alpha_m(x)m} + \sum_{k=m+1}^{\infty} \delta_{\alpha_k(x)k} \prod_{j=m}^{k-1} g_{\alpha_j(x)j}\right) -$$

$$- \left(\prod_{j=1}^{m-1} g_{\alpha_j(x_0)j}\right)\left(\delta_{\alpha_m(x_0)m} + \sum_{k=m+1}^{\infty} \delta_{\alpha_k(x_0)k} \prod_{j=m}^{k-1} g_{\alpha_j(x_0)j}\right) =$$

$$= \left(\prod_{j=1}^{m-1} g_{\alpha_j(x_0)j}\right)(C_1 - C_2) \to 0, \ m \to \infty,$$

where $0 \le C_1 = \delta_{\alpha_m(x)m} + \delta_{\alpha_{m+1}(x)[m+1]} g_{\alpha_m(x)m} + \cdots < 1$, $0 \le C_2 = \delta_{\alpha_m(x_0)m} + \delta_{\alpha_{m+1}(x_0)[m+1]} g_{\alpha_m(x_0)m} + \cdots < 1$.

Therefore,

$$\lim_{x \to x_0} f(x) = f(x_0)$$

and the function $f(x)$ is continuous at point x_0 by definition.

2. If x_0 is the Q_s^*-rational point, that is

$$x_0 = \Delta^{Q_s^*}_{\alpha_1(x)\alpha_2(x)\ldots\alpha_k(x)(0)} = \Delta^{Q_s^*}_{\alpha_1(x)\alpha_2(x)\ldots[\alpha_k(x)-1](s-1)},$$

we can use the considerations from case 1, but for the situation when x tends to x_0 from the left, it is enough to use the representation of the number x_0 with period $(s-1)$, and when x tends to x_0 from the right we should use the representation with period (0).

Theorem 21.4 *The variation of $f(x)$ is given by the formula*

$$V_0^1(f) = \prod_{n=1}^{\infty} \left(\sum_{i=0}^{s-1} |g_{in}| \right).$$

The function f has a bounded variation if and only if

$$\sum_{n=0}^{\infty} \left(1 - \sum_{i=0}^{s-1} |g_{in}| \right) < \infty.$$

Proof *Variation of function* $f(x)$ on a segment $[a; b]$ is the number

$$V_a^b(f) = \sup_T V_a^b(T; f),$$

where $V_a^b(T; f) = \sum_{k=0}^{n-1} |f(x_{k+1}) - f(x_k)|, n \in N$ and supremum is taken by all possible T-partitions of the segment $[a; b]$.

Since

$$V_1 = \sum_{i=0}^{s-1} \left| f\left(\Delta^{Q_s^*}_{i+1,(0)}\right) - f\left(\Delta^{Q_s^*}_{i(0)}\right) \right| = \sum_{i=0}^{s-1} |g_{i1}|,$$

$$V_2 = \sum_{j=0}^{s-1} |g_{j1}| \cdot \sum_{i=1}^{s-1} \left| f\left(\Delta^{Q_s^*}_{j+1,i+1,(0)}\right) - f\left(\Delta^{Q_s^*}_{j,i(0)}\right) \right| = \left(\sum_{j=0}^{s-1} |g_{j1}| \right) \cdot \left(\sum_{i=0}^{s-1} |g_{i2}| \right),$$

$$\ldots\ldots\ldots\ldots\ldots\ldots\ldots\ldots\ldots\ldots$$

$$V_n = \left(\sum_{j=0}^{s-1} |g_{j1}| \right) \cdot \left(\sum_{k=0}^{s-1} |g_{k2}| \right) \cdot \ldots \cdot \left(\sum_{i=0}^{s-1} |g_{in}| \right) = \prod_{n=1}^{\infty} \left(\sum_{i=0}^{s-1} |g_{in}| \right),$$

we get

$$V_0^1(f) \leq \overline{\lim_{n\to\infty}} V_n.$$

Moreover, it is easy to prove that

$$V_0^1(f) = \lim_{n\to\infty} V_n.$$

Let $L = \prod_{n=1}^{\infty}\left(1 - \left(1 - \sum_{i=0}^{s-1}|g_{in}|\right)\right)$.

Then $L = \infty$ if and only if $\sum_{i=0}^{s-1}\left(1 - \sum_{i=0}^{s-1}|g_{in}|\right) = \infty$. Therefore, if $\sum_{i=0}^{s-1}|g_{in}| \to 1$ as $n \to \infty$, then $V_0^1(f) = \infty$ and the function $f(x)$ is a function of unbounded variation.

The function $f(x)$ is a function of bounded variation if and only if

$$\sum_{n=0}^{\infty}\left(1 - \sum_{i=0}^{s-1}|g_{in}|\right) < \infty.$$

Corollary 21.2 *If all columns of the matrix G_s^* are the same and there are negative elements among its elements, then every function of this class is a function of unbounded variation.*

21.4 Properties of Monotonicity

Lemma 21.1 *Increment of function $\mu_f\left(\Delta^{Q_s^*}_{c_1c_2\ldots c_m}\right)$ on cylinder $\Delta^{Q_s^*}_{c_1c_2\ldots c_m}$, that is*

$$\mu_f\left(\Delta^{Q_s^*}_{c_1c_2\ldots c_m}\right) \equiv f\left(\Delta^{Q_s^*}_{c_1c_2\ldots c_m(s-1)}\right) - f\left(\Delta^{Q_s^*}_{c_1c_2\ldots c_m(0)}\right),$$

is given by the formula

$$\mu_f\left(\Delta^{Q_s^*}_{c_1c_2\ldots c_m}\right) = \prod_{i=1}^{m} g_{c_i i}. \tag{21.6}$$

Proof Since

$$f\left(\Delta^{Q_s^*}_{c_1\ldots c_m(0)}\right) = \delta_{c_1 1} + \sum_{k=2}^{m}\left(\delta_{c_k k}\prod_{i=1}^{k-1} g_{c_i i}\right),$$

$$f\left(\Delta^{Q_s^*}_{c_1\ldots c_m(s-1)}\right) = \delta_{c_1 1} + \sum_{k=2}^{m}\left(\delta_{c_k k}\prod_{i=1}^{k-1} g_{c_i i}\right) + \prod_{i=1}^{m} g_{c_i i},$$

we get

$$\mu_f\left(\Delta^{Q_s^*}_{c_1c_2\ldots c_m}\right) = f\left(\Delta^{Q_s^*}_{c_1c_2\ldots c_m(s-1)}\right) - f\left(\Delta^{Q_s^*}_{c_1c_2\ldots c_m(0)}\right) = \prod_{i=1}^{m} g_{c_i i}.$$

Corollary 21.3 *If all elements of the matrix* $\| g_{ij} \|$ *are nonnegative, then the function* $f(x)$ *is nondecreasing.*

Corollary 21.4 *The function* $f(x)$ *is constant on the cylinder* $\Delta^{Q_s^*}_{c_1c_2\ldots c_m}$ *if and only if there exists* $g_{c_k k} = 0$ *for some* $k \leq m$.

Theorem 21.5 *Function* $f(x)$ *on* $[0, 1]$:

1. *has a finite number of intervals where function is constant if the matrix* $\|g_{ij}\|$ *contains a finite number of zeros;*
2. *has an infinite number of intervals where function is constant if the matrix* $\|g_{ij}\|$ *contains an infinite number of zeros;*
3. *is piecewise monotone if there are no zeros in the matrix* $\|g_{ij}\|$ *and there are negative numbers in the finite number of columns;*
4. *is nowhere monotone if there are no zeros in the* $\|g_{ij}\|$ *matrix and there are negative numbers in the infinite number of columns.*

Proof (1) Let $\prod\limits_{i=0}^{s-1} g_{im} = 0$ and $\prod\limits_{i=0}^{s-1} g_{ij} \neq 0$ for all $j > m$.

If $\prod\limits_{i=1}^{m} g_{c_i i} = 0$, then according to Corollary 21.4 the function $f(x)$ is constant on the cylinder $\Delta^{Q_s^*}_{c_1c_2\ldots c_m}$ and there are exist at most $(s-1)^m$ such cylinders.

If $\prod\limits_{i=1}^{m} g_{c_i i} \neq 0$, then the function is not constant on the cylinder $\Delta^{Q_s^*}_{c_1c_2\ldots c_m}$ and $\prod\limits_{i=1}^{m+k} g_{c_i i} \neq 0$ for any $k \in N$. Therefore, the function $f(x)$ has no intervals of constancy belonging to the cylinder $\Delta^{Q_s^*}_{c_1c_2\ldots c_m}$.

(2) If the matrix $\|g_{ij}\|$ contains an infinite number of zeros, then it has a row that contains an infinite number of zeros. Let (g_{p, j_k}) be a sequence of zeros of the pth row. From condition (3) of the definition of the function it follows that $g_{0, j_k} \cdot g_{s-1, j_k} \neq 0$. Then the cylinders $\Delta^{Q_s^*}_{\underbrace{0\ldots0}_{j_k-1}p}$, $k = 1, 2, \ldots$ have no common ends and are intervals of monotonicity for function f.

(3) If there are no zeros among the elements of the matrix $\|g_{ij}\|$, then according to Corollary 21.4 the function f has no intervals of constancy. Suppose that there exists $q_{cm} < 0$, but $g_{ij} > 0$ for all $i \in A_s$ and $j > m$. We show that if for arbitrary cylinder of rank m there exists a cylinder of rank $(m+1)$ such that increments $\mu_f(\Delta^{Q_s^*}_{c_1c_2\ldots c_m})$ and

$\mu_f(\Delta^{Q_s^*}_{c_1c_2\ldots c_m p})$ have different plus-or-minus signs, then the function is not monotone, but if they have the same plus-or-minus signs then it is monotone.

According to the previous lemma, the increment of the function has the expression $\mu_f\left(\Delta^{Q_s^*}_{c_1c_2\ldots c_m}\right) = \prod_{i=1}^{m} g_{c_i i} \neq 0.$

Then

$$\mu_f\left(\Delta^{Q_s^*}_{c_1c_2\ldots c_m c_{m+1}\ldots c_{m+k}}\right) > 0, \forall k \in N,$$

when $\prod_{i=1}^{m} g_{c_i i} > 0$, that is the function increases on the cylinder $\Delta^{Q_s^*}_{c_1c_2\ldots c_m}$ and

$$\mu_f\left(\Delta^{Q_s^*}_{c_1c_2\ldots c_m c_{m+1}\ldots c_{m+k}}\right) < 0, \forall k \in N,$$

when $\prod_{i=1}^{m} g_{c_i i} < 0$, that is the function decreases on cylinder $\Delta^{Q_s^*}_{c_1c_2\ldots c_m}$.

Taking into account that there are s^m cylinders of the mth rank, we conclude that $f(x)$ is a piecewise monotone function.

(4) Suppose there are no zeros in the $\|g_{ij}\|$ matrix and there are negative numbers in the infinite number of columns. Suppose that there exists an interval $(a, b) \subset [0, 1]$ of the monotonicity of f. Obviously, there exists a cylinder $\Delta^{Q_s^*}_{c_1c_2\ldots c_m}$ completely contained in (a, b), and therefore it is the interval of monotonicity of f.

Since $\mu_f\left(\Delta^{Q_s^*}_{c_1c_2\ldots c_m}\right) \neq 0$, we have that cylinder $\Delta^{Q_s^*}_{c_1c_2\ldots c_m}$ is not an interval of monotonicity of the function.

Let $m + k$ be the smallest number of column containing negative elements, with $g_{i,m+k} < 0$. Then

$$\mu_f\left(\Delta^{Q_s^*}_{c_1c_2\ldots c_m \underbrace{0\ldots 0}_{k-1} 0}\right) \cdot \mu_f\left(\Delta^{Q_s^*}_{c_1c_2\ldots c_m \underbrace{0\ldots 0}_{k-1} i}\right) = \left(\prod_{j=1}^{m} g_{c_j j}\right)^2 \cdot \left(\prod_{j=1}^{k-1} g_{0j}\right)^2 g_{0,m+k} \cdot g_{i,m+k}.$$

Since $g_{0,m+k} > 0$, we see that $g_{0,m+k} \cdot g_{i,m+k} < 0$, and the function has a positive increment on one of the cylinders $\Delta^{Q_s^*}_{c_1c_2\ldots c_m \underbrace{0\ldots 0}_{k-1} 0}$ or $\Delta^{Q_s^*}_{c_1c_2\ldots c_m \underbrace{0\ldots 0}_{k-1} i}$ and it has a negative increment on the other cylinder. This contradicts to its monotonicity on the whole cylinder $\Delta^{Q_s^*}_{c_1c_2\ldots c_m}$. This contradiction proves that function is nowhere monotone.

21.5 Singular Functions of Cantor type

Lemma 21.2 *The continuous function $f(x)$ is a Cantor-type function if and only if there are an infinite number of zeros among the elements of the matrix $\|g_{ij}\|$.*

Proof The spectrum of the function f is the closure of the set

$$\{x : g_{\alpha_j(x)j} \neq 0 \quad \forall j \in N\},$$

that is, a set of Cantor type

$$C[Q_s^*, V_n] \equiv \{x : x = \Delta^{Q_s^*}_{\alpha_1 \dots \alpha_n \dots}, \alpha_n \in V_n \subset A_s \subset N\}, \quad \text{where} \quad V_n = \{k : g_{kn} \neq 0\}.$$

It is known [17] that the set $C[Q_s^*, V_n]$ is nowhere dense if and only if the inequality $V_n \neq A_s$ holds infinitely many times.

Theorem 21.6 *The function f is a singular Cantor type function if and only if the set $C[Q_s^*, V_n] = \{x : x = \Delta^{Q_s^*}_{\alpha_1\alpha_2 \dots \alpha_n \dots} \in [0, 1], \alpha_n \in V \subset N\}$, where $V_n = \{k : g_{kn} \neq 0\}$, is of zero Lebesgue measure, that is $\sum\limits_{k=1}^{\infty} W_k = \infty$, where $W_k = \sum\limits_{i:g_{ij}=0} q_{ik}$, $k \in Z$.*

Proof This statement is a corollary of the Lemma 21.2 and the known fact [30] that Lebesgue measure of the set $C[Q_s^*, V_n]$ is calculated by the formula

$$\lambda(C) = \prod_{k=1}^{\infty}(1 - W_k), \quad \text{where} \quad W_k = \sum_{i:g_{ij}=0} q_{ik}.$$

Therefore, if f is a singular Cantor type function, then $\lambda(S_f) = 0$ and therefore $\lambda(C) = 0$. If $\lambda(C) = 0$, then $\lambda(S_f) = 0$ and the function f is a singular Cantor type function.

21.6 Cantor-Type Functions That Does not Have Intervals of Monotonicity, Except the Intervals Where They Are Constant

Let $A_5 = \{0, 1, 2, 3, 4\}$ be an alphabet of a quinary numeral system and let $L_5 \equiv A_5 \times A_5 \times \cdots$ be a space of the sequences of elements of alphabet.

Let $Q_5^* = \|q_{ij}\|$ be an infinite positive stochastic matrix ($i \in A_5$, $j \in N$), where $q_{ij} > 0$, $q_{0j} + \cdots + q_{4j} = 1$, which defines Q_5^*-representation of numbers:

$$x = \beta_{\alpha_1 1} + \sum_{k=2}^{\infty} \left[\beta_{\alpha_k k} \prod_{j=1}^{k-1} q_{\alpha_j j} \right] = \Delta^{Q_5^*}_{\alpha_1(x)\alpha_2(x)\dots\alpha_k(x)\dots}, \tag{21.7}$$

where $\beta_{0k} = 0$ and

$$\beta_{ij} = \sum_{k=0}^{i-1} q_{kj}, i \in \overline{1,4}, j \in N.$$

Suppose also (ε_n) is a sequence of positive real numbers with $0 \le \varepsilon_n \le 1$; $(\overline{g_n}) = (g_{0n}, g_{1n}, g_{2n}, g_{3n}, g_{4n})$ is the sequence of vectors such that $g_{0n} = g_{4n} = \frac{2+\varepsilon_n}{4}$, $g_{1n} = g_{3n} = \frac{-\varepsilon_n}{4}$, $g_{2n} = 0, n \in N$.

Consider the function f defined by equality

$$f(x) = \delta_{\alpha_1(x)1} + \sum_{k=2}^{\infty} \left(\delta_{\alpha_k(x)k} \prod_{j=1}^{k-1} g_{\alpha_j(x)j} \right) \equiv \Delta^{G_5^*}_{\alpha_1\alpha_2\ldots\alpha_k\ldots}, \tag{21.8}$$

where $\delta_{0n} = 0$, $\delta_{1n} = \frac{2+\varepsilon_n}{4}$, $\delta_{2n} = \frac{2}{4} = \delta_{3n}$, $\delta_{4n} = \frac{2-\varepsilon_n}{4}$,

$$\delta_{[i+1]n} = \delta_{in} + g_{in} = \sum_{j=0}^{i} g_{jn}, n \in N.$$

The function $f(x)$ is continuous on the segment [0; 1]. It is

1. constant on each cylinder of the form $\Delta^{Q_5^*}_{c_1c_2\ldots c_m 2}$ as well as on the cylinders $\Delta^{Q_5^*}_{c_1c_2\ldots c_{n-1}1}$ and $\Delta^{Q_5^*}_{c_1c_2\ldots c_{n-1}3}$ if $\varepsilon_n = 0$;
2. monotone if and only if $\varepsilon_n = 0, n \in N$;
3. singular Cantor type function if the Lebesgue measure of the set $C = C[Q_5^*, \{0, 1, 3, 4\}]$ is equal to zero, i.e., $\sum_{k=1}^{\infty} q_{2k} = \infty$.

The function $f(x)$ takes all values from [0, 1], it does not have intervals of monotonicity, except the intervals where it is constant, if the inequality $\varepsilon_n \neq 0$ holds for an infinite set of values of n.

Lemma 21.3 *If* $q_{ij} = q_i = \frac{1}{5}$, *then the graph of* f *is symmetric about point* $C\left(\frac{1}{2}, \frac{1}{2}\right)$.

Proof The central symmetry of a plane with center C is analytically given by the formulas:

$$\varphi : \begin{cases} x' = 1 - x, \\ y' = 1 - y. \end{cases}$$

Let us show that $f(x) + f(x') = 1$.

Since $x' = 1 - x = \Delta^5_{[4-\alpha_1][4-\alpha_2]\ldots[4-\alpha_k]\ldots}$, but

$$f(x') = \delta_{[4-\alpha_1]1} + \delta_{[4-\alpha_2]2} g_{[4-\alpha_1]1} + \delta_{[4-\alpha_3]3} g_{[4-\alpha_1]1} g_{[4-\alpha_2]2} + \cdots$$

and

$$g_{\alpha_k k} = g_{[4-\alpha_k]k} \quad \forall k \in N,$$

we have

$$f(x) + f(x') = \left(\delta_{\alpha_1 1} + \delta_{[4-\alpha_1]1}\right) + \left(\delta_{\alpha_2 2} + \delta_{[4-\alpha_2]2}\right) g_{\alpha_1 1} + \cdots +$$

$$+ \left(\delta_{\alpha_k+1} + \delta_{[4-\alpha_{k+1}][k+1]}\right) g_{\alpha_1 1} g_{\alpha_2 2} \cdots g_{\alpha_k k} + \cdots .$$

Taking into account that $\delta_{0k} + \delta_{5k} = \delta_{1k} + \delta_{4k} = \delta_{2k} + \delta_{3k} = 1$ and $g_{jk} = g_{[4-j]k}$, we get

$$\delta_{\alpha_k k} + \delta_{[4-\alpha_k]k} = 1 - g_{[4-\alpha_k]} = 1 - g_{\alpha_k k}.$$

So

$$f(x) + f(x') = [1 - g_{\alpha_1 1}] + [1 - g_{\alpha_2 2}] g_{\alpha_1 1} + [1 - g_{\alpha_3 3}] g_{\alpha_1 1} g_{\alpha_2 2} + \cdots =$$

$$= 1 - g_{\alpha_1 1} + g_{\alpha_1 1} - g_{\alpha_1 1} g_{\alpha_2 2} + g_{\alpha_1 1} g_{\alpha_2 2} - g_{\alpha_1 1} g_{\alpha_2 2} g_{\alpha_3 3} + \cdots = 1.$$

Corollary 21.5 *The Riemann integral of the function f is equal to $\frac{1}{2}$, i.e.,*

$$\int_0^1 f(x)dx = \frac{1}{2}.$$

21.6.1 Images of Cantor Type Functions

Let all the columns of the matrix G_5^* be the same, that is, $g_{in} = g_i$, moreover $g_0 = \frac{3}{4} = g_4$, $g_1 = -\frac{1}{4} = g_3$, $g_2 = 0$, $\delta_0 = 0$, $\delta_1 = \frac{3}{4}$, $\delta_2 = \frac{2}{4} = \delta_3$, $\delta_4 = \frac{1}{4}$.

Lemma 21.4 *Under the map f, the image of a quinary cylinder is either G_5-cylinder or a point such that*

$$f(\Delta^{Q_5^*}_{c_1 c_2 \ldots c_m 2}) = \Delta^{G_5}_{c_1 c_2 \ldots c_m 2(0)} = y_0. \tag{21.9}$$

Proof 1. If the cylinder $\Delta^{Q_5^*}_{c_1 \ldots c_m}$ has at least one digit 2 in the base, i.e., $c_i = 2$ $(1 \le i \le m)$, we get $f(\Delta^{Q_5^*}_{c_1 c_2 \ldots c_m}) = \Delta^{G_5}_{c_1 c_2 \ldots c_{i-1} 2(0)}$, where $\Delta^{G_5}_{c_1 c_2 \ldots c_{i-1} 2(0)}$ is a point.

Now we prove that the image of the cylinder $\Delta^{Q_5^*}_{c_1 c_2 \ldots c_m}$, $c_i \ne 2$, $1 \le i \le m$, is a cylinder. Let $x \in \Delta^{Q_5^*}_{c_1 c_2 \ldots c_m}$, i.e., $x = \Delta^{Q_5^*}_{c_1 c_2 \ldots c_m \alpha_{m+1} \ldots \alpha_{m+k} \ldots}$, and

$$f(\Delta^{Q_5^*}_{c_1c_2\ldots c_m(0)}) = \Delta^{G_5}_{c_1c_2\ldots c_m(0)} = a,$$

$$f(\Delta^{Q_5^*}_{c_1c_2\ldots c_m(4)}) = \Delta^{G_5}_{c_1c_2\ldots c_m(4)} = b.$$

According to the property of cylinders, the function takes the largest and smallest value on the cylinder $\Delta^{G_5}_{c_1c_2\ldots c_m}$ at its endpoints, i.e.,

$$f(\Delta^{Q_5^*}_{c_1c_2\ldots c_m}) \subset [A, B],$$

where $A = \min\{a, b\}$, $B = \max\{a, b\}$.

Since $f(x)$ is a continuous function on a segment, it takes all intermediate values between A and B, that is,

$$f(\Delta^{Q_5^*}_{c_1c_2\ldots c_m}) = [A, B].$$

2. Now we prove equality (21.9). If $x = \Delta^{Q_5^*}_{c_1\ldots c_m2\alpha_{m+2}\ldots} \in \Delta^{Q_5^*}_{c_1\ldots c_m2}$, then

$$f(x) = \delta_{c_1} + \cdots + \delta_{c_m}\prod_{i=1}^{m-1} g_{c_i} + \delta_2\prod_{i=1}^{m} g_{c_i} + 0 = \Delta^{G_5}_{c_1c_2\ldots c_m2(0)} = y_0,$$

and the value of the function does not depend on the numbers $\alpha_{m+2}, \alpha_{m+3}\ldots$.

If G_5-representation of the number y uses only two digits, then we will call it a G_2-representation.

Theorem 21.7 *Under the map f, the images of Cantor type sets can be described as follows:*

(1) the image of $C_1 \equiv C[Q_5^*; \{0, 1\}]$ is a segment $\left[0, \frac{3}{4}\right]$, whose countable set of points has exactly two G_2-representation, namely:

$$\Delta^{G_2}_{c_1c_2\ldots c_m01(0)} = \Delta^{G_2}_{c_1c_2\ldots c_m11(0)}, \tag{21.10}$$

and the rest of points have a single G_2-representation;

(2) the image of $C_2 \equiv C[Q_5^*; \{1, 3\}]$ is a set of Cantor type $C_3 \equiv C[4; \{1, 2\}]$; moreover, the mapping of the set C_2 into set C_3 is bijective;

(3) the image of $C_4 \equiv C[Q_5^*; \{1, 2, 3\}]$ is the set $E = C_3 \bigcup M$, where M is a discrete subset of the set of quaternary-rational numbers, namely:

$$M = \{y : y = \Delta^4_{\alpha_1\alpha_2\ldots\alpha_{m-1}2(0)},\ \alpha_i \in \{1; 2\},\ m \in N\}.$$

Proof 1. We prove equality (21.10). Let $\varphi_m \equiv \delta_{c_1} + \sum_{k=2}^{m}\left(\delta_{c_k}\prod_{j=1}^{k-1} g_{c_j}\right)$. Then

$$\Delta^{G_2}_{c_1c_2\ldots c_m01(0)} - \Delta^{G_2}_{c_1c_2\ldots c_m11(0)} =$$

$$= \varphi_m + \delta_1 g_0 \prod_{j=1}^{m} g_{c_j} - \varphi_m - \delta_1 \prod_{j=1}^{m} g_{c_j} - \delta_1 g_1 \prod_{j=1}^{m} g_{c_j} =$$

$$= [\delta_1 g_0 - \delta_1 - \delta_1 g_1]\left(\prod_{j=1}^{m} g_{c_j}\right) = \delta_1[g_0 - 1 - g_1]\left(\prod_{j=1}^{m} g_{c_j}\right) = 0.$$

Therefore, equality (21.10) holds.

Let $I = f(C[Q_5^*; \{0, 1\}])$. Prove that $I = \left[0, \frac{3}{4}\right]$. If $x = \Delta^{Q_5^*}_{\alpha_1\ldots\alpha_n\ldots} \in C_1$, where $\alpha_n \in \{0, 1\}$, we get

$$0 < f(x) = \delta_{\alpha_1(x)} + \sum_{k=2}^{\infty}\left(\delta_{\alpha_k(x)}\prod_{j=1}^{k-1} g_{\alpha_j(x)}\right) \le g_0 = \frac{3}{4}.$$

Thus, $I \subset \left[0, \frac{3}{4}\right]$.

Let us define a cylindrical segment of rank m with base $c_1c_2\ldots c_m$ by equality

$$\Delta^*_{c_1c_2\ldots c_m} = \left[\inf \Delta^{G_2}_{c_1c_2\ldots c_m}; \sup \Delta^{G_2}_{c_1c_2\ldots c_m}\right], \text{ where } c_i \in \{0, 1\}.$$

If $\prod_{i=1}^{m} g_{c_i} \equiv D$, then it is easy to see that

$$\inf \Delta^*_{c_1c_2\ldots c_m} = \begin{cases} \Delta^*_{c_1c_2\ldots c_m(0)}, \text{ if } D > 0, \\ \Delta^*_{c_1c_2\ldots c_m1(0)}, \text{ if } D < 0, \end{cases}$$

$$\sup \Delta^*_{c_1c_2\ldots c_m} = \begin{cases} \Delta^*_{c_1c_2\ldots c_m1(0)}, \text{ if } D > 0, \\ \Delta^*_{c_1c_2\ldots c_m(0)}, \text{ if } D < 0. \end{cases}$$

Prove that for any $m \in N$

$$\left[0, \frac{3}{4}\right] = \Delta^*_0 \cup \Delta^*_1 = (\Delta^*_{00} \cup \Delta^*_{01}) \cup (\Delta^*_{11} \cup \Delta^*_{10}) = \cdots = \bigcup_{c_1=0}^{1}\bigcup_{c_2=0}^{1}\cdots\bigcup_{c_m=0}^{1}\Delta^*_{c_1c_2\ldots c_m},$$

and these formally different cylindrical segments that are part of the union do not overlap. Let us hold general considerations for the cylinder $\Delta^*_{c_1c_2\ldots c_m} = \Delta^*_{c_1c_2\ldots c_m0} \cup \Delta^*_{c_1c_2\ldots c_m1}$.

In fact, let $D > 0$. Then

$$\inf \Delta^*_{c_1c_2...c_m} = \Delta^*_{c_1c_2...c_m(0)} = \inf \Delta^*_{c_1c_2...c_m0(0)},$$

$$\sup \Delta^*_{c_1c_2...c_m0} = \Delta^*_{c_1c_2...c_m01(0)} = \Delta^*_{c_1c_2...c_m11(0)} = \inf \Delta^*_{c_1c_2...c_m1},$$

$$\sup \Delta^*_{c_1c_2...c_m1} = \Delta^*_{c_1c_2...c_m1(0)} = \sup \Delta^*_{c_1c_2...c_m}.$$

Then, for $D < 0$, we get

$$\inf \Delta^*_{c_1c_2...c_m} = \Delta^*_{c_1c_2...c_m1(0)} = \inf \Delta^*_{c_1c_2...c_m1},$$

$$\sup \Delta^*_{c_1c_2...c_m1} = \Delta^*_{c_1c_2...c_m11(0)} = \Delta^*_{c_1c_2...c_m01(0)} = \inf \Delta^*_{c_1c_2...c_m0},$$

$$\sup \Delta^*_{c_1c_2...c_m0} = \Delta^*_{c_1c_2...c_m0(0)} = \sup \Delta^*_{c_1c_2...c_m}.$$

So, by the Cantor axiom, for any $(\alpha_n) \in L_2 = A_2 \times A_2 \times \cdots$

$$\bigcap_{m=1}^{\infty} \Delta^*_{\alpha_1\alpha_2...\alpha_m} = x \in \left[0; \frac{3}{4}\right].$$

Conversely, for any $x \in \left[0; \frac{3}{4}\right]$ there exists $(\alpha_n) \in L_2$ such that

$$x \in \Delta^*_{\alpha_1\alpha_2...\alpha_m} \ \forall\, m \in N.$$

So $I = \left[0, \frac{3}{4}\right]$.

Since cylindrical segments do not overlap, points that are not endpoints of cylindrical segments have a unique G_2-representation, but endpoints have two G_2-representation, except point 0.

2. Let $x \in C_2$, i.e., $x = \Delta^{Q_5^*}_{\alpha_1\alpha_2...\alpha_n...}$, where $\alpha_n \in \{1, 3\}$, $\delta_{\alpha_n} = \frac{a_n}{4}$,

$$a_n = \begin{cases} 4 - \alpha_n, & \text{for } \alpha_n = 1, \\ 5 - \alpha_n, & \text{for } \alpha_n = 3, \end{cases} \quad \text{where } a_n \in \{3; 2\}, \ \forall n \in N. \tag{21.11}$$

Then

$$\delta_{\alpha_n} \prod_{j=1}^{n-1} g_{\alpha_j} = \frac{(-1)^{n-1} a_n}{4^n} \quad \text{and}$$

$$f(x) = \frac{a_1}{4} - \frac{a_2}{4^2} + \frac{a_3}{4^3} - \frac{a_4}{4^4} + \cdots + \frac{a_{2k-1}}{4^{2k-1}} - \frac{a_{2k}}{4^{2k}} + \cdots =$$

$$= \frac{a_1}{4} - \frac{4 - (4 - a_2)}{4^2} + \frac{a_3}{4^3} - \frac{4 - (4 - a_4)}{4^4} + \cdots =$$

$$= \frac{a_1 - 1}{4} + \frac{4 - a_2}{4^2} + \frac{a_3 - 1}{4^3} + \frac{4 - a_4}{4^4} + \cdots =$$

$$= \frac{c_1}{4} + \frac{c_2}{4^2} + \frac{c_3}{4^3} + \frac{c_4}{4^4} + \cdots = \Delta^4_{c_1 c_2 \ldots c_n \ldots},$$

where

$$c_n = \begin{cases} a_n - 1, & \text{if } n \text{ is odd}, \\ 4 - a_n, & \text{if } n \text{ is even}, \ \forall n \in N. \end{cases} \tag{21.12}$$

Since $c_n \in \{1; 2\}$, we conclude that $f(x) \in C_3$.

If $y = f(x) = \Delta^4_{c_1 c_2 \ldots c_n \ldots} \in C_3$, where $c_n \in C_3$, then

$$f^{-1}(y) = \Delta^{Q_5^*}_{\alpha_1 \alpha_2 \ldots \alpha_n \ldots},$$

where

$$\alpha_n = \begin{cases} 4 - a_n, \text{ for } a_n = 3, \\ 5 - a_n, \text{ for } a_n = 2, \end{cases} \quad \alpha_n \in \{1; 3\}, \tag{21.13}$$

$$a_n = \begin{cases} c_n - 1, \text{ if } n \text{ is odd}, \\ 4 - c_n, \text{ if } n \text{ is even}, \ \forall n \in N. \end{cases} \tag{21.14}$$

Since $\alpha_n \in \{1, 3\}$, we conclude that $x \in C_2$. Hence, $f(C_2) = C_3$.

The fact that the mapping $f : C_2 \to C_3$ is bijective follows from the uniqueness of the Q_5^*-representation of the numbers from the set C_2 and the quaternary representation of the numbers from the set C_3, and the reasoning presented above.

3. Obviously, $C_3 \bigcap M = \emptyset$. Since $C[Q_5^*; \{1, 3\}] \equiv C_2 \subset C_4$, we see that for $f(C_4) \subset E$, it is enough to show that for $x \in C_4 \setminus C_2$ we have $f(x) \in M$.

Let $\alpha_m(x) = 2$ and let $\alpha_j(x) \neq 2$ for $j < m$. Then

$$f(x) = \Delta^4_{c_1 c_2 \ldots c_{m-1} 2(0)} = \Delta^4_{c_1 c_2 \ldots c_{m-1} 1(3)} = y \in M,$$

where c_i is given by Eqs. (21.11) and (21.12).

In view of the previous reasoning, it suffices to prove that $y \in M$ implies $f^{-1}(y) \in C_4$.

Let $y = \Delta^4_{c_1 c_2 \ldots c_{m-1} 2(0)}$, where $c_i \in \{1, 2\}$. Then $f(x) = f(\Delta^{Q_5^*}_{\alpha_1 \ldots \alpha_{m-1} 2 \alpha_{m+1} \ldots})$, where α_j $(j < m)$ can be calculated by formulas (21.13) and (21.14), and α_{m+j} are arbitrary digits from set $\{1; 2; 3\}$ that belong to the set C_4. Hence, each point of the set E is a preimage of at least one point from the set C_4. Therefore, $f(C_4) = E$. The theorem is proved.

21.6.2 The Distribution of Values of f for a Given Argument Distribution

Let (τ_n) be a sequence of independent random variables that take the values $0, 1, 2, 3{,}4$ with probabilities $p_{0n}, p_{1n}, p_{2n}, p_{3n}, p_{4n}, n \in N$ respectively, i.e.,

$$P\{\tau_n = i\} = p_{in},\ i = \overline{0,4},$$

$$p_{0n} + p_{1n} + p_{2n} + p_{3n} + p_{4n} = 1,\ \forall n \in N.$$

Let $X = \Delta^{Q_5^*}_{\tau_1\tau_2\tau_3...}$ be a continuous random variable with independent quinary digits.

Theorem 21.8 ([17]) *The random variable X has the distribution of pure Lebesgue type. It is:*

1. *pure discrete if*

$$M \equiv \prod_{k=1}^{\infty} \max_i\{p_{ik}\} > 0;$$

2. *pure absolutely continuous if*

$$L \equiv \sum_{k=1}^{\infty}\sum_{i=0}^{4}\left(1 - \frac{p_{ik}}{q_{ik}}\right)^2 < \infty;$$

3. *pure singularly continuous if $M = 0$ and $L = \infty$.*

Theorem 21.9 *Let $X = \Delta^{Q_5^*}_{\tau_1\tau_2...\tau_n...}$ be a random variable such that the digits (τ_n) of its Q_5^*-representation are independent and have the distributions $P\{\tau_n = i\} = p_{in}$, $i = \overline{0,4}$, moreover,*

$$\prod_{k=1}^{\infty} \max_i\{p_{ik}\} = 0. \tag{21.15}$$

The distribution of the random variable $Y = f(X)$ is

1. *pure discrete if*

$$\sum_{n=1}^{\infty}\left[p_{2n}\prod_{k=1}^{n-1}(1 - p_{2k})\right] = 1;$$

2. *a nontrivial mixture of discrete and continuous distributions if*

$$0 < \sum_{n=1}^{\infty}\left[p_{2n}\prod_{k=1}^{n-1}(1 - p_{2k})\right] < 1,$$

moreover,

$$F_Y(x) = cF_d(x) + (1-c)F_c(x),$$

where $F_d(x) = \sum_{\Delta^{G}_{c_1\dots c_m 2(0)} = x_i < x,\ c_j \in \{0,1,3,4\}} p_{x_i}$, $p_{x_i} = P\{Y = \Delta^{G_2}_{c_1\dots c_m 2(0)}\}$;

3. *a singular distribution of Cantor type if, for any* $n \in N$, *the equalities* $p_{0n} = p_{2n} = p_{4n} = 0$ *hold.*

Proof The set of levels of the function f that contain intervals is countable and any level cannot contain more than two intervals. Hence, the event

$$P\{Y = f(\Delta^{Q_5^*}_{c_1\dots c_m 2}) = \Delta^{G_2}_{c_1\dots c_m 2(0)}\}$$

is equivalent to the event

$$X \in \Delta^{Q_5^*}_{c_1\dots c_m 2},$$

if the level $y_0 = \Delta^{G_2}_{c_1\dots c_m 2(0)}$ contains one segment or to the event

$$X \in \Delta^{Q_5^*}_{c_1\dots c_m 2} \bigcup \Delta^{Q_5^*}_{c'_1\dots c'_m 2},$$

where $f(\Delta^{Q_5^*}_{c_1\dots c_m 2(0)}) = f(\Delta^{Q_5^*}_{c'_1\dots c'_m 2(0)})$, if the level contains two intervals.

In the first case, in view of the independence of the digits of the random variable X, we get

$$P\{Y = f(\Delta^{Q_5^*}_{c_1\dots c_m 2(0)})\} = P\{X \in \Delta^{Q_5^*}_{c_1\dots c_m 2}\} \equiv \left(\prod_{i=1}^{m} p_{c_i i}\right) p_{2,m+1}.$$

In the second case, we assume that the number

$$y = f(\Delta^{Q_5^*}_{c_1\dots c_m 2(0)})$$

has two "components" y and y^*, which are formally different numbers.

Hence, $y = f\left(\Delta^{Q_5^*}_{c_1\dots c_m 2(0)}\right)$ is the atom of distribution of function f, when

$$p_{c_i i} \neq 0 \ \forall i \in \{1, 2, \dots, m\} \text{ and } p_{2,m+1} \neq 0.$$

The total mass of atoms of the distribution Y is calculated by the formula

$$S = \sum_{n=1}^{\infty}\left[p_{2n}\prod_{k=1}^{n-1}(p_{0k} + p_{1k} + p_{3k} + p_{4k})\right] = \sum_{n=1}^{\infty}\left[p_{2n}\prod_{k=1}^{n-1}(1 - p_{2k})\right].$$

Obviously, the last sum is a finite number. Moreover, it contains finitely many terms if there exists $p_{2k} = 1$, and infinitely many terms, otherwise.

If $p_{0n} = p_{1n} = p_{2n} = p_{3n} = p_{4n}$ for all $n \in N$, then $S = 1$ and the distribution is pure atomic (discrete). If all p_{in} are different and $S < 1$, then, by using the Lebesgue theorem on the structure of probability measure, we conclude that the distribution Y has atoms as well as a nontrivial continuous component. Note that $S \neq 0$, which is the result of continuous distribution X.

If $p_{0n} = p_{2n} = p_{4n} = 0$, then for any $n \in N$, the spectrum of the distribution of the random variable Y has the form of a Cantor type set $C_3 = C[4; \{1, 2\}]$ if $p_{1n}^2 + p_{3n}^2 \neq 1$ for all $n \in N$ or of its subset if the equality $p_{1m}^2 + p_{3m}^2 = 1$ is true for some m. Therefore, according to Theorem 21.7, the Lebesgue measure of the spectrum of the distribution Y is equal to zero.

The distribution of the random variable X is continuous, i.e.,

$$\prod_{n=1}^{\infty} \max_k \{p_{kn}\} = M = 0 = \prod_{n=1}^{\infty} \max\{p_{1n}, p_{3n}\}.$$

Then in view of the fact that the mapping $f : C_2 \to C_3$ Y is bijective, the distribution of Y does not have atoms. Hence, it is a singular distribution of the Cantor type. The theorem is proved.

Theorem 21.10 *Let* $X = \Delta^{Q_5^*}_{\tau_1\tau_2\ldots\tau_n\ldots}$ *be a random variable such that the digits* (τ_n) *of its* Q_5^**-representation are independent and have distributions* $P\{\tau_n = i\} = p_{in}$, $i = \overline{0, 4}$. *If* $p_{2n} = p_{3n} = p_{4n} = 0$ *for any* $n \in N$, *then the distribution of the random variable* $Y = f(X) = \Delta^{G_2}_{\eta_1\eta_2\ldots\eta_n\ldots}$, *encoded by means of the two-symbol alphabet* $A_2 = \{0, 1\}$ *in the system with zero redundancy and such that the digits* $\eta_1, \eta_2, \ldots, \eta_n, \ldots$ *of its representations are independent, is of pure Lebesgue type, and is absolutely continuous if*

$$W = \sum_{n=1}^{\infty} \left[\left(1 - \frac{4}{3} p_{0n}\right)^2 + (1 - 4p_{1n})^2 \right] < \infty$$

and singular for $W = \infty$.

Proof By Theorem 21.7, the random variable Y takes all values from the segment $[0; \frac{3}{4}]$, whose almost all points, except for the countable set, have a unique G_2-representation, and:

$$f(\Delta_0^{Q_5^*}) = \Delta_0^{G_2} = \left[0; \frac{9}{16}\right], \quad f(\Delta_1^{Q_5^*}) = \Delta_1^{G_2} = \left[\frac{9}{16}; \frac{3}{4}\right],$$

$$P_Y(\Delta_0^{G_2}) = P_X(\Delta_0^{Q_5^*}) = p_{01}, \quad P_Y(\Delta_1^{G_2}) = P_X(\Delta_1^{Q_5^*}) = p_{11}.$$

Moreover,

$$P_Y(f(\Delta^{Q_5^*}_{c_1c_2\ldots c_m i})) = P_X(\Delta^{Q_5^*}_{c_1c_2\ldots c_m i}) = p_{i[m+1]}.$$

$$F'_Y(x) = \lim_{n\to\infty} \frac{P_Y(f(\Delta^{Q_5^*}_{\alpha_1\alpha_2\ldots\alpha_n}))}{|f(\Delta^{Q_5^*}_{\alpha_1\alpha_2\ldots\alpha_n})|} = \lim_{n\to\infty} \prod_{i=1}^{n+1} \left(\frac{p_{\alpha_i i}}{|g_{\alpha_i}|}\right) = \prod_{i=1}^{\infty} \left(\frac{4p_{\alpha_i i}}{2-(-1)^{1-\alpha_i}}\right).$$

The last infinite product is convergent if and only if the series

$$\sum_{i=1}^{\infty} \ln\left(\frac{4p_{\alpha_i i}}{2-(-1)^{1-\alpha_i}}\right)$$

is convergent.

Since every point $y \in [0; \frac{3}{4}]$ is the image of a single point $x \in C[Q_5^*, \{0, 1\}]$, where $y = f(x)$, and moreover, the digits of the representations of x and y are connected by the formula

$$\alpha_i(f(x)) = \begin{cases} 0, \text{if } \alpha_i(x) = 0, \\ 1, \text{if } \alpha_i(x) = 1, \end{cases}$$

we see that the G_2-digits of the random variable Y are independent as the digits of X are independent.

Moreover,

$$P_Y\{\eta_n = 0\} = P_X\{\tau_n = 0\} = p_{0n}, \quad P_Y\{\eta_n = 1\} = P_X\{\tau_1 = 1\} = p_{1n}.$$

The G_2-representation is self-similar as the Q_2-representation of numbers is self-similar. There is an easy relation between these representations (the bijection that maps one representation into another and preserves the Lebesgue measure and the fractal Hausdorff–Besicovitch dimension). This is the basis for the conclusion: the methods of proving the purity of the distribution of the random variable Y and criterion for absolute continuity (and hence singularity) can be generalized to this case. So, by using the well-known results obtained in [17] for the Q_2-representation, we get the rest of the theorem.

21.7 Conclusion

Many physical, economic, biological, psychological, social, and other processes run unevenly over time. And this heterogeneity has different functional manifestations, which are reflected in nonlinearity, instability, irregularity of local differential properties, etc. The components of the mathematical model of such processes are Cantor functions (continuous functions with self-similar properties and a set of non-constancy of a zero or positive Lebesgue measure). Their theoretical study is conducted in different directions (a topological-metric and fractal analysis of sets of levels of functions, sets of different features, in particular, sets of non-constancy, structural analysis of properties of graphs, self-similar properties). At the same time, they are increasingly appearing in various mathematical models (digital signal processing, construction of fractal antennas), in fractal-related mathematical sections (in probability theory for the study of distributions of random variables, in theory of dynamical systems). They adequately describe transient fractal radiations, appear in problems of control of distributed systems, and the like. Therefore, the convenience of the analytical definition of such functions and their study is an urgent problem of today.

Recently, various systems of encoding for real numbers are used to solve this problem. The simplest system is the s-adic numeral system. Its generalization are Q_s-representation and Q_s^*-representation whose digits do not function as numbers and act only as indexes. We use this representation of numbers to simulate and study functions with a non-trivial local structure. Monotone singular, non-monotone singular, nowhere monotone and non-monotone singular Cantor-type functions are among them. In addition to the above-mentioned problems, for one case, we analyze the properties of functions related to the uniform and non-uniform distribution of the argument.

References

1. Agadzhanov, A.N.: Singular functions that do not have intervals of monotonicity in problems of finite control of distributed systems. Rep. Acad. Sci. **454**(5), 503–506 (2014) (in Russian)
2. Barabash, O., Sobchuk, V., Lukova-Chuiko, N., Musienko, A.: Application of Petri networks for support of functional stability of information systems. In: IEEE First International Conference on System Analysis and Intelligent Computing (SAIC), pp. 36–39 (2018)
3. Barabash, O., Kopiika, O., Zamrii, I., Sobchuk, V., Musienko, A.: Fraktal and differential properties of the inversor of digits of Q_s-representation of real number. Modern Mathematics and Mechanics. Fundamentals, Problems and Challenges, pp. 79–95. Springer International Publishing AG, New York (2019)
4. Boiko, J., Tolubko, V., Barabash, O., Eromenko, O., Havrylko, Ye.: Signal processing with frequency and phase shift keying modulation in telecommunications. TELKOMNIKA Telecommun. Comput. Electron. Control (Yogyakarta, Indonesia) **17**(4), 2025–2038 (2019)
5. Bolotov, V.N.: Transient fractal radiation. J. Eng. Phys. **70**(12), 98–101 (2000) (in Russian)
6. Dovgoshey, O., Martio, O., Ryazanov, V., Vuorinen, M.: The Cantor function. Expo. Math. **24**, 1–37 (2006)
7. Feinberg, E.A., Kasyanov, P.O., Zgurovsky, M.Z.: Convergence of probability measures and Markov decision models with incomplete information. Proc. Steklov Inst. Math. **287**(1), 96–117 (2014)
8. Kapustian, O.A., Sobchuk, V.V.: Approximate homogenized synthesis for distributed optimal control problem with superposition type cost functional. Stat. Optim. Inf. Comput. **6**(2), 233–239 (2018)
9. Kasyanov, P.O., Toscano, L., Zadoianchuk, N.V.: A criterion for the existence of strong solutions for the 3D Navier-Stokes equations. Appl. Math. Lett. **26**(1), 15–17 (2013)

10. Kravchenko, V.F., Masiuk, V.M.: New class of fractal functions in the problems of analysis and synthesis of antennas. Modern Methods of Approximation in the Theory of Antennas. Book 3 in 3 Books (2002), 72 pp. (IPRZhR, Moscow) (in Russian)
11. Mukhin, V., Loutskii, H., Barabash, O., Kornaga, Ya., Steshyn, V.: Models for analysis and prognostication of the indicators of the distributed computer systems characteristics. Int. Rev. Comput. Softw. (IRECOS) **12**(144), 247–255 (2015)
12. Musienko, A., Serdyuk, A.: Lebesgue-type inequalities for the de la Vallee Poussin sums on sets of entire functions. Ukr. Math. J. **65**(5), 709–722 (2013)
13. Musienko, A., Serdyuk, A.: Lebesgue-type inequalities for the de la Valee-Poussin sums on sets of analytic functions. Ukr. Math. J. **65**(4), 575–592 (2013)
14. Okamoto, H.: A remark on continuous, nowhere differentiable functions. Proc. Jpn. Acad. Ser. A Math. Sci. **81**(3), 47–50 (2005)
15. Okamoto, H., Wunsh, M.: A geometric construction of continuous strictly increasing singular functions. Proc. Jpn. Acad. Ser. A Math. Sci. **83**(7), 114–118 (2007)
16. Potapov, A.A.: Fractals in radiophysics and radiolocation: a sampling topology, 2nd edn. alteration. and ext. (2005), 848 pp. (University Book, Moscow) (in Russian)
17. Pratsiovitiy, M.V.: Fractal approach to the investigation of singular distributions (1998), 296 pp. (National Pedagogical Dragomanov University, Kyiv) (in Ukraine)
18. Pratsiovitiy, M.V.: Nowhere monotonic singular functions. Trans. Mykhailo Drahomanov Natl. Pedagog. Univ. Ser. 1. Phys. Math. **12**, 24–36 (2011) (National Pedagogical Dragomanov University, Kyiv) (in Ukraine)
19. Pratsiovytyi, M., Isaieva, T.: Transformations of (0; 1] preserving tails of Δ^{μ}-representation of numbers. Algebra Discret. Math. **22**(1), 102–115 (2016)
20. Pratsiovitiy, M.V., Kalashnikov, A.V.: Self-affine singular and nowhere monotonic functions related to the Q-representation of real numbers. Ukr. Math. J. **65**(3), 381–393 (2013) (in Ukraine)
21. Pratsiovitiy, M.V., Ratushnyak, S.P.: Distribution of values of a fractal function of random argument. Trans. Mykhailo Drahomanov Natl. Pedagog. Univ. Ser. 1. Phys. Math. **16**(2), 150–160 (2014) (National Pedagogical Dragomanov University, Kyiv) (in Ukraine)
22. Pratsiovitiy, M.V., Svynchuk, O.V.: Singular non-monotonic functions defined in terms of Q_s^*-representation of argument. Trans. Mykhailo Drahomanov Natl. Pedagog. Univ. Ser. 1. Phys. Math. **15**, 134–144 (2013) (National Pedagogical Dragomanov University, Kyiv) (in Ukraine)
23. Pratsiovitiy, M.V., Svynchuk, O.V.: On one family of continuous non-monotonic singular Cantor type functions with fractal properties. Trans. Mykhailo Drahomanov Natl. Pedagog. Univ. Ser. 1. Phys. Math. **17**(2), 37–48 (2015) (National Pedagogical Dragomanov University, Kyiv) (in Ukraine)
24. Pratsiovitiy, M.V., Svynchuk, O.V.: Distribution of values of one singular nonmonotonic function of Cantor type. Tr. Inst. Prikl. Mat. Mekh. **14**(2), 68–79 (2017) (in Ukraine)
25. Pratsiovytyi, M., Svynchuk, O.: Spread of values of a Cantor-type fractal continuous nonmonotone function. J. Math. Sci. **240**(3), 342–357 (2019)
26. Pratsiovitiy, M.V., Vasilenko, N.A.: Probability distribution on graphs of one class of nowhere differentiable functions. Tr. Inst. Prikl. Mat. Mekh. **26**, 159–171 (2013) (in Ukraine)
27. Pratsiovytyi, M., Vasylenko, N.: Fractal properties of functions defined in terms of Q-representation. J. Math. Anal. **7**(61), 3155–3169 (2013)
28. Pratsiovitiy, M.V., Torbin, G.M., Goncharenko, Y.V.: Contemporary problems of the singular distributions of probabilities. Sci. Notes: Collect. Sci. Artic. NPU Named After M.P. Dragomanov **42**, 18–20 (2001) (National Pedagogical Dragomanov University, Kyiv) (in Ukraine)
29. Pratsiovytyi, M., Lysenko, I., Voitovska, O.: Distribution of values of classic singular Cantor function of random argument. Random Oper. Stoch. Equ. **26**(4), 193–200 (2018)
30. Torbin, G.M., Pratsiovitiy, M.V.: Random variables with independent Q^*-signs. Random Evolutions: Theoretical and Applied Problems, pp. 95–104 (1992) (National Pedagogical Dragomanov University, Kyiv) (in Russian)
31. Zamrii, I.V., Pratsovytyi, M.V.: Singularity of the digit inversor for the Q_3-representation of the fractional part of a real number, its fractal and integral properties. J. Math. Sci. **215**(3), 323–340 (2016)

Chapter 22
Solvability Issue for Optimal Control Problem in Coefficients for Degenerate Parabolic Variational Inequality

Nina V. Kasimova

Abstract In this paper we investigate an optimal control problem in coefficients for degenerate parabolic variational inequality. Since these types of problems can exhibit the Lavrentieff phenomenon and non-uniqueness of weak solutions, there several possible statements of such problems depending on the choice of the class of admissible solutions. Here we consider the optimal control problem in the so-called class of H-admissible solutions. Using the classical approach to parabolic variational inequalities, we show that the set of admissible pairs is not empty. We prove some topological properties of the set of H-admissible solutions and show that this set possesses some compactness properties with respect to the appropriate convergence in variable spaces. Using, the direct method in Calculus of variations, we prove the theorem on the existence of H-optimal solutions.

22.1 Introduction

The purpose of this paper is to investigate optimal control problem associated with a degenerate parabolic inequality. The control is a matrix of coefficients in the main part of elliptic operator. It is well known that degenerate control problems of this type can exhibit the Lavrentieff phenomenon and non-uniqueness of solutions (see [20, 23, 24]).

Note, that optimal control problems in coefficients for PDE are not new in the literature and widely studied by many authors since this topic includes optimal shape design problems, optimization of certain evolution systems, some problems originating in mechanics and others. We could mention Butazzo and Dal Maso [3], Lions [17], Murat [19] and others.

As F. Murat showed (see [18]), in general, such problems have no solution even if the original elliptic equation is non-degenerate. It turns out that this feature is typical

N. V. Kasimova (✉)
Taras Shevchenko National University of Kyiv, Volodymyrska Street 60,
Kyiv 01601, Ukraine
e-mail: zadoianchuk.nv@gmail.com

V. A. Sadovnichiy and M. Z. Zgurovsky (eds.), *Contemporary Approaches and Methods in Fundamental Mathematics and Mechanics*, Understanding Complex Systems,
https://doi.org/10.1007/978-3-030-50302-4_22

for the majority of problems for optimal control in coefficients. So, we have to restrict our optimization problem by introducing some additional control constrains (see, for instance, [11]). An optimal control problem for a variational inequality with the so-called anisotropic p-Laplacian in the principle part of this inequality is studied in [15] and the authors showed that the original problem is well-posed and derive existence of optimal pairs. In [16] an optimal control problem associated to Dirichlet boundary value problem for non-linear elliptic equation on a bounded domain Ω is considered. The authors take the coefficient $u(x) \in L^{\infty}(\Omega) \cap BV(\Omega)$ in the main part of the non-linear differential operator as a control and in the linear part of differential operator they consider coefficients to be unbounded skew-symmetric matrix and show that, in spite of unboundedness of the non-linear differential operator, the considered Dirichlet problem admits at least one weak solution and the corresponding OCP is well-possed and solvable. In [4] the authors study the existence of optimal solutions in coefficients associated to a linear degenerate elliptic equation with mixed boundary condition where by control variable they mean a weight coefficient in the main part of the elliptic operator. The sufficient conditions of the existence of weak solutions to one class of Neumann boundary value problems (BVP) are obtained in [12], and moreover, the authors propose a way for their approximation. In [14] the existence of H-optimal solutions for optimal control problem in coefficients for degenerate variational elliptic inequalities of monotone type in the class of so-called generalized solenoidal controls was proved. In [1] the authors study the Dirichlet optimal control problem associated with a degenerate linear parabolic equation the coefficients of which they take as controls in $L^1(\Omega)$.

Here we concentrate on the optimal control problem in coefficients of a degenerate linear parabolic variational inequality in the so-called class of H-admissible solutions and prescribe the class of admissible controls which includes some div-like conditions in weighted spaces. The paper is organized as follows. In Sect. 22.2 we give the collection of preliminary results. In Sect. 22.3 we state the problem of optimal control in coefficients and prescribe the solvability of degenerate variational inequality that gives us the regularity of the original problem. Section 22.4 deals with the existence of optimal solutions to the original problem. Moreover, here we study the topological properties of the class of H-admissible solutions and show that this set possesses some compactness properties with respect to the appropriate convergence in variable spaces. And then, using the direct method in the Calculus of variations, we prove the existence of the H-optimal solutions to the original problem.

22.2 Notations and Preliminaries

22.2.1 Weighted Sobolev Spaces

Let Ω be a bounded subset of $R^N (N \geq 1)$ with a Lipschitz boundary. Let χ_E be the characteristic function of a subset $E \subseteq \Omega$ i.e. $\chi_E(x) = 1$ if $x \in E$, and $\chi_E(x) = 0$ if

$x \notin E$. The space $W_0^{1,1}(\Omega)$ is the closure of $C_0^\infty(\Omega)$ in the classical Sobolev space $W^{1,1}(\Omega)$. For any subset $E \subset \Omega$ we denote by $|E|$ its N-dimensional Lebesgue measure $\mathscr{L}^N(E)$.

Hereinafter by a weight we mean a locally integrable function ρ on R^N such that $\rho(x) > 0$ for a. e. $x \in R^N$. As a matter of fact every weight ρ gives rise to a measure on the measurable subsets of R^N through integration. This measure will also be denoted by ρ. Thus $\rho(E) = \int_E \rho dx$ for measurable sets $E \subset R^N$. We will use the standard notation $L^2(\Omega, \rho dx)$ for the set of measurable functions f on Ω such that

$$\|f\|_{L^2(\Omega,\rho dx)} = \left(\int_\Omega f^2 \rho dx\right)^{1/2} < +\infty.$$

Definition 22.1 We say that a weight function $\rho : R^N \to R_+$ is degenerate on Ω if

$$\rho + \rho^{-1} \in L^1_{loc}(R^N) \tag{22.1}$$

and the sum $\rho + \rho^{-1}$ does not belong to $L^\infty(\Omega)$.

With each of the degenerate weight functions ρ we will associate two weighted Sobolev spaces $W(\Omega, \rho dx)$ and $H(\Omega, \rho dx)$, where $W(\Omega, \rho dx)$ is the set of functions $y \in W_0^{1,1}(\Omega)$ for which the norm

$$\|y\|_\rho = \left(\int_\Omega \left(y^2 + \rho|\nabla y|^2\right) dx\right)^{1/2} \tag{22.2}$$

is finite, and $H(\Omega, \rho dx)$ is the closure of $C_0^\infty(\Omega)$ in $W(\Omega, \rho dx)$-norm. Note that due to the compact embedding $W_0^{1,1}(\Omega) \hookrightarrow L^1(\Omega)$ and estimates

$$\int_\Omega |y| dx \le |\Omega|^{1/2} \left(\int_\Omega |y|^2 dx\right)^{1/2} \le \sqrt{|\Omega|}\|y\|_\rho, \tag{22.3}$$

$$\int_\Omega |\nabla y| dx \le \left(\int_\Omega |\nabla y|^2 \rho dx\right)^{1/2} \left(\int_\Omega \rho^{-1} dx\right)^{1/2} \le C\|y\|_\rho, \tag{22.4}$$

we come to the following result (we refer to [13, 24] for the details):

Theorem 22.1 *Let $\rho : R^N \to R_+$ be a degenerate weight on Ω. Then*

1. *the spaces $H(\Omega, \rho dx)$ and $W(\Omega, \rho dx)$ are complete with respect to the norm $\|\cdot\|_\rho$;*
2. *$H(\Omega, \rho dx) \subseteq W(\Omega, \rho dx)$, and $W(\Omega, \rho dx)$, $H(\Omega, \rho dx)$ are Hilbert spaces;*
3. *$H(\Omega, \rho dx) \subset W_0^{1,1}(\Omega)$, $W(\Omega, \rho dx) \subset W_0^{1,1}(\Omega)$, and the estimate*

$$\|v\|_{W_0^{1,1}(\Omega)} \le \left(\sqrt{|\Omega|} + \left(\int_\Omega \rho^{-1} dx\right)^{1/2}\right) \|v\|_\rho$$

is valid for every element $v \in H(\Omega, \rho dx) \cup W(\Omega, \rho dx)$;
4. *the embeddings* $H(\Omega, \rho dx) \hookrightarrow L^1(\Omega)$ *and* $W(\Omega, \rho dx) \hookrightarrow L^1(\Omega)$ *are compact.*

If ρ is non-degenerate weight function, that is, ρ is bounded between two positive constants, then it is easy to verify that $W(\Omega, \rho dx) = H(\Omega, \rho dx)$. However, for a "typical" degenerate weight ρ the space of smooth functions $C_0^\infty(\Omega)$ is not dense in $W(\Omega, \rho dx)$. Hence the identity $W(\Omega, \rho dx) = H(\Omega, \rho dx)$ is not always valid (for the corresponding examples we refer to [5, 22].

We recall that by Riesz Representation Theorem the dual space $(H(\Omega, \rho dx))^*$ of weighted Sobolev space $H(\Omega, \rho dx)$ can be characterized as follows: if $g \in (H(\Omega, \rho dx))^*$ then there exist functions $g_0 \in L^2(\Omega)$ and $\mathbf{g}_1 \in L^2(\Omega, \rho dx)^N$ such that

$$\langle g, y\rangle = \int_\Omega g_0 y dx + \int_\Omega (\mathbf{g}_1, \nabla y)_{R^N} \rho dx \quad \forall y \in H(\Omega, \rho dx), \tag{22.5}$$

where by $\langle \cdot, \cdot \rangle$ we denote the duality between elements of $(H(\Omega, \rho dx))^*$ and $H(\Omega, \rho dx)$, respectively. Furthermore,

$$\|g\|_{(H(\Omega,\rho dx))^*} = inf \left\{ \left(\int_\Omega |g_0|^2 dx + \int_\Omega \|\mathbf{g}_1\|^2_{R^N} \rho dx \right)^{1/2} : \text{g satisfies (22.5)} \right\}.$$

Remark 22.1 Note that under some additional suppositions Theorem 22.1 can be specified as follows: assume that there exists $\nu \in (N/2, +\infty)$ such that $\rho^{-\nu} \in L^1(\Omega)$. Then

$$|||y||| = \left(\int_\Omega \rho |\nabla y|^2 dx \right)^2$$

is a norm defined on $H(\Omega, \rho dx)$ and it's equivalent to (22.2) and that, the embedding $H(\Omega, \rho dx) \hookrightarrow L^2(\Omega)$ is compact [6, pp 46].

To conclude this section we recall some results concerning variational triplets. Let $V_- = H(\Omega, \rho dx)$, $V = L^2(\Omega)$ and let $V_-^* = (H(\Omega, \rho dx))^*$. Let $\mathscr{X} = L^2(0, T; V_-)$. Then the dual space of $\mathscr{X}$ is $\mathscr{X}^* = L^2(0, T; V_-^*)$. For any $y \in \mathscr{X}$, let y' denotes the generalized derivative of $y(t) = y(t, \cdot)$, i.e.

$$\int_0^T y'(t)\varphi(t)dt = -\int_0^T y(t)\varphi'(t)dt \quad \forall \varphi \in C_0^\infty([0, T]).$$

Then we have the following result (see [21]):

Lemma 22.1 *Assume that there exists* $\nu \in (N/2, +\infty)$ *such that* $\rho^{-\nu} \in L^1(\Omega)$. *Then* $V_- \subseteq V \subseteq V_-^*$ *is an evolution triple, i.e. the embeddings* $V_- \hookrightarrow V \hookrightarrow V_-^*$ *are continuous, and the embedding* $V_- \hookrightarrow V$ *is compact. Moreover,* $\mathscr{W} = \{y \in \mathscr{X}, y' \in \mathscr{X}^*\}$ *equipped with the norm*

$$\|y\|_{\mathscr{W}} = \|y\|_{\mathscr{X}} + \|y'\|_{\mathscr{X}^*} := \|y\|_{L^2(0,T;H(\Omega,\rho dx))} + \|y'\|_{L^2(0,T;(H(\Omega,\rho dx))^*)}$$

is a Banach space such that

1. *the embedding* $\mathscr{W} \hookrightarrow C([0,T]; L^2(\Omega))$ *is continuous;*
2. *the embedding* $\mathscr{W} \hookrightarrow L^2(0,T; L^2(\Omega))$ *is compact.*

22.2.2 Parabolic Variational Inequalities

We present for later use some basic results in the existence theory of variational inequalities of parabolic type. We emphasize those aspects of the abstract theory which are of importance in practical applications.

Let V and H will be real Hilbert spaces, V is a dense subspace of H and

$$V \subset H \subset V^*$$

algebraically and topologically. We shall denote by $|\cdot|$ and $\|\cdot\|$ the norms in H and V, respectively, and by $(\cdot,\cdot)$ the scalar product in H and the pairing between V and its dual V^*. The norm of V^* will be denoted $\|\cdot\|_*$.

We are given a linear continuous and symmetric operator from V to V^* satisfying for some $\omega > 0$, and real α, the coercivity condition

$$(Ay, y) + \alpha|y|^2 \geq \omega\|y\|^2 \quad \text{for all } y \in V, \tag{22.6}$$

and a lower semicontinuous convex function $\phi : V \to \bar{R}$.

For $y_0 \in V$ and $f \in L^2(0,T;V^*)$, we consider the problem:
Find

$$y \in L^2(0,T;V) \cap C([0,T];H) \cap W^{1,2}([0,T];V^*)$$

such that

$$\begin{aligned}(y'(t) + Ay(t), y(t) - z) + \phi(y(t)) \quad &-\phi(z) \leq (f(t), y(t) - z)\\ &\text{a.e. } t \in (0,T) \text{ for all } z \in V\\ &y(0) = y_0.\end{aligned} \tag{22.7}$$

Here $y' = \frac{dy}{dt}$ is the strong derivative of $y : [0,T] \to V^*$.

In terms of the subgradient mapping $\partial\phi : V \to V^*$ problem (22.7) can be written as

$$\begin{aligned}y'(t) + Ay(t) + \partial\phi(y(t)) \ni f(t) \quad &\text{a.e. } t \in (0,T)\\ y(0) &= y_0.\end{aligned} \tag{22.8}$$

This is an abstract variational inequality of parabolic type. In applications to partial differential equations V is a Sobolev subspace of $H = L^2(\Omega)$, A is an elliptic differential operator on Ω and the unknown y is a function of two variables $(x, t) \in \Omega \times [0, T]$ which can be viewed as a function of t from $[0, T]$ to $L^2(\Omega)$. Then the derivative $y'(t)$ can be viewed as the partial derivative y_t of y.

In special case where $\phi = I_K$ is the indicator function of some closed convex subset K of V, i.e.

$$\phi(y) = 0 \text{ if } y \in K, \ \phi(y) = +\infty \text{ if } y \bar{\in} K, \tag{22.9}$$

the variational inequality (22.7) reduces to

$$\begin{aligned} y(t) \in K \text{ for all } t \in (0, T) \\ (y'(t) + Ay(t), y(t) - z) \leq (f(t), y(t) - z) \\ \text{a.e. } t \in (0, T) \text{ for all } z \in K \\ y(0) = y_0. \end{aligned} \tag{22.10}$$

Theorem 22.2 ([2, Theorem 4.1]) *Let $y_0 \in V$ and $f \in W^{1,2}([0, T]; V^*)$ be such that*

$$\{Ay_0 + \partial\phi(y_0) - f(0)\} \cap H \neq \emptyset. \tag{22.11}$$

Then the variational inequality (22.7) has a unique solution $y \in W^{1,2}([0, T]; V)$ $\cap W^{1,\infty}([0, T]; H)$. Moreover, the map $(y_0, f) \to y$ is Lipschitsian from $H \times L^2(0, T; V^*)$ to $C([0, T]; H) \cap L^2(0, T; V)$.

If $f \in L^2(0, T; H)$ and $\phi(y_0) < +\infty$ then (22.7) has a unique solution $y \in W^{1,2}([0, T]; H) \cap L^2(0, T; V)$ and the map $(y_0, f) \to y$ is Lipschitsian from $H \times L^2(0, T; H)$ to $C([0, T]; H) \cap L^2(0, T; V)$. Moreover, one has

$$y'(t) = (f(t) - Ay(t) - \partial\phi(y(t)))^0 \text{ a.e. } t \in (0, T), \tag{22.12}$$

where $(f(t) - Ay(t) - \partial\phi(y(t)))^0$ is the element of minimum norm in $f(t) - Ay(t) - \partial\phi(y(t))$.

Since the set of all $(y_0, f) \in V \times W^{1,2}([0, T]; V^*)$ satisfying condition (22.11) is a dense subset of $D(\phi) \times L^2(0, T; V^*)$ and the map $(y_0, f) \to y$ is Lipschitzian from $H \times L^2(0, T; V^*)$ to $C([0, T]; H) \cap L^2(0, T; V)$, we may extend it by continuity on all of $D(\phi) \times L^2(0, T; V^*)$.

Corollary 22.1 ([2, Corollary 4.1]) *For every $y_0 \in D(\phi)$ and $f \in L^2(0, T; V^*)$, (22.7) has a unique weak solution $y \in C([0, T]; H) \cap L^2(0, T; V)$.*

In particular for $\phi = I_K$, Theorem 22.2 gives

Theorem 22.3 ([2, Theorem 4.2]) *Let $y_0 \in K$ and $f \in W^{1,2}([0,T]; V^*)$ be given such that for some $\xi_0 \in H$*

$$(f(0) - Ay_0 - \xi_0, y_0 - v) \geq 0 \quad \forall v \in K. \tag{22.13}$$

Then problem (22.10) has a unique solution $y \in W^{1,2}([0,T]; V) \cap W^{1,\infty}([0,T]; H)$. If $f \in L^2(0,T; H)$ and $y_0 \in K$ then problem (22.10) has a unique solution $y \in W^{1,2}([0,T]; H) \cap L^2(0,T; V)$.

Assume in addition that for some $\omega > 0$

$$(Av, v) \geq \omega \|v\|^2 \quad \forall v \in V \tag{22.14}$$

and there exists $h \in H$ such that

$$(I + \varepsilon A_H)^{-1}(v + \varepsilon h) \in K \ for\ all\ \varepsilon > 0\ and\ all\ v \in K,\ A_H y = Ay \cap H. \tag{22.15}$$

Then $Ay \in L^2(0,T; H)$ and $y \in L^\infty(0,T; V)$.

22.2.3 Compensated Compactness Lemma in Variable Lebesgue and Sobolev Spaces

Let ρ be a weight function. We associate to ρ the space

$$X_\rho = \{\mathbf{f} \in L^2(0,T; L^2(\Omega, \rho dx)^N) | \, div(\rho \mathbf{f}) \in L^2(0,T; L^2(\Omega))\}, \tag{22.16}$$

and endow it with the norm

$$\|\mathbf{f}\|_{X_\rho} = \left(\|\mathbf{f}\|^2_{L^2(0,T;L^2(\Omega,\rho dx)^N)} + \|div(\rho \mathbf{f})\|^2_{L^2(0,T;L^2(\Omega))} \right)^{1/2}.$$

We call a sequence $\{\mathbf{f}_k \in X_\rho\}_{k\in N}$ bounded if

$$\sup_{k \in N} \|\mathbf{f}_k\|_{X_\rho} < +\infty.$$

Also let us consider the space $\mathscr{H} = \{y \in H(\Omega, \rho dx) | \, y' \in L^2(0,T; L^2(\Omega))\}$. Composing suggestions of [14, Lemma 4] and [1, Theorem 2] we obtain some refinement of the celebrated div-curl lemma of F. Murat and L.C. Tartar [19].

Lemma 22.2 *Let $\{\rho\}$ be a weight function. Let $\{\mathbf{f}_k \in L^2(0,T; L^2(\Omega, \rho dx)^N)\}_{k\in N}$ and $\{g_k \in \mathscr{H}\}_{k\in N}$ be such that $\{\mathbf{f}_k\}_{k\in \mathcal{N}}$ is bounded in the space X_ρ, $\mathbf{f}_k \rightharpoonup \mathbf{f}$ in $L^2(0,T; L^2(\Omega, \rho dx)^N)$ as $k \to \infty$, $\{g_k\}_{k\in N}$ is bounded in the space $\mathscr{H}$, $g_k \rightharpoonup g$ in $L^2(0,T; L^2(\Omega))$, and $\nabla g_k \rightharpoonup \nabla g$ in $L^2(0,T; L^2(\Omega, \rho dx)^N)$, $g_k' \rightharpoonup g'$ in $L^2(0,T; L^2(\Omega))$ as $k \to \infty$. Then*

$$\lim_{k\to\infty}\int_0^T\int_\Omega \varphi(\mathbf{f}_k,\nabla g_k)_{R^N}\rho\psi dxdt = \int_0^T\int_\Omega \varphi(\mathbf{f},\nabla g)_{R^N}\rho\psi dxdt, \tag{22.17}$$

$$\forall\varphi\in C_0^\infty(\Omega),\ \psi\in C_0^\infty(0,T).$$

22.3 Setting of the Optimal Control Problem (OCP)

Let ρ be given element of $L^1(\Omega)$ satisfying the conditions

$$0<\rho(x)\text{ a.e. in }\Omega,\ \rho^{-\nu}\in L^1(\Omega)\text{ for some }\nu\in(N/2,+\infty). \tag{22.18}$$

Then in view of the estimate

$$\int_\Omega \rho^{-1}dx \le \left(\int_\Omega \rho^{-\nu}dx\right)^{1/\nu}\left(\int_\Omega dx\right)^{1/\nu^*} = \|\rho^{-\nu}\|_{L^1(\Omega)}^{1/\nu}|\Omega|^{1/\nu^*},$$

where $\nu^*=\nu/(1-\nu)$ is the conjugate of ν, we have: $\rho^{-1}\in L^1(\Omega)$, i.e., ρ is a degenerate weight in the sense of Definition 22.1.

Let $\mathcal{K}$ be a non-empty convex closed subset of the space $L^2(0,T;H(\Omega,\rho dx)))$ such that $0\in\mathcal{K}$, y_{ad}, $f\in L^2(0,T;L^2(\Omega))$ be given elements. Consider the next OCP in coefficients for degenerate variation parabolic inequality:

$$I(U,y)=\int_0^T\int_\Omega |y-y_{ad}|^2dxdt\to\inf \tag{22.19}$$

$$\int_0^T\int_\Omega y'(v-y)dxdt +$$

$$\int_0^T\left(\sum_{i,j=1}^N\int_\Omega\left(a_{i,j}(x)\frac{\partial y}{\partial x_j}\right)\frac{\partial(v-y)}{\partial x_i}\rho dx+\int_\Omega y(v-y)dx\right)dt\ge$$

$$\ge\int_0^T\int_\Omega f(v-y)dxdt,\ \forall v\in\mathcal{K}, \tag{22.20}$$

$$U\in\mathcal{U}_{ad},\ y\in\mathcal{K},\ y'\in L^2(0,T;L^2(\Omega)), \tag{22.21}$$

$$y(0, x) = 0, \quad x \in \Omega. \tag{22.22}$$

Here

$$\mathscr{U}_{ad} = \{U = [\mathbf{a_1}, \dots, \mathbf{a_N}] \in M_2^{\alpha,\beta}(\Omega) | \, |div(\rho \mathbf{a_i})| \le \gamma_i, \text{ a.e. in } \Omega, \ \forall i = 1, \dots N\}, \tag{22.23}$$

where $\gamma = (\gamma_1, \dots, \gamma_N) \in R^N$ is a strictly positive vector, $M_2^{\alpha,\beta}(\Omega)$ $(0 < \alpha \le \beta < +\infty)$ is a set of all symmetric matrices $U(x) = \{a_{i,j}(x)\}_{1 \le i,j \le N}$ in $L^\infty(\Omega; R^N \times R^N)$ such that the following conditions are fulfilled:

$$|a_{i,j}(x)| \le \beta \text{ a.e. in } \Omega \ \forall i, j \in \{1, \dots N\}, \tag{22.24}$$

$$(U(x)(\xi - \eta), \xi - \eta)_{R^N} \ge 0 \text{ a.e. in } \Omega \ \forall \xi, \eta \in R^N, \tag{22.25}$$

$$(U(x)\xi, \xi)_{R^N} = \sum_{i,j=1}^{N} a_{i,j}(x)\xi_i\xi_j \ge \alpha|\xi|^2 \text{ a.e. in } \Omega. \tag{22.26}$$

Definition 22.2 We say that $y \in \mathscr{K}$, for which the inequality (22.20)–(22.22) takes place is called an H-solution.

For every fixed control $U \in M_2^{\alpha,\beta}(\Omega)$ let us consider the linear operator $A : H(\Omega, \rho dx) \to (H(\Omega, \rho dx))^*$ defined as

$$\langle A(y), v \rangle = \sum_{i,j=1}^{N} \int_\Omega \left(a_{i,j}(x) \frac{\partial y}{\partial x_j} \right) \frac{\partial v}{\partial x_i} \rho dx + \int_\Omega yv dx, \text{ for } v \in H(\Omega, \rho dx). \tag{22.27}$$

Hereinafter we shall suggest that the following assumption is fulfilled:

Hypothesis A. For $A : H(\Omega, \rho dx) \to (H(\Omega, \rho dx))^*$ we suggest that (22.15) is fulfilled.

Let us consider the regularity problem for OCP (22.19)–(22.22). Namely, we have such result.

Proposition 22.1 *For every control $U \in M_2^{\alpha,\beta}(\Omega)$ and every $f \in L^2(0, T; L^2(\Omega))$ there exists a unique H-solution to degenerate variation inequality (22.20)–(22.22). Moreover, $A(y) \in L^2(0, T; L^2(\Omega))$ and $y \in L^\infty(0, T; H(\Omega, \rho dx))$ having (22.14) and Hypothesis A.*

Proof Let us show that for every fixed $U \in M_2^{\alpha,\beta}(\Omega)$ the operator $A : H(\Omega, \rho dx) \to (H(\Omega, \rho dx))^*$ defined in (22.27) is such that for some $\omega > 0$ and real α_1 the coercivity condition

$$\langle A(y), y \rangle + \alpha_1 \|y\|^2_{L^2(\Omega)} = (A(y), y) + \alpha_1 \|y\|^2_{L^2(\Omega)} \ge \omega \|y\|^2_{H(\Omega,\rho dx)}, \quad \forall y \in H(\Omega, \rho dx). \tag{22.28}$$

Indeed, in view of (22.26) we have

$$\langle A(y), y\rangle = \sum_{i,j=1}^{N} \int_{\Omega} \left(a_{i,j}(x)\frac{\partial y}{\partial x_j}\right)\frac{\partial y}{\partial x_i}\rho dx + \int_{\Omega} |y|^2 dx \geq$$

$$\geq \alpha \int_{\Omega} |\nabla y|^2_{R^N} \rho dx + \int_{\Omega} |y|^2 dx \geq \min\{1, \alpha\}\|y\|^2_{H(\Omega,\rho dx)}. \tag{22.29}$$

Thus, having put $\alpha_1 := 0$, $\omega := \min\{1, \alpha\}$, we obtain (22.28).

Further, taking into account the condition (22.24) for every $v \in H(\Omega, \rho dx)$, we obtain

$$|\langle A(y), v\rangle| \leq \beta \left|\int_{\Omega} (\nabla y, \nabla v)_{R^N} \rho dx\right| + \left|\int_{\Omega} yv dx\right| \leq$$

$$\leq \beta \int_{\Omega} |(\nabla y, \nabla v)_{R^N}| \rho dx + \int_{\Omega} |yv| dx \leq$$

$$\leq \beta \|\nabla y\|^N_{L^2(\Omega,\rho dx)} \|\nabla v\|^N_{L^2(\Omega,\rho dx)} + \|y\|_{L^2(\Omega)}\|v\|_{L^2(\Omega)} \leq$$

$$\max\{\beta, 1\}\|y\|_{H(\Omega,\rho dx)}\|v\|_{H(\Omega,\rho dx)}. \tag{22.30}$$

Hence, we have that $A : H(\Omega, \rho dx) \to (H(\Omega, \rho dx))^*$ is linear continuous symmetric operator with coercivity condition (22.28). In view of Theorem 22.3 and definition and properties of Bochner integral (see [7–10]), we have that the inequality (22.20)–(22.22) admits a unique solution $y \in L^2(0, T; H(\Omega, \rho dx)) \cap L^2(0, T; L^2(\Omega))$ such that $y' \in L^2(0, T; L^2(\Omega))$. Moreover, it obviously follows from (22.29) that (22.14) takes place and taking into account *Hypothesis A*, we obtain that $A(y) \in L^2(0, T; L^2(\Omega))$ and $y \in L^\infty(0, T; H(\Omega, \rho dx))$. □

22.4 Existence of H-Optimal Solutions

Let us introduce the set of admissible pairs to the optimal control problem (22.19)–(22.22).

Definition 22.3 We say that the set Ξ_H defined as

$$\Xi_H = \{(U, y) \in U_{ad} \times L^2(0, T; H(\Omega, \rho dx))|\ y \in \mathscr{K},\ (U, y) \text{ are related by } (22.20){-}(20.22)\}$$

is the set of admissible pairs to the problem (22.19)–(22.22).

Remark 22.2 In view of Proposition 22.1 the set Ξ_H is always nonempty, and note that $\Xi_H \subset L^\infty(\Omega; R^{N\times N}) \times \mathscr{H} \subset L^\infty(\Omega; R^{N\times N}) \times L^2(0,T; H(\Omega, \rho dx))$, where $\mathscr{H} = \{v \in L^2(0,T; H(\Omega, \rho dx)) | v' \in L^2(0,T; L^2(\Omega))\}$.

Taking this observation into account, we adopt the following concept.

Definition 22.4 We say that a pair $(U^0, y^0) \in L^\infty(\Omega; R^{N\times N}) \times \mathscr{H} \subset L^\infty(\Omega; R^{N\times N}) \times L^2(0,T; H(\Omega, \rho dx))$ is an H-optimal solution to problem (22.19)–(22.22) if $(U^0, y^0) \in \Xi_H$ and $I(U^0, y^0) = \inf\limits_{(U,y)\in \Xi_H} I(U, y)$.

Our prime interest in this section deals with the solvability of optimal control problem (22.19)–(22.22) in the class of H-solutions. To begin with, we consider the topological properties of the set of admissible solutions Ξ_H to the problem (22.19)–(22.22). To do so, we introduce the following concepts:

Definition 22.5 We say that a sequence $\{(U_k, y_k) \in \Xi_H\}_{k\in N}$ is bounded if

$$\sup_{k\in N}[\|U_k\|_{L^\infty(\Omega; R^N\times R^N)} + \|y_k'\|_{L^2(0,T;L^2(\Omega))} + \|y_k\|_{L^2(0,T;L^2(\Omega))} + \|\nabla y_k\|_{L^2(0,T;L^2(\Omega;\rho dx)^N)}]$$

is finite.

Definition 22.6 We say that a bounded sequence of H-admissible solutions $\{(U_k, y_k) \in \Xi_H\}_{k\in N}$ τ-converges to a pair $(U, y) \in L^\infty(\Omega; R^{N\times N}) \times \mathscr{H}$ if

(a) $U_k \rightharpoonup U$ $*$-weakly in $L^\infty(\Omega; R^{N\times N})$;
(b) $y_k \rightharpoonup y$ in $L^2(0,T; L^2(\Omega))$;
(c) $\nabla y_k \rightharpoonup \nabla y$ in $L^2(0,T; (L^2(\Omega, \rho dx))^N)$;
(d) $y_k' \rightharpoonup y'$ in $L^2(0,T; L^2(\Omega))$.

Theorem 22.4 *Let $\rho(x) > 0$ be a degenerate weight function and let (22.14) and Hypothesis A hold true. Then for every $f \in L^2(0,T; L^2(\Omega))$ the set Ξ_H is sequentially τ-closed.*

Proof Let $\{(U_k, y_k)\}_{k\in N} \subset \Xi_H$ be a bounded τ-convergent sequence of H-admissible pairs to the problem (22.19)–(22.22) (in view of Proposition 22.1 such choice is always possible). Let (U^0, y^0) be its τ-limit. Our aim is to prove that $(U^0, y^0) \in \Xi_H$. In order to do this we have to show that $U_0 \in U_{ad}$, the validity of corresponding "limit" variational inequality and "limit" relation for initial data. Thus, we divide our proof into several steps.

Step 1. Let us show that $U^0 \in U_{ad}$. Since $\{U_k = [\mathbf{a}_{1k}, \dots, \mathbf{a}_{Nk}]\}_{k\in N} \subset U_{ad}$, it follows that $|div(\rho\mathbf{a}_{ik})| \le \gamma_i$ a.e. in Ω $\forall i = 1, \dots, N, \forall k \in N$. Passing to the limit as $k \to \infty$ in the relations

$$\int_\Omega (\mathbf{a}_{ik}, \nabla\varphi)_{R^N} \rho dx = -\int_\Omega \varphi div(\rho\mathbf{a}_{ik}) dx, \ \forall \varphi \in C_0^\infty(\Omega), \ \forall i = 1, \dots, N,$$

$$-\gamma_i \int_\Omega \varphi dx \le \int_\Omega \varphi div(\rho\mathbf{a}_{ik}) dx \le \gamma_i \int_\Omega \varphi dx, \ \forall i = 1, \dots, N, \ \forall \varphi \ge 0,$$

we may suppose that $|div(\rho \mathbf{a}_i^0)| \leq \gamma_i$ a.e. in Ω $\forall i \in \{1, \dots, N\}$ and

$$div(\rho \mathbf{a}_{ik}) \rightharpoonup div(\rho \mathbf{a}_i^0) \text{ in } L^2(\Omega) \text{ as } k \to \infty. \tag{22.31}$$

Thus, $U_k \rightharpoonup U^0 = [\mathbf{a}_1^0, \dots, \mathbf{a}_N^0]$ weakly-$*$ in $L^\infty(\Omega; R^{N\times N})$, and $U^0 \in U_{ad}$.

Step 2. Let us show that the pair (U^0, y^0) satisfies the variational inequality (22.20).

(a) Since each of the pairs (U_k, y_k) is admissible to the OCP (22.19)–(22.22), we have

$$\langle y_k', v - y_k\rangle_{L^2(0,T;H(\Omega,\rho dx))} \tag{22.32}$$
$$+\langle -div(U_k \rho(x)\nabla y_k) + y_k, v - y_k\rangle_{L^2(0,T;H(\Omega,\rho dx))} \leq$$
$$\leq \langle f, v - y_k\rangle_{L^2(0,T;H(\Omega,\rho dx))}, \ v \in \mathcal{K}. \tag{22.33}$$

Let us note that by the initial assumptions there exists a constant $C > 0$ such that

$$\|y_k\|_{L^2(0,T;L^2(\Omega))} \leq C, \ \ \|\nabla y_k\|_{L^2(0,T;(L^2(\Omega,\rho dx))^N)} \leq C,$$

$$\|y_k'\|_{L^2(0,T;L^2(\Omega))} \leq C \ \forall k \in N,$$

and taking into account that $U_k \rightharpoonup U^0$ weakly-$*$ in $L^\infty(\Omega; R^{N\times N})$, $k \to \infty$, one gets

$$div(\rho \mathbf{a}_{ik}) \rightharpoonup div(\rho \mathbf{a}_i^0) \ \text{ in } L^2(\Omega), \ \forall i = 1, \dots, N;$$

$$y_k \to y^0 \ \text{ strongly in } \ L^2(0, T; L^2(\Omega));$$

$$\nabla y_k \rightharpoonup \nabla y^0 \ \text{ in } \ (L^2(0, T; L^2(\Omega, \rho dx)))^N);$$

$$y_k' \rightharpoonup (y^0)' \ \text{ in } \ L^2(0, T; L^2(\Omega));$$

$$\{U_k \nabla y_k\}_{k\in N} \ \text{ is bounded in } \ L^2(0, T; (L^2(\Omega, \rho dx))^N).$$

Then $U_k\nabla y_k =: \boldsymbol{\xi}_k \rightharpoonup \boldsymbol{\xi}$ in $L^2(0, T; (L^2(\Omega, \rho dx))^N)$ within a subsequence.

(b) Let us now show that the sequence $\{\boldsymbol{\xi}_k\}_{k\in N}$ is bounded in X_ρ. Taking into account the property (22.14) for the operator A (see for details the proof of Proposition 22.1) and Hypothesis A, having $f \in L^2(0, T; L^2(\Omega))$, we obtain that $-div(\rho\boldsymbol{\xi}_k) + y_k \in L^2(0, T; L^2(\Omega))$ and, obviously, $div(\rho\boldsymbol{\xi}_k) \in L^2(0, T; L^2(\Omega))$ $\forall k \in N$. Further, the relation

$$\int_0^T\int_\Omega div(\rho\boldsymbol{\xi}_k)\varphi\psi dxdt = -\int_0^T\int_\Omega (\boldsymbol{\xi}_k, \nabla\varphi)_{R^N}\rho\psi dxdt \to$$

$$\to -\int_0^T\int_\Omega (\boldsymbol{\xi}, \nabla\varphi)_{R^N}\rho\psi dxdt = \int_0^T\int_\Omega div(\rho\boldsymbol{\xi})\varphi\psi dxdt,$$

$$\forall\varphi \in C_0^\infty(\Omega),\ \psi \in C_0^\infty(0,T) \text{ as } k \to \infty,$$

means that $div(\rho\boldsymbol{\xi}_k) \rightharpoonup div(\rho\boldsymbol{\xi})$ in $L^2(0,T;L^2(\Omega))$ implying $\{\boldsymbol{\xi}_k\}_{k\in N}$ is bounded in X_ρ.

(c) Having put $\mathbf{f}_k = \boldsymbol{\xi}_k$, $g_k = y_k$ for all $k \in N$, taking into account Lemma 22.2, we obtain:

$$\lim_{k\to\infty}\int_0^T\int_\Omega \varphi(\boldsymbol{\xi}_k, \nabla y_k)_{R^N}\rho\psi dxdt = \int_0^T\int_\Omega \varphi(\boldsymbol{\xi}, \nabla y)_{R^N}\rho\psi dxdt,$$

$$\forall\varphi \in C_0^\infty(\Omega),\ \psi \in C_0^\infty(0,T). \quad (22.34)$$

Further, taking into account that $div(\rho\boldsymbol{\xi}_k) \rightharpoonup div(\rho\boldsymbol{\xi})$ in $L^2(0,T;L^2(\Omega))$, and $y_k \to y^0$ in $L^2(0,T;L^2(\Omega))$, $y_k' \rightharpoonup (y^0)'$ in $L^2(0,T;L^2(\Omega))$, we pass to the limit in (22.33) as $k \to \infty$. As a result we obtain

$$\langle (y^0)', v - y^0\rangle_{L^2(0,T;H(\Omega,\rho dx))} + \langle -div(\rho\boldsymbol{\xi}) + y^0, v - y^0\rangle_{L^2(0,T;H(\Omega,\rho dx))} \geq$$
$$\geq \langle f, v - y^0\rangle_{L^2(0,T;H(\Omega,\rho dx))},\ \forall v \in \mathscr{K}. \quad (22.35)$$

d) It remains to prove that $\boldsymbol{\xi} = U^0\nabla y^0$. To do this we introduce the following scalar function

$$v(x) = (\mathbf{z}, x)_{R^N}, \quad (22.36)$$

where $\mathbf{z}$ is a fixed element of R^N. By the initial assumptions, we have

$$\int_0^T\int_\Omega \varphi(U_k(\nabla y_k - \nabla v), \nabla y_k - \nabla v)_{R^N}\rho\psi dxdt \geq 0,\ \forall\varphi \geq 0,\ \forall\psi \geq 0,$$

or, in view of (22.36), this inequality can be rewritten as

$$\int_0^T\int_\Omega \varphi(U_k(\nabla y_k - \mathbf{z}), \nabla y_k - \mathbf{z})_{R^N}\rho\psi dxdt \geq 0. \quad (22.37)$$

Now we can pass to the limit in (22.37) as $k \to \infty$ using Lemma 22.2 and having put in the statement of this Lemma $\mathbf{f}_k = U_k\nabla(y_k - v)$, $g_k = y_k - v$ for all $k \in N$. Due to the initial assumptions, we see that the preconditions of Lemma 22.2 for the sequence $\{g_k\}_{k\in N}$ obviously hold. Moreover we showed that the sequence $\{div(\rho U_k \nabla y_k)\}_{k\in N}$ is bounded in $L^2(0, T; L^2(\Omega))$. It remains only to show that the sequence $\{div(\rho U_k \mathbf{z})\}_{k\in N}$ is weakly convergent in $L^2(0, T; L^2(\Omega))$. To this end, we note that the elements $div(\rho U_k \mathbf{z})$ for all $k \in N$ are defined as follows

$$\int_\Omega (U_k\mathbf{z}, \nabla\varphi)_{R^N}\rho dx = -\int_\Omega \varphi div(\rho U_k\mathbf{z})dx, \ \forall\varphi \in C_0^\infty(\Omega), \ \forall k \in N.$$

Then, for every function $\varphi \in C_0^\infty(\Omega)$ we have

$$\int_\Omega (U_k\mathbf{z}, \nabla\varphi)_{R^N}\rho dx = \int_\Omega \left(\begin{bmatrix} (\mathbf{a}_{1k}, \mathbf{z})_{R^N} \\ \cdots \\ (\mathbf{a}_{Nk}, \mathbf{z})_{R^N} \end{bmatrix}, \nabla\varphi \right)_{R^N} \rho dx =$$

$$= \int_\Omega \sum_{i=1}^N (\mathbf{a}_{ik}(x), \mathbf{z})_{R^N} \frac{\partial\varphi}{\partial x_i}\rho dx = \int_\Omega \sum_{i=1}^N\sum_{j=1}^N a_{ij}^k(x)\frac{\partial\varphi}{\partial x_i} z_j \rho dx =$$

$$= \sum_{j=1}^N z_j \int_\Omega (\mathbf{a}_{jk}(x), \nabla\varphi)_{R^N}\rho dx = -\sum_{j=1}^N z_j \int_\Omega \varphi div(\rho\mathbf{a}_{jk})dx = J_k. \tag{22.38}$$

Then, using (22.31), we get

$$\lim_{k\to\infty} J_k = -\sum_{j=1}^N z_j \lim_{k\to\infty}\int_\Omega \varphi div(\rho\mathbf{a}_{ik})dx = -\sum_{j=1}^N z_j \int_\Omega \varphi div(\rho\mathbf{a}_j^0)dx. \tag{22.39}$$

Applying the converse transformation with (22.39) as we did it in (22.38), we come to the relation:

$$\lim_{k\to\infty}\int_\Omega \varphi div(\rho U_k\mathbf{z})dx = -\lim_{k\to\infty}\int_\Omega (U_k\mathbf{z}, \nabla\varphi)_{R^N}\rho dx =$$

$$-\int_\Omega (U^0\mathbf{z}, \nabla\varphi)_{R^N}\rho dx = \int_\Omega \varphi div(\rho U^0\mathbf{z})dx, \ \forall\varphi \in C_0^\infty(\Omega). \tag{22.40}$$

Hence, from (22.40) and boundedness of the sequence $\{div(\rho U_k\mathbf{z})\}_{k\in N}$ in $L^2(0, T; L^2(\Omega))$, we get

$$div(\rho U_k(\nabla y_k - \nabla v)) \rightharpoonup div(\rho\boldsymbol{\xi}) - div(\rho U^0\mathbf{z}). \tag{22.41}$$

Combining the property (22.41), and the fact that $U_k \nabla v = U_k \mathbf{z} \rightharpoonup U^0 \mathbf{z}$ in $L^2(0, T; (L^2(\Omega, \rho dx))^N)$ as $k \to \infty$, it is easy to see that all suppositions of Lemma 22.2 are fulfilled. Hence, passing to the limit in the inequality (22.37) as $k \to \infty$, we get

$$\int_0^T \int_\Omega \varphi(x)(\boldsymbol{\xi} - U^0\mathbf{z}, \nabla y^0 - \mathbf{z})_{R^N} \rho\psi dx dt \geq 0, \ \forall \mathbf{z} \in R^N$$

for all possible $\varphi \in C_0^\infty(\Omega)$ and $\psi \in C_0^\infty(0, T)$. After localization, we have

$$\rho(\boldsymbol{\xi} - U^0\mathbf{z}, \nabla y^0 - \mathbf{z})_{R^N} \geq 0. \tag{22.42}$$

Hence, $\boldsymbol{\xi} = U^0 \nabla y^0$ a.e. in $(0, T) \times \Omega$.

Thus, the inequality (22.35) takes the form

$$\langle (y^0)', v - y^0 \rangle_{L^2(0,T;H(\Omega,\rho dx))} + \langle -div(\rho U^0 \nabla y^0) + y^0, v - y^0 \rangle_{L^2(0,T;H(\Omega,\rho dx))} \geq$$
$$\geq \langle f, v - y^0 \rangle_{L^2(0,T;H(\Omega,\rho dx))} \ \forall v \in \mathcal{K}.$$

Step 3. In order to conclude the proof it remains to pass to the limit in the equality

$$\int_\Omega 0 \cdot \varphi dx = \lim_{t \to +0} \int_\Omega y_k \varphi dx,$$

which holds true for all $k \in N$. As a result, using that fact that $y_k \to y^0$ in $L^2(0, T; H(\Omega, \rho dx))$ as $k \to \infty$, we obtain

$$\lim_{t \to +0} \int_\Omega y^0 \varphi dx = \int_\Omega 0 \cdot \varphi dx.$$

Thus, τ-limit pair (U^0, y^0) belongs to the set Ξ_H, and this concludes the proof. □

Now we are in a position to state the existence of H-optimal pairs to the problem (22.19)–(22.22).

Theorem 22.5 *Let ρ be a degenerate weight function that satisfies (22.18) and let (22.14) and Hypothesis A hold true. Let also f, $y_{ad} \in L^2(0, T; L^2(\Omega))$ be given functions. Then the optimal control problem (22.19)–(22.22) admits at least one H-solution.*

Proof First of all we note that in virtue of Proposition 22.1 for the given function $f \in L^2(0, T; L^2(\Omega))$ and every admissible control $U \in U_{ad}$ there exists an H-solution $y = y(U, f) \in L^2(0, T; H(\Omega, \rho dx))$ such that $y' \in L^2(0, T; L^2(\Omega))$ to the problem (22.20)–(22.22).

Let $\{(U_k, y_k) \in \Xi_H\}_{k \in N}$ be an H-minimizing sequence to the problem (22.19)–(22.22): that is $\lim_{k\to\infty} I(U_k, y_k) = \inf_{(U,y)\in\Xi_H} I(U, y) < +\infty$. Hence, taking into account the definition of U_{ad} and definition of the boundedness of $\{(U_k, y_k) \in \Xi_H\}_{k\in N}$ we may assume that there exists a matrix $U^* \in L^\infty(\Omega; R^{N\times N})$ and $y^* \in L^2(0, T; H(\Omega, \rho dx))$ such that $(y^*)' \in L^2(0, T; L^2(\Omega))$ and $U_k \to U^*$ weakly-$*$ in $L^\infty(\Omega; R^{N\times N})$, $y_k \rightharpoonup y^*$ in $L^2(0, T; H(\Omega, \rho dx))$, $y_k' \rightharpoonup (y^*)'$ in $L^2(0, T; L^2(\Omega))$ as $k \to \infty$. Since Ξ_H is sequentially τ-closed, the pair (U^*, y^*) is H-admissible to the problem (22.19)–(22.22). In view of τ-semicontinuity of the cost functional we get

$$I(U^*, y^*) \le \liminf_{k\to\infty} I(U_k, y_k) = \inf_{(U,y)\in\Xi_H} I(U, y).$$

Hence, (U^*, y^*) is an H-optimal pair.

The proof is complete. □

Remark 22.3 In application a degenerate weight ρ occurs as the limit of a sequence of non-degenerate weights ρ_ε for which the corresponding "approximate" optimal control problem is solvable. Thus, naturally, it arises the question: if limit points of the family of admissible solutions to the perturbed problems appear to be H-admissible solutions to the original problem, whether all H-optimal solutions are attainable in this sense? These issues will be considered in near future articles.

References

1. Balanenko, I.G., Kogut, P.I.: H-optimal control in coefficients for Dirichlet parabolic problems. Bulletin of Dnipropetrovsk University: Series: Communications in Mathematical Modelling and Differential Equations Theory, vol. 18, pp. 45–63 (2010)
2. Barbu, V.: Optimal Control of Variational Inequalities. Pitman Advanced Publishing Program, London (1984)
3. Butazzo, G., Dal Maso, G., Garroni, A., Malusa, A.: On the relaxed formulation of some shape optimization problems. Adv. Math. Sci. Appl. **1**, 1–24 (1997)
4. Butazzo, G., Kogut, P.I.: Weak optimal controls in coefficients for linear elliptic problems. Revista Matematica Complutense **24**, 83–94 (2011)
5. Chiadó Piat, V., Serra Cassano, F.: Some remarks about the density of smooth functions in weighted Sobolev spaces. J. Convex Analysis **1**, 135–142 (1994)
6. Drabek, P., Kufner, A., Nicolosi N.: Nonlinear elliptic equations, singular and degenerate cases. University of West Bohemia, Pilsen (1996)
7. Dunford, N., Schwartz, J.T.: Linear Operators. Wiley, New York (1957)
8. Edwards, E.: Functional Analysis. Theory and Applicationas. Mir, Moscow (1969)
9. Gaevskii, Ch., Greger, K., Zacharias, K.: Nonlinear Operator Equations and Operator Differential Equations. Mir, Moscow (1978)
10. Hille, E., Phillips, R.S.: Functional Analysis and Semi-groups. FL, Moscow (1962)
11. Kapustjan, V.Ye., Kogut, O.P: Solenoidal controls in coefficients of nonlinear elliptic boundary value problems. Comput. Math. **12**, 138–143 (2010) (in Russian)
12. Kogut, P., Manzo, R., Putchenko, A.: On approximate solutions to the Neumann elliptic boundary value problem with non-linearity of exponential type. Bound. Value Probl. (2016). https://doi.org/10.1186/s13661-016-0717-1

13. Kovalevsky, A.A., Gorban, Yu. S.: Degenerate anisotropic variational inequalities with L^1-data. Preprint of IAMM of NAS (2007)
14. Kupenko, O.P.: Optimal Control Problems in Coefficients for Degenerate Variational inequalities of Monotone Type. I. Existence of optimal solutions. J. Comput. Appl. Math. **106**, 88–104 (2011)
15. Kupenko, O.P., Manzo, R.: Approximation of an optimal control problem in coefficient for variational inequality with anisotropic p-Laplacian. Nonlinear Diff. Eq. Appl. (2016). https://doi.org/10.1007/s00030-016-0387-9
16. Kupenko, O.P., Manzo, R.: On Optimal Controls in Coefficients for Ill-Posed Non-Linear Elliptic Dirichlet Boundary Value Problems. DCDS, Series B (2018) https://doi.org/10.3934/dcdsb.2018155
17. Lions, J.-L.: Optimal Control of Systems Governed by Partial Differential Equations. Springer, New York (1971)
18. Murat, F.: Un contre-exemple pour le problème de de contrôle dans les coefficients. C.R.A.S. Paris,Sér. A **273**, 708–711 (1971)
19. Murat, F.: Compacité par compensation. Ann. Sci. Norm. Sup. Pisa. **5**, 489–507 (1978)
20. Pastuhova, S.E.: Degenerate equations of monotone type: Lavrent'ev phenomenon and attainability problems. Sbornik: Mathematics **198**, 1465–1494 (2007)
21. Zeider, E.: Nonlinear Analysis and Its Applications II A and II B. Springer, New York (1990)
22. Zhikov, V.V.: On Lavrentiev phenomenon. Rus. J. Math. Phys. **3**, 249–269 (1994)
23. Zhikov, V.V.: Weighted Sobolev spaces. Sbornik: Math. **189**, 27–58 (1998)
24. Zhikov, V.V., Pastukhova, S.E.: Homogeniztion of degenerate elliptic equations. Siberian Math. J. **25**, 80–101 (2006)

Chapter 23
Group Pursuit Differential Games with Pure Time-Lag

Lesia V. Baranovska

Abstract The paper is devoted to the group pursuit differential-difference game with pure time-lag. An approach to the solution of this problem based on the method of resolving functions is proposed. For the group problem, the integral presentation of game solution based on the time-delay exponential is proposed at the first time. The guaranteed time of the game termination is found, and corresponding control law is constructed. The results are illustrated by a model example. In such game of two persons, it is possible to avoid meeting with the terminal set with any control of the pursuer. It is shown that if the pursuers are several then the pursuit game can be completed.

23.1 Introduction

In the theory of conflict-controlled processes, i.e. differential games and dynamic games, along with Isaacs ideology (see [1]), the Pontryagin's backward procedures (see [2]), and Krasovskii's extremal aiming principle (see [3]) concerning the basic equation in the theory of differential games, there exist efficient methods that can apparently be classified into a separate group. There are First Direct Method of Pontryagin and the Method of Resolving Functions of Chikrii (see [4]). They share a common principle of constructing the control of the pursuer on the basis of the Filippov–Castaing measurable choice theorem (see [5]).

In this paper, the Method of Resolving Functions is chosen as the main tool for research. An attractive side of this method is the fact that it allows us to effectively use modern technology of set-valued mappings and their selections in the substantiation of game constructions and to obtain meaningful results on their basis (see [4, 6–9]). This method was developed for the pursuit game, whose dynamics was described by

L. V. Baranovska (✉)
Institute for Applied System Analysis, National Technical University of Ukraine "Kyiv Polytechnic Institute", 37 Peremogy Ave., building 35, Kyiv 03056, Ukraine
e-mail: lesia@baranovsky.org

V. A. Sadovnichiy and M. Z. Zgurovsky (eds.), *Contemporary Approaches and Methods in Fundamental Mathematics and Mechanics*, Understanding Complex Systems,
https://doi.org/10.1007/978-3-030-50302-4_23

the system of a differential equation. The processes with fractional derivatives are studied in (see [10–16]), the nonstationary problems are considered in (see [17–21]).

It can be claimed that any real-life system has some time delay associated with it. The delay in a system may be due to one or more of he following causes: measurement of system variables, physical properties of the equipment used in the system, signal transmission. We are concerned with systems in which the time delay plays an important role. In (see [22–24] the modification of the Method of Resolving Functions for the differential-difference pursuit games is described. The scheme is based on a well-known Cauchy formula (see [25] for the differential-difference system of equations. Some modifications of the Method of Resolving Functions based on the time-delay exponential (see [26] are developed for one pursuer and one evader in (see [27–29]. The integral presentation of the game solution based on the time-delay exponential for a group is proposed for the first time. Such representation of the solution leads to a faster finding of the fundamental matrix of the system, which was difficult to do in previous cases. In contrast to the case with one pursuer, this paper considers an example where the capture of the evader is possible if there are several pursuers.

23.2 Statement of the Problem

Consider a conflict-controlled process whose evolution is described by the system of differential equations with pure time delay

$$\dot{z}_i(t) = B_i z_i(t - \tau_i) + \phi_i(u_i, v), \quad u_i \in U_i, \quad v \in V, \quad t \geq 0, \quad i = 1, \ldots, l. \tag{23.1}$$

Here $z_i(t) \in \mathbb{R}^{n_i}$, $\mathbb{R}^{n_i}$ be an Euclidean space of points $z = \left(z_1, \ldots, z_{n_i}\right)$. B_i are square constant matrices of order n_i, $\tau_i = \text{const} > 0$. The control unit is defined by the functions $\phi_i(u_i, v)$, $\phi_i : U_i \times V \to \mathbb{R}^{n_i}$, which are assumed to be jointly continuous in its variables on the direct product of nonempty compact sets U_i and V, $i = 1, \ldots, l$. The controls of the players, $u_i : R_+ \to U_i$ and $v : R_+ \to V$, $R_+ = \{t : t \geq 0\}$, are measurable functions of time.

In space $\mathbb{R}^n = \mathbb{R}^{n_1} \times \cdots \times \mathbb{R}^{n_l}$ a terminal set M^* is selected, consisting of sets M_i^*, $i = 1, \ldots, l$, each of which is cylindrical and has the form

$$M_i^* = M_i^0 + M_i, \tag{23.2}$$

where M_i^0 are the linear subspaces in $\mathbb{R}^{n_i}$ and M_i are nonempty compact sets from the orthogonal complements (L_i) of M_i^0 in $\mathbb{R}^{n_i}$.

A game to be completed if at least for one i, $i = 1, \ldots, l$ $z_i(t) \in M_i^*$ at some finite point in time. The initial conditions for system (23.1) are absolutely continuous on $[-\tau_i; 0]$ functions

$$z_i^0(t), \ i = 1, \ldots, l. \tag{23.3}$$

System state (23.1) at time t is a vector $z^t(\cdot) = \left(z_1^t(\cdot), \ldots, z_l^t(\cdot)\right)$, where $z_i^t(\cdot) = \{z(t+s),\ -\tau_i \le s \le 0\}$.

The goals of the first (pursuers, u_i) and second (evader, v) players are the opposite. The first players (l pursuers) tries to bring the trajectory of the process (23.1) to the terminal set in the shortest time, whereas the second player (one evader) tries to maximally put off the instant when the trajectory reaches the terminal set, or even avoid this meeting at all.

Let us take the side of the pursuers and assume that the evader chooses an arbitrary measurable function as a control. If the game (23.1), (23.3) occurs on an interval $[0, T]$, then we assume that the pursuers decide on its control at time t depending on the information about $z_i^0(\cdot)$ and $v^t(\cdot)$, i.e. the controls of the pursuers are either measurable functions

$$u_i(t) = u_i\left(z_i^0(\cdot),\ t,\ v^t(\cdot)\right) \in U_i, \quad t \in [0, T], \quad i = 1, \ldots, l, \tag{23.4}$$

where $v^t(\cdot) = \{v(s) : 0 \le s \le t,\ v(\cdot) \in V\}$ is the prehistory of the control of the evader up to instant t.

We say that the pursuit game can be completed from the initial state $z^0(\cdot)$ in time $T = T\left(z^0(\cdot)\right)$, if there are measurable functions such that the solution of the system (23.1) for any measurable functions $v(t),\ \ v(t) \in V,\ \ t \in [0, T]$, belongs to the corresponding set M_i^* at time $t = T$ for at least one $i,\ \ i = 1, \ldots, l$.

Denote by π_i the orthogonal projectors from $\mathbb{R}^{n_i}$ onto the subspaces $L_i,\ \ i = 1, \ldots, l$. Consider the set-valued mappings

$$W_i(t,\ v) = \pi_i X_i(B_i,\ t)\,\phi_i(U_i,\ v), \quad W_i(t) = \bigcap_{v \in V} W_i(t,\ v,), \quad i = 1, \ldots, l.$$

Here functions $X_i(B_i,\ t)$ are denoted for each $k = 1, 2, \ldots$ as a time-delay exponentials (see [30]):

$$X_i(B_i, t) = \begin{cases} \Theta, \ -\infty < t < -\tau_i;\ I,\ -\tau_i \le t < 0; \\ I + B_i \frac{t}{1!} + B_i^2 \frac{(t-\tau_i)^2}{2!} + \cdots + B_i^k \frac{(t-(k-1)\tau_i)^k}{k!}, \ (k-1)\,\tau_i \le t \le k\tau_i, \end{cases}$$

where Θ is a zero matrix, I is a unit matrix.

Condition 23.1 (*Pontryagin's condition*) The set-valued mappings $W_i(t)$ take nonempty values for all $t \ge 0,\quad i = 1, \ldots, l$.

The Pontryagin's condition and the measurable selection theorem (see [31]) imply that for any $t \ge 0$ there exists at least one selection $g_i(t)$ measurable in t such that $g_i(t) \in W_i(t),\ \ i = 1, \ldots, l$. Denote the set of such selections by G_i and introduce a function

$$\begin{aligned} &\xi_i\left(t,\ z_i^0(\cdot),\ g_i(\cdot)\right) = \\ &= \pi_i X_l(B_l, t)\, z_i^0(-\tau_i) + \\ &\int_{-\tau_i}^{0} \pi_i X_i(B_i, t - \tau_i - s)\, \dot{z}_i^0(s) ds + \int_0^t g_i(s) ds. \end{aligned}$$

Let Y be a nonempty compact set from $\mathbb{R}^n$, $0 \in Y$. Consider the Minkowski function (see [4])

$$\mu_Y(p) = \inf\{\mu \geq 0 : \quad p \in \mu Y\}$$

and the function inverse to it

$$\alpha_Y(p) = \sup\{\alpha \geq 0 : \quad \alpha p \in Y\}, \quad p \in \mathbb{R}^n.$$

Denote

$$\alpha\left(t, s, z_i^0(\,\cdot\,), m_i, v, g_i(\,\cdot\,)\right) = \alpha_{W_i(t-\tau_i-s,v)-g_i(t-\tau_i-s)}\left(m_i - \xi_i\left(t\,,\; z_i^0\,(\,\cdot\,)\,,\; g_i\,(\,\cdot\,)\right)\right)$$

for all $t \geq s \geq 0$, $v \in V$, $g_i(\,\cdot\,) \in G_i$, $m_i \in M_i$, $z_i \in \mathbb{R}^{n_i}$, $i = 1, \ldots, l$. Finally, denote

$$\alpha_i\left(t, s, z_i^0(\,\cdot\,), v, g_i(\,\cdot\,)\right) = \max_{m_i \in M_i} \alpha_i\left(t, s, z_i^0(\,\cdot\,), m_i, v, g_i(\,\cdot\,)\right), \quad i = 1, \ldots, l, \tag{23.5}$$

and then we obtain so-called resolving functions

$$\begin{gathered}\alpha_i\left(t, s, z_i^0(\cdot), v, g_i(\cdot)\right) = \sup\left\{\alpha \geq 0 :\right.\\ [W_i(t-\tau_i-s, v) - g_i(t-\tau_i-s)] \cap \\ \left.\alpha\left[M_i - \xi_i\left(t, z_i^0(\cdot), g_i(\cdot)\right)\right] \neq \emptyset\right\}.\end{gathered} \tag{23.6}$$

Consider the set

$$\begin{gathered}T = T\left(z^0\,(\,\cdot\,)\,,\; g\,(\,\cdot\,)\right) = \inf\left\{t \geq 0 :\right.\\ \left.\inf_{v \in V}\max_{i=1,\ldots,l} \int_0^t \alpha_i\left(t\,,\; s\,,\; z_i^0\,(\,\cdot\,)\,,\; v\,,\; g_i\,(\,\cdot\,)\right) ds \geq 1\right\},\end{gathered} \tag{23.7}$$

$$g\,(\,\cdot\,) = column\,(g_1\,(\,\cdot\,), \ldots, g_l\,(\,\cdot\,)), \quad g_i\,(\,\cdot\,) \in G_i.$$

Note that if $\xi_i\left(t,\; z_i^0\,(\,\cdot\,),\; g_i\,(\,\cdot\,)\right) \in M_i$ for all $i = 1, \ldots, l$, $t \geq 0$, then $\alpha_i\left(t,\; s,\; z_i^0\,(\,\cdot\,),\; v,\; g_i\,(\,\cdot\,)\right) = +\infty$ for all $s \in [0,\; t]$ and $v \in V$. In this case it is nature to set the value of the integral in relation (23.7) equals to $+\infty$, and the related inequality holds automatically. If the inequality in braces in (23.7) fails for all $t \geq 0$, we set $T\left(z^0\,(\,\cdot\,)\right) = +\infty$. Taking into account the properties of the parameters of the process (23.1), (23.3) and applying the inverse image theorems, we can show that the resolving function (23.6) is Borel measurable function in the variables s, v, $s \in [0,\; t]$, $v \in V$. The function (23.5) is jointly Borel measurable function in s, v by the theorem on the support function (see [31]) for $\xi_i\left(t,\; z_i^0\,(\,\cdot\,),\; g_i\,(\,\cdot\,)\right) \notin M_i$ for all $i = 1, \ldots, l$.

23.3 Main Results

Theorem 23.1 *Suppose that the process (23.1) satisfies the Pontryagin's condition, the sets M_i are convex for all $i = 1, \ldots, l$, and $T = T\left(z^0(\,\cdot\,)\right) < +\infty$ for the given initial state $z^0(\,\cdot\,)$ and some selection $g_i^0(\,\cdot\,) \in G_i$. Then a trajectory of the process (23.1), (23.3) can be brought to the terminal set (23.2) at the moment T by a control of the form (23.4).*

Proof Let $v(s)$, $v : [0, T] \to V$, be an arbitrary measurable function. Suppose $\xi_i\left(T, z_i^0(\,\cdot\,), g_i^0(\,\cdot\,)\right) \notin M_i$ for all $i = 1, \ldots, l$. Introduce a test function

$$h(t) = 1 - \max_{i=1,\ldots,l} \int_0^t \alpha_i\left(T, s, z_i^0(\,\cdot\,), v(s), g_i^0(\,\cdot\,)\right) ds, \quad t \geq 0.$$

It is continuous, it does not increase, and $h(\tau_i) = 1$. Since, $h(T) \leq 0$, there exists a time instant $t_* = t_*(v(\,\cdot\,))$, $0 < t_* \leq T$, such that $h(t_*) = 0$. We will call the time intervals $[0, t_*)$ and $[t_*, T]$ active and passive intervals, respectively.

Let us describe the method of controls of the pursuers on each of these intervals. Consider the set valued mapping

$$\begin{aligned} U_{1i}(s, v) = \{u_i \in U_i \,:\, \pi_i X_i(B_i, T - \tau_i - s)\phi_i(u_i, v) - g_i^0(T - \tau_i - s)\} \\ \in \alpha_i\left(T, s, z_i^0(\cdot), v(s), g_i^0(\cdot)\right)\left[M_i - \xi_i\left(T, z_i^0(\cdot), v, g_i^0(\cdot)\right)\right], \end{aligned} \tag{23.8}$$

which is measurable by the inverse image theorem. According to the measurable selection theorem (see [31]), it has at least one measurable selection $u_{1i}(s, v)$ that is compositionally measurable, particularly, $u_{1i}(s, v) = \operatorname{lex\,min} U_{1i}(s, v)$. We choose the control of the pursuers on the active time $[0, t_*)$ to be

$$u_i(s) = u_{1i}(s, v(s)), \quad i = 1, \ldots, l.$$

Consider the passive time interval $[t_*, T]$. For $s \in [t_*, T]$ and $v \in V$, we set the resolving functions $\alpha_i\left(T, s, z_i^0(\,\cdot\,), v(s), g_i^0(\,\cdot\,)\right) \equiv 0$ for all $i = 1, \ldots, l$ in expression (23.8). This yields a set-valued mapping

$$\begin{aligned} U_{2i}(s, v) = \left\{u_i \in U_i : \pi_i X_i(B_i, T - \tau_i - s)\phi_i(u_i, v) - g_i^0(T - \tau_i - s) = 0\right\}, \\ s \in [t_*, T], \; v \in V, \; i = 1, \ldots, l. \end{aligned} \tag{23.9}$$

It follows from the measurable selection theorem that the measurable mapping U_{2i} has a measurable selection: $u_{2i}(s, v) = \operatorname{lex\,min} U_{2i}(s, v)$, $i = 1, \ldots, l$.

Denote $\alpha_{i*}\left(T, s, z_{i*}^0(\,\cdot\,), v(s), g_{i*}^0(\,\cdot\,)\right) \equiv 0$ for such number i_*, as

$$\int_0^{t_*} \alpha_{i*}\left(T, s, z_{i*}^0(\,\cdot\,), v(s), g_{i*}^0(\,\cdot\,)\right) ds = 1.$$

We choose the control of i_*th pursuer on the interval $[t_*\ ,\ T]$ to be

$$u_{i*}(s) = u_{2i*}(s,\ v(s))\,,$$

and we choose the control of all others pursuers on the interval $[t_*\ ,\ T]$ to be arbitrary. When $\xi_i\left(T,\ z_i^0(\,\cdot\,),\ g_i^0(\,\cdot\,)\right) \in M_i$ for some $i = 1, \ldots, l$, we choose the control of the ith pursuer on the interval $[0, T]$ to be

$$u_i(s) = u_{2i}(s,\ v(s))\,,$$

and we choose the control of all others pursuers on this interval to be arbitrary.

Let us show that when the controls of the pursuers on the active and passive intervals are chosen in accordance with the above rules, the trajectory of system (23.1) is brought to the terminal set at time instant T at least for one $i = 1, \ldots l$ for any admissible controls of the evader.

The Cauchy formula (see [30]) for the system (23.1) implies the representation

$$\begin{aligned}\pi_i z_i(T) = \pi_i X_i(B_i, T)\, z_i^0(-\tau_i) + \int_{-\tau_i}^{0} \pi_i X_i(B_i, T - \tau_i - s)\, \dot{z}_i^0(s) ds \\ + \int_0^T \pi_i X_i(B_i, T - \tau_i - s)\, \phi_i(u_i(s), v(s)) ds.\end{aligned} \tag{23.10}$$

Consider first the case of $\xi_i\left(T,\ z_i^0(\,\cdot\,),\ g_i^0(\,\cdot\,)\right) \notin M_i$ for all $i = 1, \ldots, l$. Using relations (23.8) and (23.9), we obtain the following inclusion from (23.10):

$$\begin{aligned}\pi_i z_i(T) \in \xi_i\left(T,\ z_i^0(\,\cdot\,),\ g_i^0(\,\cdot\,)\right)\left[1 - \int_0^{t_*} \alpha_i\left(T,\ s,\ z_i^0(\,\cdot\,),\ v(s),\ g_i^0(\,\cdot\,)\right) \mathrm{d}s\right] \\ + \int_0^{t_*} \alpha_i\left(T,\ s,\ z_i^0(\,\cdot\,),\ v(s),\ g_i^0(\,\cdot\,)\right) M_i \mathrm{d}s.\end{aligned} \tag{23.11}$$

Since M_i are a convex compact sets, $\alpha_i\left(T,\ s,\ z_i^0(\,\cdot\,),\ v(s),\ g_i^0(\,\cdot\,)\right)$ are a nonnegative functions for $s \in [0\ ,\ t_*)$, $i = 1, \ldots, l$, and $\int_0^{t_*} \alpha_i\left(T,\ s,\ z_i^0(\,\cdot\,),\ v(s),\ g_i^0(\,\cdot\,)\right) \mathrm{d}s = 1$, it follows that

$$\int_0^{t_*} \alpha_i\left(T,\ s,\ z_i^0(\,\cdot\,),\ v(s),\ g_i^0(\,\cdot\,)\right) M_i \mathrm{d}s = M_i, \quad i = 1, \ldots, l.$$

Taking into account these facts, from (23.11) we obtain $\pi_i z_i(T) \in M_i$.

Suppose for some $i = 1, \ldots, l$ $\xi_i\left(T,\ z_i^0(\,\cdot\,),\ g_i^0(\,\cdot\,)\right) \in M_i$. Then, taking into account (23.9), from equality (23.10) we find that $\pi_i z_i(T) \in M_i$, or $z_i(T) \in M_i^*$.

The theorem is proved.

Corollary 23.1 *Assume that the differential-difference group pursuit game (23.1), (23.3) is linear* $(\phi_i(u_i,\ v) = u_i - v)$, *Pontryagin's condition holds, there exist a continuous positive functions* $r_i(t)\,,\ r_i : \mathbb{R} \to \mathbb{R}$, *and the numbers* $l_i \geq 0$ *such that*

$$\pi_i X_i (B_i, t) U_i = r_i (t) S_i, \quad M_i = l_i S_i, \quad i = 1, \ldots l,$$

where S_i are the unit balls centered at zero in the subspaces L_i. Then when $\xi_i \left(t,\ z_i^0 (\ \cdot\ )\ ,\ g_i (\ \cdot\ )\right) \notin M_i$, the resolving functions (23.6) are the largest roots of the quadratic equations for $\alpha_i > 0$

$$\left\| \pi_i X_i (B_i, t - \tau_i - s) v + g_i (t - \tau_i - s) - \alpha_i \xi_i \left(t,\ z_i^0 (\ \cdot\ ),\ g_i (\ \cdot\ )\right) \right\| = \\ = r_i (t - \tau_i - s) + \alpha_i l_i .$$

Proof From expression (23.6), taking into account the condition, we have that the resolving functions $\alpha_i \left(t,\ s,\ z_i^0 (\ \cdot\ ),\ v,\ g_i (\ \cdot\ )\right)$ for fixed values of arguments are the maximal numbers α_i such that

$$[r_i (t - \tau_i - s) S_i - \pi_i X_i (B_i, t - \tau_i - s) v - g_i (t - \tau_i - s)] \cap \\ \alpha_i \left[l_i S_i - \xi_i \left(t,\ z_i^0 (\ \cdot\ ),\ g_i (\ \cdot\ )\right)\right] \neq \emptyset.$$

The last expression is equivalent to the inclusion

$$\pi_i X_i (B_i, t - \tau_i - s) v + g_i (t - \tau_i - s) - \alpha_i \xi_i \left(t,\ z_i^0 (\ \cdot\ ),\ g_i (\ \cdot\ )\right) \in \\ [r_i (t - \tau_i - s) + \alpha_i l_i] S_i .$$

Due to the linearity of the left-hand side of this inclusion in α_i, the length of the vector

$$\pi_i X_i (B_i, t - \tau_i - s) v + g_i (t - \tau_i - s) - \alpha_i \xi_i \left(t,\ z_i^0 (\ \cdot\ )\ ,\ g_i (\ \cdot\ )\right)$$

is equal to the vector $[r_i (t - \tau_i - s) + \alpha_i l_i] S_i$ for the maximal value of α_i for all $i = 1, \ldots l.$.

The proof is complete.

23.4 Examples

Let the differential-difference game

$$\dot{z} (t) = bz (t - \tau) + u (t) - v (t) ,$$

$$z (t) \in \mathbb{R}^n, \quad 0 < b < 1, \quad \tau > 0,$$

be given, where the initial states

$$z (t) = z^0, \quad -\tau \leq t \leq 0,$$

$$\|u\| \leq 1, \quad \|v\| \leq 1 - b, \quad \|z^0\| = 1.$$

The terminal set $M^* = M = M_0 = \{z(t) \in \mathbb{R}^n : z = 0\}$, $L = \mathbb{R}^n$, π is the identity operator.

Consider the set-valued mapping

$$W(t, v) = \pi X(bI, t)\, \phi(U, v)$$

and verify the Pontryagin's condition:

$$W(t, v) = X(bI, t)(S - v), \quad W(t) = \{0\},$$

where S is the unit ball centered at zero in the space L, and I is the unit matrix of order n. The condition $W(t) = \{0\}$ uniquely determines the selection $g(t) = 0$.

We say that in the game from the initial state $z^0 \notin M^*$ it is possible to avoid meeting the terminal set if there exists a measurable function $v(t) \in V$, $t \geq 0$, such that $z(t) \notin M^*$ for any $t \geq 0$. It is shown in [22, 29] that if we input in $v(t) = (b-1) z^0$, $t \in [0, T]$, then $\|z(t)\| \geq 1$, that is $z(t) \notin M^*$ for any $t \geq 0$.

Thus, in such game of two persons, it is possible to avoid meeting with the terminal set with any control of the pursuer, in spite of the fact that the dynamic capabilities of the pursuer are greater. But if the pursuers are several then we can show that the pursuit game can be completed.

Consider the same pursuit game for a few pursuers and one evader:

$$\dot{z}_i(t) = bz_i(t - \tau_i) + u_i(t) - v(t),$$

$$z_i(t) \in \mathbb{R}^{n_i}, \quad 0 < b < 1, \quad \tau_i > 0, \quad i = 1, \ldots, l,$$

with the initial condition

$$z_i(t) = z_i^0, \quad -\tau_i \leq t \leq 0, \quad i = 1, \ldots, l.$$

The quasi strategies of the pursuers and the evader are

$$U_i = \{u_i \in \mathbb{R}^{n_i} : \|u_i\| \leq 1\}, \quad i = 1, \ldots, l,$$

$$V = \{v \in \mathbb{R}^{n_i} : \|v\| \leq 1 - b\},$$

respectively.

The terminal set $M_i^* = M_i = M_i^0 = \{z_i(t) \in \mathbb{R}^{n_i} : z_i = 0\}$, $L_i = \mathbb{R}^{n_i}$. π_i are the identity operators, $i = 1, \ldots, l$. As in the previous example, the Pontryagin's condition is satisfied automatically, the selections $g_i(t) = 0$, $i = 1, \ldots, l$. Then,

$$\xi_i\left(t\,,\ z_i^0\left(\,\cdot\,\right)\,,\ 0\right) =$$
$$= X_i\left(bI, t\right) z_i^0\left(-\tau_i\right) +$$
$$\int_{-\tau_i}^{0} X_i\left(bI, t - \tau_i - s\right) \dot{z}_i^0\left(s\right) ds, \quad i = 1, \ldots, l,$$

where

$$X_i\left(bI, t\right) = \begin{cases} \Theta, \ -\infty < t < -\tau_i; \\ I, \ -\tau_i \le t < 0; \\ I + bI\frac{t}{1!} + (bI)^2\frac{(t-\tau_i)^2}{2!} + \cdots + (bI)^k\frac{(t-(k-1)\tau_i)^k}{k!}, \end{cases}$$

$(k-1)\,\tau_i \le t \le k\tau_i$, Θ is a zero matrix, I is a unit matrix. Hence,

$$\left\|\xi_i\left(t\,,\ z_i^0\left(\,\cdot\,\right)\,,\ 0\right)\right\|^2 =$$
$$= \sum_{j=1}^{n}\left\{X_i\left(bI, t\right)\cdot z_{ij}\left(-\tau_i\right) + b\cdot\int_{-\tau_i}^{0} X_i\left(bI, t - \tau_i - s\right) \dot{z}_{ij}^0\left(s\right) ds\right\}^2.$$

Since $X_i\left(bI, t\right) U_i$ is a ball centered at zero and radius $X_i\left(bI, t\right)$, then we choose a positive continuous function $r_i\left(t\right) = X_i\left(bI, t\right)$, and the number $l_i = 0$. Then,

$$X_i\left(bI, t\right) U_i = r_i\left(t\right)\cdot S_i, \quad M_i = l_i S_i.$$

It follows from Corollary 23.1, with $\xi_i\left(t\,,\ z_i^0\left(\,\cdot\,\right)\,,\ 0\right) \notin M_i$, that the resolving functions are the largest roofs of the quadratic equations

$$\left\|X_i\left(bI, t - \tau_i - s\right) v - \alpha_i\cdot\xi_i\left(t\,,\ z_i^0\left(\,\cdot\,\right)\,,\ 0\right)\right\| = X_i\left(bI, t - \tau_i - s\right),$$

from which we get

$$\alpha_i\left(t, s,\ z_i^0\left(\,\cdot\,\right), v,\ 0\right) =$$
$$\frac{X_i\left(bI, t - \tau_i - s\right)}{\left\|\xi_i\left(t\,,\ z_i^0\left(\,\cdot\,\right)\,,\ 0\right)\right\|^2}\cdot\Bigg[\left(v,\ \xi_i\left(t\,,\ z_i^0\left(\,\cdot\,\right)\,,\ 0\right)\right) +$$
$$\sqrt{\left(v,\ \xi_i\left(t\,,\ z_i^0\left(\,\cdot\,\right)\,,\ 0\right)\right)^2 - \left\|\xi_i\left(t\,,\ z_i^0\left(\,\cdot\,\right)\,,\ 0\right)\right\|^2\cdot\left(\|v\|^2 - 1\right)}\Bigg]$$

The time for the completion of the group pursuit will be

$$T = T\left(z^0\left(\,\cdot\,\right), 0\right) = \min\Bigg\{t \ge 0:$$
$$\inf_{v\in V}\max_{i=1,\ldots,l}\int_0^t \alpha_i\left(t\,,\ s\,,\ z_i^0\left(\,\cdot\,\right)\,,\ v\,,\ 0\right) ds = 1\Bigg\},$$

We find sufficient conditions for the finiteness of this point in time.

Denote

$$\delta\left(z^0(\,\cdot\,)\right) = \min_{\|v|\ =1} \max_{i=1,\ldots,l} \alpha_i\left(t\,,\ s\,,\ z_i^0(\,\cdot\,)\,,\ v\,,\ 0\right).$$

Then

$$\inf_{v\in V} \max_{i=1,\ldots,l} \int_0^t \alpha_i\left(t\,,\ s\,,\ z_i^0(\,\cdot\,)\,,\ v\,,\ 0\right) ds =$$

$$\inf_{v\in V} \max_{\lambda\in\sum} \sum_{i=1}^{l} \lambda_i \int_0^t \alpha_i\left(t\,,\ s\,,\ z_i^0(\,\cdot\,)\,,\ v\,,\ 0\right) ds \geq$$

$$\inf_{v\in V} \sum_{i=1}^{l} \frac{1}{l} \int_0^t \alpha_i\left(t\,,\ s\,,\ z_i^0(\,\cdot\,)\,,\ v\,,\ 0\right) ds =$$

$$\frac{1}{l} \inf_{v\in V} \int_0^t \sum_{i=1}^{l} \alpha_i\left(t\,,\ s\,,\ z_i^0(\,\cdot\,)\,,\ v\,,\ 0\right) ds =$$

$$\frac{1}{l} \int_0^t \inf_{v\in V} \sum_{i=1}^{l} \alpha_i\left(t\,,\ s\,,\ z_i^0(\,\cdot\,)\,,\ v\,,\ 0\right) ds \geq$$

$$\frac{1}{l} \int_0^t \min_{\|v\|\leq 1} \max_{i=1,\ldots,l} \alpha_i\left(t\,,\ s\,,\ z_i^0(\,\cdot\,)\,,\ v\,,\ 0\right) ds =$$

$$\frac{t}{l}\delta\left(z^0(\,\cdot\,)\right),$$

where $\sum$ is $(l-1)$–dimensional simplex:

$$\sum = \left\{(\lambda_1, \ldots \lambda_l):\ \sum_{i=1}^{l} \lambda_i = 1,\ \ \lambda_i \geq 0\right\}.$$

Thus,

$$\inf_{v\in V} \max_{i=1,\ldots,l} \int_0^t \alpha_i\left(t\,,\ s\,,\ z_i^0(\,\cdot\,)\,,\ v\,,\ 0\right) ds \geq$$

$$\frac{t}{l}\delta\left(z^0(\,\cdot\,)\right),$$

where do we find the upper bound for the pursuit time

$$T\left(z^0(\,\cdot\,),\, 0\right) \leq \frac{l}{\delta\left(z^0(\,\cdot\,)\right)}.$$

Finiteness of time $T\left(z^0(\cdot),\, 0\right)$ is guaranteed by the condition

$$\delta\left(z^0(\cdot)\right) > 0.$$

This inequality holds if

$$\min_{\|v\|=1}\max_{i=1,\ldots,l}\left(\frac{\xi_i\left(t,\ z_i^0(\cdot),\ 0\right)}{\|\xi_i\left(t,\ z_i^0(\cdot),\ 0\right)\|},\ v\right) > 0 \tag{23.12}$$

for all $t > 0$.

To verify this, it is enough to consider cases $\|v\| = 0$ and $\|v\| = 1$. We show that the inequality (23.12) holds if it's zero space $\mathbb{R}^{n_i}$ belongs to the inner convex hull spanned by vectors

$$\frac{\xi_i\left(t,\ z_i^0(\cdot),\ 0\right)}{\|\xi_i\left(t,\ z_i^0(\cdot),\ 0\right)\|},\quad i = 1,\ldots,l,$$

$$0 \in \inf co\left\{\frac{\xi_i\left(t,\ z_i^0(\cdot),\ 0\right)}{\|\xi_i\left(t,\ z_i^0(\cdot),\ 0\right)\|}\right\}. \tag{23.13}$$

Indeed, let there be an inclusion (23.13). Since we have a convex hull of l points, this means that zero belongs to the interior of a convex polyhedron spanned by unit vectors $\frac{\xi_i\left(t,\ z_i^0(\cdot),\ 0\right)}{\|\xi_i\left(t,\ z_i^0(\cdot),\ 0\right)\|}$. Since the convex polyhedron is compact, and it's support function is

$$\max_{i=1,\ldots,l}\left(\frac{\xi_i\left(t,\ z_i^0(\cdot),\ 0\right)}{\|\xi_i\left(t,\ z_i^0(\cdot),\ 0\right)\|},\ v\right),$$

from (23.13) we obtain

$$0 < \max_{i=1,\ldots,l}\left(\frac{\xi_i\left(t,\ z_i^0(\cdot),\ 0\right)}{\|\xi_i\left(t,\ z_i^0(\cdot),\ 0\right)\|},\ v\right)$$

for all v, $\|v\| = 1$.

Since the right-hand side of this inequality is independent of l, we have

$$\min_{\|v\|=1}\max_{i=1,\ldots,l}\left(\frac{\xi_i\left(t,\ z_i^0(\cdot),\ 0\right)}{\|\xi_i\left(t,\ z_i^0(\cdot),\ 0\right)\|},\ v\right) > 0.$$

Thus, if

$$0 \in \inf co\left\{\frac{\xi_i\left(t,\ z_i^0(\cdot),\ 0\right)}{\|\xi_i\left(t,\ z_i^0(\cdot),\ 0\right)\|}\right\},\quad t \geq 0,$$

then this group pursuit problem is solvable at time $T\left(z^0(\,\cdot\,),\,0\right)$, for which the estimate

$$T\left(z^0(\,\cdot\,),\,0\right) \le \frac{l}{\delta\left(z^0(\,\cdot\,)\right)}.$$

If $t_* = t_*(v(\,\cdot\,))$ is the switching moment and zero of the control function

$$h(t) = 1 - \max_{i=1,\ldots,l} \int_0^t \alpha_i\left(T,\, s,\, z_i^0(\,\cdot\,),\, v(s),\, 0\right) ds,$$

then the control of the pursuers realizing time $T_l = T\left(z^0(\,\cdot\,),\,0\right)$ on the interval $[0\,,\; t_*)$ has the form

$$u_i(s) = v(s) - X_i^{-1}(bI, T_l - \tau_i - s) \cdot \alpha_i\left(T_l, s, z_i^0(\,\cdot\,), v, 0\right) \cdot \xi_i\left(T_l\,,\; z_i^0(\,\cdot\,)\,,\; 0\right), \quad i = 1, \ldots l,$$

and on the interval $[t_*\,,\; T_l]$ for those indices i, for which equality

$$\int_0^{t_*} \alpha_i\left(t\,,\; s\,,\; z_i^0(\,\cdot\,)\,,\; v\,,\; 0\right) ds = \max_{i=1,\ldots,l} \int_0^{t_*} \alpha_i\left(t\,,\; s\,,\; z_i^0(\,\cdot\,)\,,\; v\,,\; 0\right) ds$$

holds, the control of the pursuers has the form

$$u_i(s) = X_i^{-1}(bI, T_l - \tau_i - s) \cdot v(s)\,,$$

and for the remaining indices i the controls are arbitrary on the interval $[t_*\,,\; T_l]$.

23.5 Conclusion

An approach to the solution of the group pursuit differential-difference game with pure time-lag is proposed. In this paper, the Method of Resolving Functions is chosen as the main tool for research. Based on the above, we can conclude that in a game of two persons it is possible to avoid meeting with the terminal set, but in a case of several pursuers a group pursuit game can be completed.

The effectiveness of the Method of Resolving Functions, sufficient conditions that are easily verified, the ability to quickly build the resolution function, using the modern techniques of set-valued mappings and their selections, prove the relevance of this method for solving differential-difference games that are of great practical importance.

Acknowledgements The author is grateful to Academician Zgurovsky M.Z. for the possibility of the publication and to professor Kasyanov P.O. for assistance in publication this article.

References

1. Isaacs, R.: Differential Games: A Mathematical Theory with Applications to Warfare and Pursuit, Control and Optimization. Wiley Inc., New York (1965)
2. Pontryagin, L.S.: Selected Scientific Works, vol. 2. Nauka, Moscow (1988)
3. Krasovskii, N.N., Subbotin, A.I.: Game-Theoretical Control Problems. Springer, New York (1988)
4. Chikrii, A.A.: Conflict-Controlled Processes. Springer Science & Business Media, Berlin (2013)
5. Filippov, A.F.: Differential Equations with Discontinuous Righthand Sides. Springer Netherlands, Dordrecht (1988). https://doi.org/10.1007/978-94-015-7793-9
6. Chikrii, A.A.: An analytical method in dynamic pursuit games. Proc. Steklov Inst. Math. **271**(1), 69–85 (2010). https://doi.org/10.1134/S0081543810040073
7. Chikrii, A.A.: Multivalued mappings and their selections in game control problems. J. Autom. Inf. Sci. **27**(1), 27–38 (1995)
8. Chikrii, A.A., Rappoport, I.S., Chikrii, K.A.: Multivalued mappings and their selectors in the theory of conflict-controlled processes. Cybern. Syst. Anal. **43**(5), 719–730 (2007). https://doi.org/10.1007/s10559-007-0097-8
9. Chikrii, A.A., Rappoport, I.S.: Method of resolving functions in the theory of conflict-controlled processes. Cybern. Syst. Anal. **48**(4), 512–531 (2012). https://doi.org/10.1007/s10559-012-9430-y
10. Chikrii, A.A., Eidelman, S.D.: Generalized Mittag-Leffler matrix functions in game problems for evolutionary equations of fractional order. Cybern. Syst. Anal. **36**(3), 315–338 (2000). https://doi.org/10.1007/BF02732983
11. Chikriy, A.A., Matichin, I.I.: Presentation of solutions of linear systems with fractional derivatives in the sense of Riemann-Liouville, Caputo and Miller-Ross. J. Autom. Inf. Sci. **40**(6), 1–11 (2008). https://doi.org/10.1615/JAutomatInfScien.v40.i6.10
12. Chikrii, A.A.: Optimization of game interaction of fractional-order controlled systems. Optim. Methods Softw. **23**(1), 39–72 (2008). https://doi.org/10.1080/10556780701281309
13. Chikrii, A.A.: Game dynamic problems for systems with fractional derivatives. Springer Optimization and Its Applications, vol. 17, pp. 349–386 (2008). https://doi.org/10.1007/978-0-387-77247-9_1
14. Chikrii, A.A., Eidelman, S.D: Game problems of control for quasilinear systems with fractional Riemann-Liouville derivatives. Kibern. Sist. Anal. **6**, 66–99 (2001)
15. Chikrii, A.A., Eidelman, S.D.: Control game problems for quasilinear systems with Riemann-Liouville fractional derivatives. Cybern. Syst. Anal. **37**(6), 836–864 (2001). https://doi.org/10.1023/A:1014529914874
16. Chikrii, A.A., Eidelman, S.D.: Game problems for fractional quasilinear systems. Comput. Math. Appl. **44**(7), 835–851 (2002). https://doi.org/10.1016/S0898-1221(02)00197-9
17. Chikrii A.A.: On nonstationary game problem of motion control. J. Autom. Inf. Sci. **47**(11), 74–83 (2015). https://doi.org/10.1615/JAutomatInfScien.v47.i11.60
18. Chikrii A.A., Pepelyaev, V.A.: On the game dynamics problems for nonstationary controlled processes. J. Autom. Inf. Sci. **49**(3), 13–23 (2017). https://doi.org/10.1615/JAutomatInfScien.v49.i3.30
19. Baranovskaya, L.V., Chikrij, A.A., Chikrij, A.A.: Inverse Minkowski functionals in a nonstationary problem of group pursuit. Izv. Akad. Nauk Teor. Sist. Upr. (1), 109–114 (1997)
20. Kryvonos, I.I., Chikrii, A.A., Chikrii, K.A.: On an approach scheme in nonstationary game problems. J. Autom. Inf. Sci. **45**(11), 15–21 (2013). https://doi.org/10.1615/JAutomatInfScien.v45.i11.30
21. Baranovskaya, L.V., Chikrii, A.A., Chikrii, A.A.: Inverse Minkowski functional in a nonstationary problem of group pursuit. J. Comput. Syst. Sci. Int. **36**(1), 101–106 (1997)
22. Baranovskaya, G.G., Baranovskaya, L.V.: Group pursuit in quasilinear differential-difference games. J. Autom. Inf. Sci. **29**(1), 55–62 (1997). https://doi.org/10.1615/JAutomatInfScien.v29.i1.70

23. Baranovskaya, L.V.: A method of resolving functions for one class of pursuit problems. East.-Eur. J. Enterp. Technol. **2**(4), 4–8 (2015). https://doi.org/10.15587/1729-4061.2015.39355
24. Baranovskaya, L.V.: Method of resolving functions for the differential-difference pursuit game for different-inertia objects. Adv. Dyn. Syst. Control **69**, 159–176 (2016). https://doi.org/10.1007/978-3-319-40673-2
25. Bellman, R., Cooke, K.L.: Differential-Difference Equations. Academic, Cambridge (1963)
26. Khusainov, D.Y., Benditkis, D.D., Diblik, J.: Weak Delay in Systems with an Aftereffect. Funct. Differ. Equ. **9**(3–4), 385–404 (2002)
27. Baranovska, L.V.: On quasilinear differential-difference games of approach. J. Autom. Inf. Sci. **49**(8), 53–67 (2017). https://doi.org/10.1615/JAutomatInfScien.v49.i8.40
28. Baranovska, L.V.: Quasi-Linear Differential-Deference Game of Approach. Understanding Complex Systems, pp. 505–524 (2019). https://doi.org/10.1007/978-3-319-96755-4_26
29. Baranovska, L.V.: Pursuit differential-difference games with pure time-lag. Discret. Contin. Dyn. Syst. - B **24**(3), 1021–1031 (2019). https://doi.org/10.3934/dcdsb.2019004
30. Khusainov, D.Y., Diblik, J., Ruzhichkova, M.: Linear dynamical systems with aftereffect. Representation of Decisions, Stability, Control, Stabilization. GP Inform-Analytics Agency, Kiev (2015)
31. Aubin, J.-P., Frankovska, H.: Set-Valued Analysis. Birkhäuser, Bostom (1990)

Chapter 24
An Indirect Approach to the Existence of Quasi-optimal Controls in Coefficients for Multi-dimensional Thermistor Problem

Ciro D'Apice, Umberto De Maio, and Peter I. Kogut

Abstract The paper studies a problem of an optimal control in coefficients for the system of two coupled elliptic equations also known as thermistor problem which provides a simultaneous description of the electric field $u = u(x)$ and temperature $\theta(x)$. The coefficient b of operator div $(b(x)\nabla\theta(x))$ is used as the control in $W^{1,q}(\Omega)$ with $q > N$. The optimal control problem is to minimize the discrepancy between a given distribution $\theta_d \in L^1(\Omega)$ and the temperature of thermistor $\theta \in W_0^{1,\gamma}(\Omega)$ by choosing an appropriate anisotropic heat conductivity $b(x)$. Basing on the perturbation theory of extremal problems and the concept of fictitious controls, we propose an "approximation approach" and discuss the existence of the so-called quasi-optimal and optimal solutions to the given problem.

24.1 Introduction and Setting of the Optimal Control Problem

The aim of this paper is two-fold. The first one is to prove an existence result for the thermistor optimal control problem in coefficients with nonlinear state equations containing the p-Laplacian with variable exponent $p = p(x)$. The second one is to provide the asymptotic analysis of a special class of well-defined parametrized opti-

C. D'Apice
Dipartimento di Science Aziendali-Management e Innovation Systems, University of Salerno, Via Giovanni Paolo II, 132, Fisciano (SA), Italy
e-mail: cdapice@unisa.it

U. De Maio
Università degli Studi di Napoli "Federico II", Dipartimento di Matematica e Applicazioni "R. Caccioppoli", Complesso Monte S. Angelo, via Cintia, 80126 Napoli, Italy
e-mail: udemaio@unina.it

P. I. Kogut (✉)
Department of Differential Equations, Oles Honchar Dnipro National University, Gagarin av., 72, 49010 Dnipro, Ukraine
e-mail: p.kogut@i.ua

V. A. Sadovnichiy and M. Z. Zgurovsky (eds.), *Contemporary Approaches and Methods in Fundamental Mathematics and Mechanics*, Understanding Complex Systems, https://doi.org/10.1007/978-3-030-50302-4_24

mal control problems with fictitious controls and show that the original problem can be considered as a variational "limit"of the corresponding constrained minimization problems.

In a bounded open domain $\Omega \subset \mathbb{R}^N$, $N \geq 2$, with sufficiently smooth boundary $\partial\Omega$, we deal with the following optimization problem:

$$\text{Minimize } \left\{ J(b, u, \theta) = \int_{\Omega} |\theta(x) - \theta_d(x)|\, dx \right\} \tag{24.1}$$

subject to the constraints

$$\operatorname{div}\left(|\nabla u|^{p-2}\nabla u\right) = \operatorname{div} g \quad \text{in } \Omega, \quad u|_{\partial\Omega} = 0, \tag{24.2}$$

$$-\operatorname{div}(b\nabla\theta) = |\nabla u|^p \quad \text{in } \Omega, \quad \theta|_{\partial\Omega} = 0, \tag{24.3}$$

$$p(\cdot) = \sigma(\theta(\cdot)) \quad \text{a.e. in } \Omega, \quad b \in \mathfrak{B}_{ad}, \tag{24.4}$$

$$\mathfrak{B}_{ad} = \left\{ b \in W^{1,q}(\Omega) \;:\; m_1 \leq b(\cdot) \leq m_2 \text{ in } \Omega,\ \|\nabla b\|_{L^q(\Omega)^N} \leq \mu \right\}. \tag{24.5}$$

where m_1 and m_2 are constants such that $0 < m_1 \leq m_2 < +\infty$, $N < q < \infty$ is a given exponent, $\theta_d \in L^1(\Omega)$ and $g \in L^\infty(\Omega)^N$ are given distributions, σ is a measurable function such that $\alpha \leq \sigma(y) \leq \beta$ for all $y \in \mathbb{R}$ and the constants α and β satisfy the condition

$$1 < \alpha \leq \beta < \alpha^* = \begin{cases} +\infty, & \text{if } \alpha \geq N, \\ \frac{\alpha N}{N-\alpha}, & \text{if } \alpha < N. \end{cases} \tag{24.6}$$

Here, $\mathfrak{B}_{ad}$ stands for the class of admissible controls, where μ is a given positive value which is assumed to be large enough.

Let us begin with a short physical motivation to consider, as a control object, the boundary value problem (24.2)–(24.4). System (24.2)–(24.4) describes the coupling between the electric field with potential u and the temperature θ in a thermistor, where its heat conductivity is given by a positive function b. This model, also known as thermistor problem, is based on rational mechanics of electrorheological fluids, that takes into account the complex interactions between the electromagnetic fields and the moving liquid. In particular, the electrorheological fluids have the interesting property that their viscosity depends on the electric field in the fluid. A great deal of attention has been paid by many authors in the study of the thermistor problem during the last two decades [2–4, 18, 21, 33]. The search of the least assumptions on $\sigma(\theta)$, ensuring the (weak) solvability of the system (24.2)–(24.4), has been in the agenda of experts for decades. Earlier, existence theorems were proved only under some smallness conditions, e.g., in the case of a sufficiently small Lipschitz constant for the function $\sigma(\theta)$. For the survey of this kind of results, we refer to [36]. However, the most essential progress in the study of existence and qualitative properties of solutions to the boundary value problem (24.2)–(24.4) was achieved by Zhikov [40]. It has been shown that the solvability of these systems can be obtained in the multi-dimensional case without any smallness requirements on the function

$\sigma(\theta)$ via a regularization approach and further passing to the limit over the parameter of regularization. Another extension of the thermistor problem, which is related to the solid-state devices, can be found in [26].

However, as for the optimal control problem (24.1)–(24.6), to the best of the authors knowledge, the existence of optimal solutions for the above thermistor problem remains an open question. Only very few articles deal with optimal control for the thermistor problem in two dimensional case (see [14, 17] and references therein). There are several reasons for this:

- it is unknown whether the set of feasible points to the problem (24.1)–(24.6) is weakly closed in the corresponding functional space;
- we have no a priori estimates for the weak solutions to the boundary value problem (24.2)–(24.4) under conditions (24.6);
- the asymptotic behaviour of a minimizing sequence to the cost functional (24.1) is unclear in general;
- the optimal control problem (24.1)–(24.6) is ill-posed and relations (24.2)–(24.4) require some relaxation (see, for instance, [8]).

To circumvent the problems listed above, we propose an "indirect approach" to the solvability of the optimal control thermistor problem in coefficients. Basing on the perturbation theory of extremal problems and the concept of fictitious controls (see, for instance, [16, 19, 22, 25]), we prove the existence of so-called quasi-optimal and optimal solutions to the problem (24.1)–(24.6) and show that they can be attained by the optimal solutions of some appropriate approximations for the original optimal control problem. The main idea of our approach is based on the fact that weak solutions to the Dirichlet problem (24.2)–(24.4) can be attained through a special regularization of the exponent $p = p(x)$ and an approximation of the operator $\mathcal{A}(u) = \operatorname{div}\left(|\nabla u|^{p-2}\nabla u\right)$, using its perturbation by the $\varepsilon\Delta_\beta$-Laplacian, and the right-hand side of (24.3) by its transformation to

$$\operatorname{div}\left[\left(|\nabla u|^{\sigma(\theta)-2}\nabla u - g\right)u\right] + (g, \nabla u)_{\mathbb{R}^N}.$$

Here, by attainability of a weak solution (u, θ), we mean the existence of a sequence $\{(u_\varepsilon, \theta_\varepsilon)\}_{\varepsilon>0}$, where $(u_\varepsilon, \theta_\varepsilon)$ are the solutions of "more regular" boundary value problems, such that $(u_\varepsilon, \theta_\varepsilon) \to (u, \theta)$ in some appropriate topology as ε tends to zero.

To see these and other characteristic features of the optimisation problem (24.1)–(24.6) and our approach more clearly, we recall the well-known facts for nonlinear elliptic problems with variable exponent and discuss how each of the equations in system (24.2)–(24.4) can be interpreted. Assuming that the temperature $\theta = \theta(x)$ is known for some admissible control $b(x)$, we introduce the Sobolev-Orlicz space

$$W_0^{1,p(\cdot)}(\Omega) := \left\{u \in W_0^{1,1}(\Omega) \ : \ \int_\Omega |\nabla u(x)|^{p(x)}\,dx < +\infty\right\} \tag{24.7}$$

and equip it with the norm $\|u\|_{W_0^{1,p(\cdot)}(\Omega)} = \|\nabla u\|_{L^{p(\cdot)}(\Omega)^N}$, where $p(x) = \sigma(\theta(x))$ (for the details, we refer to Sect. 24.2.1). Here, $|\cdot|$ denotes the Euclidean norm $|\cdot|_{\mathbb{R}^N}$ in $\mathbb{R}^N$, and $L^{p(\cdot)}(\Omega)^N$ stands for the set of all measurable functions $f : \Omega \to \mathbb{R}^N$ such that $\int_\Omega |f(x)|^{p(x)}\, dx < +\infty$. It is well-known that (see, for instance, [11, 40]), unlike in classical Sobolev spaces, smooth functions are not necessarily dense in $W = W_0^{1,p(\cdot)}(\Omega)$. Hence, with variable exponent $p = p(x)$ $(1 < \alpha \le p(\cdot) \le \beta)$ it can be associated another Sobolev space,

$$H = H_0^{1,p(\cdot)}(\Omega) \text{ as the closure of the set } C_0^\infty(\Omega) \text{ in } W_0^{1,p(\cdot)}(\Omega)\text{-norm.}$$

Definition 24.1 We say that a function $u \in W_0^{1,p(\cdot)}(\Omega)$ is a weak solution of the problem (24.2) if

$$\int_\Omega \left(|\nabla u|^{p-2}\nabla u, \nabla\varphi\right)_{\mathbb{R}^N} dx = \int_\Omega (g, \nabla\varphi)_{\mathbb{R}^N}\, dx, \quad \forall \varphi \in C_0^\infty(\Omega), \tag{24.8}$$

and we say that u is the H-solution of problem (24.2), if $u \in H_0^{1,p(\cdot)}(\Omega)$ and the integral identity (24.8) holds for any test function $\varphi \in H_0^{1,p(\cdot)}(\Omega)$.

Since we can lose the density of the set $C_0^\infty(\Omega)$ in $W_0^{1,p(\cdot)}(\Omega)$ for some (irregular) variable exponents $p(x)$, it follows that a weak solution to the problem (24.2) is not unique, in general. To clarify this inference, let us associate with the Dirichlet problem (24.2) the Carathéodory vector-valued function $A_p : \Omega \times \mathbb{R}^N \to \mathbb{R}^N$ given by the rule $A_p(x,\xi) = |\xi|^{p(x)-2}\xi$. Then it is easy to see that the following strict monotonicity, coercivity, and boundedness conditions hold:

$$\left(A_p(x,\xi) - A_p(x,\varepsilon ta), \xi - \varepsilon ta\right)_{\mathbb{R}^N} > 0, \quad \forall \xi \ne \varepsilon ta, \tag{24.9}$$

$$\left(A_p(x,\xi), \xi\right)_{\mathbb{R}^N} \ge |\xi|^{p(x)} \quad \text{a.e. in } \Omega, \quad \forall \xi \in \mathbb{R}^N, \tag{24.10}$$

$$|A_p(x,\xi)|^{p'(x)} \le |\xi|^{p(x)} \quad \text{a.e. in } \Omega, \quad \forall \xi \in \mathbb{R}^N, \text{ where } p'(x) = \frac{p(x)}{p(x)-1}. \tag{24.11}$$

Typically, such conditions are referred to as $p(x)$-monotonicity conditions. Let V be a closed subspace of $W_0^{1,p(\cdot)}(\Omega)$. Then for a given $u \in V$ we determine an element $\mathcal{A}_p u \in V^*$ as follows

$$\begin{aligned}\left(\mathcal{A}_p u, v\right) &= \int_\Omega \left(A_p(x, \nabla u(x)), \nabla v(x)\right) dx \\ &= \int_\Omega \left(|\nabla u(x)|^{p(x)-2}\nabla u(x), \nabla v(x)\right) dx.\end{aligned} \tag{24.12}$$

Since

$$\left|\left(A_p(x,\nabla u),\nabla v\right)\right| \le \frac{1}{p'(x)}\left|A_p(x,\nabla u)\right|^{p'(x)} + \frac{1}{p(x)}|\nabla v|^{p(x)} \le |\nabla u|^{p(x)} + |\nabla v|^{p(x)}$$

by the Young inequality and condition (24.10), it follows that the given definition of $\mathcal{A}_p u \in V^*$ is valid and the operator $\mathcal{A}_p : V \to V^*$ is bounded. To verify that this operator is strictly monotone, coercive, and semicontinuous, we refer to Theorem 2.1 in [40]. Hence, the existence and uniqueness of a weak solution to the problem (24.2) in V-subspace is a direct consequence of monotone operator theory (see, for instance, [32]). Thus, the H-solution exists, it is unique and satisfies the energy equality

$$\int_\Omega |\nabla u(x)|^{p(x)}\,dx = \int_\Omega (g(x), \nabla u(x))_{\mathbb{R}^N}\,dx \tag{24.13}$$

which immediately follows from (24.8) and density of $C_0^\infty(\Omega)$ in $H_0^{1,p(\cdot)}(\Omega)$.

As for the second equation (24.3), its right-hand side $|\nabla u|^p$ with $p(\cdot) = \sigma(\theta(\cdot))$, a priori belongs to the space $L^1(\Omega)$. So, $f = |\nabla u|^p$ is not an element of the dual space $H^{-1}(\Omega)$ and, hence, we can not expect that the weak solution of the Dirichlet problem (24.3) belongs to $H_0^1(\Omega)$. At the same time, since the domain Ω is smooth enough and each admissible control $b \in \mathfrak{B}_{ad}$ is a Hölder continuous function, it follows from [12, Sect. 6] that the operator $-\operatorname{div}(b\nabla\theta)$ admits maximal elliptic regularity for any $b \in \mathfrak{B}_{ad}$. Hence, the distributional solution to (24.3) is unique and it coincides with the so-called duality solution (see [27, 30]).

Definition 24.2 A function $\theta \in L^1(\Omega)$ is a duality solution to problem (24.3) if

$$\int_\Omega \theta\varphi\,dx = \int_\Omega |\nabla u|^p v\,dx, \quad \forall\,\varphi \in L^\infty(\Omega),$$

where $v \in H_0^1(\Omega) \cap L^\infty(\Omega)$ is the weak solution of

$$-\operatorname{div}(b\nabla v) = \varphi \quad \text{in } \Omega \quad v = 0 \text{ on } \partial\Omega.$$

It is however imperative to note the following property of duality solutions: If $\theta \in L^1(\Omega)$ is a duality solution of (24.3), then for any sequence $\{f_n\}_{n\in\mathbb{N}} \subset L^\infty(\Omega)$ such that $f_n \to |\nabla u|^p$ strongly in $L^1(\Omega)$ and

$$\|f_n\|_{L^1(\Omega)} \le \||\nabla u|^p\|_{L^1(\Omega)} \text{ for all } n \in \mathbb{N},$$

we have $\theta_n \to \theta$ strongly in $L^1(\Omega)$ and weakly in $W_0^{1,\gamma}(\Omega)$ for all $\gamma \in [1, \frac{N}{N-1})$, where $\theta_n \in H_0^1(\Omega) \cap L^\infty(\Omega)$ is the weak solution of

$$-\operatorname{div}(b\nabla\theta_n) = f_n \quad \text{in } \Omega \quad \theta_n = 0 \text{ on } \partial\Omega.$$

The main result concerning the existence of a duality solution to the problem (24.3), can be stated as follows (see, for instance, [29, Theorems 3.3 and 4.1],[1, 5]): if Ω is a bounded domain with sufficiently smooth boundary and $|\nabla u|^{p(\cdot)} \in L^1(\Omega)$, then the Dirichlet problem (24.3) has the unique duality solution $\theta \in W_0^{1,\gamma}(\Omega)$ with $\gamma \in [1, \frac{N}{N-1})$; moreover, there exists a constant $C = C(\gamma)$ independent of $f = |\nabla u|^{p(\cdot)}$ such that

$$\|\theta\|_{W_0^{1,\gamma}(\Omega)} \leq C(\gamma)\|f\|_{L^1(\Omega)} = C(\gamma) \int_\Omega |\nabla u(x)|^{p(x)}\, dx. \tag{24.14}$$

Remark 24.1 In fact, if the datum $f = |\nabla u|^p$ is more regular, say $f \in L^{1+\delta}(\Omega)$ for some $\delta > 0$, we have the following result (see [29, Theorem 4.4]): if $|\nabla u|^p \in L^{1+\delta}(\Omega)$, $0 < \delta < \frac{N-2}{N+2}$ then the unique duality solution of (24.3) belongs to $W_0^{1,q}(\Omega)$ with $q = \frac{N(1+\delta)}{N-1-\delta} = 1 + \delta + \frac{(1+\delta)^2}{N-1-\delta}$.

The optimal control problem we consider in this paper is to minimize the discrepancy between a given distribution $\theta_d \in L^r(\Omega)$ and the temperature of thermistor θ by choosing an appropriate heat conductivity $b \in \mathfrak{B}_{ad}$.

Since for a "typical" measurable or even continuous function $\sigma(\theta)$ with properties (24.6), the set $C_0^\infty(\Omega)$ is not dense in $W_0^{1,p(\cdot)}(\Omega)$, and, hence, no uniqueness of weak solutions to (24.2)–(24.4) can be expected, the mapping $b \mapsto (u, \theta)$, where (u, θ) is a weak solution to the boundary value problem (24.2)–(24.4), can be multi-valued in general. In view of this, we introduce the set of feasible solutions to the OCP (24.1)–(24.6) as follows:

$$\Xi_0 = \left\{ (b, u, \theta, p) \;\middle|\; \begin{array}{c} b \in \mathfrak{B}_{ad},\ u \in H_0^{1,p(\cdot)}(\Omega),\ \theta \in L^1(\Omega), \\ p \in L^\infty(\Omega), \\ p(\cdot) = \sigma(\theta(\cdot)) \text{ a.e. in } \Omega, \\ u \text{ is the } H\text{-solution of (24.2)}, \\ \theta \text{ is the duality solution to (24.3).} \end{array} \right\} \tag{24.15}$$

It is clear that $J(b, u, \theta, p) < +\infty$ for all $(b, u, \theta, p) \in \Xi_0$.

Remark 24.2 The characteristic feature of the OCP (24.1)–(24.6) is the fact that a priori it is unknown whether the set Ξ_0 is nonempty. Using the assumption (24.6) and basing on a special technique of the weak convergence of fluxes to a flux, it was established in [40] that thermistor problem (24.2)–(24.4) for $b(x) = \xi$, with $\xi \in [m_1, m_2]$, and for any measurable function $\sigma(\theta)$ admits a weak solution $u \in W_0^{1,p(\cdot)}(\Omega)$. However, in this case the inclusion $u \in H_0^{1,p(\cdot)}(\Omega)$ is by no means obvious even in the case of constant function $b \in \mathfrak{B}_{ad}$ [40, Sect. 7]. Hence, the OCP (24.1)–(24.6) requires some relaxation. With that in mind we propose to consider the function $p(\cdot)$ as a fictitious control with some more regular properties and interpret the fulfilment of equality $p(\cdot) = \sigma(\theta(\cdot))$ with some accuracy.

24.1.1 Relaxation of the Original OCP

We consider the following extension of the set of feasible solutions to the original OCP. Let $k_0 > 0$ and $\tau \geq 0$ be given constants.

Definition 24.3 We say that a tuple (b, u, θ, p) is quasi-feasible to the OCP (24.1)–(24.6) if $(b, u, \theta, p) \in \widehat{\Xi}_0(\tau)$, where

$$\widehat{\Xi}_0(\tau) = \left\{ (b, u, \theta, p) \;\middle|\; \begin{array}{c} b \in \mathfrak{B}_{ad},\ u \in H_0^{1,p(\cdot)}(\Omega),\ \theta \in L^1(\Omega),\ p \in \mathfrak{S}_{ad}, \\ \|p - \sigma(\theta)\|_{L^2(\Omega)} \leq \tau, \\ u \text{ is the } H\text{-solution of (24.2)}, \\ \theta \text{ is the duality solution to (24.3)}. \end{array} \right\}, \tag{24.16}$$

$$\mathfrak{S}_{ad} = \left\{ q \in C(\overline{\Omega}) \;\middle|\; \begin{array}{c} |q(x) - q(y)| \leq \omega(|x - y|), \\ \forall x, y, \in \Omega,\ |x - y| \leq 1/2, \\ \omega(t) = k_0 / \log(|t|^{-1}), \\ 1 < \alpha \leq q(\cdot) \leq \beta \text{ in } \overline{\Omega}. \end{array} \right\}. \tag{24.17}$$

We also say that $(b^0, u^0, \theta^0, p^0) \in W^{1,q}(\Omega) \times H_0^{1,p(\cdot)}(\Omega) \times L^1(\Omega) \times C(\overline{\Omega})$ is a quasi-optimal solution to the problem (24.1)–(24.6) if

$$(b^0, u^0, \theta^0, p^0) \in \widehat{\Xi}_0(\tau) \text{ and } J(b^0, u^0, \theta^0, p^0) = \inf_{(b,u,\theta,p) \in \widehat{\Xi}_0(\tau)} J(b, u, \theta, p),$$

and this tuple is called to be optimal if $p^0(\cdot) = \sigma(\theta^0(\cdot))$ a.e. in Ω.

Remark 24.3 It is clear that $\widehat{\Xi}_0(\tau) \subset \Xi_0$ for $\tau = 0$ and, moreover, as we will see later on, the set $\widehat{\Xi}_0(\tau)$ is nonempty if only $\tau \geq \sqrt{|\Omega|}(\beta - \alpha)$. It is also worth to emphasize that the condition $p \in \mathfrak{S}_{ad}$ implies that $p(\cdot)$ has some additional regularity. Moreover, in view of the obvious relation $\lim_{t \to 0} |t|^\delta \log(|t|) = 0$ with $\delta \in (0, 1)$, it is clear that $p \in C^{0,\delta}(\Omega)$ implies $p \in \mathfrak{S}_{ad}$. Because of this $p \in \mathfrak{S}_{ad}$ is often called a locally log-Hölder continuous exponent(see [10, Definition 2.2]). Another point about benefit of the choice of the subset $\mathfrak{S}_{ad}$ is related with the following properties: (i) $\mathfrak{S}_{ad}$ is a compact subset in $C(\overline{\Omega})$ and thus provides uniformly convergent subsequences; (ii) Every cluster point p of a sequence $\{p_k\}_{k \in \mathbb{N}} \subset \mathfrak{S}_{ad}$ is a regular exponent (i.e. in this case the set $C_0^\infty(\Omega)$ is dense in $W_0^{1,p(\cdot)}(\Omega)$), which plays a key role in the situation of Propositions 24.2 and 24.3; (iii) Because of the log-Hölder continuity of an exponent $p \in \mathfrak{S}_{ad}$, the corresponding weak solution $u \in W_0^{1,p(\cdot)}(\Omega)$ to the variational problem (24.8) is such that $|\nabla u|^{(1+\delta)p(\cdot)} \in L^1(\Omega)$ for some $\delta > 0$ and satisfies the estimate

$$\int_\Omega |\nabla u(x)|^{(1+\delta)p(x)}\, dx \leq C \int_\Omega |g(x)|^{(1+\delta)p'(x)}\, dx + C, \tag{24.18}$$

where $\delta > 0$ and $C > 0$ depend only on Ω, α, N, k_0, and $\int_\Omega |g|^{p'}\,dx$. For the proof of the higher integrability of $|\nabla u|^{p(\cdot)}$, we refer to [40, Theorem 16.4] and [41, Lemma 3.3].

Remark 24.4 It is easy to show that if $u \in W^{1,p(\cdot)}(\Omega)$ is a solution to

$$\operatorname{div}(A(u)\nabla u) = \operatorname{div} g \quad \text{in } \mathcal{D}'(\Omega),$$

then

$$\big(A(u)\nabla u, \nabla u\big)_{\mathbb{R}^N} = \operatorname{div}\big((A(u)\nabla u - g)u\big) + (g, \nabla u)_{\mathbb{R}^N},$$

also in $\mathcal{D}'(\Omega)$, where

$$A(u) = |\nabla u(x)|^{p(x)-2} \quad \text{or} \quad A(u) = |\nabla u|^{p(x)-2} + \varepsilon|\nabla u|^{\beta-2}$$

and $p(\cdot)$ is a regular exponent. As a result, it allows to deduce the existence of the unique weak solution to the variational problem

$$-\operatorname{div}(b\nabla\theta) = \operatorname{div}\big((A(u)\nabla u - g)u\big) + (g, \nabla u)_{\mathbb{R}^N} \quad \text{in } \mathcal{D}'(\Omega)$$

which is also the duality solution to the Dirichlet BVP

$$-\operatorname{div}(b\nabla\theta) = \big|\big(A(u)\nabla u, \nabla u\big)_{\mathbb{R}^N}\big| \quad \text{in } \Omega, \quad \theta|_{\partial\Omega} = 0$$

(see the proof of Lemma 24.3 and Corollary 24.1).

Our main goal in this paper is to present the "approximation approach", based on the perturbation theory of extremal problems and the concept of fictitious controls. With that in mind, we make use of the following family of approximated problems

$$\begin{gathered}\text{Minimize } J_{\varepsilon,\tau}(b, u, \theta, p) = \int_\Omega |\theta - \theta_d|\,dx + \frac{1}{\varepsilon}\mu_\tau\left(\int_\Omega |p - \sigma(\theta)|^2\,dx\right)\\ \text{subject to the constraints}\end{gathered} \tag{24.19}$$

$$\operatorname{div}\big(|\nabla u|^{p(x)-2}\nabla u + \varepsilon|\nabla u|^{\beta-2}\nabla u\big) = \operatorname{div} g \quad \text{in } \Omega, \quad u|_{\partial\Omega} = 0, \tag{24.20}$$

$$\begin{aligned}-\operatorname{div}(b\nabla\theta) = {}&\operatorname{div}\big[\big(|\nabla u|^{p(x)-2}\nabla u + \varepsilon|\nabla u|^{\beta-2}\nabla u - g\big)\,u\big]\\ &+ (g, \nabla u)_{\mathbb{R}^N} \quad \text{in } \Omega, \quad \theta|_{\partial\Omega} = 0,\end{aligned} \tag{24.21}$$

$$b \in \mathfrak{B}_{ad}, \quad p \in \mathfrak{S}_{ad}. \tag{24.22}$$

Here, the function $\mu_\tau : \mathbb{R}_+ \to \mathbb{R}_+$ is defined as follows

$$\mu_\tau(s) = 0 \text{ if } 0 \le s \le \tau^2, \quad \text{and} \quad \mu_\tau(s) = s - \tau^2 \text{ if } s > \tau^2.$$

There are several principle points in the statement of approximated problem (24.19)–(24.22) that should be emphasized. The first one is related with $\varepsilon \Delta_\beta$-regularization of $p(\cdot)$-Laplacian. Though this is a standard trick in order to establish the existence of H-solution to the Dirichlet problem (24.20) with a given exponent $p(\cdot)$ (see Theorems 3.1–3.3 in [40]), however, this approach does not allow to arrive at the existence of a weak solution $(u, \theta) \in H_0^{1,p(\cdot)}(\Omega) \times L^1(\Omega)$ to the thermistor problem (24.2)–(24.4) (see Theorem 7.2 in [40]). This can be done if only the exponent $p(\cdot) = \sigma(\theta(\cdot))$ is regular, i.e. if the set $C_0^\infty(\Omega)$ is dense in $W_0^{1,p(\cdot)}(\Omega)$, and the energy density $|\nabla u(\cdot)|^{p(\cdot)}$ belongs to the space $L^{1+\delta}(\Omega)$ for some $\delta > 0$ so that the equation (24.21) holds in the sense of the distributions. With that in mind we consider the condition $p \in \mathfrak{S}_{ad}$ as an additional option for the regularization of the original OCP. The another point that should be indicated, is related with some relaxation of the equation (24.3). Namely, it is easy to see that after the formal transformations, the equation (24.3) can be transformed to the following one

$$-\operatorname{div}(b\nabla\theta) = \operatorname{div}\left[\left(|\nabla u|^{\sigma(\theta)-2}\nabla u - g\right) u\right] + (g, \nabla u)_{\mathbb{R}^N} \quad \text{in } \mathcal{D}'(\Omega). \quad (24.23)$$

The benefit of such representation and condition $p \in \mathfrak{S}_{ad}$ is the fact that, due to the estimate (24.18), the expression $\left(|\nabla u|^{\sigma(\theta)-2}\nabla u - g\right) u$ under the divergence sign in (24.23) is integrable with degree greater than 1. As follows from our further analysis, this property plays an important role in the study of OCP (24.1)–(24.6) and we consider the representation (24.23) as some relaxation of the relation (24.3).

24.1.2 Main Results

The main result of this paper is Theorem 24.1 where we claim that if the OCP (24.1)–(24.6) has a sufficiently regular feasible point, then there exist optimal solutions to the OCP and some of them are the limit as $\varepsilon \searrow 0$ of optimal solutions to (24.19)–(24.22).

Theorem 24.1 *Let Ω be an open bounded domain in $\mathbb{R}^N$ with a sufficiently smooth boundary. Assume that $\widehat{\Xi}_0(\tau) \neq emptyset$ for $\tau = 0$, i.e. there exist a heat conductivity $\widehat{b} \in \mathfrak{B}_{ad}$, an exponent $\widehat{p} \in \mathfrak{S}_{ad}$, and a weak solution to the thermistor problem* (24.2)–(24.4) $(\widehat{u}, \widehat{\theta}) \in W_0^{1,\sigma(\widehat{\theta}(\cdot))}(\Omega) \times W_0^{1,\gamma}(\Omega)$ *with $\gamma \in [1, \frac{N}{N-1})$ and $b(\cdot) = \widehat{b}(\cdot)$ such that $\widehat{\theta}$ is the duality solution to* (24.4) *and $\widehat{p} = \sigma(\widehat{\theta})$ almost everywhere in Ω. Then OCP* (24.1)–(24.6) *has a non-empty set of optimal solutions and some of them can be attained in the following way*

$$b_\varepsilon^0 \rightharpoonup b^0 \ \text{ in } W^{1,q}(\Omega), \quad u_\varepsilon^0 \rightharpoonup u^0 \ \text{ in } W_0^{1,\alpha}(\Omega), \quad (24.24)$$

$$\theta_\varepsilon^0 \rightharpoonup \theta^0 \ \text{ in } W_0^{1,\gamma}(\Omega), \quad p_\varepsilon^0 \to p^0 \ \text{ uniformly on } \overline{\Omega}, \quad (24.25)$$

as $\varepsilon \to 0$, *where* $(b_\varepsilon^0, u_\varepsilon^0, \theta_\varepsilon^0, p_\varepsilon^0)$ *are the solutions to the approximated problems* (24.19)–(24.22) *with* $\tau = \varepsilon$ *in* (24.19).

We prove this theorem below at the end of Sect. 24.4.

Remark 24.5 It is clear that the condition $\widehat{p} = \sigma(\widehat{\theta})$ in the statement of Theorem 24.1, where $\widehat{p}$ has logarithmic modulus of continuity, imposes some additional and rather special constraint on the function σ. The principle point here is the fact that this relation has to be valid for a particular function $\widehat{\theta}$ and it is not required that the function $\sigma(\theta(\cdot))$ must be at least continuous for every solution $\theta \in L^1(\Omega)$ of (24.3). It is rather delicate problem to guarantee the fulfilment of the equality $\widehat{p} = \sigma(\widehat{\theta})$ by the direct description of function σ even if we make use of the "typical" assumption (see, for instance, [14, 17, 18]): σ is a Lipschitz continuous function.

Since it is unknown whether OCP (24.1)–(24.6) is solvable or the main assumptions of Theorem 24.1 are satisfied, it is reasonable to show that this problem admits the quasi-optimal solutions and they can be attained (in some sense) by optimal solutions to special approximated problems. We prove in Sect. 24.4 the following result.

Theorem 24.2 *Let* $\left\{(b_\varepsilon^0, u_\varepsilon^0, \theta_\varepsilon^0, p_\varepsilon^0)\right\}_{\varepsilon>0}$ *be an arbitrary sequence of optimal solutions to the approximated problems (24.19)–(24.22). Assume that either there exists a constant* $C^* > 0$ *satisfying condition*

$$\limsup_{\varepsilon\to 0} \inf_{(b,u,\theta,p)\in\widehat{\Xi}_\varepsilon} J_{\varepsilon,\tau}(b,u,\theta,p) \le C^* < +\infty$$

or $\tau \ge \sqrt{|\Omega|}(\beta - \alpha)$, *where* $\widehat{\Xi}_\varepsilon$ *stands for the set of feasible solutions to the problem (24.19)–(24.22). Then any cluster tuple* $\left(b^0, u^0, \theta^0, p^0\right)$ *in the sense of convergence (24.24)–(24.25) is a quasi-optimal solution of the OCP* (24.1)–(24.6)*. Moreover, in this case the following variational property holds*

$$\lim_{\varepsilon\to 0} \inf_{(b,u,\theta,p)\in\widehat{\Xi}_\varepsilon} J_{\varepsilon,\tau}(b,u,\theta,p) = J\left(b^0, u^0, \theta^0, p^0\right) = \inf_{(b,u,\theta,p)\in\widehat{\Xi}_0(\tau)} J(b,u,\theta,p).$$

24.2 Preliminaries and Some Auxiliary Results

24.2.1 On Orlicz Spaces

Let $p(\cdot)$ be a measurable exponent function on Ω such that $1 < \alpha \le p(x) \le \beta < \infty$ a.e. in Ω, where α and β are given constants. Let $p'(\cdot) = \frac{p(\cdot)}{p(\cdot)-1}$ be the corresponding conjugate exponent. It is clear that $\beta' \le p'(\cdot) \le \alpha'$ a.e. in Ω, where β' and α' stand for the conjugates of constant exponents. Denote by $L^{p(\cdot)}(\Omega)^N$ the set of all measurable functions $f(x)$ on Ω such that $\int_\Omega |f(x)|^{p(x)}\,dx < \infty$. Then $L^{p(\cdot)}(\Omega)^N$ is a reflexive separable Banach space with respect to the Luxemburg norm (see [10, 11, 31] for the details)

$$\|f\|_{L^{p(\cdot)}(\Omega)^N} = \inf\left\{\lambda > 0 \,:\, \rho_p(\lambda^{-1} f) \le 1\right\}, \tag{24.26}$$

where

$$\rho_p(f) := \int_\Omega |f(x)|^{p(x)}\, dx.$$

The dual of $L^{p(\cdot)}(\Omega)^N$ with respect to $L^2(\Omega)$-inner product will be denoted by $L^{p'(\cdot)}(\Omega)^N$. The following estimates are well-known (see, for instance, [10, 31, 37]): if $f \in L^{p(\cdot)}(\Omega)^N$ then

$$\|f\|^\alpha_{L^{p(\cdot)}(\Omega)^N} \le \int_\Omega |f(x)|^{p(x)}\, dx \le \|f\|^\beta_{L^{p(\cdot)}(\Omega)^N}, \text{ if } \|f\|_{L^{p(\cdot)}(\Omega)^N} > 1, \tag{24.27}$$

$$\|f\|^\beta_{L^{p(\cdot)}(\Omega)^N} \le \int_\Omega |f(x)|^{p(x)}\, dx \le \|f\|^\alpha_{L^{p(\cdot)}(\Omega)^N}, \text{ if } \|f\|_{L^{p(\cdot)}(\Omega)^N} < 1,$$

$$\|f\|_{L^{p(\cdot)}(\Omega)^N} = \int_\Omega |f(x)|^{p(x)}\, dx, \text{ if } \|f\|_{L^{p(\cdot)}(\Omega)^N} = 1,$$

$$\|f\|^\alpha_{L^{p(\cdot)}(\Omega)^N} - 1 \le \int_\Omega |f(x)|^{p(x)}\, dx \le \|f\|^\beta_{L^{p(\cdot)}(\Omega)^N} + 1, \tag{24.28}$$

$$\|f\|_{L^\alpha(\Omega)^N} \le (1 + |\Omega|)^{1/\alpha}\, \|f\|_{L^{p(\cdot)}(\Omega)^N}. \tag{24.29}$$

Moreover, due to the duality method, it can be shown that

$$\|f\|_{L^{p(\cdot)}(\Omega)^N} \le (1 + |\Omega|)^{1/\beta'}\, \|f\|_{L^\beta(\Omega)^N}, \quad \beta' = \frac{\beta}{\beta - 1}, \quad \forall f \in L^\beta(\Omega)^N. \tag{24.30}$$

We make use of the following results.

Lemma 24.1 ([40], Lemma 13.3) *If a sequence $\{f_k\}_{k\in\mathbb{N}}$ is bounded in $L^{p(\cdot)}(\Omega)$ and $f_k \rightharpoonup f$ in $L^\alpha(\Omega)$ as $k \to \infty$, then $f \in L^{p(\cdot)}(\Omega)$ and $f_k \rightharpoonup f$ in $L^{p(\cdot)}(\Omega)$, i.e.*

$$\lim_{k\to\infty} \int_\Omega f_k \varphi\, dx = \int_\Omega f_k \varphi\, dx, \quad \forall \varphi \in L^{p'(\cdot)}(\Omega).$$

Lemma 24.2 *Let $\{p_k\}_{k\in\mathbb{N}} \subset \mathfrak{S}_{ad}$ and $p \in \mathfrak{S}_{ad}$ be such that $p_k(\cdot) \to p(\cdot)$ uniformly in $\overline{\Omega}$ as $k \to \infty$. If a sequence $\left\{\|f_k\|_{L^{p_k(\cdot)}(\Omega)}\right\}_{k\in\mathbb{N}}$ is bounded and $f_k \rightharpoonup f$ in $L^\alpha(\Omega)$ as $k \to \infty$, then $f \in L^{p(\cdot)}(\Omega)$.*

Proof By analogy with the proof of Lemma 24.1 (see [39, 536 p.]), to deduce the inclusion $f \in L^{p(\cdot)}(\Omega)$, it is enough to note that

$$\int_\Omega |f(x)|^{p(x)}\, dx \le \liminf_{k\to\infty} \int_\Omega |f_k(x)|^{p_k(x)}\, dx \tag{24.31}$$

by the semicontinuity of convex functionals with respect to the weak convergence in $L^\alpha(\Omega)$.

24.2.2 On the Weak Convergence of Fluxes to Flux

A typical situation arising in the study of most optimization problems and which is of fundamental importance in many other areas of nonlinear analysis, can be stated as follows: we have the weak convergence $u_k \rightharpoonup u$ in some Sobolev space $W^{1,\alpha}(\Omega)$ with $\alpha > 1$ and we have the weak convergence of fluxes $A_k(\cdot, \nabla u_k) \rightharpoonup z$ in the Lebesgue space $L^\delta(\Omega)$, $\delta > 1$, where by flux we mean the vector under the divergence sign in an elliptic equation (in our case it is $A_k(\cdot, \nabla u_k) = |\nabla u_k|^{p_k(\cdot)-2}\nabla u_k$ or $A_k(\cdot, \nabla\theta_k) = b_k(\cdot)\nabla\theta_k$). Then the problem is to show that $z = A(\cdot, \nabla u)$, although the validity of this equality is by no means obvious at this stage.

Assume that the fluxes $A_k(x, \xi)$ satisfy the following conditions:

$$A_k : \Omega \times \mathbb{R}^N \to \mathbb{R}^N \text{ are Carathéodory vector-valued functions,} \tag{24.32}$$

$$\begin{gathered}(A_k(x,\xi) - A_k(x,\varepsilon ta), \xi - \varepsilon ta)_{\mathbb{R}^N} \geq 0,\\ A_k(x,0) = 0, \text{ for a.e. } x \in \Omega \text{ and } \forall \xi, \varepsilon ta \in \mathbb{R}^N,\end{gathered} \tag{24.33}$$

$$\begin{gathered}|A_k(x,\xi)|^{\beta'} \leq C_1|\xi|^\beta + C_2,\\ \lim_{k\to\infty} A_k(x,\xi) = A(x,\xi) \text{ for a.e. } x \in \Omega \text{ and } \forall \xi \in \mathbb{R}^N.\end{gathered} \tag{24.34}$$

Let $\{v_k\}_{k\in\mathbb{N}}$ and $\{A_k(\cdot, v_k)\}_{k\in\mathbb{N}}$ be weakly convergent sequences in $L^1(\Omega)^N$, and let v and z be their weak L^1-limits, respectively. In order to clarify the conditions under which the equality $z = A(x, v)$ holds and the fluxes $A_k(\cdot, v_k)$ weakly converge to the flux $A(\cdot, v)$, we cite the following result.

Theorem 24.3 *([40], Theorem 4.6) Assume that $\{u_k\}_{k\in\mathbb{N}}$ and $\{A_k(\cdot, \nabla u_k)\}_{k\in\mathbb{N}}$ are the sequences such that conditions* (24.33)–(24.34) *hold true and*

(i) $u_k \rightharpoonup u$ *in* $W^{1,\alpha}(\Omega)$ *and* $u_k \in W^{1,\beta}(\Omega)$ *for all* $k \in \mathbb{N}$;
(ii) $\sup_{k\in\mathbb{N}} \|A_k(\cdot, \nabla u_k)\|_{L^{\beta'}(\Omega)^N} < +\infty$;
(iii) $\sup_{k\in\mathbb{N}} \|\, (A_k(\cdot, \nabla u_k), \nabla u_k)_{\mathbb{R}^N} \,\|_{L^1(\Omega)} < +\infty$;
(iv) the exponents α and β are related by the condition

$$1 < \alpha \leq \beta < \begin{cases} +\infty, & \text{if } \alpha \geq N-1,\\ \frac{\alpha(N-1)}{N-1-\alpha}, & \text{if } \alpha < N-1.\end{cases} \tag{24.35}$$

Then, up to a subsequence, the fluxes weakly converge to the flux

$$A_k(\cdot, \nabla u_k) \rightharpoonup A(\cdot, \nabla u) \quad in \;\; L^{\beta'}(\Omega)^N.$$

It is worth to note that in the case of equality $\alpha = \beta$, Theorem 24.3 becomes the well-known result of Tartar and Murat [28] also known as the div-curl Lemma.

24.3 On Approximated Optimal Control Problems in Coefficients and Their Properties

This section deals with the description of the structure of the approximated optimal control problems with respect to the original one (24.1)–(24.6) and the study of their main variational properties.

Let ε be a small parameter. Assume that the parameter ε varies within a strictly decreasing sequence of positive real numbers which converges to 0. Let $\tau \geq 0$ be a given constant. We consider the collection of approximated optimal control problems in coefficients for nonlinear elliptic equations (24.19)–(24.22). For every $\varepsilon > 0$ we denote by $\widehat{\Xi}_\varepsilon$ the set of all feasible points to the problem (24.19)–(24.22).

Definition 24.4 We say that (b, u, θ, p) is a feasible point to the problem (24.19)–(24.22) if $b \in \mathfrak{B}_{ad}$, $p \in \mathfrak{S}_{ad}$, and $u \in W_0^{1,\beta}(\Omega)$ and $\theta \in L^1(\Omega)$ are the weak solutions to the following variational problems

$$\operatorname{div}\left(|\nabla u|^{p(x)-2}\nabla u + \varepsilon|\nabla u|^{\beta-2}\nabla u\right) = \operatorname{div} g \quad \text{in } \mathcal{D}'(\Omega), \tag{24.36}$$

$$-\operatorname{div}(b\nabla\theta) = \operatorname{div}\left[\left(|\nabla u|^{p(x)-2}\nabla u + \varepsilon|\nabla u|^{\beta-2}\nabla u - g\right)u\right] + (g, \nabla u)_{\mathbb{R}^N} \text{ in } \mathcal{D}'(\Omega). \tag{24.37}$$

We begin with the following lemma reflecting the consistency of approximated optimal control problem (24.19)–(24.22).

Lemma 24.3 *Let $\theta_d \in L^1(\Omega)$ and $g \in L^\infty(\Omega)^N$ be given distributions, let $\sigma \in C(\mathbb{R})$ be a given function satisfying the conditions* (24.6), *and let τ be an arbitrary non-negative value. Then the approximated optimal control problem* (24.19)–(24.22) *is consistent for each $\varepsilon > 0$, i.e. $\widehat{\Xi}_\varepsilon \neq emptyset$.*

Proof Let us define the control functions $\widehat{b}$ and $\widehat{p}$ as follows: $\widehat{p}(\cdot) = \beta$ and $\widehat{b}(\cdot) = \xi$, where $\xi \in [m_1, m_2]$. It is clear that $\widehat{b} \in \mathfrak{B}_{ad}$, $\widehat{p} \in \mathfrak{S}_{ad}$, and the variational problem (24.36) becomes a well-posed problem for the Δ_β-Laplacian and admits a unique solution $\widehat{u}_\varepsilon \in W_0^{1,\beta}(\Omega)$. Let us show that there exists a unique element $\widehat{\theta} \in W_0^{1,\gamma}(\Omega)$ for $\gamma \in [1, \frac{N}{N-1})$, satisfying variational equality (24.37). Indeed, as follows from (24.37), we see that the identity

$$\xi \int_\Omega \left(\nabla\widehat{\theta}, \nabla\varphi\right)_{\mathbb{R}^N} dx = \int_\Omega (g, \nabla\widehat{u})_{\mathbb{R}^N} \varphi\, dx - \int_\Omega \left((1+\varepsilon)|\nabla\widehat{u}|^{\beta-2}\nabla\widehat{u} - g, \nabla\varphi\right)_{\mathbb{R}^N} \widehat{u}\, dx \tag{24.38}$$

holds for all $\varphi \in C_0^\infty(\Omega)$ and $k \in \mathbb{N}$. Since $\widehat{u}$ is a solution to (24.36), it leads us to the following relations

$$(1+\varepsilon)\int_\Omega \left(|\nabla\widehat{u}|^{\beta-2}\nabla\widehat{u}, \nabla(\varphi\widehat{u})\right)_{\mathbb{R}^N} dx = \int_\Omega (g, \nabla(\varphi\widehat{u}))_{\mathbb{R}^N}\, dx, \quad \forall\varphi \in C_0^\infty(\Omega).$$

Hence,

$$(1+\varepsilon)\int_\Omega |\nabla\widehat{u}|^\beta \varphi\,dx = \int_\Omega \left(\nabla\varphi, g-(1+\varepsilon)|\nabla\widehat{u}|^{\beta-2}\nabla\widehat{u}\right)_{\mathbb{R}^N}\widehat{u}\,dx$$
$$+\int_\Omega (g,\nabla\widehat{u})_{\mathbb{R}^N}\varphi\,dx,\quad \forall\varphi\in C_0^\infty(\Omega).$$

Combining this equality with (24.38), we arrive at the integral identity

$$\xi\int_\Omega \left(\nabla\widehat{\theta},\nabla\varphi\right)_{\mathbb{R}^N} dx = (1+\varepsilon)\int_\Omega |\nabla\widehat{u}|^\beta\varphi\,dx\quad \forall\varphi\in C_0^\infty(\Omega). \tag{24.39}$$

Thus, $\widehat{\theta}$ is the distributional solution to the Dirichlet problem

$$-\,\xi\,\mathrm{div}\left(\nabla\theta_{\varepsilon,k}\right) = (1+\varepsilon)|\nabla\widehat{u}|^\beta \text{ in } \Omega,\quad \widehat{\theta}\Big|_{\partial\Omega}=0. \tag{24.40}$$

However, since the operator $-\xi\,\mathrm{div}\,(\nabla\theta) = -\xi\Delta\theta$ admits maximal elliptic regularity on $W^{-1,p_0}(\Omega)$ for some $p_0 > N$ (see [12, Sect. 6]), it follows that distributional solution and duality solution to the problem (24.40) are the same (see [27]). Hence, by Stampacchia theorem this solution is unique in $W_0^{1,\gamma}(\Omega)$ for every $\gamma\in[1,\frac{N}{N-1})$ and it can be found via approximation of $|\nabla\widehat{u}|^\beta\in L^1(\Omega)$ by $L^\infty(\Omega)$-functions. Moreover, there exists a constant $C=C(\gamma)$ independent of $|\nabla\widehat{u}|^\beta$ such that (see [29, Theorem 4.1])

$$\int_\Omega |\nabla\widehat{\theta}|^\gamma\,dx \le C(\gamma)\||\nabla\widehat{u}|^\beta\|^\gamma_{L^1(\Omega)},\quad \forall\gamma<\frac{N}{N-1}.$$

Thus, $\widehat{\theta}\in W_0^{1,\gamma}(\Omega)$ and the tuple $(\xi,\widehat{u},\widehat{\theta},\beta)$ is a feasible point to the problem (24.19)–(24.22).

As an obvious consequence of the reasoning given in proof of Lemma 24.3 and the facts that $W^{1,q}(\Omega)\hookrightarrow C^{0,\lambda}(\overline{\Omega})$ continuously for $\lambda = 1-N/q$ and any admissible control in $\mathfrak{B}_{ad}$ can be considered as the perturbation coefficient function for $-\Delta$ which has the maximal regularity property for some $p_0 > N$ [12, Sect. 6], we can draw the following inference.

Corollary 24.1 *For given $\tau\ge 0$ and $\varepsilon>0$, let $(b,u,\theta,p)\in\widehat{\Xi}_\varepsilon$ be a feasible point to the problem* (24.19)–(24.22). *Then θ is the unique duality solution to the problem* (24.40).

The next results are crucial for our further analysis.

Lemma 24.4 *The set of fictitious controls $\mathfrak{S}_{ad}$ is convex, bounded and compact with respect to the strong topology of $C(\overline{\Omega})$.*

Proof Let $\{p_k(\cdot)\}_{k\in\mathbb{N}}\subset\mathfrak{S}_{ad}$ be an arbitrary sequence of fictitious controls. Since $\max_{x\in\overline{\Omega}}|p_k(x)|\le\beta$ and each element of this sequence has the same modulus of continuity ω, it follows that the sequence $\{p_k(\cdot)\}_{k\in\mathbb{N}}$ is uniformly bounded and equicontinuous. Hence, by Arzelà–Ascoli Theorem the set $\{p_k\}_{k\in\mathbb{N}}$ is relatively compact with respect to the strong topology of $C(\overline{\Omega})$. Taking into account that the set $\mathfrak{S}_{ad}$ is closed with respect to the uniform convergence, we can deduce: there exists an element $p \in \mathfrak{S}_{ad}$ such that, up to subsequences, $p_k(\cdot) \to p(\cdot)$ uniformly in $\overline{\Omega}$ as $k \to \infty$. The convexity and boundedness of the set $\mathfrak{S}_{ad}$ obviously follows from its definition.

Lemma 24.5 *Let $\varepsilon > 0$ be a fixed value, and let $\left\{\left(b_{\varepsilon,k}, u_{\varepsilon,k}, \theta_{\varepsilon,k}, p_{\varepsilon,k}\right)\right\}_{k\in\mathbb{N}} \subset \widehat{\Xi}_\varepsilon$ be be an arbitrary sequence of feasible points to OCP* (24.19)–(24.22). *Then the flux $A_{\varepsilon,k}(x,\xi) := |\xi|^{p_{\varepsilon,k}(x)-2}\xi + \varepsilon|\xi|^{\beta-2}\xi$ satisfies the properties* (24.33)–(24.34).

Proof The monotonicity property (24.33) is a direct consequence of the well-known inequalities:

$$\left(|\xi|^{p-2}\xi - |\varepsilon ta|^{p-2}\varepsilon ta, \xi - \varepsilon ta\right)_{\mathbb{R}^N} \geq 2^{2-p}|\xi - \varepsilon ta|^p \quad \text{for } 2 \leq p < +\infty,$$
$$\left(|\xi|^{p-2}\xi - |\varepsilon ta|^{p-2}\varepsilon ta, \xi - \varepsilon ta\right)_{\mathbb{R}^N} \geq (|\xi| + |\varepsilon ta|)^{p-2}\,|\xi - \varepsilon ta|^2 \quad \text{for } 1 < p < 2,$$

whereas the boundedness condition $(24.34)_1$ is ensured by inclusion $p_{\varepsilon,k} \in \mathfrak{S}_{ad}$,

$$|A_{\varepsilon,k}(x,\xi)| \leq |\xi|^{p_{\varepsilon,k}(x)-1} + \varepsilon|\xi|^{\beta-1} \leq (1+\varepsilon)|\xi|^{\beta-1} + 1 \quad \text{a.e. in } \Omega.$$

It remains to establish the pointwise convergence $(24.34)_2$. Due to Lemma 24.4, we can suppose that there exists an element $p_\varepsilon \in \mathfrak{S}_{ad}$ such that, up to subsequences, $p_{\varepsilon,k}(\cdot) \to p_\varepsilon(\cdot)$ uniformly in $\overline{\Omega}$ as $k \to \infty$. As a result, we have

$$\begin{aligned} A_{\varepsilon,k}(x,\xi) &:= |\xi|^{p_{\varepsilon,k}(x)-2}\xi + \varepsilon|\xi|^{\beta-2}\xi \\ &\stackrel{k\to\infty}{\longrightarrow} |\xi|^{p_\varepsilon(x)-2}\xi + \varepsilon|\xi|^{\beta-2}\xi =: A_\varepsilon(x,\xi) \quad \text{a.e. in } \Omega. \end{aligned}$$

Lemma 24.6 *Let $\left\{\left(b_{\varepsilon,k}, u_{\varepsilon,k}, \theta_{\varepsilon,k}, p_{\varepsilon,k}\right)\right\}_{k\in\mathbb{N}} \subset \widehat{\Xi}_\varepsilon$ be an arbitrary sequence. Then there exist a distribution $u_\varepsilon \in W_0^{1,\beta}(\Omega)$, an exponent $p_\varepsilon \in \mathfrak{S}_{ad}$, and a subsequence of $\left\{u_{\varepsilon,k}\right\}_{k\in\mathbb{N}}$, still denoted by the suffix (ε, k), such that*

$$\varepsilon\|u_\varepsilon\|^\beta_{W_0^{1,\beta}(\Omega)} \leq 2^{\alpha'+1}\left(\int_\Omega |g|^{\alpha'}\,dx + |\Omega|\right), \tag{24.41}$$
$$u_{\varepsilon,k} \rightharpoonup u_\varepsilon \text{ in } W_0^{1,\beta}(\Omega) \text{ as } k \to \infty. \tag{24.42}$$

Proof To prove this result, we apply the reasoning line coming from the paper [38]. Since $\left(b_{\varepsilon,k}, u_{\varepsilon,k}, \theta_{\varepsilon,k}, p_{\varepsilon,k}\right) \in \widehat{\Xi}_\varepsilon$ for each $k \in \mathbb{N}$, it follows that $u_{\varepsilon,k} \in W_0^{1,\beta}(\Omega)$ and the integral identity

$$\int_\Omega \left(|\nabla u_{\varepsilon,k}|^{p_{\varepsilon,k}-2}\nabla u_{\varepsilon,k} + \varepsilon|\nabla u_{\varepsilon,k}|^{\beta-2}\nabla u_{\varepsilon,k}, \nabla\varphi\right)_{\mathbb{R}^N} dx = \int_\Omega (g, \nabla\varphi)_{\mathbb{R}^N}\,dx \tag{24.43}$$

holds for all $\varphi \in C_0^\infty(\Omega)$. Hence, by density of $C_0^\infty(\Omega)$ in $W_0^{1,\beta}(\Omega)$, we obtain the energy equality

$$\int_\Omega \left(|\nabla u_{\varepsilon,k}|^{p_{\varepsilon,k}} + \varepsilon|\nabla u_{\varepsilon,k}|^\beta\right) dx = \int_\Omega \left(g, \nabla u_{\varepsilon,k}\right)_{\mathbb{R}^N} dx. \tag{24.44}$$

By the Young inequality, we have

$$\begin{aligned}\int_\Omega \left(g, \nabla u_{\varepsilon,k}\right)_{\mathbb{R}^N} dx &\leq \frac{2^{\alpha'}}{\beta'} \int_\Omega |g|^{p'_{\varepsilon,k}}\, dx + \frac{1}{\alpha 2^\alpha} \int_\Omega |\nabla u_{\varepsilon,k}|^{p_{\varepsilon,k}}\, dx \\ &\leq 2^{\alpha'} \int_\Omega |g|^{p'_{\varepsilon,k}}\, dx + \frac{1}{2} \int_\Omega |\nabla u_{\varepsilon,k}|^{p_{\varepsilon,k}}\, dx,\end{aligned}$$

where $\alpha' = \alpha/(\alpha - 1)$. Then it is easy to derive from (24.44) the following estimates

$$\begin{aligned}\int_\Omega \left(|\nabla u_{\varepsilon,k}|^{p_{\varepsilon,k}} + \varepsilon|\nabla u_{\varepsilon,k}|^\beta\right) dx &\leq 2^{\alpha'+1} \int_\Omega |g|^{p'_{\varepsilon,k}}\, dx \\ &\leq 2^{\alpha'+1} \left(\int_\Omega |g|^{\alpha'}\, dx + |\Omega|\right).\end{aligned} \tag{24.45}$$

Thus, the sequence $\left\{u_{\varepsilon,k}\right\}_{k\in\mathbb{N}}$ is relatively compact with respect to the weak topology of $W_0^{1,\beta}(\Omega)$. Without loss of generality, we assume that the weak convergence $u_{\varepsilon,k} \rightharpoonup u_\varepsilon$ takes place in $W_0^{1,\beta}(\Omega)$. Then it follows from (24.44) that

$$\begin{aligned}\lim_{k\to\infty} \int_\Omega \left(|\nabla u_{\varepsilon,k}|^{p_{\varepsilon,k}} + \varepsilon|\nabla u_{\varepsilon,k}|^\beta\right) dx &= \lim_{k\to\infty} \int_\Omega \left(g, \nabla u_{\varepsilon,k}\right)_{\mathbb{R}^N} dx \\ &= \int_\Omega (g, \nabla u_\varepsilon)_{\mathbb{R}^N}\, dx, \\ \varepsilon \int_\Omega |\nabla u_\varepsilon|^\beta\, dx &\leq \int_\Omega (g, \nabla u_\varepsilon)_{\mathbb{R}^N}\, dx\end{aligned} \tag{24.46}$$

because, according to the property of lower semicontinuity of $\|\cdot\|_{W_0^{1,\beta}(\Omega)}$-norm with respect to the weak convergence, we have

$$\lim_{k\to\infty} \int_\Omega |\nabla u_{\varepsilon,k}|^\beta\, dx \geq \liminf_{k\to\infty} \int_\Omega |\nabla u_{\varepsilon,k}|^\beta\, dx \geq \int_\Omega |\nabla u_\varepsilon|^\beta\, dx. \tag{24.47}$$

Thus, the estimate (24.41) immediately follows from (24.45)–(24.47).

Lemma 24.7 *Let* $\left\{\left(b_{\varepsilon,k}, u_{\varepsilon,k}, \theta_{\varepsilon,k}, p_{\varepsilon,k}\right)\right\}_{k\in\mathbb{N}} \subset \widehat{\Xi}_\varepsilon$ *be an arbitrary sequence, and let* $p_\varepsilon \in \mathfrak{S}_{ad}$ *and* $u_\varepsilon \in W_0^{1,\beta}(\Omega)$ *be such that*

$$p_{\varepsilon,k}(\cdot) \to p_\varepsilon(\cdot) \text{ uniformly in } \overline{\Omega} \quad \text{and} \quad u_{\varepsilon,k} \rightharpoonup u_\varepsilon \text{ in } W_0^{1,\beta}(\Omega) \text{ as } k \to \infty. \tag{24.48}$$

Then, up to a subsequence, we have the weak convergence of fluxes to a flux:

$$|\nabla u_{\varepsilon,k}|^{p_{\varepsilon,k}-2}\nabla u_{\varepsilon,k}+\varepsilon|\nabla u_{\varepsilon,k}|^{\beta-2}\nabla u_{\varepsilon,k} \rightharpoonup |\nabla u_\varepsilon|^{p_\varepsilon-2}\nabla u_\varepsilon+\varepsilon|\nabla u_\varepsilon|^{\beta-2}\nabla u_\varepsilon \quad in \quad L^{\beta'}(\Omega)^N. \tag{24.49}$$

Proof In order to prove the convergence (24.49), we show that all conditions of Theorem 24.3 are fulfilled. Taking into account Lemma 24.5 and the fact that $\frac{\alpha N}{N-\alpha}<\frac{\alpha(N-1)}{N-1-\alpha}$, we focus on the verification of conditions (ii)–(iii). Indeed, in view of the weak convergence $u_{\varepsilon,k}\rightharpoonup u_\varepsilon$ in $W_0^{1,\beta}(\Omega)$ and

$$\begin{aligned}
\|A_{1,k}(\cdot,\nabla u_{\varepsilon,k})\|_{L^{\beta'}(\Omega)^N} &\overset{\text{by (24.29)}}{\leq} (1+|\Omega|)^{1/\beta'}\,\|A_{1,k}(\cdot,\nabla u_{\varepsilon,k})\|_{L^{p'_{\varepsilon,k}(\cdot)}(\Omega)^N}\\
&\overset{\text{by (24.28)}}{\leq} (1+|\Omega|)^{1/\beta'}\left(\int_\Omega |A_{1,k}(x,\nabla u_{\varepsilon,k}(x))|^{p'_{\varepsilon,k}(x)}\,dx+1\right)^{1/\beta'}\\
&= (1+|\Omega|)^{1/\beta'}\left(\int_\Omega |\nabla u_{\varepsilon,k}(x)|^{p_{\varepsilon,k}(x)}\,dx+1\right)^{1/\beta'}\\
&\leq (1+|\Omega|)^{1/\beta'}\left(\int_\Omega |\nabla u_{\varepsilon,k}(x)|^{\beta}\,dx+|\Omega|+1\right)^{1/\beta'} < +\infty,
\end{aligned} \tag{24.50}$$

$$\|A_{2,k}(\cdot,\nabla u_{\varepsilon,k})\|_{L^{\beta'}(\Omega)^N} = \left(\int_\Omega |\nabla u_{\varepsilon,k}(x)|^{\beta}\,dx\right)^{1/\beta'}, \tag{24.51}$$

where

$$\begin{aligned}
&A_{\varepsilon,k}(x,\nabla u_{\varepsilon,k}) := A_{1,k}(x,\nabla u_{\varepsilon,k})+\varepsilon A_{2,k}(x,\nabla u_{\varepsilon,k}),\\
&A_{1,k}(x,\nabla u_{\varepsilon,k}) := |\nabla u_{\varepsilon,k}|^{p_{\varepsilon,k}(x)-2}\nabla u_{\varepsilon,k},\quad A_{2,k}(x,\nabla u_{\varepsilon,k}) := |\nabla u_{\varepsilon,k}|^{\beta-2}\nabla u_{\varepsilon,k},
\end{aligned}$$

we see that the fluxes $A_{\varepsilon,k}(\cdot,\nabla u_{\varepsilon,k})$ are bounded in $L^{\beta'}(\Omega)^N$. To check the condition (iii) of Theorem 24.3, it is enough to apply the estimate (24.41) and note that

$$\begin{aligned}
\|\left(A_{\varepsilon,k}(\cdot,\nabla u_{\varepsilon,k}),\nabla u_{\varepsilon,k}\right)_{\mathbb{R}^N}\|_{L^1(\Omega)} &= \int_\Omega |\nabla u_{\varepsilon,k}|^{p_{\varepsilon,k}}\,dx+\varepsilon\int_\Omega |\nabla u_{\varepsilon,k}|^{\beta}\,dx\\
&\leq (1+\varepsilon)\sup_{k\in\mathbb{N}}\left(\int_\Omega |\nabla u_{\varepsilon,k}|^{\beta}\,dx\right)+|\Omega|.
\end{aligned} \tag{24.52}$$

Thus, the weak convergence of fluxes to a flux (24.49) follows from Theorem 24.3.

Lemma 24.8 *Let $p_\varepsilon\in\mathfrak{S}_{ad}$ and $u_\varepsilon\in W_0^{1,\beta}(\Omega)$ be as in Lemma 24.7. Then u_ε is the unique weak solution to the Dirichlet problem*

$$\operatorname{div}\left(|\nabla u|^{p_\varepsilon(x)-2}\nabla u+\varepsilon|\nabla u|^{\beta-2}\nabla u\right)=\operatorname{div} g \quad in \quad \Omega,\quad u|_{\partial\Omega}=0. \tag{24.53}$$

Proof Since $\{(b_{\varepsilon,k}, u_{\varepsilon,k}, \theta_{\varepsilon,k}, p_{\varepsilon,k})\}_{k\in\mathbb{N}}$ are feasible points to OCP (24.19)–(24.22), it follows that the integral identity (24.43) holds true for all $\varphi \in C_0^\infty(\Omega)$ and for all $k \in \mathbb{N}$. Then (24.49) implies that that

$$\begin{aligned}\int_\Omega \left(|\nabla u_\varepsilon|^{p_\varepsilon(x)-2}\nabla u_\varepsilon + \varepsilon|\nabla u_\varepsilon|^{\beta-2}\nabla u_\varepsilon, \nabla\varphi\right)_{\mathbb{R}^N} dx \\ = \lim_{k\to\infty}\int_\Omega \left(|\nabla u_{\varepsilon,k}|^{p_{\varepsilon,k}(x)-2}\nabla u_{\varepsilon,k} + \varepsilon|\nabla u_{\varepsilon,k}|^{\beta-2}\nabla u_{\varepsilon,k}, \nabla\varphi\right)_{\mathbb{R}^N} dx \\ = \int_\Omega (g, \nabla\varphi)_{\mathbb{R}^N}\, dx, \quad \forall\,\varphi \in C_0^\infty(\Omega). \end{aligned} \tag{24.54}$$

Hence, u_ε is a weak solution to the boundary value problem (24.53). In view of the strict monotonicity of the operator $\mathcal{A}_\varepsilon : W_0^{1,\beta}(\Omega) \to W^{-1,\beta'}(\Omega)$, given by the equality

$$(\mathcal{A}_\varepsilon u, v) = \int_\Omega \left(|\nabla u|^{p(x)-2}\nabla u + \varepsilon|\nabla u|^{\beta-2}\nabla u, \nabla v\right)_{\mathbb{R}^N} dx, \quad \forall\, v \in W_0^{1,\beta}(\Omega), \tag{24.55}$$

this solution is unique.

The next results are based on the compactness property of the set of controls $\mathfrak{B}_{ad}$ with respect to the uniform topology of $C(\overline{\Omega})$. For other types of admissible controls in coefficients, we refer to [6–9, 15, 23]. We recall that in accordance with Morrey's inequality and Sobolev embedding theorem, there exists a constant C, depending only on $q > N$, Ω, and N, such that [13, Sect. 5.6]

$$\|y\|_{C^{0,\lambda}(\overline{\Omega})} \le C\|y\|_{W^{1,q}(\Omega)}, \quad \forall\, y \in W^{1,q}(\Omega) \text{ with } \lambda = 1 - N/q. \tag{24.56}$$

So, we may always suppose that if $y \in W^{1,q}(\Omega)$, then y is in fact a Hölder continuous function, after possibly being redefined on a set of measure zero. Moreover, each uniformly bounded set in $W^{1,q}(\Omega)$-norm is relatively compact with respect to the uniform topology of $C^{0,\lambda}(\overline{\Omega})$.

Since the set $\{b \in W^{1,q}(\Omega) : m_1 \le b \le m_2, \text{ in } \Omega\}$ is closed with respect to the pointwise convergence, the following property of the class of admissible controls $\mathfrak{B}_{ad}$ holds true.

Proposition 24.1 *For any given $m_1 > 0$, $m_2 > m_1$, and $\mu > 0$, the set $\mathfrak{B}_{ad}$ is nonempty, uniformly bounded, convex, and sequentially compact with respect to the uniform topology of $C^{0,\lambda}(\overline{\Omega})$.*

Lemma 24.9 *Let $\{(b_{\varepsilon,k}, u_{\varepsilon,k}, \theta_{\varepsilon,k}, p_{\varepsilon,k})\}_{k\in\mathbb{N}} \subset \widehat{\Xi}_\varepsilon$ be a sequence such that $b_{\varepsilon,k} \rightharpoonup b_\varepsilon$ in $W^{1,q}(\Omega)$ and $\theta_{\varepsilon,k} \rightharpoonup \theta_\varepsilon$ in $W_0^{1,\gamma}(\Omega)$ for some $\gamma \in [1, \frac{N}{N-1})$. Then we have*

$$\lim_{k\to\infty}\int_\Omega b_{\varepsilon,k}\left(\nabla\theta_{\varepsilon,k}, \nabla\varphi\right)_{\mathbb{R}^N} dx = \int_\Omega b_\varepsilon(\nabla\theta_\varepsilon, \nabla\varphi)_{\mathbb{R}^N}\, dx, \quad \forall\varphi \in C_0^\infty(\Omega). \tag{24.57}$$

Proof Since $b_{\varepsilon,k} \to b_\varepsilon$ in $C(\overline{\Omega})$ and the sequence $\{b_{\varepsilon,k}\}_{k\in\mathbb{N}}$ is bounded in $L^\infty(\Omega)$, we infer $b_{\varepsilon,k} \to b_\varepsilon$ strongly in $L^r(\Omega)$ for every $1 \le r < +\infty$. In particular, $b_{\varepsilon,k}\nabla\varphi \to b_\varepsilon\nabla\varphi$ in $L^{\gamma'}(\Omega)$ with $\gamma' = \frac{\gamma}{\gamma-1}$ and $\nabla\theta_{\varepsilon,k} \rightharpoonup \nabla\theta_\varepsilon$ in $L^\gamma(\Omega)^N$. Hence, it is immediate to pass to the limit and to deduce (24.57).

The next lemma reveals some compactness properties of the set of feasible solutions $\widehat{\Xi}_\varepsilon$.

Lemma 24.10 *Let* $\left\{\left(b_{\varepsilon,k}, u_{\varepsilon,k}, \theta_{\varepsilon,k}, p_{\varepsilon,k}\right)\right\}_{k\in\mathbb{N}} \subset \widehat{\Xi}_\varepsilon$ *be a sequence of feasible points for OCP* (24.19)–(24.22). *Then there exist* $\delta > 0$ *and distributions* $p_\varepsilon \in \mathfrak{S}_{ad}$, $u_\varepsilon \in W_0^{1,\beta}(\Omega)$, $\theta_\varepsilon \in W_0^{1,\gamma}(\Omega)$ *for* $\gamma \in [1, \frac{N}{N-1})$, *and a control* $b_\varepsilon \in \mathfrak{B}_{ad}$ *such that* $|\nabla u_\varepsilon|^\beta \in L^{1+\delta}(\Omega)$ *and, up to subsequences,*

$$p_{\varepsilon,k}(\cdot) \to p_\varepsilon(\cdot) \text{ uniformly in } \overline{\Omega}, \quad u_{\varepsilon,k} \rightharpoonup u_\varepsilon \text{ in } W_0^{1,\beta}(\Omega), \tag{24.58}$$

$$b_{\varepsilon,k} \rightharpoonup b_\varepsilon \text{ in } W^{1,q}(\Omega), \quad \theta_{\varepsilon,k} \rightharpoonup \theta_\varepsilon \text{ in } W_0^{1,\gamma}(\Omega), \tag{24.59}$$

where θ_ε *is a weak solution to the Dirichlet problem*

$$-\operatorname{div}(b_\varepsilon\nabla\theta) = \operatorname{div}\left[\left(|\nabla u_\varepsilon|^{p_\varepsilon(x)-2}\nabla u_\varepsilon + \varepsilon|\nabla u_\varepsilon|^{\beta-2}\nabla u_\varepsilon - g\right)u_\varepsilon\right] + (g, \nabla u_\varepsilon)_{\mathbb{R}^N}, \tag{24.60}$$

$$\theta|_{\partial\Omega} = 0 \tag{24.61}$$

satisfying the integral identity

$$\int_\Omega (b_\varepsilon\nabla\theta_\varepsilon, \nabla\varphi)_{\mathbb{R}^N}\,dx = \int_\Omega \left(|\nabla u_\varepsilon|^{p_\varepsilon(x)} + \varepsilon|\nabla u_\varepsilon|^\beta\right)\varphi\,dx, \quad \forall\varphi \in C_0^\infty(\Omega). \tag{24.62}$$

Moreover, if $N > 2$ *then* $\theta_\varepsilon \in W_0^{1,q}(\Omega)$ *for*

$$q = \frac{N(1+\delta)}{N-1-\delta} = \frac{N}{N-1}\left(1 + \frac{N\delta}{N-1-\delta}\right) \quad \textit{provided } \delta \in \left(0, \frac{N-2}{N+2}\right).$$

Proof To begin with, we note that the convergence $(24.59)_1$ and inclusion $b_\varepsilon \in \mathfrak{B}_{ad}$ is a direct consequence of Proposition 24.1, whereas (24.58) follows from Lemmas 24.4–24.6. In view of the initial assumptions, the distribution $\theta_{\varepsilon,k}$ is the weak solution to (24.21) for each $k \in \mathbb{N}$, i.e. the identity

$$\int_\Omega \left(b_{\varepsilon,k}\nabla\theta_{\varepsilon,k}, \nabla\varphi\right)_{\mathbb{R}^N}dx = \int_\Omega \left(g, \nabla u_{\varepsilon,k}\right)_{\mathbb{R}^N}\varphi\,dx - \int_\Omega \left(|\nabla u_{\varepsilon,k}|^{p_{\varepsilon,k}(x)-2}\nabla u_{\varepsilon,k} + \varepsilon|\nabla u_{\varepsilon,k}|^{\beta-2}\nabla u_{\varepsilon,k} - g, \nabla\varphi\right)_{\mathbb{R}^N} u_{\varepsilon,k}\,dx \tag{24.63}$$

holds for all $\varphi \in C_0^\infty(\Omega)$ and $k \in \mathbb{N}$. Hence, by analogy with the proof of Lemma 24.3, we deduce that the integral identity

$$\int_{\Omega} \left(b_{\varepsilon,k}\nabla\theta_{\varepsilon,k}, \nabla\varphi\right)_{\mathbb{R}^N} dx = \int_{\Omega} \left(|\nabla u_{\varepsilon,k}|^{p_{\varepsilon,k}(x)} + \varepsilon|\nabla u_{\varepsilon,k}|^{\beta}\right) \varphi\, dx \tag{24.64}$$

holds for every $\varphi \in C_0^{\infty}(\Omega)$. Hence, $\theta_{\varepsilon,k}$ is the distributional solution to the Dirichlet problem

$$-\operatorname{div}\left(b_{\varepsilon,k}\nabla\theta_{\varepsilon,k}\right) = |\nabla u_{\varepsilon,k}|^{p_{\varepsilon,k}(x)} + \varepsilon|\nabla u_{\varepsilon,k}|^{\beta} \text{ in } \Omega, \quad \theta_{\varepsilon,k}\big|_{\partial\Omega} = 0. \tag{24.65}$$

Let us show that this fact implies the boundedness of the sequence $\left\{\theta_{\varepsilon,k}\right\}_{k\in\mathbb{N}}$ in $W_0^{1,\gamma}(\Omega)$ for $\gamma \in [1, \frac{N}{N-1})$. Indeed, since the distributional solution and duality solution to the problem (24.65) are the same, it follows that $\theta_{\varepsilon,k}$ is the duality solution to (24.65) and by Stampacchia theorem this solution is unique, $\theta_{\varepsilon,k}$ belongs to $W_0^{1,\gamma}(\Omega)$, for every $\gamma \in [1, \frac{N}{N-1})$, and there exists a constant C independent of $f_{\varepsilon,k}$ and k such that

$$\int_{\Omega} |\nabla\theta_{\varepsilon,k}|^{\gamma}\, dx \le C\|f_{\varepsilon,k}\|^{\gamma}_{L^1(\Omega)}.$$

In fact, the constant C depends only on N, γ, and m_1. Since

$$\begin{aligned}\int_{\Omega} |f_{\varepsilon,k}|\, dx &= \int_{\Omega} \left(|\nabla u_{\varepsilon,k}|^{p_{\varepsilon,k}(x)} + \varepsilon|\nabla u_{\varepsilon,k}|^{\beta}\right) dx \\ &\overset{\text{by (24.52)}}{\le} \left((1+\varepsilon)\sup_{k\in\mathbb{N}}\left(\int_{\Omega} |\nabla u_{\varepsilon,k}|^{\beta}\, dx\right) + |\Omega|\right),\end{aligned}$$

it follows that

$$\|\theta_{\varepsilon,k}\|_{W_0^{1,\gamma}(\Omega)} \le C^{\frac{1}{\gamma}} \left((1+\varepsilon)\sup_{k\in\mathbb{N}}\left(\int_{\Omega} |\nabla u_{\varepsilon,k}|^{\beta}\, dx\right) + |\Omega|\right), \quad \forall k \in \mathbb{N}. \tag{24.66}$$

Thus, the weak convergence $(24.59)_2$ immediately follows from estimate (24.66). It remains to show that the limit function θ_{ε} is a weak solution to the problem (24.60) and satisfies the identity (24.62). With that in mind, we note that

$$\begin{aligned}&|\nabla u_{\varepsilon,k}|^{p_{\varepsilon,k}-2}\nabla u_{\varepsilon,k} + |\nabla u_{\varepsilon,k}|^{\beta-2}\nabla u_{\varepsilon,k} - g \\ &\qquad \rightharpoonup |\nabla u_{\varepsilon}|^{p_{\varepsilon}-2}\nabla u_{\varepsilon} + \varepsilon|\nabla u_{\varepsilon}|^{\beta-2}\nabla u_{\varepsilon} - g \text{ in } L^{\beta'}(\Omega)^N, \\ &u_{\varepsilon,k} \to u_{\varepsilon} \text{ in } L^{\beta}(\Omega), \\ &\left(g, \nabla u_{\varepsilon,k}\right)_{\mathbb{R}^N} \to (g, \nabla u_{\varepsilon})_{\mathbb{R}^N} \text{ in } L^{\beta}(\Omega)\end{aligned}$$

by Lemma 24.7 and compactness of the embedding $W_0^{1,\beta}(\Omega) \hookrightarrow L^{\beta}(\Omega)$. Hence,

$$\begin{aligned}&\left(|\nabla u_{\varepsilon,k}|^{p_{\varepsilon,k}-2}\nabla u_{\varepsilon,k} + \varepsilon|\nabla u_{\varepsilon,k}|^{\beta-2}\nabla u_{\varepsilon,k} - g\right) u_{\varepsilon,k} \\ &\qquad \rightharpoonup \left(|\nabla u_{\varepsilon}|^{p_{\varepsilon}-2}\nabla u_{\varepsilon} + \varepsilon|\nabla u_{\varepsilon}|^{\beta-2}\nabla u_{\varepsilon} - g\right) u_{\varepsilon} \text{ in } L^1(\Omega).\end{aligned}$$

Taking these facts and Lemmas 24.4 and 24.9 into account, we can pass to the limit in (24.63) as $k \to \infty$. As a result, we obtain

$$\int_\Omega (b_\varepsilon \nabla\theta_\varepsilon, \nabla\varphi)_{\mathbb{R}^N}\, dx = \int_\Omega (g, \nabla u_\varepsilon)_{\mathbb{R}^N}\, \varphi\, dx - \int_\Omega \left(|\nabla u_\varepsilon|^{p_\varepsilon-2}\nabla u_\varepsilon + \varepsilon|\nabla u_\varepsilon|^{\beta-2}\nabla u_\varepsilon - g, \nabla\varphi\right)_{\mathbb{R}^N} u_\varepsilon\, dx, \quad (24.67)$$

i.e.

$$-\operatorname{div}(b_\varepsilon\nabla\theta_\varepsilon) = \operatorname{div}\left[\left(|\nabla u_\varepsilon|^{p_\varepsilon(x)-2}\nabla u_\varepsilon + \varepsilon|\nabla u_\varepsilon|^{\beta-2}\nabla u_\varepsilon - g\right) u_\varepsilon\right] + (g, \nabla u_\varepsilon)_{\mathbb{R}^N}$$

in the sense of distributions. In order to establish the integral identity (24.62), it is enough to observe that $u_\varepsilon \in W_0^{1,\beta}(\Omega)$ is the unique weak solution to the Dirichlet problem (24.53) (see Lemma 24.8) and apply then the transformations, that we used in the proof of Lemma 24.3, to the identity (24.67). As for the inclusions $|\nabla u_\varepsilon|^\beta \in L^{1+\delta}(\Omega)$ for $N \ge 2$ and $\theta_\varepsilon \in W_0^{1,q}(\Omega)$ for $N > 2$ and $q = \frac{N(1+\delta)}{N-1-\delta}$, we should apply the arguments of the higher integrability of the gradient (see Remark 24.3 for the details) and Theorem 4.4 in [29].

To conclude this section, we give the existence result for the approximated OCP (24.19)–(24.22).

Theorem 24.4 *Let $\theta_d \in L^1(\Omega)$ and $g \in L^\infty(\Omega)^N$ be given distributions, let σ be a given measurable function satisfying the conditions* (24.6), *and let τ be an arbitrary non-negative value. Then optimal control problem* (24.19)–(24.22) *admits at least one solution for each $\varepsilon > 0$.*

Proof Since the set $\widehat{\Xi}_\varepsilon$ is nonempty (see Lemma 24.3) and the cost functional $J_{\varepsilon,\tau}$ is bounded from below on $\widehat{\Xi}_\varepsilon$, it follows that there exists a minimizing sequence

$$\left\{\left(b_{\varepsilon,k}, u_{\varepsilon,k}, \theta_{\varepsilon,k}, p_{\varepsilon,k}\right)\right\}_{k\in\mathbb{N}} \subset \widehat{\Xi}_\varepsilon$$

to problem (24.19)–(24.22), i.e.

$$\inf_{(b,u,\theta,p)\in\widehat{\Xi}_\varepsilon} J_{\varepsilon,\tau}(b,u,\theta,p) = \lim_{k\to\infty}\left[\int_\Omega |\theta_{\varepsilon,k}(x) - \theta_d(x)|\, dx + \frac{1}{\varepsilon}\mu_\tau\left(\int_\Omega |p_{\varepsilon,k}(x) - \sigma(\theta_{\varepsilon,k}(x))|^2\, dx\right)\right] < +\infty.$$

Hence, in view of Lemma 24.6 and definition of the sets $\mathfrak{B}_{ad}$ and $\mathfrak{S}_{ad}$, the minimizing sequence $\left\{\left(b_{\varepsilon,k}, u_{\varepsilon,k}, \theta_{\varepsilon,k}, p_{\varepsilon,k}\right)\right\}_{k\in\mathbb{N}}$ is bounded in $W^{1,q}(\Omega) \times W_0^{1,\beta}(\Omega) \times W^{1,\gamma}(\Omega) \times C(\overline{\Omega})$. From (24.42) and Lemmas 24.4 and 24.8–24.10, we deduce the existence of a subsequence, that we denote in the same way, and a tuple $\left(b_\varepsilon^0, u_\varepsilon^0, \theta_\varepsilon^0, p_\varepsilon^0\right) \in \widehat{\Xi}_\varepsilon$ such that $b_{\varepsilon,k} \rightharpoonup b_\varepsilon^0$ in $W^{1,q}(\Omega)$, $u_{\varepsilon,k} \rightharpoonup u_\varepsilon^0$ in $W_0^{1,\beta}(\Omega)$,

$\theta_{\varepsilon,k} \rightharpoonup \theta_\varepsilon^0$ in $W_0^{1,\gamma}(\Omega)$, and $p_{\varepsilon,k}(\cdot) \to p_\varepsilon^0(\cdot)$ uniformly in $\overline{\Omega}$. Then taking into account the compact embedding $W_0^{1,\gamma}(\Omega) \hookrightarrow L^q(\Omega)$ for all $q \in [1, \frac{N}{N-2})$, we have $\theta_{\varepsilon,k}(\cdot) \to \theta_\varepsilon^0(\cdot)$ in $L^1(\Omega)$. So, we can suppose that $\theta_{\varepsilon,k}(\cdot) \to \theta_\varepsilon^0(\cdot)$ almost everywhere in Ω. Hence, in view of the fact that $\mu_\tau \in C_{loc}(\mathbb{R}_+)$, we have

$$\liminf_{k\to\infty} \int_\Omega |p_{\varepsilon,k}(x) - \sigma(\theta_{\varepsilon,k}(x))|^2\, dx \geq \int_\Omega |p_\varepsilon^0(x) - \sigma(\theta_\varepsilon^0(x))|^2\, dx.$$

As a result, we finally infer

$$\inf_{(b,u,\theta,p)\in\widehat{\Xi}_\varepsilon} J_{\varepsilon,\tau}(b,u,\theta,p) = \lim_{k\to\infty} J_{\varepsilon,\tau}\left(b_{\varepsilon,k}, u_{\varepsilon,k}, \theta_{\varepsilon,k}, p_{\varepsilon,k}\right) \geq J_{\varepsilon,\tau}\left(b_\varepsilon^0, u_\varepsilon^0, \theta_\varepsilon^0, p_\varepsilon^0\right).$$

Thus, $\left(b_\varepsilon^0, u_\varepsilon^0, \theta_\varepsilon^0, p_\varepsilon^0\right)$ is a solution of the approximated OCP (24.19)–(24.22).

As a consequence of Theorem 24.4, Lemma 24.10, and Corollary 24.1, it is worth to emphasize the following property of optimal solutions.

Corollary 24.2 *If $\left(b_\varepsilon^0, u_\varepsilon^0, \theta_\varepsilon^0, p_\varepsilon^0\right) \in \Xi_\varepsilon$ is an optimal solution to the approximated OCP (24.19)–(24.22), then $\theta_\varepsilon^0 \in W_0^{1,\gamma}(\Omega)$ with $\gamma \in [1, \frac{N}{N-1})$ is a unique duality solution to the Dirichlet problem*

$$-\operatorname{div}\left(b_\varepsilon^0 \nabla \theta_\varepsilon^0\right) = |\nabla u_\varepsilon^0|^{p_\varepsilon^0(x)} + \varepsilon |\nabla u_\varepsilon^0|^\beta \text{ in } \Omega, \quad \theta_\varepsilon^0\big|_{\partial\Omega} = 0 \tag{24.68}$$

where u_ε^0 is the unique weak solution to boundary value problem (24.20) with $p = p_\varepsilon^0$.

Remark 24.6 As follows from the proof of Lemma 24.7 and Theorem 24.3, the existence result for approximated OCPs (24.19)–(24.22) in the form of Theorem 24.4 remains valid even if we omit the condition (24.22) on the parameters α and β because in this case, instead of Theorem 24.3, we can apply the celebrated div-curl Lemma of Tartar and Murat [28].

24.4 Asymptotic Analysis of the Approximated OCP (24.19)–(24.22) as $\varepsilon \to 0$

Our main intention in this section is to show that quasi-optimal solutions to the OCP (24.1)–(24.6) can be attained (in some sense) by optimal solutions to the approximated problems (24.19)–(24.22). In order to do it, we do not use the concept of variational convergence of constrained minimization problems (see [20, 22, 24]) but rather apply the direct analysis to the study of asymptotic behaviour of optimal solutions for OCPs (24.19)–(24.22) as $\varepsilon \to 0$.

Proposition 24.2 *Let $\left\{(b_\varepsilon^0, u_\varepsilon^0, \theta_\varepsilon^0, p_\varepsilon^0) \in \widehat{\Xi}_\varepsilon\right\}_{\varepsilon>0}$ be a sequence of optimal solutions to the approximated problems (24.19)–(24.22) when the small parameter $\varepsilon > 0$*

varies in a strictly decreasing sequence of positive numbers which converges to 0. *Assume there exists a constant* $C^* > 0$ *such that*

$$\limsup_{\varepsilon\to 0} \inf_{(b,u,\theta,p)\in\widehat{\Xi}_\varepsilon} J_{\varepsilon,\tau}(b,u,\theta,p) \le C^* < +\infty. \tag{24.69}$$

Then there is a subsequence of $\left\{(b_\varepsilon^0, u_\varepsilon^0, \theta_\varepsilon^0, p_\varepsilon^0)\right\}_{\varepsilon>0}$, *still denoted by the suffix* ε, *and distributions* $b^0 \in \mathfrak{B}_{ad}$, $p^0 \in \mathfrak{S}_{ad}$, $u^0 \in W_0^{1,\alpha}(\Omega)$, *and* $\theta^0 \in W_0^{1,\gamma}(\Omega)$ *such that*

$$b_\varepsilon^0 \rightharpoonup b^0 \ \text{in } W^{1,q}(\Omega), \quad u_\varepsilon^0 \rightharpoonup u^0 \ \text{in } W_0^{1,\alpha}(\Omega), \tag{24.70}$$

$$\theta_\varepsilon^0 \rightharpoonup \theta^0 \ \text{in } W_0^{1,\gamma}(\Omega), \quad p_\varepsilon^0(\cdot) \to p^0(\cdot) \ \text{uniformly on } \overline{\Omega}, \tag{24.71}$$

$$\|p^0 - \sigma(\theta^0)\|_{L^2(\Omega)} \le \tau. \tag{24.72}$$

Proof Having applied the estimates (24.41) and (24.45) to elements of the sequence of optimal solutions, we obtain

$$\begin{aligned}
\int_\Omega |\nabla u_\varepsilon^0|^\alpha\,dx &\overset{\text{by (24.29)}}{\le} (1+|\Omega|)\,\|\nabla u_\varepsilon^0\|^\alpha_{L^{p_\varepsilon^0(\cdot)}(\Omega)^N}\\
&\overset{\text{by (24.28)}}{\le} (1+|\Omega|)\left(\int_\Omega |\nabla u_\varepsilon^0|^{p_\varepsilon^0}\,dx + 1\right)\\
&\overset{\text{by (24.45)}}{\le} (1+|\Omega|)\left(2^{\alpha'+1}\int_\Omega |g|^{(p_\varepsilon^0)'}\,dx + 1\right)\\
&\le (1+|\Omega|)\left(2^{\alpha'+1}\left(\int_\Omega |g|^{\alpha'}\,dx + |\Omega|\right) + 1\right),
\end{aligned} \tag{24.73}$$

$$\varepsilon\int_\Omega |\nabla u_\varepsilon^0|^\beta\,dx \le 2^{\alpha'+1}\left(\int_\Omega |g|^{\alpha'}\,dx + |\Omega|\right). \tag{24.74}$$

Hence, the sequence $\left\{u_\varepsilon^0\right\}_{\varepsilon>0}$ is relatively compact with respect to the weak topology of $W_0^{1,\alpha}(\Omega)$. Therefore, we can suppose that the convergence $(24.70)_2$ is valid. Moreover, in view of estimates (24.73)–(24.74), it is easy to conclude that the right-hand side of the equation (24.68) is L^1-bounded. In view of (24.66), the distributional solution θ_ε^0 to this equation is obeyed to the estimate

$$\begin{aligned}
\|\theta_\varepsilon^0\|_{W_0^{1,\gamma}(\Omega)} &\le C(\gamma)\int_\Omega \left(|\nabla u_\varepsilon^0|^{p_\varepsilon^0(x)} + \varepsilon|\nabla u_\varepsilon^0|^\beta\right)dx\\
&\overset{\text{by (24.73)–(24.74)}}{\le} C(\gamma)2^{\alpha'+2}\left(\int_\Omega |g|^{\alpha'}\,dx + |\Omega|\right).
\end{aligned}$$

Thus, there exists a distribution $\theta^0 \in W_0^{1,\gamma}(\Omega)$ satisfying the property $(24.71)_1$. Therefore, up to a subsequence, we have the pointwise convergence

$$\sigma(\theta_\varepsilon^0) \to \sigma(\theta^0) \ \text{a.e. in } \Omega \ \text{as } \varepsilon \to 0. \tag{24.75}$$

By Lemma 24.4 the family $\{p_{\varepsilon,k}\}_{k\in\mathbb{N}} \subset \mathfrak{S}_{ad}$ is relatively compact in $C(\overline{\Omega})$. Hence, there exists an element $p^0 \in \mathfrak{S}_{ad}$ such that, up to a subsequence, $p_\varepsilon^0 \to p^0$ uniformly in Ω and, in view of (24.75), the sequence $\{p_\varepsilon^0 - \sigma(\theta_\varepsilon^0)\}_{\varepsilon>0}$ weakly converges in $L^2(\Omega)$ to $p^0 - \sigma(\theta^0)$ as $\varepsilon \to 0$. Taking this fact into account, we have

$$\liminf_{\varepsilon\to 0} \|p_\varepsilon^0 - \sigma(\theta_\varepsilon^0)\|^2_{L^2(\Omega)} \geq \|p^0 - \sigma(\theta^0)\|^2_{L^2(\Omega)}. \tag{24.76}$$

At the same time, the condition (24.69) implies that

$$\mu_\tau\left(\int_\Omega |p_\varepsilon^0(x) - \sigma(\theta_\varepsilon^0(x))|^2\,dx\right) \leq \varepsilon C^*. \tag{24.77}$$

Hence, $\|p_\varepsilon^0 - \sigma(\theta_\varepsilon^0)\|_{L^2(\Omega)} \leq \tau$ for ε small enough. Combining this fact with (24.76), we arrive at the desired property (24.72).

It remains to note that the existence of a control $b^0 \in \mathfrak{B}_{ad}$ with property $(24.70)_1$ is a direct consequence of Proposition 24.1.

The next step of our analysis is to show that the tuple $(b^0, u^0, \theta^0, p^0)$ is a quasi-feasible point to the original OCP (24.1)–(24.6).

Proposition 24.3 *Assume the condition* (24.69) *holds. Let*

$$(b^0, u^0, \theta^0, p^0) \in W^{1,q}(\Omega) \times W_0^{1,\beta}(\Omega) \times W_0^{1,\gamma}(\Omega) \times \mathfrak{S}_{ad},$$

with $\gamma \in [1, \frac{N}{N-1})$*, be a cluster tuple (in the sense of convergence (24.70)–(24.71)) of a given sequence of optimal solutions* $\{(b_\varepsilon^0, u_\varepsilon^0, \theta_\varepsilon^0, p_\varepsilon^0) \in \widehat{\Xi}_\varepsilon\}_{\varepsilon>0}$*. Then* $(b^0, u^0, \theta^0, p^0)$ *is an element of* $\widehat{\Xi}_0(\tau)$ *and*

$$\int_\Omega |\nabla u_\varepsilon^0|^{p_\varepsilon^0}\,dx + \varepsilon\int_\Omega |\nabla u_\varepsilon^0|^\beta\,dx \to \int_\Omega |\nabla u^0|^{p^0}\,dx, \tag{24.78}$$

$$\int_\Omega |\nabla u^0|^{p^0}\,dx = \int_\Omega \left(g, \nabla u^0\right)_{\mathbb{R}^N}\,dx, \tag{24.79}$$

$u^0 \in H_0^{1,p^0(\cdot)}(\Omega)$ *is the unique H-variational solution of the Dirichlet problem*

$$\operatorname{div}\left(|\nabla u|^{p^0-2}\nabla u\right) = \operatorname{div} g \quad in\ \Omega, \quad u|_{\partial\Omega} = 0, \tag{24.80}$$

and $\theta^0 \in W_0^{1,\gamma}(\Omega)$ *is the unique duality solution to the Dirichlet problem with* L^1*-data*

$$-\operatorname{div}\left(b^0\nabla\theta\right) = |\nabla u^0|^{p^0} \quad in\ \Omega, \quad \theta|_{\partial\Omega} = 0. \tag{24.81}$$

Proof In order to conclude the energy equality (24.79), we apply the following reasoning. At the first step, let us show that distribution u^0 is a weak solution to the problem (24.80). With that in mind we have to pass to the limit as $\varepsilon \to 0$ in integral identity

$$\int_\Omega \left(|\nabla u_\varepsilon^0|^{p_\varepsilon^0(x)-2} \nabla u_\varepsilon^0 + \varepsilon |\nabla u_\varepsilon^0|^{\beta-2} \nabla u_\varepsilon^0, \nabla \varphi \right)_{\mathbb{R}^N} dx = \int_\Omega (g, \nabla\varphi)_{\mathbb{R}^N}\, dx, \quad (24.82)$$

where $\varphi \in C_0^\infty(\Omega)$.

Taking into account estimate (24.74), we see that the sequence $\left\{\varepsilon |\nabla u_\varepsilon^0|^{\beta-2} \nabla u_\varepsilon^0\right\}_{\varepsilon>0}$ is bounded in $L^{\beta'}(\Omega)^N$, and in addition, for any vector-valued function $\psi \in C_0^\infty(\Omega)^N$, we have (see Lemma 3.8 in [40])

$$\begin{aligned}
\int_\Omega \left(\varepsilon |\nabla u_\varepsilon^0|^{\beta-2} \nabla u_\varepsilon^0, \psi\right)_{\mathbb{R}^N} dx &\le \varepsilon \int_\Omega |\nabla u_\varepsilon^0|^{\beta-1} |\psi|\, dx \\
&\le \varepsilon \left(\int_\Omega |\nabla u_\varepsilon^0|^\beta\, dx \right)^{1/\beta'} \|\psi\|_{L^\beta(\Omega)^N} \\
&\le \varepsilon^{1-\frac{1}{\beta'}} \left(\varepsilon \int_\Omega |\nabla u_\varepsilon^0|^\beta\, dx \right)^{1/\beta'} \|\psi\|_{L^\beta(\Omega)^N} \\
&\overset{\text{by (24.74)}}{\le} C \varepsilon^{1-\frac{1}{\beta'}} \to 0.
\end{aligned}$$

Hence,

$$\varepsilon |\nabla u_\varepsilon^0|^{\beta-2} \nabla u_\varepsilon^0 \rightharpoonup 0 \text{ in } L^{\beta'}(\Omega)^N. \quad (24.83)$$

It remains to show the weak convergence of fluxes to a flux:

$$|\nabla u_\varepsilon^0|^{p_\varepsilon^0-2} \nabla u_\varepsilon^0 \rightharpoonup |\nabla u^0|^{p^0-2} \nabla u^0 \quad \text{in} \quad L^{\beta'}(\Omega)^N. \quad (24.84)$$

Since $\frac{\alpha N}{N-\alpha} < \frac{\alpha(N-1)}{N-1-\alpha}$, $u_\varepsilon^0 \rightharpoonup u^0$ in $W_0^{1,\alpha}(\Omega)$,

$$\begin{aligned}
\int_\Omega \left(|\nabla u_\varepsilon^0|^{p_\varepsilon^0-1} \right)^{\beta'} dx &\overset{\text{by (24.50)}}{\le} (1+|\Omega|) \left(\int_\Omega |\nabla u_\varepsilon^0|^{p_\varepsilon^0}\, dx + 1 \right) \\
&\overset{\text{by (24.45)}}{\le} (1+|\Omega|) \left(2^{\alpha'+1} \left(\int_\Omega |g|^{\alpha'}\, dx + |\Omega| \right) + 1 \right) < +\infty,
\end{aligned}$$

and the sequence $\left\{ |\nabla u_\varepsilon^0|^{p_\varepsilon^0} \right\}_{\varepsilon>0}$ is L^1-bounded (see estimate (24.45)), the convergence property of fluxes (24.84) follows from Theorem 24.3. Thus, in view of the properties (24.83)–(24.84), the limit passage in (24.82) as $\varepsilon \to 0$ immediately leads to the relation

$$\int_\Omega \left(|\nabla u^0|^{p^0-2} \nabla u^0, \nabla\varphi \right)_{\mathbb{R}^N} dx = \int_\Omega (g, \nabla\varphi)_{\mathbb{R}^N}\, dx, \quad \forall \varphi \in C_0^\infty(\Omega). \quad (24.85)$$

Since the inclusion $u^0 \in W_0^{1,p^0(\cdot)}(\Omega)$ is guaranteed by Lemma 24.2 and convergence $(24.70)_2$, it follows that u^0 is a weak solution to the boundary value problem (24.80).

Taking into account the fact that $u^0 \in W_0^{1,p^0(\cdot)}(\Omega)$ and $\omega(t) = k_0/\log(|t|^{-1})$ is a modulus of continuity of the exponent $p^0 \in \mathfrak{S}_{ad}$, it follows that the set $C_0^\infty(\Omega)$ is dense in $W_0^{1,p^0(\cdot)}(\Omega)$ (see [40, Theorem 13.10]). Hence, we can consider $\varphi = u^0$ in the identity (24.85) as a test function. As a result, we arrive at the energy equality (24.79). The fact that $u^0 \in W_0^{1,p^0(\cdot)}(\Omega)$ is the unique variational solution to the Dirichlet problem (24.80) follows from the strict monotonicity of the nonlinear operator

$$\mathcal{A}_{p^0} : W_0^{1,p^0(\cdot)}(\Omega) \to \left(W_0^{1,p^0(\cdot)}(\Omega)\right)^*$$

given by the equality (24.12).

As for the property (24.78), we have

$$\begin{aligned}\lim_{\varepsilon\to 0}\left(\int_\Omega |\nabla u_\varepsilon^0|^{p_\varepsilon^0}\,dx + \varepsilon\int_\Omega |\nabla u_\varepsilon^0|^\beta\,dx\right) &\overset{\text{by (24.82)}}{=} \lim_{\varepsilon\to 0}\left(\int_\Omega \left(g, \nabla u_\varepsilon^0\right)_{\mathbb{R}^N}dx\right)\\ &\overset{\text{by (24.70)}_2}{=} \int_\Omega \left(g, \nabla u^0\right)_{\mathbb{R}^N}dx \overset{\text{by (24.79)}}{=} \int_\Omega |\nabla u^0|^{p^0}\,dx.\end{aligned}$$

It remains to establish the relation (24.81). To this end, we have to pass to the limit in integral identity (24.67) as $\varepsilon \to 0$. By Sobolev Embedding Theorem, the conditions (24.6) and the weak convergence $(24.70)_2$ imply the strong convergence $u_\varepsilon^0 \to u^0$ in $L^\beta(\Omega)$. Combining this fact with the properties (24.83) and (24.84), we obtain

$$\int_\Omega \left(|\nabla u_\varepsilon^0|^{p_\varepsilon^0-2}\nabla u_\varepsilon^0 + \varepsilon|\nabla u_\varepsilon^0|^{\beta-2}\nabla u_\varepsilon^0, \nabla\varphi\right)_{\mathbb{R}^N}dx \to \int_\Omega \left(|\nabla u^0|^{p^0-2}\nabla u^0, \nabla\varphi\right)_{\mathbb{R}^N}dx,$$
$$\int_\Omega \left(g, \nabla u_\varepsilon^0\right)_{\mathbb{R}^N}dx \to \int_\Omega \left(g, \nabla u^0\right)_{\mathbb{R}^N}dx,$$

and $b_\varepsilon^0\nabla\theta_\varepsilon^0 \rightharpoonup b^0\nabla\theta^0$ in $L^1(\Omega)^N$ by Lemma 24.9. Thus, the limit passage in (24.67) leads to the equality

$$-\operatorname{div}\left(b^0\nabla\theta^0\right) = \operatorname{div}\left[\left(|\nabla u^0|^{p^0-2}\nabla u^0 - g\right)u^0\right] + \left(g, \nabla u^0\right)_{\mathbb{R}^N} \quad \text{in } \mathcal{D}'(\Omega). \tag{24.86}$$

Since the set $C_0^\infty(\Omega)$ is dense in $W_0^{1,p^0(\cdot)}(\Omega)$, we can apply the transformations, that we used in Lemma 24.3, to show that

$$\operatorname{div}\left[\left(|\nabla u^0|^{p^0-2}\nabla u^0 - g\right)u^0\right] + \left(g, \nabla u^0\right)_{\mathbb{R}^N} = |\nabla u^0|^{p^0}$$

in the sense of distributions. Taking into account the fact that $b^0 \in C^{0,\lambda}(\overline{\Omega})$ and arguments from the proof of Lemma 24.10, it can be shown that $\theta^0 \in W_0^{1,\gamma}(\Omega)$ is the duality solution of the boundary value problem (24.81). The uniqueness of θ^0 follows from the results of L^1-theory of elliptic equations [29].

Proposition 24.4 *Let* $\tau \geq \sqrt{|\Omega|}(\beta - \alpha)$*. Then there exists a constant* $C^* > 0$ *such that estimate* (24.69) *holds true.*

Proof For an arbitrary $\xi \in [m_1, m_2]$, we set $\widehat{b}(\cdot) = \xi$. Then due to the fixed point principle, it was shown in [40, p.495] that the system

$$\operatorname{div}\left(|\nabla u|^{\sigma(\theta)-2}\nabla u + \varepsilon|\nabla u|^{\beta-2}\nabla u\right) = \operatorname{div} g \quad \text{in } \Omega, \quad u|_{\partial\Omega} = 0, \tag{24.87}$$

$$-\xi\,\Delta\theta = |\nabla u|^{\sigma(\theta)} + \varepsilon|\nabla u|^{\beta} \quad \text{in } \Omega, \quad \theta|_{\partial\Omega} = 0 \tag{24.88}$$

has at least one solution

$$(u_\varepsilon, \theta_\varepsilon) \in W_0^{1,\beta}(\Omega) \times \left(W^{2,1+\delta}(\Omega) \cap W_0^{1,1+\delta}(\Omega)\right)$$

for all $\varepsilon > 0$ with some positive $\delta > 0$. Moreover, there exists a constant $C > 0$ such that

$$\sup_{\varepsilon>0} \|u_\varepsilon\|_{W_0^{1,\alpha}(\Omega)} \leq C, \quad \sup_{\varepsilon>0} \|\theta_\varepsilon\|_{W^{2,1+\delta}(\Omega)} \leq C \quad \text{for some } \delta > 0, \text{ and}$$

$$u_\varepsilon \rightharpoonup u \text{ in } W_0^{1,\alpha}(\Omega) \text{ and } \theta_\varepsilon \rightharpoonup \theta \text{ in } W^{2,1+\delta}(\Omega) \text{ as } \varepsilon \to 0, \tag{24.89}$$

where (u, θ) satisfies (in the sense of distributions) the following relations

$$\operatorname{div}\left(|\nabla u|^{\sigma(\theta)-2}\nabla u\right) = \operatorname{div} g,$$

$$-\xi\,\operatorname{div}(\nabla\theta) = \operatorname{div}\left[\left(|\nabla u|^{\sigma(\theta)-2}\nabla u - g\right)u\right] + (g, \nabla u)_{\mathbb{R}^N}.$$

We set $\widehat{p}_{\varepsilon,\lambda} = T_\lambda(\sigma(\theta_\varepsilon))$, where T_λ are smoothing operators satisfying the properties

$$T_\lambda(\sigma(\theta_\varepsilon)) \to \sigma(\theta_\varepsilon) \quad \text{in } L^2(\Omega) \text{ as } \lambda \to 0, \tag{24.90}$$

$$\begin{gathered} \alpha \leq T_\lambda(\sigma(\theta_\varepsilon)) \leq \beta, \ T_\lambda(\sigma(\theta_\varepsilon)) \in C^1_{loc}(\mathbb{R}) \\ \forall\, \varepsilon \in (0, \lambda(\varepsilon)) \text{ with some } \lambda(\varepsilon) > 0. \end{gathered} \tag{24.91}$$

As an example of such operators, we can take the following one

$$T_\lambda(\sigma(\theta_\varepsilon)) = \max\left\{\alpha, \min\left\{\sigma(\theta_\varepsilon) * \rho_\lambda, \beta\right\}\right\},$$

where $\{\rho_\lambda\}_{\lambda>0}$ is any rescaled family of smooth mollifiers such that $\operatorname{supp} \rho_\lambda \in B(0, \lambda)$.

Let $(\widehat{u}_{\varepsilon,\lambda(\varepsilon)}, \widehat{\theta}_{\varepsilon,\lambda(\varepsilon)}) \in W_0^{1,\beta}(\Omega) \times W_0^{1,1+\delta}(\Omega)$ be a unique solution to the system

$$\operatorname{div}\left(|\nabla u|^{p_{\varepsilon,\lambda(\varepsilon)}(x)-2}\nabla u+\varepsilon|\nabla u|^{\beta-2}\nabla u\right)=\operatorname{div} g \quad \text{in } \Omega, \quad u|_{\partial\Omega}=0, \tag{24.92}$$

$$-\xi \operatorname{div}(\nabla\theta)=|\nabla u|^{p_{\varepsilon,\lambda(\varepsilon)}(x)} \quad \text{in } \Omega, \quad \theta|_{\partial\Omega}=0. \tag{24.93}$$

We note that, in view of estimate (24.14) (see also the proof of Lemma 24.3), the sequence $\left\{\widehat{\theta}_{\varepsilon,\lambda(\varepsilon)}\right\}_{\varepsilon>0}$ is bounded in $W_0^{1,\gamma}(\Omega)$ for all $\gamma\in[1,\frac{N}{N-1})$. Since

$$\begin{aligned}
&|\nabla\widehat{u}_{\varepsilon,\lambda(\varepsilon)}|^{\widehat{p}_{\varepsilon,\lambda(\varepsilon)}}\\
&\quad=\operatorname{div}\left[\left(|\nabla\widehat{u}_{\varepsilon,\lambda(\varepsilon)}|^{\widehat{p}_{\varepsilon,\lambda(\varepsilon)}-2}\nabla\widehat{u}_{\varepsilon,\lambda(\varepsilon)}+\varepsilon|\nabla\widehat{u}_{\varepsilon,\lambda(\varepsilon)}|^{\beta-2}\nabla\widehat{u}_{\varepsilon,\lambda(\varepsilon)}-g\right)\widehat{u}_{\varepsilon,\lambda(\varepsilon)}\right]\\
&\qquad+\left(g,\nabla\widehat{u}_{\varepsilon,\lambda(\varepsilon)}\right)_{\mathbb{R}^N}
\end{aligned}$$

in the sense of distributions (see Lemma 24.10), it follows that

$$\left(\widehat{b},\widehat{u}_{\varepsilon,\lambda(\varepsilon)},\widehat{\theta}_{\varepsilon,\lambda(\varepsilon)},\widehat{p}_{\varepsilon,\lambda(\varepsilon)}\right)\in\widehat{\Xi}_\varepsilon \quad \text{for all } \varepsilon>0.$$

Then by Lemmas 24.8–24.10 and definition of the set $\mathfrak{B}_{ad}$, the sequence

$$\left\{\left(\widehat{b},\widehat{u}_{\varepsilon,\lambda(\varepsilon)},\widehat{\theta}_{\varepsilon,\lambda(\varepsilon)},\widehat{p}_{\varepsilon,\lambda(\varepsilon)}\right)\right\}_{\varepsilon>0}$$

is bounded in $W^{1,q}(\Omega)\times W_0^{1,\alpha}(\Omega)\times W_0^{1,\gamma}(\Omega)\times C(\overline{\Omega})$ and

$$\|\widehat{p}_{\varepsilon,\lambda(\varepsilon)}-\sigma(\widehat{\theta}_{\varepsilon,\lambda(\varepsilon)})\|_{L^2(\Omega)}\le(\beta-\alpha)\sqrt{|\Omega|} \quad \forall\,\varepsilon>0.$$

Since $\tau\ge(\beta-\alpha)\sqrt{|\Omega|}$ and $\sup_{\varepsilon>0}\|\widehat{\theta}_{\varepsilon,\lambda(\varepsilon)}\|_{L^\gamma(\Omega)}<+\infty$, it follows from definition of the function μ_τ that

$$\begin{aligned}
\limsup_{\varepsilon\to0}\inf_{(b,u,\theta,p)\in\widehat{\Xi}_\varepsilon}J_{\varepsilon,\tau}(b,u,\theta,p)&\le\limsup_{\varepsilon\to0}J_{\varepsilon,\tau}\left(\widehat{b},\widehat{u}_{\varepsilon,\lambda(\varepsilon)},\widehat{\theta}_{\varepsilon,\lambda(\varepsilon)},\widehat{p}_{\varepsilon,\lambda(\varepsilon)}\right)\\
&=\limsup_{\varepsilon\to0}\int_\Omega|\widehat{\theta}_{\varepsilon,\lambda(\varepsilon)}(x)-\theta_d(x)|\,dx<+\infty.
\end{aligned}$$

Summing up Propositions 24.2 and 24.3, we are led to the following conclusion: the fulfilment of (24.69) or $\tau\ge(\beta-\alpha)\sqrt{|\Omega|}$ suffices to claim that any cluster tuple (in the context of convergence (24.70)–(24.71)) of the sequence of optimal solutions $\left\{(b_\varepsilon^0,u_\varepsilon^0,\theta_\varepsilon^0,p_\varepsilon^0)\in\widehat{\Xi}_\varepsilon\right\}_{\varepsilon>0}$ is a quasi-feasible point to the original OCP (24.1)–(24.6). We are now in a position to prove our main result.

Theorem 24.5 *Let $\left\{(b_\varepsilon^0,u_\varepsilon^0,\theta_\varepsilon^0,p_\varepsilon^0)\in\widehat{\Xi}_\varepsilon\right\}_{\varepsilon>0}$ be an arbitrary sequence of optimal solutions to the approximated problems (24.19)–(24.22). If the condition* (24.69) *holds true, then any cluster tuple $\left(b^0,u^0,\theta^0,p^0\right)$ is a quasi-optimal solution of the OCP* (24.1)–(24.6). *Moreover, in this case the following variational property holds*

$$\lim_{\varepsilon\to0}\inf_{(b,u,\theta,p)\in\widehat{\Xi}_\varepsilon}J_{\varepsilon,\tau}(b,u,\theta,p)=J\left(b^0,u^0,\theta^0,p^0\right)=\inf_{(b,u,\theta,p)\in\widehat{\Xi}_0(\tau)}J(b,u,\theta,p). \tag{24.94}$$

Proof As follows from Propositions 24.2–24.4, the set of feasible points $\widehat{\Xi}_0(\tau)$ is nonempty. Let us assume that there exists a tuple $\left(\widehat{b}, \widehat{u}, \widehat{\theta}, \widehat{p}\right)$ in $\widehat{\Xi}_0(\tau)$ such that

$$J\left(\widehat{b}, \widehat{u}, \widehat{\theta}, \widehat{p}\right) < J\left(b^0, u^0, \theta^0, p^0\right). \tag{24.95}$$

By definition of the set $\widehat{\Xi}_0(\tau)$, we have $\widehat{b} \in \mathfrak{B}_{ad}$, $\widehat{p} \in \mathfrak{S}_{ad}$, and $\|\widehat{p} - \sigma(\widehat{\theta})\|_{L^2(\Omega)} \leq \tau$. We define the sequence $\left\{(\widehat{b}_\varepsilon, \widehat{u}_\varepsilon, \widehat{\theta}_\varepsilon, \widehat{p}_\varepsilon)\right\}_{\varepsilon>0}$ as follows

$$\begin{aligned} &\widehat{p}_\varepsilon \to \widehat{p} \text{ in } C(\overline{\Omega}) \text{ as } \varepsilon \to 0, \quad \widehat{b}_\varepsilon \equiv \widehat{b}, \quad \text{and} \quad \widehat{p}_\varepsilon \in \mathfrak{S}_{ad}\ \forall \varepsilon > 0, \\ &\|\widehat{p}_\varepsilon - \sigma(\widehat{\theta})\|_{L^2(\Omega)} \leq \tau + \frac{\varepsilon^2}{2} \quad \text{for } \varepsilon \in (0, \tau) \text{ small enough,} \end{aligned} \tag{24.96}$$

and each of the pairs $(\widehat{u}_\varepsilon, \widehat{\theta}_\varepsilon)$ is the corresponding solution to the boundary value problems (24.20)–(24.21) with $b = \widehat{b}_\varepsilon$ and $p = \widehat{p}_\varepsilon$. Since each of these problems admit the unique solution in $W_0^{1,\beta}(\Omega) \times W_0^{1,\gamma}(\Omega)$ satisfying the a priori estimates (24.41) and (24.66), it follows that the tuples $(\widehat{b}_\varepsilon, \widehat{u}_\varepsilon, \widehat{\theta}_\varepsilon, \widehat{p}_\varepsilon)$ are feasible points to the corresponding approximated OCPs (24.19)–(24.22), i.e. $(\widehat{b}_\varepsilon, \widehat{u}_\varepsilon, \widehat{\theta}_\varepsilon, \widehat{p}_\varepsilon) \in \widehat{\Xi}_\varepsilon$ for all $\varepsilon > 0$. Moreover, in view of Remark 24.3 and estimates

$$\begin{aligned} \|\widehat{u}_\varepsilon\|^\alpha_{W_0^{1,\alpha}(\Omega)} &\overset{\text{by (24.29), (24.28)}}{\leq} (1 + |\Omega|)\left(\int_\Omega |\nabla \widehat{u}_\varepsilon|^{\widehat{p}_\varepsilon}\, dx + 1\right) \\ &\overset{\text{by (24.45)}}{\leq} (1 + |\Omega|)\left(2^{\alpha'+1}\left(\int_\Omega |g|^{\alpha'}\, dx + |\Omega|\right) + 1\right), \\ \|\widehat{\theta}_\varepsilon\|_{W_0^{1,\gamma}(\Omega)} &\overset{\text{by (24.41), (24.66)}}{\leq} C(\gamma) 2^{\alpha'+1}\left(\int_\Omega |g|^{\alpha'}\, dx + |\Omega|\right), \end{aligned}$$

the sequence $\left\{(\widehat{u}_\varepsilon, \widehat{\theta}_\varepsilon)\right\}_{\varepsilon>0}$ is bounded in $W_0^{1,\alpha}(\Omega) \times W_0^{1,\gamma}(\Omega)$. Hence, by analogy with Propositions 24.2 and 24.3, it can be shown that

$$\widehat{u}_\varepsilon \rightharpoonup \widehat{u} \text{ in } W_0^{1,\alpha}(\Omega) \quad \text{and} \quad \widehat{\theta}_\varepsilon \rightharpoonup \widehat{\theta} \text{ in } W_0^{1,\gamma}(\Omega), \tag{24.97}$$

where $(\widehat{u}, \widehat{\theta})$ is a weak solution to the boundary value problem (24.2)–(24.3) with $p = \widehat{p}$ and $\widehat{\theta}$ is the duality solution to (24.3). Since this solution is unique (by the strict monotonicity and the log-Hölder continuity of the exponent $\widehat{p}$) it is not necessary to pass to a subsequence in (24.97). Moreover, since the embedding $W^{1,\gamma}(\Omega) \hookrightarrow L^\gamma(\Omega)$ is compact, we can suppose that there exists a subsequence $\left\{\widehat{\theta}_{\delta(\varepsilon)}\right\}_{\delta(\varepsilon)>0}$ of $\left\{\widehat{\theta}_\varepsilon\right\}_{\varepsilon>0}$ such that $\delta(\varepsilon) \to 0$ as $\varepsilon \to 0$, $\delta(\varepsilon) \leq \varepsilon$, and $\widehat{\theta}_{\delta(\varepsilon)} \to \widehat{\theta}$ almost everywhere in Ω. Hence, by boundedness of the sequence $\left\{\sigma(\widehat{\theta}_\varepsilon)\right\}_{\varepsilon>0}$ and Lebesgue theorem, we can suppose that $\sigma(\widehat{\theta}_{\delta(\varepsilon)}) \to \sigma(\widehat{\theta})$ strongly in $L^2(\Omega)$ as $\varepsilon \to 0$ and

$$\|\sigma(\widehat{\theta}_{\delta(\varepsilon)}) - \sigma(\widehat{\theta})\|_{L^2(\Omega)} \leq \frac{\varepsilon^2}{2} \quad \text{for } \varepsilon \text{ small enough.} \tag{24.98}$$

As a result, (24.96) and (24.98) imply that

$$\begin{aligned}\|\widehat{p}_{\delta(\varepsilon)} - \sigma(\widehat{\theta}_{\delta(\varepsilon)})\|_{L^2(\Omega)} &\le \|\widehat{p}_{\delta(\varepsilon)} - \sigma(\widehat{\theta})\|_{L^2(\Omega)} + \|\sigma(\widehat{\theta}_{\delta(\varepsilon)}) - \sigma(\widehat{\theta})\|_{L^2(\Omega)}\\ &\le \tau + \delta(\varepsilon)^2 \le \tau + \varepsilon^2\end{aligned}$$

for ε small enough. Hence, there exists $\varepsilon_0 > 0$ such that

$$0 \le \mu_\tau \left(\int_\Omega |\widehat{p}_{\delta(\varepsilon)}(x) - \sigma(\widehat{\theta}_{\delta(\varepsilon)}(x))|^2 \, dx \right) \le \varepsilon^2, \quad \forall \varepsilon \in (0, \varepsilon_0). \tag{24.99}$$

Taking this fact into account, we get

$$\begin{aligned}J\left(b^0, u^0, \theta^0, p^0\right) &\overset{\text{by (24.71)}}{=} \liminf_{\varepsilon\to 0} \int_\Omega |\theta^0_\varepsilon(x) - \theta_d(x)| \, dx\\ &\le \liminf_{\varepsilon\to 0} J_{\varepsilon,\tau}\left(b^0_\varepsilon, u^0_\varepsilon, \theta^0_\varepsilon, p^0_\varepsilon\right)\\ &= \liminf_{\varepsilon\to 0} \inf_{(b,u,\theta,p)\in\widehat{\Xi}_\varepsilon} J_{\varepsilon,\tau}(b, u, \theta, p)\\ &\le \liminf_{\varepsilon\to 0} J_{\varepsilon,\tau}\left(\widehat{b}_{\delta(\varepsilon)}, \widehat{u}_{\delta(\varepsilon)}, \widehat{\theta}_{\delta(\varepsilon)}, \widehat{p}_{\delta(\varepsilon)}\right)\\ &= \liminf_{\varepsilon\to 0} J\left(\widehat{b}_{\delta(\varepsilon)}, \widehat{u}_{\delta(\varepsilon)}, \widehat{\theta}_{\delta(\varepsilon)}, \widehat{p}_{\delta(\varepsilon)}\right)\\ &\quad + \lim_{\varepsilon\to 0} \frac{1}{\varepsilon}\mu_\tau\left(\int_\Omega |\widehat{p}_{\delta(\varepsilon)} - \sigma(\widehat{\theta}_{\delta(\varepsilon)})|^2 dx\right)\\ &= J\left(\widehat{b}, \widehat{u}, \widehat{\theta}, \widehat{p}\right).\end{aligned} \tag{24.100}$$

Coming into conflict with (24.95), we conclude: relation (24.100) holds true only as an equality which immediately yields (24.94). Thus, $\left(b^0, u^0, \theta^0, p^0\right)$ is a quasi-optimal solution of the OCP (24.1)–(24.6).

It remains to discuss the question about optimal solutions to the OCP (24.1)–(24.6) in the sense of Definition 24.3. As follows from Theorem 24.5, it suffices to show that $\Xi_0 \neq emptyset$ and consider instead of the approximated problems (24.19)–(24.22) the following ones (with $\tau = \varepsilon$ in the approximation cost functional)

$$\begin{aligned}\text{Minimize } & J_\varepsilon(b, u, \theta, p)\\ &= \int_\Omega |\theta(x) - \theta_d(x)| \, dx + \frac{1}{\varepsilon}\mu_\varepsilon\left(\int_\Omega |p(x) - \sigma(\theta(x))|^2 \, dx\right)\\ &\text{subject to the constraints (24.20)–(24.22).}\end{aligned} \tag{24.101}$$

As a result, the validity of the main Theorem 24.1 immediately follows from the following result.

Theorem 24.6 *Let Ω be an open bounded domain in $\mathbb{R}^N$ with a sufficiently smooth boundary. Assume that $\widehat{\Xi}_0(\tau) \neq emptyset$ for $\tau = 0$, i.e. there exist a control $\widehat{b} \in \mathfrak{B}_{ad}$, an exponent $\widehat{p} \in \mathfrak{S}_{ad}$, and a weak solution to the thermistor problem* (24.1)–(24.6) $(\widehat{u}, \widehat{\theta}) \in W_0^{1,\sigma(\widehat{\theta}(\cdot))}(\Omega) \times W_0^{1,\gamma}(\Omega)$ *with $b(\cdot) = \widehat{b}(\cdot)$ such that $\widehat{\theta}$ is the duality solution to* (24.6) *and $\widehat{p} = \sigma(\widehat{\theta})$ almost everywhere in Ω. Then the OCP* (24.1)–(24.6) *has a non-empty set of optimal solutions and some of them can be attained (in the sense of convergence (24.70)–(24.71)) by solutions $(b_\varepsilon^0, u_\varepsilon^0, \theta_\varepsilon^0, p_\varepsilon^0)$ to the approximated problem* (24.101).

Proof To begin with, let us note that, in view of the initial assumptions and Remarks 24.3 and 24.4 the set Ξ_0, given by (24.15), is nonempty. To get the solvability of the original OCP (24.1)–(24.6), we pass to its perturbation in the form of the family of approximated problem (24.101). Due to Theorem 24.4, each of the problems (24.101) has a nonempty set of solutions. Let $(b_\varepsilon^0, u_\varepsilon^0, \theta_\varepsilon^0, p_\varepsilon^0)$ be optimal tuples to the approximated problems (24.101). As follows from Proposition 24.2, compactness of this sequence with respect to the convergence (24.70)–(24.71) can be guaranteed by the condition (24.69). To show that the estimate (24.69) holds true with some $C^* > 0$, we will closely follow the proof-line of Theorem 24.5. As a result, it can be constructed a sequence $\left\{(\widehat{b}_\varepsilon, \widehat{u}_\varepsilon, \widehat{\theta}_\varepsilon, \widehat{p}_\varepsilon) \in \widehat{\Xi}_\varepsilon\right\}_{\varepsilon>0}$ such that

$$\begin{gathered}
\widehat{p}_\varepsilon \to \widehat{p} \text{ in } C(\overline{\Omega}) \text{ as } \varepsilon \to 0, \quad \widehat{b}_\varepsilon \equiv \widehat{b}, \quad \text{and} \quad \widehat{p}_\varepsilon \in \mathfrak{S}_{ad}\ \forall \varepsilon > 0, \\
\|\widehat{p}_\varepsilon - \sigma(\widehat{\theta})\|_{L^2(\Omega)} \leq \varepsilon/2 \quad \text{for } \varepsilon > 0 \text{ small enough}, \\
\widehat{u}_\varepsilon \rightharpoonup \widehat{u} \text{ in } W_0^{1,\alpha}(\Omega) \quad \text{and} \quad \widehat{\theta}_\varepsilon \rightharpoonup \widehat{\theta} \text{ in } W_0^{1,\gamma}(\Omega) \quad \text{for some } \delta > 0, \\
\sigma(\widehat{\theta}_\varepsilon) \to \sigma(\widehat{\theta}) \quad \text{strongly in } L^2(\Omega) \text{ as } \varepsilon \to 0.
\end{gathered}$$

Hence, applying the arguments of the proof of Theorem 24.5, we can conclude that there exists $\varepsilon_0 > 0$ and a subsequence $\{\delta(\varepsilon)\}$ of $\{\varepsilon\}$ such that

$$\|\sigma(\widehat{\theta}_{\delta(\varepsilon)}) - \sigma(\widehat{\theta})\|_{L^2(\Omega)} \leq \frac{\varepsilon}{2} \quad \text{for all } \varepsilon < \varepsilon_0$$

and

$$\begin{aligned}
\|\widehat{p}_{\delta(\varepsilon)} - \sigma(\widehat{\theta}_{\delta(\varepsilon)})\|_{L^2(\Omega)} &\leq \|\widehat{p}_{\delta(\varepsilon)} - \sigma(\widehat{\theta})\|_{L^2(\Omega)} + \|\sigma(\widehat{\theta}_{\delta(\varepsilon)}) - \sigma(\widehat{\theta})\|_{L^2(\Omega)} \\
&\leq \frac{\varepsilon}{2} + \frac{\varepsilon}{2}, \quad \forall \varepsilon \in (0, \varepsilon_0).
\end{aligned}$$

Therefore,

$$\mu_{\delta(\varepsilon)}\left(\int_\Omega |\widehat{p}_{\delta(\varepsilon)}(x) - \sigma(\widehat{\theta}_{\delta(\varepsilon)}(x))|^2\, dx\right) = 0, \quad \forall \varepsilon \in (0, \varepsilon_0). \tag{24.102}$$

Thus,

$$\limsup_{\varepsilon\to 0} \inf_{(b,u,\theta,p)\in\widehat{\Xi}_{\delta(\varepsilon)}} J_{\delta(\varepsilon)}(b,u,\theta,p) \le \limsup_{\varepsilon\to 0} J_{\delta(\varepsilon)}(\widehat{b}_{\delta(\varepsilon)},\widehat{u}_{\delta(\varepsilon)},\widehat{\theta}_{\delta(\varepsilon)},\widehat{p}_{\delta(\varepsilon)})$$
$$= \limsup_{\varepsilon\to 0} \int_\Omega |\widehat{\theta}_{\delta(\varepsilon)} - \theta_d|\,dx = C^* < +\infty.$$

As a result, Propositions 24.2 and 24.3 imply the existence of a cluster tuple

$$\left(b^0,u^0,\theta^0,p^0\right) \in W^{1,q}(\Omega)\times W_0^{1,\beta}(\Omega)\times W_0^{1,\gamma}(\Omega)\times\mathfrak{S}_{ad},$$

with $\gamma=\frac{Nr}{N+r}$, of the sequence $\left\{(\widehat{b}_\varepsilon,\widehat{u}_\varepsilon,\widehat{\theta}_\varepsilon,\widehat{p}_\varepsilon)\right\}_{\varepsilon>0}$ in the sense of convergence (24.70)–(24.71) such that (see (24.95) and (24.100))

$$\begin{aligned}
J\left(b^0,u^0,\theta^0,p^0\right) &= \liminf_{\varepsilon\to 0}\int_\Omega |\theta^0_\varepsilon-\theta_d|\,dx = \liminf_{\varepsilon\to 0}\int_\Omega |\theta^0_{\delta(\varepsilon)}-\theta_d|^r\,dx\\
&= \liminf_{\varepsilon\to 0} J_{\delta(\varepsilon)}\left(b^0_{\delta(\varepsilon)},u^0_{\delta(\varepsilon)},\theta^0_{\delta(\varepsilon)},p^0_{\delta(\varepsilon)}\right)\\
&= \liminf_{\varepsilon\to 0}\inf_{(b,u,\theta,p)\in\widehat{\Xi}_{\delta(\varepsilon)}} J_{\delta(\varepsilon)}\left(b,u,\theta,p\right)\\
&\ge \liminf_{\varepsilon\to 0}\inf_{(b,u,\theta,p)\in\widehat{\Xi}_{\varepsilon}} J_{\varepsilon}\left(b,u,\theta,p\right)\\
&\ge \liminf_{\varepsilon\to 0}\int_\Omega |\theta^0_\varepsilon-\theta_d|\,dx = J\left(b^0,u^0,\theta^0,p^0\right).
\end{aligned}$$

Taking into account the strong convergence $\widehat{p}_{\delta(\varepsilon)}\to\widehat{p}=\sigma(\widehat{\theta})$ in $C(\overline{\Omega})$, we finally deduce that $\left(b^0,u^0,\theta^0,p^0\right)\in\Xi_0$. In order to show that $\left(b^0,u^0,\theta^0,p^0\right)$ is an optimal solution to the original problem, we can assume the converse statement and apply the arguments of the proof of Theorem 24.5.

Remark 24.7 In the general case, the question about non-emptiness of the set of feasible solutions Ξ_0 to the OCP (24.1)–(24.6) remains open even for the smooth functions σ. At the same time, if we consider, instead of equation (24.3), its relaxation in the form (24.23), Theorem 7.2 in [40] says that the relaxed version of the thermistor problem (24.2),(24.4),(24.23) admits a solution $(u,\theta)\in W_0^{1,\sigma(\theta(b))}(\Omega)\times W_0^{1,\gamma}(\Omega)$ for any $\gamma\in[1,\frac{N}{N-1})$, $b\in\mathfrak{B}_{ad}$, and any continuous function $\sigma(\theta)$ satisfying the condition (24.6) with $\beta<+\infty$ if $\alpha\ge N-1$, and $\beta<\frac{\alpha(N-1)}{N-1-\alpha}$ if $\alpha<N-1$. Moreover, in this case we have $u\in W_0^{1,\alpha}(\Omega)$. However, as it was mentioned in Remark 24.2, in this case the inclusion $u\in H_0^{1,p(\cdot)}(\Omega)$ is by no means obvious. In order to circumvent this artefact, we can apply Theorem 7.2 in [40] to a function $\sigma(\theta)$ that has a logarithmic modulus of continuity. Then $W_0^{1,p(\cdot)}(\Omega)=H_0^{1,p(\cdot)}(\Omega)$ with $p:=\sigma(\theta(b))$, i.e. the H-solution coincide with the W-solution, and, therefore, the tuple (b,u,θ,σ) is a feasible solution to the modified OCP (24.1), (24.2), (24.23), (24.4), (24.6) for any admissible control $b\in\mathfrak{B}_{ad}$. Thus, its solvability can be established by analogy with Theorem 24.6.

Acknowledgements This research was partially supported by National Research Foundation of Ukraine (Grant No. 2020.02/0066).

References

1. Alkhutov, Y.A., Kondratev, V.A: Solvability of the Dirichlet problem for second-order elliptic equations in a convex region. Differ. Uravn. **28** (5), 806–818 (1992)
2. Antontsev, S.N., Chipot, M.: The thermistor problem: existence, smoothness, uniqueness, blowup. SIAM J. Math. Anal. **25**, 1128–1156 (1994)
3. Antontsev, S.N., Rodrigues, J.F.: On stationary thermo-rheological viscous flows. Ann. Univ. Ferrara **52**(1), 19–36 (2006)
4. Baranger, J., Mikelić, A.: Stationary solutions to a quasi-Newtonian flow with viscous heating. Math. Models Methods Appl. Sci. **5**(6), 725–738 (1995)
5. Boccardo, L., Gallouet, T.: Nonlinear elliptic and parabolic equations involving measure data. J. Funct. Anal. **87**, 149–169 (1989)
6. Buttazzo, G., Kogut, P.I.: Weak optimal controls in coefficients for linear elliptic problems. Revista Matematica Complutense **24**, 83–94 (2011)
7. Casas, E., Kogut, P.I., Leugering, G.: Approximation of Optimal Control Problems in the Coefficient for the p-Laplace Equation. I. Convergence Result. SIAM J. Control Optim. **54**(3), 1406–1422 (2016)
8. D'Apice, C., De Maio, U., Kogut, O.P.: On shape stability of Dirichlet optimal control problems in coefficients for nonlinear elliptic equations. Adv. Differ. Equ. **15**(7–8), 689–720 (2010)
9. D'Apice, C., De Maio, U., Kogut, O.P.: Optimal control problems in coefficients for degenerate equations of monotone type: Shape stability and attainability problems. SIAM J. Control Optim. **50**(3), 1174–1199 (2012)
10. Cruz-Uribe, D.V., A. Fiorenza, A.: Variable Lebesgue Spaces: Foundations and Harmonic Analysis. Birkhäuser, New York (2013)
11. Diening, L., Harjulehto, P., Hästö, P., Růžiĉka, M.: Lebesgue and Sobolev Spaces with Variable Exponents. Springer, New York (2011)
12. Disser, K., Kaiser, H.C., Rehberg, J.: Optimal Sobolev regularity for linear second-order divergence elliptic operators occurring in real- world problems. SIAM J. Math. Anal. **47**(3), 1719–1746 (2015)
13. Evans, L.C.: Partial Differential Equations. AMS, New York (2000)
14. Hömberg, D., Meyer, C., Rehberg, J., Ring, W.: Optimal control for the thermistor problem. SIAM J. Control Optim. **49**(5), 3449–3481 (2010)
15. Horsin, T., Kogut, P.I.: Optimal L^2-control problem in coefficients for a linear elliptic equation. Math. Control Relat. Fields **5**(1), 73–96 (2015)
16. Horsin, T., Kogut, P.I.: On unbounded optimal controls in coefficients for ill-posed elliptic Dirichlet boundary value problems. Asymptot. Anal. **98**, 155–188 (2016)
17. Hrynkiv, V.: Optimal boundary control for a time dependent thermistor problem. Electr. J. Differ. Equ. **2009**(83), 1–22 (2009)
18. Howison, S.D., Rodrigues, J.F., Shillor, M.: Stationary solutions to the thermistor problem. J. Math. Anal. Appl. **174**(2), 573–588 (1993)
19. Kogut, P.: On approximation of an optimal boundary control problem for linear elliptic equation with unbounded coefficients. Discret. Contin. Dyn. Syst. - Ser. A **34**(5), 2105–2133 (2014)
20. Kogut, P.: Variational S-convergence of minimization problems. Part I. Definitions and basic properties. Problemy Upravleniya I Informatiki (Avtomatika) **5**, 29–42 (1996)
21. Kogut, P., Kupenko, O.: Approximation Methods in Optimization of Nonlinear Systems, De Gruyter Series in Nonlinear Analysis and Applications 32. Walter de Gruyter GmbH, Berlin (2019)

22. Kogut, P., Leugering, G.: Optimal control problems for partial differential equations on reticulated domains. Approximation and Asymptotic Analysis. Series: Systems and Control. Birkhäuser Verlag, Boston (2011)
23. Kogut, P., Leugering, G.: Matrix-valued L^1-optimal control in the coefficients of linear elliptic problems. J. Anal. Appl. (ZAA) **32**(4), 433–456 (2013)
24. Kogut, P., Leugering, G.: S-homogenization of optimal control problems in Banach spaces. Math. Nachr. **233–234**, 141–169 (2002)
25. Kogut, P., Manzo, R., Putchenko, A.: On approximate solutions to the Neumann elliptic boundary value problem with non-linearity of exponential type. Bound. Value Prob. **2016**(1), 1–32 (2016)
26. Kuttler, K.L., Shillor, M., Fernandez, J.R.: Existence for the thermoviscoelastic thermistor problem. Differ. Equ. Dyn. Syst. **16**(4), 309–332 (2008)
27. Meyer, C., Panizzi, L., Schiela, A.: Uniqueness criteria for the adjoint equation in state-constrained elliptic optimal control. Numer. Funct. Anal. Optim. **32**(9), 983–1007 (2011)
28. Murat, F.:, Compacite par compensation. Ann. Scuola Norm. Sup. Pisa. Cl. Sci. Fis. Math. **5**, 481–507 (1978)
29. Orsina, L.: Elliptic Equations with Measure Data. Sapienza University of Rome (2011)
30. Prignet, A.: Remarks on existence and uniqueness of solutions of elliptic problems with right-hand side measures. Rend. Maht. Appl. VII. **15**(3), 321–337 (1995)
31. Rădulescu, V.D., Repovš, D.D.: Partial Differential Equations with Variable Exponents. Variational Methods and Qualitative Analysis. CRC Press, New York (2015)
32. Roubíček, T.: Nonlinear Partial Differential Equations with Applications. Birkhäuser, Bosnot (2013)
33. Růžička, M.: Electrorheological Fluids: Modeling and Mathematical Theory. Lecture Notes in Mathematics, vol. 1748. Springer, Berlin (2000)
34. Stampacchi, G.: Equations Elliptiques du Second Ordre a Coefficients Discontinus. Les Presses de L'Universite de Montreal (1966)
35. Zhikov, V.V.: On some variational problems. Russ. J. Math. Phys. **5**(1), 105–116 (1997)
36. Zhikov, V.V.: Existence theorem for some pairs of coupled elliptic equations. Russ. Acad. Sci. Dokl. (Math.) **77**(1), 80–84 (2008)
37. Zhikov, V.V.: Solvability of the three-dimensional thermistor problem. Proc. Steklov Inst. Math. **281**, 98–111 (2008)
38. Zhikov, V.V.: On the technique for passing to the limit in nonlinear elliptic problems. Funct. Anal. Appl. **43**(2), 96–112 (2009)
39. Zhikov, V.V.: On the weak convergence of fluxes to flux. Russ. Acad. Sci. Dokl. (Math.) **81**(1), 58–62 (2010)
40. Zhikov, V.V.: On variational problems and nonlinear elliptic equations with nonstandard growth conditions. J. Math. Sci. **173**(5), 463–570 (2011)
41. Zhikov, V.V., Pastukhova, S.E.: Improved integrability of the gradients of solutions of elliptic equations with variable nonlinearity exponent. Math. Sb. **199**(12), 19–52 (2008)

CPSIA information can be obtained
at www.ICGtesting.com
Printed in the USA
LVHW081147301120
673012LV00020B/151

9 783030 503017